Early Precambrian Basic Magmatism

EARLY PRECAMBRIAN BASIC MAGMATISM

edited by

R.P. HALL and D.J. HUGHES

Department of Geology
Portsmouth Polytechnic

Blackie
Glasgow and London

Published in the USA by
Chapman and Hall
New York

Blackie & Son Limited.
Bishopbriggs, Glasgow G64 2NZ
and
7 Leicester Place, London WC2H 7BP

Published in the USA by
Chapman and Hall
a division of Routledge, Chapman and Hall, Inc.
29 West 35th Street, New York, NY 10001-2291

British Library Cataloguing in Publication Data

Early Precambrian basic magmatism.
1, pre-Cambrian strata. Magmatism
I. Hall, R.P. II. Hughes D.J.
551.71

ISBN-13: 978-94-010-6666-2   e-ISBN-13: 978-94-009-0399-9
DOI: 10.1007/978-94-009-0399-9
For the USA, the International Standard Book Number is

0-412-02061-0

Typesetting by Thomson Press (India) Ltd., New Delhi

Preface

Basic magmatic rocks make up approximately three-quarters of the crust of the present day Earth. Because we can observe and study the volcanic products of present day tectonic regimes comprehensively, we can shed light on ancient tectono-magmatic provinces, and thereby deduce the petrogenesis and evolution of the oldest basic rocks. This is the primary objective of this book.

The book was conceived in order to provide a comprehensive review of the basic rocks produced during the first half of the Precambrian, i.e. the Archaean and early Proterozoic, to about 1.8 Ga years ago. Two major questions are addressed. First, what basic magmas were generated during the early Precambrian: were these magmas globally uniform, and to what extent were prevailing tectonic controls and compositions analogous to those of the present day? Clearly, this can be answered only by bringing together fundamental information about all relevant basic magmatic events. Second, is there any systematic temporal variation in the nature of basic suites, and what implications might such variations have on our interpretations of early Earth history? Are there important differences between early Archaean, late Archaean, Proterozoic and modern basic magmatic suites? The book uses two approaches to address these questions. Early chapters examine the fundamental characteristics of these basic rocks, whilst later chapters assess regional distribution and development by providing an overview of each major early Precambrian craton.

There is considerable evidence for the rapid growth of continental crust towards the end of the Archaean (c. 2.5 Ga). This was a diachronous event and associated basic magmatic activity also responded, but at different rates and in different ways throughout the world. For this reason we decided to include the early Proterozoic under our umbrella term 'early Precambrian'.

Despite summarising the wealth of data and the intensity of geological research on early Precambrian basic rocks, this book probably raises as many new questions as have been answered. Nonetheless, we hope that it represents a comprehensive account which will provide further stimulus for future work into the occurrence, character, origin and development of all types of early Precambrian basic rocks, and that ultimately this will lead to a better understanding of early Earth history.

This book is a contribution to IGCP projects 217 (Proterozoic geochemistry), 257 (Precambrian mafic dyke swarms) and 280 (The oldest rocks on Earth).

RPH
DJH

Contents

7 Lunar magmatism   136
S.B. SIMON

8 Mineralisation associated with early Precambrian basic magmatism   157
S. ROBERTS, R.P. FOSTER and R.W. NESBITT

PART II   REGIONAL SYNTHESES

9 Early Precambrian basic rocks of the USA   191
G.L. SNYDER, R.P. HALL, D.J. HUGHES and K.R. LUDWIG

10 Early Precambrian basic rocks of the Canadian Shield   221
P.C. THURSTON

17 Early Precambrian basic rocks of South America   379

K.R. WIRTH, E.P. OLIVEIRA, J.H. SILVA SÁ and J. TARNEY

Contributors

M.J. Bickle Department of Earth Sciences, University of Cambridge, Downing Street, Cambridge CB2 3EQ, UK.

T.S. Brewer Department of Mining Engineering, University of Nottingham, University Park, Nottingham NG7 2RD, UK.

A.C. Cattell Department of Geology, University of Exeter, North Park Road, Exeter EX4 4QE, UK.

K.C. Condie Department of Geoscience, New Mexico Institute of Mining and Technology, Socorro, New Mexico. 87801, USA.

R.P. Foster Department of Geology, University of Southampton, Southampton SO9 5NH, UK.

R.P. Hall Department of Geology, Portsmouth Polytechnic, Burnaby Road, Portsmouth PO1 3QL, UK.

C.J. Hatton Institute for Geological Research on the Bushveld Complex, University of Pretoria, Hillcrest, Pretoria, South Africa.

D.J. Hughes Department of Geology, Portsmouth Polytechnic, Burnaby Road, Portsmouth PO1 3QL, UK.

B.-M. Jahn Centre Armoricain d'Etude Structurale des Socles, Institut de Géologie, Université de Rennes, 35042 Rennes Cedex, France.

K.R. Ludwig U.S. Geological Survey, Federal Center, Denver, Colorado 80225, USA.

R.W. Nesbitt Department of Geology, University of Southampton, Southampton SO9 5NH, UK.

E.P. Oliveira Department of Geology, University of Leicester, University Road, Leicester LE1 7RH, UK.

T.C. Pharaoh British Geological Survey, Keyworth, Nottingham NG12 5GG, U.K.

A.C. Purvis Pontifex and Associates, 26 Kensington Road, Rose Park 5067, South Australia.

S. Roberts Department of Geology, University of Southampton, Southampton SO9 5NH, UK.

J.H. Silva Sá Department of Geology, University of Leicester, University Road, Leicester LE1 7RH, UK.

S.B. Simon   South Dakota School of Mines and Technology, 501 E. St. Joseph Street, Rapid City, South Dakota 57701-3995, USA.
(present address: Department of Geophysical Sciences, University of Chicago, 5734 South Ellis Avenue, Chicago, Illinois 60637, USA)

H.S. Smith   Department of Geochemistry, University of Cape Town, Rondebosch, Cape, South Africa.
(present address: Rio Diamond CC, PO Box 280, Kleinsee, Cape, South Africa)

G.L. Snyder   U.S. Geological Survey, Box 25046, M.S.913, Federal Center, Denver, Colorado 80225, USA.

J. Tarney   Department of Geology, University of Leicester, University Road, Leicester LE1 7RH, UK.

R.N. Taylor   Department of Geology, University of Southampton, Southampton SO9 5NH, UK.

P.C. Thurston   Ontario Geological Survey, 77 Grenville Street, Toronto, M7A 1W4, Canada.

G. Von Gruenewaldt   Institute for Geological Research on the Bushveld Complex, University of Pretoria, Hillcrest, Pretoria, South Africa.

B.L. Weaver   School of Geology and Geophysics, University of Oklahoma, 830 Van Vleet Oval, Norman, Oklahoma 73019, USA.

K.R. Wirth   Department of Geological Sciences, Cornell University, Ithaca, New York, 14853-1504, USA.

Symbols and Abbreviations

Rock types

BIF	banded iron-formation
CAB	calc-alkali basalt
IAT	island are tholeiite
HMB	high-Mg basalt
MORB	mid-ocean ridge basalt (N-MORB: normal; E-MORB: enriched; T-MORB: transitional)
OIB	ocean island basalt
SHMB	siliceous high-Mg basalt
STPK	spinifex-textured peridotitic komatiite
VAB	volcanic arc basalt

Minerals

Ab	albite
An	anorthite
En	enstatite
Fs	ferrosilite
Fo	forsterite
Mt	magnetite
Ol	olivine
Pl	plagioclase
Px	pyroxene
X_{Cr}^{spinel}	ionic $Cr/(Cr + Al)$ of spinel
X_{Mg}^{px}	ionic $Mg/(Mg + Fe)$ of pyroxene

Geochemistry

Ga	thousand million years (10^9 years)
Ma	million years (10^6 years)
AFM	Alkali–Fe–Mg wt% oxide proportions
CHUR	chondritic uniform reservoir
ε_{Nd}	Nd epsilon parameter: $10^4 \times [(Nd_i/Nd_{CHUR}^t) - 1]$, where Nd_{CHUR}^t is the $^{143}Nd/^{144}Nd$ ratio of CHUR at the time of formation of the rock
$Fe_2O_3^*$, FeO^*	all Fe expressed as Fe_2O_3, FeO
fO_2	oxygen fugacity
GPa	giga pascal (10 kbars)
HFSE	high field strength elements (Nb, Ti, P, Zr)
LILE	large ion lithophile elements (K, Rb, Sr, Ba, LREE)
La_N	La value normalised to La in average chondrite
$K_{D(PGE)}^{sulphide}$	distribution coefficient (PGE in sulphide phase)
KREEP	K-, REE-, and P-rich lunar rocks
mg'	Mg number, $Mg/(Mg + Fe)$
Nd_i	initial $^{143}Nd/^{144}Nd$ ratio
NME	normalised multi-element plot (spider diagram)
P	pressure (GPa)
PGE	platinum group elements (Ru, Rh, Pd, Re, Os, Ir, Pt)
ppm	parts per million
REE	rare-earth elements (H: heavy; L: light; M: middle)
REE_N	chondrite-normalised REE values
Sr_i	initial $^{87}Sr/^{86}Sr$ ratio
T	temperature (°C)
T_p	potential temperature

1 Introduction: basic magmatism and crustal evolution

R.P. HALL and D.J. HUGHES

The bulk of the crust of the Earth comprises, and probably always has comprised, igneous rocks which are basic or basic derivative in character, and a vast amount of scientific effort has gone into understanding their petrogenesis over the past fifty years. Why, then, distinguish the basic magmatism of the early Precambrian for particular attention? In a sense the reasons are obvious and simple. At that time, the Earth was a hotter body than now, more capable of producing higher temperature magmas. The brittle plate tectonics which control modern magma genesis were possibly not so prevalent, and entirely different processes may have operated. The present differentiation between oceanic and continental crust and their attendant lithospheres was probably less well defined. All of these factors influenced and contributed to differences in the composition, style and volume of basic magmatism in the early Precambrian compared with what we see today. Understanding the mechanisms and details of the various processes must bring a clearer insight into the evolution of the mantle and thus to the processes which produce basic magmas – both old and modern. It also contributes to our knowledge of how and when continental crust differentiated and developed its modern characteristics, and to how and why much of the metallogenesis of the early Earth took place.

The text has two distinct parts: Chapters 2 to 8 present a thematic view of various aspects of early basic magmatism, whereas the remaining nine chapters are regional syntheses, cataloguing the distribution and petrological characteristics of basic magmatism in the major Precambrian cratons of the world (Figure 1.1). The second part of the review is not balanced or fully comprehensive – but nor is the available information. For example, perhaps one of the most glaring omissions is the lack of reference to the Aldan Shield of eastern Siberia (Kazansky and Moralev, 1981; Bibikova, 1984). In some chapters, where the particular emphasis is on the Archaean alone rather than including the early Proterozoic, again this generally reflects the balance of available information.

What is the significance of 'Early Precambrian' rather than, say, 'Archaean' in the title? This simply reflects a pragmatic approach to what the geological record presents. For a variety of good reasons, early Precambrian stratigraphers can happily place the Archaean–Proterozoic boundary at or about 2.5 Ga (depending on the craton). However, discussions on the temporal geochemical changes in basalts (rather than komatiites, which are almost exclusively Archaean) would be pointlessly restricted by choosing 2.5 Ga as a limit (Condie, Chapter 3). Similarly, taking this specific age as an upper limit would make discussions about major layered intrusions rather difficult as, for example, the Bushveld Complex (2.05 Ga) would have to be excluded, although this complex is clearly related to processes which became significant in the late Archaean

Figure 1.1   Sketch map showing the distribution of the major early Precambrian shields of the world (cross-hatching: Archaean; stipple: early Proterozoic). Map compiled from Condie (1982) and authors of Chapters 9 to 17.

(Hatton and Von Gruenewaldt, Chapter 4; Roberts *et al.*, Chapter 8). There is much evidence to suggest that the Archaean–Proterozoic transition (as an interval of time, rather than the 2.5 Ga boundary) was a period of profound magmatic and tectonic change, marked by a rapid increase in the rate of continental crustal growth, by the inception of modern plate-tectonic processes and by a clear change in the character of basic magmatism. The production of silica-poor ultramafic komatiites stopped and the newly thickened and extended continental crust was intruded by large volumes of more silicic noritic magmas which, when mixed with contemporaneous tholeiitic magmas, formed the major layered intrusive complexes that characterise this period of time. So we are flexible; 'Early Precambrian' has no particular limit although it broadly corresponds to the Archaean and early Proterozoic (> 1.75 Ga).

1.1   Early thermal and magmatic history of the Earth

Arndt (1987) commented on the interpretation of the geochemistry of komatiites, pointing out the problems in unravelling the effects of contamination and the consequent difficulties in determining the nature of source compositions. He used the apposite phrase 'through a dirty window' to describe our view of the Earth's early mantle as revealed by the geological record. In reality the situation is worse than that. The windows are indeed dirty, but they also only look out on part of the view. There are many aspects of the early history of the Earth of which we have virtually no view at all. For example, from the time of condensation or aggregation from the solar nebula to about 4.3 Ga we have no record whatsoever. Indeed, we know vastly more about the magmatic history of the Moon than we do of the Earth between 4.6 and 3.8 Ga, which was the period of highest terrestrial planetary heat production, when the effects of decay of the highly energetic short-lived nuclides ^{244}Pu and, more particularly, ^{26}Al (Figure 1.2), combined with gravitational heating.

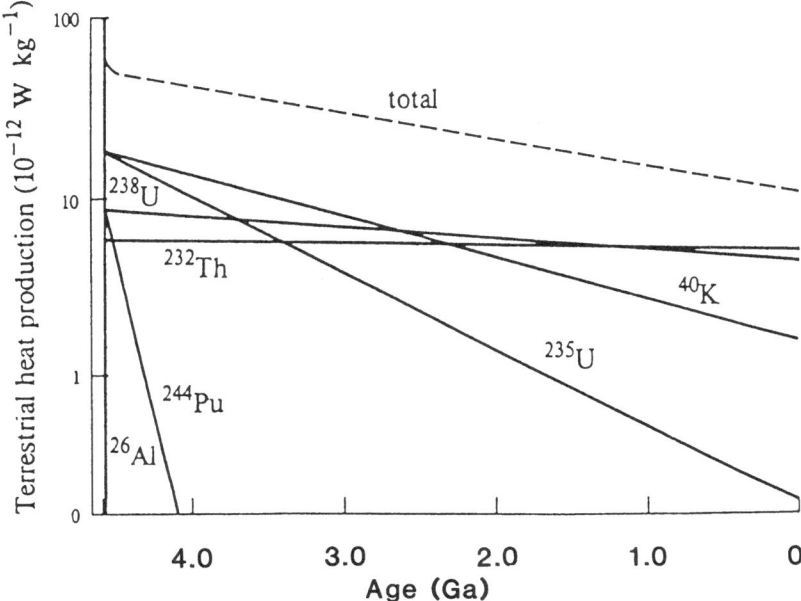

Figure 1.2 Heat production from the significant radioactive nuclides available at the Earth's formation. K, Th and U values extrapolated from present crust and mantle abundances. Al and Pu values estimated from nucleosynthesis theory using stable isotopes in the Allende meteorite (after O'Nions *et al.*, 1978).

Simon (Chapter 7) shows that the Moon provides us with a very clear example of how basic magmatism might have evolved under these conditions. Intense early heating led to the formation of a lunar magma ocean at around 4.6 Ga, the fractionation and solidification of which produced the bedrocks of the lunar highlands. Bombardment by residual nebula debris produced impact melts and highland regolith material between 4.6 and 3.8 Ga and, finally, the mare basalts formed by partial melting of heterogeneous cumulates of the former magma ocean. However, although this is an example of how basic magmatism developed in a terrestrial planet, it is very unlikely to be an appropriate model for the Earth's early history because of the intrinsic geochemical differences between the two planets.

Bickle (Chapter 6) points out that the evidence against substantial early Precambrian mantle degassing, combined with the lack of large-scale fractionation of many trace elements in the mantle, mitigates strongly against there ever having been a major terrestrial mantle melting event, a magma ocean (cf. Nisbett and Walker, 1982). However, the geological record suggests that at least part of the upper mantle may have undergone significant melting during the first 100 Ma of the Earth's history. Similarly, the total heat supply and the nature of the Moon's accretion and fractionation did not allow it to continue as a dynamic system after about 3 Ga, whereas the Earth was by this time clearly producing a wide range of basic magmas from the largely unfractionated silicate mantle which had efficiently separated from the core. In contrast, the last major event on the Moon was very different, producing mare basalts by partial melting of a highly fractionated source. The one aspect that must have been common to both planets was intense early meteoritic bombardment and cratering. But this event had largely finished by 3.8 Ga and any record of it has been lost on the Earth, save for the evidence of the occasional late major impact, of which the Sudbury event (1.85 Ga) was possibly one (Hall and Hughes, Chapter 5).

1.2   Early Precambrian oceans and greenstone belts

Cattell and Taylor (Chapter 2) rightly emphasise the predominance and importance of komatiites and related rocks in the Archaean record. As both they and Bickle (Chapter 6) point out, the early Archaean (pre-3.0 Ga) equivalents of ocean basins must have been dominated by ultramafic lavas which would have led to a rapid overturn in (at least) upper mantle material and must have been the major source of terrestrial heat loss. However, we now see almost none of these ancient oceanic rocks. Perhaps the only candidates are the 3.4–3.1 Ga Malene supracrustal rocks in southern West Greenland (Hall et al., Chapter 11) where komatiitic and tholeiitic metavolcanics occur with abundant slices and sheared fragments of harzburgite. However, these rocks are isoclinally folded together with slightly younger acid gneisses and are generally at amphibolite facies. There are no komatiitic ophiolites, no unequivocal sections through the Earth's early oceanic crust – but perhaps we could hardly expect that.

Of course, what we do find is that the early Precambrian komatiites, komatiitic basalts and tholeiites occur mainly in so-called greenstone belts. There is considerable discussion in the following chapters (and much more elsewhere: see Condie, Chapter 3 and Smith, Chapter 16, Table 16.4) about the nature and significance of greenstone belts. Were they entirely ensialic? Were they equivalent to back-arc basins with incipient oceanic crust? Was true subduction involved? The extent to which greenstone belts represent contracted minor oceans is the subject of much controversy, and Smith (Chapter 16) comments on recent suggestions that the Barberton greenstone belt in the Kaapvaal Craton of South Africa might be an ophiolite. Without doubt there are greenstone belts and greenstone belts, but all too often fragmentary preservation and metamorphic overprinting obscures their significance. There is certainly hope in the better preserved examples and, for example, Thurston (Chapter 10) shows that in the Archaean Superior Province in Canada it is possible to recognise four distinct lithological associations within the different greenstone belt assemblages: shallow platform sequences, oceanic sequences, arc volcanism and pull-apart basins. However, one of the major problems which still remains in interpreting virtually every greenstone belt is the nature of their original base. Recent geophysical evidence suggests that greenstone belts may extend at most down to 6 km (Smith, Chapter 16). Critical basal contacts are all too often obscured, sheared or are simply tectonic. Perhaps the only unequivocal example of a greenstone belt in unconformable contact with underlying gneissic basement is the 2.6 Ga Belingwe (now Mberengwa) greenstone belt in Zimbabwe (Smith, Chapter 16). However, it seems clear from geochemical evidence that association with continental crustal rocks of some sort is a prerequisite for most greenstone sequences. It is interesting to note that the oldest preserved rocks of all, those at Isua in West Greenland (3.8 Ga; Moorbath and Taylor, 1981), comprise a sediment-dominated supracrustal sequence and, indeed, one which has no komatiites among its volcanics (Gill et al., 1981). The 4.3–4.1 Ga zircons from the Yilgarn Block in Western Australia (Froude et al., 1983) were presumably derived from acid differentiates of some sort. Thus, although the nature and size of the early Archaean 'continent' at Isua is totally unknown and, as Moorbath (1983) points out, 'four zircons do not make a continent', evolved continental crustal material was clearly being produced at or about 4.0 Ga and was available to support the earliest greenstone belts.

The oldest greenstone belts (> 3.0 Ga) are only well preserved in the 3.55 Ga Warrawoona Group in the Pilbara Block in Western Australia (Purvis, Chapter 14) and in the 3.46 Ga Barberton greenstone belt (Smith, Chapter 16). In these, a familiar stratigraphic sequence is found in which basal komatiites are followed by tholeiitic basalts, their derivatives and various sediments. In other cratons, the oldest belts are fragmented at best. Remnants of komatiite–tholeiite sequences occur in the 3.6–3.3 Ga Minnesota River gneiss terrane in the USA (Snyder et al., Chapter 9), in a variety of > 3.0 Ga sequences in the Zimbabwe and Gabon Cratons in Africa (Smith, Chapter 16), in the Yilgarn Block in Australia (Purvis, Chapter 14) and possibly in the 3.4–3.0 Ga Sargur Group of the Dharwar Craton in India (Weaver, Chapter 15) and the > 3.5 Ga Imataca Complex in the Venezuelan part of the Amazonian Craton in South America (Wirth et al., Chapter 17). Elsewhere, such as in the 3.5 Ga Qianxi Group in Hebei Province of eastern China (Jahn, Chapter 13) and in the > 3.2 Ga portions of the Singhbhum and Aravalli Cratons in India (Weaver, Chapter 15), amphibolite enclaves in gneisses are all that can be recognised. The early geological record is hardly a comprehensive one on which to base modelling for the > 3.0 Ga Earth.

After 3.0 Ga the geological record expands dramatically. Virtually every early Precambrian craton has extensive greenstone belt terranes of between 3.0 and 1.8 Ga, often with very well preserved stratigraphies. Many Archaean ones, such as the 2.7 Ga Abitibi Belt in the Superior Province of Canada, are dominated by komatiites in their lower portions (Thurston, Chapter 10), whereas early Proterozoic ones, such as the 2.1 Ga Jatulian succession in northern Norway (Brewer and Pharaoh, Chapter 12), are predominantly tholeiitic. To a large extent this variation is time dependent. There was a marked decline in the rate of komatiite production with the approach of the Archaean–Proterozoic boundary and after 2.5 Ga komatiitic rocks are very rare in most greenstone belts. However, they are not totally absent. For example, both the 2.1 Ga Rio Itapicuru greenstone belt in the São Francisco Craton and the 2.1 Ga Inini belt in the Amazonian Craton of South America contain komatiitic rocks (Wirth et al., Chapter 17).

The nature of the associated derivative rocks in the stratigraphically higher parts of greenstone belts also appears to have changed with time. Older sequences tend to have evolved members with 'arc-like' characteristics, whereas younger ones tend to have more obviously calc-alkaline members (Condie, Chapter 3). There is a general progression towards the bimodal basalt–rhyolite association which characterises so many Proterozoic greenstone belts (Pharaoh et al., 1987). However, this is neither a simple progression nor a universal rule. There are many instances of Archaean greenstone belts with extensive, highly evolved components. For example, many of the Dharwar greenstone belts in India have metavolcanic sequences ranging from basalt through andesite to rhyolite, with only a rare komatiitic component (Weaver, Chapter 15).

The rather enigmatic term 'greenstone belt' sometimes becomes less than appropriate for some of the more well-preserved sequences of basic volcanic rocks, particularly in the late Archaean and early Proterozoic. For example, the basalts and komatiites of the Hamersley Basin in Western Australia (Purvis, Chapter 14) and the Ventersdorp

Supergroup in South Africa (Smith, Chapter 16) were clearly erupted onto fully stabilised crust and are the equivalents of within-plate flood basalts. Elsewhere in the early Proterozoic, basaltic rocks occur in the earliest complete and unequivocal examples of modern plate-tectonic style orogenic belts. Of these, the 2.0–1.8 Ga Wopmay Orogen on the western side of the Archaean Slave Province in Canada is perhaps the best known (Hoffman, 1980), although there are many others, for example those in the 2.0–1.75 Ga Svecofennian of Scandinavia (Brewer and Pharaoh, Chapter 12).

Many of these changes are reflected much more subtly in the geochemistry of the basic rocks and Cattell and Taylor (Chapter 2) and Condie (Chapter 3) in particular discuss the nature of temporal geochemical variation in the basic rocks of the early Precambrian. In general, komatiites fall into two groups: those which are Al-depleted and those which are Al-undepleted. 3.0–2.5 Ga komatiites (Munro-type) tend to be Al-undepleted but are LREE-depleted. However, although > 3.0 Ga komatiites tend to be Al-depleted, the temporal pattern falls down rapidly. Of the two well-preserved early Archaean examples, those from Barberton are not LREE-depleted whereas those of similar age from Pilbara are strongly depleted. However, it is emphasised that this might well point to the paucity of statistically meaningful information for the > 3.0 Ga Earth rather than to the lack of real temporal change.

A range of siliceous high-magnesium basalts (SHMB) which often show LREE enrichment and negative ε_{Nd} begin to appear as flows and minor intrusions in late Archaean greenstone belts. The best documented of these occur in the Yilgarn Block in Western Australia (Cattell and Taylor, Chapter 2, and Purvis, Chapter 14). Cattell and Taylor do not distinguish these rocks from komatiitic basalts and they favour a simple model for their origin in which komatiitic magma is contaminated by continental crustal material. With Purvis, they point to the evidence provided by the presence of 3.5 Ga zircons, presumed to represent an old sialic crustal contaminant, in the 2.7 Ga flows at Kambalda in the Yilgarn Block. However, Purvis discusses the difficulties in applying the crustal contamination model universally and suggests the possibility of a contaminated lithospheric source in some cases.

Tholeiitic basalts are widespread in every early Precambrian volcanic environment. However, Condie (Chapter 3) points out that what we see in the record is very unlikely to represent statistically what was erupted, because of inevitable preservational bias. In particular, early Precambrian basalts with oceanic (MORB) and within-plate basalt characteristics are rare. Most early Precambrian basalts have a subduction-zone component. Those from the Archaean tend to have island arc affinities whereas those from the Proterozoic tend to be similar to modern calc-alkaline suites. If the geological record is at all statistically reliable, this geochemical variation presumably reflects the nature of the greenstone belts in which most of these basalts occur and the increasing maturity of the continental crust with which they are associated.

1.3 Continental crust and the major basic intrusions

The rate of growth of continental crust has been the subject of much speculation. Some authors have preferred an early, rapid growth model, with continental crust broadly achieving its present extent prior to the earliest geological record, whereas others prefer a slower model, with close to the present crustal extent not being reached until the Archaean–Proterozoic transition (Figure 1.3). Much of this is speculation indeed.

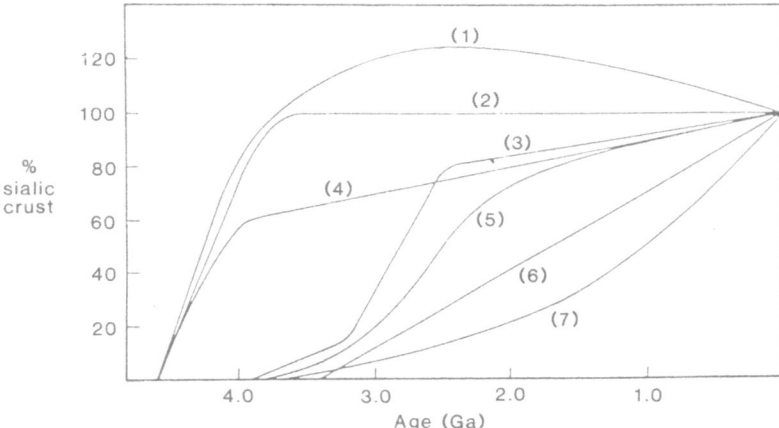

Figure 1.3 A selection of crustal growth models. 1: Fyfe (1978); 2: Armstrong (1981); 3: McLennan and Taylor (1982); 4: Reymer and Schubert (1984); 5: Veizer and Jansen (1979); 6: Hurley (1968); 7: Hurley and Rand (1969) (after Taylor and McLennan, 1985).

Without doubt, the earliest rocks have continental crustal affinities. Those from Isua in West Greenland (3.8 Ga) are perhaps less certain (Nutman *et al.*, 1984). The 3.8 Ga rocks from the Western Gneiss Belt in the Yilgarn Block in Western Australia, which also contain 4.3–4.0 Ga zircons (Froude *et al.*, 1983) are more convincing. Even so, the geological evidence is very thin on which to invoke global-scale continents.

What the geological record does show is that with the proliferation of preserved late Archaean greenstone belts there is a similar increase in the volume of related granitoids of the tonalite–trondhjemite–granodiorite (TTG) association (e.g. Brewer and Pharaoh, Chapter 12; Jahn, Chapter 13). The way in which these are interpreted in relation to adjacent greenstone belts depends upon the choice of model, but the age relationships are in little doubt. The significance of this as far as basic magmatism is concerned is that TTG granitoids and greenstone belt volcanics and sediments accreted during the late Archaean and early Proterozoic to form thick, rigid continental crust. This crust was then able to support the voluminous injections of basic magmas represented by the late Archaean–early Proterozoic large layered intrusions and extensive dyke swarms (Hatton and Von Gruenewaldt, Chapter 4; Hall and Hughes, Chapter 5). The large intrusions, such as the 2.7 Ga Stillwater Complex in Montana, USA, the 2.45 Ga Great Dyke in Zimbabwe, and the 2.05 Ga Bushveld Complex in South Africa, are described by Hatton and Von Gruenewaldt in terms of complex magma mixing models involving at least an ultramafic and a basaltic component. Hall and Hughes describe associations of dykes, such as the BN and MD swarms in West Greenland, in which discrete norite and tholeiite dykes occur. All of these authors discuss whether the distinctive siliceous, high-Mg and yet high-K and high-LREE signature of the ultramafic component is derived predominantly from crustal assimilation or, as they prefer, from a contaminated mantle source.

This major magmatic event, which does not seem to be represented in intrusions much younger than 2.0 Ga, probably marks the real transition between the basic magmatism of the early Precambrian and that of modern times.

Part I
General Aspects

2 Archaean basic magmas

A.C. CATTELL and R.N. TAYLOR

2.1 Distribution of Archaean basic volcanic suites

This chapter discusses Archaean basic magmatism, concentrating in particular on komatiite, komatiitic basalt and tholeiitic basalt suites. Archaean tholeiites are chemically distinct from their more modern counterparts and they are the most abundant magma type represented in greenstone belts, and thus they warrant significant attention. Several magma types are not discussed in this chapter since, although they were erupted during the Archaean, they are commonplace in more recent assemblages. These basalt types include those from typical continental bimodal suites (e.g. Bowen *et al.*, 1986), calc-alkaline (high-Al) basalts (e.g. Barley *et al.*, 1984) and alkaline basalts (e.g. Cooke and Moorhouse, 1969). Also excluded from this account is a discussion of the parental magmas of Archaean gabbro–anorthosite complexes (Weaver *et al.*, 1981; Windley *et al.*, 1981), which are described in Chapter 4. Representative analyses of rocks discussed in this chapter are presented in Table 2.1.

2.1.1 *Greenstone belts*

Mafic and ultramafic lavas form a significant proportion of most low-grade Archaean supracrustal successions (greenstone belts). In some belts such as the 3.5 Ga Barberton belt of South Africa (Anhaeusser, 1971), such lavas are concentrated towards the base of the succession, whilst in others such as the Abitibi belt of Ontario, Canada (Jensen and Pyke, 1982) they occur sporadically throughout the supracrustal belt succession. Associated intrusions are commonly derived from magmas of similar composition to the mafic and ultramafic lavas of these belts (Williams and Hallberg, 1973; Naldrett and Mason, 1968; McRae, 1969; Weaver *et al.*, 1981).

The common feature of mafic–ultramafic successions in greenstone belts is the wide variety of magmas represented within relatively thin stratigraphic sequences. The three predominant lava types (komatiite, komatiitic basalt and tholeiitic basalt) frequently occur together and each type is often divisible into two or more sub-types on the basis of geochemistry. Two well-studied greenstone belt examples are used below to illustrate the major features of typical mafic–ultramafic volcanic sequences.

The 2.7 Ga succession at Kambalda, Western Australia, has been intensively studied because of the widespread association of Ni sulphide mineralisation (Gresham and Loftus-Hills, 1981). The Kambalda sequence is deformed and metamorphosed, but igneous structures and textures are preserved. The lower part of the Kambalda mafic–ultramafic succession consists of over 2000 m of pillowed and massive tholeiitic basalts and minor intercalated sediments. The metabasalts comprise a chemically homogeneous series of tholeiites characterised by flat REE patterns and chondritic Ti-Zr-Y

Table 2.1 Representative whole rock analyses

	1	2	3	4	5	6	7	8	9	10	11	12	13	14
SiO_2	46.62	47.59	55.94	50.95	48.45	51.60	50.58	47.17	52.14	49.45	47.25	54.08	53.67	45.75
TiO_2	0.78	0.32	0.84	1.20	0.33	0.55	0.81	0.50	0.76	2.45	0.63	0.37	0.19	0.68
Al_2O_3	5.46	3.37	10.15	11.69	7.81	11.01	15.64	7.58	12.89	13.51	16.27	14.73	12.85	12.42
Fe_2O_3	14.78	15.03	10.26	13.41	10.43	10.66	11.33	12.78	11.55	18.28	12.14	8.79	8.49	12.74
MnO	0.25	0.18	0.17	0.26	0.18	0.19	0.18	0.25	0.19	0.29	0.21	0.15	0.15	0.19
MgO	23.45	31.99	9.58	7.57	24.18	15.61	5.96	21.70	9.77	5.99	6.48	9.40	13.47	16.46
CaO	8.49	3.37	9.60	12.46	8.17	8.72	11.24	9.22	9.71	6.79	15.14	10.75	9.75	10.45
Na_2O	0.06	0.09	3.14	2.21	0.39	1.34	3.62	0.19	2.48	2.66	1.18	1.56	0.97	1.17
K_2O	0.02	0.05	0.17	0.10	0.03	0.28	0.60	0.14	0.20	0.20	0.33	0.14	0.45	0.02
P_2O_5	0.07	0.07	0.11	0.15	0.01	0.04	—	0.03	0.09	0.36	—	0.02	0.01	0.06
Sc	—	—	—	—	22	32	47	38	45	50	—	42	42	31
V	208	102	188	301	152	202	293	204	262	566	241	252	217	260
Cr	4100	5020	576	619	3113	2019	375	1327	733	8	424	397	800	1250
Ni	1450	1450	128	156	1267	468	177	354	172	52	209	129	350	720
Sr	12	9	22	138	31	54	188	76	151	75	69	80	88	58
Y	11	4	13	20	9	15	21	14	18	38	28	12	9	14
Zr	41	17	56	90	20	57	51	34	59	149	26	24	19	30
Nb	—	—	2	6	<0.9	3	1.2	2.6	3.7	—	3	<1	<1	0.5
La	1.03	3.38	5.50	9.32	—	7.46	2.56	1.12	5.15	21.3	0.7	1.61	1.41	0.56
Ce	2.39	—	14.09	20.38	2.04	14.09	7.28	3.22	12.28	—	2.9	3.18	2.50	2.10
Sm	1.23	0.95	2.25	3.44	0.74	1.78	2.09	1.12	2.18	4.91	1.4	0.18	0.28	1.46
Eu	0.32	0.36	0.71	1.12	—	0.58	0.79	0.49	0.76	1.82	0.54	0.35	0.12	0.58
Gd	1.90	1.21	2.83	3.89	1.03	2.12	2.84	1.62	2.66	6.17	2.0	—	—	1.93
Yb	0.96	0.57	1.61	2.07	0.79	1.41	2.15	1.08	1.77	4.10	2.48	1.18	0.91	1.36
Lu	0.14	0.08	0.25	0.31	0.12	0.21	0.31			0.53	0.45	0.20	0.15	0.21

All analyses recalculated to 100% volatile free, total Fe as Fe_2O_3

1. Komatiite, 3.5 Ga, Talga Talga, Pilbara (Gruau, 1987)
2. Komatiite, 3.5 Ga, Onverwacht (Jahn et al., 1982)
3. Komatiite, 3.5 Ga, Onverwacht (Jahn et al., 1982)
4. Tholeiitic basalt, 3.5 Ga, Onverwacht (Jahn et al., 1982)
5. Komatiite, 2.7 Ga, Kambalda (Arndt and Jenner, 1986)
6. Komatiitic basalt, 2.7 Ga, Kambalda (Arndt and Jenner, 1986)
7. Tholeiitic basalt, 2.7 Ga, Kambalda (Arndt and Jenner, 1986)
8. Komatiite, 2.7 Ga, Newton Township (Cattell, 1987)
9. Komatiitic basalt, 2.7 Ga, Newton Township (Cattell, 1987)
10. Fe-rich tholeiitic basalt, 2.7 Ga, Newton Township (Cattell, 1985)
11. Tholeiitic amphibolite, >3.0 Ga?, Malene Series, Buksefjorden, Westsouthern Greenland (Chadwick, 1981)
12. Olivine basalt, Troodos ophiolite, Cyprus, Sample X.35 (c)
13. Olivine basalt, Troodos ophiolite, Cyprus, Sample X.63 (a)
14. Komatiite, Gorgona Island, Colombia, Sample 48 (Dietrich et al., 1981)

proportions (Redman and Keays, 1985). The only variation is a slight decrease in MgO and increase in incompatible element abundances in the uppermost lavas. The middle part of the sequence consists of up to 1000 m of komatiite lava flows, grading up section from thick (10–100 m) flows dominated by massive, olivine-porphyritic lava, to thin flows with well developed olivine spinifex zones (Gresham and Loftus-Hills, 1981). The komatiites all show consistent incompatible element characteristics, having depleted LREE, flat HREE and chondritic Al_2O_3/TiO_2 (c. 20) (Sun and Nesbitt, 1978). Subtle chemical variations may be due to contamination during thermal erosion of underlying basement/sediments, although this interpretation remains controversial (Huppert et al., 1984; Claoué-Long and Nesbitt, 1986; Groves et al., 1986; Arndt and Jenner, 1986). Komatiitic basalts form the upper 600 m of the Kambalda succession. Although all of the komatiitic basalts are LREE-enriched, the lavas in the uppermost part of the sequence demonstrate the greatest degree of LREE enrichment (Arndt and Jenner, 1986).

The 2.7 Ga mafic–ultramafic metavolcanic sequence at Newton Township, Ontario, is part of the Abitibi greenstone belt and contains a more varied set of magma types than occur at Kambalda. As at Kambalda, the mafic–ultramafic sequence commences with tholeiitic basalts, but at Newton most of the metatholeiites are highly differentiated and Fe-rich. These tholeiitic flows are overlain by more Mg-rich tholeiitic basalts, indistinguishable from those at Kambalda. Above the tholeiitic flows is a complex succession of komatiites and komatiitic basalts and, in contrast to the situation at Kambalda, the two lava types are interlayered as a sequence of at least eight units; five of komatiite and three of komatiitic basalt. As an additional complication there are two chemical varieties of komatiites, one similar to the komatiites from Kambalda (LREE-depleted, flat HREE, $Al_2O_3/TiO_2 = 20$), and a second, HREE- and Al-depleted type (Cattell and Arndt, 1987). The komatiitic basalts share the HREE and Al characteristics of the latter type, but are LREE-enriched (Cattell, 1987). Subtle geochemical differences between the sub-types are ascribed to magma-mixing. The combined succession of mafic and ultramafic lavas at Newton Township is less than 2000 m thick, within which at least five, and probably more, distinctly different magma types are represented. However, such diversity within a relatively restricted stratigraphic and, presumably, temporal range does not seem to be unusual. Other examples of highly complex basic lava sequences include those at Munro Township, Ontario (Arndt et al., 1977; Arndt and Nesbitt, 1982, 1984; Canil, 1987) and Diemals, Western Australia (Nesbitt et al., 1984). Detailed work continues to bring to light increasing complexity in sequences originally thought to have been relatively simple and homogeneous (Redman and Keays, 1985).

2.1.2 High-grade gneiss terranes

Most studies of Archaean mafic volcanic suites have been concentrated on those occurring in low-grade greenstone belts, since their better preservation affords more precise interpretations. However, igneous structures are also occasionally well preserved in mafic metavolcanic rocks in high-grade terranes, and work on the geochemistry of these rocks and on comparisons with less metamorphosed suites indicates that the high-grade mafic metavolcanic suites are compositionally similar to those found in greenstone belts. This is illustrated by a brief description of one such well preserved high-grade Archaean supracrustal suite.

The middle Archaean Malene metavolcanic amphibolites in the Archaean craton of southern West Greenland contain well preserved pillow structures. These amphibolites are usually tectonically interleaved with older and younger orthogneisses and therefore their original setting is obscure. Field relationships locally suggest that some may have been erupted on to sialic basement (Chadwick and Nutman, 1979) while elsewhere their primitive chemistry, association with harzburgite and field relationships which indicate that they represent the oldest crustal component of the gneiss terrane have been interpreted as indicating an oceanic origin (Hall *et al.*, 1987a). Their relationship to middle Archaean amphibolite dykes which intruded the early Archaean sialic crust is also not entirely clear. In some areas the dykes and supracrustal amphibolites are chemically similar (Chadwick, 1981), whilst in other areas many of the supracrustal amphibolites are markedly more primitive (Hall *et al.*, 1987a; Gill and Bridgwater, 1979). The Malene metavolcanics are geochemically similar to greenstone belt tholeiitic basalts and komatiites. The komatiitic amphibolites are similar to komatiites from late Archaean greenstone belts (section 2.2) except that they have flat rather than depleted chondrite-normalised LREE patterns (Hall *et al.*, 1987a). The tholeiitic amphibolites either have flat REE patterns typical of flows in late Archaean greenstone belts (section 2.3.1) or are LREE-depleted (Chadwick, 1981; Hall *et al.*, 1987a). The tholeiitic amphibolites share with their greenstone belt counterparts the high Cr and Ni contents typical of Archaean tholeiitic magmas (section 2.3.1). Since they are distinct from modern basalt types in this repect, their geochemistry alone cannot be used to assign a tectonic setting to the Malene tholeiitic amphibolites.

2.2 Komatiites

The term komatiite was introduced twenty years ago by Viljoen and Viljoen (1969a) to describe ultramafic lavas from the Barberton greenstone belt of South Africa. The name was derived from the Komati River and has been widely accepted by other authors and is now in common use. However, with the subsequent recognition of various types around the world, the definition of komatiite is still unclear. A review and discussion of the evolution of komatiite terminology is presented by Arndt and Nisbet (1982a, b). In this work the term komatiite is used for volcanic rocks with greater than 18% MgO (calculated anhydrous), derived from a liquid with greater than 18% MgO; the rider is added in order to exclude basaltic cumulates.

A diagnostic feature of many, though not all, komatiite lava flows is the presence of olivine spinifex texture (Nesbitt, 1971), and a great deal has been published on the significance of this texture (e.g. Nesbitt, 1971; Donaldson, 1976, 1982; Arndt, 1986; Turner *et al.*, 1986). Komatiite lavas are not typically spinifex-textured throughout, but commonly form layered flows with spinifex-textured tops and olivine-phyric bases. In an early description of the well exposed komatiite lava flows of Munro Township, Ontario, Pyke *et al.* (1973) interpreted the basal porphyritic layer as an early formed olivine cumulate, which incorporated all the olivine phenocrysts from the magma and allowed subsequent growth of the olivine spinifex zone down from the flow top into a crystal-free liquid. This interpretation was recently questioned by Arndt (1986) and Turner *et al.* (1986), who concluded that komatiite liquids would flow turbulently and therefore retain olivine phenocrysts in suspension until after the spinifex zone had developed. This interpretation is rather different from an older 'spinifex-first' model (Lajoie and Gelinas, 1978) which envisaged continued flow through the base whilst the spinifex zone was growing from the flow top.

Although studies of komatiites have concentrated on the more spectacular spinifex-textured rocks, somewhat less than half of most komatiite successions consist of flows of this type (e.g. Gresham and Loftus-Hills, 1981). Many komatiite lava flows are thin (< 5 m) and olivine–phyric throughout, the only layering being a slight concentration of olivine crystals toward the centre of the flows (Arndt et al., 1977). Arndt (1986) suggested that these lavas had cooled considerably and crystallised extensively prior to cessation of flow, such that the effective viscosity of the partially crystallised magma was too high to allow convection or crystal settling. The increase in effective viscosity during flow in such lavas could have been sufficient to lead to a transition from turbulent to laminar flow, resulting in a concentration of olivine crystals in the centre of the flows by flow differentiation.

The basal parts of some komatiite sequences are marked by very thick (> 20 m) flows which are almost entirely olivine–phyric (Gresham and Loftus-Hills, 1981; McQueen, 1981; Green and Naldrett, 1981; Cattell and Arndt, 1987). In some localities the thick basal flows contain economic Ni sulphide deposits. The sulphide occurs at the base of the flows, grading upward from massive through network to disseminated ore as the proportion of sulphide to silicate (mostly olivine or its alteration products) decreases. The ore is richest in the lowermost flows although lesser deposits do occur in overlying flows (Chapter 8). In a few cases ore is found in the spinifex zones at the top of flows, apparently introduced by downward melting from the base of the overlying flow (Groves et al., 1986).

The only common phenocryst phase in komatiites is olivine, which is typically in the range Fo_{94}–Fo_{90} (Table 2.2). The equant phenocrystic olivines in these rocks are unzoned or have a very thin rim of slightly less magnesian olivine, while spinifex-textured olivines are continuously zoned from magnesian cores to less magnesian margins (Arndt, 1986). Magnesian chromite appears on the liquidus in the less magnesian komatiites, but does not occur as a phenocryst phase. Olivine-phyric komatiites are depleted in Cr relative to spinifex-textured komatiites, and therefore did not accumulate chromite with olivine (cf. komatiitic basalts, section 2.4). Sprays or parallel arrays of Al-rich magnesian augite needles occur in the interstices of olivine spinifex zones, and augite also occurs as stubby grains or occasionally as oikocrysts in the more slowly-cooled olivine-phyric rocks.

Komatiites are the products of liquids which had greater than 18% MgO (Arndt and Nisbet, 1982b), but although the MgO content of komatiite liquids is thus of prime importance, it is difficult to determine because aphyric flow tops are almost invariably altered and olivine spinifex-textured samples may contain cumulus olivine (Bickle, 1982; Arndt, 1986; Cattell and Arndt, 1987). Bickle (1982) used olivine compositions to infer liquids with 24–30% MgO for several komatiite suites, while Cattell and Arndt (1987) calculated liquid compositions with 18–22% MgO for komatiites from Newton Township, and Arndt (1986) calculated a liquid composition of 28% MgO for a flow from Munro Township, Ontario. An upper limit of 32% MgO seems likely for komatiite magmas.

Many early papers on komatiite geochemistry deal at length with CaO–Al_2O_3–TiO_2 relationships. This is because the type komatiites from Barberton had high CaO/Al_2O_3 ($\geqslant 1$) and this was included in early komatiite definitions (Viljoen and Viljoen, 1969a; Brooks and Hart, 1974; Arndt et al., 1977). Subsequent work has shown that such komatiites are in fact unusual and that most komatiites have CaO/Al_2O_3 of between 0.8 and 1. Higher CaO/Al_2O_3 appears to be due to Al depletion, which leads also to low Al_2O_3/TiO_2 values ($\leqslant 14$) (Nesbitt et al., 1979). Komatiites with a strong

Table 2.2 Representative mineral analyses

	1 ol	2 ol	3 ol	4 chr	5 cpx	6 cpx	7 cpx	8 cpx	9 cpx
SiO_2	41.8	41.8	40.6	—	49.00	51.60	50.60	48.65	49.03
TiO_2	—	—	—	0.41	0.68	0.35	0.41	1.09	0.66
Al_2O_3	—	—	—	9.41	7.30	3.10	3.05	2.53	1.56
FeO	5.65	7.16	12.00	26.54	7.20	9.48	8.21	17.58	24.17
MnO	—	—	—	0.43	0.00	0.21	0.11	0.33	0.55
MgO	52.10	50.90	46.9	7.92	14.70	17.70	15.96	14.92	9.56
CaO	0.23	0.22	0.16	—	20.40	17.20	19.69	14.37	13.45
Na_2O	—	—	—	—	0.17	0.13	0.69	0.56	0.44
Cr_2O_3	0.19	0.26	0.21	55.97	0.46	—	0.52	0.21	0.00
NiO	0.40	0.41	0.32	—	—	—	—	—	—
Total	100.40	100.80	100.20	100.68	99.91	99.90	99.20	100.25	99.41
Fo	94.1	92.7	87.5	—	—	—	—	—	—
En	—	—	—	—	44.0	50.0	46.0	42.5	29.1
Fs	—	—	—	—	12.1	15.0	13.3	28.1	41.3
Wo	—	—	—	—	43.9	34.9	40.7	29.4	29.5

1. Centre of phenocryst in chilled komatiite flow top (Arndt, 1986)
2. Centre of spinifex blade, komatiite flow (Arndt, 1986)
3. Margin of spinifex blade, komatiite flow (Arndt, 1986)
4. Core of equant grain, B-zone of komatiite flow (Cattell and Arndt, 1987)
5. Skeletal grain from olivine spinifex zone, komatiite flow (Arndt et al., 1977)
6. Subhedral prismatic grain, komatiitic basalt (Arndt et al., 1977)
7. Oikocryst, massive tholeiitic basalt
8. Core of groundmass grain, Fe-rich tholeiitic basalt
9. Rim of groundmass grain, Fe-rich tholeiitic basalt

Total iron expressed as FeO.
ol: olivine; chr: chromite; cpx: clinopyroxene.

depletion in Al are restricted to sequences older than 3.0 Ga (Nesbitt et al., 1982), while most komatiites younger than 3.0 Ga have Al_2O_3/TiO_2 in the range 18 to 22 (Figure 2.1). Almost all the komatiites shown in Figure 2.1 that have Al_2O_3/TiO_2 between 14 and 18 come from the Lower Komatiites unit at Newton Township (Cattell and Arndt, 1987).

Ni contents of komatiites fall continuously from roughly 1500 ppm at 30% MgO to 600 ppm at 18% MgO, due to olivine fractionation. The partition coefficient for Cr between olivine and liquid is less than unity and therefore Cr contents rise from roughly 2500 ppm at 30% MgO to a maximum of around 3000 ppm at 25% MgO, after which chromite crystallisation progressively depletes the liquid in Cr. Cattell and Arndt (1987) calculated that olivine and chromite crystallise together in the proportions 50:1, and pointed out that olivine spinifex-textured lava may contain accumulated chromite, thereby producing Cr-enriched samples which do not represent liquids.

The concentrations of incompatible and partially compatible elements in komatiites reflect the two-fold subdivision into Al-depleted and Al-undepleted types. Komatiites with low Al contents also contain low concentrations of the trace elements V, Sc, Ga, Y and the HREE (Nesbitt and Sun, 1976; Sun and Nesbitt, 1978a, b; Nesbitt et al., 1979; Jahn et al., 1982; Smith and Erlank, 1982; Gruau et al., 1987). Comparison with chondritic abundances suggests that these elements are depleted to decreasing degrees in the order Al > V > Y, HREE (Yb > Gd) > Sc > Ga. The small group of komatiites

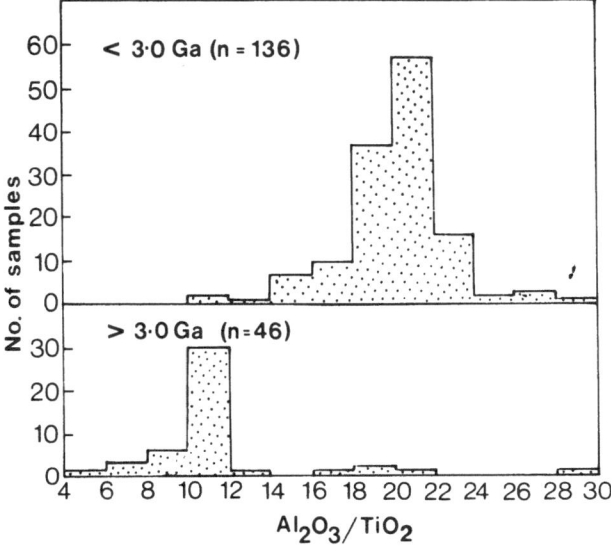

Figure 2.1 Histogram of Al_2O_3/TiO_2 in komatiites (MgO > 18%). Multiple analyses of individual flows are not included. The > 3.0 Ga data set is strongly biased by samples from the Barberton greenstone belt of South Africa.

that show mild Al depletion also show mild depletion in these elements (Cattell and Arndt, 1987). The low Al content of many pre-3.0 Ga komatiites is explained by the removal of garnet from the source prior to melting, although the exact mechanism is not well understood (Ohtani, 1984). The other elements depleted together with Al are all trivalent and relatively compatible in garnet.

Besides the differences related to Al depletion, komatiites also show variation in the relative abundances of the more incompatible elements (Ti, Zr, Nb, LREE). Ti/Zr varies between 60 and 150 in komatiites, though mostly within the range 80–130 (Figure 2.2). There is a slight increase in Ti/Zr from the early to late Archaean rocks, but the significance of this is a little obscure. As would be expected from various elemental equivalences in basic rocks (Zr = Sm, Ti = Eu*) suggested by Sun *et al.*, (1979), there is a rough though imperfect inverse correlation between Ti/Zr and the degree of LREE depletion. Most late Archaean komatiites are LREE-depleted, but while the early Archaean komatiites from Barberton are not LREE-depleted, those of similar age from Pilbara are strongly depleted and hence it is unlikely that there is a general temporal change in the depletion of komatiites or their source. Good quality data are not available for the least compatible elements, either because of analytical difficulties at the low concentrations present (e.g. Nb) or because of clear evidence of post-eruptive mobility (e.g. U, K, Rb, Cs), but it seems likely that these elements would have been even more strongly depleted than the LREE in komatiites relative to chondritic (or 'primordial mantle') abundances.

Platinum group element (PGE) abundances are high in komatiites, indicating that an immiscible sulphide liquid which would efficiently scavenge all the PGE from a silicate magma was not present at any stage in their petrogenesis (Keays, 1982;

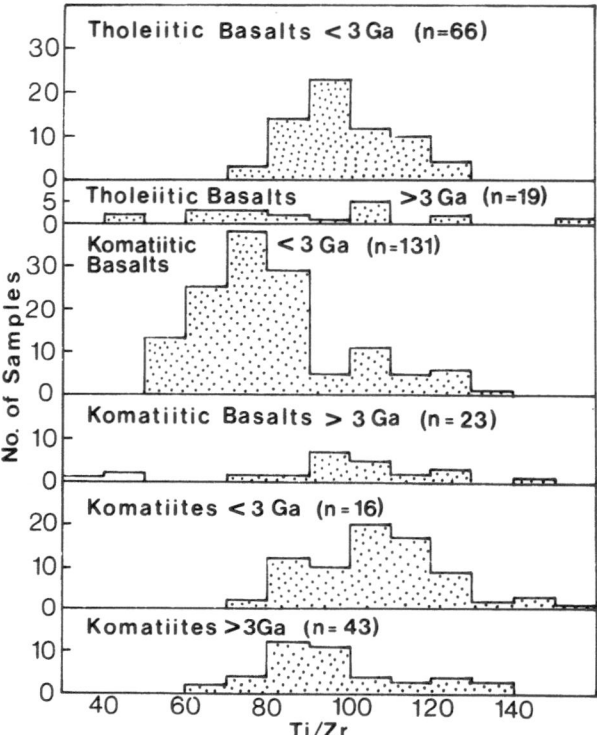

Figure 2.2 Histogram of Ti/Zr relationships in Archaean komatiites, komatiitic basalts and tholeiitic basalts. > 3.0 Ga data sets are strongly biased towards the Barberton greenstone belt samples. The small group of > 3.0 Ga samples with Ti/Zr ≤ 50 are from Barberton and Pilbara, and are also characterised by $Al_2O_3/TiO_2 \geq 20$ and $Gd_N/Yb_N < 1$.

Crockett and McRae, 1986). Pd/Ir ratios are extremely low relative to those found in basaltic rocks, reflecting the incorporation of refractory Ir into the large-percentage melts produced during komatiite genesis (Keays, 1982). During fractionation, there is a general increase in Pd and a decrease in Ir (and Os and Ru) in the liquid. The compatibility of Ir is puzzling, since its chemistry suggests that it would be highly incompatible in olivine, the only observed phenocryst (Barnes *et al.*, 1985). However, Brugmann *et al.* (1987) argued that Ir should be compatible in olivine, and that its partitioning into olivine would explain Ir variation in komatiite suites. They also argued that the main problem in accounting for PGE distribution in komatiite suites is to explain why Pd and Pt do not partition strongly in olivine.

2.3 Tholeiitic basalts

Archaean tholeiitic basalts occur most commonly as pillowed lava flows, although massive or weakly layered flows are also common. Volcaniclastic material is rare and restricted to breccias and hyaloclastites, which are produced by break up during flow and thermal contraction, rather than by explosive expansion of entrained volatiles. The flows are usually mildly amygdaloidal. Dimroth *et al.* (1978) have investigated the structure of the basaltic lavas and have discussed possible palaeogeographic interpretations. Except where they are exceptionally thick (e.g. Theo's Flow, Munro Township; Arndt, 1977a), tholeiitic flows lack the compositional layering which is characteristic of their komatiitic basalt and komatiite counterparts.

Archaean tholeiitic basalts are typically only sparsely porphyritic. Rapidly cooled lavas commonly contain a few percent of plagioclase and augite microphenocrysts, often with quenched overgrowths (Gelinas and Brooks, 1974), while more slowly cooled, massive lavas often have an ophitic texture, except in high-Fe flows where textures tend to be equigranular (Cattell, 1985). Olivine occurs only rarely. Arndt *et al.* (1977) noted its absence from the rather Fe-rich tholeiites of Munro Township, and Gelinas and Brooks (1974) recorded it in only one sample from sections in Quebec (their Mg-rich tholeiites would here be grouped with the komatiitic basalt series). It should be noted that in most cases the phenocrysts are inferred from pseudomorphs and therefore the identification of ferromagnesian phases (e.g. olivine, augite, orthopyroxene, pigeonite) is often uncertain. Calcic pyroxenes are commonly the only preserved minerals, both as phenocrysts and groundmass grains, and also as oikocrysts in slowly cooled lavas. The clinopyroxene compositions reflect those of the host basalts, ranging from $En_{45}Fs_{14}Wo_{41}$ in magnesian tholeiites (mg' = 0.58) to $En_{39}Fs_{26}Wo_{35}$ in Fe- rich tholeiites (mg' = 0.37), while extreme zoning in grains in the latter lavas leads to pyroxene compositions at least as Fe-rich as $En_{17}Fs_{53}Wo_{30}$ (Cattell, unpub. data).

The major element chemistry of Archaean tholeiitic basalts is broadly similar to that of modern low-K tholeiites (Condie, 1985). The most significant difference is the relatively high Fe content of Archaean lavas, matched inevitably by a high Mn content. In this context, high Fe refers to high Fe at a particular MgO content in fairly primitive lavas, and not to the strongly fractionated Fe-rich tholeiitic basalts common to many Archaean sequences. The high Fe content leads to problems in comparing Archaean and modern basalts, since Archaean basalts at the same degree of fractionation as their modern counterparts (as measured by mg') and with equivalent Al contents have 1.11 to 1.23 times greater concentrations of Fe and Mg (Condie, 1985). The high Fe and Mg contents give rise to the crystallisation sequence: olivine (+ chromite), (Cr-bearing) augite, augite + plagioclase, augite + plagioclase + Fe-Ti oxide. Some orthopyroxene may also be present. This sequence is recorded most clearly in layered tholeiitic sills (e.g. McRae, 1969). One consequence of the late crystallisation of plagioclase is that mg' values decline rapidly relative to the increase in incompatible element abundances. The net effect is that Archaean basalts have relatively low incompatible element concentrations at a particular mg' value. This accounts for the lower concentrations of TiO_2 and other incompatible elements in Archaean basalts (Condie, 1985). Any difference in incompatible element concentrations inherited from the source is superimposed on this effect. However, incompatible element *ratios* are unaffected.

Strongly fractionated tholeiitic basalts are common in many Archaean volcanic sequences (Arndt *et al.*, 1977; Naldrett and Turner, 1977). Fe-Ti oxides typically appear late in the fractionation history of Archaean tholeiitic basalts, allowing the attainment of very high Fe contents ($> 20\%$ $Fe_2O_3^*$) (Naldrett and Turner, 1977). The high Fe contents of these evolved lavas are due to a combination of high-Fe parental magmas and the suppression of Fe-Ti oxide crystallisation by low oxygen fugacities (fO_2).

The concentrations of the compatible elements Ni, Cr and Co are higher in Archaean than modern tholeiitic basalts (Nesbitt and Sun, 1976; Gill, 1979; Condie, 1985), and the available data suggest that their enrichment is most marked in early Archaean rocks (Condie, 1985). Ni and Cr levels decrease markedly from Mg-rich to Fe-rich tholeiitic basalts, as first olivine + chromite, and then calcic pyroxenes crystallise. V, Sc and Ga show partially compatible behaviour in Archaean tholeiitic basalt suites, reflecting their incorporation into augite, and leading to steadily increasing Ti/V, Ti/Sc and Ti/Ga (Redman and Keays, 1985; Nesbitt *et al.*, 1984; Cattell, unpub. data). The compatibility of V, although lower than that of Sc and Ga, is sufficient to suggest high V^{3+}/V^{5+} and therefore low fO_2 conditions during fractionation (Shervais, 1982), in accord with the late onset of Fe oxide crystallisation noted above. Once Fe-Ti oxides start to crystallise, V is highly compatible and its concentration declines rapidly, leading to a sharp rise in Ti/V values and a reversal of the gradual increase in V/Sc (Arndt and Nesbitt, 1982; Cattell, unpub. data).

The incompatible elements Ti, Zr, Y and Nb are present in Archaean tholeiitic basalts in near-chondritic proportions. Nb may be slightly depleted (Sun, 1984), but this is not well constrained because of the lack of precise Nb data. Although Ti is partially compatible in calcic pyroxenes (estimated crystal–liquid Ti partition coefficient (K_D) values are 0.4 for both Mg- and Fe-rich Archaean tholeiites) there is no decrease in Ti/Y and Ti/Zr with decreasing mg'. The inference must be that Zr and Y also partition into calcic pyroxene, and that their K_D values are roughly equivalent to that of Ti.

There is an important group of Fe-rich tholeiites which have a rather different trace element chemistry. These rocks are depleted in Y relative to Zr and Ti, and have roughly chondritic Ti/Zr values. Their chemistry indicates derivation by the fractionation of a Mg-rich tholeiite magma with a similar depletion in Y, but such compositions are not found amongst the erupted lavas (Arndt and Nesbitt, 1982). Most Archaean tholeiitic basalts have unfractionated REE distributions (Figure 2.3), with overall REE concentrations about ten times chondritic values in the less evolved samples (Sun and Nesbitt, 1978a; Hawkesworth and O'Nions, 1977). High-Fe tholeiites have REE concentrations of twenty to thirty times chondrite. Those Fe-rich tholeiites which are depleted in Y are also strongly depleted in the HREE, and typically have upwardly convex LREE patterns (Figure 2.3; Arndt and Nesbitt, 1982; Cattell, 1985).

Mg-rich tholeiitic basalts have high PGE abundances (e.g. 9–32 ppb Pd), while less magnesian tholeiites have relatively low abundances (1–5 ppb Pd) (Redman and Keays, 1985). These data suggest that Archaean tholeiitic magmas became saturated in sulphur within the compositional range of erupted liquids.

2.4 Komatiitic basalts

Many Archaean mafic–ultramafic volcanic sequences contain a suite of Mg- and Si-rich basalts with characteristic pyroxene spinifex textures. Such lavas are referred to in this chapter as komatiitic basalts (Arndt and Nisbet, 1982b). The suitability of this name will not be discussed here, but it should be noted that the name does not necessarily indicate a genetic link between these rocks and komatiites. Alternative names used for essentially the same rocks include basaltic komatiite, clinopyroxene (spinifex-textured) komatiite, siliceous high-magnesian basalt (SHMB) and spinifex-textured basalt.

Komatiitic basalts occur as lava flows of extraordinary diversity and complexity. Their most common and distinctive feature is coarse acicular pyroxene (pyroxene

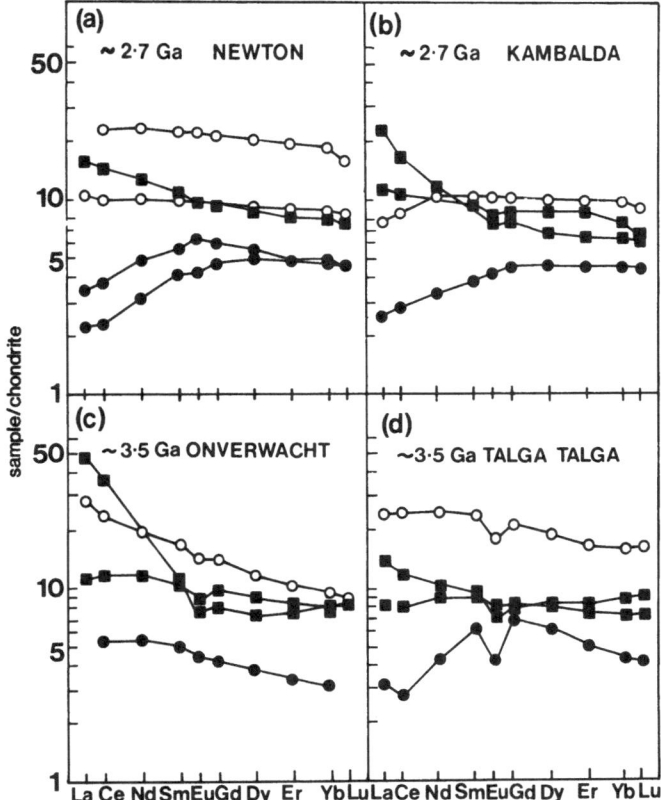

Figure 2.3 Chondrite-normalised rare-earth element diagrams for komatiites (filled circles), komatiitic basalts (filled squares) and tholeiitic basalts (open circles) from Newton Township, Ontario (Cattell, 1985, 1987; Cattell and Arndt, 1987), Kambalda, Western Australia (Arndt and Jenner, 1986), Onverwacht, South Africa (Jahn *et al.*, 1982) and Talga Talga, Pilbara Block, Western Australia (Gruau *et al.*, 1987).

spinifex texture), but this may be lacking in more mafic flows which more closely resemble komatiites. This latter type of flow may be olivine-phyric throughout or may have poorly developed olivine spinifex texture near to the flow top (Arndt *et al.*, 1979). Komatiitic basalt flows commonly show extreme compositional and textural layering, and although these characteristics are best known from very thick (> 100 m) flows (Arndt, 1977a) they are shared by relatively thin (*c.* 10 m) flows (Arndt *et al.*, 1979; Nisbet *et al.*, 1977; Cattell, 1985). The lower parts of many such flows are cumulates, whilst the upper spinifex-textured zones represent crystallisation downward from the flow top. Thicker flows may also have a central gabbroic portion. Less mafic lavas show little if any compositional layering, and occur as pillowed or massive flows (Arndt *et al.*, 1979). Komatiitic basalt flow tops are usually aphyric, although their textures are often masked by alteration.

In the more mafic of these lavas, the flow tops are often underlain by a zone containing randomly oriented spinifex-textured olivine. Downward through olivine spinifex zones, the interstitial acicular pyroxenes gradually become coarser and the spinifex-textured olivines finer, until the lava grades into a pyroxene spinifex-textured rock. The coarse pyroxene needles may be either randomly oriented or elongate perpendicular to the flow top, giving rise to a 'stringbeef' texture (Arndt and Fleet, 1979). The coarse pyroxenes are zoned from highly magnesian pigeonite cores to augite rims, and they are typically set in a groundmass of plagioclase–augite intergrowths (Arndt and Fleet, 1979). The pyroxenes in the thicker flows become larger and less elongated, giving a gabbroic texture (Arndt, 1977a) which may have a granophyric mesostasis (Cattell, 1985). Whole-rock compositions become progressively less mafic downward from the flow top, indicating the evolution of the liquid composition as crystallisation advanced down into the flow. The lower parts of layered flows commonly comprise olivine-, augite-, and bronzite-cumulates. Experimental work indicates that the erupted liquid compositions would not fractionate to the orthopyroxene saturation surface, and the occurrence of bronzite as a cumulus phase is somewhat puzzling, and has led to the erection of complex models to explain its presence (Kinzler and Grove, 1985).

Komatiitic basalts were erupted with MgO contents of between 7 and 15%. *In situ* fractionation gave rise to evolved gabbroic centres in some flows, which have as little as 5% MgO, while mafic cumulates in layered flows have 18–35% MgO. Although chemical discriminants to distinguish these komatiitic basalts from tholeiites have been proposed (e.g. Arndt *et al.*, 1977), none seems universally applicable. In general, komatiitic basalts tend to be relatively Si-rich and Al-poor when compared to tholeiitic basalts of similar mg' value. The low Al content of komatiitic basalts is reflected by the late crystallisation of plagioclase, which in turn leads to fairly constant Al_2O_3/TiO_2 during fractionation, accompanied by subdued Fe enrichment. The onset of plagioclase fractionation is only observed when MgO values drop below 7.5%, at which point Al_2O_3/TiO_2 rapidly declines (e.g. Arndt and Nesbitt, 1982; Cattell, 1987). Prior to plagioclase crystallisation, Al_2O_3/TiO_2 is almost chondritic (20) in most komatiitic basalt suites (Jenner and Arndt, 1982; Arndt and Nesbitt, 1982), but is lower in Al- and HREE-depleted suites (Cattell, 1987; Jahn *et al.*, 1982; Nesbitt *et al.*, 1984).

Cr and Ni contents are around 1500–2000 ppm and 400–500 ppm respectively in the most magnesian komatiitic basalts (MgO = 15%) and fall to 100 and 50 ppm respectively in rocks with MgO of around 7.5% (Arndt and Nesbitt, 1982; Cattell, 1987; Redman and Keays, 1985). Olivine cumulates at the base of layered flows are enriched in Cr, indicating that chromite also settled out with olivine, and that it was on the liquidus of the most magnesian samples of this series (Cattell, 1985). Co concentrations fall from only 70 to 50 ppm as MgO falls from 15 to 7.5% (Redman and Keays, 1985).

The low Al_2O_3/TiO_2 values of some komatiitic basalt suites is matched by depletion of V, Sc, Y and the HREE, as it is in komatiite suites. V and Sc show compatible behaviour in komatiitic basalt suites, presumably due to their incorporation into clinopyroxene. This compatibility leads to slight increases in Ti/V and Ti/Sc with fractionation, and a blurring of the boundary between Al-depleted and undepleted suites.

A notable chemical feature of many komatiitic basalts is a marked enrichment in the LREE and Zr, giving rise to low Ti/Zr ratios (Figure 2.2) (Sun and Nesbitt, 1978a; Arndt and Jenner, 1986; Barley, 1986; Cattell, 1987). The degree of enrichment does not vary systematically with the degree of fractionation within individual komatiitic basalt suites, but it does vary significantly from one suite to another. Accurate Nb data are difficult to obtain at the concentrations present in these rocks (*c.* 1–5 ppm) but most studies indicate that Nb is not relatively enriched in komatiitic basalts (Nesbitt *et al.*, 1984; Barley, 1986; Cattell, 1987).

Redman and Keays (1985) analysed Pd and Ir in komatiitic basalts from Western Australia. The contents of both these elements are relatively high, indicating that the magmas were not S-saturated, that is, an immiscible sulphide liquid had not stripped out the chalcophile PGE. Mean Pd/Ir values in komatiitic basalts are high (40) compared with komatiites (Pd/Ir = 6) and slightly high compared with Archaean tholeiitic basalts (Pd/Ir = 22) (Redman and Keays, 1985; Keays 1982).

2.5 Petrogenesis

2.5.1 *Komatiites*

Several models of komatiite genesis have evolved over the past twenty years as more chemical and physical data have become available. The various models are reviewed in this section, before a discussion of how recent advances in related fields may shed light on the question of komatiite generation.

There is a simple recipe for making komatiites: induce > 50% melting of the mantle and then extract and erupt the resultant liquid. The flaw in this model is that melt will tend to escape from the source long before 50% melt is produced, and commonly before even 5% melt is produced (McKenzie, 1985). Various authors have devised schemes to overcome this difficulty. A group of models proposing the involvement of more than one stage of melting has been put forward, the komatiitic liquid being produced in these models by the last melting episode. Since melt extraction renders the mantle more refractory and Mg-rich, the percentage of melting required to produce a magnesian, komatiitic liquid is significantly reduced. The prior melting episodes may be unrelated earlier events (Weaver and Tarney, 1979) or an earlier part of the same event (Arndt, 1977b). Most komatiites appear to have been derived from mantle from which melts had previously been extracted. Their marked LREE depletion leaves no room for doubt on this point, but it seems unlikely that the komatiite mantle source could be so refractory as to produce low percentage (< 5%) melts of komatiite composition.

The driving force which separates liquid and solid during melting is gravity, buoyant liquids being driven upward relative to dense residual solid. Since liquids are more compressible than solids, the density of liquids increases more rapidly with increasing pressure than the density of solids, unless the solid inverts to a denser structure. Therefore, with increasing pressure, the density contrast between liquid and solid declines. This is particularly true for minerals such as olivine which are relatively incompressible. Both theoretical and experimental work indicates that olivine may be less dense than silicate liquids under high pressure, and therefore a liquid within an olivine-dominated residue would not tend to rise, or might even sink (Nisbet and Walker, 1982; Agee and Walker, 1988). If liquid–solid separation is prevented, then high percentage (komatiitic) melts may be produced. Clearly, the crux of the matter is whether or not olivine is denser than the natural liquid compositions at the appropriate mantle depths. This is the subject of recent controversy (Herzberg, 1987).

Several recent komatiite generation models have been based on the observation that low percentage mantle melts become increasingly magnesian with increasing pressure (Takahashi and Scarfe, 1985; Herzberg and O'Hara, 1985; Herzberg and Ohtani, 1988).

Recent improvements in high-pressure experimental apparatus and research techniques have shown that the trend towards increasingly magnesian melts continues until at 5 GPa (50 kb) initial melts are komatiitic (Takahashi and Scarfe, 1985). If komatiites do reflect very high pressure melting, then the problems discussed above do not apply, since only small degrees of melting are then necessary. Furthermore, the wide ranges in composition of komatiites may be due in part to their derivation from small degree or 'pseudo-invariant' melts produced over a large pressure range (Herzberg and O'Hara, 1985; Herzberg and Ohtani, 1988). Although initial liquids produced at great depth (> 150 km) may thus be komatiitic, the inferred trace element compositions for such liquids are far removed from those of erupted komatiites (Bickle and Nisbet, 1986; Nisbet, 1987). The abundances of incompatible trace elements in komatiites are typically one to three times those of estimated primordial mantle. Initial or low-percentage melts should have far higher concentrations of incompatible elements, unless all estimates of partition coefficients are not applicable to the high P and T conditions under consideration.

Much light has been shed recently on basalt genesis by the analysis of melt generation and extraction, and by the careful consideration of what melts arriving at the base of the crust actually represent (Ahren and Turcotte, 1979; McKenzie, 1984; Thompson, 1987; McKenzie and Bickle, 1988). Basalts are generated from upwelling portions of the mantle which begin melting after intersecting the solidus. Melting continues as the mantle rises until the liquid fraction can escape upward. This probably occurs when the melt fraction is still small, perhaps only 2% (McKenzie, 1985). On continued ascent, melting continues and more melt escapes upward, and so on. It is still not clear whether melt escape is episodic (batch melting) or continuous (Rayleigh melting). This process occurs simultaneously at different levels in the melting zone, and thus melts from varying depths are available at the same time. These different melts must mix to a greater or lesser degree, leading to liquids with an average composition of melt extracted from the melting zone, which does not necessarily correlate with the composition of any individual partial melt, nor with any liquid that could be generated in a laboratory melting experiment. Although all the liquids are low-degree melts, the mantle becomes progressively refractory as it passes up through the melting zone and, thus, those melts derived from the upper part of the zone are strongly depleted. The net effect is that the average melt is not significantly different from a high-percentage melt, the main differences resulting from the higher pressure of equilibration of a large proportion of the average melt. This result is of profound importance to komatiite petrogenesis models. Given sufficiently high temperatures (a problem in itself), an upwelling column of mantle can yield a liquid whose composition is close to that of a 50% mantle melt, yet the amount of melt which exists in any part of the melting column is never more than, say, 2%. Producing komatiites in this way reconciles the chemical evidence in favour of high degrees of melting with the physical evidence that such high degrees are implausible.

Producing a komatiite in such a way can be illustrated by taking, for example, komatiite lavas with c. 25% MgO, which are common, especially in the late Archaean. The amount of melting of mantle peridotite required to give a magma with 25% MgO can be calculated from the curves given by McKenzie and Bickle (1988) as roughly 39%. As Thompson (1987) pointed out, the source of MORB (and also of LREE-depleted komatiites) is more refractory than most of the samples used in the experiments which constrain these melting curves. The estimated degree of melting ought to be somewhat

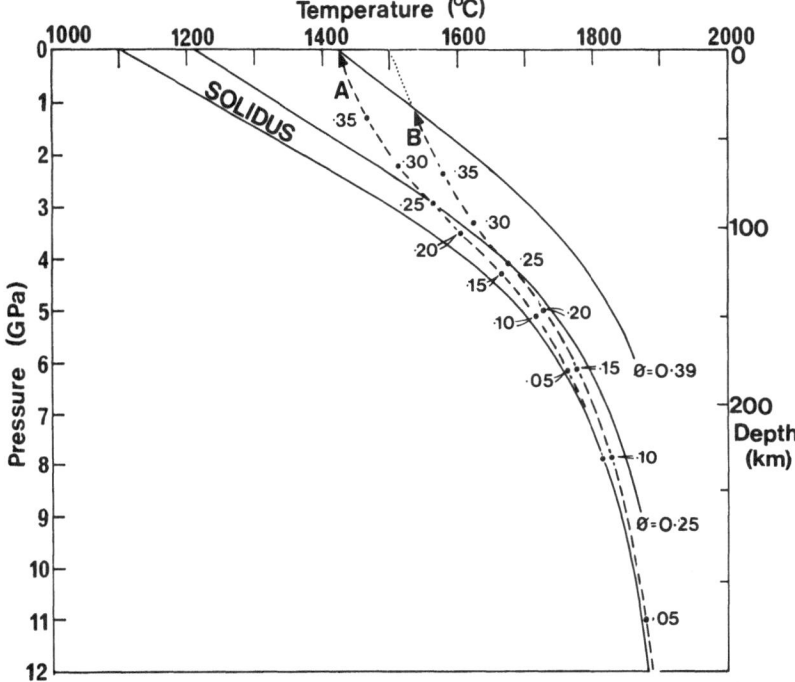

Figure 2.4 Pressure–temperature diagram illustrating the generation of komatiite melts by decompression. Solidus and melt fraction curves constructed using expressions 18 to 22 of McKenzie and Bickle (1988). The illustrated melting paths were constructed assuming that for ascending mantle under the appropriate P–T conditions, the melt fraction (ϕ) increases by approximately 0.1 as T falls by 100°C (McKenzie, 1984). Path A indicates the unlikely and limiting situation where melting continues to the surface. Path B represents melts arriving at the surface at 1500°C. In both cases, melt segregation occurs at $\phi = 0.39$, at which point the melt will have a typical komatiitic composition (MgO = 25%).

less than 39%, although since the starting material is more refractory, the temperature required is not significantly different. For simplicity this complication is ignored, since extensive data are not available for melts generated from refractory peridotite. The top of a melting column which yields a 25% MgO komatiite should therefore lie somewhere along the 39% melt fraction curve of the source (Figure 2.4). The P–T ascent paths shown on Figure 2.4 are chosen arbitrarily, but it is clear that to produce a komatiite with 25% MgO requires a mantle potential temperature (T_p) of at least 1770°C, and probably somewhat more. The generation of even more magnesian komatiite liquids requires higher mantle T_p values. Bickle (Chapter 6) argues that a T_p greater than 1700°C cannot be sustained in the upper mantle. The high temperature required for komatiite generation implies that komatiites were generated in unusually hot parts of the mantle, such as in hot rising jets. The relations shown on Figure 2.4 indicate that melting in such a jet commences at depths of 250 km or greater, if komatiites are to be generated. As noted above, even the initial melts are komatiitic at such depths. Figure 2.5 illustrates the liquid evolution in a hot rising jet that begins melting at

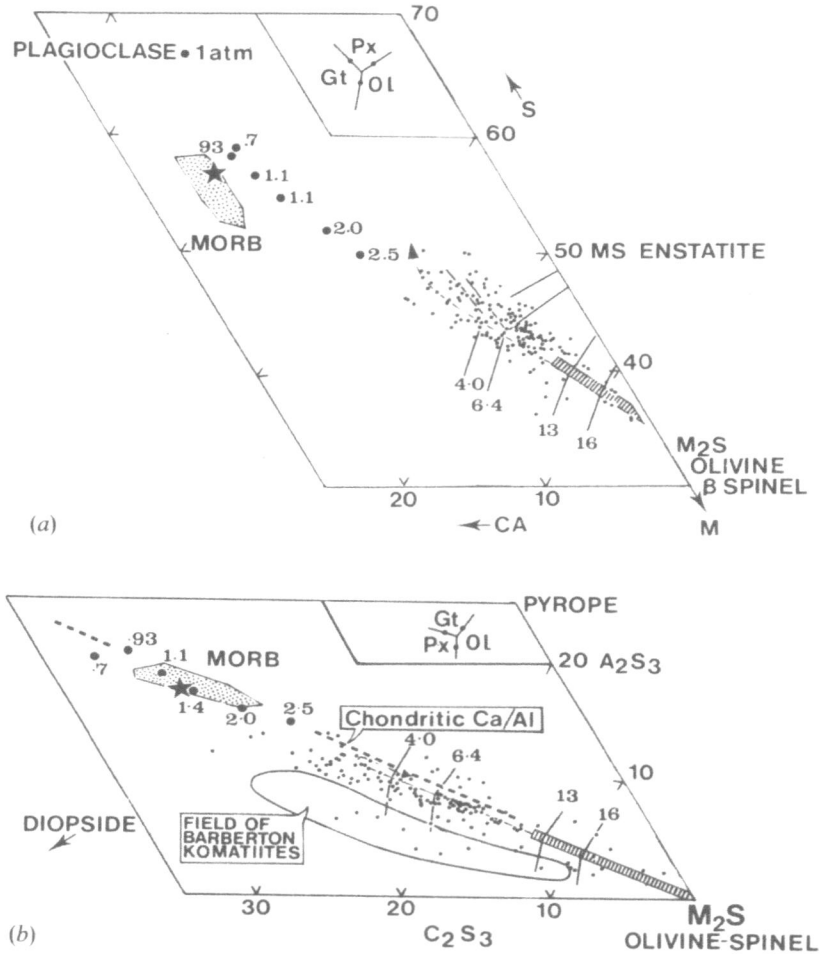

Figure 2.5   Phase relations during komatiite generation in the upper mantle (after Herzberg and Ohtani, 1988) in the system CMAS, shown as (a) projection from diopside on to the plane CA–M–S, and (b) projection from enstatite on to the plane C_2S_3–A_2S_3–M_2S. Small dots are komatiite compositions; line shaded field: asthenospheric peridotites; stippled field: MORB; star: average MORB; dashed line: locus of melt compositions generated in ascending mantle (see text for discussion). Pressure given in GPa.

500 km (case B on Figure 2.4). The combination of declining pressure and increasing degrees of melting mean that melts generated throughout the melting column are komatiitic, at least in terms of elements used to construct the plots. Incompatible element concentrations are very high in the melts generated near the base of the melting column and decline rapidly upward.

Initial melts are in equilibrium with all four main mantle phases (olivine, orthopyroxene, garnet and clinopyroxene), and therefore lie at the junction of the various phase volumes. Further up through the melting column, melts are generated from successively more refractory mantle at successively lower pressures. For simplicity it is assumed that no phases are completely consumed until 25% melting is reached, and that the phase relations for rather fertile peridotites on which the phase boundaries of Figure 2.5 are based are applicable to the more refractory (high Mg/Fe) compositions which would occur within the melting zone.

The composition of the liquids produced in the upper part of the melting column depends on whether clinopyroxene or garnet is exhausted first. The two phases probably disappear almost simultaneously, at least under the chosen conditions. In this case, the liquid is then saturated only with olivine and orthopyroxene, and liquid compositions trend back toward the olivine apex (Figure 2.5a) as melting continues and orthopyroxene and olivine are progressively consumed. If either garnet or clinopyroxene were consumed significantly before the other, then the liquid would be displaced away from the consumed phase on the opposite phase boundary. For example, if garnet were consumed first, the liquid would lie on the clinopyroxene–olivine phase boundary, shifting with it as pressure fell. Once clinopyroxene was consumed, then the liquid would trend back toward the olivine apex.

The trends of liquids produced within the melting zone are shown on the diopside projection on Figure 2.5b to illustrate olivine–orthopyroxene relationships. On both projections komatiites plot within the field of predicted liquid compositions. The Al-depleted komatiites from Barberton plot away from the main body of komatiites, on the diopside side of the trend of pseudo-invariant compositions. The retention of garnet in the residue of melting after the consumption of clinopyroxene is not a possible mechanism to produce the Al-depleted lavas, since progressive melting of a garnet-bearing, clinopyroxene-free residue can only lead to Ca/Al ratios progressively less than chondritic, rather than to the elevated Ca/Al of the Barberton Al-depleted lavas. Progressive melting of a garnet-free, clinopyroxene-bearing residue would produce liquids with high Ca/Al, but these would not have the observed komatiitic low Al/Ti and depleted HREE patterns. The source of Al-depleted komatiites must itself have been depleted in Al and HREE prior to komatiite generation (section 2.6.2).

The above discussion is a simplified assessment of komatiite generation in the light of recent ideas concerning melt generation and segregation (McKenzie, 1984; McKenzie and Bickle, 1988). The discussion deliberately avoids complications of the three-dimensional nature of the melting zone (O'Hara, 1985) and thermal implications of the inferred mantle T_p (see Chapter 6), in order to stress the point that the apparent conflict between komatiite chemistry and the physics of melt segregation is resolved if the dynamic nature of the melting process is realised.

2.5.2 Komatiitic basalts

There are two conflicting hypotheses to explain the origin of komatiitic basalts. One assumes that there is a genetic link between them and komatiites, while the other suggests that there is no link and that komatiitic basalts reflect separate mantle melts, perhaps petrogenetically analogous to modern boninites. The former hypothesis is tested in this section and the similarities between komatiitic basalts and their recent analogues are reviewed in section 2.7.

The major element compositions of komatiitic basalts are exactly matched by the products of olivine fractionation of komatiite liquid, as clearly demonstrated by differentiated horizons within thick komatiite flows (Cattell, 1987). However, the incompatible element concentrations in many komatiitic basalts are higher than those in fractionated komatiites, and the relative abundance of the incompatible elements are often dramatically different in the two groups. For example, komatiites are LREE-depleted while komatiitic basalts typically show varying degrees of LREE enrichment (Figure 2.3). These features indicate that komatiitic basalts cannot simply be fractionated komatiites (Sun and Nesbitt, 1978a).

The geochemical differences between komatiites and komatiitic basalts can be reconciled by the komatiites being contaminated by continental crust during differentiation, to give komatiitic basalt compositions, and given the very high temperatures of komatiite liquids this is certainly a feasible process (Huppert and Sparks, 1985). The assimilation of continental crust would explain the LREE enrichment of the komatiitic basalts, their lack of Nb enrichment (Cattell, 1987; Sun, 1984), their enriched isotopic character (Arndt and Jenner, 1986), and perhaps their relatively siliceous nature. Clear evidence that some form of contamination occurred came from the discovery of zircon xenocrysts up to 3.4 Ga old in a 2.7 Ga komatiitic basalt from Kambalda. This komatiitic basaltic liquid clearly digested old felsic crust during its ascent (Compston *et al.*, 1986). Further evidence for crustal contamination comes from the correlation between Nd-isotopic enrichment in komatiitic basalts and the age of the basement at the time of their eruption; < 200 Ma old basement in the case of the Abitibi belt in Newton Township, Ontario where ε_{Nd} is slightly positive (Cattell, 1987) and 700 Ma old basement at Kambalda, Western Australia, where ε_{Nd} in the komatiitic basalts is markedly negative (Chauvel *et al.*, 1985). More data are required for komatiitic basalts from other areas to confirm this correlation.

The most significant tectonic inference to be drawn from this interpretation is that komatiitic basalts, together with the stratigraphically juxtaposed komatiites and tholeiites, were erupted through and on to continental crust. Discrepancies between the chemical evolution of the komatiites and the komatiitic basalts could be explained by the komatiites having escaped the effects of contamination by their ascent through conduits lined with the frozen komatiite from previous eruptions (Nisbet, 1987).

2.5.3   *Tholeiitic basalts*

Compared with modern tholeiites with similar MgO contents, Archaean tholeiites are enriched in Fe, Cr, Co and Ni (section 2.3; Figure 2.6). Our understanding of these differences and of the origin of these basalts is to a large extent dependent on our understanding of how modern tholeiitic basalts are generated. New light has been shed on modern basalt generation by a better understanding of how melting and melt extraction occur (section 2.6.1).

These new petrogenetic ideas have been applied to explain the relatively high Ni content of Archaean tholeiites (Arndt, 1989). The rationale behind the explanation is simple. Melts generated within a rising column of mantle will show a positive correlation between Ni and Mg, both elements increasing with depth and the degree of melting. However, the melt compositions do not show a linear relationship between Ni and Mg. Ni increases more rapidly than increasing MgO. The net effect is that mixtures of melts will always lie on the Ni side of the correlation curve defined by simple batch melts. Additionally, the degree of Ni enrichment of average or 'pooled' melt compositions increases with increasing depth of the onset of melting. Since the somewhat

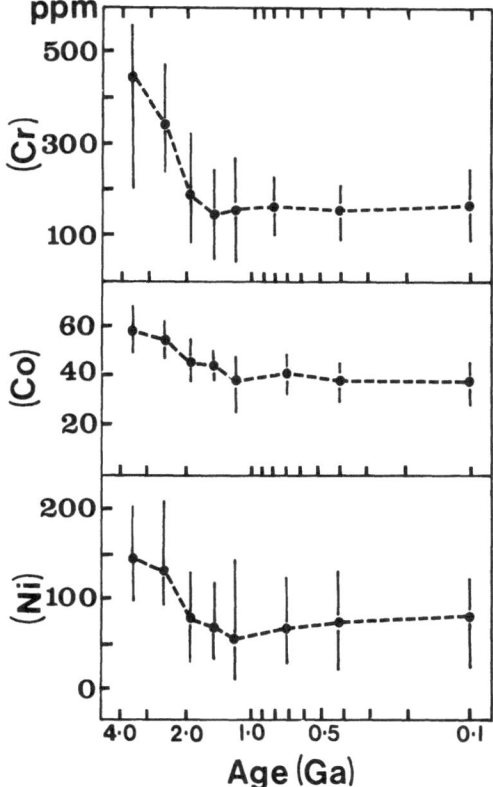

Figure 2.6 Ni, Co and Cr contents of basalts with respect to age based on data grouped into the eight age categories 3.8–3.0, 3.0–2.5, 2.5–1.6, 1.6–1.2, 1.2–0.9, 0.9–0.6, 0.6–0.2, and 0.2–0.06 Ga (after Condie, 1985). Data normalised to a mg' value of 0.6. Vertical bars indicate range of values about the mean.

hotter Archaean mantle would begin to melt at greater depth, melts richer in Ni relative to MgO would be generated. This enrichment in Ni relative to Mg would not be lost during subsequent olivine fractionation (Arndt, 1989). Several authors have noted that deeper melting leads to higher Fe at a particular mg', thus explaining the high Fe of Archaean tholeiites (Gill, 1979; Condie, 1985).

The model outlined above for Archaean tholeiites is similar to that for komatiites. The difference is that komatiites were generated from an unusually high T_p mantle which commenced melting at greater depth. If Archaean tholeiites were generated from an ascending mantle column which intersected the solidus at roughly 5 GPa (Arndt, 1989), then a T_p of 1575°C is implied, compared with 1770°C for komatiite generation (section 2.5.1). It seems probable that Archaean tholeiitic basalts represent the normal products of decompressive melting of roughly average T_p mantle, and are thus the nearest equivalents of modern MORB. However, an exact parallel cannot be drawn since most preserved Archaean tholeiites were probably erupted through continental crust (section 2.6) although they usually show little evidence of contamination.

2.5.4   Archaean 'oceanic' basic magmatism

At the present day, basic volcanism in ocean basins dwarfs the volume of basic magmas erupted on the continents. Various theoretical considerations, and particularly the need for the Earth to dissipate its greater radiogenic heat production, imply even greater ocean crust production during the Archaean (Bickle, 1978). However, unequivocal examples of Archaean oceanic crust are extremely rare. Nearly all greenstone belt lavas appear to represent magmas erupted through continental crust. Although the Archaean oceanic crust is not, therefore, directly observable, some comments can be made about its character. It must have been basic or even ultrabasic, and therefore represents by far the most important site of Archaean basic–ultrabasic magmatism, despite the lack of its preservation in the geological record. It is stressed that in discussing Archaean basic lavas only the anomalous c. 1% erupted onto (or near) the continents rather than the 99% erupted to make the ephemeral Archaean oceanic crust can be examined.

Arguments concerning the nature of Archaean plate tectonics are beyond the scope of this contribution. The high radiogenic heat production in the Archaean probably led to a more intense plate tectonic regime than is operative at present, leading inevitably to the necessity of subducting younger (and thus hotter and more buoyant) oceanic crust (Bickle, 1978). Arndt (1983) and Nisbet and Fowler (1983) suggested that the difficulty of subducting young hot oceanic crust would be overcome if the crust had a bulk komatiitic composition. Bickle (1986; Chapter 6) points out that a more ultramafic crust, produced by higher degrees of partial melting, would also be thicker than modern oceanic crust. The major difference between the chemistry of lavas from Archaean oceanic crust compared with their continental counterparts would obviously be the total lack of a crustal contamination chemical component. It is difficult to envisage how fragments of Archaean oceanic crust would be recognised given only these negative criteria. It is unfortunate that the products of the most extensive and voluminous basic magmatism, those in Archaean oceanic regions, remain hypothetical.

2.6   The Archaean mantle

2.6.1   Mantle evolution

Samples of sub-continental lithospheric mantle of Archaean age are available as xenoliths and as inclusions in diamonds, brought to the near surface by kimberlitic volcanism (Richardson et al., 1984). However, no such samples of the Archaean asthenosphere are available, and therefore its composition cannot be measured directly, but must instead be inferred from the compositions of its partial melts, Archaean basic and ultrabasic magmas. The upper mantle sampled by present day basaltic volcanism is strongly depleted in those elements which preferentially enter a melt, and these elements are now concentrated in the continental crust. The mantle is also depleted, relative to chondritic meteorites, in the siderophile elements, which have been partitioned into the Earth's core. The Archaean basic–ultrabasic volcanic record stretches from roughly 3.8 Ga to 2.5 Ga, and should allow us to answer at least two fundamental questions about the evolution of the mantle. Firstly, had the formation of the Earth's core finished by the time of the eruption of the earliest basic–ultrabasic volcanics, or did it continue through the Archaean? Secondly, when did the astheno-spheric mantle become depleted in the more incompatible elements?

The first question can be answered by using the approach adopted by Sun (1982, 1984, 1987), which is to look at the relative abundances of pairs of elements which behave in a similar way during petrogenetic processes giving rise to basic–ultrabasic magmas, i.e. they show similar compatibility in silicate minerals, but which behave in contrasting ways during processes related to core formation in that they have very different silicate–metal partition coefficients (K_D). The ratios of such element pairs should have changed dramatically whilst core formation was in progress, but would then remain fixed since they are not modified by subsequent magmatic events. One of the most illuminating pairs of elements are the minor elements Ti and P. Both are incompatible during melting processes and P strongly partitions into a Fe-rich metallic phase whilst Ti does not. The Ti/P ratio in modern basalts is 10 ± 1 compared to a solar abundance ratio of < 1, the difference reflecting the strong partitioning of P into the Earth's core. If the core had formed prior to 3.8 Ga, then Archaean basic–ultrabasic magmas should also have Ti/P of c. 10. If the core were still forming during the Archaean then the basic–ultrabasic magmas would have Ti/P < 10, and increasing Ti/P with time through the Archaean until the formation of the core was complete. The data collected by Sun (1987) show that all Archaean basic–ultrabasic magmas have Ti/P $= c. 10$, indicating that core formation was indeed complete by 3.8 Ga.

The depletion in the upper mantle of the more incompatible elements is more difficult to interpret. Consider the element pair Sm–Nd. Sm is slightly more compatible than Nd during mantle partial melting, leading to melts with Sm/Nd less than that of the original source, and a residual solid with Sm/Nd greater than that of the source. The extraction of melts therefore leads to an increase in Sm/Nd in the mantle. The evolution of mantle Sm/Nd can be monitored using Nd isotopes. ^{147}Sm decays to ^{143}Nd and thus the ratio ^{143}Nd/^{144}Nd in an erupted magma reflects the time-integrated Sm/Nd of its source. The Nd isotopic composition is conventionally normalised to the Nd isotopic composition of the bulk Earth (estimated from chondrites), and deviations from the bulk Earth are expressed in ε_{Nd} units where 1 ε_{Nd} unit represents a deviation of one part in ten thousand, and positive values indicate ^{143}Nd/^{144}Nd greater than bulk Earth.

Modern MORB have Sm/Nd ratios greater than bulk Earth and strongly positive ε_{Nd} values ($+8$ to $+12$), indicating that the upper mantle from which they were produced had low Sm/Nd for a significant time. These data can be interpreted as representing a single depletion (i.e. melting) episode at roughly 2.0–1.5 Ga, or the integrated sum of many depletion events throughout geological time. The latter interpretation is supported by the Nd isotopic compositions of Archaean and Proterozoic ultrabasic and basic volcanics. ε_{Nd} values show gradually increasing positive values through time (Jacobsen, 1988) and, significantly, the rate of increase of ε_{Nd} also increases through time, implying that there is a gradual increase in Sm/Nd in the source. Even the earliest known basic volcanics, the 3.8 Ga metavolcanic amphibolites from Isua, West Greenland, have ε_{Nd} of $+2$, suggesting that melt extraction from the mantle had occurred for a significant period prior to 3.8 Ga (Hamilton et al., 1983). These data answer the second question, concerning the depletion of the upper mantle in the more incompatible elements. This has clearly been a continuous process from before the eruption of the oldest preserved basic volcanics to the present day.

2.6.2   Mantle heterogeneity

Most, although not all, Archaean tholeiitic basalts have Sm/Nd slightly greater than chondritic values, and therefore close to the Sm/Nd of the upper mantle at the time, as inferred from the Nd isotope evolution curve. Tholeiitic basalts could therefore be regarded as partial melts of average Archaean mantle, with little or no change in Sm/Nd during melting or subsequent fractionation. In contrast, virtually all Archaean komatiites have Sm/Nd ratios well above chondritic (i.e. they are strongly LREE-depleted), the only major exception being the 3.5 Ga komatiites from the Barberton greenstone belt (Sun and Nesbitt, 1978; Jahn et al., 1982). The ε_{Nd} of these LREE-depleted, high Sm/Nd komatiites are similar to, or only slightly greater than, the ε_{Nd} of associated tholeiitic basalts (Zindler, 1982; Cattell et al., 1984; Chauvel et al., 1985; Machado et al., 1986). The greater compatibility of Sm compared to Nd means that the komatiite source must have had as high, or even higher, Sm/Nd than the komatiites themselves. Note that any crustal contamination would also lower Sm/Nd, again meaning that the Sm/Nd of the komatiites must represent the minimum possible Sm/Nd of their source. While the Nd isotopic composition of the komatiites is therefore consistent with their derivation by melting of average Archaean upper mantle, their Sm/Nd ratios show that their source is far more depleted than average Archaean upper mantle. The combination of chemical heterogeneity and relative isotopic homogeneity in Archaean basic–ultrabasic magmas suggests that the Archaean mantle was heterogeneous, but that the heterogeneities were short lived, presumably re-mixing and homogenising before large-scale isotopic variations could develop. The ε_{Nd} and Sm/Nd relations of komatiites, as compared to tholeiitic basalts and estimated average Archaean mantle values, suggest that the upper mantle was being depleted by several episodes of melt extraction. These melting episodes produced chemically depleted (high Sm/Nd) mantle volumes, but these volumes were re-mixed with less depleted mantle before significant isotopic variations could develop. Komatiites were generated within the short-lived depleted mantle volumes produced by melt extraction episodes.

A second type of mantle heterogeneity is suggested by the existence of two different komatiite types, namely Al-depleted and Al-undepleted (section 2.2; Figure 2.1). The low Al content of the former type is matched by depletion in Ga, Sc, V, Y and, especially, the HREE. Since all of these elements partition into garnet, most authors have assumed that garnet plays a key role in producing the different komatiite types. More precisely, they envisage no role for garnet in the generation of Al-undepleted komatiites but a key role in the generation of Al-depleted types.

The Al-depletion cannot be due to residual garnet in the komatiite source, since the Ca/Al of the liquid is buffered by residual clinopyroxene and garnet at roughly the chondritic ratio (Figure 2.5), and if clinopyroxene is consumed first, then the Ca/Al of the liquid will fall as the remaining garnet is consumed. If garnet is consumed before clinopyroxene, then liquids with increasing Ca/Al will be produced as the remaining clinopyroxene is consumed, but these liquids will have chondritic Al/Ti. The only way to produce Al-depleted komatiite liquids with their characteristically high Ca/Al and low Al/Ti is to remove garnet from the source prior to the melting episode. The garnet removal process is presumably gravity-driven and exploits the large density contrast between garnet and other mantle silicates (Ohtani, 1984).

It is possible to date the garnet separation event using the Lu–Hf isotope system, since Lu partitions strongly into garnet whilst Hf does not. Preliminary data from the

3.5 Ga old Barberton komatiites suggest that they were produced from a source region which had been poor in garnet (i.e. had low Lu/Hf) for a significant period prior to eruption, but a lack of correlation between Lu/Hf and Hf isotopic composition suggests further garnet extraction during the komatiite-generating event itself (Gruau *et al.*, 1986; Sun, 1987). More Hf isotopic data are needed to define the garnet segregation event, but an interim conclusion is that the Archaean mantle was heterogeneous with respect to garnet.

2.7 Phanerozoic analogues of Archaean mafic volcanism

The tectonic setting of Archaean volcanism cannot be determined directly. However, recent mafic volcanics have geochemical and petrological features which are often unique to their particular tectonic setting (e.g. Pearce and Cann, 1973; Shervais, 1982). This section compares and contrasts mafic–ultramafic volcanism of the Archaean and the Phanerozoic in the light of recent geochemical data and developments in petrogenetic understanding.

2.7.1 *Komatiites*

Komatiites younger than the Precambrian have only been discovered on the small island of Gorgona, off the Pacific coast of Colombia. The komatiitic rocks on Gorgona are associated with a sequence of plutonic microgabbros, troctolites, wehrlites and dunites dated at *c.* 86 Ma, and a volcanic sequence of picrites and ultramafic pyroclastics dated at > 25 Ma (Aitken and Echeverria, 1984). Like its Archaean counterparts, the extrusive part of the Gorgona basic igneous complex includes extensive tholeiitic basalt lavas (section 2.7.3). The Gorgona komatiite flows are between 1.5 and 6 m thick with well-defined polyhedrally jointed chilled upper surfaces. Lower sections of the flows contain spinifex and microspinifex textures, but basal olivine cumulates, which are common in Archaean komatiites, have not been found on Gorgona (Echeverria, 1980). The spinifex-textured rocks comprise parallel-linked skeletal olivines surrounded by magnesiochromite in a matrix of augite and plagioclase. This matrix assemblage contrasts with Archaean komatiites, in which Al remains in the glass and plagioclase does not crystallise. The crystallisation of plagioclase may reflect the higher Al content of the Gorgona komatiites (or slower cooling rates).

The Gorgona komatiites have MgO contents in the range 13–24% (volatile free), with an average of 18.6% (Aitken and Echeverria, 1984). Their high MgO contents and olivine spinifex textures demonstrate that most of the Gorgona ultramafic lavas are komatiites, at least according to the criteria of Arndt and Nisbet (1982b). However, these authors suggested a minimum MgO content of 18% for komatiites, which coincides with a natural gap in MgO contents in many Archaean komatiite–komatiitic basalt sequences. There is no such chemical gap in the Gorgona lavas, and no textural or mineralogical changes to indicate any significant difference in the genesis of the various members of the suite. The Gorgona komatiite suite seems to represent a series of liquids produced by olivine-dominated fractionation (Aitken and Echeverria, 1984).

Compared with Archaean komatiites of similar MgO content, the Gorgona komatiites have relatively high Al_2O_3 and low SiO_2. The CaO/Al_2O_3 and Al_2O_3/TiO_2 ratios in the Gorgona komatiites are close to chondritic values, and they therefore

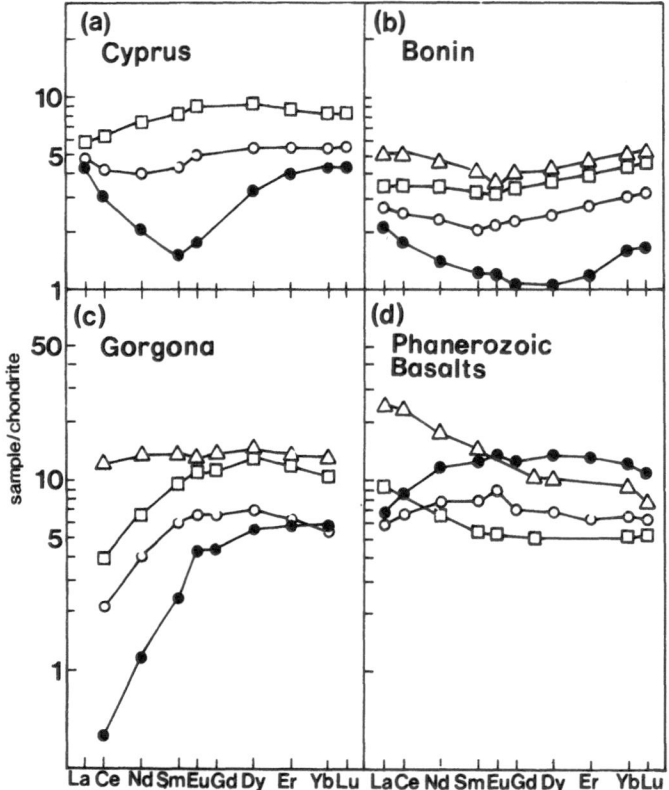

Figure 2.7 Chondrite-normalised REE distributions in various types of Phanerozoic mafic lavas. (a) three olivine basalts from Cyprus (Taylor, 1988; Taylor and Nesbitt, 1988); (b) two boninites (circles), an andesite (squares) and a dacite (triangles) from Bonin Island (Nesbitt and Taylor, unpubl. data); (c) komatiite (open circles), tuff-breccia (filled circles), T-tholeiite (triangles) and K-tholeiite (squares) from Gorgona Island (Aitken and Echeverria, 1984; Echeverria and Aitken, 1986; (d) typical calc-alkaline basalt (triangles), N-MORB (filled circles), island arc tholeiite (open circles) and low-Ti tholeiite (squares) from, respectively, Jorullo volcano, Mexico (Luhr and Carmichael, 1985), Yap trench, Western Pacific (Sun *et al.*, 1979), Hahajima, Ogasawara arc (Nesbitt and Taylor, unpubl. data) and Manam volcano, Papua New Guinea (Johnson *et al.*, 1985).

resemble Archaean Al-undepleted komatiites (section 2.2). However, the Gorgona komatiites have a marked LREE depletion (Figure 2.7) which is a characteristic of the Archaean Al-depleted type (section 2.2).

A sequence of ultramafic pyroclastic rocks (tuff breccias) is intruded by picrites at the southern end of Gorgona Island. The age relations between the ultramafic pyroclastic rocks and the komatiite lava sequence are not clear, and the pyroclastics may be much younger (Echeverria and Aitken, 1986). The pyroclastic rocks and picrites are chemically similar to the komatiite lavas, although they are slightly more magnesian ($MgO = 25\%$) and slightly richer in SiO_2 and Al_2O_3. The pyroclastic rocks and picrites may represent liquids derived by re-melting the source of the komatiites, a process which would explain their extreme LREE depletion (Figure 2.7) (Echeverria and Aitken, 1986). However, the Pb isotopic compositions of the two groups are subtly different, the komatiite flows containing less radiogenic lead, implying their derivation from a different source (Dupré and Echeverria, 1984).

Ultramafic pyroclastic rocks also occur in the Precambrian. Archaean examples are found near the Scotia Ni sulphide deposit in the Yilgarn Block of Australia (Page and Shumulian, 1981; Stolz and Nesbitt, 1981) and at Ruth Well in the Pilbara Block (Nisbet and Chinner, 1981). The Scotia pyroclastic sequence consists of layers 0.1–14 m thick of lapilli tuff which contain both lithic (komatiite, dunite) and altered vitric clasts. The tuffs at Scotia are associated with spinifex-textured komatiite lavas, and are interpreted as the proximal products of sub-aqueous explosive volcanism (Page and Shumulian, 1981). Ultramafic pyroclastic rocks also occur in the early Proterozoic greenstone belts of Lapland (Saverikko, 1983, 1985). The komatiitic sequences here are unusual in that virtually all the volcanics are pyroclastic. The komatiitic pyroclastic rocks are similar to modern basaltic equivalents and represent various facies around large central volcanic complexes.

2.7.2 Komatiitic basalts

Similarities between Archaean komatiitic basalts and certain Phanerozoic primitive basalts have been noted by various authors (e.g. Sun and Nesbitt, 1978b; Cameron et al., 1979; Hickey and Frey, 1982; Malpas and Langdon, 1984). The closest analogues to komatiitic basalts are the boninitic lavas of the western Pacific margin and certain ophiolitic basalts. The boninitic series shows the crystallisation sequence olivine, orthopyroxene, clinopyroxene, plagioclase, which distinguishes it from modern MORB-type tholeiites which precipitate feldspar before clinopyroxene, and from Archaean tholeiites which do not have orthopyroxene on the liquidus. Many of the recent ophiolitic lavas from the Troodos Massif in Cyprus are akin to boninites as they show the early crystallisation of pyroxene. In the Troodos extrusives the initial pyroxene to crystallise is calcic, although in certain primitive high-silica variants it is preceded by orthopyroxene.

The most magnesian Troodos lavas are layered flows which have olivine-phyric bases (up to 55% modal olivine) and grade up into sparsely olivine-phyric lava tops. These layered flows clearly indicate the accumulation of olivine phenocrysts, although virtually identical material is also found in discrete pillows with little evidence of crystal settling. These 'picrite pillows' contain equant olivine grains separated by a meshwork of acicular aluminous augite set in a glassy matrix, a texture similar to that found in thin unlayered komatiite flows (section 2.2). The pillows were initially thought of as representing ultramafic liquids (Gass, 1958), but they are now considered to be the products of continued movement at the leading edge of an already differentiated flow, or an intratelluric accumulation of olivine (Taylor, 1987).

Certain parallels can be drawn between the chemistry of Phanerozoic boninites and ophiolitic basalts, and Archaean komatiitic basalts (Sun et al., 1989). The most significant feature is the shared combination of high Mg and high Si, which is the cause of the distinctive early appearance of orthopyroxene. As would be expected for such magnesian lavas, Cr and Ni concentrations are also high. The most significant differences in the chemistry of the Phanerozoic and Archaean lavas are the very low concentrations of incompatible trace elements in modern boninites and ophiolitic

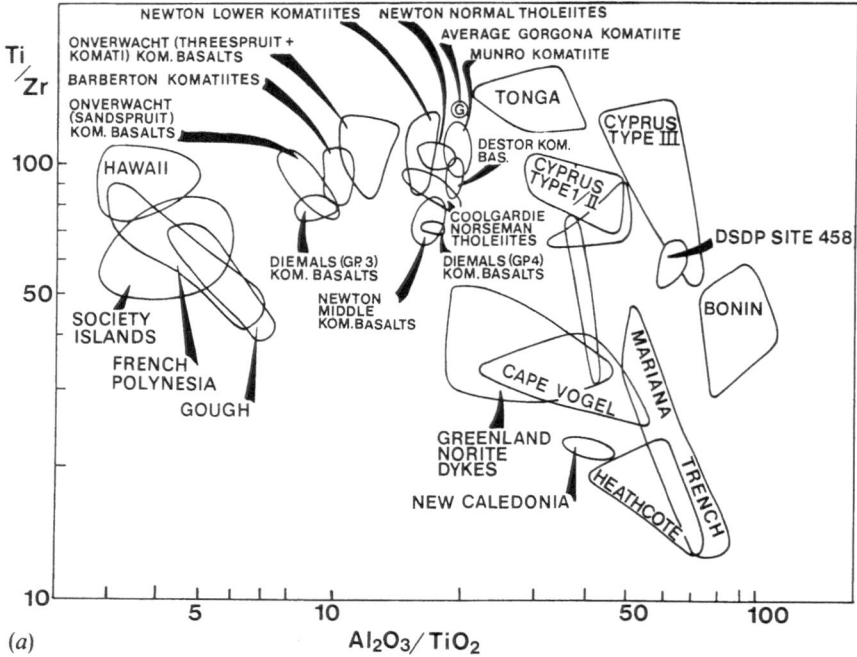

(a)

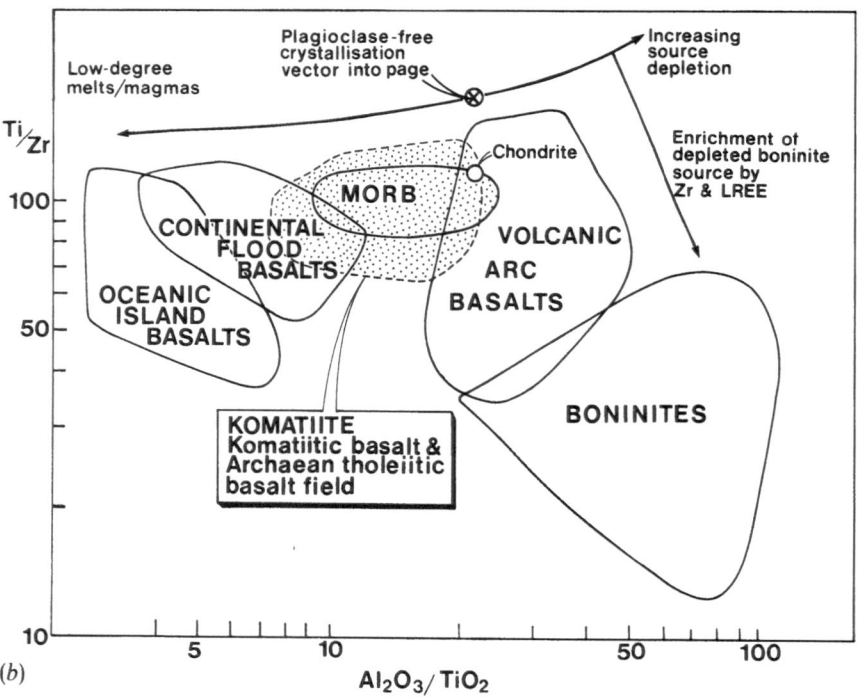

(b)

lavas. These are well illustrated by their REE profiles and Figure 2.7 shows the REE concentrations of two to three times those of chondrite for a boninite (13% MgO), and for an ophiolitic basalt from Troodos (9.3% MgO) with REE concentrations of four to six times chondrite. In contrast, a komatiitic basalt from Newton Township, Ontario, with 9.77% MgO, has REE concentrations between seven and sixteen times chondrite (Figure 2.3). The low concentrations of incompatible trace elements in boninitic and ophiolitic lavas are matched by low Ti contents but not by low Al and Ca contents. The very low Ti contents of boninites and ophiolitic lavas lead to very high Al_2O_3/TiO_2 ratios, compared with the chondritic (or lower) ratios found in komatiitic basalts (Figure 2.8). Such low incompatible element concentrations in the Phanerozoic Si- and Mg- rich lavas indicate their derivation from a source which was impoverished in these elements by a previous melting episode (Sun and Nesbitt, 1978b). Both recent boninites and Archaean komatiitic basalts show LREE enrichment superimposed on the effects of any previous melting episode in their source region, and the role of contamination by continental crust in producing the LREE enrichment of komatiitic basalts was discussed in section 2.5.2. Since boninites are found in intra-oceanic island arcs, their LREE enrichment can not be due to contemporaneous continental crust, but must be a feature of their source. Could the same be true for Archaean komatiitic basalts?

Boninites often display upwardly concave, U-shaped or V-shaped chondrite-normalised REE patterns (Figure 2.7). The simplest explanation of such complex REE distributions is that the source had previously been melted, and was strongly depleted in the LREE relative to the HREE. A component rich in LREE may have been added to this source prior to the boninite-generating event (Sun and Nesbitt, 1978b). Similar explanations have been offered to explain the similar REE patterns found in Ti-poor ophiolitic lavas (Cameron et al., 1983; McCulloch and Cameron, 1983; Cameron, 1985). The REE patterns for the Troodos low-Ti lavas (Figure 2.7) at first sight seem almost randomly variable, and it is not obvious that the various lavas are genetically linked. However, Taylor and Nesbitt (1988) showed that these REE patterns can be explained by a variably-depleted source region having been contaminated to varying degrees by a homogeneous, LREE-rich component. The LREE-rich component also introduced Zr (and Hf), but not significant Ti, P or Y. Boninites thus have very low

---

Figure 2.8 Ti/Zr versus Al_2O_3/TiO_2 for (a) various suites of komatiites, komatiitic basalts, tholeiites, low-Ti arc and ophiolitic lavas, boninites, and ocean island basalts, and (b) discrimination diagram highlighting the differences between Phanerozoic volcanic suites and komatiitic suites. Data sources:
Komatiitic provinces: Hallberg (1972), Gill (1979), Dietrich et al. (1981), Jahn et al. (1982), Ludden and Gelinas (1982), Nesbitt et al. (1982), Smith and Erlank (1982), Arndt and Nesbitt (1984) and Cattell (1985).
Volcanic arc basalts and low-Ti ophiolitic basalts: Ewart et al. (1973), Gill (1976), Arculus (1978), Mann (1983), Johnson et al. (1985), Luhr and Carmichael (1985), Crawford et al. (1986), Myers et al. (1986), Sakuyama and Nesbitt (1986), Volpe et al. (1987), Groves et al. (1988), Nixon (1988), Taylor (1988, 1989).
Oceanic basalts: Dostal et al. (1982), Chen and Frey (1985), Le Roex (1985), Liotard et al. (1986).
Continental flood basalts: McDougal (1976), Cox and Hawkesworth (1984), Fodor et al. (1985), Hooper (1985).
MORB: Blanchard et al. (1976), Gill (1979), Sun et al. (1979).
Boninites: Dietrich et al. (1978), Meijer (1980), Hickey and Frey (1982), Cameron et al. (1983), Walker and Cameron (1983), Crawford and Cameron (1985), Hall and Hughes (1987), Bloomer and Hawkins (1988), R. W. Nesbitt and R. N. Taylor (unpubl. data).

Ti/Zr ratios of 15 to 60 (Hickey and Frey, 1982; Hall and Hughes, 1987), compared with a chondritic value of 110 (Figure 2.8).

The enrichment of the LREE and Zr seems, at first sight, to indicate a similar genesis for komatiitic basalts, low-Ti ophiolitic basalts and boninites. However, in detail the observed enrichments are different. The degree of LREE enrichment in the Troodos lavas correlates with the degree of previous source depletion, as measured by, for example, Al_2O_3/TiO_2 values. This is not coincidental. The LREE-rich component which is added to the depleted source is best explained as a water-rich fluid, perhaps carrying a fraction of silicate melt, which acts as a flux. The more depleted the source, the greater is the amount of flux needed to cause melting (Taylor and Nesbitt, 1988). Such correlations are not found in Archaean komatiitic basalts.

2.7.3   Tholeiitic basalts

Tholeiitic basalt lavas are found in close association with komatiites and komatiitic basalts in almost all Archaean greenstone belts (section 2.1). The apparently unique Phanerozoic komatiite sequence of Gorgona Island is also overlain by tholeiitic basalts, enabling a direct comparison between Archaean and Phanerozoic tholeiite–komatiite associations to be made.

Aitken and Echeverria (1984) divided the Gorgona tholeiites into two types (K-type and T-type), on the basis of their REE profiles (Figure 2.7c) and Nd and Sr isotopic compositions. The K-tholeiites have REE profiles parallel to those of the Gorgona komatiites (Figure 2.7c), suggesting that the two lava types may be cogenetic. The major element data are also consistent with the hypothesis that the komatiites are parental to the K-tholeiites, the fractionating phases being first those observed in the komatiites (olivine) and then those in K-tholeiites (augite and plagioclase). However, Aitken and Echeverria (1984) suggested that the K-tholeiites represent simply a smaller degree of melting of the komatiite source, although the mg' values of the K-tholeiites are too low for this explanation to be viable. The T-tholeiites lack the marked LREE depletion of the komatiites and K-tholeiites (Figure 2.7c), and their markedly lower $^{143}Nd/^{144}Nd$ ratios show that they were derived from a different mantle source (Aitken and Echeverria, 1984). The T-tholeiites from Gorgona are similar to most Archaean basalts (section 2.3), while the K-tholeiites have only rare Archaean counterparts such as the LREE-depleted variety of amphibolites in the Archaean craton of West Greenland (Chadwick, 1981).

Many authors have noted the general similarities between Archaean and Phanerozoic low-K tholeiites (Nesbitt and Sun, 1976; Gill, 1979; Condie, 1985). However, no modern tholeiite corresponds exactly with the Archaean examples. Archaean tholeiites are enriched in Ni, Cr and Fe when compared with all modern basalts of similar mg' value, and they also show subtle differences in their incompatible trace element concentrations (Gill, 1979; Condie, 1985). Some of these geochemical differences are illustrated in Figure 2.6. Unfortunately, the unique trace element character of the Archaean tholeiites means that the assigning of these ancient basic volcanic rocks to a specific tectonic setting on the basis of their trace element chemistry is not reliable.

2.8   Concluding remarks

In the light of the geochemical characteristics of Archaean mafic–ultramafic volcanic suites discussed in this chapter, it can be seen that few, if any, products of Phanerozoic volcanism equate exactly with Archaean rocks. This in itself perhaps indicates a secular change in petrogenetic systems. Such a change could have been influenced by several factors which are not necessarily mutually exclusive, and include (i) variations in the thermal regime under which melting occurred, (ii) a temporal change in the chemistry of the upper mantle reflecting depletion events and (iii) evolution of the tectonic setting of magmatism through time.

Komatiites represent the products of high-degree partial melting of the mantle, and those preserved in greenstone belts were probably generated in sub-continental mantle. Associated komatiitic basalts are predominantly derivative magmas produced by interaction of komatiitic magma with continental crustal material. The virtual absence of komatiitic lavas from the Phanerozoic record reflects a secular decline in the temperature of the upper mantle. In the Archaean, komatiites could be generated only in hot rising plumes (section 2.5.1; Chapter 6), while under present global thermal conditions even plumes with temperatures 200°C above ambient mantle values are only capable of producing basalts (McKenzie and Bickle, 1988). At the present state of our knowledge, it appears that Archaean komatiitic basalts and Phanerozoic boninites are only superficially similar. However, there is no reason why boninitic melts should not have been generated in the early Precambrian, and it would be dangerous to assume that all Archaean high-Si high-Mg lavas are contaminated komatiites.

3 Geochemical characteristics of Precambrian basaltic greenstones

K.C. CONDIE

3.1 Introduction

One of the most fundamental questions concerning early Precambrian basic magmas is whether or not they are geochemically the same as those produced in equivalent tectonic settings at the present day. However, the geochemical classification of ancient basic igneous suites on the basis of inferred tectonic setting, and the interpretation of ancient tectonic settings based on geochemical criteria are, to a large extent, iterative but circular processes. One of the most significant differences between basic metavolcanic suites of the Archaean and those of post-Archaean times is the abundance of Archaean komatiitic suites. There are few younger analogues with which to compare these suites directly and therefore little can be achieved in terms of identifying temporal changes in these types of magmas. On the other hand, basaltic volcanic rocks have been present throughout the geological record.

This chapter catalogues some of the secular changes in basaltic volcanic rocks, the tectonic settings in which they might have formed in comparison to the compositions of modern basaltic rocks from different environments and the implications for their source mantle compositions. Many subtle changes in basaltic geochemical compositions with time have been reported previously, and are dealt with in this review simply by reference to previous work. Other discriminant geochemical criteria, and in particular trace element ratios discovered more recently, are documented in more detail.

The understanding of secular changes in the composition of basalts is important both to studies of mantle evolution and also in constraining crustal evolution (Glikson, 1971; Sun and Nesbitt, 1977; Condie, 1985, 1989a). The only well documented geochemical change in Archaean basalts compared to post-Archaean basalts is the progressive decrease in Ni, Co, Cr and related elements (Gill, 1979; Condie, 1985). Geochemical studies of Precambrian and Phanerozoic basalts clearly indicate that the mantle is heterogeneous on various scales and that these heterogeneities have been present in the order of 1.0 Ga (Allègre *et al.*, 1982; Zindler and Hart, 1986; Shirey and Hanson, 1986). Furthermore, major differences in incompatible element contents of basalts produced in different tectonic settings suggest a close relationship between mantle source composition and tectonic setting. Three factors must be addressed in accurately identifying the changes in source mantle composition with time using geochemical data from basalts: (i) the effects of alteration and metamorphism, which tend to mobilise some incompatible elements; (ii) the effects of magma–crystal equilibria imposed on basaltic magmas during their formation; and (iii) comparison of basalts of different ages from similar tectonic settings. The first two factors can be minimised by comparing ratios of incompatible elements that are relatively immobile during secondary processes. Elements such as the rare earth elements (REE), Ti, Y, Hf, Zr, Nb, Ta and Th have proved to be most successful in this respect.

The problem of identifying ancient tectonic settings is not so easily overcome. However, to ignore this factor as some previous studies have done, can only lead to misinterpretations of compositional changes in basalts with time. The average compositions of Precambrian basalts of a given age are meaningless, as are average compositions of modern basalts, if they are not grouped according to tectonic setting (Condie, 1989a). One way to minimise the tectonic setting effect is to compare basalts of different ages from the same lithological association. Dickinson (1971) showed that modern lithological assemblages reflect their tectonic settings and similar assemblages occur in the Phanerozoic and Proterozoic (Kröner, 1981; Condie, 1982; Pharaoh et al., 1987), and some are also recognisable in Archaean successions (Condie, 1989b). Of the various tectonic settings in which we know basalts occur at the present day, island arc systems, oceanic crust (ophiolites) and continental rifts (including mafic dyke swarms) are documented to about 2.0 Ga ago (e.g. Halls and Fahrig, 1987; Pharaoh et al., 1987; Kontinen, 1987). The greenstone assemblage, which appears to be characteristic of ancient arc systems, is found in both Archaean and post-Archaean terranes. Although Archaean mafic dyke swarms are also common, only one or two minor Archaean continental rift assemblages have been recognised (Saverikko, 1988), and clearly developed ophiolitic suites with sheeted dykes have not as yet been described from Archaean terranes.

The term greenstone belt has been used in a variety of contexts, but in this chapter it is restricted to *volcanic-dominated* submarine supracrustal assemblages (Table 3.1). This definition includes most supracrustal assemblages described from the Archaean and many Proterozoic assemblages. Greenstone belt assemblages are similar to modern arc-related assemblages deposited in various fore-arc, intra-arc, and back-arc basins. Archaean greenstone belts were at one time thought to reflect a distinct tectonic setting limited to the Archaean. However, as we learn more about Proterozoic greenstone successions, the apparently distinctive character of Archaean greenstone sequences begins to fade. Numerous early Proterozoic greenstone successions are lithologically indistinguishable from Archaean greenstones (Barager et al., 1977; Watters and Armstrong, 1985; Condie 1986a; Pharaoh et al., 1987) and clearly show that the 'greenstone' tectonic setting spanned the Archaean–Proterozoic boundary.

Table 3.1 Lithological assemblages of greenstone belts

Rock Types	Abundance (%)
Mafic volcanics*	20–65
Intermediate to felsic volcanics	5–30
Volcaniclastic sediments (incl. greywackes)	10–50
Chert and banded iron-formation (BIF)	$\leqslant 3$
Carbonate	0–1
Quartzite, arkose, conglomerate, non-volcanic shale	0–10
Hypabyssal intrusives	5–20

*Includes up to 50% komatiite in some Archaean successions

In this chapter, Precambrian basalts of various ages from greenstone assemblages are compared geochemically to evaluate more fully their reliability in monitoring mantle evolution, and to see if they record significant petrogenetic changes across the Archaean–Proterozoic boundary. Results indicate that greenstone assemblage basalts probably sample only a very small proportion of the mantle at any given time, and that they should not be used to record changes in the composition of large volumes of the mantle. The compositional differences between late Archaean and early Proterozoic basalts appear to reflect the rapid growth of the continents towards the end of the Archaean.

3.2　Classification of greenstone basalts

Analyses have been selected for this comparative study from published and un-published chemical data on greenstone basalts. Samples described as porphyries or cumulates have not been included. Since few post-Archaean komatiites are known, komatiites and basaltic komatiites also are not considered in this study although the absence of these high-Mg rocks from post-Archaean suites is, of course, one of the most fundamental of secular changes in greenstone belt metavolcanic assemblages. Basaltic rocks considered here include samples with 48–54% SiO_2, 15–21% $(CaO + MgO)$, > 12% Al_2O_3 and ≤ 15% MgO. Each chemical analysis has been passed through a succession of alteration 'screens' described by Condie (1985). However, less than 10% of the analyses have been rejected as having been severely geochemically altered. Because of the strong coupling between Zr and Hf, Nb and Ta, and Th and U in mafic volcanic rocks, Hf, Ta and U are calculated, if not reported in published analyses, as $Hf = Zr/37$; $Ta = Nb/17$; $U = Th/3$ (Jochum et al., 1986; Hofmann, 1986; O'Nions, 1987). Ratios of relatively immobile elements (such as Ti/Zr, Nb/Y, Hf/Ta, Th/Yb) are used to fine-tune rock classification (Condie, 1989a).

The greenstone basalts have been grouped into four age categories for comparative purposes: > 3.0 Ga, 3.0–2.5 Ga, 2.5–1.5 Ga and 1.5–0.5 Ga. Most published geochem-ical data are from greenstone successions of the two intermediate age groups and thus, the geochemical results may be more representative for these two categories. Only five successions are represented in the > 3.0 Ga group. These are Isua, southern West Greenland; Ancient Gneiss Complex, South Africa; Barberton, South Africa; Pilbara, Western Australia and Holenarasipur, southern India. Most data for the 3.0–2.5 Ga group are from late Archaean (2.8–2.6 Ga) greenstone belts in Canada, Western Australia and Scandinavia. Data in the 2.5–1.5 Ga group are chiefly from Scandinavia and southwestern USA, and those in the 1.5–0.5 Ga group are dominantly from the Pan-African terranes of North Africa and Arabia, and from North America and Scandinavia. The study includes over 1000 complete or partial chemical analyses, some of which are unpublished analyses by the author and his associates. Also included are a large number of published major element and trace element analyses from the Pilbara succession in Western Australia (Glikson and Hickman, 1981; Glikson et al., 1986). The Pilbara data include volcanics from both the > 3.0 Ga and 3.0–2.5 Ga age groups and thus provide a means of characterising secular compositional changes within a given geographic area (Condie, 1989a).

Five geochemical 'screens' have been used sequentially to classify greenstone basalts according to trace element ratios (Condie, 1989a). A summary of this classification is given in Table 3.2. Two of the most striking features of Precambrian basalts are the

Table 3.2 Classification of basalts from Precambrian greenstone successions in terms of modern basalt types

	> 3.0 Ga		3.0–2.5 Ga		2.5–1.5 Ga		1.5–0.5 Ga	
	n	%	n	%	n	%	n	%
IAB	15	29	104	67	49	18	9	20
CABI	21	40	26	17	120	44	25	54
CABC	13	25	15	10	96	35	10	22
N-MORB	3	6	10	6	5	2	2	4
WPB, T-MORB, E-MORB	0	0	0	0	3	1	0	0
Total	52		155		273		46	

n: number of samples; IAB: island arc basalt; CABI, CABC: calc–alkaline basalt from island arcs and continental-margin arcs, respectively; MORB: mid-ocean ridge basalt (N-, T-, E-: normal, transitional, enriched); WPB: oceanic within-plate basalt

absence of basalts that have the geochemical characteristics of modern within-plate basalts (WPB) and T-, E-MORB basalts in all groups apart from the 2.5–1.5 Ga group (where they are exceedingly rare), and the rarity of basalts with N-MORB geochemical characteristics in greenstone basalts of any age. Almost all Precambrian greenstone basalts have trace element distributions similar to those in modern basalts associated with island arcs. It is unlikely that the lack of rocks with N-type MORB compositions is due to the enrichment of incompatible elements during alteration or other secondary processes, because the geochemical screens used to identify N-MORB compositions are based on ratios of relatively immobile elements. In addition, if sampling is representative, late Archaean successions (3.0–2.5 Ga) are dominated by basalts with island arc basalt (IAB) geochemical characteristics. Basalts with geochemical features similar to calc-alkaline basalts from island arcs (CABI) dominate in the 1.5–0.5 Ga category. The absence or sparsity of basalts that exhibit WPB or MORB-type geochemical characteristics from Precambrian greenstone successions of all ages supports inferences based on lithological assemblages, which indicate that greenstone successions (as previously defined) do not represent remnants of oceanic crust, continental rifts or cratons. Portions of some Archaean greenstone sequences have recently been described as possible ophiolites (Helmstaedt et al., 1986; de Wit et al., 1987). The best candidates for Archaean ophiolites are those successions where komatiites are important, such as the lower part of the Barberton succession in South Africa. Based on the geochemical data (Tables 3.2 and 3.3), it is apparent that if some greenstones are ophiolites, they probably represent remnants of immature arcs or back-arc basins and not large ocean basins. The oldest major occurrences of basaltic rocks with MORB or WPB geochemical affinities are those reported from Proterozoic mafic dykes, ophiolites and rift associations (Pharaoh et al., 1987; Kontinen, 1987; Halls and Fahrig, 1987).

3.3 Geochemical characteristics

Average concentrations of selected major and trace elements and average element ratios for Precambrian greenstone basalts from each of the four age categories are summarised in Table 3.3. Average mg' values (Mg/(Mg + Fe)) in basalts of all ages are similar, falling in the range of 0.6–0.65. If komatiites were included in the statistical

Table 3.3 Average selected element concentrations and ratios in Precambrian greenstone basalts

	> 3.0 Ga			3.0–2.5 Ga			2.5–1.5 Ga			1.5–0.5 Ga		
	$\bar{x}$	σ	n	$\bar{x}$	σ	n	$\bar{x}$	σ	n	$\bar{x}$	σ	n
Sr	152	10	52	122	10	155	264	8	282	246	26	43
Ba	147	14	38	204	47	77	320	77	242	151	23	14
K_2O (%)	0.43	0.04	51	0.33	0.03	180	0.69	0.05	180	0.77	0.15	54
Rb	12	1	46	9.1	1	141	21	1.8	263	15	3.2	36
Th	1.0	0.3	19	0.60	0.08	41	1.4	0.07	180	1.6	0.5	12
Pb	4.1	1	49	3.6	1.1	252	5.3	1.1	16	7.6	1.3	28
La	7.4	1	38	5.5	0.6	70	11	0.5	205	13	2.5	19
Ce	19	1.8	47	15	1.5	80	27	1.2	205	31	4.4	19
Nd	11.2	2.0	27	9.6	1.0	80	15	0.7	195	18.2	2.0	12
Sm	3.0	0.23	27	2.8	0.16	80	3.8	0.14	195	4.2	0.84	12
Eu	1.0	0.1	27	0.97	0.05	65	1.3	0.05	195	1.3	0.16	12
Tb	0.56	0.07	14	0.68	0.07	27	0.72	0.02	182	0.75	0.11	12
Yb	2.3	0.22	21	2.5	0.14	73	2.7	0.1	197	2.8	0.38	12
TiO_2 (%)	0.99	0.07	53	0.97	0.03	192	1.19	0.03	288	1.10	0.10	57
Zr	81	6	54	65	3	181	92	3	273	92	10	57
Hf	2.2	0.18	44	1.8	0.62	53	2.7	0.09	204	2.4	0.91	57
Y	21	1	53	23	2	179	27	1	261	27	1.4	37
Ta	0.31	0.02	42	0.30	0.03	70	0.26	0.02	140	0.24	0.07	15
Nb	5.2	0.03	44	5.1	0.82	76	4.4	0.43	140	4.0	1.0	22
P_2O_5 (%)	0.11	0.07	52	0.12	0.02	139	0.19	0.06	265	0.25	0.03	51
V	244	13	28	314	28	93	289	7.1	140	253	26	14
Cr	446	98	41	406	29	273	169	16	266			
Ni	155	26	52	186	15	175	83	7	250	121	14	15
mg′	0.63			0.65			0.60			0.62		
La/Yb	3.2	0.45	21	2.1	0.25	70	4.2	0.21	186	4.6	0.92	12
Sm/Nd	0.27	0.05	25	0.29	0.03	80	0.25	0.01	190	0.23	0.04	12
Eu/Eu*	0.96	0.10	27	0.93	0.08	65	0.99	0.07	195	0.91	0.11	12
Zr/Y	3.8	0.26	50	2.8	0.28	176	3.5	0.16	260	3.4	0.41	57
Ti/Zr	81	6.5	51	102	5.1	176	84	3.3	273	86	10	57
Ti/Y	283	22	52	253	25	179	264	12	253	254	28	37
Ti/V	24	1.9	28	19	2.0	93	25	0.8	135	26	3.1	14
Hf/Ta	7.1	0.7	40	7.0	2.1	52	11	3.6	135	10	3.7	15
Th/Ta	3.2	0.9	19	2.0	0.26	40	5.4	2.5	140	6.7	2.1	12
Th/Yb	0.43	0.15	18	0.27	0.04	41	0.50	0.25	176	0.57	0.18	12
Ta/Yb	0.13	0.01	21	0.13	0.01	70	0.10	0.09	140	0.09	0.03	10
Zr/Nb	16	1.1	44	14	2.2	75	21	2.1	137	23	5.7	22
Nb/La	0.7	0.09	36	1.0	0.16	70	0.41	0.04	140	0.31	0.08	15
(Nb/U)	16	4.8	19	25	3.3	40	9.4	1.0	138	7.5	2.3	12
Th/Nb	0.19	0.06	19	0.12	0.02	40	0.32	0.03	138	0.4	0.12	12
Ce/Nb	3.7	0.41	43	2.9	0.47	76	6.1	0.50	186	7.8	1.9	19
Ce/Pb	4.6	1.1	45	4.2	1.3	78	5.1	1.1	16	4.1	0.70	18
Hf/Th	2.2	0.90	19	3.0	0.15	40	2.0	1.0	140	1.5	0.42	12
K/Rb	297	32	46	301	30	138	273	24	173	426	89	36
Rb/Sr	0.08	0.01	46	0.08	0.01	140	0.08	0.01	260	0.06	0.01	33

Concentrations in ppm unless noted otherwise; σ: one standard deviation of the mean; $\bar{x}$: ratios are mean values of individual samples; n: number of samples; mg′ $MgO/(MgO + FeO)$ molecular ratio with $Fe_2O_3/FeO = 0.15$; Eu/Eu* = (Eu/0.069)/(Sm/0.181) − ([Sm/0.181 − Tb/0.047]/3); Nb/U = (Nb/Th) × 3

analysis, the Archaean averages would of course exceed 0.65. Archaean greenstone basalts have higher contents of Cr, Ni, Co, Fe and platinum-group elements (PGE) than their post-Archaean equivalents (Glikson, 1971; Condie, 1985; Redman and Keays, 1985). Although the large-ion lithophile elements (LILE) are known to be mobile during alteration and metamorphism, the fact that the mean values of K, Rb, Sr, Ba and Th in Precambrian basalts have similar magnitudes to those in modern arc-related basalts suggests that losses and gains of these elements are averaged-out when the mean values from greenstone successions are considered. Mean K/Rb and Rb/Sr ratios range from 273 to 426 and 0.06 to 0.08 respectively, and do not show significant temporal variation (Table 3.3). Th and Hf contents of the Proterozoic groups are higher than those of the Archaean groups, and mean Hf/Th ratios are somewhat lower in Proterozoic basalts. The high Hf/Th ratio (3.0) in the 3.0–2.5 Ga group reflects the low mean Th value (0.6 ppm) of this group. In terms of the Th/Yb and Ta/Yb ratios considered by Pearce (1983), most greenstone basalts fall in or near modern arc-related basalt fields (Figure 3.1). Th/Yb ratios are lower and Ta/Yb ratios somewhat higher in Archaean than in Proterozoic basalts. These differences are apparent in both mean values and overall sample distribution. Relatively few Precambrian samples fall in the modern WPB or N-MORB fields in terms of these elements.

Average chondrite-normalised REE patterns of greenstone basalts are LREE-enriched with very small or negligible negative Eu anomalies (mean Eu/Eu* = 0.91–0.99) (Table 3.3) (Condie, 1989a). Proterozoic basalts show a greater degree of LREE enrichment (mean La_N/Yb_N = 4–5) than Archaean ones (mean La_N/Yb_N = 2–3) and also have greater overall REE contents. Less than 10% of greenstone basalts show LREE depletion and none has the LREE enrichment typical of modern within-plate

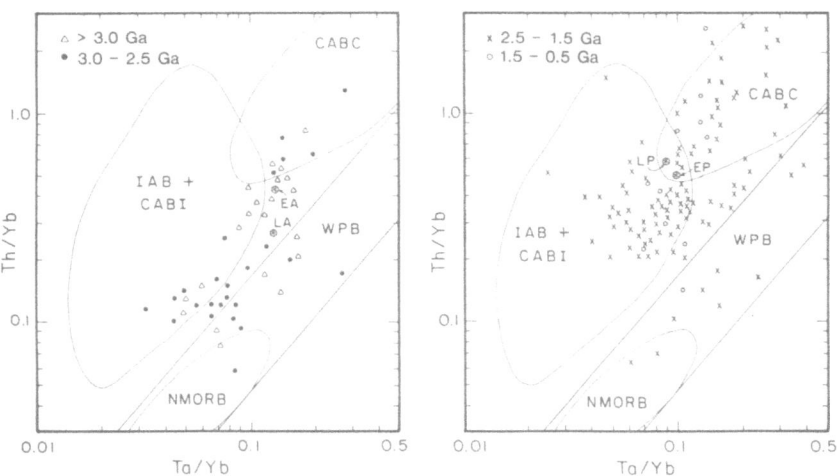

Figure 3.1   Th/Yb versus Ta/Yb diagrams showing the distribution of Precambrian greenstone basalts with respect to these elements. Modern basalt fields (after Pearce, 1983): WPB: within-plate basalt; IAB: island arc basalt; CABI and CABC: calc-alkaline basalts from oceanic and continental margin arcs respectively; N-MORB: normal mid-ocean ridge basalts; EA, LA, EP and LP: mean early and late Archaean and early and late Proterozoic basalts respectively.

basalts. In general, the late Archaean greenstone basalts appear to exhibit the least fractionated REE patterns.

Archaean basalts also have lower contents of Ti, Zr, Y, Hf and P than their Proterozoic counterparts (Figures 3.2, 3.3). Zr is especially low in the 3.0–2.5 Ga group, accounting for the high mean Ti/Zr and low Zr/Y values (102 and 2.8 respectively) in this group. The > 3.0 Ga group has the lowest Ti/Zr and highest Zr/Y ratios of any of the Precambrian groups (81 and 3.8 respectively), although the mean Ti/Y ratios, exhibit only limited variation (250–280; Table 3.3), and again the > 3.0 Ga rocks have the highest values. Ti–V distributions from all age groups scatter around the boundary between arc and WPB–MORB fields on the diagram proposed by Shervais (1982). Mean Ti/V ratios range from 24–26 except for the 3.0–2.5 Ga group which has a mean Ti/V value of 19. Mean Ta and Nb values are slightly higher in Archaean than Proterozoic basalts (Table 3.3), and the low Zr/Nb and Hf/Ta ratios in the Archaean groups reflect these differences. Low Th/Ta, Th/Nb, La/Ta, Ce/Nb and high Nb/La and Nb/U ratios in Archaean basalts are due to both Th, U and La depletion and slight relative Nb and Ta enrichment. These element ratios in all Precambrian greenstone basalts are similar to ratios characteristic of modern subduction-related basalts.

The mean incompatible element contents of basalts from each age category from Table 3.3 are shown on a MORB-normalised multi-element (NME) diagram in Figure 3.4. Overall, incompatible elements are relatively enriched in the Proterozoic basalts compared to the Archaean ones. The arc-like geochemical signatures of these patterns are striking. Particularly noteworthy is the negative Nb–Ta anomaly, a feature characteristic of basalts produced at modern convergent plate margins (Pearce, 1983), but also found in many continental within-plate basalts (CWPB) (Duncan, 1987; Marsh, 1987). Analytical errors generally result in higher concentrations of Nb and Ta, and hence the Nb–Ta anomalies in Figure 3.2 should be considered a minimum. The enrichment of LREE and Th relative to Nb and Ta is considered to reflect a subduction-zone geochemical component (SZC) and appears to reflect crustal contamination or/and production of magmas in a mantle wedge that has been enriched in LILE and LREE during devolatilisation of a descending plate (Gill, 1981; Pearce, 1983).

More than 90% of the chemical analyses for which Ta or Nb are available from Precambrian greenstone basalts are characterised by an SZC. The chief exceptions are basalts from some Archaean successions where komatiite is an important component. These basalts exhibit rather flat MORB-normalised element distributions from Th to Yb which reflect their depleted mantle sources.

3.4 Precambrian mantle sources

To recognise changes in mantle composition using element distributions in greenstone basalts it is necessary to evaluate the effects of varying degrees of melting and of fractional crystallisation on basaltic magmas. As previously mentioned, these effects can be minimised by considering incompatible element ratios, although open-system fractional crystallisation with small degrees of leakage can cause some incompatible element ratios to change significantly (Pankhurst, 1977). To avoid this problem, element ratios are selected which, for moderate degrees of open-system fractional crystallisation (< 50%), leakage of 1–20%, and a fractionating assemblage of olivine + clinopyroxene + plagioclase ± magnetite, change by factors generally of less than 2

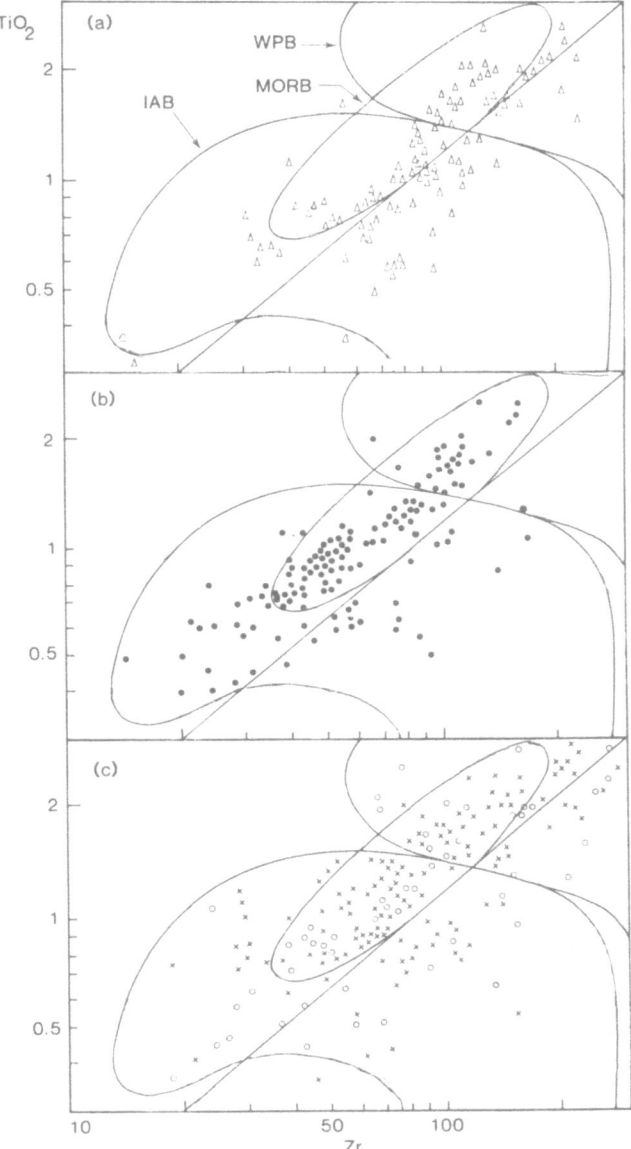

Figure 3.2   Plot of TiO_2 (%) versus Zr (ppm) for (a) early Archaean (> 3.0 Ga), (b) late Archaean (3.0–2.5 Ga), and (c) early Proterozoic (2.5–1.5 Ga; crosses) and late Proterozoic (1.5–0.5 Ga; open circles) greenstone basalts. Fields of modern island arc basalts (IAB), MORB and within-plate basalts (WPB) from Pharaoh and Pearce (1984). The diagonal line (Ti/Zr = 70) separates primitive (to the left) from evolved (to the right) members of these suites. Most of the early Archaean data in (a) are from the Pilbara Block, Western Australia. Data from sources mentioned in the text. (Redrawn from Condie, 1989a.)

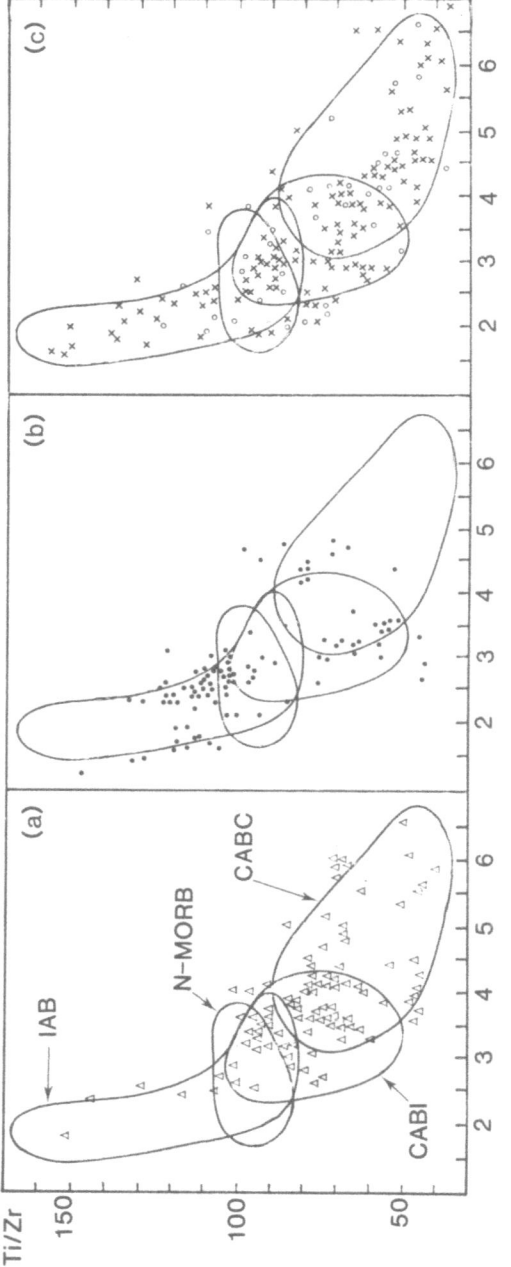

Figure 3.3   Plot of Ti/Zr versus Zr/Y for (a) early Archaean, (b) late Archaean, and (c) early and late Proterozoic greenstone basalts compared to the fields of modern **IAB**, **MORB** and calc-alkaline basalts. Symbols as in Figure 3.2. (Redrawn after Condie, 1989a.)

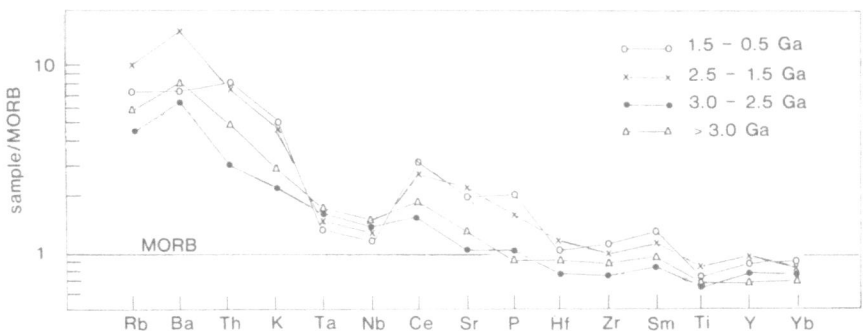

Figure 3.4 MORB-normalised multi-element plot of average early Archaean, late Archaean, early Proterozoic and late Proterozoic greenstone basalts (Table 3.3), normalised to MORB values quoted by Pearce (1983).

(Condie, 1985, 1989a). Differences in these element ratios between the four age groups of greenstone basalts should chiefly reflect differences in their mantle source compositions. The results indicate an overall enrichment in Proterozoic mantle sources as indicated by high Th/Nb, Zr/Nb, U/Nb, Ce/Nb, Hf/Ta, Th/Ta and La/Yb ratios and low Ti/Zr, Nb/La, Zr/Y, Hf/Th and Ta/Yb ratios (Table 3.3). A surprising anomaly characterises the 3.0–2.5 Ga group, which has Th/Nb, Ce/Nb, Th/Ta and Zr/Nb ratios that are lower and Ti/Zr, Nb/U, and Hf/Th ratios that are higher than the > 3.0 Ga group. A similar trend is observed for Ti/Zr and Zr/Y ratios between the early and late Archaean Pilbara data (Condie, 1989a). If sampling is representative for the > 3.0 Ga group, such differences indicate that late Archaean mantle sources were more depleted than corresponding early Archaean sources.

Rb/Sr and Sm/Nd ratios of basalts are important in monitoring these ratios in mantle sources, and thus in constraining the $^{87}Sr/^{86}Sr$ and $^{143}Nd/^{144}Nd$ ratios of the mantle with time. Sm/Nd ratios of greenstone basalts (0.23–0.29) fall in the range of modern arc-related basalts, while Rb/Sr ratios (⩾ 0.06) are higher than those of arc-related basalts (0.02–0.04) (Table 3.3) (Condie, 1989a). Because the samples in all four greenstone basalt age groups were selected on the basis that they appear to exhibit the same degree of fractionation, it is likely that the low Sm/Nd ratios in the Proterozoic samples indicate more enriched mantle sources than those of the Archaean basalts. However, the fact that ε_{Nd} values for Archaean greenstone basalts are chiefly positive (Fletcher and Rosman, 1982; Condie, 1986b; Shirey and Hanson, 1986) indicates depleted mantle sources. Thus, the low Sm/Nd ratios in those basalts that do reflect enriched sources must be produced by short-lived enrichment events at or near the time of magma production. It is likely that these low ratios reflect an SZC produced at, or not more than 100 Ma before, the time of magma production. The anomalously high Rb/Sr ratios in greenstone basalts compared to modern arc-related basalts are probably caused by the remobilisation of Rb or/and Sr during alteration and metamorphism.

Although Sr, Pb and Nd isotopes are most definitive in characterising mantle sources (Zindler and Hart, 1986), recent studies indicate that Ce/Nb, Nb/U, Th/Nb and Ce/Pb ratios in basalts are also sensitive to mantle composition (Hofmann *et al.*, 1986;

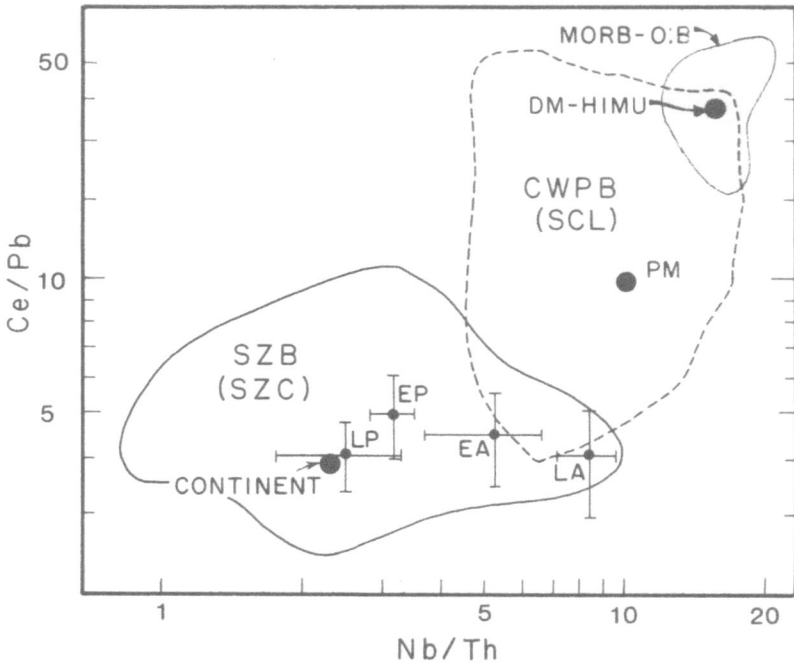

Figure 3.5 Ce/Pb versus Nb/Th diagram for modern and Precambrian greenstone basalts. Precambrian values are mean values with one standard deviation (Table 3.3). OIB: ocean island basalts; SZB: subduction zone related basalts; SZC: subduction-zone component; CWPB: continental within-plate basalts; DM: depleted mantle; HIMU; high U/Pb (Th/Pb) mantle; PM: primordial mantle (Wood, 1979); SCL: sub-continental lithosphere. Other symbols as in Figure 3.1. Average upper continent composition from Taylor and McLennan (1985).

Saunders *et al.*, 1988). The differences between modern basalts from various tectonic settings in terms of these ratios are shown in Figures 3.5 and 3.6. The major mantle types recognised from Nd, Pb and Sr isotopic studies (Zindler and Hart, 1986) are also shown. Oceanic island basalts (OIB) and N-MORB exhibit a limited distribution and are indistinguishable from each other in terms of their Ce/Pb and Nb/U ratios (Figure 3.5). They are, however, distinguished by their respective Ce/Pb and Th/Nb ratios (Figure 3.6). Subduction zone related basalts (SZB) (which carry an SZC) are enriched in U, Th, Ce and Pb relative to Nb. Continental within-plate basalts (CWPB) generally overlap SZB or fall between SZB and OIB and N-MORB in terms of these ratios. CWPB also carry a SZC, although sometimes only weakly developed. These basalts are probably derived from the sub-continental lithosphere (SCL) (Hawkesworth *et al.*, 1983; Marsh, 1987), which acquired a SZC during earlier periods of subduction. Estimates of primordial mantle (PM) and continental crust composition fall in either SCL or SZC fields. Average compositions of Archaean and Proterozoic greenstone basalts plot in the SZC fields on Figures 3.5 and 3.6. The lack of basalts with MORB or OIB characteristics is striking. In addition, Archaean basalts clearly exhibit lower Ce/Nb and Th/Nb ratios and higher Nb/U ratios than Proterozoic basalts. However,

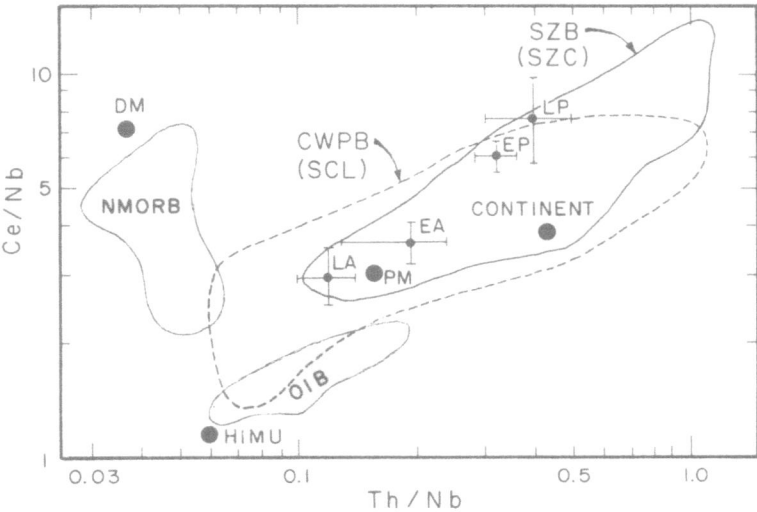

Figure 3.6 Ce/Nb versus Th/Nb diagram for modern and Precambrian basalts. Symbols as in Figures 3.1 and 3.5.

as with other previously mentioned trace element ratios, a reversal in the degree of LILE enrichment appears to occur in going from early to late Archaean basalts.

3.5   Discussion

3.5.1   *Selective preservation of lithological assemblages*

Both the geological and geochemical data suggest that Precambrian greenstone successions formed in island arc-like environments, and hence that some form of plate tectonics was operative on the Earth by 3.8 Ga (Condie, 1989a, b). Although several lines of evidence suggest that many Archaean greenstones formed in rifts on or near older continental crust (Thurston *et al.*, 1985; Nisbet, 1986; Compston *et al.*, 1986; Kröner *et al.*, 1988), the geochemical evidence suggests that such rifts must have been associated with subduction, representing perhaps intra-arc or back-arc basins. Thus in terms of our understanding of mantle reservoirs, it is possible that greenstone basalts only sampled a relatively small volume of the mantle (mantle wedges?) at any given period of time.

Precambrian oceanic crust must have been at least as widespread as modern oceanic crust. Why then are there no greenstones that contain significant quantities of basalt with MORB-like compositions? This could reflect inadequate sampling of rocks with MORB compositions in greenstone sequences, the alteration or crustal contamination of Precambrian MORB so that they resemble arc basalts, or the selective preservation of arc-like greenstone successions. Although the sample population and number of greenstone localities considered so far may be inadequate to identify and characterise early Archaean tectonic settings, this does not seem to be a problem for the late

Archaean successions. The sample population for the early Proterozoic sequences is probably representative, although the geographic distribution may not be so adequate (most come from the southwestern USA and Scandinavia). Although the number of samples from late Proterozoic greenstone belts is not great (Table 3.2), it may be representative in that the compositional range is similar to that found in Phanerozoic arc successions. The problem of element remobilisation during secondary processes has been minimised by considering only relatively immobile elements. Although LILE and LREE can be added to MORB during deep-sea alteration, the elements and element ratios selected to characterise MORB are among the least mobile during these processes (Condie *et al.*, 1977; Hajash, 1984; Condie, 1989a). It is highly unlikely that all of the relatively immobile elements were enriched in just the right proportions to make greenstone basalts compositionally similar to arc basalts, if they originally represented MORB-type lavas. The possibility of MORB contamination by continental crust is not so easily discounted. Nd isotopic data, however, suggest that most Precambrian greenstone basalts are derived from depleted mantle sources (Shirey and Hanson, 1986) with little if any contamination by significantly older continental crust.

The selective preservation of lithological assemblages from certain tectonic settings is an important constraint in assessing the relative areal distribution of Precambrian tectonic regimes. For instance, Phanerozoic oceanic regimes, which include oceanic crust, island arcs and related basins, and passive continental margins are rapidly recycled with estimated half-lives of 30 to 80 Ma. Continental regimes (cratonic basins) on the other hand have half-lives of 300 Ma or more (Veizer, 1988). Hence, it is not surprising that MORB (ophiolites) and OIB are not readily preserved in the Precambrian record. Likewise, continental rift basalts may have been rapidly eroded from the continents as continents rose isostatically during cratonisation and widespread Precambrian mafic dyke swarms (Halls and Fahrig, 1987) may be the preserved plumbing systems of Precambrian rifts. The apparently selective preservation of Precambrian greenstones which should have rapid recycling rates similar to or greater than Phanerozoic arcs requires an explanation. Most Precambrian greenstones, and especially late Archaean ones, may owe their widespread preservation to underplating by tonalite–trondhjemite–granodiorite (TTG) suite material which prevented them from being subducted. However, why Archaean oceanic crust and oceanic islands are not occasionally preserved by obduction or addition to accretionary prisms is not clear. In any case, it is probably not valid to estimate the original areal distributions and importance of specific Precambrian tectonic settings based on the proportions of various preserved lithological assemblages.

3.5.2 *Early and late Archaean greenstone basalts*

It is important to know if the differences in composition between early and late Archaean greenstone basalts (Table 3.3) represent simply a preservational and sampling bias, or reflect real differences in mantle source compositions between the early and late Archaean. The few preserved successions which are > 3.0 Ga old contain basalts (and andesites) similar in composition to basalts from modern continental-margin arcs. The composition of late Archaean basalts, on the other hand, is suggestive of more immature arc systems in which IAB dominate, and these rocks also reflect more depleted mantle sources (Figures 3.5 and 3.6). Perhaps large numbers of these IAB-dominated arcs did also exist in the early Archaean but were not preserved by TTG underplating. Extensive TTG plutonism in the late Archaean may have been responsible for the widespread preservation of relatively immature arcs, whereas in the early Archaean plutonism may have been somewhat more localised in the few continental-margin arcs that existed. Alternatively, early Archaean basalts may contain a large number of crustally contaminated komatiites. Due to their high temperatures, komatiitic magmas could have become contaminated with continental crust and compositionally might look like basalts (Huppert and Sparks, 1985; Arndt and Jenner, 1986; Condie and Crow, 1989). Early Archaean greenstones do appear to have formed on or near to continental crust (Kröner and Todt, 1983; Boak and Dymek, 1989).

A third possibility is that the differences may reflect the progressive extraction of large volumes of continent from the upper mantle during the Archaean. In this case, the > 3.0 Ga mantle would be less depleted or undepleted relative to the late Archaean mantle. This needs to be tested further with Nd isotopic studies.

3.5.3   The Archaean–Proterozoic boundary

Inadequate sampling or preservational biases are probably not responsible for the compositional differences between Archaean and Proterozoic greenstone basalts (Table 3.3). It is difficult to escape the conclusion that Archaean greenstone basalts are dominantly tholeiites with geochemical compositions characteristic of relatively depleted SZC-mantle sources, whereas Proterozoic basalts are probably derived from more enriched SZC-mantle sources. The abundance of IAB-like basalts in Archaean greenstones may reflect relatively large degrees of melting at shallow depths in response to steeper geotherms in the Archaean.

The relative enrichment of siderophile elements in Archaean basalts has been attributed to two different causes, namely the enrichment of these elements in Archaean mantle sources (Glikson, 1971; Gill, 1979), and larger degrees of partial melting in the Archaean upper mantle as a result of steeper geotherm (Nesbitt and Sun, 1976; Condie, 1985). However, because (rare) Phanerozoic komatiites have similar enrichments of Ni, Cr and Co to Archaean komatiites, a secular change in siderophile element contents of the mantle seems unlikely. Other arguments against changes in mantle composition with time are reviewed by Condie (1985). Larger degrees of melting in the Archaean mantle should result not only in increased siderophile element contents of basalts, but also in elevated MgO contents (Gill, 1979; Arndt, 1989). Hence, when basalts with comparable MgO contents or mg′ numbers are compared, both Archaean and post-Archaean basalts should have similar siderophile element contents. Thus, the siderophile element enrichment without MgO enrichment in Archaean basalts is a serious obstacle to the melting model. Arndt (1989) has recently proposed a third possibility to explain these differences between Archaean and post-Archaean basalts. This is that many Archaean basalts are the products of mixing of komatiitic and basaltic magmas. This model is attractive in that variable mixing ratios can account for the wide range in Ni contents of Archaean basalts at specific MgO contents or mg′ values (Condie, 1985).

The enrichment of incompatible elements in Proterozoic relative to Archaean greenstone basalts seems to necessitate an increased role of continental crust in Proterozoic magma production. This enrichment may reflect an increased amount of SZC in mantle wedges or/and the contamination of basaltic magmas with continental crust. Both mechanisms involve continental crust. An increased SZC in mantle wedges in the Proterozoic could result from an increase in the amount of subducted terrigenous sediment or an increase in the amount of incompatible elements in subducted pelagic sediments and altered oceanic crust. However, the absence of HREE depletion in early Proterozoic greenstone basalts does not favour an increase in the amount of subducted continental sediment. Because Archaean granitoids typically show HREE depletion (Martin, 1986), sediments derived from them should also exhibit HREE depletion, and equally, if these sediments were subducted and contributed to the SZC, the derivative contaminated basalts should show signs of HREE depletion. An increase in the volume of continental crust in the late Archaean would have increased the rate of supply of incompatible elements to the oceans (via weathering and erosion), which would in turn have increased the contribution of these elements to pelagic sediments and to altered MORB-type basalts. The net result would have been an increase in the amount of incompatible elements subducted and subsequently carried into mantle wedges by escaping volatiles. Various lines of evidence for the rapid growth of the continents in the late Archaean support this explanation for the increased incompatible element contents in Early Proterozoic basalts (Nelson and De Paolo, 1985; Taylor and McLennan, 1985; Condie, 1986b).

Mean values of Hf/Ta and Nb/Th ratios in Precambrian greenstone basalts and in basalts from mafic dyke swarms in the southern part of the Superior Province in

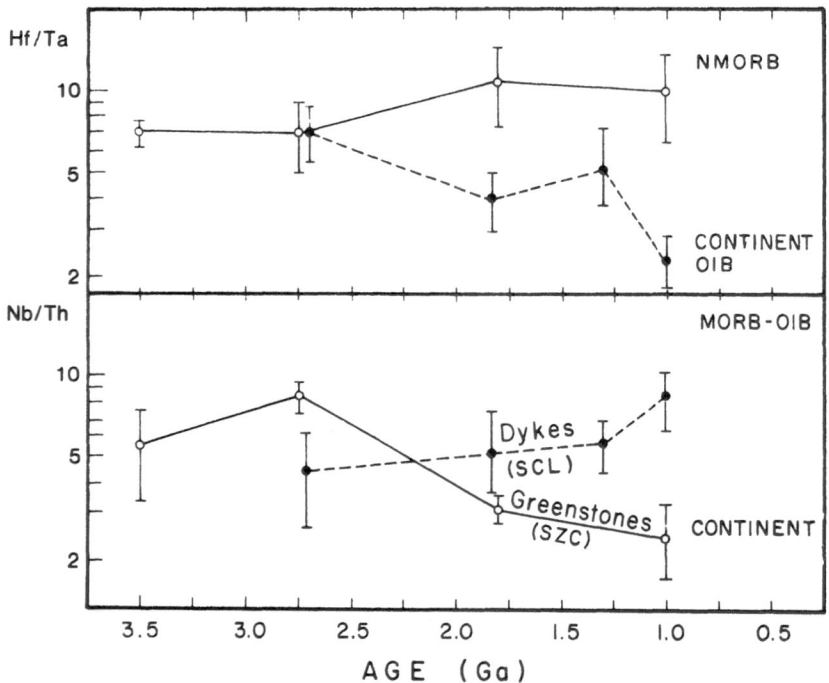

Figure 3.7 Hf/Ta and Nb/Th ratios in Precambrian basalts (open circles) and Superior Province mafic dykes (dots) plotted against time. Each point is a mean value with one standard deviation (vertical bar). Data from Table 3.3 and Condie *et al.* (1987).

Canada (Condie *et al.*, 1987) are shown in Figure 3.7 (see Chapter 10). The latter are continental within-plate basalts probably derived from the sub-continental lithosphere (SCL) beneath the Superior Province. The greenstone basalts show an increase in the Hf/Ta ratio and a decrease in the Nb/U ratio after the end of the Archaean. The dykes, on the other hand, show a decrease in Hf/Ta ratio and a slightly increasing Nb/Th (and Nb/U) ratio for the same time period. The increase in Hf/Ta and decrease in Nb/Th ratios in the greenstone basalts may reflect an increased SZC due to late Archaean continental growth, with Nb and Ta preferentially retained in descending slabs, perhaps in minor phases (Pearce, 1983). The opposite behaviour of these ratios in the Precambrian dykes may be caused by the contamination of the basaltic magmas by continental crust or by mixing with OIB mantle sources (Figure 3.7). The increase in Nb/Th in the youngest dykes, however, favours the mixing of SCL and OIB mantle sources, a trend which may have increased in intensity with time. This mixing, in turn, may reflect the interaction of mantle plumes with the SCL beneath the Superior Province. In any case, the Hf/Ta and Nb/Th ratios in basalts appear to be sensitive indicators of Precambrian mantle sources of both subduction zone related basalts and continental within-plate basalts.

3.6 Conclusions

1. Due to selective preservation in the geological record, it is unlikely that Precambrian basaltic volcanic assemblages are representative of the original areal proportions of Precambrian basalts from different tectonic settings.
2. Basalts with island-arc geochemical affinities are dominant in Archaean greenstone assemblages while those with calc–alkaline affinities are more abundant in Proterozoic greenstones. Basalts with MORB or within-plate geochemical characteristics are rare in Precambrian greenstone sequences of all ages.
3. Most Precambrian greenstone basalts, regardless of age, possess a subduction zone geochemical component similar to that in their basaltic counterparts from modern arc systems.
4. The preserved Precambrian greenstone basalts appear to be derived from mantle sources that occupy a very small volume of the total mantle (i.e., mantle wedges).
5. Archaean greenstone basalts are enriched in siderophile elements relative to post-Archaean basalts. This enrichment may reflect varying degrees of mixing of komatiitic and basaltic magmas during the Archaean.
6. If not caused by a preservational bias, early Archaean greenstone basalts (> 3.0 Ga) reflect either a less depleted mantle source than late Archaean greenstone basalts (3.0–2.5 Ga), or a greater proportion of crustally contaminated komatiites in the early Archaean.
7. Proterozoic greenstone basalts are derived from relatively enriched mantle sources compared to Archaean sources, a feature which may be due to increased incompatible element contents of subducted pelagic sediments and altered oceanic crust in the early Proterozoic. These element enrichments appear to reflect rapid growth of the continents in the late Archaean.
8. Hf/Ta and Nb/Th ratios may be sensitive indicators of secular changes in source composition of both subduction zone related basalts and continental within-plate basalts.

4 Early Precambrian layered intrusions

C.J. HATTON and G. VON GRUENEWALDT

4.1 Introduction: intrusion and crystallisation mechanisms

Large-scale layered basic intrusions are a major phenomenon in the late Archaean and early Proterozoic. The most important formed in three periods: the Stillwater Complex of southern Montana, USA at 2.7 Ga, the Great Dyke of Zimbabwe, the Jimberlana intrusion of Western Australia, and the intrusions of the Tornio–Narankavaara belt of the Baltic Shield at 2.45 Ga, and the Bushveld and Molopo Farms complexes in the Kaapvaal craton, South Africa at 2.05 Ga. Earlier Archaean major anorthosite complexes are also widespread (Table 4.1). The 2.7 Ga and 2.05 Ga intrusions were emplaced as sills and it will be argued that they may be related to major orogenies. The 2.45 Ga intrusions are canoe-shaped in form and may be related to rifting. All the intrusions have a broadly similar lithostratigraphy, a lower ultramafic unit and an upper mafic unit. Orthopyroxenite is a major component of the ultramafic units and may indicate a distinctive high-Si, high-Mg parental magma. High K, LREE and Zr contents in these magmas raise questions about an upper crustal component, either assimilated during magma ascent or incorporated into the source mantle by the earlier subduction of sediments.

Early studies of layered intrusions assumed that they formed by a process of crystal fractionation, and that the layering was produced by a combination of gravity settling and convection of a single parental magma. This model was elaborated upon by Hess (1960) and Wager and Brown (1968) to explain rhythmic successions of gravity stratified layers and homogeneous cumulates, which are characteristic features of large-scale layered intrusions. From this hypothesis a 'cumulate terminology' for the description of the igneous rocks in layered intrusions has evolved (Wager *et al.*, 1960; Jackson, 1967; Irvine, 1982). More recently, the traditional concept of gravity settling has been questioned. Jackson (1961) postulated a theory of bottom-crystallisation and variable depth convection to account for the repetitive stratigraphy in the ultramafic sequences of the Stillwater Complex and other layered intrusions (Jackson, 1970). Density calculations of fractionated basaltic liquids (Bottinga and Weil, 1970) and experimental studies on crystal buoyancy (Campbell *et al.*, 1978) showed that plagioclase is unlikely to settle through the late stage of Fe-rich liquids of layered intrusions. Other indications that crystal settling is not as important as previously suggested were summarised by Irvine (1979, 1980), who concluded that the formation of cumulates involved very little crystal settling and that the fractionated crystals either grew *in situ* or were transported to their position at the base of the magma chamber by magmatic density currents. The first alternative, referred to as bottom growth, is now widely acclaimed as an important process in layered intrusions, and the concept has been extended and applied not only to low density minerals like plagioclase but also to phases such as magnetite which are denser than any reasonable magma composition (Cawthorn and McCarthy, 1980). The problems associated with gravity settling have led to the exploration of alternative processes by which layering and differentiation might have developed. One of these is double-diffusive convection, a process illustrated experimentally by Turner and Chen (1974) and applied to layered intrusions by McBirney and Noyes (1979) and Irvine (1980). This process is thought to have operated in magma chambers where a stratification of liquids has developed in response to the addition of magma with a composition, temperature and density different to that already in the chamber.

The concept of periodic replenishment of the magma chamber with fresh, undifferentiated magma was postulated to account for the cyclicity of layering in the Bay of Islands Complex, and the Rhum (see Emeleus, 1987) and Muskox layered intrusions (see Irvine, 1980). This model explains the disproportionately large amount of ultramafic rocks by suggesting that the new influx displaces some of the existing differentiated liquid upwards in the chamber and to the surface as volcanic fissure eruptions. Although in earlier papers it was suggested that the new and the already partially differentiated magmas mixed, numerous recent fluid dynamic modelling experiments of magma chambers (e.g. Huppert and Sparks, 1980; Huppert et al., 1984; Sparks and Huppert, 1984) have shown that periodically replenished magma chambers develop a gravitational stratification of horizontal liquid layers of upward decreasing densities. Another aspect which has gained prominence in recent years is the possibility that more than one magma type must have been involved in the formation of the large Precambrian layered intrusions. In the case of the Bushveld Complex the observed variations in initial $^{87}Sr/^{86}Sr$ (Sr_i) ratios (Kruger and Marsh, 1982; Sharpe, 1985) present irrefutable evidence for multiple magma intrusions, and Todd et al. (1982) and Barnes and Naldrett (1985) reached a similar conclusion for the Stillwater Complex from petrogenetic and geochemical considerations respectively.

Early Precambrian layered intrusions are notable for their extensive chromitite and platiniferous layers, which are now widely believed to have formed by the interaction between different magma types (Irvine et al., 1983; Sharpe and Irvine, 1983; Campbell et al., 1983; Naldrett et al., 1986; Hatton et al., 1986). Major deposits of chromitite and platiniferous layers are found in the Bushveld, Great Dyke and Stillwater intrusions, and minor platinum and chromite deposits occur in several smaller early Precambrian intrusions. Apparently, major chromitite layers are a unique feature of early Precambrian layered intrusions. Younger intrusions contain either very thin (e.g. Rhum) or no (Skaergaard) chromitite layers. The whole question of the origins of igneous layering has been reviewed by Parsons (1987).

In this chapter we highlight those features which may be common to all early Precambrian layered intrusions, in order to gain some insight into the processes which gave rise to their emplacement and to the spectacular layering contained in them. We focus on the Bushveld and Stillwater complexes and major dyke-like intrusions (the Great Dyke of Zimbabwe, those in northern Finland and the Jimberlana intrusion of Western Australia) because these are the largest, best preserved and most intensively studied. Anorthosite complexes, layered ultramafic sills and some intensely deformed early Archaean complexes are compared with these major intrusions. For each intrusion the location, age, intrusion form, lithology and parental magma composition (where known) are described. These topics are summarised in Tables 4.1 and 4.2, and

Table 4.1 Early Precambrian layered intrusions

Intrusion	Province	Age* (Ga)	Areal extent (km)	Thickness (km)	Intrusion form	Lithologies	Refs.
Manfred	Y	3.73	3 × 2		A	An, gab	22, 35
Selukwe	Z	>3.5	10 × 10	1	dS	Px, dun	15, 44, 45
Barberton	K	3.54	(120 × 100)	1.3	dSs	Dun, hb, px	7
Julimar	Y	A	(80 × 30)		dSs	Gab, nor	22
Donnelly R.	Y	A	(60 × 40)		dSs	Serp, dun	22
Sittampundi	I	3–3.2	20 × 20	1.8	A	An, gab, ec	46
Messina	L	3.15	(300 × 60)	1	A	An, gab	43, 9
Holenarasipur	I	3.095	20 × 3		A	Serp, gab, an	28
Fiskenaesset	G	2.8–3	(200 × 15)		A	An, gab, px, per	34, 53, 57
Usushwana	K	2.87	(60 × 50)		DS	px-gabbro	23, 26
Dore Lake	S	A	43 × 6	6.2	A	An	6
Shawmere	S	A	(40 × 20)		A	An	42
Bad V. Lake	S	2.7	30 × 4		A	Gab, an	8
Windimurra	Y	>2.6	85 × 35	5	A	Gab, an	2, 24, 38
Stillwater	M	2.7	55 × 8	6.5	S	Br, hb; gbnr, an	30, 37, 40
Mushaba	Z	2.7	(80 × 30)	4	dS	Dun, hb; gab, nor	45, 55
Rooiwater	K	>2.65	55 × 10	7.5	S	Gab, an	48
Bird River	S	2.75	20 × 1	0.7	S	Per, gab	25, 36
Abitibi	S	A	(130 × 15)	1	Ss	Per, cp, gab	31
Donaldson	S	A	(140 × 20)		dS	Px, hb	16
Munni Munni	P	A	26 × 9	5.5	S	Per, cp, gab	17
Mt Sholl	P	A	20 × 5	2		Serp, gab	24
E. Goldfields	Y	A	(220 × 150)	2	Ss	Per, br; nor, gab	54
Mt. Kilkenny	Y	A	8 × 5	0.6	S	Tr, gab	27
Barrambie	Y	A	21 × 0.4	1	S	An, gab	24
Orissa State	I	A			dS	Serp, px	45
Goiás	SF	A			dS	Hb, px, gab	45
Bahia	SF	A			dS	Gab, an	45

Great Dyke	Z	2.46	550 × 8	3	D	Br, hb; gbnr	39, 56
Jimberlana	Y		200 × 2		D		
Binneringie	Y	2.37	320 × 3		D	Gab, nor	32
Tornio-Narankavaara belt	B	2.44	300 × 100	3	D	Gbnr	29
Bushveld	K	2.05	300 × 200	6	S	Br, per; gbnr	49
Molopo Farms	K	2.05	130 × 190	3.5	S	Hb, br; gbnr	20, 50
Fox River	S	P	250 × 3	2.2	S	Dun; pl. lh	41
McIntosh	HC	P	14.5 × 5	1.5	S	Tr, gab	24
Panton	HC	P	11 × 3.2	1	S	Ub, gab	24
Lamboo	HC	P	10 × 6.5		S	Ub, gab	24

*Where no isotopic age estimates are available ages are given as A: Archaean and P: Proterozoic. **Provinces:** B: Baltic; G: Greenland; HC: Hall's Creek; I: India, K: Kaapvaal; L: Limpopo; Ma: Malagasy; M: Montana; P: Pilbara; S: Superior; SF: São Francisco; Y: Yilgarn; Z: Zimbabwe.

Lithological abbreviations: an: anorthosite; gab: gabbro; px: pyroxenite; dun: dunite; ec: eclogite; serp: serpentinite; per: peridotite; br: bronzitite; hb: harzburgite; gbnr: gabbronorite; nor: norite; pl. lh: plagioclase lherzolite; tr: troctolite. **Intrusion form abbreviations:** D: dyke, S: sill; DS: intrusion exhibiting both dyke and sill-like form; dS: deformed sill (deformation more severe than tilting); dSs: deformed sills; A: anorthosite complex (deformed sills).

References (see also Figures 4.3 and 4.7): 1 Abbott and Ferguson (1965); 2 Ahmat and de Laeter (1982); 3 Alapieti and Piirainen (1984); 4 Alapieti and Lahtinen (1986); 5 Alapieti et al. (1979); 6 Allard (1970, 1976); 7 Anhaeusser (1985); 8 Ashwal et al. (1983); 9 Barton et al. (1979); 10 Buchanan (1975); 11 Cameron (1978, 1980, 1982); 12 Campbell (1970); 13 Campbell et al. (1977); 14 Coertze (1970); 15 Cotterill (1979); 16 Dillon-Leitch et al. (1986); 17 Donaldson (1974); 18 Engelbrecht (1985); 19 Furgason (1977); 20 Gould et al. (1986); 21 Grobler and Whitfield (1970); 22 Harrison (1984); 23 Hegner et al. (1984); 24 Hoatson (1984); 25 Hulbert et al. (1988); 26 Hunter (1970); 27 Jaques (1976); 28 Kutty et al. (1984); 29 Lahtinen (1985); 30 Lambert et al (1985); 31 MacRae (1969); 32 McCall and Peers (1971); 33 Molyneux (1974); 34 Myers (1985); 35 Myers (1988); 36 Ohnenstetter et al. (1986); 37 Page et al. (1985); 38 Parks and Hill (1986); 39 Prendergast (1987); 40 Raedeke et al. (1985); 41 Scoates and Eckstrand (1986); 42 Simmons et al. (1980); 43 Sohnge et al. (1948); 44 Stowe (1987a); 45 Stowe (1987b); 46 Subramaniam (1956); 47 Van der Merwe (1976); 48 Vearncombe et al. (1987); 49 Von Gruenewaldt et al. (1985); 50 Von Gruenewaldt et al. (1988); 51 Von Gruenewaldt (1973); 52 Vuorelainen et al. (1982); 53 Weaver et al. (1981); 54 Williams and Hallberg (1973); 55 Wilson (1968); 56 Wilson (1982); 57 Windley (1973)

Table 4.2 Possible parental magma and bulk compositions of layered intrusions

	Komatiitic	High magnesium basalt									Tholeiitic			Anorthositic		
	1 Barb	2 Abi	3 MtM	4 Jimb	5 Binn	6 Koil	7 EGD	8 Stw	9 Blw	10 Ble	11 B2	12 B3	13 Mtk	14 Mess	15 BVL	16 Fisk
SiO_2	47.62	51.54	51.91	53.87	54.13	54.26	53.03	50.40	55.63	56.32	49.94	51.51	49.47	50.81	50.23	48.57
TiO_2	0.36	1.25	0.41	0.41	0.60	0.47	0.55	0.54	0.36	0.33	0.68	0.40	0.49	1.11	0.56	0.57
Al_2O_3	3.97	9.71	12.40	13.16	16.51	15.26	11.05	13.74	12.72	11.43	16.21	16.32	16.89	14.07	19.75	16.26
FeO^*	11.78	13.54	9.70	9.16	6.76	7.73	9.35	11.73	8.77	9.42	11.96	9.01	10.07	11.66	8.11	10.00
MnO	0.19	0.21	0.17	0.17	0.15	0.13	0.14	0.21	0.09	0.18	0.20	0.16	0.19	0.26	0.16	0.20
MgO	29.36	11.38	15.70	11.83	8.44	10.09	15.67	11.41	12.39	13.14	6.95	8.37	8.19	9.11	6.94	9.47
CaO	6.82	9.68	8.10	8.57	9.76	8.99	7.63	10.20	6.99	6.45	11.61	11.64	12.87	10.12	11.53	13.07
Na_2O	0.00	2.42	1.21	2.14	2.30	2.93	1.78	1.35	2.02	1.75	2.15	2.35	1.57	2.33	2.46	1.54
K_2O	0.00	0.27	0.34	0.68	1.34	0.14	0.69	0.27	1.03	0.90	0.15	0.20	0.13	0.51	0.18	0.30
P_2O_5	0.00	0.00	0.06	0.00	0.00	0.00	0.11	0.15	0.00	0.07	0.14	0.03	0.12	0.02	0.08	0.03

All iron is given as FeO and all analyses are normalised to 100%. 1 (Barb): Barberton komatiite equivalent to parental magma of intrusions in the Barberton Mountain Land (Anhaeusser, 1985). 2 (Abi): estimated bulk composition of the Dundonald sill, representative of intrusions in the Abitibi area (MacRae, 1969). 3 (MtM): bulk composition of the Mt Monger sill, representative of intrusions in the eastern Goldfields area, Yilgarn Block (Williams and Hallberg, 1973). 4 (Jimb): estimated bulk composition of the Jimberlana dyke, Yilgarn Block (Campbell, 1977). 5 (Binn): chilled bronzite gabbro from Murdunna Hill, representative of magma parental to the Binneringie dyke, Yilgarn Block (McCall and Peers, 1971). 6 (Koil): chilled margin in the Koillismaa intrusion, Finland (Alapieti, 1982). 7 (EGD): chill from the East Dyke, representative of magma parental to the Great Dyke (Wilson, 1982). 8 (Stw): high-Mg gabbronorite from the Nye basin, possibly representative of magma parental to the Ultramafic Series of the Stillwater Complex (Helz, 1985). 9 (Blw): orthopyroxenitic sill from the southwestern Bushveld, representative of magma parental to the lower zone (Davies et al., 1980). 10 (Ble): average composition of orthopyroxenitic sills from the southeastern Bushveld, representative of magma parental to the lower zone (Harmer and Sharpe, 1985). 11: average composition of B2 tholeiitic sills from the southeastern Bushveld, possibly representative of magma added during the formation of the lower critical zone (Harmer and Sharpe, 1985). 12: average composition of B3 tholeiitic sills from the southeastern Bushveld, possibly representative of magma added during the formation of the upper critical zone. 13 (MtK): bulk composition of the tholeiitic Mt Kilkenny sill, Yilgarn (Jaques, 1976). 14 (Mess): estimated parental magma to the Messina intrusion, Limpopo belt (Barton et al., 1979). 15 (BVL): bulk composition of the Bad Vermilion Lake anorthosite complex (Ashwal et al., 1983). 16 (Fisk): average composition of amphibolites bordering the Fiskenaesset complex (Weaver et al., 1981)

Figure 4.1 Locality map showing the distribution of early Precambrian layered intrusions in relation to Archaean cratons.

extended commentary is confined to the major intrusions. The global distribution of the major layered intrusions is shown in Figure 4.1.

4.2 The Bushveld Complex

4.2.1 *Regional setting*

The Bushveld Complex is located on the northern margin of the Kaapvaal craton of southern Africa (Figure 4.2). It lies largely within the confines of the Transvaal basin (Figure 4.3) and was emplaced into a thick succession of volcanic and sedimentary rocks of the Transvaal Sequence. Both the intrusive complex and its host rocks are little deformed, and rocks generally dip less than 20° towards the centre of the basin. A notable exception to this is along the Thabazimbi–Murchison lineament where dips steepen to more than 60°. To the north of this lineament the Bushveld rocks transgress the enclosing sediments so that the larger part of the exposed Potgietersrus lobe (Figure 4.3) rests on Archaean basement. Little is known of the structure along the northernmost margin of this lobe, except for the intense deformation along the Palala shear zone in the Villa Nora area, which ceased before 1.7 Ga (Barton and McCourt, 1983).

Relationships in the Bushveld Complex correlate with those in the Molopo Farms Complex in southern Botswana (Gould *et al.*, 1986; Von Gruenewaldt *et al.*, 1988). This complex is also situated at the edge of the Kaapvaal craton and intrudes rocks of Transvaal Sequence age. The Kgomodikae lineament, a western continuation of the Thabazimbi–Murchison lineament, transects the complex. To the north the Molopo rocks have steep dips, are intensely faulted, and transgress northwards onto basement rocks, while to the south the complex generally comprises shallower dipping and less deformed rocks (Von Gruenewaldt *et al.*, 1988).

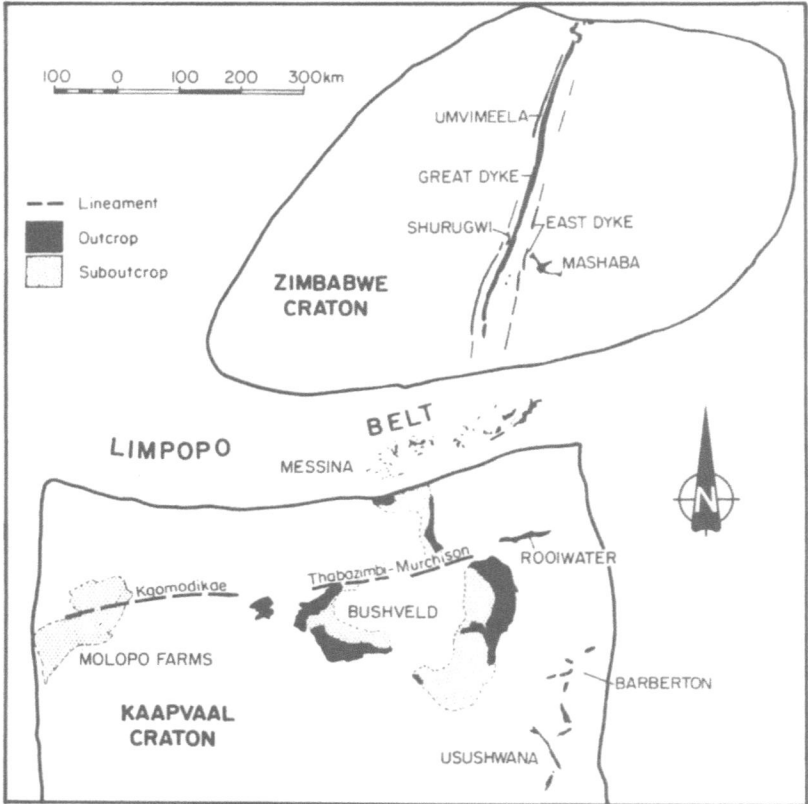

Figure 4.2 Zimbabwe craton and northern portion of Kaapvaal craton with localities of layered intrusions. Suboutcrop areas indicate inferred distribution of layered rocks beneath Proterozoic and Phanerozoic cover rocks.

4.2.2 *Intrusion form*

The structure of the Bushveld intrusion is best elucidated by the eastern sections where good exposure has been complemented by extensive geophysical studies. The thickness of the sequence is estimated to be about 5 km (Hattingh, 1980; Molyneux and Klinkert, 1978), with a maximum composite thickness of 9 km (Von Gruenewaldt, 1973). An axis of maximum thickness runs slightly west of the N–S trending outcrop (Hattingh, 1980). The intrusion gradually thins and is bent upward to either side of this axis. The overall geometry of the intrusion corresponds to several overlapping conical sills. The apices of the cones coincide with gravity highs situated along the axis of maximum thickness and have been interpreted as the sites of feeders to the complex (Sharpe and Snyman, 1980). Palaeomagnetic data suggest that the layering was originally horizontal (Gough and van Niekerk, 1959; Hattingh, 1986). Several anticlinal structures in the floor of the complex separate lower zone lithologies in a number of basins, each with its own succession of ultramafic rock types.

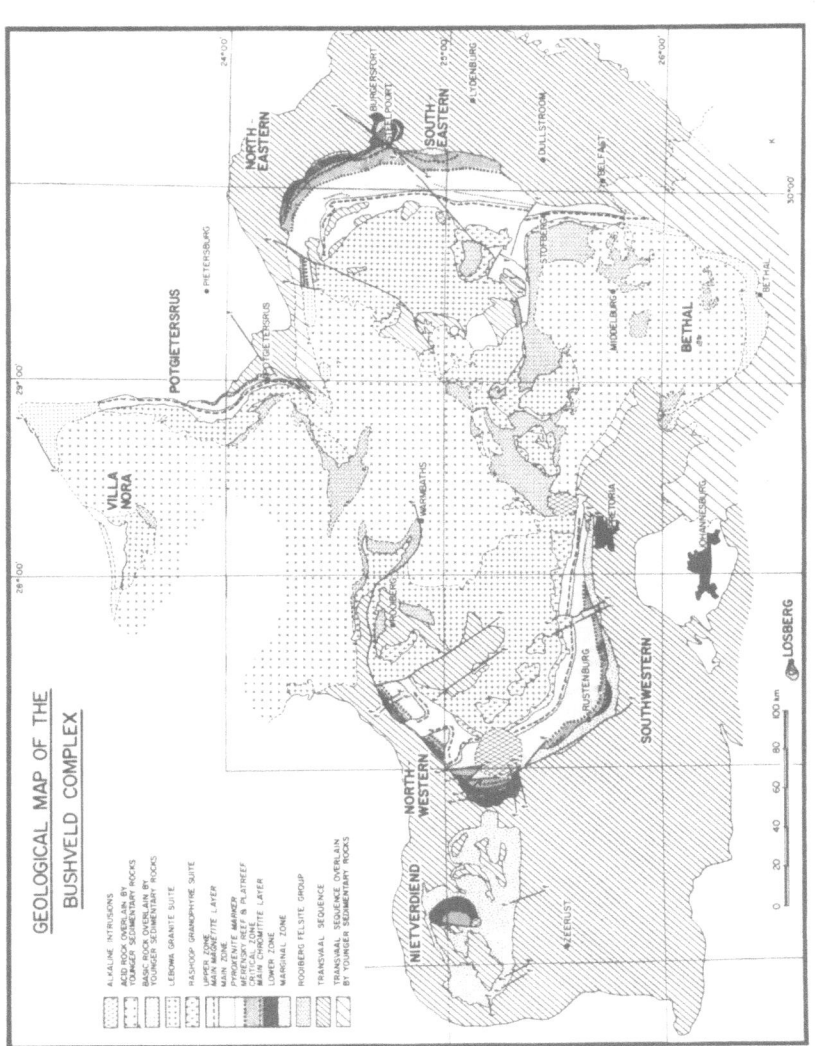

Figure 4.3 Geological map of the Bushveld Complex. Descriptions of specific portions of the Complex may be found in references listed in Table 4.1; ref. nos 1, 10, 11, 14, 18, 19, 33 and 51.

In the western and northern sections of the complex the structure is not as well known, but the available evidence is consistent with the original formation of shallow conical sills centred on gravity highs. A recent deep vibroseis survey in the southwestern Bushveld, however, shows that the complex does not extend to any great depth, but is truncated abruptly by a structure which has been inferred as an E–W trending feeder dyke (Du Plessis and Kleywegt, 1987). Although a fault-bounded truncation cannot be ruled out at this stage, the present configuration clearly suggests an asymmetrical shape of the complex in this area.

The complex is overlain by a thick succession of pre-Bushveld acid volcanics (Twist, 1985), which in turn are overlain by unconformity-bounded sequences of redbeds, devoid of any volcanics. There is no indication of the presence of volcanic rocks related to the layered rocks of the intrusive complex. Density relations indicate that additions of magma remained in the chamber and did not cause the displacement of any partially differentiated magma from the chamber as volcanic eruptions (Sharpe, 1985). This is substantiated by ample evidence that new magma influxes were accommodated by the lateral expansion of the magma chamber, such as the successively larger outcrop areas of the successive zones of the complex (Figure 4.4), and the pronounced transgressions of main zone rocks by the upper zone in the northwestern sector of the western compartment (Coertze and Schumann, 1962) and in the northern compartment (Van der Merwe, 1976).

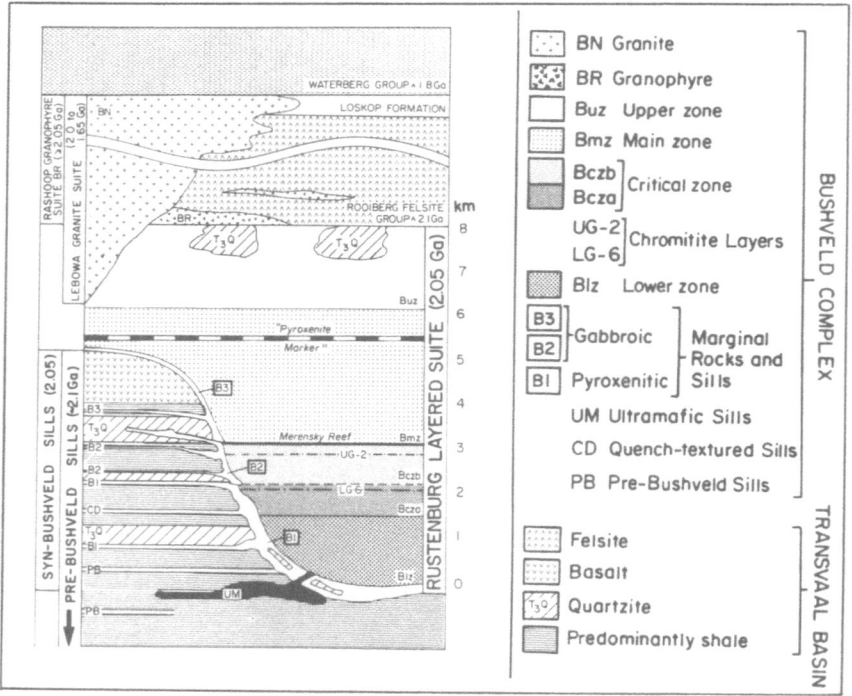

Figure 4.4 Schematic column illustrating intrusive events associated with the Bushveld Complex.

4.2.3 *Lithology*

The nature and the succession of rock types within the three major compartments of the Bushveld Complex are broadly similar, although in detail there are distinct differences between them. This suggests that similar magmas were supplied to these magma chambers and that they evolved largely independently of one another but that magmatic processes within them were very similar. Whether these three chambers were interconnected is open to speculation, but from the striking similarities of especially the upper critical zone and the lower part of the main zone in the eastern and western compartments, it is difficult to envisage how they could have crystallised unless the magma chambers were interconnected.

The layered mafic rocks, known as the Rustenburg Layered Suite, are subdivided on the basis of characteristic rock types and mineral assemblages into lower, critical, main and upper zones (Figure 4.5). The lower zone in the northeastern and northwestern sectors consists of a repeated succession of bronzitites, harzburgites and dunites while in the southeastern and southwestern sectors olivine-bearing lithologies are absent and the lower zone consists of bronzitites and norites. The lower zone in these areas contains no significant chromite concentrations (Vermaak, 1970; Cameron, 1978). In contrast, the lower zone in the Potgietersrus area contains a cyclic succession of dunite, harzburgite and pyroxenite, which hosts a number of chromitite layers (Hulbert and Von Gruenewaldt, 1985). All chromitite layers in the eastern and western Bushveld are hosted by the critical zone, which is subdivided into a lower subzone dominated by feldspathic pyroxenite, and an upper subzone in which plagioclase-rich cumulates (norites, leuconorites and anorthosites) alternate with feldspathic pyroxenite and chromitite layers. This zone is the most economically important (Von Gruenewaldt, 1977) and hosts the well documented platiniferous Merensky Reef (Vermaak, 1976; Kinloch, 1982) and UG-2 chromitite layer (McLaren and De Villiers, 1982; Gain, 1985), and other exploitable chromitite layers, most notably the LG-6 and MG-1 layers (Hatton and Von Gruenewaldt, 1987). The overlying main zone is, in comparison, a fairly homogeneous sequence of gabbronorites and gabbros, followed by the lithologically diverse upper zone where magnetite, Fe-rich olivine and apatite appear as cumulus phases at different levels (Figure 4.5) (Von Gruenewaldt, 1973).

This package of layered rocks is separated from the floor rocks by the marginal zone. Here, a variety of different fine-grained lithologies have been classified on the basis of field relations, lithology and geochemistry into the B1, B2, and B3 types (Sharpe, 1981) (Figure 4.4). The B1 group consists of quench-textured pyroxenites and norites and is marginal to the lower zone and lower part of the critical zone. The B2 and B3 types, on the other hand, are gabbroic and are marginal to the upper critical and main zones (Sharpe, 1985). These marginal zone rocks have equivalents as sills in the floor-rocks of the complex, in which Sharpe (1984) has also distinguished a suite of basic sills of pre-Bushveld age (Figure 4.4). The sills are especially voluminous adjacent to the southeastern sector where their total thickness within the sedimentary sequence adds up to more than 2 km, with individual sills as much as 600 m thick.

4.2.4 *Mineralogy*

The layered rocks of the Bushveld Complex display the most complete products of crystal fractionation of any of the layered intrusions. Olivines range in composition

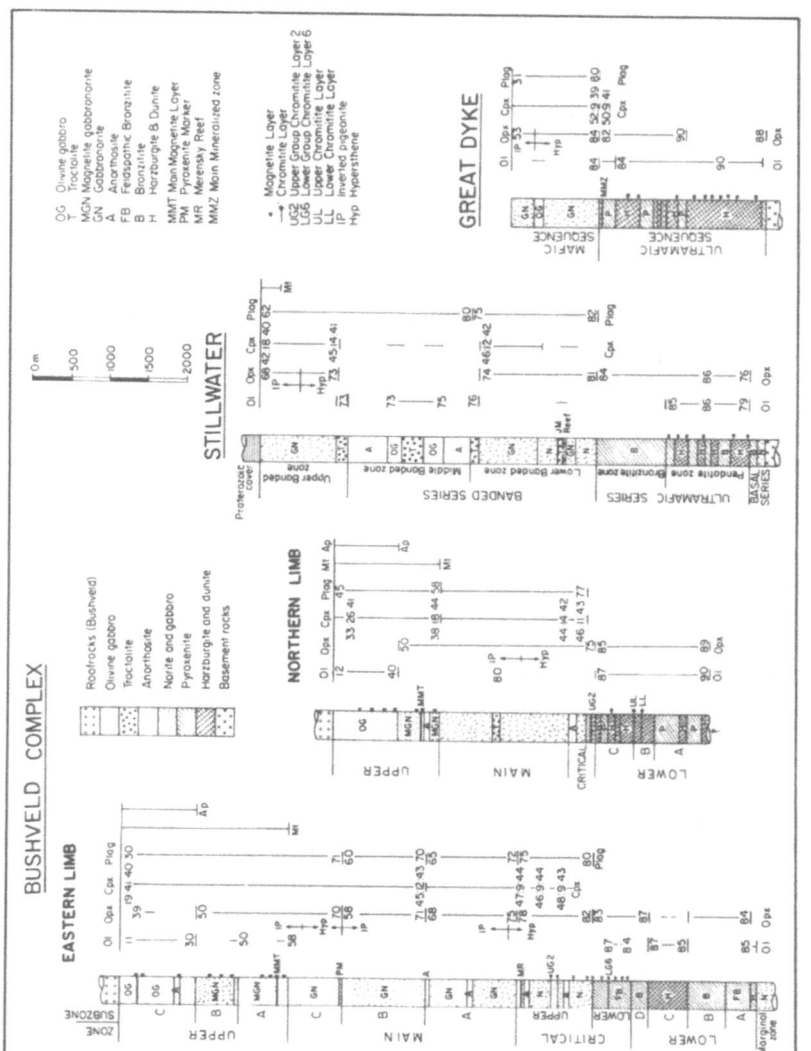

Figure 4.5 Stratigraphic column through the Bushveld, Stillwater and Great Dyke complexes. Columns are aligned along the contact between the lower, dominantly ultramafic and upper, dominantly mafic parts of the intrusions. Distribution of cumulus minerals and their compositions (mol.% Fo for olivine; mol.% En for orthopyroxene; atomic %Mg, Fe and Ca for clinopyroxene and mol.% An for plagioclase) is shown to the right of the columnar sections.

from Fo_{90} in the lower zone south of the Potgietersrus lobe to fayalite at the top of the intrusion. Orthopyroxene varies from En_{90} at the base of the lower zone to En_{38} near the top of the upper zone, and plagioclase ranges from An_{80} at the base of the upper critical subzone to An_{30} at the roof (Figure 4.5).

Orthopyroxene dominates the lower and lower critical zone lithologies and constitutes on average more than 50% of these rocks. Olivine makes up almost 40% of the lower zone rocks of the Potgietersrus compartment, less than 10% in similar rocks of the northeastern and northwestern Bushveld, but is absent in the southeastern and southwestern sectors of the complex. Plagioclase is mostly an interstitial component of lower and lower critical zone rocks, but it appears abruptly as a cumulus phase in an anorthosite at the base of the upper critical zone and from there upwards it is the dominant mineral in the layered sequence, constituting between 50 and 60% of the upper critical zone, more than 65% of the main zone and between 50 and 60% of the upper zone. Clinopyroxene is a minor, intercumulus constituent of most of the lower and lower critical zone rocks, and although it appears as a cumulus phase some distance from the base of the upper critical zone it remains, apart from a few layers, a minor constituent of the rocks. It is an abundant cumulus constituent of the gabbroic main and upper zones where it constitutes between 15 and 18% of the rocks.

4.2.5 Parental magma

Wager and Brown (1968) envisaged the crystallisation of the Bushveld Complex from a single magma, but subsequent isotopic investigations have shown that at least three magma types were present. Across the boundary between the lower critical and upper critical zones Sr_i values change from 0.705 to 0.706 (Hatton et al., 1986; Reichhardt, in prep.) and from the upper critical to the main zone Sr_i increases to 0.7085 (Kruger and Marsh, 1982; Sharpe, 1985). The isotopic evidence that new magma was added before the upper critical zone formed accords well with the mineralogical changes, which could be explained by the addition of plagioclase-saturated magma at this level. Sharpe and Irvine (1983) have shown by experiment that mixing between olivine- and plagioclase-saturated magmas resulted in the presence of chromite only on the liquidus of the mixed magma, and this hypothesis of magma addition also accounts for the chromitite layers of the upper critical zone. The presence of chromitite layers in the lower critical zone also provides circumstantial evidence for the addition of new magma during crystallisation of this unit.

Information as to the compositions of the various magma types is obtained from the marginal border groups and sills (B1, B2 and B3) associated with the complex (Sharpe, 1981; Harmer and Sharpe, 1985). The B1 rocks contain abundant orthopyroxene and experiments on these rocks (Cawthorn and Davies, 1983; Sharpe and Irvine, 1983) show that their compositions are appropriate for the parental magma to the lower zone. The Sr_i ratios of B1 rocks range between 0.702 and 0.704. B2 and B3 rocks have plagioclase to pyroxene ratios of 2:1, with orthopyroxene and clinopyroxene present in roughly equal amounts, and Sr_i ratios of between 0.706 and 0.707. These rocks cannot be representative of magmas added during the formation of the main zone (since Sr_i reaches 0.7085 in the main zone). Instead, it is considered that B2 and B3 were added during crystallisation of the lower critical and upper critical zones. The magma added during crystallisation of the main zone is clearly not represented in the marginal rocks, but approximate mass balance calculations indicate that its composition is that of a high-Al basalt. Representative compositions of the parental magmas are given in Table 4.2.

From a profile of Sr_i through the main zone, Sharpe (1985) concluded that the added main zone magma was more dense than the magma already in the chamber, so that the resident magma was displaced upwards. After crystallisation of the main zone, the upper zone formed from the displaced magma. Since the densities of B2 and B3 magma are greater than that of B1, Hatton (1988) suggested that upward displacement of resident magma by new magma was a general feature in the Bushveld Complex evolution. The liquidus temperatures of B2 and B3 are less than that of B1 (Sharpe and Irvine, 1983; Cawthorn and Davies, 1983), so that new magma would have cooled the overlying B1 magma and caused the crystallisation of orthopyroxene. Since the density of orthopyroxene is substantially greater than that of magma, portions of B1 magma which contained more than about 5% orthopyroxene crystals would have become more dense than underlying B2 or B3 magma, and would have collapsed into it. This process provides a mechanism for the mixing between B1 and B2/B3 magmas and the consequent formation of chromitite layers.

4.3   The Stillwater Complex

4.3.1   *Regional setting*

The Stillwater Complex lies in the northern part of the Beartooth Mountains, in one of several segments of Archaean crust of the Wyoming province (Figure 4.6). Its emplacement at about 2.7 Ga (De Paolo and Wasserburg, 1979; Lambert *et al.*, 1985) was into a stable environment, as suggested by the continuously traceable thin cumulate layers (Page and Zientek, 1985). The basal contact of the complex is exposed

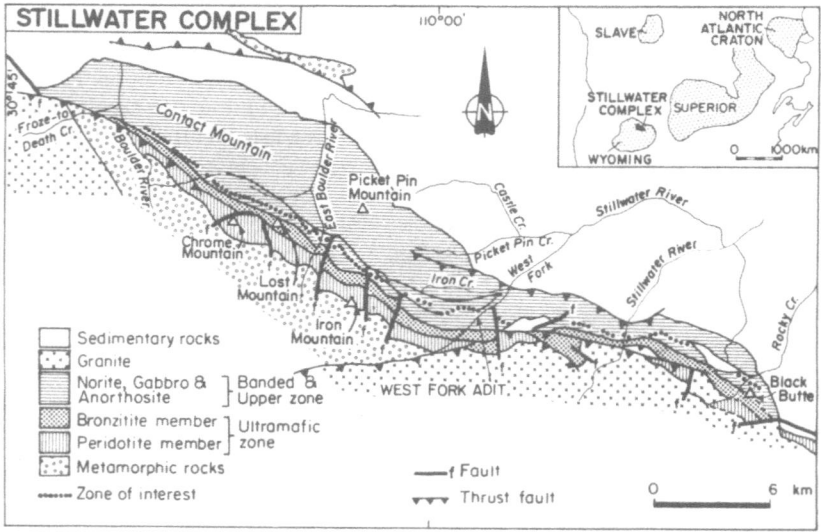

Figure 4.6   Geological map of the Stillwater Complex.

along its southern edge where it is seen to intrude into isoclinally folded 3.12 Ga metasediments which were metamorphosed to a variety of different hornfelses by the complex. Metamorphic mineral assemblages indicate a maximum depth of emplacement of about 12 km (Labotka, 1985). The transgressive nature of the basal contact of the complex across these sediments suggests its possible emplacement along a fault zone or unconformity (Page and Zientek, 1985). Palaeomagnetic data suggest that the Stillwater block is allochthonous and that the intrusion and tectonic emplacement of the intrusive complex, possibly along a wrench fault associated with the Nye–Boulder lineament, occurred in rapid succession (Mogk and Geissman, 1984). 'Attachment' to the much larger Beartooth block is considered to have taken place at 2.7 Ga during the emplacement of post-Stillwater Complex quartz-monzonite plutons, along the contact zone of the Stillwater and Beartooth blocks (Page and Zientek, 1985).

4.3.2 Intrusion form

The Stillwater Complex is exposed as a 48 km long, steeply dipping wedge of layered mafic and ultramafic rocks lying along a persistent high-gravity zone which coincides with the faulted front of the Beartooth Mountains and the Nye–Boulder structural zone (Kleinkopf, 1985). This wedge, which is between 5.5 and 6.5 km wide, is considered to be the upturned distal edge of a sill-like lopolith centred to the northeast beneath a cover of younger sedimentary rocks. This is indicated by magnetic and gravity data according to which the complex has a synformal shape and extends for more than 2500 km^2 to the northeast of its present outcrop position (Kleinkopf, 1985), as well as by numerous xenoliths of Stillwater-type cumulate rocks in dacitic rocks of late Cretaceous intrusive centres situated between 8 and 12 km north of the Stillwater Complex (Brozdowski, 1985).

4.3.3 Lithology

The exposed layered rocks of the Stillwater Complex attain a maximum thickness of about 6.5 km and are subdivided into the Basal, Ultramafic and Banded series (Figure 4.5). The Basal Series is a laterally persistent, but locally heterogeneous unit consisting of a basal norite zone that contains disseminated to massive Ni-Cu sulphides at several localities, and an overlying bronzitite zone. The Ultramafic Series overlies the Basal Series, but locally lies directly on basement. This Ultramafic Series is divided into a lower peridotite zone, consisting of up to twenty-one cyclic units, which comprise the mineralogical sequence olivine, olivine + chromite, olivine + orthopyroxene and orthopyroxenite, and an upper bronzitite zone of laminated and size-graded bronzite cumulates (Jackson, 1970).

The appearance of cumulus plagioclase in the sequence marks the base of the Banded Series, a 4.7 km thick succession of norites, gabbros, anorthosites and troctolites, grouped into a lower, a middle and an upper series. Olivine occurs as a cumulus mineral at five intervals, referred to simply as olivine-bearing zones (OBI to OBV) by McCallum et al. (1980), and as troctolite anorthosite zones (TAZI to TAZV) by Todd et al. (1982). The important platiniferous 'J-M Reef' occurs within the first olivine-bearing zone (OBI), which is itself a complicated cyclic sequence of norite, gabbro, anorthosite, troctolite and peridotite layers (Barnes and Naldrett, 1985, 1986).

The uppermost rocks of the complex are magnetite-bearing pigeonite gabbros which have equivalents at the base of the upper zone in the Bushveld Complex. From this and mass balance considerations, Hess (1960) estimated that a 2–3 km thick sequence of ferrogabbroic and ferrodioritic rocks were eroded prior to the deposition of the overlying Proterozoic sediments, and speculated that a hidden zone of magnetite-rich rocks is preserved beneath the angular unconformity to the north.

4.3.4 *Mineralogy*

Compositional variations of the major mineral constituents are, apart from the less refractory components, similar to those displayed by the Bushveld Complex (Figure 4.5). Orthopyroxene and olivine constitute on average about 74 and 23% respectively of the ultramafic zone lithologies (Raedeke and McCallum, 1985). Plagioclase appears abruptly as a cumulus phase at the base of the Banded Series of which it constitutes about 69% whereas cumulus clinopyroxene makes its appearance some 250 m higher in the sequence. Orthopyroxene and clinopyroxene are present in roughly equal proportions in this series, although their modal concentrations vary widely. Olivine is not an abundant constituent of the Banded Series, apart from the significant concentrations which occur in the thin troctolite and peridotite layers of the five 'olivine-bearing' zones.

4.3.5 *Parental magma*

As yet, no systematic studies of variation in isotopic ratio with height have been undertaken in the Stillwater Complex, so that the evidence for multiple magma injection is less firm than that for the Bushveld Complex. However, there are several lines of evidence that two different parental magmas were involved in the evolution of this complex. On the basis of variable Sr_i and Nd_i ratios, Lambert *et al.* (1985) postulated that the Stillwater Complex crystallised from at least two geochemically distinct, LREE-enriched magma types, namely a Mg- and Si-rich magma parental to the Ultramafic Series, and an Al-rich magma parental to the olivine- and plagioclase-rich intervals of the Banded Series. This agrees with the findings of Todd *et al.* (1982) who showed the difficulties in explaining the abundance of plagioclase-rich cumulates in the Banded Series by a simple process of fractional crystallisation of the high-Mg magma that gave rise to the Ultramafic Series, and who postulated that a basaltic magma with plagioclase alone on the liquidus invaded the Stillwater magma chamber at some stage after the formation of the Ultramafic Series. Furthermore, Barnes and Naldrett (1985) demonstrated a distinct increase in the Sr content of plagioclase across the OBI sequence, a feature only accounted for by the influx of a compositionally different magma. High-Mg dykes and sills in the footwall of the complex and in the Beartooth Mountains in close proximity of the complex are possible representatives of the magma responsible for the Ultramafic Series (Longhi *et al.*, 1983; Helz, 1985; Lambert *et al.*, 1985).

As in the Bushveld Complex, the mixing of two compositionally distinct magmas is clearly of great petrogenetic significance. Irvine *et al.* (1983) proposed that a complex mixing process between a high-Mg magma enriched in platinum-group elements (PGE), and a high-Al magma enriched in sulphur, gave rise to the olivine-bearing zones and to the J-M Reef.

4.4 Dyke-like layered intrusions

4.4.1 Great Dyke of Zimbabwe

The Great Dyke of Zimbabwe is one of the world's unique geological features, extending for 550 km NNE across the entire Zimbabwe craton (Figure 4.2). The intrusion varies in width from 4 to 12 km and, together with its associated satellite dykes the Umvimeela dyke to the west and the East dyke, was intruded c. 2.46 Ga ago into Archaean granites, gneisses and associated greenstone belts of the Zimbabwe craton (Wilson, 1982). The northern extremity of the Great Dyke extends into the Zambesi mobile belt, where it underwent intensive deformation during an orogeny about 500 Ma ago, parts of the dyke having become dislodged, folded and rotated (Wiles, 1968). In the south, the dyke rocks terminate just short of the Tuli-Sabi shear zone of the Limpopo mobile belt, although the southern satellite dykes continue for a further 80 km into the mobile belt, where they were deformed some 1950 Ma ago (Jones et al., 1975).

Worst (1960) interpreted the Great Dyke as a fault-bounded remnant of four contiguous canoe-shaped layered complexes. These four complexes, named from north to south the Musengezi, Hartley, Selukwe and Wedza complexes, all consist of a thick succession of ultramafic rocks capped in the central areas by gabbroic rocks which were considered to overlie the feeders to each of the complexes. A recent gravity survey (Podmore and Wilson, 1987) indicates that a single deep dyke-like feature underlies the complex over most of its length and not only at the central zones, but revealed a noticeably shallower succession within the dyke near the junction of the Hartley and Selukwe complexes, leading to a subdivision of the dyke into southern and northern chambers (Prendergast, 1987).

The stratigraphy of all four sections is broadly similar. In each a thick, cyclically layered sequence of ultramafic rocks is capped by a comparatively homogeneous succession of gabbroic rocks (Figure 4.5). The ultramafic sequence can be divided into a lower cyclic succession of predominantly serpentinised dunite and chromitite layers, and an upper cyclic succession in which the cycles consist of a basal chromitite layer, followed by dunite or harzburgite which grades upward into olivine bronzitite and bronzitite at the top (Worst, 1960; Wilson, 1982). This sequence is best developed in the 2000 m thick Hartley Complex where fourteen such cyclic units occur (Wilson, 1982). The uppermost cyclic unit is remarkably similar in all four complexes, and differs from the underlying cycles in that cumulus clinopyroxene appears in a websterite layer at the top and because the platiniferous Main Mineralized Zone is located in a bronzitite directly below this websterite.

The thickness of the preserved overlying gabbroic rocks varies considerably from one complex to the next, but in the Hartley complex it is estimated to be 1150 m. Here, the mafic sequence commences with an olivine gabbro which grades into norite, the dominant rock type, and to magnetite- and quartz-bearing pigeonite gabbros at the top (Wilson and Prendergast, 1987).

Although the compositions of the minerals of the Great Dyke are very similar to those encountered in the large sheet-like intrusions, the proportions of the cumulus phases in the ultramafic rocks are reversed. Olivine constitutes about 70% of the ultramafic sequence where it varies in composition from Fo_{90} to Fo_{84} (Wilson, 1982).

Orthopyroxene is largely restricted to the upper half of this sequence and changes in composition from En_{90} near the base to En_{82} in the websterite layer at the top of the first unit. Plagioclase is the dominant mineral in the mafic sequence, where it ranges from An_{80} at the base to An_{31} at the highest exposed level of the intrusion. Orthopyroxene, the next most abundant mineral, varies from En_{84} to En_{53} over the same interval (Wilson and Prendergast, 1987). The highly fractionated nature of the plagioclase and orthopyroxene, together with the presence of magnetite, interstitial quartz and apatite led these authors to believe that the highest exposed level in the Hartley Complex must be very close to the original roof of the intrusion.

A liquid with about 15 wt% MgO and similar in composition to the chill of one of the satellite dykes (Table 4.2) is in closest agreement with the observed and modelled crystallisation sequences and compositions in the ultramafic sequence, and hence considered to be the parental magma of the Great Dyke (Wilson, 1982). This is in broad agreement with Hughes (1976) who postulated the parental liquid to be a magnesian basalt on the basis of circumstantial evidence.

4.4.2 Finnish layered intrusions

A belt of more than twenty early Proterozoic layered intrusions extends for more than 300 km eastward from Tornio in northern Finland into the Soviet Union. These layered intrusions were emplaced 2.44 Ga ago, soon after cratonisation of the late Archaean crust, as a single elongate belt which subsequently has been tectonically sliced into several separate blocks during the Svecokarelian orogeny (Alapieti and Piirainen, 1984; Gaál, 1985). The intrusions are classed into three groups: those of Tornio–Kemi–Penikat; those in the Narkaus area, including the Suhanko and Kontijarvi bodies; and several intrusions of the Koillismaa Complex (Figure 4.7). All these intrusions are surrounded by granitic rocks of the Archaean basement, or located at the junction between the basement and supracrustal metasediments and volcanics of the Kemi and Koillismaa schist belts. Gaál (1985) proposed that these schist belts evolved in the failed arm of a triple junction which formed during the breaking up of Archaean continental crust, and that the layered intrusions were emplaced in response to an initial rifting event. Alapieti and Lahtinen (1986) have also suggested that this belt of layered intrusions may be related to an aulacogen. The Koitelainen and some smaller, related layered intrusions in Finnish Lapland are of the same age and considered to belong to the same magmatic event (Lahtinen, 1985; Mutanen et al., 1987).

Although the layered intrusions of northern Finland are considered to have originally constituted a single belt generated from four intrusion centres and emplaced along a deep seated tensional fracture zone (Gaál, 1985), subsequent dislocation into several separate blocks has complicated the reconstruction of their original form. Their general geometry seems to be that of Great Dyke-like canoe-shaped structures linked along the length of the entire belt (Gaál, 1985; Alapieti and Lahtinen, 1986). This is demonstrated for the Koillismaa complex by Alapieti and Piirainen (1984) who have postulated that the western part of this complex is a broad synformal feature, connected to the narrow canoe-shaped Narankavaara intrusion by a hidden dyke, on the basis of a prominent gravity high which extends for more than 80 km from the western part of the complex through the Narankavaara intrusion into the Soviet Union. The dyke is 3 km wide but tapers gradually with depth and represents a feeder to the Koillismaa complex.

Figure 4.7 Distribution of layered intrusions (black) in northern Finland in relation to Svecokarelian plutonic rocks (line shading), volcanogenic and sedimentary rocks (stippled) and pre-Svecokarelian basement (hatched shading). Descriptions of specific portions of the Tornio–Narankavaara belt may be found in the references listed in Table 4.1, ref. nos. 3, 4, 5, 29, 45 and 52.

In general terms the Finnish intrusions are ultramafic near their bases and become mafic or anorthositic at their tops. The lithologies vary significantly along strike. In the Koillismaa complex, the Narankavaara intrusion, which is situated within the feeder dyke, is predominantly ultramafic, while the western sheet-like intrusions are mafic and contain magnetite-rich differentiates which have been mined for their V content. Chromitite layers occur in the ultramafic parts of the Tornio, Kemi and Penikat intrusions (Lahtinen, 1985), whilst the Penikat intrusion also contains well developed megacyclic units with anorthosite layers and thin layers with disseminated sulphides enriched in PGE (Alapieti and Lahtinen, 1986). Disseminated Ni-Cu-PGE sulphides are also a feature of the gabbroic marginal rocks of the intrusions in the Narkaus area and the western part of the Koillismaa complex (Vuorelainen *et al.*, 1982; Alapieti and Piirainen, 1984; Lahtinen, 1985).

The Finnish intrusions contain typical layered intrusion mineral assemblages, comprising orthopyroxene, olivine, plagioclase and clinopyroxene. Compositional investigations in the Koillismaa complex show them to be well fractionated, varying from En_{88} to En_{57} for Ca-poor pyroxene, Fo_{88} to Fo_{69} for olivine and An_{78} to An_{64} for plagioclase (Alapieti and Piirainen, 1984).

The Koitelainen layered intrusion seems to be different from the other Finnish intrusions in that Fe-rich chromitite layers occur in close proximity to magnetite-bearing gabbroic rocks high in the intrusion, and in that disseminated PGE-enriched sulphides also predominate in late differentiates. The close association of chlorapatite with these platiniferous sulphides suggests that a high chlorine content in this intrusion caused the PGE to remain in solution in the magma until an advanced stage in the evolution of the intrusion (Mutanen *et al.*, 1987).

Fine-grained, chilled marginal rocks that could be representative of the parental magma of the layered intrusions have been encountered in the Koillismaa (Alapieti and Piirainen, 1984) and Penikat (Alapieti and Lahtinen, 1986) intrusions. Their compositions are tholeiitic basaltic, and are characterised by relatively high MgO and low TiO_2 contents (Table 4.2). Although repeated influxes of parental magma are considered important to explain features such as the megacyclic variation in these intrusions (Alapieti and Lahtinen, 1986), there is no indication as yet that parental magmas of contrasting composition were involved in the evolution of the Finnish layered intrusions.

4.4.3 *Jimberlana intrusion*

The Jimberlana intrusion is part of the easterly trending Widgiemooltha dyke suite, emplaced 2.37 Ga ago in the southern part of the Yilgarn Block of Western Australia (Campbell *et al.*, 1970) (Figure 4.8). The Jimberlana intrusion extends for 180 km and has an average width of 1.5 km, but widens at seven points along its length into canoe-shaped complexes which are joined by a connecting dyke. These complexes have very steep, funnel-shaped cross-sections and contain cumulate layers which are flat in their central portions and steepen towards their edges (McClay and Campbell, 1976).

The overall intrusion has been divided into (i) a lower series of five macrorhythmic units, each consisting of olivine cumulates at the base and bronzite cumulates at the top, overlain by a thick layer of plagioclase–augite–hypersthene cumulates; (ii) an upper layered series which rests unconformably on the lower series and consists of several macrorhythmic units of olivine and bronzite cumulates overlain by plagioclase–augite–hypersthene cumulates, as well as a granophyric layer at the top; and (iii) the steeply dipping, reversed sequence of the marginal layered series below the lower series, consisting of plagioclase–augite–hypersthene cumulates at the base, overlain by bronzite cumulates which in turn are overlain by olivine cumulates.

Information on parental liquid compositions is limited, but field relationships and petrographic evidence suggest that the upper layered series crystallised from a major

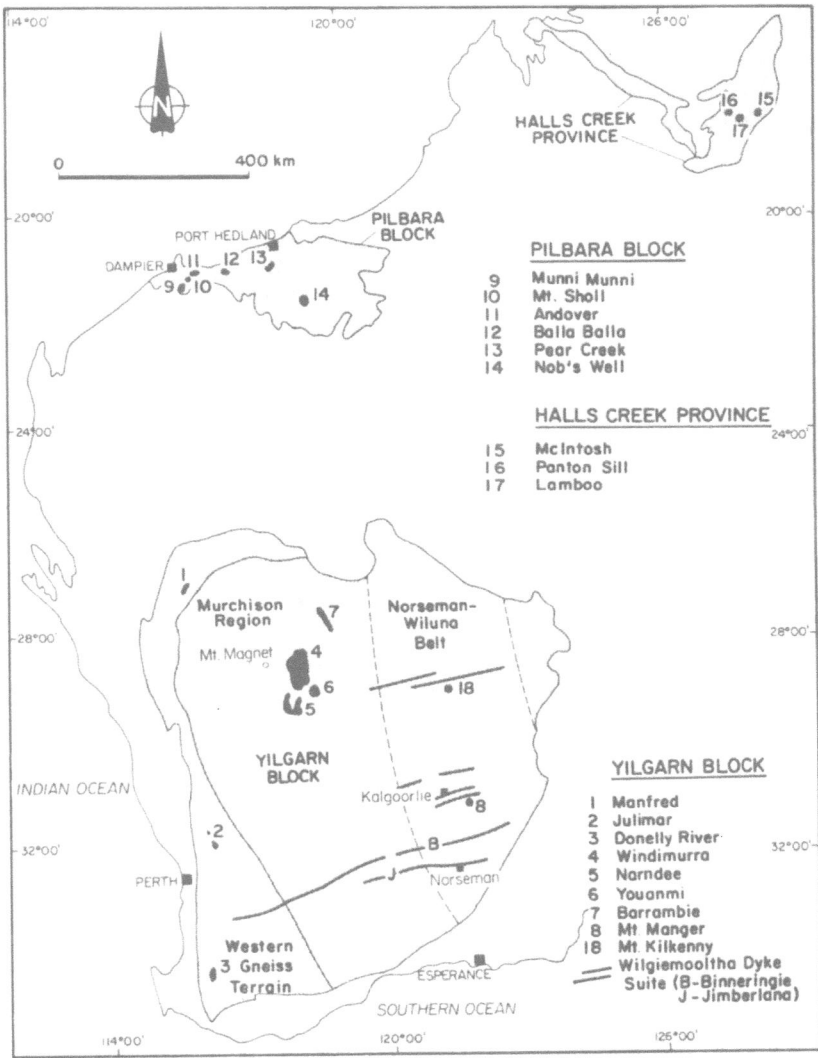

Figure 4.8 Map of Western Australia showing Yilgarn, Pilbara and Halls Creek cratons, and associated layered intrusions. Many of the intrusions shown have not yet been described in any detail, and are not listed in Table 4.1.

new pulse of magma which entered the chamber during the final stages of crystallisation of the lower layered series. A bronzite-rich gabbro dyke, which cuts the lower layered series, has a composition very close to bulk composition of the upper layered series (Table 4.2) and is considered by Campbell (1977) to be a possible feeder dyke to this series.

4.5   Archaean anorthosite complexes

Archaean anorthosite complexes are known from a number of localities throughout the world. Their distinguishing features are the presence of coarse (up to 30 cm), equidimensional euhedral to subhedral, calcic (An_{80-95}) plagioclase megacrysts in a finer grained mafic matrix (Ashwal *et al.*, 1983). Preserved igneous layering clearly indicates parallels with other layered intrusions. In one of the earliest studies Subramaniam (1956) compared the chromitite layers of the Sittampundi Complex of India with the Dwars River chromitites of the Bushveld Complex. Archaean anorthosite complexes range in age from the early Archaean (3.7 Ga) Manfred Complex of Western Australia, through the 3.2–2.8 Ga Fiskenaesset, Sittampundi, Messina (South Africa) and Holenarasipur (India) complexes, to the 2.7 Ga Bad Vermilion Lake Complex of the Superior Province (Table 4.1).

All of the Archaean anorthosite complexes have undergone severe deformation and metamorphism to amphibolite or granulite facies, so that their original form is difficult to ascertain. Weaver *et al.* (1981) proposed that the Fiskenaesset anorthosite complex of southern West Greenland was tectonically emplaced into the tonalitic gneisses with which it is now associated, although cross-cutting relationships suggest that the gneisses are younger. The Messina anorthosite complex of South Africa was originally sill-like (Sohnge *et al.*, 1948).

Although amphibole is present in all the Archaean complexes, it was never a primary magmatic phase (Barton *et al.*, 1979; Weaver *et al.*, 1981) and probably crystallised initially as a postcumulus mineral in the cumulate pile (Weaver *et al.*, 1981). In the ultramafic layers olivine, orthopyroxene and clinopyroxene were the primary cumulus minerals, and in the mafic layers plagioclase and clinopyroxene clearly predominated (Myers 1985; Weaver *et al.*, 1981; Kutty *et al.*, 1984; Simmons *et al.*, 1980). Spinel-rich layers are found in many anorthosite complexes; chromitite in Sittampundi (Subramaniam, 1956), magnetite in Dore Lake and Bad Vermilion Lake (Allard, 1976; Ashwal *et al.*, 1983) and both chromitite and magnetite in Fiskenaesset and Messina (Myers, 1985; Sohnge *et al.*, 1948).

With regard to the petrogenesis of Archaean anorthosite complexes, the fusion of a depleted mantle under hydrous conditions has been proposed by Weaver *et al.* (1981), and the high water pressure inferred from the presence of amphibole may be related to crystallisation under moderately high pressure conditions (Mullan and Bussell, 1977). The parental magmas are considered to be tholeiitic (Barton *et al.*, 1979; Simmons *et al.*, 1980; Weaver *et al.*, 1981), although the estimated bulk composition of an anorthosite complex is considerably more aluminous than tholeiitic magma. This has been explained on the basis that anorthosite complexes represent cumulates in a magma chamber from which volcanics with typical greenstone tholeiite compositions were being erupted (Barton *et al.*, 1979; Weaver *et al.*, 1981; Simmons *et al.*, 1980; Ashwal *et al.*, 1983).

4.6   Other layered intrusions

A large number of other layered intrusions of different size and age, too numerous to described individually, are known (Table 4.1). Many of these are sill-like ultramafic or ultramafic to mafic intrusions in greenstone belts, and have been interpreted as hypabyssal equivalents of the associated komatiitic and tholeiitic basalts. The better known differentiated sills of this type are the different types of intrusions in the Barberton Mountain Land (Viljoen and Viljoen, 1970; Anhaeusser, 1985), in the Cape Smith–Wakeham Bay thrust fold belt of Quebec also known as the Ungava trough (Lamothe *et al.*, 1987), in the Labrador trough of Quebec and Newfoundland (Clark, 1987), in the Abitibi area of Ontario (MacRae, 1969) and small intrusions around Kalgoorlie in the Norseman–Willoona belt of the Yilgarn block in Western Australia (Williams and Hallberg, 1973). Several other layered intrusions are also known which are not necessarily associated with greenstone belt magmatism. Brief reference is made here only to the better documented and larger intrusions.

4.6.1 *Kaapvaal craton*

The Rooiwater Complex (> 2.65 Ga) intrudes supracrustal rocks of the Murchison schist belt in the Kaapvaal craton (Vearncombe *et al.*, 1987). It differs from most other Precambrian layered intrusions in that it contains no ultramafic component in its exposed 7.5 km thick sequence. Metagabbro and subordinate layers of anorthosite dominate the succession, which also contains thick titanomagnetite layers near the top. The gabbroic succession is overlain by quartz gabbro and hornblende granite. Vearncombe *et al.* (1987) considered these three rock suites to have been co-magmatic and to represent differentiates from the same basic magma source, which was relatively felsic compared to other layered intrusions.

The Usushwana Complex in Swaziland and adjoining South Africa is an approximately 2 km thick body with a 750 m pyroxenite horizon overlain by gabbro and gabbronorite. Younger granitic plutons have disrupted the intrusion, but it appears to have comprised two parallel dyke-like bodies linked by a gabbroic sill (Hunter, 1970). Hegner *et al.* (1984) have presented isotopic evidence for a substantial crustal component in the magmas which gave rise to this complex at around 2.87 Ga.

4.6.2 *Yilgarn Block*

Fragments of layered intrusions are found throughout the Western Gneiss Terrain of the Yilgarn Block of Western Australia (see Chapter 14). The Manfred intrusion is the largest segment of a > 3.7 Ga complex near Mt Narryer (Myers and Williams, 1985), and the Donnelly River Complex and the Julimar Complex are collective names for fragments of layered intrusions found in the southern and central portions respectively of the Western Gneiss Terrain. On the basis of Pt/Pd and Cu/Ni ratios, the Julimar Complex has been related to the Noril'sk intrusion and the associated tectonic environment of rifting close to the margins of a stable craton (Harrison, 1984).

The Windimurra intrusion is the largest of a number of Archaean layered complexes in the Murchison granite–greenstone province of the Yilgarn Block (Figure 4.8). It covers an area of 2345 km² and represents a deformed and partly metamorphosed remnant of a larger mass emplaced at about 3.07 Ga (Ahmat and de Laeter, 1982). Gravity data suggest the complex to have been a flat-lying tabular mass not more than 5 km thick, surrounded on all sides by major shear zones. Rhythmically layered anorthositic gabbros and troctolites, which dominate the succession, define a basin-shaped structure with an apparent stratigraphic thickness of 13 km. Localised upfaulted blocks contain serpentinised ultramafic rocks and a few thin chromitite layers. Magnetite-bearing gabbroic rocks, including several magnetitite layers, dominate the uppermost part of the succession, which has a transgressive contact with the underlying layered rocks (Ahmat, 1986).

Crystallisation sequences in contrasting mafic and plagioclase-rich macrolayers in the Windimurra intrusion suggest that two magma-types were involved in the evolution of the complex (Ahmat, 1986). One was a high-Al, silica-saturated type and the other a more mafic silica-undersaturated type. Low Sr_i ratios (0.701) indicate primitive source rocks of both parental magmas and that there was no significant crustal contamination component. The complex is considered to have been emplaced in an extensional environment, such as a rift or spreading centre.

The largest of the associated intrusions is the Narndee Complex (Figure 4.8), which consists of a lower 1250 m thick cyclic sequence of peridotite, orthopyroxenite and melanorite, overlain by a 2500 m thick upper cyclic sequence of peridotite, olivine norite and gabbronorite (S. Olissoff, personal communication, 1987).

4.6.3 Pilbara Block

Many of the Archaean layered intrusions in the Pilbara Block have been grouped by Fitton et al. (1975) into the Millindinna Complex, which they believed to be a single extensive complex emplaced along a regional unconformity between distinct Archaean successions. However, Hickman (1983) has argued that the component fragments are located at various stratigraphic levels within the Pilbara Supergroup, indicating more than one intrusive event. The best known of these intrusions is the 32 km² Munni Munni Complex (Figure 4.8), which was emplaced 2.8 Ga ago into 3.0 Ga old granitic rocks. This complex consists of a lower 1850 m thick cyclic succession of ultramafic rocks overlain by a comparatively homogeneous gabbroic sequence in excess of 3000 m thick (Donaldson, 1974; Hoatson, 1984). Disseminated platiniferous sulphides occur concentrated in a websterite a few metres below the contact between the mafic and ultramafic rocks. Other layered intrusions in this region are indicated on Figure 4.8.

4.6.4 Superior Province

The best documented differentiated sill of the Archaean greenstone belts in the Superior Province of Canada is the Bird River Sill of southern Manitoba (Hulbert et al., 1988). This sill consists of a 200 m thick ultramafic series of peridotites and dunites with six groups of chromitite layers near the top, overlain by a 400 m thick series of gabbros and quartz gabbros which are unconformably overlain by mafic volcaniclastic rocks.

The largest of the layered sills of this province is undoubtedly the Fox River Sill, a Proterozoic stratiform 2.5 km thick mafic to ultramafic intrusion which extends discontinuously for 250 km (Scoates and Eckstrand, 1986). It is predominantly ultramafic, the cumulates displaying a variety of cyclic units of variable composition. It was emplaced into a sequence a few hundred metres thick of sandstones and siltstones overlain by komatiitic and tholeiitic basalts. These volcanic rocks are considered to have erupted from the Fox River Sill magma at different times during its crystallisation history.

4.7 Tectonic setting and petrogenesis

Early Precambrian layered intrusions do not appear to be distributed randomly in time. Relatively short-lived structural events at approximately 3.6, 2.7, 1.9, 1.1 and

0.6 Ga have been recognised throughout the world (e.g. Sutton, 1963) and many early Precambrian basic intrusions appear to have formed just prior to collisional events at around 3.6, 2.7 and 1.9 Ga. The Selukwe (Shurugwi) complex is preserved in a thrust sheet formed during a 3.5 Ga collisional event (Stowe, 1968, 1987a; Cotterill, 1979). Several intrusions including the Stillwater, Mashaba and Rooiwater complexes formed at c. 2.7 Ga, and of these the Stillwater Complex has been related to a late Archaean collisional orogeny (Mueller et al., 1985; Mogk et al., 1988). The Bushveld and the related Molopo Farms complexes formed at 2.05 Ga and they may be related to orogenic events preceding the collision of the Zimbabwe and Kaapvaal cratons at c. 1.9 Ga (Watkeys, 1983).

Other basic intrusions appear to be related to rifting events. The Great Dyke of Zimbabwe, the Widgiemooltha dyke swarm of Western Australia and the Finnish layered intrusions formed in a very narrow time span (2.37–2.46 Ga) during the rifting of late Archaean continents (Gaál, 1985). The Finnish intrusions appear to have intruded sediments which were deposited in an aulacogen formed during the rifting event (Gaál, 1985). The 2.9 Ga Usushwana Complex of southern Africa formed in a similar setting and may also be related to a rifting event.

Layered intrusions may therefore be related to repeated cycles of continental collision and rifting; 3.5 Ga collision–2.9 Ga rifting–2.7 Ga collision–2.4 Ga rifting–2.0 Ga collision. The most coherent theoretical framework for the interpretation of long-term regularities in geological processes is the extended Wilson cycle, which holds that the cycle of continental rifting–continental collision is repeated through the geological record. Although certain authors have argued that modern plate tectonic processes only began at c. 1 Ga (Kröner, 1977, 1984; Baer, 1977), detailed investigations of early Archaean provinces such as the Barberton greenstone belt (de Wit, 1982; Jackson et al., 1987) and the Shurugwi complex (Stowe, 1968) have led to interpretations which can readily be framed in the context of modern plate tectonic processes (Dewey and Spall, 1975). On balance it would appear that the gross features of the plate tectonic process, as expressed in the Wilson cycle, may be discerned as far back as the geological record is preserved (e.g. Windley, 1981).

With the exception of hypotheses related to astrophysical forces (Williams, 1973), changes in the mantle convection system are generally viewed as the driving mechanism for long term orogenic episodicity (Runcorn, 1965; Sutton, 1963; Anderson, 1982; Pavoni, 1981, 1985). Earlier theories related episodic mantle convection to growth of the core (Runcorn, 1965) or to thermal blanketing by continents (Pavoni, 1981, 1985; Anderson, 1982). Recently, attention has shifted to the thickness of the D'' layer at the base of the mantle as the regulator of mantle convection (Stacey and Loper, 1983; Loper and McCartney, 1986; Courtillot and Besse, 1987). Episodicity in mantle convection may be related to the thermal relaxation time which is, in part, governed by the square of the thickness of the D'' layer (Loper and McCartney, 1986). There is some uncertainty concerning the value of this parameter (Courtillot and Besse, 1987), but if the D'' layer is on average 150 km thick, as suggested by modelling of seismic data (Dziewonski and Anderson, 1981), then the thermal relaxation time could lie in the range appropriate to a 500 to 900 Ma episodicity in global orogeny.

To drive a recurrent process of collison and rifting, the mantle convection system must undergo some rearrangement between cycles. A coherent framework in which this rearrangement takes place has been presented by Pavoni (1981, 1985), according to whom the mantle convects by hemispherical upward motion focussed at two antipodal points on the equator. At present, these points are the African and the Pacific spreading centres at 0°N, 10°E and 0°, 170°W respectively (Kanasewich, 1976; Le Pichon and Huchon, 1984). Convecting mantle ascends at these centres then spreads radially away, and at a radial distance of about 60° begins to descend. The two largest, approximately circular plates, the African and the Pacific, are located above the spreading centres, while the smaller, elliptical Antarctic, South American, North American, Eurasian and Indian plates lie in the transpolar girdle between the larger plates (Kanasewich, 1976). Pavoni (1981) proposed that the accumulation of continents in the transpolar girdle produces a thermal blanketing effect which disturbs and finally interupts the mantle convection system. A new convection system is then established, with the poles again located at antipodal points on the equator, but at 90° from the previous poles. The theory thus predicts that supercontinents form at regular intervals, but inevitably break up since the formation of a supercontinent is inherently self-destructive as a result of the thermal anomaly which builds up below it.

An impressive array of features on the modern Earth is explained by such a hemispherical mantle convection system; these include the distribution of heights on the residual geoid (Crough and Jurdy, 1980; Hager, 1984) and on the lower mantle equivalent geoid (Dzienowski, 1984) and the pattern of global heat flow (Chapman and

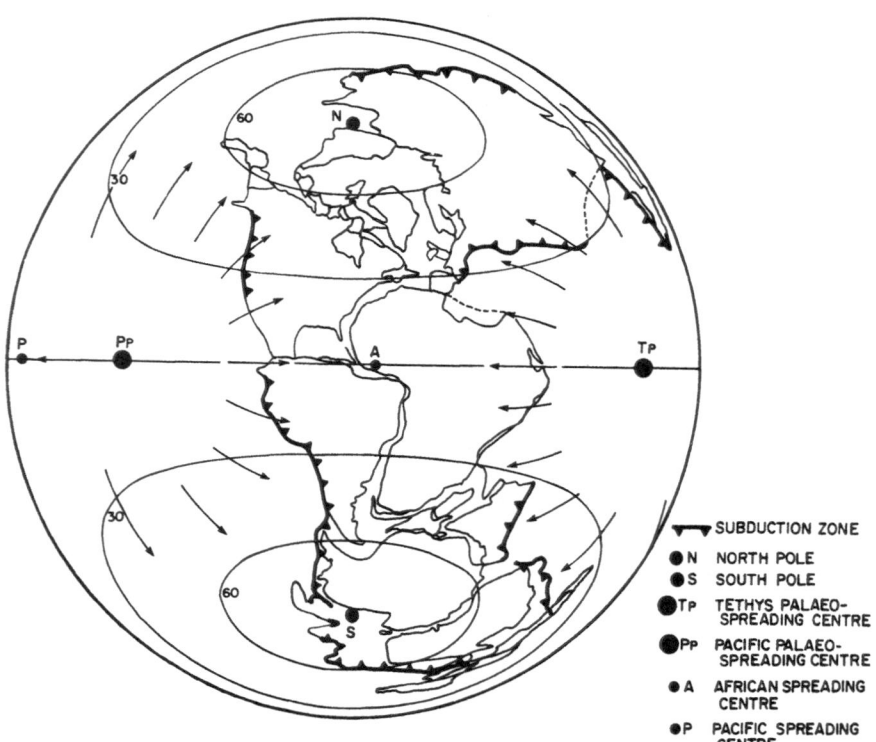

Figure 4.9 Mesozoic supercontinent with bounding subduction zones and localities of spreading centres.

Pollack, 1975). Furthermore, the Mesozoic supercontinent as reconstructed by Hay *et al.* (1981) lies on a transpolar girdle (Figure 4.9) as predicted by Pavoni (1981). An interesting feature of the Mesozoic supercontinent is that it is almost entirely surrounded by subduction zones (Figure 4.9), as might be expected if the mantle convection cells were descending along its margins. Slabs penetrate the 670 km discontinuity (Creager and Jordan, 1984) and may descend to the core–mantle boundary (Loper, 1985). Since descent is essentially vertical (Loper, 1985) the slabs would tend to accumulate directly below the supercontinent. Heat flux from the core into these accumulated slabs eventually leads to their destabilisation and subsequent ascent, counter to the flow of the descending slabs. Consequently the convection system is rearranged, along the lines proposed by Pavoni (1981).

A further important feature of a model involving a supercontinent surrounded by subduction zones is that eroded sediment might be carried below the supercontinental margins in the subduction zones. The detection of [10]Be anomalies in recent island arc volcanics (Brown *et al.*, 1982; Tera *et al.*, 1986), and seismic profiles of subducted oceanic crust (Hilde, 1983; Kvenvolden and Von Huene, 1986) indicate that the proportion of sediment that is subducted rather than scraped off at the plate margin, is very high. The possibility that sediment may be subucted below continents has particular relevance for Precambrian layered intrusions.

Evidence that an upper crustal component is present in the parental magma to layered basic intrusions has been obtained from the Bushveld (Hamilton, 1977; Sharpe, 1981; Harmer and Sharpe, 1985), Stillwater (Longhi *et al.*, 1983) and Usushwana (Hegner *et al.*, 1984) intrusions. Although it has been proposed that this upper crustal component was incorporated during ascent of the magma through the crust (Longhi *et al.*, 1983; Sparks, 1986), features such as the lack of correlation between Rb/Sr and Sr_i ratios (Harmer and Sharpe, 1985) and narrow ranges in Ti/Zr ratios (Hatton and Sharpe, 1988; Hegner *et al.*, 1984) suggest instead that the upper crustal signature reflects the unusual mantle source from which the parental magma was derived (Hamilton, 1977). The suggested mechanism for the generation of the unusual mantle source is sediment subduction (Hatton and Sharpe, 1989).

4.8 Concluding remarks

A plausible proposal for a general relation between the emplacement of layered intrusions and long term orogenic cycles is as follows. During the assembly of a supercontinent, sediment is subducted below the continental margin and contaminates the subcontinental lithosphere. As ocean basins close, the ridge (mid-oceanic spreading centre) is also subducted below the continent. Ridge subduction ultimately leads to a strong upwelling of the asthenosphere (Damon, 1983) and this in turn results in the melting of the overlying, sediment-contaminated, subcontinental lithosphere. Similarly, during the break-up of a supercontinent, the asthenosphere also ascends beneath the lithosphere. In this situation the result is the melting of lithosphere which may have been contaminated by subducted sediment. Such a contaminated lithosphere is the source of the high-Si, high-Mg boninitic rocks associated with the Bushveld (Sharpe, 1981; Davies *et al.*, 1980), Stillwater (Longhi *et al.*, 1983), Great Dyke (Wilson, 1982), Jimberlana (Campbell, 1977), Binneringie (McCall and Peers, 1971) and Koillismaa (Alapieti, 1982) intrusions (Table 4.2). In terms of magma genesis, the basic process of melting of lithosphere by upwelling asthenosphere is relatively independent of whether the upwelling was associated with ridge subduction or continental break-up. The intrusion form, however, does appear to be related to the overall tectonic environment. The Bushveld and Stillwater complexes, which we suggest formed during supercontinent assembly, are dominantly sill-like in form, while the Great Dyke, Widgiemooltha dyke swarm and Finnish intrusions formed during continental rifting and are predominantly dyke-like.

5 Noritic magmatism

R.P. HALL and D.J. HUGHES

5.1 Introduction: high-Mg basic rock types

Over the past two decades much effort among petrologists has been devoted to the problems of magnesium-rich basic magmatic suites. In the early Precambrian these suites comprise komatiites, komatiitic basalts (previously known as basaltic komatiites), siliceous high-Mg basalts (SHMB), norites and the parental magmas of the noritic portions of major layered, Bushveld-type intrusions ('U'-type magmas). However, of these, the petrology of norites has not been given a great deal of attention. Indeed, norites are mentioned only fleetingly, or not at all, in most petrology texts. Norites are mafic rocks consisting essentially of orthopyroxene and plagioclase, with varying amounts of olivine (olivine-norite) and clinopyroxene (augite-norite, gabbronorite). Thus, they are distinctive and easily recognised. The relative paucity of petrological data for norites results principally from the fact that they occur most obviously as cumulates in large layered intrusions. However, the abundance of early orthopyroxene suggests that they are not derived simply from tholeiitic magmas, in which orthopyroxene is not a significant low pressure liquidus phase. Norites warrant a little more emphasis here than they have been afforded previously because they represent a considerable volume of mafic magmatism during the early Precambrian, and they are distributed world wide.

Since the recognition of komatiites in the Archaean of South Africa twenty years ago (Viljoen and Viljoen, 1969a–c, 1982), similar Archaean metavolcanics have been identified in virtually every ancient craton (Arndt and Nisbet, 1982b) and even high-grade, amphibolite facies Mg-rich basic rocks have been recognised as originally komatiitic extrusive lavas having apparently undergone more or less isochemical metamorphism (e.g. Hall et al., 1987a). Unaltered komatiitic lavas have a distinctive mineralogy and chemistry which are not found, except in rare instances, in modern magnesian basaltic suites. They are characterised mineralogically by their 'spinifex-textured' (Nesbitt, 1971), randomly oriented sheaves of quenched olivine and clinopyroxene rods and plates within a glassy matrix (Figure 5.1a) (Donaldson, 1982). Plagioclase is clearly not a liquidus phase and is apparent only normatively in the (formerly) glass matrix.

Komatiites are geochemically distinctive in that they have low SiO_2 ($< 50\%$), TiO_2 ($< 0.9\%$) and alkali contents ($K_2O < 0.5\%$), and high MgO ($> 18\%$), Cr (> 500 ppm) and Ni (> 250 ppm) (Brooks and Hart, 1974; Arndt et al., 1977; Arndt and Nisbet, 1982b). Their high clinopyroxene:plagioclase ratios are mirrored by their relatively high CaO/Al_2O_3 values (> 0.9) compared with typical modern tholeiitic basalts ($CaO/Al_2O_3 = 0.5$–0.8) (Figure 5.2). Komatiitic basalts (MgO = 18–9%) are fre-

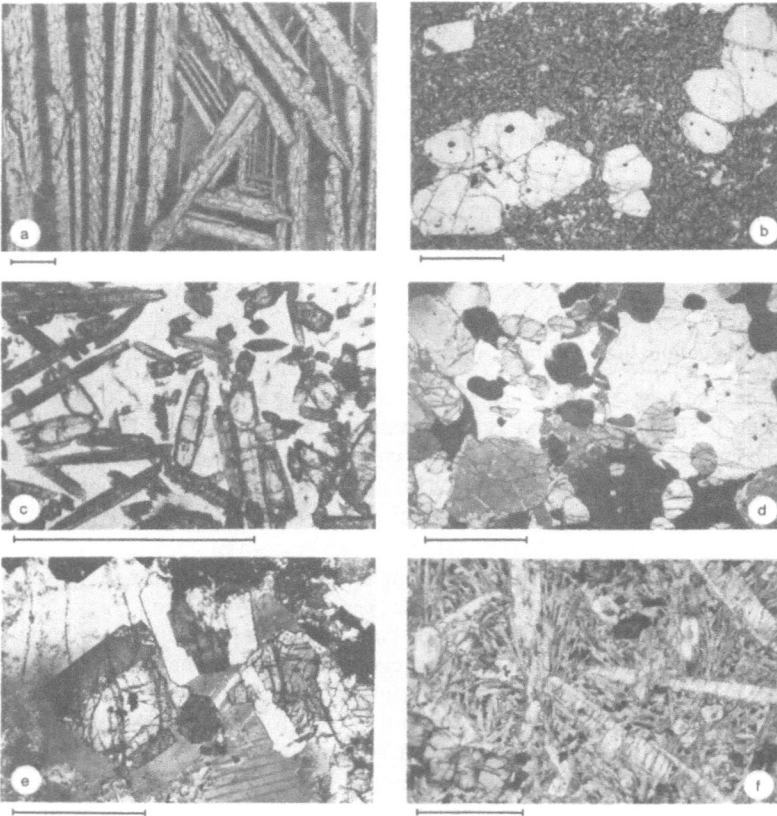

Figure 5.1   Photomicrographs of high-Mg rock types discussed in the text. Scale bar represents 1 mm. (a) Spinifex-textured komatiite (ppl). Skeletal olivine blades with interstitial pyroxene and devitrified glass. Abitibi Belt, Munro Township, Ontario. (b) Picritic basalt (ppl). Abundant olivine phenocrysts in a plagioclase, clinopyroxene and magnetite matrix. Mauna Loa. (c) Boninite (ppl). Ca-poor pyroxene phenocrysts, some of which are sandwiched between calcic pyroxene, in a glassy matrix. Bonin Island, Western Pacific (sample kindly supplied by R.N. Taylor). (d) Plagioclase-harzburgite (xpl). Large bronzite (En_{83-76}) crystals enclose smaller olivines (Fo_{80}), with interstitial diopside and plagioclase (An_{80-47}). Sample WY13, Cherry Creek sill, central Laramie range, Wyoming, USA (see Figure 9.8, p. 206). (e) Augite-norite (xpl). Bronzite chadacrysts sandwiched between augite and enclosed by plagioclase oikocrysts. The bronzites are weakly concentrically zoned to thin hypersthene rims, and the augites are also zoned to more Fe-rich compositions away from the orthopyroxene contacts (Figure 5.3). BN dyke sample RPH 156023, southern West Greenland. (f) Quenched BN dyke, southern West Greenland (ppl). Phenocrysts of olivine, bronzite and, mainly, elongate pigeonites mantled by augite, in a matrix of dendritic plagioclase, more Fe-rich clinopyroxenes and magnetite (Figure 5.3).

quently not simply the fractionated products of a more Mg-rich komatiitic parent. In many instances, their high LREE contents suggest a mechanism other than the simple fractionation of the komatiite liquidus phases (olivine and clinopyroxene) and often invoke a significant continental, or subduction zone-related contamination component (SZC; see also Chapter 3).

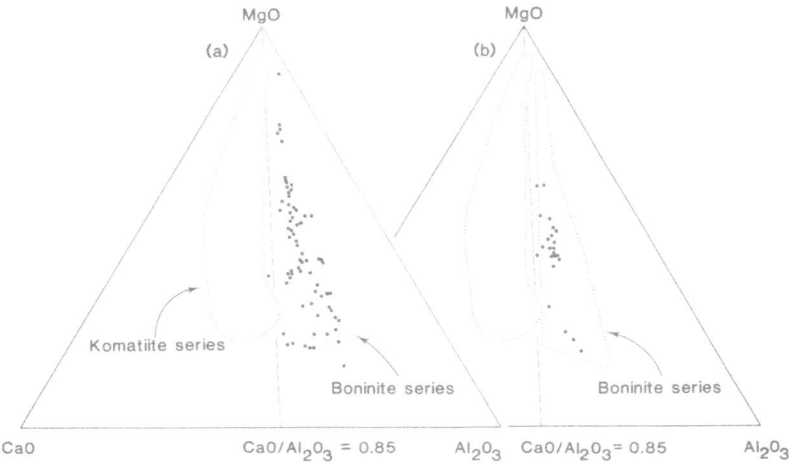

Figure 5.2 MgO:CaO:Al$_2$O$_3$ proportions in (a) boninite series (boninites and high-Mg andesites; 70 points) compared to the field of the komatiite series, and (b) BN norite dykes of southern West Greenland. Komatiites generally have a relatively high CaO/Al$_2$O$_3$ ratio. Komatiite field taken from Arndt *et al.* (1977), Nisbet *et al.* (1977), Auvray *et al.* (1982), Binns *et al.* (1982) and Viljoen *et al.* (1982). Boninite series data from Dietrich *et al.* (1978), Kuroda *et al.* (1978), Cameron *et al.* (1979, 1983), Jenner (1981), Hickey and Frey (1982), Tatsumi and Ishizaka (1982) and Walker and Cameron (1983).

The tectonic setting of many komatiites remains uncertain. The Archaean Earth was presumably dominated by basic–ultrabasic oceanic crust (Nisbet, 1987), and yet most komatiitic basalts appear to have been erupted through sialic crust. It has been proposed that they formed in environments equivalent to modern mid-ocean ridges, island arcs and continental rift systems. Experimental modelling has demonstrated that high-Mg komatiitic lavas would be capable of digesting crustal material (Huppert *et al.*, 1984; Huppert and Sparks, 1985) to give rise to the contaminated compositions encountered in many komatiitic basalts (Arndt and Nesbitt, 1984; Nisbet, 1984; Arndt and Jenner, 1986), although, paradoxically, many komatiites apparently associated with continental crust show no evidence of significant contamination (Claoue'-Long and Nesbitt, 1985; Arndt *et al.*, 1986). While it is generally agreed that komatiites are ultimately derived from high degrees of partial melting of relatively fertile mantle (see Chapters 2 and 6), the composition of the mantle source is somewhat obscured by the higher level modifications to the komatiitic magmas. Therefore, we view the mantle through a 'dirty window' (Arndt, 1986).

Modern high-Mg basic lavas comprise two main types, namely picrites and boninites, characterised respectively by their high olivine and orthopyroxene contents (McKenzie *et al.*, 1982). Both of these magnesian suites differ significantly from Archaean komatiites, picrites having low CaO/Al$_2$O$_3$ ratios (Hall *et al.*, 1987a) and olivine-phyric textures (Figures 5.1b), and boninites being recognised by their relatively high silica contents and Ca-poor pyroxene-phyric nature (Figures 5.1c, 5.2). The origin of modern boninites (and high-Mg andesites) is also quite different to that of komatiites. These rocks appear to be derived from the fluid-induced melting of the refractory, depleted, but metasomatically replenished harzburgitic portion of litho-spheric mantle above the subducted slab in a primitive arc system, contaminated by fluids derived from the entrapped, largely sedimentary veneer material (Hickey and Frey, 1982; Cameron *et al.*, 1983). However, despite their recent petrological pro-minence (Crawford, 1989), modern boninites are volumetrically insignificant and restricted in their distribution (Cameron *et al.*, 1983).

Another type of high-Mg basic rock suite became prominent in the late Archaean and early Proterozoic. The rocks of this suite are different from those mentioned above in that they occur principally as plutonic and hypabyssal intrusions, constituting a series of layered complexes and dyke swarms of high-Mg noritic composition, or at least with a large noritic component. The major layered noritic intrusions include the 2.7 Ga Stillwater Complex of Montana, USA (Czamanske and Zientek, 1985), the 2.46 Ga Great Dyke of Zimbabwe (Nisbet, 1982; Podmore and Wilson, 1987), the 2.4 Ga Jimberlana intrusion of Western Australia (Campbell, 1977) and the 2.1 Ga Bushveld Complex of South Africa (Von Gruenewaldt *et al.*, 1985). These and numerous other minor intrusions are reviewed by Hatton and Von Gruenewaldt (Chapter 4). Volumetrically, these noritic intrusions collectively far exceed what we see preserved of the early Precambrian high-Mg volcanic suites (see Figure 4.1, p. 61).

The most obvious and well-documented feature of these large-scale intrusions is their extensive and complex mineralogical layering (Wager and Brown, 1968; Parsons, 1987). The coarse cumulus assemblages leave little doubt that individual samples cannot represent liquid compositions. Parental magma compositions are therefore only estimated by calculating bulk compositions for the intrusions or by examining fine-grained, homogeneous satellite basic dykes or sills which may have been feeders or offshoots of the complexes. In recent years numerous high-Mg basic dyke swarms have come to light. Several varieties of noritic dykes occur in the Beartooth Mountains of southern Montana, USA, near to the Stillwater Complex (Longhi *et al.*, 1983; Helz, 1985; Hall *et al.*, 1987b), and fine-grained noritic ('micropyroxenitic') dykes also occur near to the margins of the Bushveld Complex (Sharpe, 1981; Sharpe and Hulbert, 1985). Noritic dyke swarms have been recognised in Western and South Australia (Hallberg, 1986; Fletcher *et al.*, 1987; Mortimer *et al.*, 1988), Antarctica (Sheraton *et al.*, 1987; Keuhner, 1989), Scotland (Tarney and Weaver, 1987a, b), the Bighorn Mountains, Wyoming, USA (Hall *et al.*, 1987b), Greenland (Hall and Hughes, 1987) and South America (Sial *et al.*, 1987).

Despite the obviously widespread distribution of large- and small-scale late Archaean or early Proterozoic noritic intrusions, equivalent extrusive rocks are not so apparent. This could, of course, be due either to their not having formed or to their not having been preserved. It is quite possible that the norite dyke swarms in particular could have reached the surface and fed lava flows. If the deeper, layered intrusive complexes are presently exposed, it is perhaps unlikely that the corresponding volcanics would also be preserved in the same portion of the craton. They would have simply been eroded from the top of the continental crust through which they passed in the dyke fissures and onto which they were erupted.

In this chapter we shall highlight the similarities and the differences between the ancient and modern siliceous high-Mg volcanic suites and these late Archaean and early Proterozoic magnesian noritic intrusives, and to examine whether the distinctive petrology of the norites is simply a function of their tectonic setting. Because of the problems in overcoming the varied petrological processes operative within large-scale intrusions, and since these complexes have been dealt with in the previous chapter, we concentrate mainly on the noritic dykes.

5.2 Mineralogy

Norites are by no means uniform rocks but comprise a variety of types which are related by the partial melting, fractionation, accumulation and contamination processes which affect all basic magmas. The commonest norites occur in layered intrusions (see Chapters 4 and 8) and comprise varying proportions of orthopyroxene, plagioclase, clinopyroxene and olivine. Most fall into the category of orthopyroxene (bronzite) cumulates, often with interstitial or poikilitic plagioclase and clinopyroxene, although there is still controversy about the origin and significance of the cumulus layering in the large-scale intrusions (Irvine, 1980; McBirney, 1985; Huppert et al., 1986, 1987). Similar poikilitic textures to those in the layered intrusions also occur in smaller, essentially unlayered intrusions such as the Quad Creek intrusion of the southern Beartooth mountains of Montana (Eckelman and Poldervaart, 1957; Prinz, 1964) and the Cherry Creek and Tony Ridge plagioclase-harzburgite intrusions of the central Laramie range, Wyoming, USA (Figure 5.1d) and in smaller dykes, such as those in western North America, Greenland (Figure 5.1e) and Scotland. While concentric zoning of minerals has been described in some noritic dykes (e.g. Hall and Hughes, 1987), and ascribed in some instances to the crystallisation of inclined magma sheets (Tarney and Weaver, 1987a; cf. Huppert et al., 1987), most of the dykes are not layered, except for the presence of thin chilled margin rocks, and they do not appear to have any significant cumulus component, but probably represent an essentially *in situ* crystallisation sequence.

The crystallisation sequence of norites is usually very different from that shown by contemporary tholeiitic intrusions. The swarm of early Proterozoic norite dykes and another of tholeiitic dolerites in southern West Greenland, for example (Chapter 11), are geochemically as well as petrographically distinct. Isotopic data suggest that both swarms are of approximately the same absolute age, c. 2.1 Ga (Bridgwater et al., 1985; Kalsbeek and Taylor, 1985). The so-called 'MD' (meta dolerite) dykes are commonly deuterically altered to uralitic amphibole-rich assmblages, but cross-cutting relationships locally demonstrate that these dykes are younger than the unaltered norites. They are overwhelmingly subophitic or ophitic rocks, in which complex clinopyroxene assemblages crystallised together with, or slightly after, plagioclase (Hall et al., 1985, 1986, 1988a). The norite dykes on the other hand have the crystallisation sequence olivine–orthopyroxene–clinopyroxene–plagioclase (Hall and Hughes, 1986, 1987). Similar types of contrasting crystallisation sequences in noritic and doleritic dykes have also been noted in North America (Hall et al., 1987b) and Scotland (Weaver and Tarney, 1981a).

The orthopyroxene chadacrysts in the Greenlandic norites comprise mainly weakly-zoned bronzite (En_{88-70}) with thin Fe-rich (hypersthene, En_{60}) rims (Figure 5.3a). These orthopyroxenes are often mantled by epitaxial overgrowths of augite (Hall et al., 1985; cf. Tarney, 1969), zoned to more Fe-rich compositions away from the contact with the Ca-poor pyroxene (Figures 5.1e, 5.3a). The volcanic analogue of this texture occurs in modern boninites, which commonly comprise phenocrysts of Ca-poor pyroxene sandwiched between calcic pyroxene, within a glassy matrix (Figure 5.1c) (Cameron et al., 1979; Wood, 1980; Natland, 1982). Olivine compositions range from

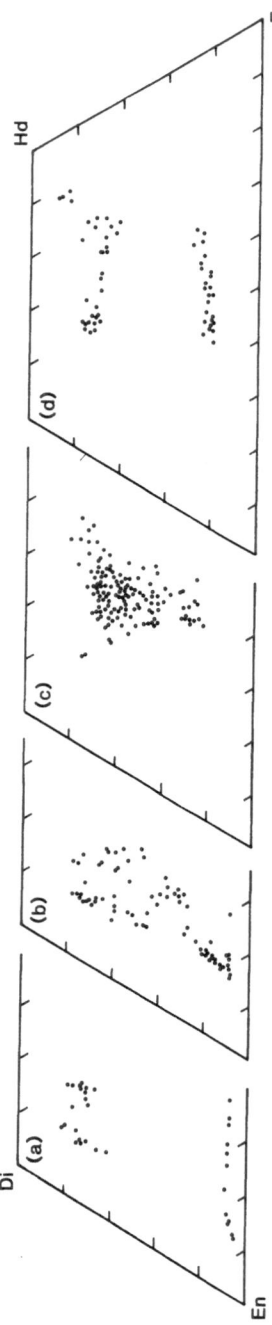

Figure 5.3 Pyroxene assemblages in (a) BN norite (sample 156023), (b) quenched BN norite (sample 155686) early Proterozoic dykes from southern West Greenland, and (d) Mesozoic dolerite (sample TA5) from the Mt Wellington sill, Tasmania, in terms of their mol. proportions of $MgSiO_3$–$(Ca, Mg)SiO_3$–$(Ca, Fe)SiO_3$–$FeSiO_3$ (En–Di–Hd–Fs). Data from Hall and Hughes (1987) and Hall *et al.* (1985, 1986, 1988a).

Fo_{85} to Fo_{65}. These olivines and composite pyroxenes are typically enclosed by large plagioclase oikocrysts, weakly zoned (labradorite–andesine) from their centres to their rims. In some samples, the olivine and orthopyroxene primocrysts occur within a more finely grained groundmass of dominantly dendritic plagioclase and clinopyroxene, and one thin dyke comprises a quenched assemblage of skeletal or subhedral olivines (Fo_{90}), orthopyroxenes (En_{85}) and, predominantly, pigeonites mantled by augite, all within a fine-grained dendritic plagioclase matrix (Figure 5.1f)(Hall and Hughes, 1986, 1987). The composite pigeonite–augite crystals are also Mg-rich ($X_{Mg} = 0.85$), while the groundmass pyroxenes are more Fe-rich augites and subcalcic augites ($X_{Mg} = 0.70$) (Figure 5.3b). Dykes with similar quench textures and mineralogy have been recognised in micronoritic dykes in the Wyoming Province, USA (Longhi et al., 1983; Hall et al., 1987b) and as 'micropyroxenitic' sills at the margins of the Bushveld Complex (Cawthorn et al., 1981; Sharpe, 1981; Sharpe and Hulbert, 1985). While no equivalent extrusive rocks have been identified, these quenched dykes give a good indication of the type of lava that would develop from the parental magmas of these high-level noritic intrusions.

The concentric Fe-enrichment zoning trend leaves little doubt about the pyroxene crystallisation sequence in the norites. Late-stage calcic pyroxene intergrown with the plagioclase is the most Fe-rich pyroxene. However, the nearby tholeiitic dolerites show even more complex sequential pyroxene crystallisation histories, but in this case shown by extreme Ca as well as Mg/Fe variation (Figure 5.3c). The pyroxene compositions vary greatly from one dyke to another and the variation is largely dependent on the degree of prior plagioclase crystallisation and oxygen fugacity (fO_2)-controlled Fe-Ti oxide precipitation (Hall et al., 1986, 1988a). The degree of complexity in these hypabyssal rock pyroxenes is clearly significant in evaluating the emplacement and post-emplacement mechanisms in minor intrusions, for if it reflects a dynamic (turbulent/laminar flow) crystallisation regime (e.g. Huppert and Sparks, 1985; Delany, 1987), it brings into question to what extent any sample taken from an intrusion corresponds to the composition of the liquid from which it was derived. However, in the case of the Greenlandic noritic dyke swarm, the chemistry of the partially quenched dyke is probably close to that of the magma composition (Hall and Hughes, 1987) and the fact that the coarser 'cumulate' norites have the same composition indicates that they have probably not been severely modified by crystal accumulation.

Such considerations become even more significant in layered intrusions. Despite the post-cumulus processes in such intrusions (Sparks et al., 1985; Hunter, 1987) cryptic variation in mineral chemistry has for many years been taken to reflect the progressive evolution of a large-scale crystallising body of magma (Wager and Brown, 1968). However, while the progressive upward Fe-enrichment trend of ferromagnesian phases (and hence whole-rock chemistry) does clearly reflect the magmatic differentiation history, individual samples can themselves mirror this evolution trend in the apparently rare cases where it is not obscured by subsolidus re-equilibration. For example, in the case of the Mesozoic (Karoo) dolerite sills of Tasmania, pyroxene compositions of individual small samples, and indeed grains, can show chemical zoning on the scale of around 50% of the evolution trend of the entire intrusion (Figure 5.3d) (Hall et al., 1988b). This cannot reflect sequential crystallisation during the gravitational settling of a grain through a magma body. It is unrealistic to argue that all of the grains passed through half of a chemically layered magma body. The disequilibrium crystallisation trends clearly reflect, in part at least, in situ crystallisation and the evolution of residual liquid on a very small scale, perhaps in 'cells' (Nwe, 1975). It is equally possible that the norite dykes crystallised more or less *in situ* rather than having been derived from the gross accumulation of grains settling or rising into a more Fe-rich liquid, and that samples of the dykes do reflect the parental magma composition.

Most norites in layered intrusions are clearly controlled by petrogenetic effects such as double diffusive convection, gravity stratification, turbulence, magma replenishment, post-cumulus and re-equilibration processes, depending on the attitude and scale of the intrusion (Irvine, 1980; Sparks *et al.*, 1985; Huppert *et al.*, 1987; Hunter, 1987), and it therefore remains difficult to find samples from a layered noritic intrusion whose composition could be interpreted as that of the parental magma. The close spatial and temporal association of norite and dolerite dykes – for example in Greenland and Scotland (Hall *et al.*, 1978b; Tarney and Weaver 1987b) – is mirrored in many of the large layered intrusions. Two (or more) different magma types have been postulated as having been emplaced together and mixed to form the lower noritic and upper gabbroic portions of complexes such as Bushveld and Stillwater (Todd *et al.*, 1982; Irvine and Sharpe, 1982; Barnes and Naldrett, 1986). The two end-member rocks which most closely reflect these liquids are magnesian norite and gabbro, with the crystallisation sequences olivine–bronzite–augite–plagioclase, and plagioclase–olivine–augite–bronzite respectively. These different magmas gave rise to associations of ultramafic and anorthositic cumulates and are referred to as 'U'- and 'A'-type magmas respectively (Irvine and Sharpe, 1982). This is clearly analogous to the apparently contemporaneous intrusion elsewhere of contrasting noritic and doleritic dyke swarms. The petrological data suggest that the only significant difference between the norite and dolerite dyke swarms, and the complexly layered noritic–gabbroic Bushveld-type intrusions, is their style of emplacement.

5.3 Geochemistry

The geochemistry of the early Proterozoic noritic (BN) dykes of Greenland is very distinctive and markedly different to that of normal tholeiitic basic (doleritic–gabbroic) intrusions (Hall and Hughes, 1987). Moreover, the chemical characteristics of the Greenlandic norites appear to be common to many other noritic intrusive suites. The chemistry of these norites is described here as these dykes constitute a fairly uniform suite and presumably reflect relatively consistent petrogenetic processes, whereas equivalent dykes in, for example, the Wyoming craton are rather more diverse and may have resulted from more than one parental magma, or from a magma which has undergone the influence of a variety of petrogenetic processes (Helz, 1985; Longhi *et al.*, 1983).

Most of the BN dykes are Mg-rich (14–18% MgO). A few have up to 22% MgO and rare granophyric augite-norites have as little as 6% MgO (Hall and Hughes, 1987; see Figure 11.5, p. 264). Cr and Ni contents are correspondingly high (5000–200 ppm, and 1300–60 ppm respectively). Fe contents are low (Fe_2O_3* from 12 to 9%), and mg' ($Mg/(Mg + Fe)$) values are high, decreasing from *c.* 0.8 to 0.6 (Hall *et al.*, 1987b). For such magnesian rocks, their silica contents are also high, increasing from 49 to 57% SiO_2 with decreasing MgO. Their CaO contents are generally low, increasing from *c.* 6 to 9%, and Al_2O_3 contents increase from 9 to 16% ($CaO/Al_2O_3 = 0.55$–0.65; Figure 5.2). Consequently, normative *hy:di* ratios are high (*c.* 1.5–4) compared to most tholeiites ($\leqslant 1$), in accordance with the modal mineralogy. Sc and V contents are also

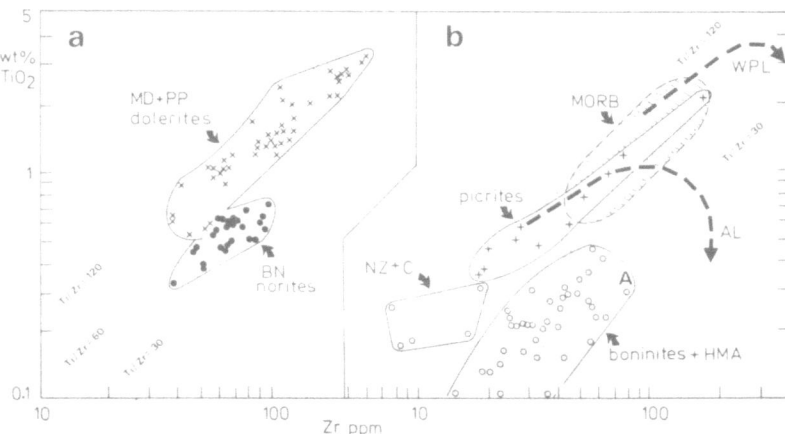

Figure 5.4 TiO_2 (wt %) versus Zr (ppm) in (a) early Proterozoic BN norites and MD and PP dolerites of southern West Greenland, and (b) boninites and high-Mg andesites (HMA) and average Bushveld 'micropyroxenite' sill (A) of Harmer and Sharpe (1985). Field of MORB and trends of within plate lavas (WPL) and arc lavas (AL) from Pearce (1980) and Pearce and Norry (1979). Boninite series data from Dietrich et al. (1978), Wood (1980), Jenner (1981), Cameron et al. (1983) and Nelson et al. (1984). Picrite data from Clarke (1970) and Ramsay et al. (1984). Diagram reproduced from Hall and Hughes (1987).

uniformly low (c. 25 ppm and 175 ppm respectively), compared to those of the neighbouring tholeiitic dolerites (20–50 ppm, and 200–500 ppm respectively) (see Figures 11.5, 11.6, pp. 264, 265).

Concentrations of all of the HFS elements (Ti, Zr, Nb, P) are low, as would be expected for such magnesian rocks. TiO_2 and Zr contents range from 0.3 to 0.7% and 100 to 400 ppm respectively. Ti/Zr ratios are relatively low (40–60) compared to tholeiitic picritic and basaltic rocks (Figure 5.4). Similarly, Y contents are relatively low (< 15 ppm) and Zr/Y values correspondingly high (Figure 5.5). Ti:Zr:Y proportions are more closely akin to those in calc-alkaline basalts than tholeiitic ones (Hall et al.,

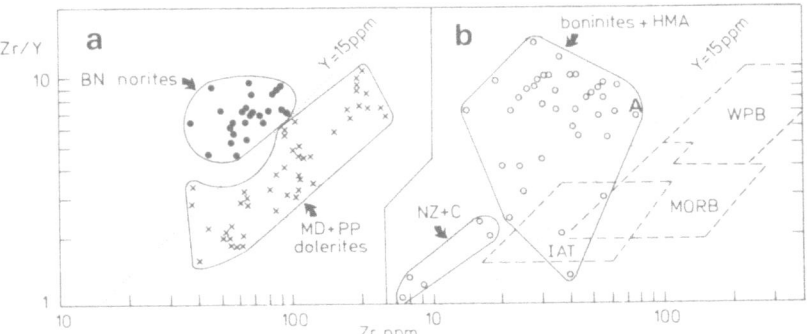

Figure 5.5 Zr:Y relationships in (a) BN and MD dykes, southern West Greenland and (b) boninite series rocks and average Bushveld quenched norite sill. Data sources as in Figure 5.4. Diagram reproduced from Hall and Hughes (1987).

1987b). Nb contents are consistently low (*c.* 2 ppm), and anomalously so compared to trace element proportions in primitive tholeiitic basalts (Figure 5.6).

Perhaps the most distinctive aspect of the geochemistry of the BN norites, considering their magnesian character, is their high LILE concentrations (Table 5.1). K, Rb, Ba and the LREE (La, Ce and Nd) are all markedly higher than in other Mg-rich suites (komatiites, picrites). Average values for these elements in the norites are 0.75% K_2O, 15 ppm Rb, 200 ppm Ba, 12 ppm La, 30 ppm Ce and 15 ppm Nd. These elements are enriched by between 2 (Nd) and 20 (Ba) times their concentrations in MORB (Figure 5.6; see also Figure 11.5, p. 264), and so far as the petrogenesis of these norites is concerned, it is significant that there is a marked contrast between the levels of these elements in the norites and the most magnesian (MgO = *c.* 10%) of the neighbouring tholeiitic dolerite dykes (K_2O = 0.2%, Rb = 5 ppm, Ba = 50 ppm, La = 3 ppm). The MD dykes are typical continental tholeiitic dolerites with mild LILE enrichment compared to MORB (Figure 5.6). The HREE (and Y) contents of the norites are relatively low with respect to MORB, and consequently La_N/Lu_N values are markedly high (*c.* 4.4–8), in contrast to the only mildly LREE-enriched MD dolerites (La_N = 10 − 50; La_N/Lu_N = 2 − 4).

The contrast in geochemistry between the BN noritic and MD doleritic dykes in West Greenland raises the question of why two such chemically different swarms of mafic dykes should have been developed so close together at virtually the same time, and also lends direct support to the hypothesis that two contrasting basic magmas may have mixed to produce the diversity of lithologies which comprises major layered intrusions such as Bushveld and Stillwater.

5.4 Petrogenesis

5.4.1 *Komatiites, SHMB, 'U'-type magmas, norites and boninites*

Norite dykes such as those in southern West Greenland appear, at first approximation at least, to be mineralogically and geochemically the holocrystalline equivalents of siliceous high-Mg basalts (SHMB), which themselves may represent crustally contaminated komatiitic magmas, which are almost exclusively Archaean (Arndt and Jenner, 1986; Barley, 1986; Cattell, 1987). Equally, they appear to be the unfractionated analogue of certain layered rocks in major intrusions, derived from 'U'-type magmas (which ultimately give rise to ultramafic cumulates). A modern siliceous Mg-rich volcanic suite with which these rocks have been compared are olivine-boninites (Cawthorn *et al.*, 1981; Cameron *et al.*, 1983; Campbell, 1985; Sharpe and Hulbert, 1985; Sparks, 1986; Hall and Hughes, 1987; Barnes, 1989; Sun *et al.*, 1989). Because the volcanic analogue of norites is in question, their parental magmas will be referred to as noritic. The principal petrogenetic processes which may be responsible for, or at least influential in, the formation of these distinctive mafic rocks are evaluated in the following sections. It is emphasised at the outset that different petrogenetic pathways could, and probably do, lead to these similar suites and that geological evidence for the evolution of noritic magmas in one instance may not be valid elsewhere.

5.4.2 *Chemical mobility and alteration*

It is well known that some elements are relatively mobile and others less so during alteration and metamorphism, and consequently the chemical classification of even

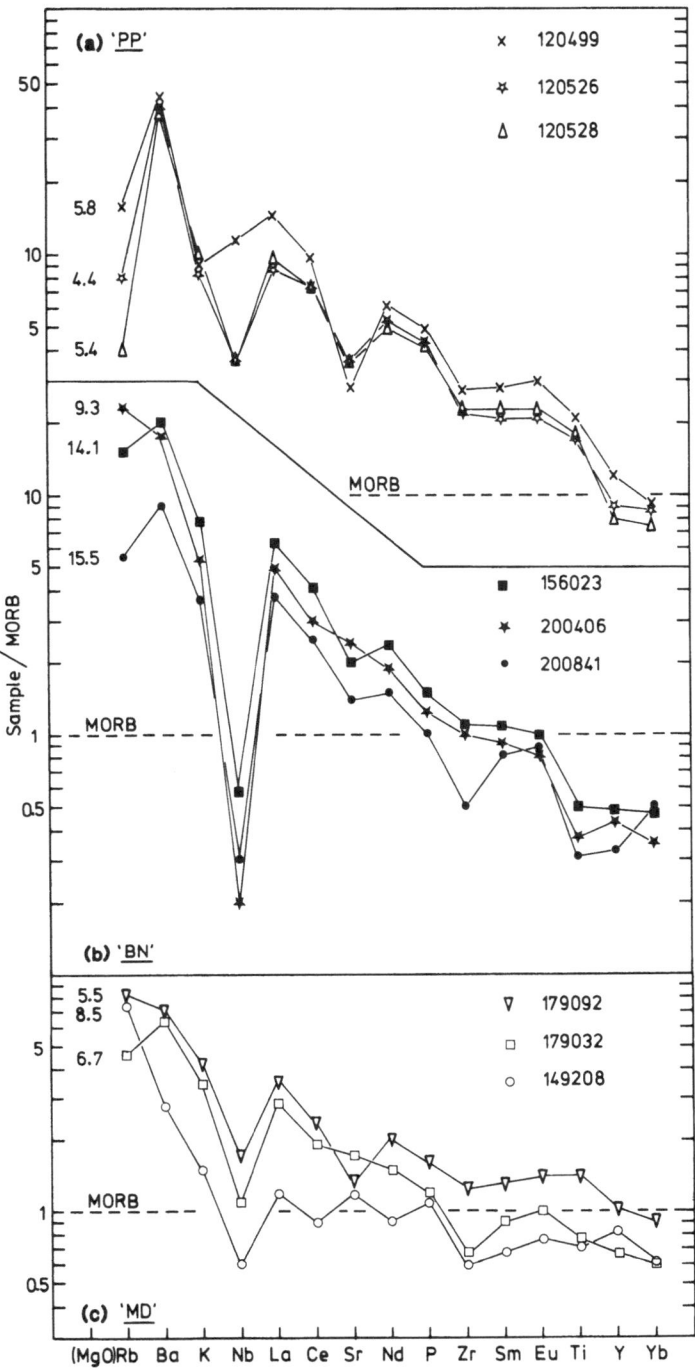

Figure 5.6 MORB-normalised multi-element (MORB-NME) plot of representative (a) PP (plagioclase-phyric) dolerite, (b) BN noritic and (c) MD dolerite dykes, southern West Greenland. The PP and BN dyke patterns have a similar slope, but note the contrast in MgO contents and the lack of a negative Nb anomaly in the PP dykes. The MD dolerites have flatter (MORB-like) patterns, apart from slight enrichment of the LILE and LREE, typical of continental tholeiites. Data from Hall *et al.* (1985, 1987b, 1987c).

Table 5.1 Geochemistry of noritic and associated rock types

	1 BN	2 EG	3 SC	4 BH	5 BT	6 NA	7 VH	8 PL	9 BU	10 CKOM	11 SHMB	12 BON	13 KREEP
SiO_2	51.60	52.81	51.14	51.17	49.33	55.23	53.90	51.67	56.07	51.60	54.59	54.31	49.10
TiO_2	0.56	0.41	0.67	0.58	0.95	0.98	0.71	0.75	0.34	0.55	0.55	0.23	1.72
Al_2O_3	12.06	11.09	14.4	12.48	14.04	12.42	11.84	13.70	11.47	11.01	11.40	10.62	16.85
Fe_2O_3	2.05	9.24	10.95	2.00	2.40	2.00	1.22	10.70	—	—	10.78	1.83	—
FeO	7.80	—	—	7.86	9.60	9.11	9.10	—	9.53	10.66	—	6.25	9.61
MnO	0.18	0.15	0.18	0.19	0.18	0.17	0.16	0.17	0.18	0.19	0.20	0.17	—
MgO	15.02	14.83	12.91	13.11	11.28	7.71	9.95	10.41	12.96	15.61	11.63	12.65	9.39
CaO	7.82	7.01	8.04	9.79	8.66	8.58	8.63	9.80	6.68	8.72	7.97	7.16	10.52
Na_2O	1.92	1.22	2.08	1.16	2.37	1.68	1.82	1.98	1.68	1.34	2.32	1.37	0.64
K_2O	0.82	0.74	0.98	0.63	0.81	1.19	0.96	0.77	0.80	0.28	0.47	0.32	0.61
P_2O_5	0.10	0.07	0.09	0.14	0.24	0.12	0.10	0.11	0.07	0.04	0.06	0.03	0.58
Cr	1862	1727	1031	1080	771	512	771	1218	1240	2019	1464	983	2430
Ni	624	626	459	310	443	151	221	276	295	468	258	279	229
Rb	14	19	20	13	10	54	39	26	30	6	14	7	14
Sr	228	244	195	112	345	161	133	180	158	54	133	124	179
Y	11	9	12	13	14	22	17	22	13	15	15	7	—
Zr	70	64	78	40	93	128	91	98	77	57	68	32	—
Nb	1	2	2	<1	2	7	4	6	4	3	4	1	—
Ba	220	312	315	128	428	295	242	310	270	131	227	41	770
La	12	13.4	12	5.3	15.6	21	17	16	15	7	11	2.2	75
Ce	27	28.2	26	9.6	28.6	39	30	33	30	14	17	4.9	191
Nd	15	12.9	12.2	6.2	15.5	—	13	16	12	7	9	2.7	114
Sm	2.5	2.2	2.6	1.9	3.5	—	2	3.5	2.7	1.8	2.1	0.7	32.1
Eu	0.75	0.62	0.76	0.65	1.23	—	0.6	1.07	0.71	0.58	0.61	0.25	2.5
Yb	1.3	0.98	1.4	1.46	1.64		1.5	2.3	1.14	1.41	1.32	0.8	20

1 Average of twenty-three BN dykes, SW Greenland (Hall and Hughes, 1987); 2 Norite dyke, SE Greenland (Hall et al., 1989b); 3 Average of five Scourie norite dykes, NW Scotland (Weaver and Tarney, 1981); 4,5 Qenched dykes, northern Bighorn and southern Beartooth mountains, Wyoming, USA (WY102, 108; authors' unpublished data); 6,7 Average of eight Napier Complex and eleven Vestfold Hills dykes, Antarctica (Sheraton et al., 1987; Kuehner, 1989; 8 Average of eight Port Lincoln norite dykes, South Australia (Mortimer et al., 1988); 9 Average of nine Bushveld 'micropyroxenite' sills (Sharpe and Hulbert, 1985; Ba data interpolated from Cawthorn, 1983); 10 Contaminated komatiite, Kambalda, Western Australia (sample 3, Arndt and Jenner, 1986); 11 Average of four SHMB, Negri Volcanics, Pilbara Block, Western Australia (Sun et al., 1989); 12 Average of ten boninites (Cameron et al., 1983); 13 Average of five lunar KREEP samples (Simon, Chapter 7). Average Cr and Ni values interpolated from Simon et al. (1985).

recent basic volcanic suites has been restricted to the relatively immobile elements (e.g. Pearce and Cann, 1973; Pharaoh and Pearce, 1984; Winchester and Floyd, 1977; Floyd and Winchester, 1978). In metamorphosed volcanic or intrusive basic suites, the most suspect of elements are the large-ion lithophile elements (LILE: Ba, Rb, K), which are characteristically high in the relatively siliceous and magnesian rocks under consideration. However, many of the noritic intrusive suites are virtually free from secondary metasomatic or metamorphic effects. Indeed, it is a characteristic feature of these intrusive complexes and dyke swarms that they retain their igneous mineral assemblages and textures. The relatively deep-seated layered complexes have subsolidus assemblages resulting simply from their depth of intrusion rather than subsequent metamorphic effects. Quenched dykes such as those in southern West Greenland (Hall and Hughes, 1986, 1987) and at the margin of the Bushveld and Stillwater complexes retain pristine high-temperature magnesian assemblages (Sharpe, 1981; Longhi et al., 1983; Hall et al., 1987b). Thus, while LILE remobilisation cannot be ignored in metamorphosed volcanic and intrusive suites (e.g. Gill and Bridgwater, 1979; Zeck et al., 1983; Rajamani et al., 1985; Stahle et al., 1987), and although LILE contents do not always correlate with metamorphic grade (Condie and Allen, 1984), this is clearly not a significant problem in evaluating the petrogenesis of the major late Archaean and early Proterozoic intrusive norite suites.

5.4.3 Fractional crystallisation

The norite and dolerite dyke swarms in southern West Greenland comprise two geochemically distinct groups. The first and most obvious question is whether the two swarms were derived from the same primary magma and the distinctive composition of the norites resulted from fractionation of a primitive tholeiite, which is now represented by the earliest of the neighbouring dolerite dykes. If this were the case, then the compositional difference between the two types could have been effected in one of two ways: either the norites (MgO = 16%) would have to be ferromagnesian mineral (orthopyroxene and olivine) cumulates derived from a less magnesian (MgO = 10%), early MD-type tholeiitic magma, or the MD dolerites must have resulted from the separation of these phases from a noritic magma.

The presence of a quenched variety of the noritic dykes, with the same geochemistry as the coarser dykes, renders the idea of their compositions being due to crystal accumulation highly unlikely. Additionally, some of the most significant petrochemical differences between the norites and dolerites are in elements which are not compatible in the potentially cumulus phases. Most significantly, K, Rb, Ba and the LREE are richer in the more magnesian norites (Figures 5.6, 11.5). These elements cannot have been increased in a basic magma by the accumulation of olivine and orthopyroxene, nor can the separation of these phases from a noritic liquid have reduced the levels of these elements to those encountered in the dolerite dykes. The only remaining explanation is that the norite and dolerite dyke swarms cannot be related by fractional crystallisation or crystal accumulation, but that they were derived from different magmas.

Noritic dykes also occur in Scotland (Scourie dykes: Tarney and Weaver, 1987b, c), Western Australia (Jimberlana, Binneringie and associated Widgiemooltha dykes: Parker et al., 1987), South Australia (Port Lincoln dykes: Mortimer et al., 1988), Antarctica (Vestfold Hills and Napier Complex dykes: Sheraton et al., 1987; Kuehner,

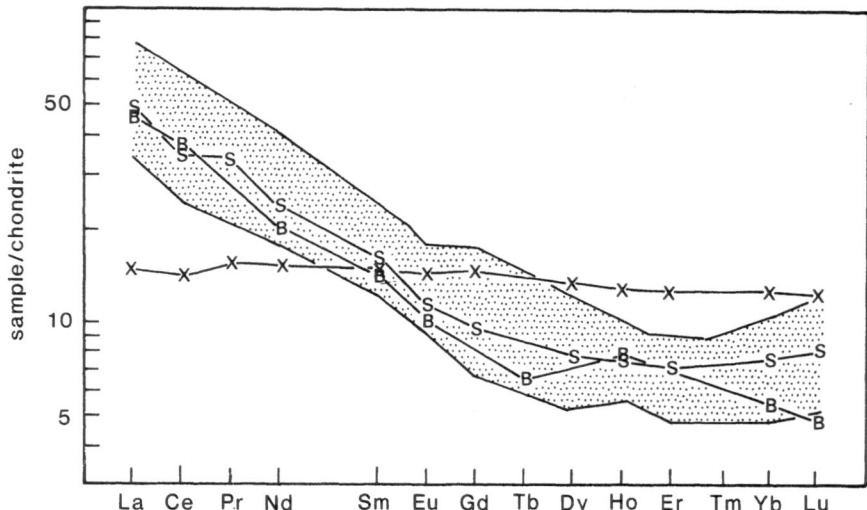

Figure 5.7 Chondrite-normalised REE distribution patterns of average Bushveld 'micro-pyroxenite' sills (B), a Scourie norite dyke, northwest Scotland (S), and an early MD dolerite dyke, southern West Greenland (X), compared to the range of BN norite dykes, southern West Greenland (stippled field). Bushveld data from Sharpe and Hulbert (1985). REE abundances in noritic dykes elsewhere in the world are shown in Section II of the book, and can be interpreted from Figure 5.8.

1989), South America (Uauá dykes: Chapter 17), North America (Stillwater satellite, Beartooth and Bighorn dykes: Longhi *et al.*, 1983; Helz, 1985; Hall *et al.*, 1987b), and as sills marginal to the Bushveld Complex in South Africa (Cawthorn *et al.*, 1981; Sharpe and Hulbert, 1985). Most share many of the geochemical characteristics of those in Greenland (Figures 5.7, 5.8) and many also include quenched varieties. Therefore, so far as fractionation is concerned, the same arguments can be applied to these swarms: they cannot have been derived simply by crystal separation from, or accumulation in, a tholeiitic magma. They must represent noritic magmas.

5.4.4 *Degree of partial melting*

The Mg content of the norite dykes (MgO up to 20%) indicates that they must have crystallised from a highly magnesian liquid, presumably derived by relatively high degrees of mantle partial melting. High degree partial melting would also account for the low levels of incompatible elements, such as Ti, Zr and Y (Figures 5.4, 5.5, Table 5.1). However, higher degrees of melting of the source which was capable of producing the near-chondritic, tholeiitic dolerite magmas (Figure 5.6) cannot have produced magmas which, while rich in Mg, Ni and Cr, were also richer in Si, K, Rb, Ba, La and Ce. The linking of the parental norite and dolerite dyke magmas by varying degrees of partial melting of the same source is as untenable as a common fractionation model. Such a mechanism is clearly at odds with the relatively high LILE and LREE concentrations (ten to twenty times MORB values), compared to the MORB-like, or continental tholeiitic trace element characteristics of the most Mg-rich of the dolerite dykes of the same region (Figure 5.6).

5.4.5 Magma mixing

The above considerations suggest that the penecontemporaneous early Proterozoic doleritic and noritic dykes of southern West Greenland, and many other parts of the world, were probably not linked to a common source or common parental magma by any straightforward closed-system petrogenetic process. It is most likely that they are derived from separate, geochemically distinct mafic magmas, one a tholeiitic magma (MgO = 10%) which gave rise to primitive and progressively evolved Fe-rich dolerites (Figure 5.6; Chapter 11), and a second magma rich in MgO (c. 20%), SiO_2 (c. 50%), Cr, Ni, Ba, La and Rb, but low in Ti, Zr, Y and the HREE. On a large scale, the composite noritic–gabbroic rocks of the Stillwater and Bushveld complexes have been ascribed to the mixing of similar discrete tholeiitic and noritic magmas, respectively dubbed 'A'-type and 'U'-type magmas (Sharpe, 1981; Irvine and Sharpe, 1982; Sharpe et al., 1986; Todd et al., 1982). Could any of the smaller intrusions, such as the Greenlandic dykes, also show the effects of magma mixing? The noritic and doleritic dyke swarms of southern West Greenland provide clear evidence for the simultaneous intrusion of two contrasting types of mafic magma, broadly corresponding to the Bushveld and Stillwater 'U'- and 'A'-type magmas. In general, there is little evidence for mixing between these two magmas in the dyke swarms.

The mixing between komatiitic and tholeiitic magmas sometimes invoked to account for komatiitic basalt suites (Arndt and Nesbitt, 1984; Cattell and Arndt, 1987) does not seem appropriate in explaining the composition of the BN dykes, as nowhere is there evidence of a possible pre-mixing, komatiitic component as dykes, despite the fact that elsewhere in the world such mixed magmas appear to reach the surface to occur as flows. However, one set of the MD dykes (rare 'MD2' dykes) includes members which have a remarkably varied assemblage of olivines (Fo_{70-35}), orthopyroxenes (En_{80-55}), augites ($X_{Mg} = 90-55$), pigeonites ($X_{Mg} = 70-60$) and relatively Fe-rich, subcalcic augites (Hall et al., 1986). This generation of dykes was intruded between the earliest ('MD1') dolerite and BN norite dykes, and the youngest ('MD3') dolerites which have the most variable calcic pyroxene assemblage (Figure 5.3) (Hall et al., 1985, 1988a). One such dyke has a whole-rock chemistry and mineralogy falling half-way between the typical noritic and doleritic dykes (Figures 5.9, 5.10).

Mixed magmas in the 2.7 Ga Stillwater Complex probably gave rise to the PGE-rich J-M reef (Irvine et al., 1983; Naldrett and Barnes, 1987) and, in the 2.1 Ga Bushveld Complex, to the famous Merensky reef (Irvine and Sharpe, 1981; Campbell et al., 1983). It is possible that the 20 m wide MD2 dyke is petrologically equivalent to these critical horizons in the large layered intrusions and that it perhaps represents the mixing of two different magmas rather than the crustal contamination of a tholeiitic magma as suggested by Kalsbeek and Taylor (1985).

5.4.6 Crustal contamination

The geochemical similarity between the quenched and cumulus-textured norite dykes in Greenland, and the occurrence of similar quenched sills beneath the Bushveld Complex (Sharpe, 1981; Sharpe and Hulbert, 1985; Cawthorn et al., 1983) and in the Beartooth and Bighorn mountains in the USA (see Chapter 9), leave little doubt that these noritic dykes and sills were derived from magmas which were not significantly different from the compositions of the dyke samples themselves. This is clearly not true

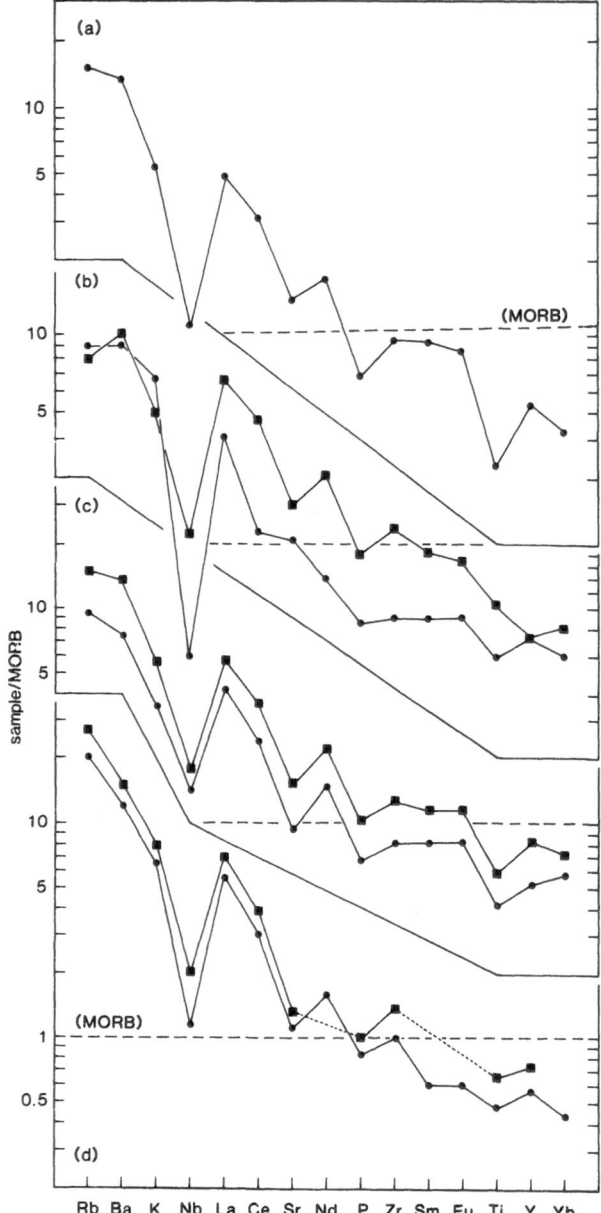

Figure 5.8 MORB-NME plots for norite dykes and sills. (a) Average of nine quenched 'micropyroxenite' sills (MgO = 13%) marginal to the Bushveld Complex (Sharpe and Hulbert, 1985). (b) Two samples of *c.* 2.4 Ga Scourie norite dykes, northwest Scotland. Samples L639 (dots) and L480 (squares) have MgO contents of 10% and 18.2% respectively (Weaver and Tarney, 1981a). (c) Two samples of the *c.* 1.6 G Port Lincoln norite dykes, South Australia. Samples B58 (dots) and B16 (squares) have MgO values of 15.9% and 7.7% respectively (Mortimer *et al.*, 1988). (d) Average of eleven Vestfold Block (dots; MgO = 10%) and eight Napier complex (squares; MgO = 7.7%) 2.4 Ga norite dykes, Antarctica (Sheraton, 1987). REE data interpolated from Kuehner (1989). (e) Two quenched norite (filled circles and squares; MgO = 15.1% and 11.3% respectively)

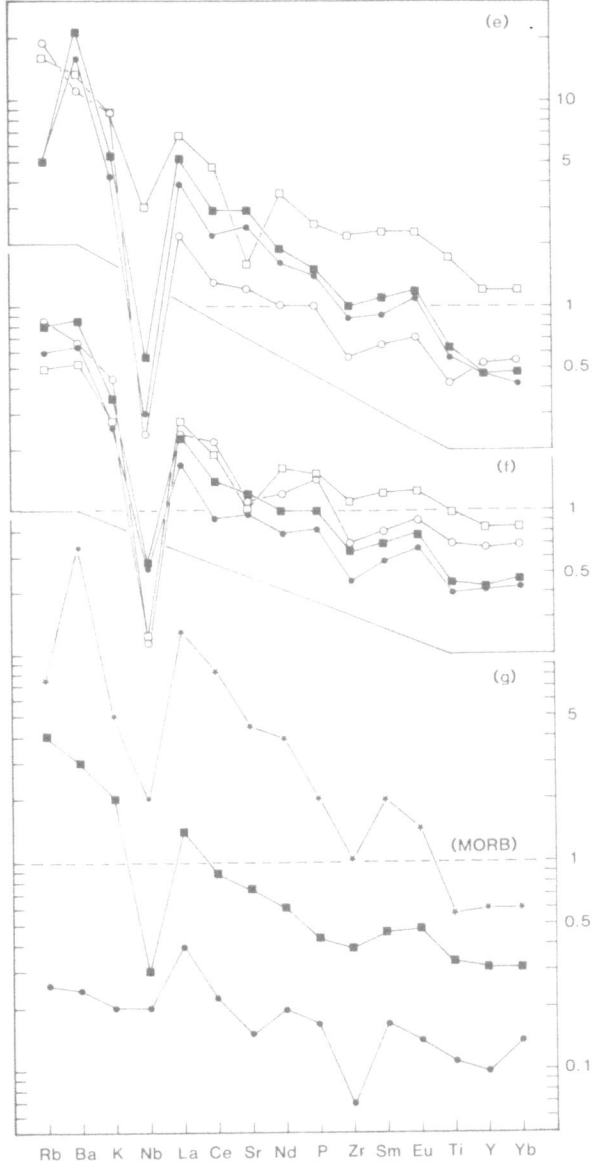

and two dolerite (open circles and squares; MgO = 8.5% and 4.6% respectively) dykes from the southern Beartooth Mountains, Wyoming craton, USA (Hall *et al.*, 1987b) and authors' unpublished data). (f) Two quenched norite (filled circles and squares; MgO = 11.2% and 9.7% respectively) and two dolerite (open symbols; MgO = 7.5% and 6.8% respectively) dykes from the Bighorn mountains, Wyoming, USA (Hall *et al.*, 1987b) and authors' unpublished data). (g) Norite dykes from the central Laramie range (squares) and Hartville uplift (asterisks), Wyoming, USA (MgO = 14.7% and 6.7% respectively). Dots show the composition of a harzburgite sill (MgO = 33%) (see Chapter 9). Data from Snyder (1984), Hall *et al.* (1987b) and authors' unpublished data.

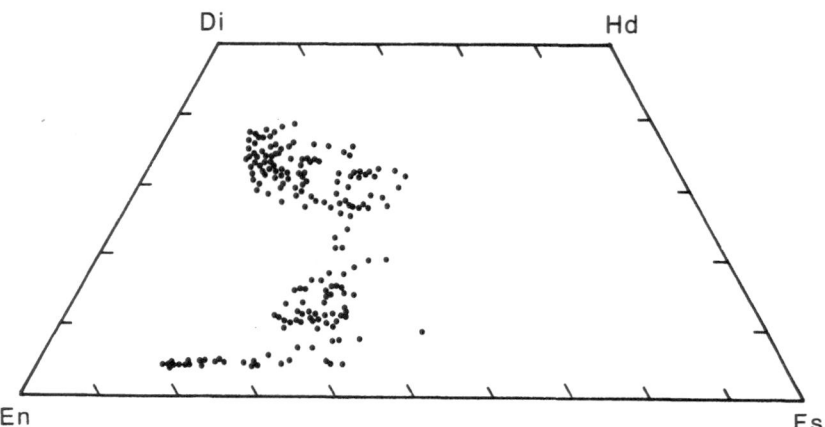

Figure 5.9 Complex pyroxene assemblage in a sample of a second generation (MD2) early Proterozoic dolerite dyke, southern West Greenland (sample 120432), comprising orthopyroxenes (En_{80-57}), pigeonites ($X_{Mg} = 0.62-0.44$), subcalcic augites and augites ($X_{Mg} = 0.85-0.5$). Replotted from Hall *et al.* (1985, 1986). Individual pyroxene 'grains' comprise the entire range of compositions shown by the sample, reflecting a complex crystallisation sequence involving the partial resorption of early orthopyroxenes and calcic pyroxenes, and possibly reflecting the mixing of noritic and tholeiitic magmas.

for samples of larger-scale layered intrusions which are subject to a variety of closed- and open-system magma chamber influences (O'Hara, 1977; Irvine *et al.*, 1983; McBirney, 1985). It is also clear that the major mafic intrusive complexes of the world, all late Archaean or early Proterozoic in age, have a significant noritic component. The fact that these magmas formed in a continental environment may of course be why they are so abundant and well preserved (see Chapter 1). But, were the magmas noritic *because* they formed there? Were they simply the result of the contamination of basic magma by continental crust?

There are many lines of evidence which suggest that a crustal contamination model is perhaps the simplest and most effective means of accounting for the distinctive composition, and, in particular, the crustal signature (high LILE and LREE) of some 'noritic' suites. To produce a magma with *c.* 20% MgO, the initial, uncontaminated magma would clearly need to have been even more magnesian. This can simply be calculated by the degree of contamination required to raise the level of the sialic components to the observed level. Komatiite is an obvious candidate for a primary magnesian magma. Consideration of the thermal properties of such low-viscosity magnesian magmas has provided a clear indication of the potential for crustal erosion and contamination during their passage through and onto the crust (Huppert *et al.*, 1984; Huppert and Sparks, 1985; Sparks, 1986). Many suites of komatiites and, more particularly, komatiitic basalts which have a crustal contamination component have been recognised (e.g. Gariepy *et al.*, 1984; Arndt, 1986; Arndt and Jenner, 1986; Compston *et al.*, 1986; Edwards and Nisbet, 1986; Cattell, 1987). Zircon xenocrysts within the metavolcanics at Kambalda, Western Australia, range in age from 2.7 to 3.5 Ga (Compston *et al.*, 1986) and provide some of the strongest evidence for the crustal contamination of komatiitic magmas (Arndt and Jenner, 1986). However, the geological evidence for thermal erosion is not always so clear (Claoué-Long and

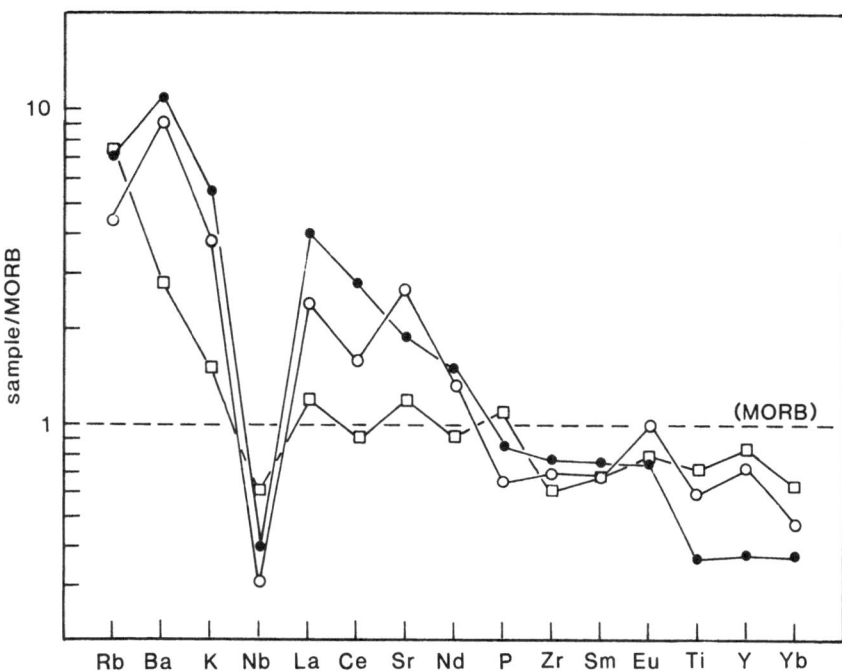

Figure 5.10 MORB-NME plot for MD2 bronzite-dolerite dyke 120432 (open circles; MgO = 7.7%) compared to another of the earliest and most primitive MD dykes (149208; open squares; MgO = 8.5%) and an average of twenty-three BN norite dyke samples (dots; eleven samples for Sm, Eu and Yb and thirteen samples for Nb; average MgO = 15.85; range 22–5.7%). Sample 120432 has a complex orthopyroxene and clinopyroxene assemblage (Figure 5.9) and has geochemical characteristics which suggest the mixing of the BN noritic and the primitive MD tholeiitic magmas. Data from Hall *et al.* (1985, 1986) and Hall and Hughes (1987).

Nesbitt, 1985; Groves *et al.*, 1986) and the occurrence of uncontaminated komatiites which have apparently passed through continental crust together with contaminated ones demonstrates that contamination need not always occur (Arndt *et al.*, 1986). The interpretation of Sm–Nd isotopic systematics has also proved difficult in trying to resolve the relationship of contaminated komatiitic lavas to basement, even whether they form belts which are younger or older than 'basement' gneisses (Cattell *et al.*, 1984; Claoué-Long *et al.*, 1984; Chauvel *et al.*, 1985).

Some of the most direct evidence for crustal contamination occurs in the 1.85 Ga Sudbury intrusion (Krogh *et al.*, 1982) in which strongly fractionated REE distributions ($La_N = 60–200$; $La_N/Lu_N = 10–20$) and PGE abundances indicate contamination by silica-rich material (Kuo and Crocket, 1979; Naldrett *et al.*, 1982; Dressler *et al.*, 1987). This is one of the rare instances in which marginal country rocks can be seen to be strongly disrupted and deformed, and the Sudbury structure has been interpreted by some workers to be the result of a meteorite impact (Dietz, 1964; Dence, 1972). However, the Sudbury norites are relatively Mg-poor (MgO = 6%) and are different from those of most other intrusions in several important respects. There is little doubt that they are heavily contaminated, but they are the exception rather than the rule.

There are several difficulties in applying a crustal contamination model to explain the compositions of most norites. Suites of contaminated komatiites and lesser magnesian basalts can often be modelled in terms of the combined effects of the degree of partial melting, fractional crystallisation and contamination (Arndt and Jenner, 1986; Cattell and Arndt, 1987; Sun *et al.*, 1989). This is not the case with the noritic and tholeiitic dykes of, for example, Greenland and North America (Longhi *et al.*, 1983; Hall *et al.*, 1987b). If norites are developed by the contamination of high-Mg basaltic (komatiitic?) magmas, then they should perhaps all carry the same chemical signature. The MORB-normalised multi-element (NME) patterns for the Greenlandic norites are not the same in detail as those for the North American norites (Figure 5.8). Norite dykes in the Beartooth Mountains of the Wyoming craton have similar REE distributions to those in neighbouring dolerite dykes (see Chapter 9) and similar MORB-NME patterns (Figure 5.8). It is unlikely that the LILE-enrichment of the doleritic (not Si-enriched) dykes is due, solely at least, to a contamination component (although there may be an equal partial contamination component in the norites and dolerites). If more magnesian magmas have a greater potential for the digestion of crustal material (Huppert *et al.*, 1984; Huppert and Sparks, 1985), the most Mg-rich intrusions might be expected to be the most LILE-enriched. The most magnesian dolerites in southern West Greenland, however, do not have a significant crustal component. With rare exceptions (Figure 5.6), there is a marked contrast between the most magnesian dolerities and the norites. The dolerite dykes which show the strongest crustal signature are the least magnesian, plagioclase-phyric ('PP') dykes (Figure 5.6).

A set of discordant sills in the Laramie range of the Wyoming craton, USA (Snyder, 1984; see Chapter 9), comprise mainly plagioclase-harzburgites (Figure 5.1d). These rocks have a 'dish'-shaped (upward concave) chondrite-normalised REE pattern (see Figure 9.9) with $La_N \geqslant 3.5$, $La_N/Gd_N = 2.5$ and $Gd_N/Lu_N = 2$. This REE distribution is similar to that encountered in many boninites (Hickey and Frey, 1982; Cameron *et al.*, 1983; Nelson *et al.*, 1984; Barnes, 1989). These differences may, of course, reflect the combination of variations in the degree of contamination, fractionation and the composition of the contamination component (sialic crust or sediments). In the case of komatiitic magmas, contamination may occur either during ascent through the crust or by the erosion of sediments at the surface (Huppert *et al.*, 1984; Huppert and Sparks, 1985). However, contaminated komatiites do not appear to show the same level of a contamination component as that required by a contamination model for the norites (Figure 5.11). While they do have fractionated REE patterns, they are not so enriched in, for example, Rb and K, and show no sign of a negative Nb anomaly, distinctive of the MORB-NME plots for many of the norites (Figure 5.8). The degree of contamination by granulite or amphibolite facies gneissic crust (Weaver and Tarney, 1981b) required to produce the levels of Rb, Ba, K and La in the norites is extremely high (Figure 5.12). Assuming that the supposed uncontaminated magma had primitive tholeiitic (N-MORB type) elemental proportions, a contamination component of over 50% is required. If the average upper crustal or total crustal compositions estimated by Taylor and McLennan (1985) are used as a contaminant, and if the original basic magma was more magnesian than average MORB (as it must have been to give norites with around 16% MgO), with consequently lower incompatible element concentrations, then the dergee of contamination would have to have been even higher (*c.* 75%?). The crustal contamination model becomes a little suspect in the light of these estimates.

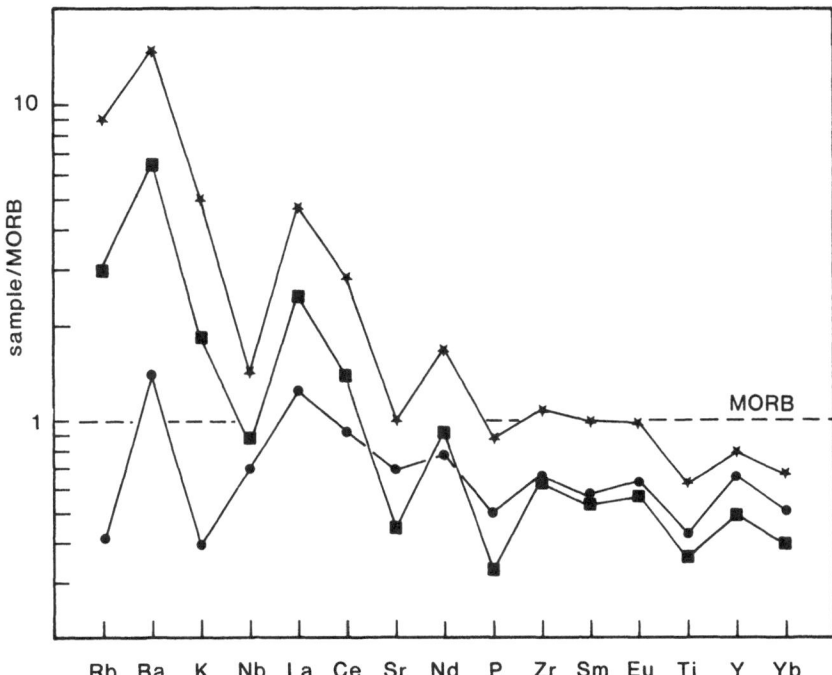

Figure 5.11 MORB-NME plot for three samples of crustally contaminated komatiites from Kambalda, Western Australia. Data are from samples 21, 3 and 11 of Arndt and Jenner (1986), which have 12.9, 15.6 and 5.5% MgO respectively.

Another problem involves the magnitude of the negative Nb anomaly in the NME plot. This is estimated by interpolating a theoretical Nb value (Nb*) between the normalised K and La data. The average Nb*/Nb value for the Greenlandic norites is *c.* 10, which cannot be accounted for by the addition of a granulite or amphibolite facies gneiss contamination component, since the average Nb*/Nb values of these rocks are lower (4.8 and 7 respectively) (Figure 5.12). The 'PP' doleritic dykes with the strongest crustal signature have only a small negative Nb anomaly (Figure 5.6). The same is true for many of the noritic and doleritic dykes of Scotland (Tarney and Weaver, 1987a, b) and North America (Figure 5.8). The Nb*/Nb values are not so high in the noritic dykes of South Australia or Antarctica (Nb*/Nb = 2.6–5; Figure 5.8). However, Mortimer *et al.* (1988) and Kuehner (1989) concluded from trace element and, in particular, Sr isotope constraints (Collerson and Sheraton, 1986) that the 1.6 Ga noritic dykes of Port Lincoln, South Australia, and the 2.4 Ga dykes of the Vestfold Hills, Antarctica, respectively were also not the products of sialic crustal contamination.

Several fundamental geological questions arise from models which attempt to explain the genesis of norites by contamination of primitive, komatiitic (?) magmas. One is whether an intrusion as large as the Bushveld Complex (*c.* 150 000 km^2) could have become uniformly contaminated by ingested crustal material (Willemse and Viljoen, 1970; Hamilton, 1977; Cawthorn and Davies, 1983; Sharpe, 1985). Another is

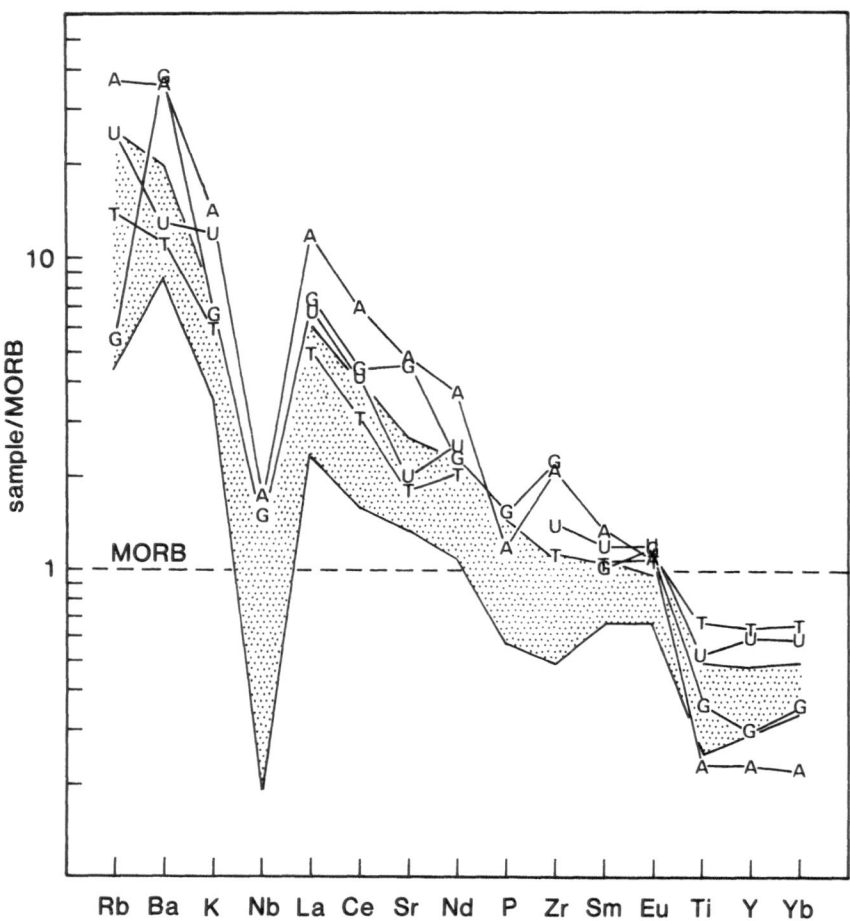

Figure 5.12   MORB-NME plot showing the compositions of amphibolite (A) and granulite (G) facies gneisses (Weaver and Tarney, 1981b) and estimated upper (U) and total (T) crust (Taylor and McLennan, 1985) compared to the field of the BN norite dykes of southern West Greenland (stippled). No Nb or P data are available for U and T. The degree of crustal contamination of a magnesian MORB-like parental magma required to obtain the compositions of the norites appears to be unrealistically high.

the disparity between the ages of preserved noritic and komatiitic suites (Figure 5.13). If any komatiitic magma will partly digest the continental crust through which it passes, why then are not more komatiite suites severely contaminated (Arndt *et al.*, 1986)? Conversely, if in such a model many magmas do not become contaminated, as seems to be the case with the abundant late Archaean komatiitic metavolcanic suites, why are *all* the widespread early Proterozoic high-Mg intrusive suites apparently so very strongly modified by a crustal component, and why are there no extrusive komatiitic analogues of these rocks among the abundant early Proterozoic volcanic suites (Pharaoh *et al.*,

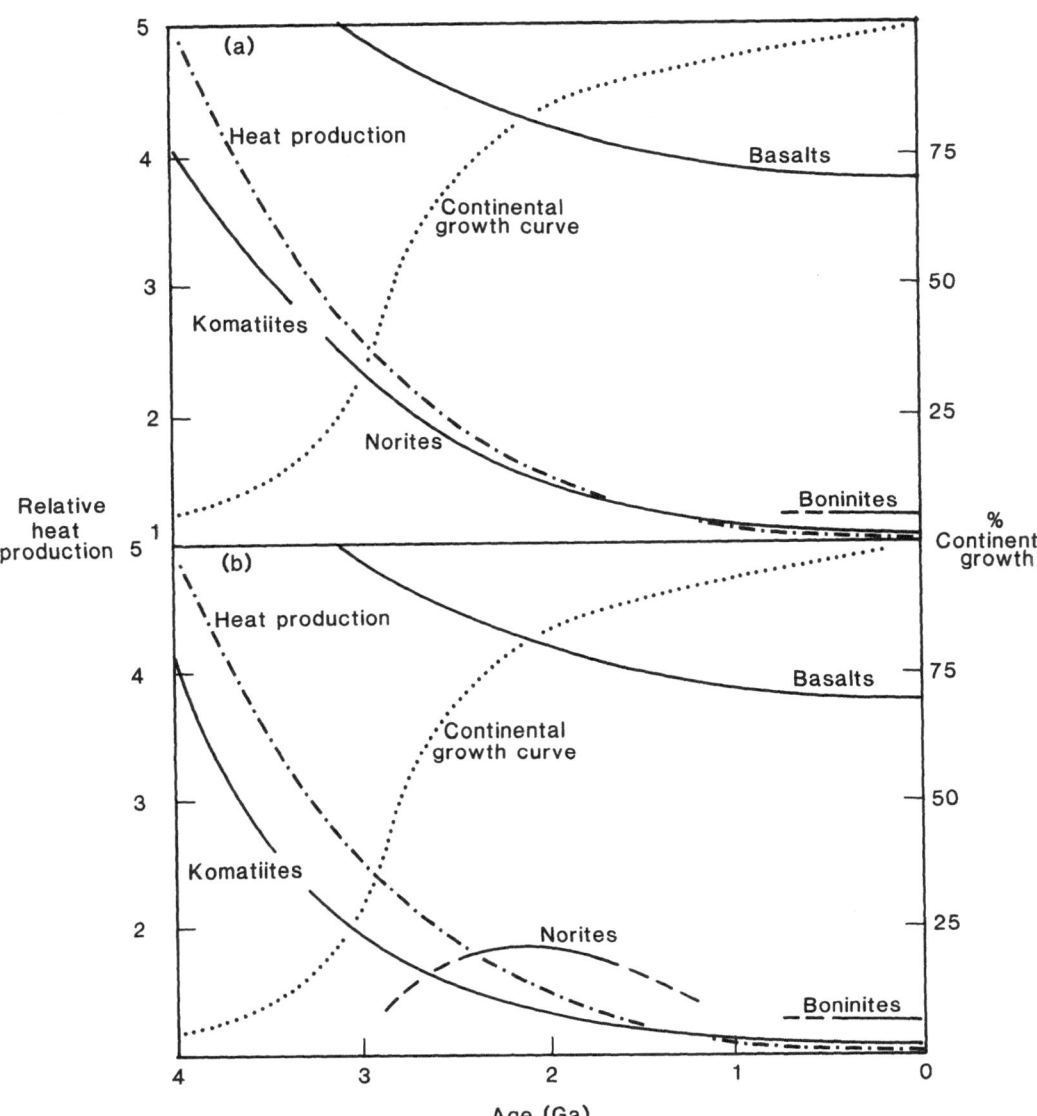

Figure 5.13 Relative rates of production (schematic) of basalts, komatiites, norites and boninites with respect to time, continental growth (dotted lines), and relative global heat production (dash-dot lines). In (a), norites are interpreted as continued, late Archaean–early Proterozoic komatiitic liquids, contaminated by crustal material. Boninites are an independent suite of rocks not generated (or recognised) in the Precambrian. In (b) (the preferred model), komatiite generation decreases towards the end of the Archaean and norites represent a different suite of magmas generated beneath, and in response to, the rapid development of continental crust in the late Archaean, by the partial melting of depleted, harzburgitic mantle, replenished during metasomatism by LILE- and LREE-enriched hydrous fluids. In this sense, they are more closely related to boninite generation. The relatively high basalt production rate is the same in both models. Schematic relative heat production curve adapted from O'Nions et al. (1978). Crustal growth curve adapted from Veizer and Janson (1979) and McLennan and Taylor (1982).

1987)? Komatiitic and noritic suites appear to be distinct chemically, temporally and in their tectonic settings (Figure 5.13).

5.4.7 Mantle metasomatism

The combination of high Mg, Cr and Ni, and high Si, LILE and LREE is clearly not one which can be achieved from the advanced melting of primitive fertile mantle. If this signature is not the result of contamination of basic magma, then it perhaps simply reflects the composition of the source mantle. The early history of the Earth must have involved the formation of a vast amount of komatiitic crust, both as primordial oceanic crust (Arndt, 1983; Nisbet, 1987) and as lavas which passed through early sialic basement. This is turn must have significantly depleted large portions of the mantle. Whether the crust and mantle were recycled in the early Precambrian in a way similar to the mechanisms operative at the present day (Hofmann and White, 1982; Pearce, 1983; Ringwood, 1982; Sun, 1984; Nakamura et al., 1985; McKenzie, 1986) is not known, but extensive depleted, harzburgitic mantle, residual from the prolonged episodes of komatiite formation, must have been present in the lithosphere beneath the rapidly developing, thickening continents at the end of the Archaean.

In places where Archaean oceanic crust and continental nucleii have been tectonically interleaved, such as the Archaean craton of West Greenland (see Chapter 11), large slices of harzburgitic material are preserved together with pillow-structured komatiitic and tholeiitic metavolcanics and remnants of anorthositic layered intrusions (Hall et al., 1987a). Similar material subducted below the newly thickened continents would perhaps have become contaminated by hydrous (granitic?) fluids rich in LILE and LREE, derived either from the destabilisation of material from the base of the continental crust, or from trapped sediments at the Archaean–Proterozoic boundary equivalent of a destructive plate margin. It may be significant that the centre of the BN noritic dyke swarm in Greenland, for example, is at the northern margin of the Archaean craton, near to the boundary with the early Proterozoic (Nagssugtoqidian) mobile belt. In a limited way, this model of the contamination of harzburgitic mantle by metasomatic fluids derived from sediments or recycled continental crust is akin to the boninite generation model (Hickey and Frey, 1982; Cameron et al., 1983; Crawford, 1989). However, boninites are somewhat different from the BN-type norites in that they are even more Si-rich (Table 5.1), but do not show such a high level of LILE and LREE contents as the norites (Figure 5.14). This presumably reflects the contrast in the sediment-derived fluids involved in boninite genesis compared to the lower crust-derived (granitic?) fluids which were responsible for the metasomatism of the mid-Precambrian mantle in the genesis of the norites.

Sr isotope compositions also tend to support a mantle metasomatism model. The initial $^{87}Sr/^{86}Sr$ ratio (Sr_i) of virtually all noritic intrusives is high. The BN norite dykes of Greenland have Sr_i values of around 0.704 (Bridgwater et al., 1985). The Bushveld marginal sills have Sr_i values of 0.704–0.707 (Hamilton, 1977; Davies et al., 1980; Harmer and Sharpe, 1985; Sharpe, 1985) and the Port Lincoln norites have values of c. 0.706 (Mortimer et al., 1988). If these high values are not due directly to magma contamination then they must surely reflect mantle enrichment. The somewhat lower Sr_i values (c. 0.702) of the Vestfold Hills and Napier Complex norite dykes of Antarctica

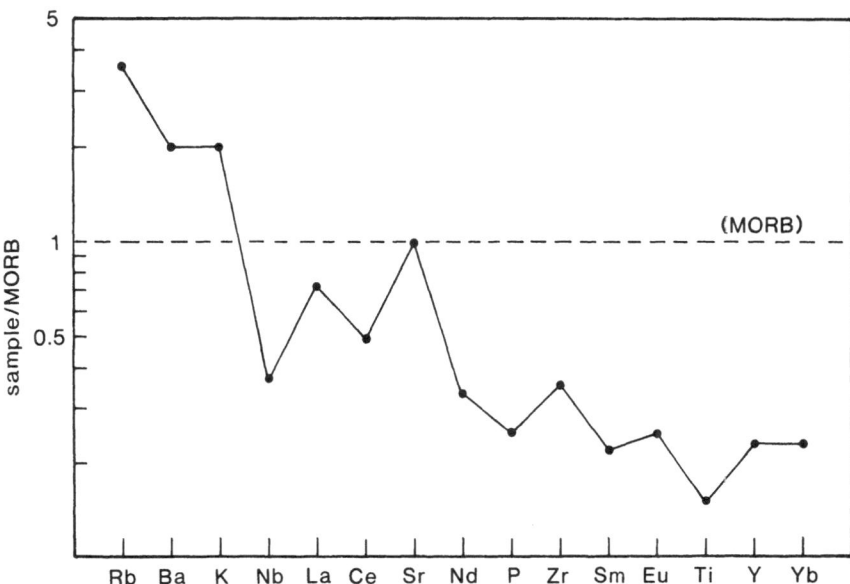

Figure 5.14   MORB–NME plot for boninite (average of ten boninite analyses given by Cameron *et al.*, 1983).

have equally been interpreted as signifying the lack of a crustal contamination component (Collerson and Sheraton, 1986; Kuehner, 1989). Isotopic studies on the 2.7 Ga Stillwater complex (DePaolo and Wasserberg, 1979; Simmons and Lambert, 1982) also suggest an Sr_i value of 0.702, interpreted as incorporating a source with a history of Rb/Sr enrichment (Lambert *et al.*, 1985), although isotopic compositions do not resolve the differences between mixed parental magmas (e.g. Ryder and Spettel, 1985). Modern boninitic lavas, which share the same gross chemical characteristics as the hypabyssal noritic rocks (Table 5.1), also have high but variable Sr_i ratios (0.704–0.707), attributed to variation in contamination of both their source and parental magmas (and to secondary Rb and Sr mobility) (Hickey and Frey, 1982; Cameron *et al.*, 1983; Nelson *et al.*, 1984; Crawford, 1989). The significance of the isotopic data is not clear-cut.

5.4.8   *KREEP: a lunar analogue?*

If there is one line of irrefutable evidence that K and REE-rich noritic rocks can be generated without the involvement of contamination by continental crust, it is that provided by the occurrence of KREEP on the Moon (Ryder, 1987b; see Chapter 7). KREEP basalts are characteristically rich in K, REE and P (from which the acronym KREEP is derived), and also in Rb and Ba, despite the fact that they are magnesian rocks (Table 5.1). They are relatively more siliceous than the lunar Mg-suite norites (Figure 7.5, p. 143) and have bronzite (En_{85-65}) as the main pyroxene (Figure 7.9, p. 153). They are rare and only occur as rock fragments but they are significant in that they are considered to have been derived from the residual melt of a Moon-wide magma ocean, after the separation of the anorthositic highland crust and the Mg-suite dunitic to troctolitic and gabbroic rocks. In this respect (alone), they have petrogenetic affinities with terrestrial norites derived from the depleted harzburgitic mantle residue left from the early extraction of the Archaean komatiitic–basaltic crust.

In detail, of course, terrestrial noritic rocks and KREEP are very different. The enrichment of all incompatible elements in KREEP is due simply to the failure of these elements to become incorporated into any of the early lunar crustal components. Their REE contents are extremely high and slightly fractionated ($La_N = c. 200$; $La_N/Yb_N = c. 2$; Figure 7.1, p. 138). K_2O, Rb and Ba values are around 0.6% and 15 and 800 ppm respectively. Sr contents are not notably high (c. 180 ppm) and, together with the negative Eu anomalies, reflect the earlier separation of the lunar highland anorthosites. P_2O_5 contents are around 0.6%.

The KREEP rocks (breccias and lavas) were formed at around 3.9 Ga (Nyquist *et al.*, 1975) and there has, of course, been no reworking of the lunar crust during which the incompatible elements could be rehoused in derived, terrestrial-type lithologies. However, the significant point about KREEP petrogenesis in the context of the terrestrial norites is that it does not involve a crustal contamination component. KREEP also formed by the melting of a residual layer, but one enriched in incompatible elements simply by primordial differentiation rather than by secondary, metasomatic enrichment.

5.5 Concluding remarks

It has now long been recognised that the geochemical and mineralogical similarities between Bushveld 'U'-type magmas, boninites, SHMB and norite dykes are unlikely to be accidental, and the petrogenetic links between these different suites have been discussed by many authors (e.g. Cawthorn and Davies, 1982; Longhi *et al.*, 1983; Walker and Cameron, 1983; Sharpe and Hulbert, 1985; Hall and Hughes, 1987; Barnes, 1989; Sun *et al.*, 1989; Crawford, 1989). Barnes (1989) highlighted the relative similarities and dissimilarities, and pointed out significant geochemical differences between, in particular, SHMB, 'U'-type magmas and boninites.

The difference in tectonic setting between the continental intrusive suites (Bushveld-type layered complexes and Greenlandic BN-type dyke swarms) and arc-related boninites hardly needs emphasising. What is in question is whether essentially similar petrogenetic processes have operated in the formation of these different suites. As a first approximation, SHMB are a half-way stage between the two, but are clearly derived predominantly from contaminated komatiitic magmas, which are not part of boninite genesis. Uncontaminated komatiites are commonly associated with these SHMB suites. SHMB differ geochemically from modern boninites in that they possess, for example, higher TiO_2 contents (0.3–0.8%; compared to 0.1–0.5% in boninites) and consequently lower Zr/Ti, V/Ti and Y/Ti values (Figure 5.15). No extrusive equivalents of the parental magmas of the noritic intrusions have as yet been recognised. This could, of course, simply reflect the fact that they have been eroded from the surface, beneath which the preserved and exposed intrusions were formed. However, komatiitic and associated SHMB suites appear to peter out at the end of the Archaean (Figure 5.12), while noritic intrusions become widespread at this time (indeed the intrusion of the Great Dyke has been used to *define* the Archaean–Proterozoic

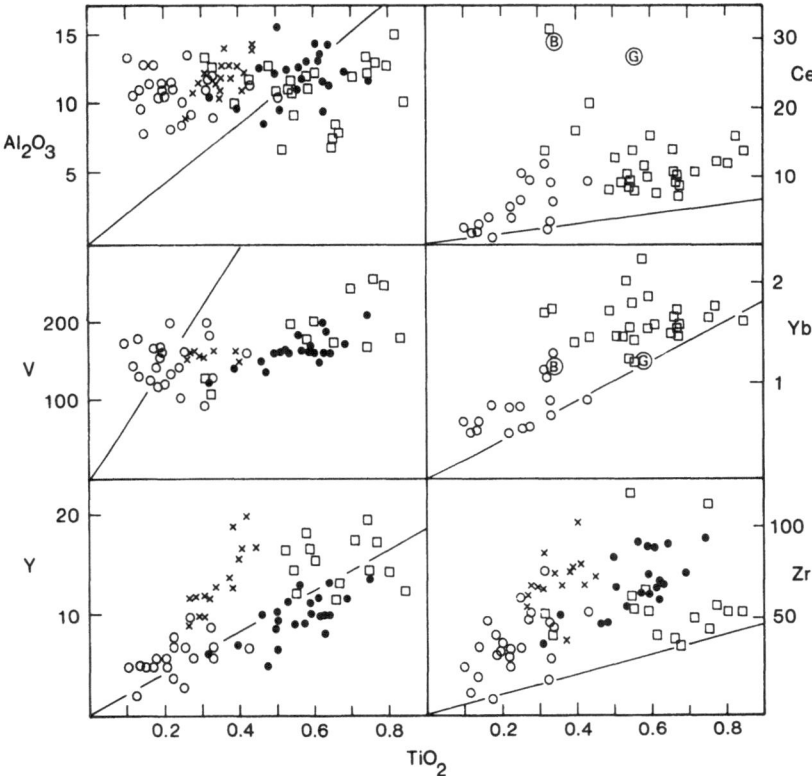

Figure 5.15 Al_2O_3 (%) and V, Y, Ce, Yb and Zr (ppm) versus TiO_2 (%) contents in the BN norite dykes of southern West Greenland (dots) relative to Bushveld 'micropyroxenite' sills (crosses, boninites (open circles) and SHMB (open squares)). Average values for the Bushveld (B) and Greenlandic (G) rocks are plotted for Ce and Yb. Diagram replotted from Barnes (1989) with authors' BN dyke data.

transition in the Zimbabwe craton; Nisbet, 1982). Are the Archaean–Proterozoic transition and the komatiitic–noritic magmatism transition then just coincidental? Is this change in high-Mg suites with the major episode of crustal thickening also coincidental (Figure 5.12), or do they signify fundamentally different forms of high-Mg magmatism, in response to different conditions? Modern style plate tectonics and associated basic volcanism are clearly recognisable from the early Proterozoic (Kröner, 1981; Condie, 1982b; Pharaoh et al., 1987). Boninitic magmas continue to be formed at the present day, from the remelting of replenished and metasomatised harzburgitic mantle wedges. It seems more likely that the noritic magmas of the latest Archaean and early Proterozoic developed in response to similar petrogenetic processes to those producing boninites, rather than by a complete, and apparently unique, contamination of magnesian magmas ponded in the continental crust. Virtually uncontaminated tholeiitic basic magmas were, after all, intruded at the same time (e.g. the earliest MD dolerite dykes of Greenland; Hall and Hughes, 1987).

Geochemically, the norites appear to fall half-way between the compositions of boninites and those of SHMB (Figure 5.13). The differences between them and SHMB do not appear to be simply the effect of advanced contamination. The extremely high degree of contamination required to give rise to the levels of LILE and LREE in the norites virtually precludes this. The fact that the norite suites appear in the geological record only after the major late Archaean crust-forming event presumably signifies that it was only at this stage that the crust was stable enough to accommodate such large-scale intrusions, and that it was *due* to the thickening of the crust that the high-Mg magmas were produced by the melting of the enriched sub-continental lithosphere. Magmas produced from such a metasomatised harzburgitic source would not, of course, be exempt from crustal contamination during magma ponding at either lower or upper crustal levels. Equally, differences in the degree of partial melting and subsequent fractionation would also have strongly influenced the contamination potential of noritic magmas intruded at higher levels in the crust (or erupted onto the surface). Thus, several petrogenetic processes appear to be able to give rise to mafic magmas with noritic compositions.

6 Mantle evolution

M.J. BICKLE

6.1 Introduction

An understanding of early Precambrian tectonic and magmatic processes cannot be separated from an understanding of the corresponding processes operating in the early mantle. Much of our knowledge is based on studies of the present mantle but the early Precambrian geological record, and in particular the sensitivity of the magmatic record to the thermal and chemical state of the mantle, places constraints on mantle evolution. These constraints, in turn, provide critical tests of models based on observations of the present-day mantle. Therefore, in this chapter the discussion of the early mantle is intimately interwoven with that of the nature and processes of the present day.

Both the thermal and chemical state of the mantle will have undergone secular changes over the lifetime of the Earth. The thermal state reflects the balances between the initial state, decreasing radiogenic heat production and convective heat loss. The Earth has been cooling since the early Precambrian, with radiogenic heat production decreasing by about a factor of 3 since 3.5 Ga and a factor of 2 since 2.5 Ga. The chemistry of the mantle results from the processes that have led to the gross chemical divisions within the Earth of core, crust and mantle. The mantle itself can probably be divided into two distinct chemical reservoirs. The secular changes in mantle chemistry which reflect continued crustal growth with time have received much attention in attempts to monitor crustal growth rates and the size of the mantle reservoir depleted by the extraction of crustal components. Interaction between distinct geochemical reservoirs within the mantle may also have affected the chemical and isotopic evolution of the mantle.

The continental crust contains an imperfect record of the chemical and thermal state of the mantle over most of the Earth's history. Mafic igneous rocks more or less directly reflect mantle chemistry and temperature. Relatively rare xenoliths of mantle materials are present in some magmatic environments. The thermal state of crustal rocks reflects convection-driven heat loss processes from the mantle and their chemical state is a result of the poorly understood crust-forming processes operating ultimately on mantle materials. The geological record can only be related to the mantle through an understanding of how mantle processes relate to the surface tectonics which mould the geological record. Our imperfect understanding of the governing mantle convection processes is based on knowledge of the physical properties of mantle materials, numerical modelling of mantle convection and on surface-based geophysical observations. The first part of this chapter begins with a brief review of the results of the theoretical modelling since this guides interpretations of the geophysical manifestations of convection. The structure and behaviour of the lithosphere are then reviewed as it is the substantial viscosity change across the lithosphere that imparts the distinctive plate tectonic style to geological processes on Earth. Geoid anomalies and the gravity field which contain critical information on the scales of mantle convection are discussed briefly, followed by an outline of the processes which control melting in the present-day mantle, an understanding of which is crucial for interpreting the magmatic record.

6.2 Mantle processes

6.2.1 *Convection*

It is generally accepted that most of the mantle convects and that plate-tectonics is one surface manifestation of this convection. There is much less agreement on the scale(s)

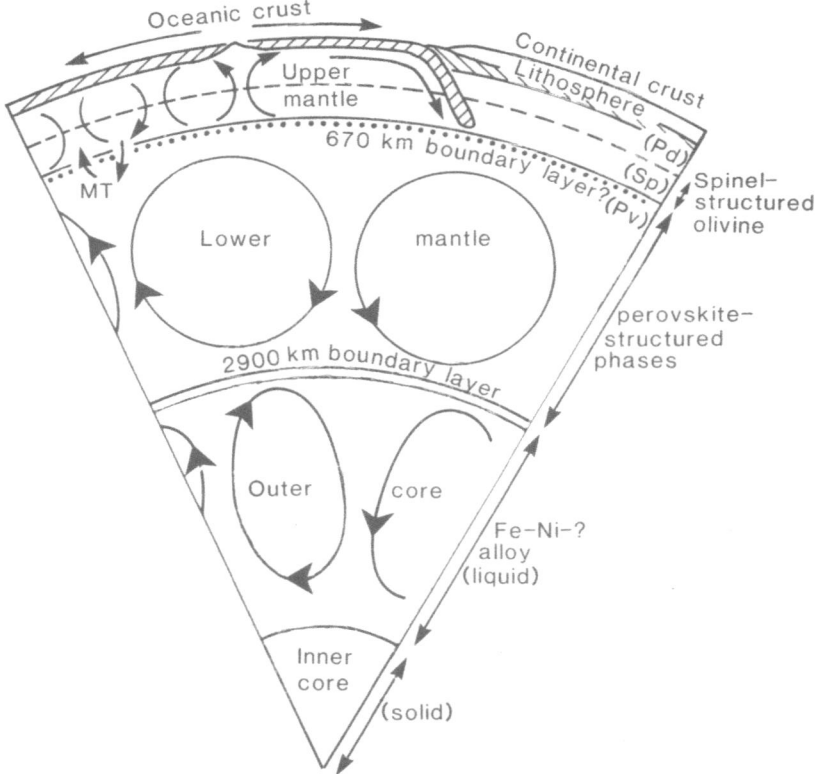

Figure 6.1   Structure of the Earth in terms of chemical layering, convection and phase transitions (Pd: peridotite; Sp: spinel-structured olivine; Pv: perovskite-structured phases). The existence of a boundary layer at 670 km is contentious. Note the relatively small plate-scale convection in the upper mantle, but that the scale of convective circulation in the lower mantle is speculative. The convection in the outer core is linked to the thermal structure of the lower mantle (Gubbins and Richards, 1986). The rate of mass transfer (MT) between the upper and lower mantle is also controversial.

and geometry of mantle convection, and in particular how convection in strongly temperature dependent and non-Newtonian fluids differs from that in constant viscosity fluids on which most of the modelling has been carried out. A crucial point is whether the whole mantle convects or whether the mantle is layered with independently convecting upper and lower mantles separated by a boundary layer region at about 670 km (Figure 6.1). Whether or not the mantle is layered has major implications for both its thermal and chemical evolution during Earth history, and these implications may be tested against the early Precambrian geological record. A two-layered mantle has a substantially longer thermal time constant than a mantle convecting as a whole and would be expected to show less temperature variation with time. The upper layer of a two-layer system would be expected to show a more marked chemical depletion resulting from crust formation than a convectively mixed whole-mantle system.

The possibility, or indeed the probability, that the mantle is layered also has implications for scales on which the mantle convects. Many convection cell systems have aspect ratios (width/height) in the order of unity. This observation led early workers on mantle convection to relate the 10^3 to 10^4 km wide lithospheric plates to whole mantle convection in cells $c.$ 3×10^3 km deep (e.g. Turcotte and Oxburgh, 1967). Aspect ratios may be increased by strongly variable viscosities or certain boundary conditions. For example, constant thermal gradient conditions on the upper boundary could give rise to convection cells with aspect ratios of $c.$ 10 comparable to plate scale convection restricted to the upper mantle. Under these circumstances, with the approximately constant thermal gradient boundary condition imposed by the rigid lithospheric plate capping the system, large aspect ratio cells would be established with smaller wavelength perturbations manifest as a series of unsteady rising and sinking hot and cold 'plumes' (e.g. Hewitt et al., 1980; Figure 6.2). There is a certain amount of observational evidence to suggest that this style of convection does occur in the upper mantle, and that the lower mantle convects largely independently. This evidence includes (1) the cessation of subduction-related earthquakes at a depth of about 650 km, (2) geophysical observations related to current mantle dynamics and heat loss, (3) geochemical observations related to the size and number of well-mixed geochemical reservoirs within the mantle and (4) heat loss rate from the mantle and evidence for the variation of internal temperature with time.

6.2.2 The lithosphere

The gravitational and thermal signals of the smaller scale convective motions are damped at the surface by the lithosphere, which is the cold and rigid upper boundary to the plate-scale convection system. The lithosphere is also a major control on melting in the mantle and has a substantial influence on the heat loss processes from the Earth. Understanding the thermal and mechanical state of the lithosphere is therefore crucial in interpreting the deeper structure of the Earth, the evolution of convective processes with time and the control on magmatism and tectonics.

One of the early successes of the recognition of plate tectonic processes was explaining the decrease in heat flow and increase in ocean depth away from spreading ridges in terms of simple conductive cooling from the surface (e.g. Parsons and Sclater, 1977). Such a model predicts that heat flow should decrease inversely with the square-root of time ($1/\sqrt{t}$), and depth increase proportionally to $\sqrt{t}$, the depth increase

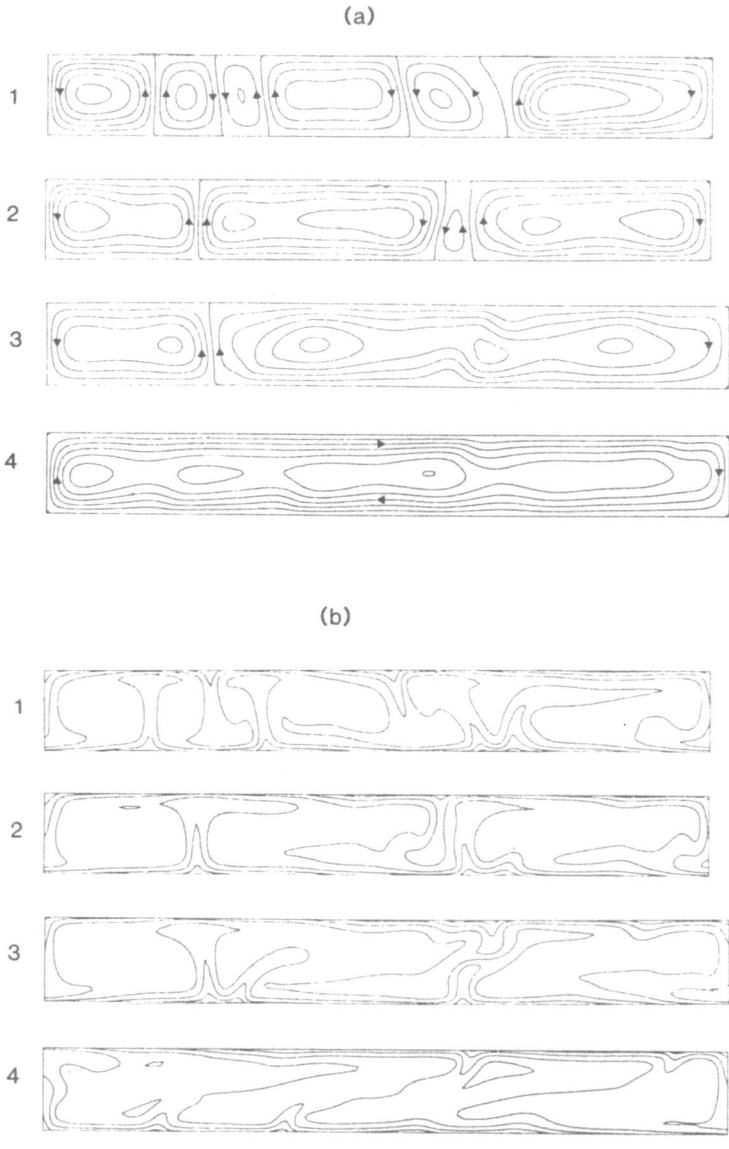

Figure 6.2   Theoretical upper mantle-type convection driven by heat supplied from below and with heat flux fixed on both boundaries, showing the relationship between (a) material stream lines and (b) temperature contours with respect to time. (1) to (4) represent a time period of the order of the age of the Earth. Note that (4) is from a separate experiment than sequence (1) to (3). (Redrawn from Hewitt *et al.*, 1980.)

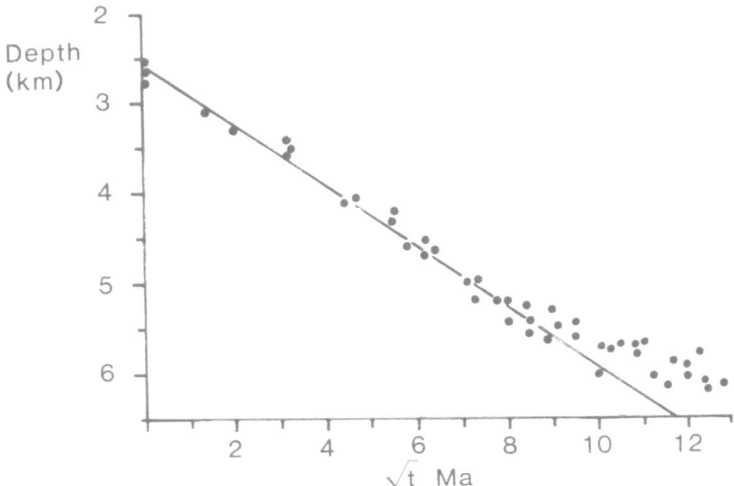

Figure 6.3 Depth of oceanic crust plotted against the square-root of age ($\sqrt{t}$), corrected for sediment loading. Solid line shows square-root relationship for crust younger than 80 Ma ($\sqrt{t} = 9$ Ma). (After Bickle, 1978.)

being a consequence of isostatic compensation of the older cooler and therefore denser lithosphere. The fit of the depth observations to the $\sqrt{t}$ variation is particularly impressive for crust up to c. 80 Ma but older crust shows significant deviations from the theoretical curve (Figure 6.3). Early plate models (e.g. McKenzie, 1967) arbitrarily, although in retrospect correctly, assumed that lower boundary conditions were represented by a horizontal isotherm which gave a good fit to the deviations from the $\sqrt{t}$: depth variation. More recently it has been suggested that the instabilities related to the smaller scale of upper mantle convection transfer heat to the base of the lithosphere at a rate which becomes significant for oceanic lithosphere older than about 80 Ma. Parsons and McKenzie (1978) and Houseman and McKenzie (1982) analysed the stability of cooling lithosphere in detail and suggested the subdivisions illustrated in Figure 6.4 which have important implications for the preservation of geochemical anomalies as well as for mantle melting. The modelled lithosphere comprises an upper, cooler, rigid portion within which heat transfer is only by conduction (the mechanical boundary layer) and a lower portion which acts as a thermal boundary layer to the small-scale upper mantle convection. This boundary layer becomes unstable at about 80 Ma and in lithosphere older than this it will cool and detach at a rate which, although slower than that of the underlying vigorously convecting mantle, will be sufficient to ensure that this part of the lithosphere is well mixed and will not accumulate geochemical anomalies with respect to the upper mantle reservoir.

The structure of the continental lithosphere is less well known, being largely hidden from geophysical observations by the overlying crust. Its thickness is controversial. Jordan (1978) has argued from lateral variations of mantle seismic velocity evidence that the thickness of the continental lithosphere is in excess of 400 km. However, the thermal time constant for the subsidence of the few well studied continental

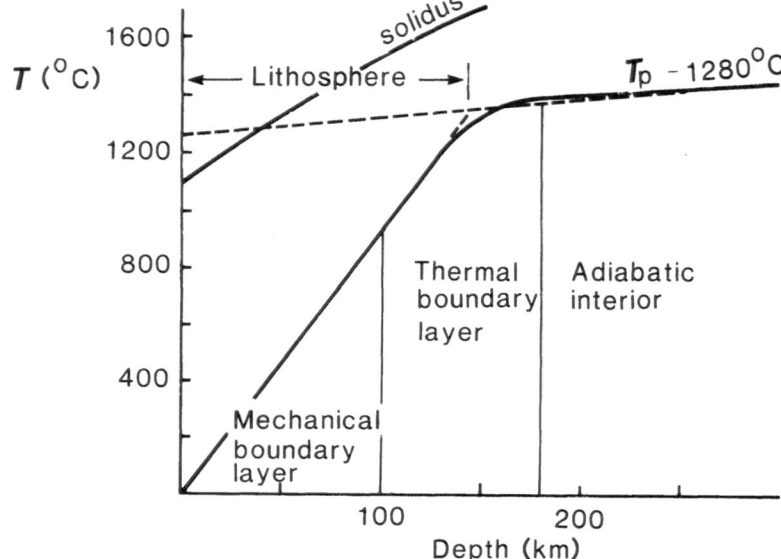

Figure 6.4 Relationship of the mechanical boundary layer, thermal boundary layer and adiabatic interior to the lithosphere and upper mantle, in terms of temperature and depth, for a mantle with a potential temperature (T_p) of 1280°C and a mechanical boundary layer thickness of 100 km. The upper mantle solidus is shown for comparison. (Redrawn from McKenzie and Bickle, 1988.)

sedimentary basins is similar to that of oceanic lithosphere, implying similar lithosphere thicknesses (e.g. McKenzie, 1978). Estimates of the thermal structure from surface heat flow estimates also predict thicknesses of c. 150 km (Sclater et al., 1980) although these estimates are very sensitive to the choice of heat-producing element distributions (Pollack and Chapman, 1977). The discovery that South African diamonds contain garnet and clinopyroxene inclusions with c. 3.3 Ga Sm–Nd ages (Richardson et al., 1984) implies mantle sources which were part of stable lithosphere at depths of 150 to 200 km from the Archaean to the Cretaceous.

The possibility that continental lithosphere might be stabilised by compositional differences has been suggested by a number of authors (O'Hara et al., 1975; Oxburgh and Parmentier, 1977; Jordan, 1978; Sleep, 1979; Bickle, 1986). According to these models, partial melting leaves a residue which is up to 0.05 gm/cm³ less dense, due to the elimination of the denser aluminous phase and preferential extraction of iron. However, density profiles modelled for possible Archaean melting regimes (Figure 6.5) demonstrate that such density stabilisation is marginal (Bickle, 1986). If the Archaean or younger continental lithosphere were a stabilised depleted residue, this would have important implications for magmatism in continental areas as well as insulating the Archaean crust from higher mantle temperatures.

6.2.3 Geoid anomalies and convection scales

The only immediate, widely available observations which could verify the postulated small-scale convective circulation of rising and sinking plumes are geoid, ocean depth

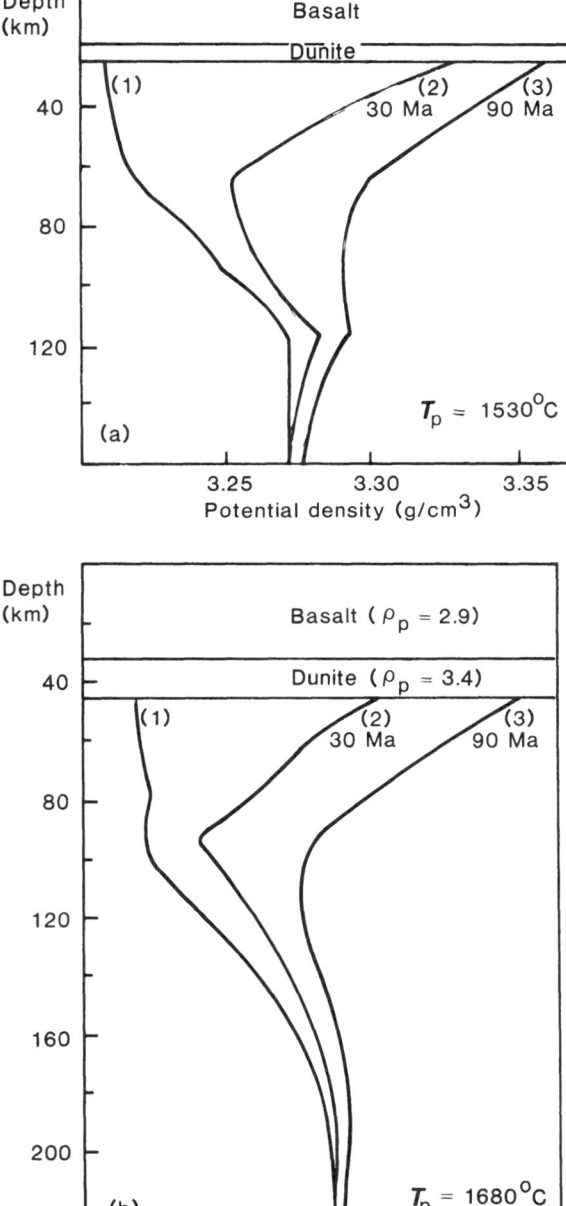

Figure 6.5 Thickness of oceanic crust and potential density (ρ_p) structure of depleted lithosphere for mantle T_p of (a) 1530°C and (b) 1680°C. Potential density calculated at pressure of solidus intersection. Curve (1) is calculated at solidus intersection pressure and temperature to illustrate composition–density differences. Curves (2) and (3) calculated for lithosphere cooled for 30 and 90 Ma respectively, the 90 Ma thermal gradient being approximately that under stable continental crust. (After Bickle, 1986.)

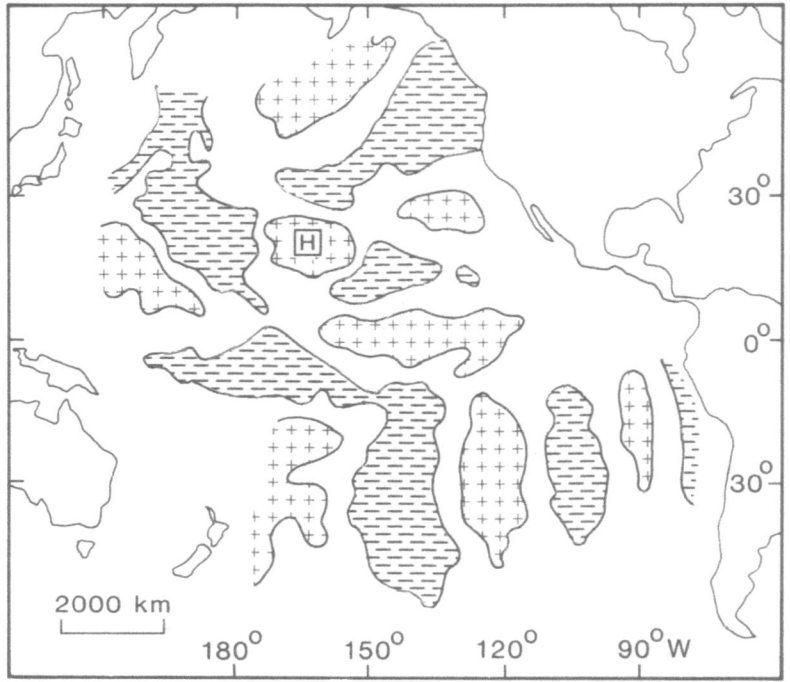

Figure 6.6   Residual geoid anomalies in the Pacific Ocean. ' + ' symbol: positive anomalies (2 to 6 m); ' — ' symbol: negative anomalies (− 2 to − 6 m); H: Hawaii. (Redrawn from Watts and Davies, 1981.)

and gravity variations on scales greater than 100 km in the ocean basins (e.g. Watts *et al.*, 1985). Figure 6.6 illustrates the deviations of the sea-surface geoid from a reference geoid in the Pacific ocean (Watts and Davies, 1981). Positive deviations relate to upwelling regions and negative deviations to downwelling regions. The spacing at *c.* 1000 km is consistent with small-scale convection within the upper mantle, and the anomaly distribution is unrelated to recognisable plate tectonic circulation. Some upwelling plumes are clearly associated with magmatism, the Hawaiian system being a good example. In this, as in most cases, the relative velocity of the lithospheric plate to the plume is such that no detectable heat flow anomaly would be expected at the surface. The Cape Verde Islands plume, which has fortuitously moved with the overlying eastern Atlantic–African plate for *c.* 50 Ma, represents an exception and the expected positive heat flow anomaly of *c.* 20% has been detected by Courtney and White (1986).

6.2.4   *Magmatism*

The ocean lithosphere depth–age relation (Figure 6.3) and geoid and gravitational anomalies (Figure 6.6) support a model for a relatively vigorous and unsteady mantle convection (Figure 6.2) in which vertical temperature gradients within the convecting region are near-adiabatic ($c.\,0.6°C/km$) with lateral temperature variations of about $\pm\,200°C$. The magnitude of the thermal anomalies generated within boundary layers is limited by the conductive time constant for the layers and the rate at which the thermal anomaly is dissipated by the forming of a rising or sinking plume. Because the adiabatic temperature variation across the mantle is of the same order as the lateral variations it is useful to describe mantle temperature in terms of its potential temperature (T_p), that is, its temperature extrapolated adiabatically to the surface as a convenient reference.

The superposition of both cold downwelling plumes and hot rising plumes across spreading ridges is consistent with lateral temperature variations relating principally to small-scale convective instabilities and demonstrates that ridges are not, in general, associated with hot rising plumes. Ridges are implicitly passive features mostly underlain by mantle of constant potential temperature. The thermal state at spreading ridges is special for two reasons. First, the mantle is rising sufficiently quickly ($>1\,mm/a$) for conductive cooling to be negligible in contrast to the surrounding plates, and second, melting occurs where mantle rises above the depth at which its adiabatic gradient crosses the dry solidus, and the melting-buffered adiabatic gradient is steeper than the equivalent solid adiabat.

McKenzie and Bickle (1988) have discussed the physical constraints on mantle melting. If normal potential temperature mantle ($T_p = c.\,1280°C$) wells up adiabatically, it intercepts the dry solidus at a depth of about 45 km (Figure 6.7). As it rises above this depth the mantle will start to melt, and the adiabatic gradient becomes rather steeper, reflecting the latent heat of melting. The melt fraction continues to increase until the mantle stops rising. At ridges, the amount of melt produced is controlled only by the T_p for mantle of given composition, the degree of melting for any

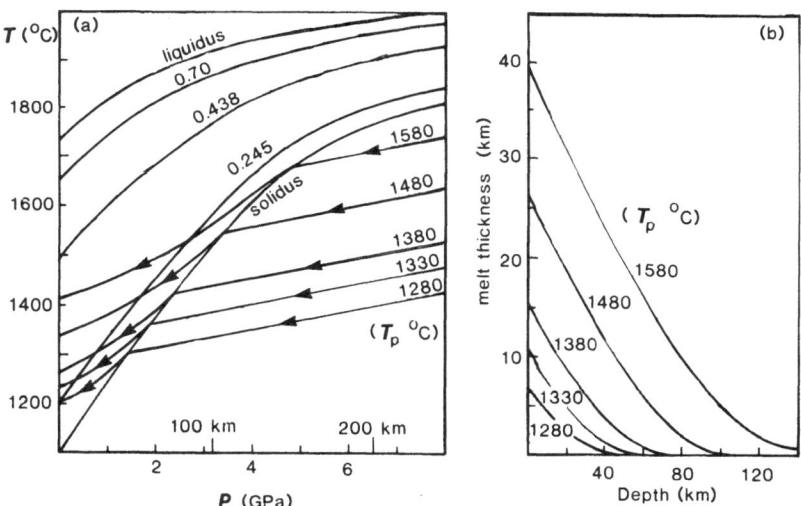

Figure 6.7 (a) Paths of adiabatic mantle upwelling labelled with potential temperatures (T_p) and calculated assuming that melt and solid rise together. Melt fraction contours shown sub-parallel to solidus and liquidus. (b) Thickness of partial melt (oceanic crust) developed in relation to T_p at a given depth. (Redrawn after McKenzie and Bickle, 1988.)

packet of mantle being determined by its final height within what becomes lithosphere by conductive cooling. McKenzie and Bickle (1988) showed, by parameterising melt fraction as a function of pressure (P) and temperature (T) from experimental petrological data, that $c.$ 7 km of oceanic crust would be produced by melting mantle with a T_p of $c.$ 1280°C (Figure 6.7b). The consistent thickness of oceanic crust is a measure of the relatively uniform upper mantle temperature and structure, and magmatism is mainly restricted to ridges because the thickness of most of the lithosphere is greater than the critical $c.$ 45 km depth required for the initiation of dry melting.

Away from the ridges mantle melts are only likely to be produced where extension thins the lithosphere to less than 45 km, in hot upwelling plumes which start to melt at greater depths (1580°C T_p mantle melts at $c.$ 140 km), or as H_2O or CO_2-rich melts at temperatures below the dry solidus as alkalic or calc-alkalic magmas. Calc-alkaline magmatism results from fluids being introduced into the mantle at subduction zones. The significance of hot upwelling plumes is well illustrated by the increase in oceanic crustal thickness in aseismic ridges such as that adjacent to Iceland for which McKenzie and Bickle (1988) estimate that melting of $c.$ 1480°C T_p mantle is necessary to give the 25 km crustal thickness (Figure 6.7b).

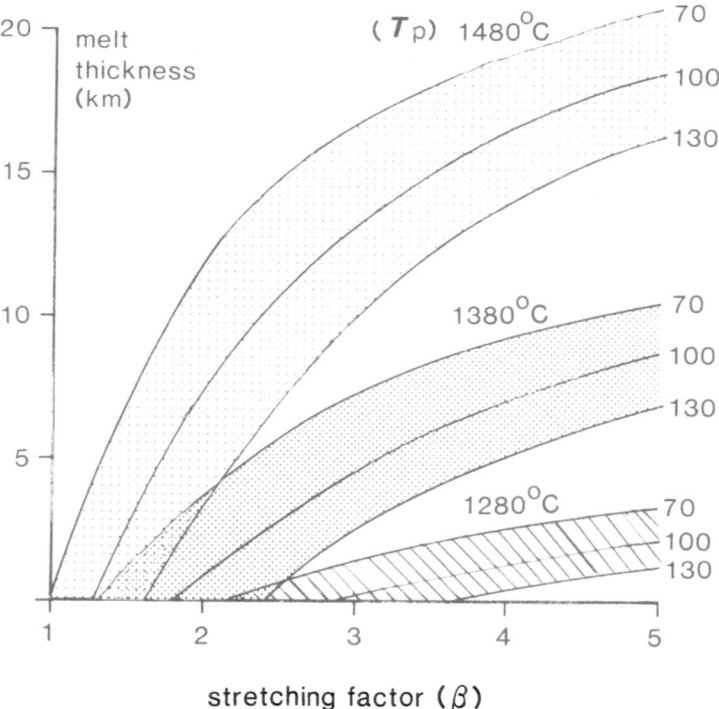

Figure 6.8 Melt thickness as a function T_p and thickness (70, 100 and 130 km) of mechanical boundary layer, and generated by extension of the continental lithosphere (β = stretching factor). (Redrawn after McKenzie and Bickle, 1988.)

The other profound implication of melting due to adiabatic upwelling is that melts separate from the residual mantle mineralogies (as so called 'primary melts') over a range of depths. The physics of melt segregation processes in upwelling mantle is complex and difficult to model. Klein and Langmuir (1987) and McKenzie and Bickle (1988) independently showed that a relatively simple batch-melting model, in which melt moves with the adjacent solid during adiabatic ascent, was capable of explaining the compositional range of MORB, allowing for subsequent high level fractionation.

McKenzie and Bickle (1988) also calculated the melt volume and composition as a function of the extension of continental lithosphere (Figure 6.8). The depth of melting and the amount of melt produced depend on the thickness of the lithosphere and more particularly on the thermal gradient in the basal boundary layer. Figure 6.9 shows that for small and moderate degrees of extension, melting is confined to the thermal boundary layer and underlying mantle. The permanent part of the lithosphere, termed the mechanical boundary layer, only melts with relatively large amounts of thinning ($\beta < 2.5$) and never contributes a volumetrically significant fraction of the total melt even at high mantle potential temperatures. This is not surprising as the cold lithosphere is not a promising location for melting. It has important geochemical implications

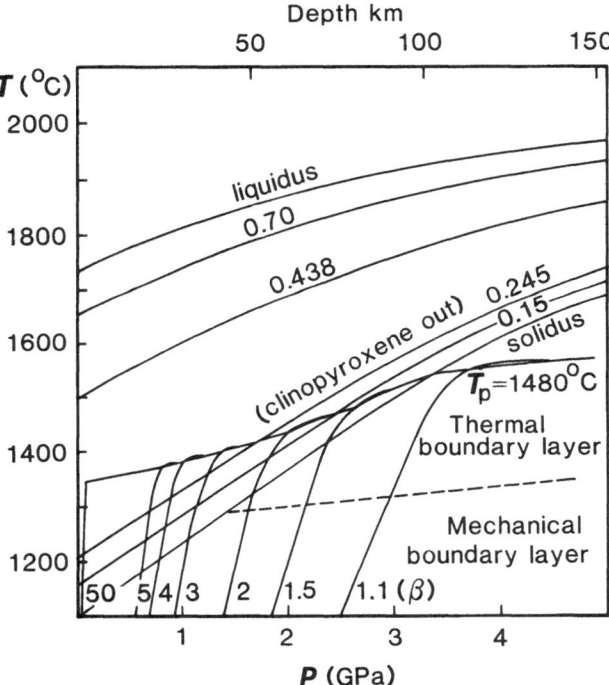

Figure 6.9 Thermal structure of variably stretched continental lithosphere for an original mechanical boundary layer thickness of 100 km and a mantle T_p of 1480°C. Curves are calculated for stretching factors (β) of 1.1 to 50. Dashed line shows loci of boundary between mechanical and thermal boundary layers. Note that only after $\beta > 2.5$ will stretching have been sufficient for the intersection of the geotherm with the mantle solidus to take place within the mechanical boundary layer. (Redrawn after McKenzie and Bickle, 1988.)

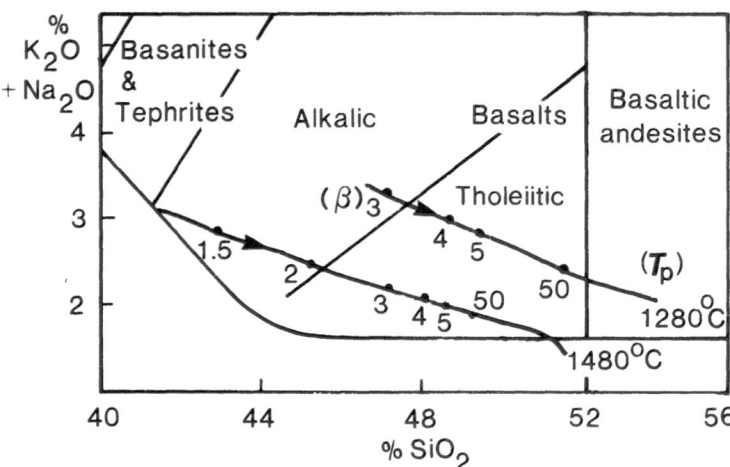

Figure 6.10   Alkalis versus silica diagram showing the composition of first melts according to the model outlined in Figure 6.9 for varying degrees of continental lithospheric extension (β from 1.1 to 50) for an original mechanical boundary layer thickness of 100 km and for mantle T_p of 1480°C and 1280°C. The melt thicknesses generated for these degrees of extension are shown in Figure 6.8. Lithological fields after Cox et al. (1979). Alkali basalt–tholeiite divide from MacDonald and Katsura (1964). (Redrawn after McKenzie and Bickle, 1988.)

because the continental lithosphere represents a long-lived isolated reservoir within which isotopic anomalies may be generated although melts passing through the continental lithosphere may be contaminated by melting of wall rock (e.g. Sparks, 1986).

Figure 6.10 illustrates the loci of primary melt compositions produced by lithospheric extension (McKenzie and Bickle, 1988). The main trend is from alkali-enriched basalts at low degrees of partial melting to tholeiitic basalts at higher degrees of melting. Which side of the important alkali basalt:tholeiite basalt divide the magmas fall is controlled largely by differing degrees of melting.

6.3   Thermal evolution of the mantle

6.3.1   *Geological evidence*

The discussion above establishes that mantle potential temperature is a prime control on magma genesis. The interpretation that mid-ocean ridges produce about 7 km of melt provides our most precise estimate of the present upper mantle potential temperature (c. 1280°C) with an uncertainty of c. 30°C related to the entropy of melting and the amount of heat dissipated by upward moving melt (McKenzie, 1984a; McKenzie and Bickle, 1988).

It is likely that the mantle was hotter in the past, reflecting the greater radiogenic heat production and stored heat from the initial accretion of the Earth and formation of the core. The restriction of magnesian komatiitic lavas to the Archaean is direct evidence for a hotter Archaean mantle, although aspects of komatiite petrogenesis are problematical as discussed in Chapter 2 and below. In fact, there may be no precise relationship between maximum komatiite eruption temperature and average mantle temperature if komatiites are produced in hot rising plumes. The most robust geological constraint on mantle temperature would be oceanic crustal thickness or average composition, but the only record of pre-Jurassic oceanic crust is preserved in ophiolite complexes which are invariably tectonised, of uncertain petrogenesis and even, in the early Precambrian, of dubious identity.

The other main observable constraints on the Earth's thermal evolution are the present-day heat loss and heat production rates, the ancient crustal thermal record preserved in metamorphic rocks, and the lithospheric $P-T$ data indicated by mantle xenoliths, and particularly Archaean diamonds (Richardson et al., 1984).

The excess of present-day heat loss over radiogenic heat production is a measure of how fast the Earth is cooling. Total global heat loss is about 4.2×10^{13} W (Sclater et al., 1980) whereas total radiogenic heat production, based on a chondrite Earth model and a K/U ratio of 1.27×10^4, is about 2×10^{13} W (O'Nions, 1987). It follows that about half the estimated heat lost is stored heat, and that the Earth must be cooling with a relatively long thermal time constant of $c.\,1\,Ga$.

Another, somewhat problematical, record of the thermal state of the mantle is derived from continental lithospheric thermal gradient evidence. Continental thickness and the $P-T$ range of granulite and amphibolite facies metamorphism appear to have changed little throughout the geological record (Perkins and Newton, 1981; Boak and Dymeck, 1982; Bickle, 1978; Burke and Kidd, 1978; Windley, 1977). These data have been interpreted to indicate a similar range of lithospheric thermal gradients back through time. This is a paradox, given the expectation and evidence for higher mantle temperatures in the past, which would logically result in higher lithospheric gradients (England and Bickle, 1984; Richter, 1985). Numerous solutions to this paradox have been proposed including (i) the noting that metamorphic conditions invariably represent perturbations from average thermal conditions and may be buffered by additional factors such as crustal melting (e.g. Morgan, 1985, 1986); (ii) the suggestion that the increased global heat loss in the past is partitioned unequally between the small scale convective component conducted through the lithosphere and the plate tectonic component lost by plate creation and subduction (Bickle, 1978, 1986; Burke and Kidd, 1978); and (iii) the suggestion that older continental lithosphere is relatively thick and refractory and insulates the continents (e.g. Jordon, 1978; Bickle, 1986). The surprising discovery that some diamonds are Archaean in age (Richardson et al., 1984) provides corroboration that some early Precambrian continental lithosphere was around 200 km thick and was thus characterised by relatively low geothermal gradients.

6.3.2 The thermal state: convection models

An alternative approach to evaluating the Earth's internal thermal evolution is to attempt to quantify the relationship between heat loss and internal temperature by numerical modelling of a convecting body crudely analogous to the Earth. Although in principle this method is essentially independent of the geological record, nearly all published studies are constrained by the geological evidence that surface temperatures of the Earth have lain between 0 and 100°C since 3.8 Ga and that a hotter mantle is implied by the abundance of Archaean komatiites. Precise modelling of global thermal histories is frustrated by the complexities of the Earth's heat loss processes and by the complications of modelling convection in a non-Newtonian fluid with a very temperature-sensitive viscosity.

Oceanic and continental regimes represent regions of very different heat loss mechanisms. Two major mechanisms of global heat transport must be considered in thermal modelling, namely a plate-tectonic scale circulation, and small-scale convection transporting heat to the base of both oceanic and continental plates (Bickle, 1978; Richter, 1985). The partition of heat loss between these two mechanisms is likely to have varied in the past, yet few models explicitly discuss such regionalisation of convective regimes. Richter (1984a, b, 1985) did present such modelling of the Archaean thermal regime and concluded that unless the Archaean lithosphere is stabilised by a mechanism in addition to its thermal state, the small-scale convective regime would have penetrated the melt region, leading to a vertical recycling of continental lithosphere under voluminous basalt. The preservation of Archaean crust from 3.8 Ga is evidence that such recycling was at least limited in extent. However, the applicability of the parameterisation adopted by Richter may be questioned for two main reasons. First, it is not clear how good is the parameterisation of plate velocities when the resistive forces to plate motions may be largely generated in relatively cool regions of anomalous viscosity. Second, the parameterised calculations of the sort adopted by Richter and in a number of other studies do not model well the effect of the very temperature-sensitive mantle viscosity. The parameterised calculations model the effect of temperature on viscosity by comparing models with internal viscosity constant over the model and run at different viscosities appropriate to some function of average temperature. McKenzie and Weiss (1980) and Christensen (1984) have shown from numerical experiments that this is not a good approximation and for similar models (whole-mantle convection) the variable viscosity model of Christensen (1985) predicts global internal temperature changes double those of Richter's (1984a, b) parameterised calculations. The reason for this discrepancy is that it is the relatively cool, more viscous upper boundary layer that controls the rate of heat loss. The temperature within this boundary layer falls well below the interior temperature used for scaling the parameterised models (Christensen, 1985). Computational power and the lack of a proper formulation for mantle viscosity are fundamental limitations on current models of global thermal evolution.

The probability that the mantle convects as two layers separated by a boundary at a depth of c. 670 km (Figure 6.1) is an additional uncertainty of considerable significance to global thermal evolution. A layered convection system is far less efficient at transporting heat than a single system. By reviewing thermal time constants for mantle convection models, Richter (1984) estimated that for a variable viscosity mantle, whole-mantle convection would result in a heat loss:heat production ratio of c. 1.1:1, whereas a two-layer mantle would have the observed ratio of 2:1.

6.3.3 Initial thermal state and 'magma oceans'?

A less certain control on mantle thermal evolution is its initial thermal state. Both initial accretion and the core formation could have released sufficient energy to completely melt the Earth. Given the very low diffusivities for gases in the solid Earth, the extent of degassing is probably a measure of the extent to which mantle material was cycled through the melt region at the surface of the primordial Earth. Evidence for degassing of at least part of the Earth comes from rare-gas systematics. Xenon isotopic compositions place the most severe constraint on the timing of this degassing. Modern MORB has a $^{129}Xe/^{130}Xe$ value of $c.$ 6.8 which is significantly enriched compared with the atmospheric value of 6.48 and the primordial value of 6.3 (Staudacher and Allègre, 1982). The enrichment in MORB is related to the decay of ^{129}I to ^{129}Xe in a mantle reservoir already depleted in Xe by degassing, which must have taken place within $c.$ 100 Ma of the formation of the Earth as ^{129}I has a half-life of only 17 Ma. Similarly, ^{36}Ar is depleted in MORB. The degassing can only have been partial because, compared to the atmosphere, MORB and especially some ocean island volcanoes are characterised by a substantial excess of ^{3}He. Ocean-island magmas are also characterised by a substantial enrichment in ^{36}Ar over MORB.

Kato et al. (1988a, b) cite high apparent crystal–liquid partition coefficients (K_D) for Mg- and Ca-perovskite, phases stable at depths greater than $c.$ 670 km, as further evidence that early mantle melting was limited in extent. Agee and Walker (1989) question the experimental validity of these K_D determinations. If Kato et al.'s values are correct, then if these phases were in equilibrium with a largely molten upper mantle, this mantle would be characterised by severe, unobserved depletions in what are incompatible elements at lower pressures. These data are significant in terms of a recent suggestion (Stolper et al., 1981; Nisbet and Walker, 1982) that basaltic and komatiitic liquids, being more compressible, become denser than the solid mantle at depths between 150 and 300 km. Recent work has not yet directly confirmed this density inversion, although Agee and Walker (1988a) predict that mantle olivines would float in a coexisting komatiitic liquid at pressures in excess of 8 GPa. Nisbet and Walker (1982) speculated that early in the Earth's history mantle temperatures were high enough to stabilise a major magma ocean at depths greater than the liquid–solid density reversal. Ohtani (1984) and Walker (1986) also speculated that the apparent near coincidence of mantle solidus and liquidus at pressures greater than 15 GPa was due to the upper mantle being effectively a eutectic composition having crystallised from a magma ocean. Further improvement in the understanding of the role of melting in the early mantle must await better constraints on the very high pressure phase relations of mantle materials, data which are currently being actively sought by a number of research groups.

To conclude, if the initial mantle was largely molten we would expect to find evidence of both substantial degassing and large-scale fractionation of many trace elements, neither of which is observed. The evidence for partial degassing suggests that at least the upper mantle was cycled through a melt region within $c.$ 100 Ma, implying substantially more vigorous convection and higher temperatures than in the present upper mantle, which is cycled through the melt region in about 1 Ga.

6.4 Melting in the early Precambrian mantle

6.4.1 Melt production at higher temperatures

The early Precambrian mantle was almost certainly hotter than the present-day mantle. The various convection models suggest temperature differences of up to 400°C and, as discussed below, potential temperatures higher than 1700°C would be precluded by copious melting. This temperature difference would have had important effects on magmatism. Volcanicity would have been more voluminous, higher degrees of partial melting would have occurred and melting would have been initiated at greater depths than in analogous Phanerozoic magmatic environments. To a first order of approximation early Precambrian geology reflects these predictions. Mafic volcanic units are prominent in most supracrustal successions, particularly in Archaean greenstone belts. The mafic successions are also distinctive in containing a higher proportion of high-Mg basalts (komatiitic basalts) as well as komatiites (MgO > 20%) not seen in successions younger than the Archaean. This is easily explained as a consequence of the higher Archaean mantle temperatures as illustrated in Figure 6.7. At the present day, the extension of continental lithosphere by a factor of two over 'normal' mantle ($T_p = 1280°C$) is predicted not to elevate mantle temperatures above the dry solidus and in Phanerozoic sedimentary basins such as the North Sea volcanicity is limited to very minor development of alkali basalt magmas (Dixon et al., 1981; Figures 6.9, 6.10). However, similar amounts of extension over a 1480°C T_p mantle is predicted to produce between 3 and 12 km of basalts spanning the tholeiite–alkali basalt divide, which is what is observed on the margins of the northeast Atlantic where continental rifting took place above the Iceland thermal plume (White et al., 1987). In the Archaean, where average mantle potential temperatures may have been in the region of 1400°C to 1600°C, any extension would have led to melting and it has been suggested that the mafic volcanic phases of Archaean greenstone belts were formed in a similar extensional environment (McKenzie et al., 1980; Bickle and Eriksson, 1982), the mafic volcanism being more voluminous because of the higher mantle temperature.

6.4.2   Heat loss by melting

The more voluminous melting over a larger depth range in the early Precambrian would have significantly affected the thickness of the oceanic crust, the structure and stability of the lithosphere and the efficiency of the heat loss processes. The existence of early Precambrian oceanic crust and the evidence for and against the operation of plate tectonics in the Archaean has been widely debated (e.g. Kröner, 1981) but the relatively low continental thermal gradients (see section 6.3.1) require that heat must have been lost preferentially in oceanic or analogous regions (Bickle, 1978, 1986; Burke and Kidd, 1978). If the 66% of global heat lost through plate creation and subduction (Sclater et al., 1980) were lost instead by conduction through the lithosphere in the Archaean, crustal thermal gradients would have been greater by an order of magnitude, allowing for the higher radiogenic heat production. This is incompatible with what we know about the thermal state of Archaean crust.

Figure 6.5 illustrates the relationship between oceanic crustal thickness and the density structure of the underlying lithosphere calculated for mantles with a T_p of 1530°C and 1690°C (Bickle, 1986). Oceanic crustal thicknesses of up to 40 km are suggested, with depletion due to melting extending down to between 120 and 200 km. In Phanerozoic lithosphere the 45 km thickness of depleted mantle is significantly less than the thickness of the conductively cooled lithosphere (100–150 km) and Bickle (1986) speculated that the density reduction in early Precambrian lithosphere might have been sufficient to stabilise it against small-scale convective instability but not against subduction, particularly where subduction was aided by the conversion of thick oceanic crust to eclogite.

The more extensive magmatism of the early Precambrian would also have had a profound effect on global heat loss, an effect not incorporated in the global thermal evolution model discussed above. The reason for volcanicity being ignored is that the processes operating within the top 100 km of the mantle and within the upper boundary layer were thought to be insignificant on a mantle scale. However, the effect of melting and efficient melt transport to the surface where it cools rapidly, is to provide a geologically instantaneous cooling, the heat being removed from the mantle as latent heat of melting. At present with a 1280°C T_p mantle the heat so transported in this way is a negligible fraction of the plate creation and subduction heat flux. However, with a 1530°C T_p mantle, the magma-transported heat is approximately 30% and with a 1690°C mantle 50% of that lost by a 45 Ma plate (Bickle, 1986). This heat is lost virtually instantaneously, drastically reducing the time constant for cooling of the lithosphere. The inclusion of the magma heat transport component and allowance for higher mantle temperatures invalidates the general conclusion from convective modelling that feedback from the high internal temperatures with faster convective circulation and subduction of younger, thinner lithosphere leads to a drastic decrease in the efficiency of the plate creation–subduction–heat loss mechanism (e.g. England and Bickle, 1984; Richter, 1985). An important implication of heat loss calculations is that the rate of melt production from the mantle during the Archaean must have been between ten and twenty times that at present. The other major consequence of melting is that the increasing slope of the solidus with depth effectively precludes average mantle potential temperatures any greater than 1700°C as copious melting would rapidly cool at least the upper few hundred kilometres.

6.4.3 Komatiites

Komatiites are prime evidence of higher mantle temperatures in the early Precambrian. The more magnesian komatiites, erupted with MgO contents between 25 and 33%, would have had liquidus temperatures of between 1500°C and 1650°C (Bickle, 1978) although there is some dispute as to whether komatiites more magnesian than 25% MgO are cumulus olivine-enriched (Elthon, 1978). Uniform upwelling of a 1600°C T_p mantle in mid-ocean ridge environments would produce 50 km of melts with an average temperature of only about 1450°C (Figure 6.7). The proposal that the Archaean oceanic crust was thin and dominated by komatiite (Arndt, 1983; Nisbet and Fowler, 1983) and the claim that the thin komatiite-dominated sequence in the Barberton greenstone belt was oceanic crust (De Wit et al., 1987) seem unlikely given this constraint, although it should be remembered that thicknesses inferred from most younger ophiolite complexes are unrepresentative of the oceanic crust (Moores and Jackson, 1974).

Alternatively, komatiites may have been produced by the extension of the continental lithosphere, possibly above hot upwelling plumes. Such a setting is consistent with the widespread evidence for their contamination with small amounts of crustal material (e.g. Arndt, 1986). Figure 6.7 suggests that the melting of normally fertile mantle in such a setting is unlikely to have produced melts much in excess of 1500°C. It is also possible that komatiites were produced from already depleted mantle which would have given smaller amounts of melt of more magnesian composition, although these melts would have been severely depleted in incompatible elements. Another mechanism for komatiite genesis is limited eutectic melting at depth in equilibrium with an olivine–pyroxene–garnet residue (O'Hara et al., 1975). High pressure experimental work has shown that primary eutectic melts become progress-

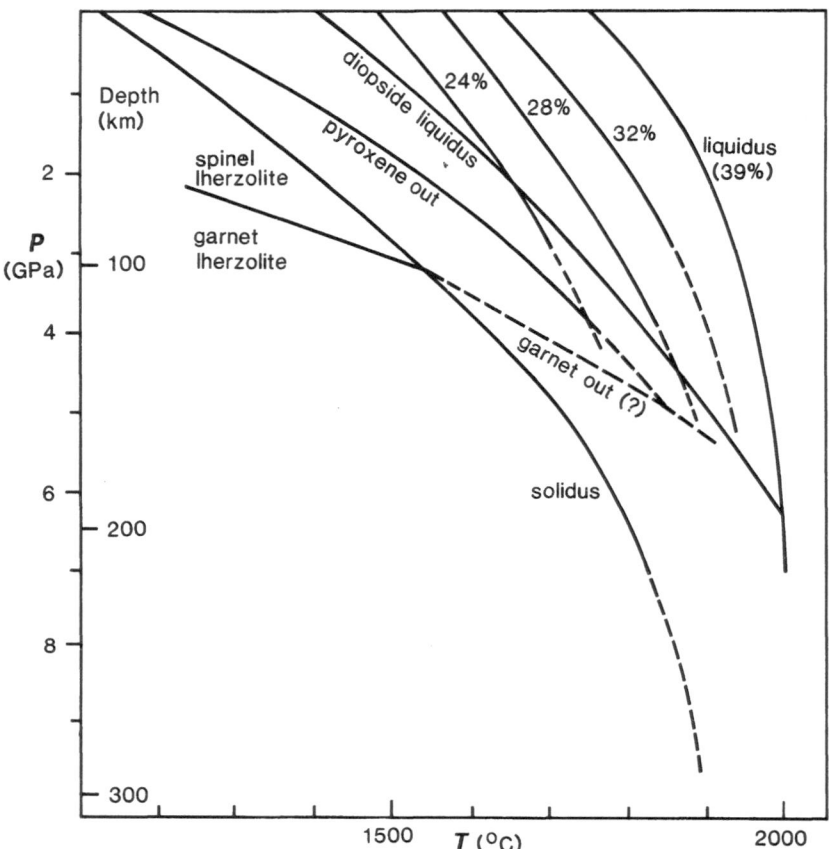

Figure 6.11   Schematic partial melting relations of komatiites as a function of P and T and MgO content (%), based on available experimental work (redrawn after Bickle, 1986). Projected dashed lines are extrapolated.

ively more magnesian with increasing pressure, and extrapolation of available experimental work suggests that eutectic liquids with komatiitic MgO contents would be produced at between 6 and 10 GPa (200 to 300 km) and at temperatures between 1670°C (25% MgO) and 1830°C (32% MgO) (Figure 6.11). However, if komatiites are produced by low degrees of melting, which would overcome the thermal problems of generating voluminous melts, then these komatiites should be proportionally enriched in the elements incompatible in the residue.

The geochemistry of komatiites is surprisingly consistent with their being derived from a mantle exhibiting about the same degrees of depletion as the present-day source for MORB and by high degrees of melting (Chapter 2). Thus, models which require their genesis either from already strongly depleted mantle or as low-degree eutectic melts have to explain the relatively consistent incompatible trace element concentrations present in amounts to be expected in 50% melts of a slightly depleted mantle source with a dunite residue. At present, there is no satisfactory model which explains all the geochemical and thermal aspects of komatiite genesis.

6.5 Chemical evolution of the early mantle

6.5.1 Chemical reservoirs

The main chemical divisions within the Earth are its crust, mantle and core and it is probable that the mantle contains at least two discrete geochemical reservoirs. Rare-gas data provide the most compelling evidence for the existence of long-lived separated geochemical reservoirs within the mantle (O'Nions, 1987). The ^{129}Xe anomaly as well as the Ar isotopic composition of the atmosphere discussed previously (see section 6.3.2) strongly suggest early degassing of at least the upper mantle. However, the ^{3}He anomalies associated with certain seamounts and with MORB, the discrepancy between the ^{4}He flux from the mantle and that estimated from radioactive decay (O'Nions and Oxburgh, 1983) and the enhanced ^{36}Ar flux from some ocean islands, all imply the existence of a long-lived reservoir of primordial gases. It is plausible on geophysical grounds that this reservoir is the lower mantle, although the geochemical data alone do not conclusively reveal its source.

6.5.2 The growth of geochemical reservoirs

The constancy of siderophile element concentrations in basic magmas throughout the geological record is consistent with the formation of the Earth's core prior to 3.8 Ga (e.g. Newsom et al., 1986). In contrast, the continental crust has had continual additions from the mantle and much geochemical study has been directed at assessing the amount and rate of crustal growth and the nature of the processes involved. In addition, much effort has gone into documenting isotopic and chemical variations within the mantle, as sampled by volcanic rocks and xenoliths. In this respect it should be noted that magmatism at mid-ocean ridges (25 km^3/a) elevates about 250 km^3/a of mantle above the dry solidus and at this rate processes the entire upper mantle in c. 1 Ga. MORB is most likely to be a representative sample from this upper mantle.

6.5.3 Mantle chemistry: primitive mantle and crustal depletion

The compositions of the bulk crust and mantle source of MORB are shown in Figure 6.12, normalised to hypothetical primordial mantle values (Wood et al., 1979). The elements are arranged in order of decreasing incompatibility into MORB-type melts from left to right. This order is not perfectly known and the desire to present smooth patterns may obscure anomalous fractionation (e.g. Nb), related to crust-forming processes (Hoffman et al., 1986). The choice of primitive mantle composition is a second critical assumption inherent in this diagram. The refractory elements (e.g. Th, U, Sm, Nd, Ca, Al, Ti) are assumed to be present in chondritic porportions allowing for the 1.5 times enrichment in the silicate Earth following core formation. For some of the critical elements isotopic constraints are available, of which the most significant is the isotopic co-variation of Sr with Nd in modern magmas (Jacobsen and Wesserburg, 1979; O'Nions et al., 1979). The principal observation from present-day MORB chemistry is that the upper mantle source region is depleted in those elements most enriched in the bulk crust (Figure 6.12).

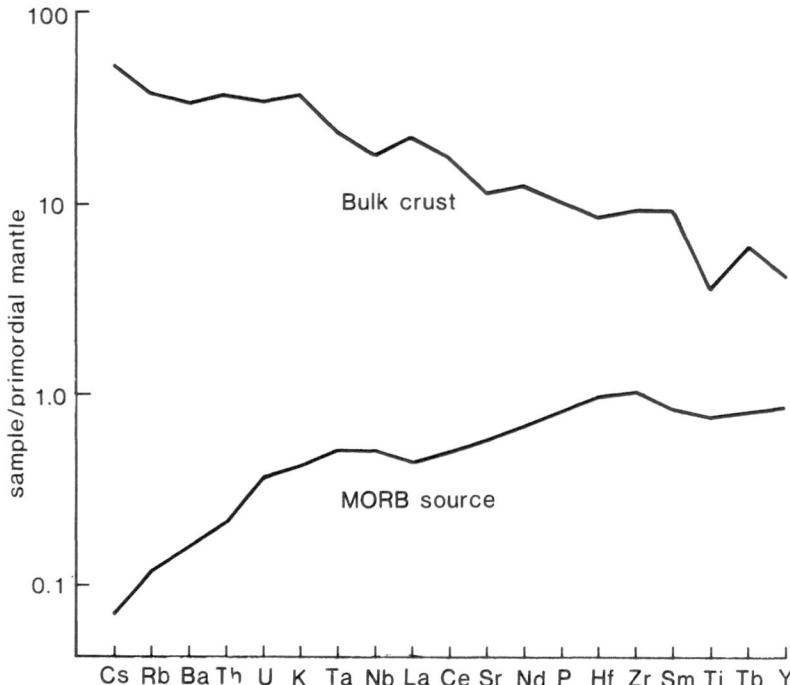

Figure 6.12   Incompatible element concentrations of bulk crust (Taylor and McLennan, 1985) and the modern MORB mantle source normalised to primitive mantle values (Wood *et al.*, 1979). Elements arranged from left to right in order of decreasing incompatibility in MORB.

Mass-balance calculations on the isotopically constrained element concentrations of Sm, Nd, Rb and Sr indicate that the degree of depletion observed in the MORB source would be attained if the crust was extracted from only 0.3 to 0.5 of the whole mantle (Jacobsen and Wesserburg, 1979; O'Nions *et al.*, 1979; De Paolo, 1980; O'Nions, 1987). Davies (1981) questioned the robustness of this conclusion to uncertainties in the data, but in doing so invoked rather high concentrations of these elements in the continental lithosphere and extended the estimated initial isotopic Nd composition of the Earth to its feasible limit. The ^{40}Ar budget is relatively well constrained given the estimate of total K in the Earth, derived from the observed K/U ratio of 10^4 and chondritic U contents. The ^{40}Ar concentration of the atmosphere requires the extraction of Ar from between 0.45 to 0.65 of the whole mantle (Galer *et al.*, 1989). The upper mantle (27% by mass) is an obvious candidate for this depleted portion of the whole mantle.

6.5.4   *The timing of crustal growth*

Another important question concerns the timing of continental crustal growth. Rb–Sr and Sm–Nd isotopic systems have again provided the most straightforward approach and a number of estimates in the region of *c.* 2.0 Ga have been made for the mean age of the continental crust (e.g. Jacobsen and Wasserburg, 1979; O'Nions *et al.*, 1979;

Goldstein *et al.*, 1984). Implicit in this estimate that the mean age of the crust is about half that of the Earth are the conclusions that the crust has grown, on average, approximately uniformly through geological time and that the mantle, or rather the depleted portion of the mantle, should show progressive depletion with time. Strangely, there is no evidence of such progressive depletion from the geological record. The oldest well preserved granitoid crustal rocks have concentrations of the most incompatible elements similar to those in the youngest additions to the crust, even for elements whose concentrations have been depleted in the mantle by an order of magnitude (e.g. Bickle *et al.*, 1983). The Nd isotopic evolution of volcanic rocks does not show the increase in ε_{Nd} with time expected from a progressive depletion. Rather, it fits a linear evolution, but a linear evolution not consistent with present estimates of the upper mantle Sm/Nd ratio (Figure 6.13). There are limitations on the use of these controls on upper mantle composition in the past because, for example, the relation between crustal composition and the ultimate source mantle composition is known poorly at best, and the claim that komatiites are a sample of the depleted mantle is not proven. Taylor and Mclennan (1985) have monitored the variation in fine grained clastic sediment composition with time as a measure of changes in crustal composition. They concluded that there was a

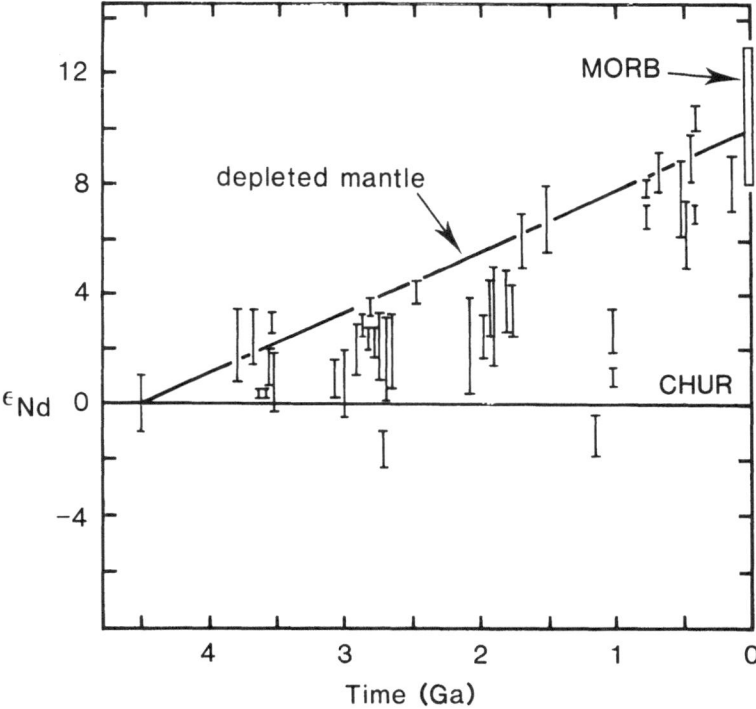

Figure 6.13 Initial Nd isotopic compositions of samples of Phanerozoic and Precambrian continental crust and mafic igneous rocks. The samples with the highest positive ε_{Nd} compositions at a given time are assumed to lie closest to contemporary mantle values (cf. depleted mantle trend of Goldstein *et al.*, 1984). Samples with negative ε_{Nd} values are assumed to be mixed with older crust which evolves to negative ε_{Nd} with time. Redrawn after Galer *et al.* (1989).

substantial change in average upper crustal composition at the end of the Archaean, with the Archaean crust being on average much more mafic. However, they thought that this difference reflects the preponderance of mafic lavas in the Archaean rather than fundamental differences in the granitic component of Archaean crust.

6.5.5   U–Th–Pb systematics: crustal or lower mantle recycling?

The problem revealed in modelling the apparently simpler Rb–Sr and Sm–Nd isotope systematics were predated by those produced by modelling U–Th–Pb isotopes. U, Th and Pb are even more incompatible than Sm and Nd and therefore more depleted in the mantle and more affected by external inputs. MORB Pb exhibits distinct excesses of radiogenic ^{206}Pb (from ^{238}U decay) and ^{208}Pb (from ^{232}Th decay) over the values which would be expected if the source had maintained bulk Earth U–Th–Pb proportions. These excesses could be explained by a secular increase in U/Pb and Th/Pb ratios in the MORB source over Earth history, but there is no evidence of preferential Pb loss to the crust and the siderophile element distributions indicate that the formation of the core was very early, as discussed previously. Armstrong (1968) suggested that the high MORB ^{206}Pb and ^{208}Pb contents reflect recycling of relatively radiogenic crust into the mantle, a theme enlarged upon by Armstrong (1981), Russell and Birney (1974) and Zartman and Doe (1981) among others.

Pb isotopic systematics have recently been fundamentally reinterpreted by Galer and O'Nions (1985). They showed that the upper mantle Th/U ratio, interpreted from MORB ^{208}Pb/^{206}Pb, is about 3.8, close to the bulk Earth estimates (c. 3.9), whereas ^{232}Th/^{230}Th activity ratios from MORB samples correspond to Th/U of around 2.5 in their upper mantle sources. The ^{208}Pb/^{206}Pb ratio reflects the long-term average Th/U ratio for the source of the Pb whereas the ^{232}Th/^{230}Th activity ratio in MORB should reflect the present Th/U ratio in the upper mantle. To resolve this paradox, Galer and O'Nions (1985) calculated that the Pb sampled by MORB can only have been resident in the MORB source, presumably the well mixed upper mantle, for about 600 Ma, otherwise the observed radiogenic ^{208}Pb/^{206}Pb ratio would be detectably less than that of the bulk Earth reflecting the lower Th/U ratio in the upper mantle. The implications of this are that Pb must be cycled through the upper mantle relatively rapidly from an external source with a near-bulk Earth Th/U ratio, and that the upper mantle must be well mixed on a time scale considerably less than 600 Ma. Galer and O'Nions (1985) rejected crustal recycling models mainly because they are 'mechanistically difficult to envisage', and such models require Th/U ratios of around 3.3 in the depleted mantle, well above the value of c. 2.5 derived from ^{232}Th/^{230}Th activity measurements in MORB samples. They postulated that it is the lower mantle which is cycled through the upper mantle at a relatively low rate, but one sufficient to buffer the upper mantle Pb isotopic composition. However, it is difficult to exclude crustal recycling because the bulk compositions of the crust and continental lithosphere are so poorly known.

Galer et al. (1989) have carried out a comprehensive review of the geochemical consequences of the entrainment of lower mantle into the upper mantle. They consider three separate reservoirs: the lower mantle, the upper mantle and the continental lithosphere. Their modelling illustrates some important points about geochemical evolution due to the interaction between multiple reservoirs, although as the number of reservoirs is increased, it becomes easier to explain any given set of geochemical observations. Galer *et al.* (1989) reject models based only on the formation and recycling of crust or the growth of the crust from the undepleted reservoir. They claim that a model involving the cycling of lower mantle into the upper mantle with a corresponding depleted return flow is most consistent with geochemical observations. Elemental behaviour is strongly dependent on the enrichment factor in the, as yet, poorly constrained crust-forming melts. The most incompatible elements (U, Th, Pb, Cs, Rb, K) are rapidly depleted from the upper mantle in a time corresponding to their residence time (e.g. 600 Ma for U, 400 Ma for Th) (Galer and O'Nions, 1985). At times much longer than their residence times the concentrations of these elements will stabilise at a steady state value in which the upper mantle concentration reflects a balance between the flux in and out of the lower mantle and the flux into the continental crust (Figure 6.14). One line of evidence in favour of such behaviour is that many late Archaean komatiites exhibit Th-depletions similar to those of MORB (Brevart *et al.*, 1986). For the less incompatible elements (e.g. Sm, Nd and Sr), the residence times are much longer and the concentrations of these elements in the upper mantle have probably decreased throughout Earth history (Figure 6.14). However, the rate of decrease and the total change are too limited to be detectable in the magmatic record.

Thus, it can be concluded that the concentrations of the more incompatible elements will also be somewhat depleted in the lower mantle and that the concentrations of the less incompatible elements in the upper mantle will reduce systematically with time. The Sr and Nd radiogenic isotope ratios are controlled both by the change in parent–daughter ratios in the upper mantle and by inputs of lower mantle material. The

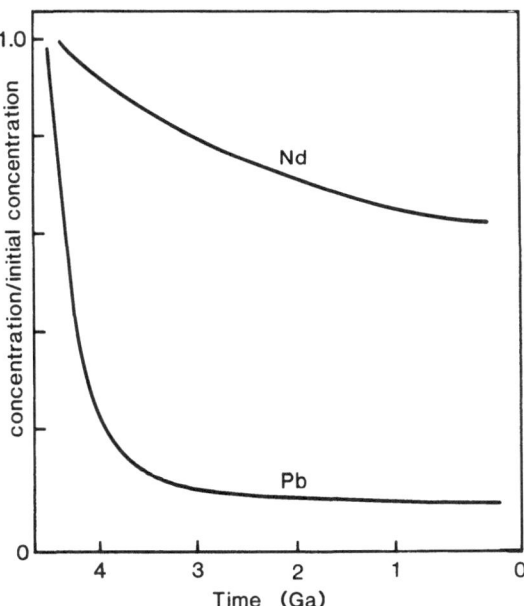

Figure 6.14 Variation of Pb (strongly incompatible) and Nd (moderately incompatible) in the upper mantle with respect to time, due to the cycling of lower mantle through the upper mantle and continuous crustal growth through time. Note that Pb with a residence time of 0.6 Ga in the upper mantle reaches a steady state in less than 1 Ga. Redrawn from Galer *et al.* (1989).

U–Th–Pb isotopic compositions provide compelling evidence for buffering of the upper mantle composition by the cycling from one or more external reservoirs. The remaining uncertainty is the precise recognition of the nature of the reservoir(s) given the poor constraints of the bulk composition of the continental lithosphere.

6.5.6   *Crust formation*

Some aspects of crust formation are still enigmatic. Continental crustal rocks are fractionated in terms of element pairs such as Ce–Pb and Nb–U, which apparently remain unfractionated in all other magmatic environments. Hofmann *et al.* (1986) cited Ce–Pb and Nb–U as element pairs with apparently identical ratios and therefore partition coefficients in a wide range of MORB and OIB. The MORB, OIB and therefore presumably upper mantle Ce/Pb ratio of 25 ± 5 and Nb/U ratio of 47 ± 10 differ from crustal ratios of 4 and 10 respectively and, in the opposite sense, from primitive mantle ratios of 9 and 30. This is consistent with depletion of crustal material in Ce and Nb with respect to Pb and U, and Hofmann *et al.* (1986) suggested that this was the result of an early crust-forming event. Galer *et al.* (1989) would explain the relatively depleted but consistent Ce/Pb and Nb/U ratios as a function of their short residence times and buffering by lower mantle recycling. If this is so, it suggests that subduction zones are the only sites at which the continental crust-like fractionation is generated. Magmatism above rising plumes and the generation of MORB are ruled out by Hofmann *et al.*'s (1986) data. If OIB isotopic anomalies are due to the melting of detached continental lithosphere (McKenzie and O'Nions, 1983) then significant continental growth by the extension of the crust and the resultant melting (e.g. McKenzie, 1984b) would seem unlikely, as fractionated Ce/Pb and Nb/U ratios would be expected in the lithospheric residua. Conversely, Ellam and Hawkesworth (1988) have argued that the Rb/Sr ratios of mantle additions to crust above present-day subduction zones, the other possible site of crustal growth, are too low to create average crustal values, but that Archaean mantle additions had substantially higher Rb/Sr ratios. However, there is an obvious problem in identifying and discriminating between mafic mantle-derived additions, a more fractionated mantle addition and the contamination or re-melting of pre-existing crust in both modern subduction and Archaean calc-alkaline environments.

    Hofmann *et al.* (1986) and Ellam and Hawkesworth (1988) favour early growth of a crust enriched in incompatible elements. However, this would require recycling of bulk crust through the upper mantle to explain the relatively young age of crustal material implied by the Sm–Nd and Rb–Sr isotopic systematics. Recycling of crust would mean that fractionated melts must have been extracted from the mantle throughout Earth history. The lack of Ce/Pb and Nb/U fractionation in OIB and MORB source material from the melting of Ce- and Nb-enriched residues is surprising in this context. The mechanism and chemistry of crustal growth is still the outstanding geochemical problem. The detailed comparison of early Precambrian calc-alkaline magmas with their modern counterparts will be one essential component of the solution.

6.6   Conclusions

Our understanding of mantle evolution over Earth history is based primarily on its 'end state' as understood imperfectly today. However, the Precambrian geological record places some important thermal and geochemical constraints on its evolution. Major points are:

1. The mantle convects on at least two scales, one related to plate tectonics and one to mantle plumes as mapped by geoid anomalies in the oceans. A number of lines of evidence support this model. These include the termination of subduction-related earthquakes at a depth of c. 650 km, the thermal time constant of the Earth estimated from present-day heat loss, and geochemical arguments for the existence of at least two geochemical reservoirs within the mantle. Principal among these geochemical arguments are the evidence for early degassing of the present upper mantle but preservation of significant primordial ^{3}He and ^{36}Ar. This and liquid–perovskite K_D data may indicate that any substantial early melting was restricted to the upper mantle.

2. The mantle melts where adiabatic upwelling brings mantle material above the solidus. Melt volumes and compositions are predictable for given magmatic environments and mantle temperatures.

3. Thermal models for mantle evolution indicate a secular decrease in upper mantle potential temperature by several hundred degrees since the early Precambrian. Temperatures 200–300°C hotter than today are indicated by the eruption of Archaean komatiitic lavas with maximum liquidus temperatures in the range 1500–1650°C. Maximum potential temperature is anyway limited to c. 1700°C, as at higher temperatures copious melting to great depths would rapidly cool the mantle, if present estimates of the shape of the mantle solidus are approximately correct.

4. Mass-balance calculations indicate that the crust was largely formed by extraction from the upper mantle, although U–Th–Pb and other very incompatible element concentrations in the upper mantle are now buffered by cycling of lower mantle material or bulk-crust through the upper mantle. Evidence for sources as depleted as MORB from the Precambrian magmatic record and the REE patterns, Th/U ratios and Nd isotopic compositions of komatiitic magmas all support this interpretation.

5. Heat loss budgets in the Archaean, together with the evidence for moderate continental thermal gradients and the thick lithosphere, imply a substantial heat loss in ocean-like regions. Higher mantle temperatures would have resulted in much more voluminous magmatism, particularly in the oceans where the production of oceanic crust at rates ten to twenty times those of today would have transported a major fraction of global heat loss. Such variation in magmatic and possible crust-forming rates has not been considered in current geochemical models.

6 The precise mechanisms responsible for crustal growth and the fractionation of elements into the crust are still enigmatic. Material has been continually added to the crust throughout Earth history, but the determination of the most likely crustal mass versus time curve is frustrated by uncertainty over the relationship of the residence times of the key isotopic tracers compared with the major element constituents. The early Precambrian geological record implies little change in crustal composition and particularly the relative fractionations of the incompatible elements since 3.8 Ga.

7 Lunar magmatism

S.B. SIMON

7.1 Introduction

Over the past two decades, our understanding of the formation and geology of the Moon has greatly increased. In the minds of many scientists the Moon has evolved from a distant object of curiosity to a natural laboratory for planetary geology. Because of the lack of tectonic recycling and chemical weathering, fresh rocks more than 4.0 Ga old still exist on the Moon and these rocks yield invaluable information about the early history of the solar system that is unavailable on Earth. Furthermore, understanding the Moon's formation and early evolution helps increase our knowledge of the Earth, because the two bodies appear to have formed close together in space and time.

We have only recently approached a consensus on the origin of the Moon. Compositional similarities, including oxygen isotopic signatures, indicate that the Earth and Moon are somehow related, and argue against the capture of the Moon by the Earth. However, compositional differences such as the Moon's relative depletion in volatiles and enrichment in refractory elements (Wood, 1986) preclude the formation of the Moon directly from an unaltered bulk Earth composition. Through improved understanding of the dynamics of the early solar system and the growth of the planets, mainly with advances in computational capabilities (Ryder, 1987a), the 'collisional ejection' hypothesis has emerged as the favoured model for the origin of the Moon (Hartmann, 1986; Wood, 1986). Simply put, this model involves the impact of a Mars-sized body into the Earth, producing a disk of material from which the Moon formed, containing components of the impactor and the Earth. The model accounts for the compositional similarities and significant differences between the Earth and Moon, and also the high angular momentum of the Earth–Moon system.

The early differentiation of the Moon provides an additional constraint on models of formation, as it requires an energetic process (Wood, 1986) which led to the melting and differentiation of the outer part of the Moon. This event led to the formation of the feldspathic lunar crust (highlands), and presumably to mafic cumulates at depth. The highlands contain a second important group of lithologically varied mafic rocks collectively known as the Mg-suite whose origin is less certain.

From the time of its formation until about 3.8 Ga ago, the Moon and most of the other planets and moons in the solar system experienced a period of high meteorite flux. With planets sweeping up debris, cratering rates at 4.6 Ga approached 10^{10} times the present rate, declining to 100 times the present rate by 3.8 Ga (Hartmann, 1980). Evidence of this early epoch in planetary formation can be seen in the densely cratered highland surfaces of the Moon, Mercury and many of the moons of Mars, Jupiter and Saturn. The Earth presumably experienced equally intense cratering at this time, but here active geological processes have erased that part of the record. The large multi-ringed basins on the Moon testify to the size and force of some of the projectiles.

This period of intense bombardment is something of a mixed blessing for geologists. The impacts thoroughly brecciated and mixed many of the rocks of the ancient crust, and melted, buried or obliterated many others and reset isotopic clocks. On the positive side, the impacts excavated deep-seated rocks that would otherwise be inaccessible and distributed rock fragments laterally away from their places of origin. This last point is very important with respect to lunar sample studies. Although only nine localities on the Moon have been directly sampled, we have samples of many different rock types due to transport of 'exotic' materials by meteoritic impacts. Thus, a single sample of lunar soil, itself a regolith of impact-derived debris, may contain a variety of rock types from different localities. Remote sensing techniques have also been used to identify the limit of outcrops of sampled lithologies (e.g. Head *et al.*, 1978), further extending our coverage. Thus, despite the somewhat limited data base, there is now a fairly good understanding of lunar geology and petrology. However, remote sensing has also indicated that there are other basalt types which have not as yet been sampled (Pieters, 1978).

Another important aspect of the basin-forming events is that they thinned and cracked the underlying crust, allowing mare basalts to reach the surface. Their dark, smooth surfaces looked like seas to the early European astronomers, who called the basins *maria* – Latin for seas. Although they occupy impact basins, lunar mare basalts are endogenous rather than impact melts. They are too basic to be formed from remelting of the crust, and their clast-free textures are also inconsistent with an impact origin. The flows have relatively few craters because their eruption post-dates the time of high meteorite flux. Most simplistically, the rocks of the highlands, in which feldspathic rocks predominate, and the mare basalts comprise the two major lunar lithological units. A third, less abundant group is KREEP (K-, REE- and P-rich) basalts. These three suites will be considered in this chapter.

7.2 Lunar highland rocks

The returned lunar samples have provided much evidence which supports the concept of an early Moonwide magma ocean or 'magmasphere' (Shirley, 1983) soon after the formation of the Moon. The complete melting of the outer 200 to 500 km of the Moon has been proposed to account for this magmasphere. Possible heat sources include energy derived from the large number of impacts, and the decay of short-lived radionuclides. However, pristine monomict (with no meteoritic contamination) highland rocks are *c.* 4.5–4.0 Ga old. This range of ages indicates their crystallisation within a few hundred million years of the formation of the Moon, not enough time for large-scale internal heating to have become significant. The lunar highland crust contains on average around 75 modal % plagioclase and mostly comprises cumulate rocks. The most plausible way to account for such a large volume of plagioclase is not by the generation of anorthositic magma, but by the flotation of plagioclase from an anhydrous basaltic melt.

In the reducing environment of the Moon, Eu occurs as Eu^{2+}, enhancing its substitution into plagioclase and fractionating it from the other REE. The ancient lunar anorthosites, therefore, have strong positive Eu anomalies and most other lithologies have complementary negative Eu anomalies (Figure 7.1) indicating the prior separ-

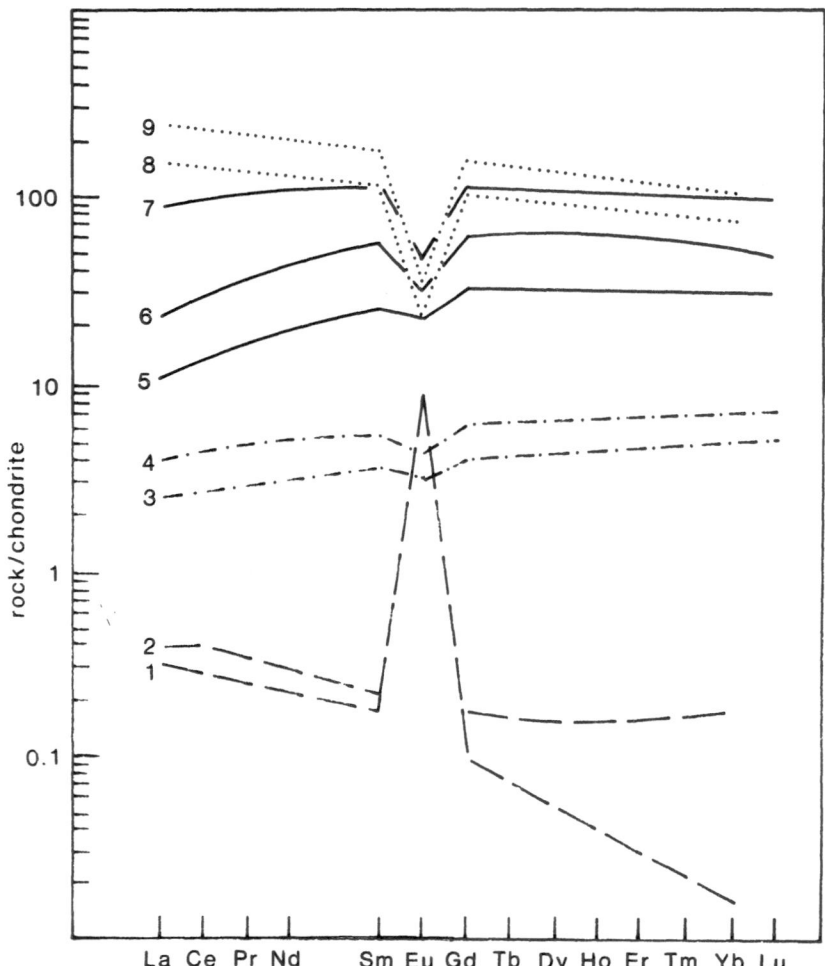

Figure 7.1 Chondrite-normalised REE abundances in representative lunar ferroan anorthosites (1, 2), VLT (3, 4), high-Ti (5, 6, 7) and KREEP (8, 9) basalts. From Taylor (1982). Sample numbers: 1: 60055; 2: 61016; 3: 70007; 4: 78526; 5: 70017; 6: 75055; 7: 10049; 8: 14310; 9: 15386.

ation of plagioclase from their sources. Warren (1985) elaborates on these and other arguments for a lunar magma ocean (or primordial differentiation event) and discusses other models. The data do not necessarily require a *single* magma ocean, but at least virtually moonwide, simultaneous magmatism. For example, the model of Shirley (1983) considers a partially molten magma ocean (*c*. 20% melt), and Walker (1983) proposes a serial magmatism model, which does not call for a magma ocean.

Walker's (1983) theory accounts for the complementary Eu anomalies of anorthosites and mare basalts by plagioclase fractionation in magma chambers. According to this model, the magmas fractionated until they reached plagioclase saturation. After plagioclase fractionation, the introduction of a fresh magma pulse produced a replenished magma with a negative Eu anomaly due to the prior plagioclase separation, but which did not have plagioclase on the new liquidus. However, a problem with this theory is that it requires that the mare basalt eruptions consistently occurred after the addition of fresh magma and before a return to plagioclase saturation (Warren, 1985).

7.2.1 Anorthosites

The evolution and probable products of a lunar magma ocean can now be considered. A schematic mineralogical and selected chemical profile through the ocean is shown in Figure 7.2. Olivine was the first phase to crystallise, followed by plagioclase, pigeonite and augite (Solomon and Longhi, 1977). Plagioclase did not nucleate until the magma ocean was at least 50% crystallised (Solomon and Longhi, 1977; Longhi, 1977). The crystallisation of Mg-rich olivine effected a decrease in the mg' $(Mg/(Mg + Fe))$ of the melt, thereby increasing its density and causing plagioclase to float (Walker and Hays, 1977), whereas it would not float on a hydrous terrestrial magma (Taylor, 1982). The resulting feldspar cumulates contain Ca-rich plagioclase ($c.$ An$_{95}$) and relatively Fe-rich olivines and pyroxenes, and are referred to as ferroan anorthosites (Dowty $et\ al.$, 1974). These anorthosites exhibit a wide range of X_{Mg} in mafic silicates without a

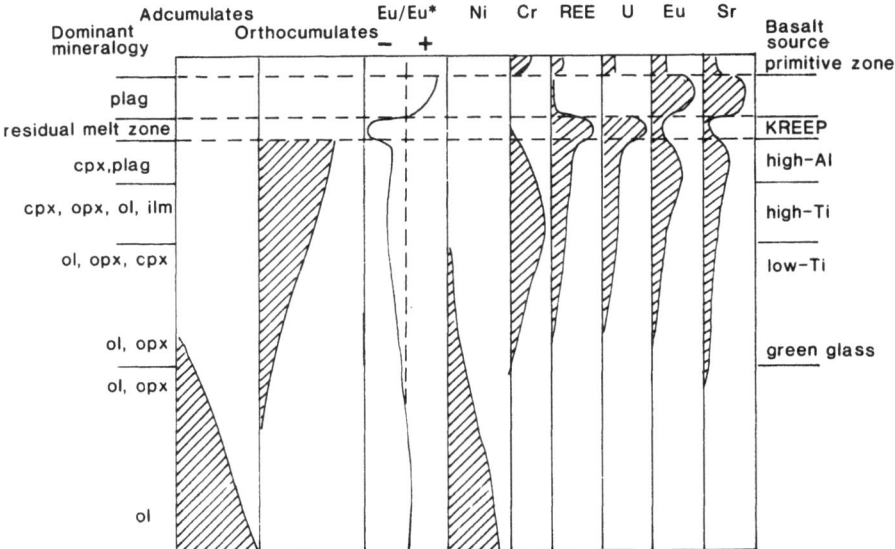

Figure 7.2 Schematic petrological and chemical section through the lunar magma ocean at 4.4 Ga showing idealised relative abundances of Ni, Cr, REE, U, Eu and Sr and the possible sources of different types of subsequent basaltic magmas. Eu* is the inferred Eu value for no Eu anomaly interpolated from Sm and Gd data and Eu/Eu* indicates the size of the positive or negative Eu anomaly. The upper, plagioclase-dominated unit is the ferroan anorthosite, while the various lower ultramafic units comprise the Mg-suite. Section not drawn to scale represents a thickness in the order of 1000 km. After Taylor (1982).

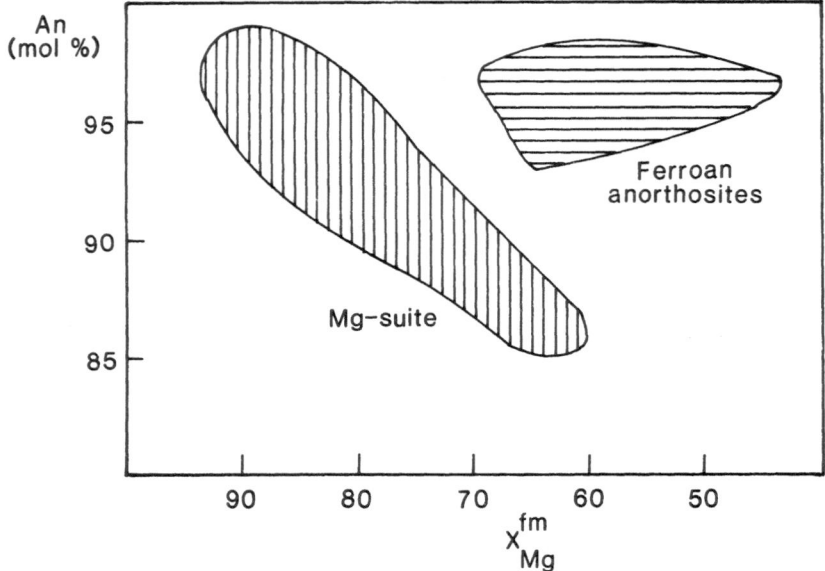

Figure 7.3   Plot of Mg/(Mg + Fe) of mafic minerals (X_{Mg}^{fm}) versus mole % An content of plagioclase showing the contrast between the mineralogy of the two major types of endogenous lunar highland rocks, the ferroan anorthosites and the Mg-suite. After Papike and Simon (1984).

corresponding variation in plagioclase composition (Figure 7.3). Orthopyroxene is generally in the range En_{55-65} and olivine is in the range Fo_{40-66} and most commonly is around Fo_{60}. The large amount of plagioclase relative to the mafics may have buffered the plagioclase composition (Raedeke and McCallum, 1980). Alternatively, the anorthosites may simply reflect the very low Na_2O contents of the parent liquids (Ryder, 1982).

Most lunar anorthosites have cataclastic textures, although the degree of recrystallisation ranges from zero to 100%. Two examples are shown in Figure 7.4 (a,b). Norman and Ryder (1979) estimated that the original plagioclase grain size was probably about 5 mm. Minor augite is present as well as orthopyroxene in many samples and in a suite of Apollo 16 anorthosites studied by Dixon and Papike (1975) the pyroxene compositions indicate equilibration at below 800°C. Exsolution lamellae, however, are rare. Common accessory phases in the ferroan anorthosites are ilmenite, Fe metal, chromite and silica (Norman and Ryder, 1979).

Major element abundances are controlled by the predominance of plagioclase. The anorthosite analyses given in Table 7.1 are virtually identical to microprobe analyses of the constituent anorthite feldspars.

Low REE abundances and strong positive Eu anomalies (relative to chondrites) are typical of ferroan anorthosites (Figure 7.1). Their low trace element abundances and shock histories make them difficult to date accurately. Ages of 4.2–4.0 Ga are typically obtained, but these are thought to have been reset by shock, and they probably do not represent crystallisation ages. For further reading see Dixon and Papike (1975), Norman and Ryder (1979) and James (1980).

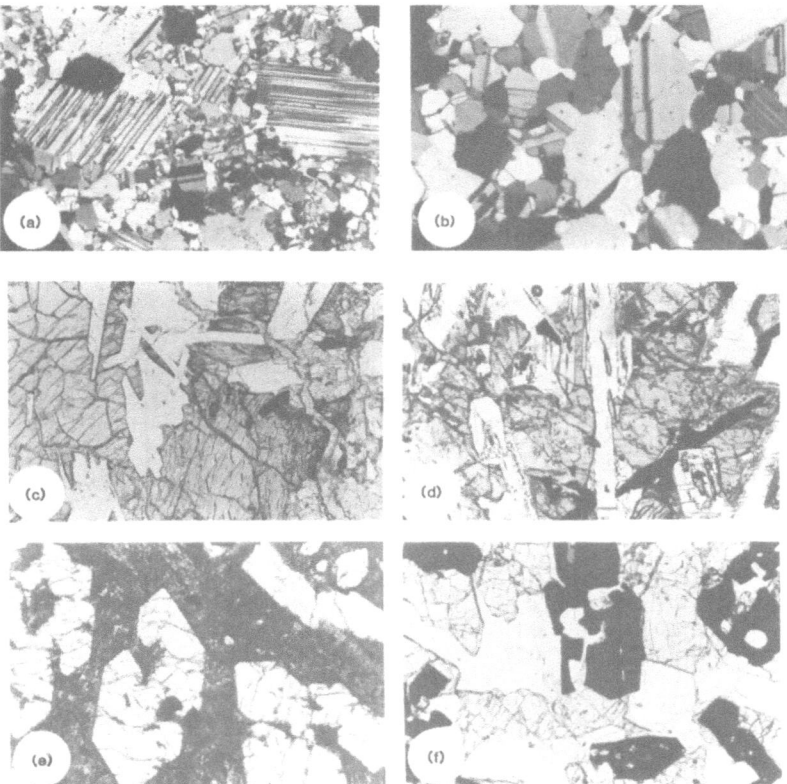

Figure 7.4 Photomicrographs showing the textures of lunar anorthosites (a, b, cross polarised light) and mare basalts (c to f, plane polarised light). a: partially recrystallised anorthosite 15223; b: recrystallised anorthosite 72704; c: A-11 (Apollo 11) low-K, high-Ti basalt 10047 showing plagioclase (white), clinopyroxene (grey), ilmenite (black) and cristobalite (mottled); d: A-12 low-Ti pigeonite basalt 12021; e: A-15 rapidly cooled low-Ti pigeonite basalt 15499 consisting of pyroxene phenocrysts in a fine-grained groundmass; f: A-17 high-Ti basalt 70035 showing plagioclase (white), pyroxene (grey) and ilmenite (black). All fields are 2 mm across.

7.2.2 *Mg-suite*

The average anorthite content of plagioclase in the ferroan anorthosites varies little despite a considerable range in X_{Mg} of the mafic silicates (Figure 7.3). However, another suite of lunar highland rocks known collectively as the Mg-suite includes a variety of lithologies with a much wider range of mineral compositions. Ferromagnesian mineral compositions extend to higher values of X_{Mg} (0.9–0.6) and plagioclase compositions range from An_{98} to An_{85}. This suite includes dunites, gabbroic to troctolitic anorthosites and various intermediate and basic rock types. Norites are probably the most abundant among the returned samples. New rock types are still being discovered, such as the Mg-anorthosites (Lindstrom *et al.*, 1984) and Mg-gabbronorites (James

Table 7.1 Compositions of selected lunar highland rocks*

Rock Type:	Anorthosite	Anorthosite	Gabbroic Anorthosite	Norite	Troctolite	Dunite
Sample No.	15415	60055	68415	78235	76535	72417
wt%						
SiO_2	44.1	44.3	45.5	49.5	42.9	39.8
TiO_2	0.02	—	0.32	0.16	0.05	0.03
Al_2O_3	35.5	34.0	28.6	20.9	20.7	1.3
FeO	0.23	0.34	4.25	5.05	5.0	11.9
MnO	—	0.10	0.06	0.08	0.07	0.11
MgO	0.09	0.33	4.38	11.8	19.1	45.4
CaO	19.7	19.0	16.4	11.7	11.4	1.1
Na_2O	0.34	0.34	0.41	0.35	0.20	0.013
K_2O	—	0.01	0.06	0.061	0.03	0.002
Cr_2O_3	—	0.005	0.10	0.23	0.11	0.34
ppm						
Ni	1.0	1.9	180	12	44	160
Rb	0.22	—	1.7	0.92	0.24	0.045
Sr	173	—	182	—	115	8.2
Zr	—	—	72	—	24	—
Ba	6.3	11	76	80	33	4.1
La	0.12	0.13	6.8	—	1.51	0.15
Ce	0.33	0.27	18.3	9.2	3.8	0.37
Nd	0.18	—	10.9	5.4	2.3	—
Sm	0.046	0.040	3.09	1.49	0.61	0.080
Eu	0.81	0.76	1.11	1.03	0.73	0.061
Gd	0.05	—	3.78	—	0.73	—
Dy	0.044	—	4.18	2.26	0.80	0.11
Er	0.019	—	2.57	1.47	0.53	—
Yb	—	0.035	2.29	1.64	0.56	0.074
Lu	—	0.004	0.34	0.24	0.08	0.012

*From Taylor (1982)

and Flohr, 1983). The bulk compositions of these rocks (Table 7.1) reflect their lower plagioclase and higher mafic mineral contents relative to the anorthosites. Trace element contents are higher, and typical REE abundances are approximately ten times chondrite values (Taylor, 1982).

Early classifications of the Mg-suite were based on modal mineralogy, using the three mineralogical end-members plagioclase (anorthosite), pyroxene (pyroxenite) and olivine (dunite) (Prinz and Keil, 1977; Stöffler et al., 1980), but because of the small sample size and generally coarse-grained character of these rocks, mineral compositions are now preferred, especially in distinguishing the ferroan from the Mg-suite rocks (Figure 7.3).

7.2.3   Petrogenesis

The different relationships between ferromagnesian mineral and feldspar compositions and, importantly, the gap between them indicate that the ferroan anorthosites and the

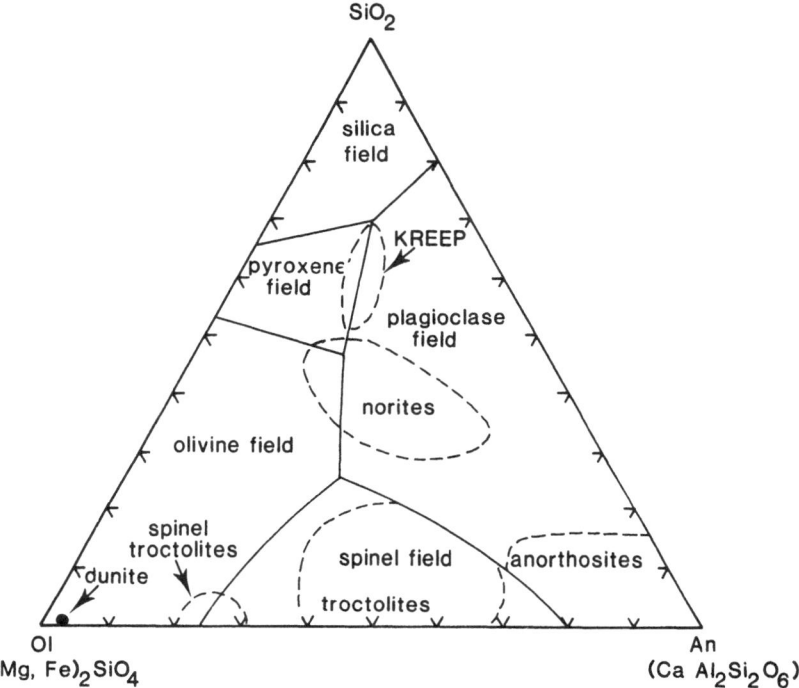

Figure 7.5 Compositions of pristine highland rocks in terms of their normative olivine–anorthite–silica components. After Taylor (1982).

Mg-rich suite could not have crystallised from the same magma (Warner *et al.*, 1976; Longhi, 1980). Raedeke and McCallum (1980) observed similar trends in samples from the Stillwater Complex, but without the compositional gap. The variety of endogenous highland lithologies also argues for more than one magma. The compositions of some of the major lithologies are indicated in Figure 7.5, and representative analyses are given in Table 7.1. Although the norites appear to form a relatively uniform suite (Figure 7.5), the Apollo 17 norites can be divided into two groups (James, 1980; James and Flohr, 1983), Mg-gabbronorites and Mg-norites, and Lindstrom (1984) also recognised alkali gabbronorites. An even wider variety of lithologies occurs along the Ol-An join (Figure 7.5), from dunite through troctolite, spinel-troctolite, troctolitic anorthosite and anorthosite.

Most workers now agree that the Mg-suite rocks formed from intrusions that post-date the formation of the anorthositic crust, but the source of the melts is uncertain. According to the model of James (1980), partial melting began in a primitive mantle after crystallisation of the magma ocean was complete except for a discontinuous KREEP layer (see section 7.4) between the mafic cumulates and the anorthosite (Figure 7.2). Thus, some of the rising magmas required a KREEP component, while others did not. After intrusion into the anorthositic crust, these Mg-rich magmas formed layered plutons (Figure 7.6a), which were then exposed and brecciated, granulated and melted by meteoritic bombardment (Figure 7.6b). The variable

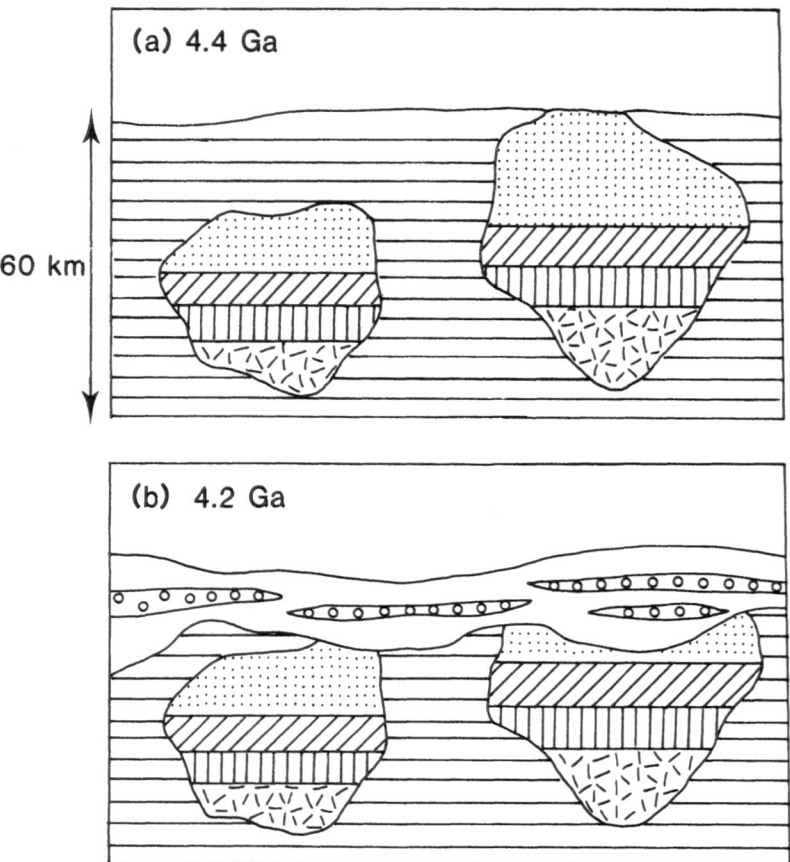

Figure 7.6   Model of James (1980) for the derivation of Mg-suite lithologies in layered plutons (irregular-shaped horizontally banded variously ornamented bodies) within anorthositic crust (horizontal line shading). a: At 4.4 Ga the plutons are intact. b: By 4.2 Ga surface rocks have been mixed, melted and granulated by impacts. A layer of breccias (circle shading) and melt rocks (unshaded) overlies the anorthositic crust and Mg-suite plutons. After James (1980) and Papike and Simon (1984).

assimilation of a KREEP component and the mixing of lithologies from layered plutons could account for most of the observed compositional diversity. In contrast, Binder (1980) suggested their source to be the cumulates from the magma ocean. However, Warren (1985) proposed that the most likely place for melting would be at the mantle–cumulate interface, the rationale being that the earliest cumulates from the magma ocean should be the most Mg-rich and less dense than later rocks. Thus, they should tend to rise and melt as they enter lower pressure regimes.

7.3   Mare basalts

Although mare basalts comprise only about 1% of the lunar crust and 17% of the surface area (Head, 1976), they provide much information that is necessary for an understanding of lunar thermal history, geochemistry and the nature of the lunar interior. The basalt flows constitute smooth, relatively sparsely cratered terrains. The A-11 (Apollo 11), A-12, L-16 (Luna 16) and L-24 sites are all on basalt, and Apollo 15 and Apollo 17 landed on mare basalts near to highland terrains.

The chemistry and mineralogy of mare basalts have been reviewed by Papike *et al.* (1976, 1981) and S.R. Taylor (1982). As with the highland Mg-suite rocks, despite the large amount of petrological data, the origin of the mare basalts remains uncertain. The compositions of representative mare basalts are listed in Table 7.2. Different types of basalts have been discovered at each site and are named by their collection locality and distinguishing features (A-11, low-K, A-12 olivine etc.). Generally, they are divided into three groups on the basis of their Ti contents (Figure 7.7). High-Ti basalts ($TiO_2 = 8$–13%) occur at the A-11 and A-17 sites, low-Ti basalts ($TiO_2 > 1.5$, $< 5.5\%$) at A-12 and A-15 sites, and very low-Ti (VLT) basalts ($TiO_2 < 1.5\%$) at A-17 and L-24 sites. The wide gap between the low- and high-Ti groups is probably merely a function of sampling, particularly in view of known unsampled basaltic units (Pieters, 1978) and it is likely that there is a continuum of Ti contents.

The basalts have low K_2O contents comparable to those of terrestrial low-K tholeiites (Taylor, 1982). The K is present in late-stage Si-rich mesostases or in rare K-feldspar grains. The low K and Rb contents are indicative of the Moon's overall depletion in volatiles relative to the Earth. In contrast, the mare basalts are somewhat enriched in non-volatile incompatible elements such as the REE, Sr and Ba. This is consistent with the origin of the basaltic liquids as partial melts, the compatible elements having been retained in the source and the incompatible elements having readily entered the melt. Zr is also incompatible and because of its ionic charge $(4 +)$ is directly associated with Ti. Therefore, the high-Ti basalts (A-11 and A-17) have far higher Zr contents than the VLT basalts. Ni and V contents are generally low relative to terrestrial basalts (Taylor, 1982), due either to a Moonwide depletion or to the retention of these elements in deep-seated ultramafic rocks.

Chondrite-normalised REE abundances in representative mare basalts are illustrated in Figure 7.1 (Taylor, 1982). The lower-Ti basalts have lower REE concentrations and slightly flatter patterns. Negative Eu anomalies are observed in all types of mare basalts and are taken to indicate prior plagioclase fractionation of the source. In this way the mare basalts provide strong evidence in support of an early magma ocean and the derivation of the basalts from the resultant mafic cumulates, or at least from a non-primitive source.

7.3.1 *Petrography and mineralogy*

Representative photomicrographs of different basalt types are shown in Figure 7.4. The samples represent a variety of compositions, formation conditions and cooling histories. Quickly cooled samples have small skeletal olivine or pyroxene phenocrysts in a glassy matrix, whereas slowly cooled basalts have well formed crystals up to several centimetres long. Textures cover the full range for basaltic rocks, from vitrophyric to ophitic (Lofgren, 1981). Experimental studies of mare basalts have been reviewed by Kesson and Lindsley (1976) and Wyllie *et al.* (1981). Crystallisation temperature and gas fugacity data are summarised by Haggerty (1981).

Typical modal proportions for various types of lunar basalts are given in Table 7.3. In all cases, pyroxene is more abundant than feldspar, and opaque phases (mostly

Table 7.2 Compositions of lunar mare basalts*

	A-11 Low K	A-11 High K	A-12 Olivine	A-12 Pigeonite	A-12 Ilmenite	A-15 Olivine	A-15 Pigeonite	A-17 Group A	A-17 Group B	L-16 High-Al	L-24 VLT	A-17 VLT	A-14 Feldspathic
Sample No.	10003	10049	12009	12021	12022	15555	15597	75055	70215	L-16/81	24174	78526	14053
wt%													
SiO_2	39.76	41.00	45.03	46.68	42.77	44.57	47.98	40.60	37.79	43.80	46.82	48.8	46.4
TiO_2	10.50	11.30	2.90	3.53	4.85	2.10	1.80	10.79	12.97	4.90	1.00	0.69	2.64
Al_2O_3	10.43	9.5	8.59	10.78	9.08	8.69	9.44	9.67	8.85	13.65	12.58	10.0	13.6
FeO	19.80	18.7	21.03	19.31	21.75	22.53	20.23	18.01	19.66	19.35	20.46	17.9	16.8
MnO	0.30	0.25	0.28	0.26	0.25	0.29	0.30	0.29	0.27	0.20	0.24	0.30	0.26
MgO	6.69	7.03	11.55	7.39	11.01	11.36	8.74	7.05	8.44	7.05	6.65	11.8	8.48
CaO	11.13	11.0	9.42	11.38	9.47	9.40	10.43	12.35	10.74	10.40	12.49	9.4	11.2
Na_2O	0.40	0.51	0.23	0.31	0.38	0.27	0.32	0.43	0.36	0.33	0.10	0.06	—
K_2O	0.06	0.36	0.06	0.07	0.07	0.09	0.06	0.08	0.05	0.15	0.02	0.02	0.10
Cr_2O_3	0.25	0.32	0.55	0.40	0.56	0.61	0.48	0.27	0.41	0.28	0.16	0.69	—
ppm													
Ni	—	—	52	16	42	42	30	2	3	—	—	—	75
Rb	0.62	6.2	1.04	1.14	0.74	0.45	1.13	0.58	0.36	1.58	—	—	2.19
Sr	161	161	95.6	129	143	84.4	111	191	121	437	110	—	98
Zr	309	—	107	123	180	76	—	272	192	218	50	—	215
Ba	108	330	60	71.1	55	32.2	52	76.2	56.9	—	50	—	146
La	15.5	28.8	6.1	—	—	—	4.86	6.26	5.22	—	2.87	1.2	13.0
Ce	47.2	82.8	16.8	19.8	17.4	8.06	13	21.5	16.5	—	8.6	—	34.5
Nd	40.0	62.8	16	14.4	14.4	6.26	9.3	23.9	16.7	—	7.0	—	21.9
Sm	14.4	22.3	4.53	4.84	5.38	2.09	3.09	10.05	6.69	—	2.10	—	6.56
Eu	1.81	2.29	0.94	1.116	1.26	0.688	0.84	2.09	1.37	—	0.83	1.0	0.21
Gd	19.5	29.3	5.2	6.59	7.71	2.9	4.4	15.7	10.4	—	—	0.30	8.59
Dy	12.9	33.4	7.13	7.86	9.37	3.27	4.51	18.1	12.2	—	2.9	—	10.5
Er	13.6	30.9	3.6	4.53	5.42	1.7	1.9	10.72	7.4	—	—	2.0	6.51
Yb	13.2	20.2	3.74	4.12	5.69	1.45	2.13	9.79	7.04	—	2.00	1.4	6
Lu	1.0	—	0.55	0.64	—	—	0.301	—	1.03	—	0.31	0.23	—

*From Papike et al. (1981)

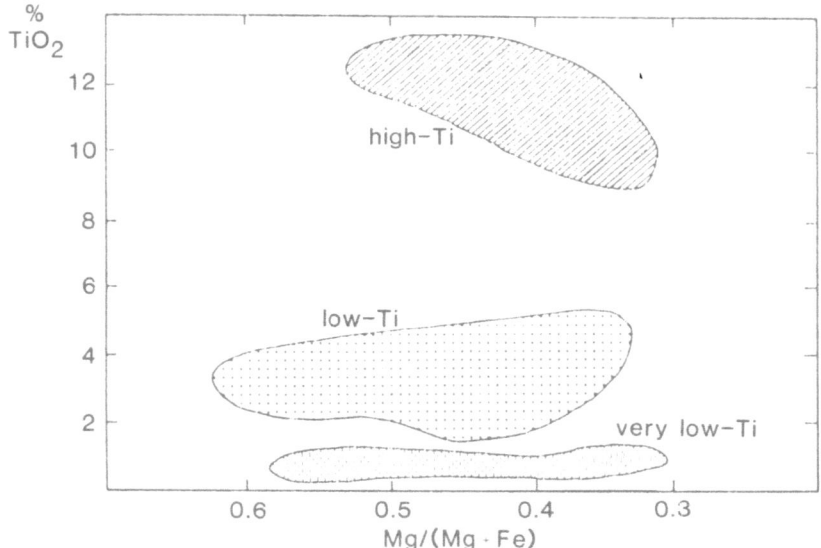

Figure 7.7 TiO$_2$(wt %) versus atomic Mg/(Mg + Fe) in mare basalts. After Papike and Simon (1984).

Table 7.3 Modal mineralogy of lunar basalts

Mission	Type	Pyroxene	Feldspar	Olivine	Opaque Minerals
L-24	VLT	60	34	4	2
A-17	VLT	62	32	5	1
A-12	olivine	54	19	20	7
A-15	olivine	63	24	7	6
A-12	pigeonite	69	21	1	9
A-15	pigeonite	62	34	–	4
A-12	ilmenite	61	26	4	9
A-11	low-K	51	32	2	15
A-17	high-Ti	52	33	–	15
A-17	VHT	48	23	5	24
L-16	high-Al	52	41	–	7
A-14	high-Al	54	43	–	3

VLT = very low-Ti; VHT = very high-Ti.
After Taylor (1982).

ilmenite) increase with increasing TiO$_2$ content. Most minerals are zoned; the compositional ranges of pyroxenes, olivines and plagioclases are summarised in Figure 7.8. Mare basalts do not contain Mg-rich orthopyroxene (found in Mg-suite plutonic rocks), but they have a very wide range of clinopyroxene compositions, extending to the 'forbidden' (Lindsley and Munoz, 1969) region of the hedenbergite–

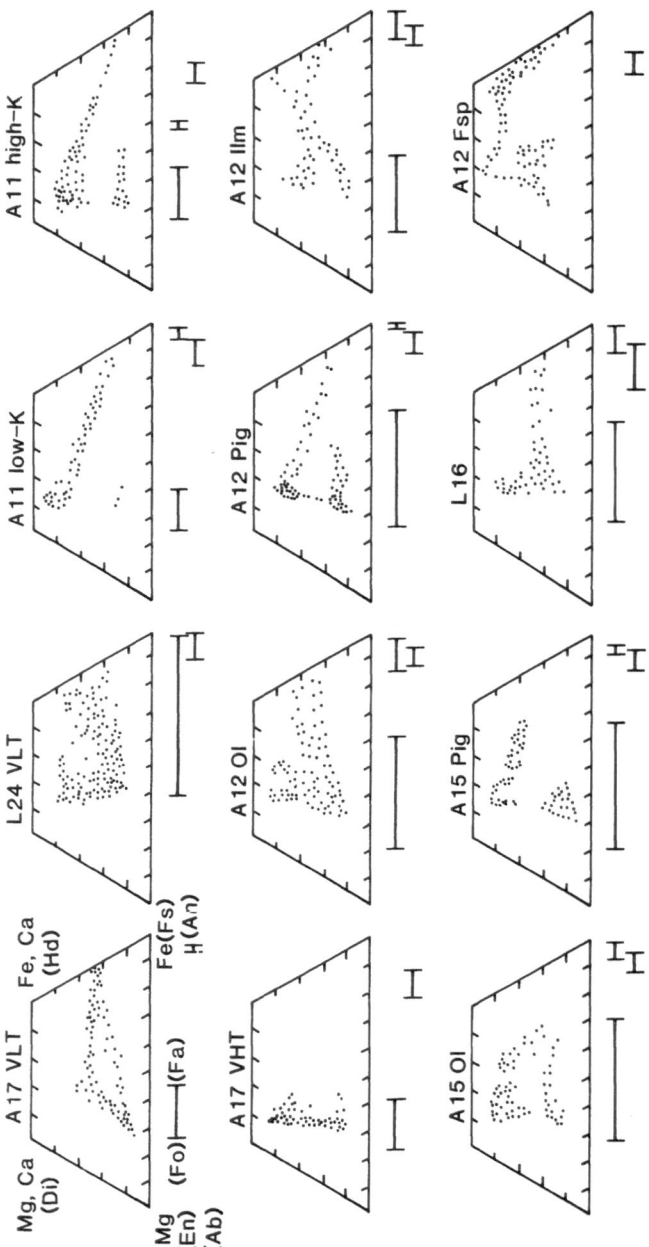

Figure 7.8 Mineral compositions in the major types of basaltic rocks retrieved from Apollo (A11, A12, A15, A17) and Luna (L16, L24) missions. Pyroxene compositions plotted in terms of their ionic Mg, (Mg, Ca), (Fe, Ca) and Fe (En, Di, Hd, Fs) proportions. Bars below the pyroxene quadrilaterals indicate ranges of olivine (upper bars) and plagioclase (lower bars) compositions. Basalt types: VLT: very low-Ti; VHT: very high-Ti; Ol: olivine; Pig: pigeonite; Ilm: ilmenite; Fsp: feldspathic. After Papike and Vaniman (1978).

ferrosilite join as the pyroxenoid pyroxferroite (Chao et al., 1970). Bence and Papike (1972) showed that in addition to the variation in quadrilateral components (Ca, Mg and Fe), Al, Ti and Cr abundances also make pyroxenes effective recorders of basalt crystallisation histories.

Olivine compositions range from Fo_{80} to fayalite. Some mare basalts have up to 36 modal % olivine (Papike et al., 1976) which always occurs as an early phase. Mare basalt feldspar compositions (Figure 7.8) have a much narrower compositional range (An_{99-75}) and are generally more calcic than their counterparts in terrestrial basalts (Papike, 1981). For example, ocean floor basalt plagioclase compositions range from An_{90} to An_{25}.

A wide variety of opaque phases is found in mare basalts. Ilmenite is the most common, and occurs in all basalt types. In most samples it is almost end-member $FeTiO_3$. MgO contents range up to 6 wt % (Papike et al., 1976) and probably reflect the bulk mg' of the parental liquid. Ilmenite compositions are also affected by whether it precipitates as an early (high-Ti basalts) or late phase (low-Ti basalts). Unlike the terrestrial variety, lunar ilmenite contains no Fe^{3+}.

A mineral associated with ilmenite in high-Ti basalts is armalcolite $((Fe_{0.5}Mg_{0.5})Ti_2O_5)$, named after the three Apollo 11 astronauts (ARMstrong, ALdrin, and COLlins) and first discovered in Apollo 11 samples (Anderson et al., 1970; Haggerty et al., 1970). It is an intermediate solid solution of $FeTi_2O_5$ and $MgTi_2O_5$. Because it tends to react with liquid to form magnesian ilmenite, it occurs only in relatively quickly cooled samples and is commonly mantled by ilmenite. Most of the Ti in armalcolites is Ti^{4+}, but Ti^{3+} is also present (Papike et al., 1989). Terrestrial armalcolite, recognised after the return of the lunar samples, has no Ti^{3+}.

Spinels are the second most abundant opaque minerals in mare basalts, some having as much as 12 modal % spinel. Their compositions can be approximately represented by the ternary system Fe_2TiO_4 (ulvöspinel)–$FeCr_2O_4$ (chromite)–$FeAl_2O_4$ (hercynite). Other cations include Mg, Mn, V and Zr. Spinel crystallisation trends begin with chromite, and then move toward ulvöspinel as FeO and TiO_2 contents increase and Al_2O_3, MgO and Cr_2O_3 decrease (Papike et al., 1989). Some basalts have spinels with titanian chromite cores and chromian ulvöspinel rims, with a sharp contact between the two, indicating an interruption of spinel crystallisation. For more details on the oxide mineralogy of mare basalts see Papike et al., (1976) and El Goresy (1976).

The only sulphide of any importance in mare basalts is troilite, essentially pure FeS. Its modal abundance is always less than 1% (Papike et al., 1989). Native Fe is also present in many mare basalts, indicative of the extremely low oxygen fugacities (below Fe-Wu) under which these rocks formed.

7.3.2 Volcanic glasses

In addition to the basalts described above, lunar mare volcanism is also manifested in the form of volcanic glass. Concentrations of glass beads were found at the Apollo 15 and Apollo 17 sites and through the study of soil and soil breccia samples from all the sampling sites, twenty-five varieties of volcanic glass have been identified (Delano, 1986). The compositions of a representative group are given in Table 7.4. They provide important clues to lunar basalt petrogenesis because they are more primitive than crystalline basalts, and have generally undergone less fractionation and crustal contamination than the basalts (Delano, 1986). Also, glass beads exhibit surficial

Table 7.4 Representative analyses of pristine lunar volcanic glasses*

	1	2	3	4	5	6	7	8	9	10	11	12
SiO_2	48.0	43.9	46.0	45.3	44.1	42.9	40.8	38.5	38.8	35.6	33.4	34.0
TiO_2	0.26	0.39	0.55	0.66	0.97	3.48	4.58	9.12	9.30	13.8	16.4	16.4
Al_2O_3	7.74	7.83	9.30	9.60	6.71	8.30	6.16	5.79	7.62	7.15	4.60	4.60
Cr_2O_3	0.57	0.39	0.58	0.40	0.56	0.59	0.41	0.69	0.66	0.77	0.84	0.92
FeO	16.5	21.9	18.2	19.6	23.1	22.1	24.7	22.9	22.9	21.9	23.9	24.5
MnO	0.19	0.24	0.21	0.26	0.28	0.27	0.30	n.a	0.29	0.25	0.30	0.31
MgO	18.2	16.9	15.9	15.0	16.6	13.5	14.8	14.9	11.6	12.1	13.0	13.3
CaO	8.57	8.44	9.24	9.40	7.94	8.50	7.74	7.40	8.55	7.89	6.27	6.90
Na_2O	n.d.	n.d.	0.11	0.27	n.d.	0.45	0.42	0.38	0.39	0.49	0.05	0.23
K_2O	n.d.	n.d.	0.07	0.04	n.d.	n.d.	0.10	n.d.	n.d.	0.12	0.12	0.16

*Arranged in order of increasing Ti content. n.a. = not analysed; n.d. = not detected; 1: A-15 green C; 2: A-16 green; 3: A-14 green; 4: A-17 VLT; 5: A-14 green A; 6: A-15 yellow; 7: A-14 yellow; 8: A-17 orange (74220-type), 9: A-17 orange; 10: A-15 red; 11: A-12 red; 12: A-14 black. After Delano (1986)

enrichments in many volatile elements, indicating that eruption was associated with a vapour phase. The glasses are thought to have been erupted in fire-fountaining events.

7.3.3 Age of mare basalts

Ages of mare basalts are important not only for determining the chronology of lunar basaltic volcanism, but also in defining the minimum ages for the large basins that contain the flows, and to calibrate relative ages obtained from crater counts.

Taylor (1982) showed that the ages of the major mare basalt units that have been sampled range from 3.86 ± 0.7 Ga to 3.08 ± 0.5 Ga, with some high-Al basalts from Apollo 14 at about 4.0 Ga. Dating of basalt clasts extracted from Apollo 14 breccias has extended the range of basalt ages to older than 4.2 Ga (Taylor *et al.*, 1983; Dasch *et al.*, 1987). However, it is not known how extensive basaltic volcanism was at that time. These basalts formed before the great bombardment ended, so the basins that these flows probably occupied have been obscured by later basins and ejecta. Although there are older samples and younger (unsampled) units, if the returned samples are an accurate indication, the main epoch of mare basalt volcanism was from 3.9 to 3.0 Ga.

7.3.4 Origin of mare basalts

Although there is no clear understanding of the origin of mare basalts, two theories that have been previously considered can now be ruled out. From their compositions we know that they are not impact melts of the crust (Taylor, 1982) and from their compositional diversity and negative Eu anomalies, a previously fractionated source is indicated, ruling out a primitive (chondritic) source of bulk moon composition.

A major unresolved controversy is their depth of origin. Based on volcanic glass data and experimentally determined phase equilibria, Delano (1980) concluded that all mare liquids were produced at depths between 400 and 500 km. However, Binder (1982, 1985) has proposed that the basaltic magmas were produced at depths of less than 200 km. This somewhat complicated model calls for a laterally heterogeneous source region, resulting from convection in the magma ocean. The early formed density-stratified rocks then partially melted to produce basaltic magmas whose contrasting Ti contents reflect those of the sources.

No simple model for mare basalt petrogenesis explains all the data. It is most likely that a heterogeneous cumulate source was melted at depths between 150 and 500 km. No mare basalts have been identified as having been derived from a primitive, unfractionated source. All bear the trace element imprint (e.g. negative Eu anomaly) of the early lunar differentiation event (Papike and Simon, 1984), including the 4.2 Ga clast described by Taylor *et al.* (1983). The presence of a negative Eu anomaly in such an ancient rock supports the theory that mare basalts were derived from Eu-depleted cumulate sources, and indicates that mare volcanism began soon after these sources formed.

7.4 KREEP basalts

Models for the lunar magma ocean incorporate three major products: anorthositic crust, mafic cumulates and a highly fractionated residuum between the two. The incompatible element-rich residuum is known by the acronym KREEP (for K, REE

Table 7.5 Bulk compositions of KREEP basalts

	A-12	A-15	A-15	A-16	A-16
Sample No.	12033	15382	15386	60315	62235
wt%					
SiO_2	48.2	52.4	50.83	46.84	47.05
TiO_2	2.0	1.78	2.23	1.39	1.19
Al_2O_3	15.5	17.8	14.77	17.24	18.88
FeO	10.6	8.6	10.55	8.86	9.45
MgO	7.7	7.1	8.17	13.81	10.16
CaO	10.9	9.9	9.71	10.50	11.60
Na_2O	0.6	0.96	0.73	0.51	0.42
K_2O	1.1	0.57	0.67	0.39	0.33
P_2O_5	0.8	0.55	0.70	0.48	0.39
ppm					
Rb	18.3	16.1	18.5	9.80	9.32
Sr	195	195	187	156	161
Ba	1207	793	837	445	568
La	105	79.5	83.5	45.5	60.1
Ce	265	212	211	113	153
Nd	149	127	131	71.3	94.3
Sm	40.5	35.2	37.5	20.1	27.1
Eu	3.11	2.77	2.72	1.89	2.03
Gd	50.0	42.9	45.4	23.8	32.2
Dy	—	45.7	46.3	26.3	35.0
Er	—	28.1	27.3	15.5	21.2
Yb	—	24.0	24.4	14.0	18.7
Lu	—	3.43	—	—	—

Major element data from Meyer (1977).
Trace elements from Hubbard and Gast (1971): 12033; Hubbard *et al.*, (1973): 15382, 60315, 62235; Hubbard *et al.* (1974): 15386.

and P). Although it has not been recognised as a rock type, it is a fragmental component of many lunar rocks and soils. Most KREEPy samples are polymict breccias, but igneous samples known as KREEP basalts have been recognised at most of the landing sites. Reviews of KREEP basalt are presented by Irving (1977), Meyer (1977), Dymek (1986) and Ryder (1987b).

Compositions of KREEP basalts are given in Table 7.5, which shows their enrichment in incompatible elements relative to other lunar rock types. Their REE abundances (Figure 7.1) are 100 to 200 times those of average chondrite, with distinct negative Eu anomalies, and the samples are also strongly enriched in Ba and Rb. Based on Rb-Sr and Sm-Nd model ages of c. 4.4 Ga (Lugmair and Carlson, 1978) and its relatively uniform composition (Warren and Wasson, 1979), the KREEP component is thought to have originated in the magma ocean.

In contrast with the mineralogy of the mare basalts, KREEP basalts have bronzite as the dominant pyroxene (Figure 7.9). In Apollo 15 (A-15) samples the orthopyroxene is zoned to more Fe-rich pigeonite, which is followed by Fe-rich, Ca-poor augite later in the crystallisation sequence.

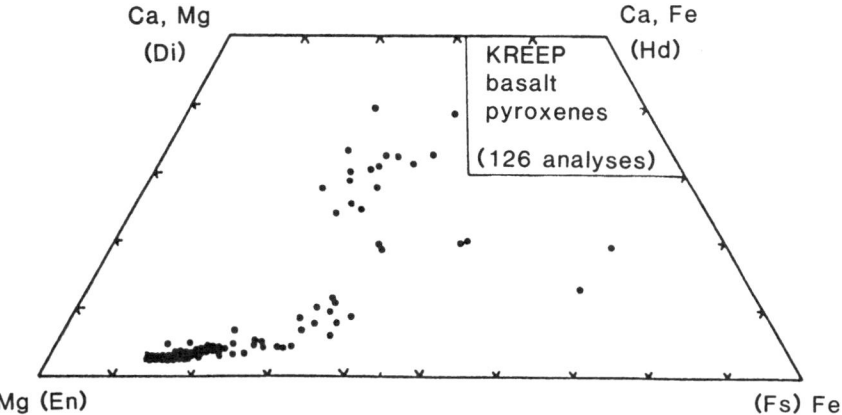

Figure 7.9 Pyroxene compositions in A-15 KREEP basalt. Note the predominance of Ca-poor pyroxenes in contrast to the pyroxene assemblages in the various other types of basaltic rocks shown in Figure 7.8. After Simon *et al.* (1988).

Plagioclase generally occurs as subhedral laths with compositions in the range An_{92}–An_{70}. Most grains are between $50\,\mu m$ and $500\,\mu m$, averaging about $200\,\mu m$ (Meyer, 1977). Despite the relatively K-rich bulk compositions (Table 7.5), the plagioclase does not have a significant Or component, except for the most sodic grains. Instead, in most samples, the K and other incompatible elements tend to concentrate in the Si-rich mesostasis. Fe-rich pyroxene, Ba-K feldspar and REE-rich phosphates (whitlockite and apatite) also occur in the mesostasis.

Although KREEP gives a model age of 4.4 Ga, the A-15 KREEP basalts give crystallisation ages of about 3.9 Ga and have been interpreted as impact melts, possibly related to the event that formed the Imbrium basin (Nyquist *et al.*, 1975; Taylor, 1982). However, there is a lack of meteoritic contamination in these basalts and recent petrographic observations indicate a two-stage cooling history (Ryder, 1987b), making a strong case for a volcanic origin for the A-15 KREEP basalts.

Another problem concerning the petrogenesis of the KREEP basalts is the interpretation of apparently contradictory results regarding the degree of partial melting required for their production (Dymek, 1986). This stems from their relatively primitive mineral compositions (Mg-rich orthopyroxene and Ca-rich plagioclase) which indicate high degrees of partial melting, combined with their incompatible element-rich bulk compositions which suggest small degrees of partial melting. It is possible that they originate from high degree partial melts of Mg-rich cumulates and picked up their strong KREEP component from the KREEP-rich residual melt layer of the magma ocean in areas where this is concentrated between the anorthosites and mafic cumulates.

7.5 Comparison with terrestrial magmatism

Comparing lunar and terrestrial basic magmatism is perhaps best done with an understanding of the similarities and differences of the bulk compositions of the Earth and Moon, because the bulk characteristics strongly affect the mineral and rock

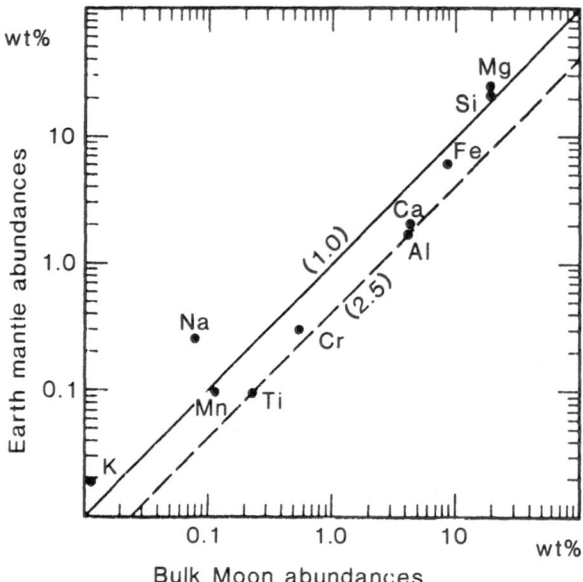

Figure 7.10 Comparison of major element abundances in the bulk Moon and primitive terrestrial mantle. Note the relative lunar depletion in Na and K and enrichment in Ti, Al and Ca. After Taylor (1982). Solid and dashed lines indicate bulk Moon/terrestrial mantle abundances of 1 and 2.5, respectively.

compositions. As one might surmise from the previous discussion of the major lunar rock types, the Moon is relatively depleted in Na and K and enriched in Ti, Al and Ca compared to the Earth (Figure 7.10). For the major elements in general the Moon is depleted in volatile elements and enriched in refractory elements. The lunar depletion in Na and K and enrichment in Ca accounts for the abundance of near end-member anorthite in the rocks, with only rare fractionation to more sodic compositions.

With respect to mare basalts, the lunar enrichments in Ti, Cr and Fe are important, as are their lowered oxidation states. Some Ti^{4+} is reduced to Ti^{3+}, some Cr from Cr^{3+} to Cr^{2+}, and Fe is present as Fe^{2+} and Fe^0, but not Fe^{3+}. These differences obviously affect their geochemical behaviour. Also, according to the data compiled by Taylor (1982), the Moon is uniformly enriched in many trace elements, including U, Th and the REE, but relatively depleted in Rb. Models for the formation of the Moon must account for all of the above compositional features.

7.5.1 *Anorthosites*

Terrestrial anorthosites are generally concordant intrusive plutons, commonly surrounded by gabbroic rocks (Basu and Pettingill, 1983). Most massif-type anorthosites are between 1.4 and 1.1 Ga old, whereas lunar anorthosites are c. 4.4 Ga old. Terrestrial anorthosites have a much wider range of grain sizes, from a few millimetres up to at least 30 cm (Ashwal *et al.*, 1983), than do lunar anorthosites and also have much wider ranges in mineral compositions. For example, most plagioclase in massif-type anorthosites is between An_{35} and An_{65} (Basu and Pettingill, 1983), and Ashwal et al. (1983) report a range from An_{92} to albite (including some metamorphic feldspar) in the 2.7 Ga Bad Vermilion Lake anorthosite complex in Ontario. The primary plagioclase composition is rather calcic (c. An_{80}) and typical of Archaean anorthosites (Ashwal et al., 1983) but not as calcic as that of the lunar ferroan anorthosites (An_{94-97}). This reflects the lunar depletion in Na.

Terrestrial anorthosites are thought to form by crystal accumulation (e.g. Basu and Pettingill, 1983; Ashwal et al., 1983), but mechanisms for their separation from their parent magmas are not well understood. Lunar anorthosites are thought to be adcumulates (e.g. Morse, 1982), grown downward at the underside of the crust at the top of the magma ocean (Figure 7.2). However, the mafic cumulates that would be complementary to the ferroan anorthosites have not been recognised. Also, as stated by Morse (1982): 'Despite its apparent simplicity, the process of generating and sustaining a low-density crust by crystal flotation turns out on close examination to be not so simple.' Efficient plagioclase separation and a remote cooling regime producing laterally-flowing supercooled magma are required.

7.5.2 Basalts

The basis for comparing lunar and terrestrial basalts is that both are endogenous melts and many terrestrial basalts, like the mare basalts, are extruded during fissure-type eruptions. This includes the voluminous ocean floor basalts erupted at mid-ocean ridges (MORB). Papike and Bence (1978) have compared mare basalts with MORB. Their data show that MORB has a wider range of mg' values and higher Al, Na and K contents than mare basalts. The Al depletion in the mare basalts probably reflects the nature of their source whereas the relative alkali depletion reflects planetary characteristics. On the other hand, mare basalts are relatively enriched in Ti and Cr. The Cr enrichment may be due to reduction of Cr^{3+} to Cr^{2+}. The latter is a large ion and tends to enter the melt more readily, an important factor during the formation of basaltic liquids. Schreiber and Haskin (1976) showed that pyroxene/liquid distribution coefficients for Cr decrease with decreasing partial pressure of oxygen.

Unlike mare basalts (described above), terrestrial basalts have no consistent negative Eu anomaly. This signifies the lack of plagioclase fractionation from the terrestrial basalt sources, and if plagioclase fractionation did occur on the Earth, it was not as predominant as it was on the Moon (Bence et al., 1980). A summary of the silicate mineralogy of planetary (terrestrial, lunar, meteoritic) basalts is given by Papike (1981). To a large extent, mineral compositions reflect bulk rock compositions. Mare basalt feldspars are restricted to calcic compositions (An_{95-75}), whereas terrestrial basalts have plagioclase as sodic as An_{20}. In contrast, mare basalt olivine can be end-member fayalite, but is generally between Fo_{80} and Fo_{40}. Archaean basalts have the most magnesian olivines (Fo_{95}) of the known planetary basalt suite (Papike, 1981). However, much younger terrestrial basalts also have magnesian olivines (Fo_{90}), showing that unevolved sources are not restricted to Archaean times. These compositions indicate a magnesian source, especially when compared to that of the mare basalts, whose most Mg-rich olivines are around Fo_{80}.

Not surprisingly, terrestrial basalt pyroxenes also tend to be more magnesian than mare pyroxenes (Papike and Bence, 1978; Papike, 1981). In terms of the quadrilateral components (Figure 7.8) mare basalt pyroxenes include pigeonites, augites and Fe-rich compositions extending to the Hd-Fs join (e.g. Papike *et al.*, 1976). On the other hand, many terrestrial basalt suites including modern MORB and Archaean basalts have mainly high-Ca pyroxenes, the latter suite being dominated by diopsidic pyroxene.

7.6 Concluding remarks

The lunar rock types described above are consistent with a very early Moonwide differentiation event. The exact form of this event, whether a single magma ocean or numerous contemporaneous events, is not known. The depth to which melting occurred is also uncertain.

Isotopic systems closed 4.4 Ga ago, indicating solidification of the magma ocean by that time. Impact resetting of these systems and ages of impact-induced melt sheets show that the Moon underwent heavy meteoritic bombardment from the time of its formation until 3.8 Ga ago, as debris left over from the formation of the solar system was swept up. Simply by virtue of its proximity in space and its stronger gravity, the Earth must also have been affected by this bombardment. Recent falls of lunar meteorites (e.g. Laul *et al.*, 1983) suggest that if a relatively recent lunar impact (less than 3.0 Ma; Tuniz *et al.*, 1983) could deliver a meteorite from the Moon, then the giant basin-forming impacts of 3.8 Ga ago could have delivered a large amount of lunar material to the Earth at that time.

One other point worth consideration is whether a terrestrial planet should be expected to form a magma ocean and produce a thick crust (i.e. the Earth is anomalous?) or a relatively thin crust is to be expected (i.e. the Moon is anomalous?). Warren (1985) discussed the possibility of the existence of a lunar-type magma ocean on the other terrestrial planets and concluded that only Mercury has any indication of an anorthositic crust. Based on studies of the Earth and inferences from basaltic meteorites, Walker (1983) suggested that a major lunar magma ocean is a planetary anomaly.

While there is no evidence for a terrestrial crust-forming magma ocean, models of magma oceans below the Earth's surface have been proposed. According to Walker (1983) an early terrestrial magma ocean would be buried by its own compressibility and the Earth's strong gravity. Nisbet and Walker (1982) have suggested that Archaean komatiites formed from leaks in a buried magma ocean and Andersen (1981) proposed a magma ocean origin for eclogite at depths of between 300 and 670 km.

These models, although devised for the Earth, would probably not have been possible without the knowledge we have gained from lunar samples. The lunar rocks correspond to a portion of the Earth's geological record that has been erased. Without them, our ideas of planetary formation and early evolution would be even more vague. By studying the Moon, we have been able to advance our understanding of the early Earth.

Acknowledgements

J.J. Papike and F.G. Gibb are thanked for their comments on an earlier version of this chapter.

8 Mineralisation associated with early Precambrian basic magmatism

S. ROBERTS, R.P. FOSTER and R.W. NESBITT

8.1 Introduction

The Archaean hosts a wide variety of mineralisation including both precious and base metals, Cr, Li, asbestos and semi-precious stones. Deposits of Ni, Cr, the platinum group elements (PGE), Cu, Au and asbestos are found in direct association with mafic and ultramafic magmatism. This chapter first considers Archaean metallogenesis associated with volcanic suites and reviews the classical association between Fe–Ni sulphide mineralisation and komatiitic suites and is followed by a discussion of the role of mafic and ultramafic rocks in the extensive occurrence of gold mineralisation in Archaean volcanic suites. A review of the mineralisation associated with basic intrusive complexes is also presented, focusing on the chromite and PGE mineralisation of the major early Precambrian layered intrusions, and in particular the Bushveld, Stillwater and Great Dyke complexes.

8.2 Nickel mineralisation: Archaean volcanic suites

Although Fe–Ni sulphides are found associated with mafic and ultramafic rocks within both Archaean and lower Proterozoic basic rocks, it is the older rock successions which are volumetrically and economically more important. Within the Archaean, the main sulphide mineralisation is found in two different associations. The first association is characterised by Cu-rich ores typically found in layered intrusions (e.g. Purvis *et al.*, 1972; Ross and Travis, 1981) and has analogues in younger intrusive gabbroic occurrences (e.g. the Duluth gabbro, Lynn Lake, Giant Mascot). The second association is between Ni-sulphides and komatiites and apart from isolated occurrences in the Lower Proterozoic this ore type appears to be restricted to the Archaean (Ross and Travis, 1981). As well as differences in the nature of their host rocks, the two associations are also distinguished by the nature of the sulphides, the gabbro association having high $Cu/(Cu + Ni)$ ratios ($ > 0.6$), and the komatiite-hosted sulphides low ratios ($ < 0.3$). The reader is referred to Naldrett (1981) for descriptions of younger examples of Ni mineralisation.

8.2.1 *The komatiite–sulphide association*

The association of Fe–Ni–Cu sulphides with highly magnesian volcanic rocks (komatiites) is found locally in areas of some Archaean greenstone belts. Such deposits were originally recognised in Canada and Zimbabwe but attracted little attention because they were small and their association with extrusive ultramafics was not recognised. The recognition and naming of komatiites in South Africa (Viljoen and Viljoen, 1969a) closely followed the discovery and descriptions of Ni sulphide mineralisation in ultramafic rocks of Western Australia (Woodall and Travis, 1969). Two decades of exploration have confirmed Western Australia as the most economically important region for this type of mineralisation (e.g. Gresham and Loftus-Hills, 1981; Marston et al., 1981). Other important regions include Zimbabwe (e.g. Williams, 1979; Chimimba and Ncube, 1986) and Canada (Green and Naldrett, 1981) but these are much smaller and economically less significant.

It has been suggested on the basis of examples from the Yilgarn Block that the komatiite–sulphide association can occur both in plutonic and volcanic associations (e.g. Marston et al., 1981). This subdivision is based on the presence or absence of volcanic features in some of the overlying, spatially-related komatiites, since, for the most part, the sulphides are always found within massive dunites. In komatiites, volcanic features include spinifex-textured olivines (Nesbitt, 1971), ultramafic hyaloclastite (Stolz and Nesbitt, 1981) and pillow flows (Viljoen and Viljoen, 1969a). For reasons which will become evident, the absence of such criteria is by no means conclusive evidence that a deposit is non-volcanic. For the sake of simplicity, this section will concentrate on deposits showing unequivocal volcanic-style characteristics.

8.2.2   Typical greenstone belts

Archaean greenstone belts essentially consist of metamorphosed volcano-sedimentary accumulations of ultramafic, basaltic and felsic igneous rocks, interlayered with sediments (e.g. Condie, 1981; Ayres et al., 1985). The sediments range from pyritic black shales, through volcaniclastics and clastics to chemical sediments (principally banded iron-formations). In any given greenstone belt, the nature and abundance of both igneous and sedimentary rock types vary enormously. This variation is also reflected in the associated mineralisation, with the ultramafic-dominated belts having the greatest potential for Ni mineralisation, whilst the felsic-dominated areas have a greater Cu–Zn–Au potential (e.g. parts of the Abitibi Belt, Canada). However, it is important to note that the Barberton greenstone belt, which is the type area for komatiites, has no known occurrences of major Fe–Ni sulphides.

Superimposed on these volcano-sedimentary piles is a complex history of metamorphism, deformation and granitoid emplacement. In any one belt, the metamorphism varies from lower greenschist to upper amphibolite facies. The komatiitic rocks are commonly made up of assemblages of serpentine, chlorite, tremolite, carbonate and talc, and unless original textures are preserved, the recognition of komatiites can be extremely difficult. In turn, deformation and metamorphism have major effects on the nature and distribution of the Fe–Ni mineralisation. In some cases (perhaps all) it would appear that these processes have been responsible for upgrading the concentration of the metal content of the ore body.

8.2.3   Kambalda and Scotia deposits

The Eastern Goldfields region of Western Australia contains the major deposits of the komatiite-hosted Fe–Ni association. By far the greatest concentration of these deposits

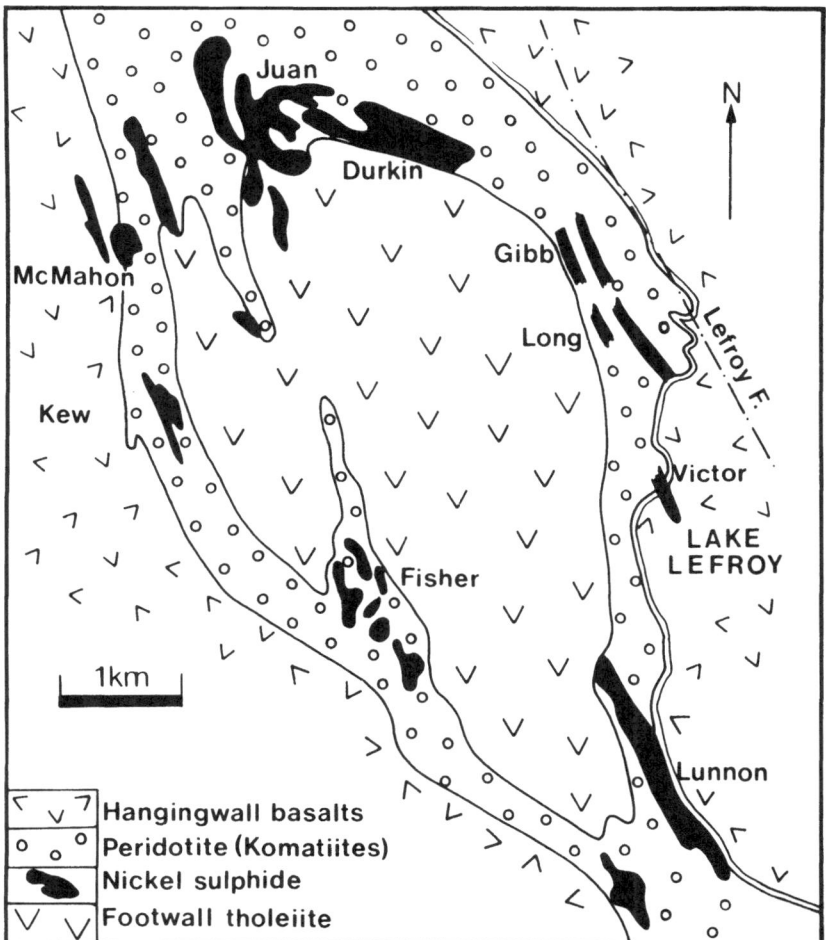

Figure 8.1 Simplified geology of the Kambalda Dome, Western Australia (after Keays *et al.*, 1981). The position of the named sub-surface sulphide shoots (which occur at the contact of the tholeiitic basalts and the overlying komatiites) is represented on the sketch map by projection to the surface. Granite intrusions omitted from the map for clarity.

occurs in the Kambalda Dome, south of Kalgoorlie. For comparison, the small deposit at Scotia (about 60 km north of Kambalda) is also described in this section.

The Kambalda Dome (Figure 8.1) has been described in several publications (e.g. Gresham and Loftus-Hills, 1981; Marston and Kay, 1980; Lesher *et al.*, 1984). It consists of a pile of tholeiitic basalts over 2 km thick of which the base is not exposed, overlain by over 1 km of komatiitic flows which in turn are overlain by high-Mg basalts, the so-called komatiitic basalts. Within the pile there are more than twenty important Ni sulphide deposits containing over 30 million tonnes of ore with an average grade of 3.6 wt% Ni. More than 80% of the ore shoots occur at the base of the komatiitic pile, at the footwall contact between the komatiites and tholeiites. About 17% of the deposits occur in a hanging wall position, almost always associated with

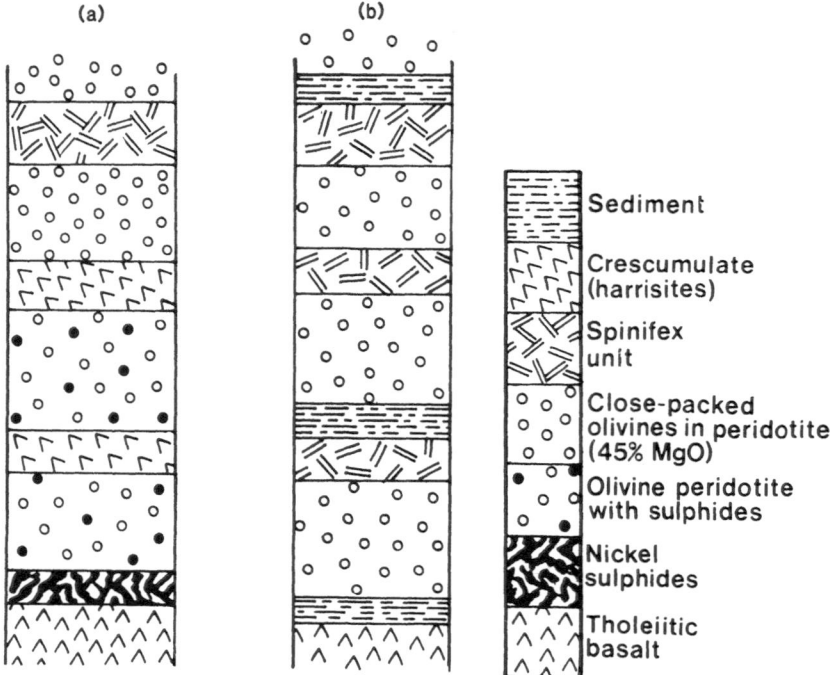

Figure 8.2   Two schematic sections through the Kambalda komatiite sequence (after Gresham and Loftus-Hills, 1982). (a) shows the positions of the Ni sulphides, the zone of sulphide-enriched komatiite (non-economic) and the zones of crescumulate olivine. These latter zones may provide evidence for a multi-stage emplacement of the flow (lava tube?). Note the absence of sediments. (b) shows a section away from the ore zone which is characterised by thin alternating sediments and flows.

footwall deposits and generally at the base of the second komatiitic flow unit. A minor percentage of the ore is found in tectonically transposed positions.

Since the greatest number and largest tonnage of ore occurrences are found at the base of the komatiite sequence, much research has been carried out on the exact nature of the host ultramafic and its contact with the underlying rocks. Interpretations of the host ultramafic (e.g. Lesher *et al.*, 1984; Cowden, 1988) suggest that it is a multistage volcanic complex, which acted as a long-lived feeder channel. These interpretations are based on textural variations through the flow in which alternations of cumulus and crescumulate olivines are found (Figure 8.2). Ross and Hopkins (1975) and Gresham and Loftus Hills (1981) described the massive ore as being confined to embayments or depressions at the contact. These structures are generally narrow (< 300 m) and are as much as 2.5 km long, with a direction and plunge of elongation roughly parallel to the regional fabric. Most researchers have interpreted the troughs as being of primary origin but Cowden (1988) believes that they result from polyphase deformation and suggests that the ore was originally deposited on a planar surface. Whatever the exact cause of the depressions, there is a remarkable regularity in the stratigraphic position of the massive ores which must be accounted for in any genetic model. Figure 8.3 shows a schematic cross-section through a typical Kambalda deposit. The whole assemblage

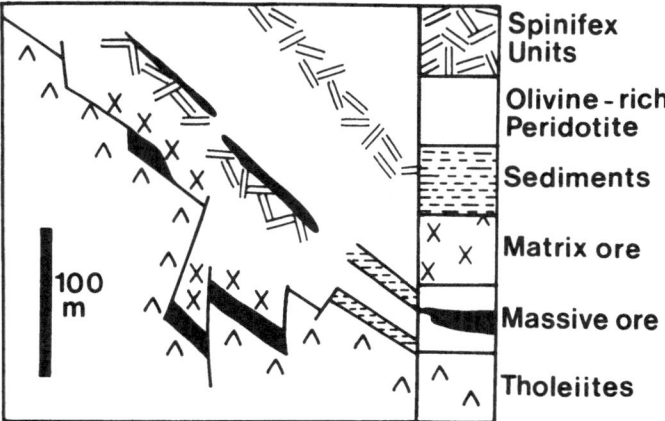

Figure 8.3 Schematic cross section through a typical komatiite-hosted Ni sulphide ore body at Kambalda (after Cowden, 1988). The ore bodies occur in trough structures, the origin of which is considered to reflect (a) original sea floor topography, (b) assimilation by hot komatiitic flows, or (c) a post-ore structure. The sketch shows the antipathy between sulphide and sediment.

has been deformed into a domal structure (Figure 8.1) which is cut by by granite and quartz porphyry intrusions.

In general, the principal ores can be classified as massive, matrix and disseminated types (e.g. Gresham, 1982). The massive ores consist of over 80% sulphides, matrix ores contain 40–80% sulphide and disseminated ores 10–40%. The ore zones themselves rarely exceed 10 m thick and more commonly are 1–3 m. Mineralogically, they are dominated by pyrrhotite and pentlandite with pyrite and minor chalcopyrite and spinel. Three other ore types are much less abundant. Breccia ore consists of a mixture of country rock fragments and massive ore and was apparently formed by the mechanical incorporation of the silicate rocks into the sulphides. This appears to have occurred by the ductile flow of the sulphides during deformation and is manifest by the pervasive foliation found in the sulphides. Blebby ore is known from a few localities and comprises a mixture of Ni-rich sulphide and Mg-rich silicates. Finally, spinifex ores are found in some hanging-wall positions where massive sulphide is in contact with spinifex-textured komatiites of the underlying unit (see Groves *et al.*, 1986). This ore type consists of a mixture of coarse-grained bladed/skeletal olivine and massive sulphide.

An important feature of the deposits is the variability in the grade of the ore. Major variations in S/Ni ratios occur between individual ore bodies and even within the same ore body (Gresham and Loftus-Hills, 1981). Average S/Ni ratios vary from over 4 to less than 2 with the high tenor massive ores (> 15 wt% Ni) having ratios of less than 2.5. Some massive ores are pyrrhotite-rich and as a consequence Ni values drop to as low as 8 wt%. Explanations for this range of S/Ni ratios include contamination by sulphidic metasediments, metamorphic overprinting, fractionation of the host silicate liquid, variations in the proportion of silicate to sulphide liquid (the so-called 'R factor' of Campbell and Naldrett, 1979) and variations in oxygen fugacity ($f O_2$). This problem is reviewed by Cowden and Woolrich (1987).

Interflow units of sulphidic metasediments form an important component of the lower part of the komatiitic section (Bavinton, 1981). The sediments appear to be a mixture of mafic and felsic detrital material, and consist of albite, quartz, chlorite and amphibole, plus pyrite and pyrrhotite. Although S contents are high (> 5%), Ni and Cu are always low (< 0.1% combined). The sediments are thin (< 5 m) and rarely extend for more than 500 m. A particularly important feature of the metasediments is their spatial distribution in relation to the Ni ores. Ross and Hopkins (1975) and Bavinton (1981) have pointed out that although the sediments are found within the same stratigraphic level as the Ni sulphides, they are almost entirely absent from ore zones. Indeed, there appears to be a zone between the sediments and the ores which is free of both components. This curious spatial relationship has important implications in terms of genetic models.

The Scotia deposit (Page and Schmulian, 1981; Stolz and Nesbitt, 1981) is 65 km north of Kalgoorlie and thus far is the only deposit found within an elongate ultramafic belt over 15 km in strike length. The deposit is remarkable because of the preservation of textures and mineralogy within the associated ultramafic rocks and could be taken as a 'type' example of the komatiite–Ni sulphide association. The Scotia deposit, in common with the Kambalda deposits, occurs in a footwall depression which confines the ore to a narrow trough. There is a strong tectonic control on the distribution of the ore body since the orientation of the trough coincides with a strong mineral lineation in the footwall basalts Like the Kambalda deposits, the footwall sequence is dominated by tholeiitic basalts, but at Scotia there is a laterally continuous sediment separating the ore from the basalts. This sediment is made up of epiclastic debris from a felsic volcanic centre and represents an important break between tholeiitic and komatiitic volcanism. The massive ore which overlies the sediment is about 0.8 m thick and passes upwards through matrix and disseminated ore, with the total ore zone being about 7 m thick.

An important feature of the Scotia deposit is the nature of the host peridotite, which is about 30 m thick. It is clear that this peridotite contains anomalous sulphide and Ni concentrations with values of over 0.5 wt% Ni immediately above the economic zone. Textural studies demonstrate the absence of a spinifex component and the unit is capped by sulphidic black slates and some altered quartz-rich clastics. Stolz and Nesbitt (1981) have suggested that the flow top of the host peridotite may have been eroded prior to the sediment deposition. It is important to note that the host peridotite appears to be confined between two sediment horizons and there does not appear to be the mutually exclusive ore–sediment relationship found at Kambalda. Above the hanging wall sediment there is a sequence almost 1 km thick of komatiites which are made up of alternating thin (a few metres) and thick (up to 100 metres) flows, all with excellent flow tops and many showing well-preserved ultramafic hyaloclastites and olivine crescumulates.

8.2.4 *Summary and genetic models*

The genesis of these komatiite-hosted Ni sulphide deposits has provoked major debate and a variety of models have been proposed. Before reviewing these it is worth noting the major characteristics of a 'type' deposit, bearing in mind that any plausible model must be able to explain all of these characteristics. Typical komatiite-hosted Ni sulphide deposits show the following features.

1. They occur in greenstone belts containing komatiite-dominated assemblages. The mineralisation almost always occurs at the base of the komatiite sequence. Since the majority of these komatiite piles are interpreted as sequences of lava flows, this means that the sulphides are associated with the base of the first (and more rarely the second) flow.

2. The presence of one sulphide ore body generally indicates the presence of others at the same stratigraphic level. These may be equally economically important, as in the original discovery area of the Kambalda Dome, or consist of lower tonnage, lower-grade deposits which are not economically viable (as in most other areas).

3. The contacts between the footwall and the mineralisation are sharp, with massive sulphides (generally a few metres thick) passing upwards through network or matrix sulphides which themselves give way to disseminated ore. The grade of the massive ore varies greatly both within and between deposits, and is controlled by the sulphide mineral assemblages of the ore bodies which in turn reflect their S/Ni ratios.

4. The ore bodies commonly appear to lie within major embayments on the footwall, although there is debate about this point (Cowden, 1988). The embayments can be over 1 km long, tens of metres deep and up to 300 m wide.

5. Many deposits appear to be situated in the thickest part of the komatiite sequence.

6. The ore bodies are almost invariably overlain by massive meta-dunite which hosts the network and disseminated mineralisation. The host dunites, which commonly reach 50–100 m thick, may show evidence of multistage emplacement (e.g. Lesher *et al.*, 1984) and may be capped by further massive dunite, spinifex-textured units or sediments (Gresham and Loftus-Hills, 1981; Stolz and Nesbitt, 1981).

7. Where hanging wall sulphides occur in subsequent komatiite flows they are found vertically (in a stratigraphic sense) above the footwall ore bodies and always at the base of flows which are stratigraphically close to the base of the volcanic pile (Gresham and Loftus-Hills, 1981).

8. In some cases, there appears to be a mutually exclusive relationship between sediments and ore zones.

In attempting to derive a coherent model for these deposits, the ultimate factor is the origin of the S and the point at which S saturation occurs. The solubility of S in silicate melts is very low (a maximum of *c*. 0.2 wt%) and is critically dependent on a number of factors (Haughton *et al.*, 1974). In general, the solubility decreases with decreasing temperature and decreasing Fe/Mg ratio of the silicate liquid. This means that the high liquidus temperatures of komatiites give them the potential to carry higher than average dissolved S, but their low Fe/Mg ratio acts in the opposite way. It is thus likely that as the komatiite liquid cooled, S saturation would rapidly be attained and the S would appear in an immiscible liquid. This immiscibility mechanism is fundamental to almost all models of komatiite-hosted nickel deposits.

Early models proposed that the S was mantle-derived and that sulphide saturation was reached either in mantle magma chambers or during the passage of the magma through the crust (Figure 8.4). In such models, the dense sulphide liquid rapidly settled

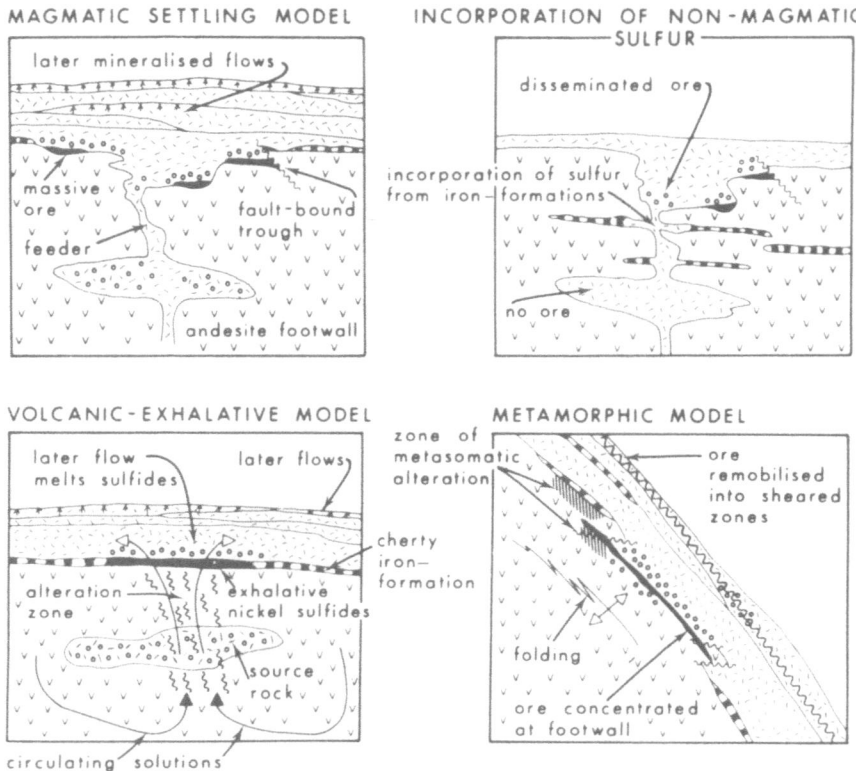

Figure 8.4 Various models for the formation of Fe-Ni sulphide mineralisation associated with komatiites (after Green and Naldrett, 1981).

to the base of the flow to be followed by olivine crystals which were either intratelluric or cumulus in origin. Such models therefore envisaged a crystalline ultramafic rock floating on a pool of molten sulphide. The 'billiard ball' model of Usselmann *et al.* (1979) attempted to model the buoyancy of this assemblage and suggested that as the pile of olivine crystals accumulated, the lower part of the pile would be forced down into the molten sulphide producing the matrix ore zone.

Models involving mantle-derived sulphides have been criticised on the basis of the large density discrepancy between the sulphide (SG = *c*. 4.0) and the carrier liquid (SG = *c*. 2.7) which would make the transport of previously segregated sulphide liquids by silicate liquids extremely difficult. This problem is perhaps overcome by particularly vigorous eruption rates which would disperse the sulphide as fine droplets but which would not explain why the sulphides were erupted in the first ultramafic flow. Alternative models are based on crustal derivation of S, either by a volcanic-exhalative model (Lusk, 1976) or by the contamination of the komatiite liquid by sulphidic sediments (e.g. Hopwood, 1981). Komatiitic liquids have a very low viscosity, high density, high liquidus temperature and high heat content (Huppert *et al.*, 1984). The net effect is that a komatiitic liquid would be turbulent and therefore capable of maintaining liquidus temperatures at its base. This behaviour contrasts with basaltic liquids which display laminar flow and develop basal chill zones. Huppert *et al.* (1984) suggested that because of this turbulent flow, komatiites would have been capable of melting and assimilating the footwall rocks over which they flowed, and thus, if the basal komatiite flowed over sulphidic sediments, S would be incorporated into the silicate liquid. This contamination would produce immediate S saturation and the formation of an immiscible sulphide liquid. Given the flow turbulence and the very strong tendency of Ni to enter sulphide liquid in preference to silicate liquid, the net effect would be a Ni-enriched sulphide liquid.

This model has many attractive features. It explains the trough structures as thermal erosion features and the antipathetic sediment–sulphide distribution at Kambalda as the result of assimilation of the sediment. However, the model has been criticised by Claoue-Long and Nesbitt (1985) on the basis of the lack of field evidence for the phenomenon and, more particularly, the presence of komatiites in contact with sediments without evidence of melting. These objections are overcome if contamination occurs on the way to the surface, for which there are two lines of evidence. The first is the presence of xenocrystic zircons in the Kambalda komatiitic basalts (Compston *et al.*, 1985), indicating that not only have they passed through continental crust but also that there has been large-scale assimilation. The second is that such assimilation would produce a major change in the chemistry of the komatiite and may therefore be the mechanism by which komatiitic basalts are produced (Barley, 1987).

If crustal assimilation by ascending komatiitic magmas is the process of Ni sulphide genesis, several problems still remain. The first is the sporadic distribution of sulphide deposits in the Archaean, despite the very widespread occurrences of komatiites, sulphidic sediments and komatiitic basalts. The second concerns the mechanism by which dense sulphides can be transported to the surface within a low-viscosity silicate liquid. It is clearly necessary to model contamination processes in komatiites and komatiitic basalts particularly in relation to sulphur and metal species.

8.3 Gold mineralisation in volcanic suites

8.3.1 *Gold production from basic rocks*

Gold is mined from Archaean terranes in many countries and regions, the most important of these being the Yilgarn Block in Australia, the Superior Province of Canada, the Brazilian Shield of South America and the Zimbabwe and Kaapvaal cratons in southern Africa. Lesser, but nevertheless important, quantities of gold are also recovered from small cratons and Archaean supracrustal remnants in West Africa and the USA and some economic potential exists in northern Europe. The Archaean craton of North China is a major source of gold, but unlike the other ancient terranes, the majority of the gold deposits here were formed during late Mesozoic deformation and anatexis of the Precambrian crust (Sang and Ho, 1987).

Despite the dominance of granitoid rocks in most Archaean terranes, the bulk of all Au production has invariably come from the volcano-sedimentary sequences which constitute the greenstone belts. Structurally-controlled lode-type deposits are most

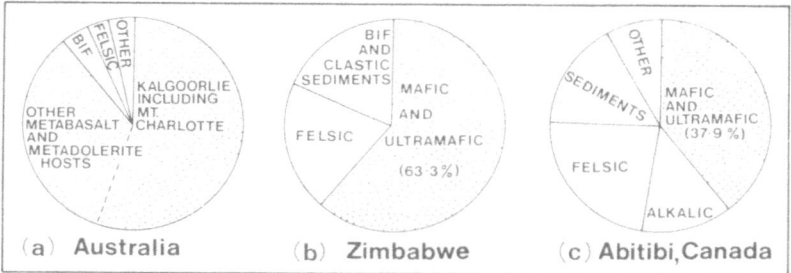

Figure 8.5 Proportions of Archaean Au deposits associated with different host rocks. Most deposits are structurally controlled. Note that (a) and (b) indicate Au production but (c) incorporates only the percentages of mines. Data sources: (a) Groves *et al.* (1985); (b) Foster and Wilson (1984); (c) Colvine *et al.* (1984, after Hodgson, 1983).

important and these are most commonly hosted by mafic and ultramafic rocks (Figure 8.5).

The importance of basic host rocks in Western Australia is greatly biased by the huge production from the Kalgoorlie deposits hosted by the Golden Mile differentiated tholeiitic dolerite (Groves *et al.*, 1985). Nevertheless, significant production has also come from deposits in other basalt-dominated sequences (Figure 8.5). Mafic volcanics are important host rocks to major Au deposits in the Abitibi greenstone belt of Canada but Hodgson (1983) and Colvine *et al.* (1984) have emphasised that lode deposits have been discovered and mined in a very wide range of rock types. The dominance of extrusive mafic rocks in most greenstone belts undoubtedly introduces a bias to the Au production statistics if taken at face value. However, the consistent association of Au and mafic rocks clearly indicates an important link between Au metallogeny and the products, if not the processes, of basaltic magmatism. In Zimbabwe, 61% of the lode gold (excluding BIF-hosted deposits) has come from mafic sequences and more than half of this has been recovered from tholeiitic basalts of the late Archaean Basaltic Units (Foster *et al.*, 1986).

8.3.2 Lode deposits

The auriferous lodes range from sulphide-impregnated ductile shear zones, through complex shear zone vein arrays, to massive extensional quartz veins (Table 8.1). The sense of displacement in the vertical plane is commonly reversed, and there is evidence that some lodes were generated or modified by thrusting (e.g. the Antelope Mine, Zimbabwe: Phaup, 1932; Vearncombe *et al.*, 1988), although a prominent strike-slip component is usually evident. Fe sulphides (commonly pyrite, less commonly pyrrhotite) together with arsenopyrite, base metal sulphides and scheelite are usually evident. The gold occurs as free grains in quartz and occasionally in carbonate minerals, as fine inclusions in the sulphide minerals, and possibly in solid solution in some sulphide minerals, notably pyrite, pyrrhotite and arsenopyrite (Henley, 1975; Cabri, 1987).

Alteration assemblages in the basaltic wallrocks commonly comprise ferruginous carbonates (siderite or ankerite) and either K-mica or albite, passing outwards from the lode into assemblages dominated by chlorite + calcite (Table 8.1; Colvine *et al.*, 1984).

High-Mg rocks also often exhibit evidence of intense carbonatisation, developing magnesite halos within 1–5 m of many lodes and a wider envelope of talc–carbonate (commonly calcite).

Fluid inclusion and stable isotope studies have revealed a fairly consistent depositional temperature regime of 450–250°C and pressures of 1–3 kb (Fyfe and Kerrich, 1984; Smith et al., 1984; Ho et al., 1985; Ho, 1987). The hydrothermal fluids were generally low salinity (< 4 wt% equiv NaCl), CO_2-rich, near-neutral aqueous fluids from which the Au precipitated in response to chemical exchange between fluid and wallrocks, increases in pH due to protonation of silicate minerals, cooling of the fluids and changes in the buffer capacities of the fluids (Foster et al., 1979; Groves et al., 1985; Ho, 1987; Foster, 1988).

8.3.3   Crustal setting

Most Archaean Au lode deposits are related directly or indirectly to major crustal lineaments which tapped aqueous fluids released by prograde greenschist to amphibolite facies metamorphism of the volcano-sedimentary sequences (Kerrich and Fryer, 1979; Colvine et al., 1984; Hodgson, 1986). These zones of dislocation channelled Au-bearing fluids into the upper regions of the Earth's crust where the sub-amphibolite facies P–T–X environment was conducive to Au precipitation. The temporal coincidence between late Archaean magmatism, hydrothermal activity and cratonisation may even have been linked to CO_2 degassing of the mantle, granulitisation of the lower crust, and the transfer of CO_2–H_2O fluids, Au and silicate magmas along the structural conduits into the middle and upper crust (Colvine et al., 1988).

The same crustal dislocations were responsible for ensialic rifting and the development of the late Archaean (2.7 Ga) supracrustal komatiitic–tholeiitic–calc-alkaline volcanic successions and derived sediments that are evident in most Archaean terranes. Such rift-type sequences are particularly important hosts for a range of Au and other deposit types (Groves and Batt, 1984; Foster, 1985), although the lode-forming hydrothermal activity was synchronous with at least one later phase of compressional tectonism. Oblique compression during the closure of a greenstone basin has been suggested for the highly productive Norseman–Wiluna Belt of the Yilgarn Block in Western Australia (Groves, 1988).

8.3.4   Role of basic magmatism

From the foregoing brief review it is evident that Archaean Au lode deposits are not exclusively hosted by basic rocks, nor is their genesis linked in any apparently unique way to basic magmatic processes. Nevertheless, there is a clear spatial and temporal link between the evolution of basalt-dominated greenstone sequences and gold metallogeny. To understand this link it is necessary to determine whether the basic rocks merely played a passive, albeit highly effective, role during the structural and hydrothermal evolution of the lodes or whether magmatic processes were more directly responsible for the transfer and/or concentration of gold in the Earth's crust.

Passive role: chemical and structural controls   The solubility of Au in low-salinity hydrothermal fluids at temperatures up to at least 350°C is controlled by thio-complexing, with $Au(HS)_2^-$ being the most stable aqueous species at near-neutral pH

Table 8.1 Characteristics of major Archaean lode-type gold deposits hosted predominantly by mafic rocks

Deposit	Gold production	Lode-type	Ore minerals	Host rocks	Alteration (P – proximal; D – distal)	Reference
Yilgarn, Australia						
Golden Mile (Kalgoorlie)	~1200	Steeply dipping brittle-ductile shear-zone sets and tensional fracture set	Pyrite, chalcopyrite; minor arsenopyrite; late tellurides	Tholeiitic dolerite sill	(P) muscovite + pyrite → ankerite/siderite → (D) chlorite + calcite ± ankerite	Groves et al. (1985), Phillips (1986)
Sons of Gwalia (Leonora)	>89	Quartz veins and lenses in shear zone	Pyrite, chalcopyrite, pyrrhotite, arsenopyrite	Tholeiitic dolerite	(P) Muscovite + pyrite → (D) carbonate	Louthean (1983), Groves et al. (1985), Phillips (1985)
Mararoa-Crown (Norseman)	>69	Folded laminated quartz veins	Pyrite, galena; minor tellurides	Tholeiitic basalt	Biotite + carbonate and chlorite + carbonate	
Abitibi, Canada						
Hollinger (Timmins) (Schumacker)	>594	Veins, stockworks, sinuous lodes	Pyrite, pyrrhotite, galena, sphalerite	Mafic lavas; minor felsic flows, pyroclastics and quartz-albite porphyry	(P) Sericite + ankerite → chlorite + ankerite → (D) chlorite + calcite + epidote + actinolite (quartz and albite ubiquitous)	Hodgson and MacGeehan (1982), Smith et al. (1954)
Kerr-Addison (Larder Lake)	>320	Stockwork, ladder veins, and schistose replacement zones	Pyrite, scheelite, minor arsenopyrite	High-Mg basalts, tholeiitic basalts; minor albitite dykes, clastic sediments, syenitic intrusions	(P) Ankerite → albite → dolomite + muscovite → (D) chlorite + dolomite/calcite	Kishida and Kerrich (1987), Colvine et al. (1984), Hodgson and MacGeehan (1982),

Sigma (Val d'Or)	> 110	Subvertical veins and breccias in ductile shear zone; flat veins	Pyrite, pyrrhotite; minor chalcopyrite, sphalerite, galena, tellurides, scheelite	Intrusive dolerite plug and andesitic flows; minor feldspar porphyry dykes	(P) Calcite →calcite + white mica → (D) calcite →white mica + chlorite	Robert and Brown (1986a,b)
Zimbabwe						
Cam and Motor (Kadoma)	> 146	Steeply dipping veins, shear zone vein arrays and stockworks	Pyrite, arsenopyrite, stibnite, sphalerite, scheelite	Tholeiitic basalt/andesite; high-Mg rocks; dolerite intrusions; minor clastic sediments.	(P) Quartz + ankerite; serpentine in high-Mg rocks	Wiles (1957), Collender (1964), Foster et al. (1986)
Phoenix (Kwekwe)	~ 105	Quartz-veins; minor stockworks and siliceous impregnations	Pyrite, stibnite, arsenopyrite, sphalerite, galena	Dunite-peridotite intrusive complex	(P) Magnesite (breunnerite) ± fuchsite→(D) talc + carbonate	Foster et al. (1986), Macgregor (1932), Harrison (1970),
Dalny* (Kadoma)	> 50	Steeply dipping ductile-brittle shear zone	Pyrite, arsenopyrite galena, chalcopyrite, sphalerite, scheelite	Tholeiitic basalt/andesite flows	(P) Pyrite →ankerite → white mica + chlorite →albite →(D) chlorite + calcite	Foster et al. (1986), Carter (1988)

*Total production will have now exceeded the latest published statistics

(Seward, 1973, 1984). The destabilisation of such complexes and precipitation of gold can be achieved by an increase or decrease in fluid pH due to wallrock alteration reactions of the type (Fyfe and Kerrich, 1984):

$$3NaAlSi_3O_8 + K^+ + H^+ \rightleftharpoons KAl_3Si_3O_{10}(OH)_2 + SiO_2 + 3Na^+ \qquad (8.1)$$

(albite)   (muscovite)

$$3(Mg, Fe)_4Al_4Si_2O_{10} + 6Ca_2Al_3Si_3O_{12}(OH) + 6SiO_2 + 24CO_2 + 10K^+ \rightleftharpoons$$

(aluminous chlorite)   (epidote)

$$10KAl_3Si_2O_{10}(OH)_2 + 12Ca(Mg, Fe)(CO_3)_2 + 10H^+ \qquad (8.2)$$

(muscovite)   (ferroan dolomite)

Even more importantly, the hydrothermal introduction of S-bearing complexes leads to sulphidation of Fe species in the wallrocks and the concomitant precipitation of Fe sulphides and Au:

$$2Au(HS)^{2-} + FeO \rightarrow 2Au^0 + FeS_2 + 2HS^- + H_2O \qquad (8.3)$$

(silicate)

$$2Au(HS)^{2-} + FeTiO_3 \rightarrow 2Au^0 + FeS_2 + TiO_2 + 2HS^- + H_2O \qquad (8.4)$$

(ilmenite)   (rutile)

The association of Au with such alteration products is well documented (Foster et al., 1979; Phillips, 1985, 1986; Robert and Brown, 1986a, b) and the feasibility of such reactions leading to Au precipitation has been confirmed thermodynamically (Neall, 1987). These and other reactions can also be redox-driven with reduction of the fluids being a particularly important mechanism, although under certain circumstances even oxidation can lead to precipitation of gold (Henley, 1984).

The nature and extent of fluid–rock interaction is dictated by the response of basic and other rocks to deformation, and this depends on a number of complex parameters, the more important being strain rate, prevailing grade of metamorphism, and the competence characteristics of associated rock types which may or may not preferentially take up the strain. However, these factors cannot explain the basalt–gold association on anything other than a local scale. Nonetheless, high-Mg rocks are particularly prone to hydrothermal alteration. The significant volume increase that accompanies serpentinisation (Coleman, 1971) can lead to enhanced permeability within volcanic sequences. More importantly, the introduction of CO_2-rich fluids can generate talc which promotes ductile behaviour in the high-Mg rocks and this in turn may facilitate fluid access and thus increase the potential for generating lode gold deposits.

Active role   Basic magmas may have contributed directly to Au metallogeny in one of three ways: (i) by magmatic transfer of Au across the mantle–crust boundary and into upper crustal regions; (ii) concentration by magmatic processes within the Earth's crust; (iii) concentration by hydrothermal processes related to the magma's thermal flux and site of emplacement. Ancient and modern mafic rocks, both volcanic and intrusive, tend to be slightly enriched in Au compared to their felsic counterparts, although the actual Au content rarely exceeds 10 ppb, compared with a crustal average of 1–2 ppb (Tilling et al., 1973; Crocket, 1974). Au tends to be slightly enriched in the mafic silicate minerals (Tilling *et al.*, 1973; Crocket, 1974) and thus there is considerable potential for hydrothermal Au mobilisation during prograde metamorphism. However, the highest Au contents are almost invariably present within accessory Fe oxide and sulphide minerals (Kwong and Crocket, 1978; Saager *et al.*, 1982; Keays, 1984), the sulphide phases being particularly vulnerable to hydrothermal breakdown.

There is little available information to indicate the Au content of the Archaean mantle. Consideration of core–mantle and hence metal–silicate fractionation processes intuitively suggest that Au and other siderophile elements would have partitioned into the molten core leaving a severely depleted mantle. However, Au contents of mantle rocks (10 ppb) are higher than those predicted by modelling of such partitioning processes, a factor attributed by Arculus and Delano (1981) to the accretion of a chondritic component subsequent to core segregation. Partial melting of the Archaean mantle seems to have led to slight Au enrichment (1.5–3 times) in komatiitic liquids (Brugmann *et al.*, 1987; see Table 8.2). The primary Au content of many komatiitic spinifex-textured flows is generally low (Table 8.2) and on the basis of Pd/Au ratios, Keays (1984) argued that considerable quantities of Au must have been removed during seawater alteration of the volcanic rocks. Nevertheless, exceptionally high Au contents have been reported from some unaltered komatiitic flows in the Mberengwe (previously 'Belingwe') greenstone belt of Zimbabwe (Table 8.2).

A possible link between calc-alkaline lamprophyre dykes and Archaean (and younger) Au lode deposits has recently been noted (McNeil and Kerrich 1986; Rock *et al.*, 1987, 1988). Such rocks almost certainly originated from the mantle and their coincidence with auriferous lode deposits at the very least suggests that magmas and fluids utilised the same conduit. High Au contents (0.1 to > 1000 ppb, arithmetic mean = 87 ppb) are evident in a number of these lamprophyres and Rock *et al.* (1988) have suggested that Au may have been transferred metasomatically from the Earth's core into the upper mantle prior to the intrusion of the lamprophyres into the crust. However, subsequent contamination by hydrothermal processes in the crust cannot be discounted (McNeil and Kerrich, 1986).

The chalcophile characteristics of gold dictate that it partitions into the sulphide phase, together with Cu, Ni and the PGE, whenever an immiscible Fe–S–O liquid separates within a silicate magma. Most mafic magmas, including MORB-type basalts, become S-saturated early in their evolution and so are effectively stripped of much of their gold. However, the high-temperature (up to 1600°C), high-Mg komatiitic magmas only attained S-saturation immediately prior to or during their ascent through the Earth's crust (Keays, 1984). Thus their Au was transported into subvolcanic and even volcanic environments as an integral component of the magma-entrained immiscible sulphide melt and subsequently accumulated with Cu–Ni–Fe–S minerals in the basal portions of sills and flows. Total Au contents of the crystalline sulphide assemblages are of the order of 100 to 1000 ppb, with much of the precious metal directly associated with either pyrite or chalcopyrite (Table 8.2). Such Au contents are often of economic significance to Ni mining operations.

Larger intrusive bodies, although not necessarily formed from high-Mg liquids, also exhibit S-saturation phenomena and the accumulation of important Cu–Ni–Au–PGE assemblages. The 2.5 Ga Great Dyke of Zimbabwe is regarded by Wilson (1979) as an abortive attempt to form a greenstone belt. Its parental liquid contained around 15% MgO (Wilson, 1982) yet it hosts a number of Cu–Ni sulphide horizons which are enriched in Au (Wilson and Prendergast, 1987). The younger (2.0 Ga) Bushveld

Table 8.2 Au content of Archaean komatiite-hosted magmatic sulphides for spinifex-textured komatiites and magmatic sulphide samples

Location	Sample type	Au content (ppb)	Key mineralogy
Spinifex-textured komatiites			
Kambalda (W. Australia)	—	0.63–25.1	—
Mt Clifford (W. Australia)	—	0.18–2.96	—
Freds Flow (Abitibi)	—	0.80–5.94	—
Barberton (S. Africa)	—	0.10–0.58	—
Mberengwe (Zimbabwe)	—	1.50–56.2	—
Magmatic sulphides			
W. Australia (Kambalda)			
Lunnon	Carted ore	1009–7215	po, pe, (py, cp)
Hunt	Carted ore	4308–4857	po, pe, (py, cp, mt)
Juan	Carted ore	399–686	po, pe, (py, cp, mt)
Fisher	Carted ore	2613–6286	po, pe, (cp, mt)
Long	Carted ore	351–588	po, pe, (py, cp, mt)
Mt Edwards	Carted ore	420	po, pe, (py, cp, mt)
McMahon	Carted ore	278–453	po, pe, (py, cp)
Ken	Carted ore	656	po, pe, (cp)
Durkin	Carted ore	570–669	po, pe, (cp, mt)
Jan	Carted ore	258–705	po, pe, (py, cp)
Carnilya	Carted ore	270–543	po, pe, (py, cp)
Nepean	Carted ore	277	po, pe, (py, cp)
Nepean	pyrrhotite	14.2–33.5	
Nepean	pentlandite	53.9–241	
Nepean	pyrite	83.1–1161	
Nepean	chalcopyrite	301–885	
Canada (Abitibi)			
Dundonald	DIS in interflow breccia and peridotite	495	pe, po, hz
Texmont	DIS in komatiitic flows	681	pe, po
Hartz	DIS, NET	67	po, pe
Alexo main pit	DIS, NET, MAS	301	po, pe
Alexo, small pit	DIS	76	po, pe
Zimbabwe			
Shangani	MAS	56	po, (pe)
	DIS in flow	312	pe, (po)
Trojan	MAS	56	po, pe, cp
	DIS in flow	176	po, pe, cp
Epoch	MAS	155	mi, py, vp
	DIS in altered sill	695	?

Komatiite data from Crocket and MacRae (1986); Kambalda data from Cowden *et al.* (1986); mineral data from Keays *et al.* (1981); Abitibi data from Barnes and Naldrett (1987); Zimbabwe data from Viljoen and Bernasconi (1979), Naldrett (1981), Baglow (1986) and Chimimba and Ncube (1986).
DIS: disseminated sulphides; MAS: massive sulphides; NET: net-textured sulphides. Mineralogy: dominant minerals; hz: heazlewoodite; mi: millerite; pe: pentlandite; po: pyrrhotite; py: pyrite; vp: violarite-polydymite; minor minerals (in parentheses): cp: chalcopyrite; mt: magnetite.

Complex in South Africa is a source of PGE and associated Au, and a link between sulphide immiscibility and the separation of chromite is becoming evident (McCarthy et al., 1984; Lee and Fesq, 1986). These and similar intrusive bodies are considered later in this chapter.

The emplacement of high-Mg magmas into middle and upper levels of the early Precambrian crust may well have represented an important intermediate step in Archaean gold metallogeny. The extrusion of komatiitic magmas, bearing an Au-enriched immiscible sulphide phase, onto the sea floor would have led to the rapid thermal cracking of the flows and the leaching of gold from the reactive sulphide accumulations by hot sea water (Keays, 1984). Pillowed flows would have been particularly susceptible to such leaching processes (Keays and Scott, 1976). Such processes may well account for the presence of Au-rich interflow sediments in Archaean volcanic sequences (Bavinton and Keays, 1978; Kerrich et al., 1981) and would also explain the anomalously low Au contents in many komatiitic rocks (Keays, 1984; Table 8.2).

Low concentrations of Au in komatiite-hosted Ni sulphide ores have been reported from a number of Ni deposits and can almost invariably be attributed to synmetamorphic hydrothermal leaching, Au commonly being concentrated in adjacent quartz veins (Keays et al., 1981; Cowden et al., 1986; Barnes and Naldrett, 1987). One such vein currently being mined for Au at the Hunt Mine, Kambalda, Western Australia, occurs immediately beneath Ni–Cu sulphide ore (Phillips and Groves, 1984). The lode terminates upwards against a talc–magnesite–chlorite schist containing Fe–Ni–Cu sulphides and it is probable that at least some gold was derived from magmatic sulphides in the volcanic sequence. Thus Au-enriched sulphide phases in high-Mg volcanics and inter-flow sediments in some instances probably contributed significantly to the genesis of individual Au deposits and to the hydrothermal Au budget of certain greenstone belts.

8.3.5   Gold and tholeiitic versus calc-alkaline magmatism: an Archaean enigma?

In more thoroughly studied cratons it is evident that the majority of the lode gold deposits were formed during late Archaean (c. 2.7 Ga) low- to medium-grade regional metamorphism, deformation and synkinematic tonalitic–trondhjemitic–granodioritic (TTG) magmatism (Colvine et al., 1984, 1988; Foster, 1985). This event marked the culmination of komatiitic → tholeiitic → calc-alkaline volcanicity which can be attributed, in Zimbabwe at least, to prolonged partial melting of a homogeneous peridotite source (Hawkesworth and O'Nions, 1977) and emplacement into an intra-continental rift environment (Nisbet et al., 1981). The ensuing metamorphism, deformation and crustal de-watering overprinted the complete, late Archaean volcano-sedimentary succession in Zimbabwe, yet Au lode deposits are abundant in tholeiitic basalt sequences but are almost completely absent from the 4 km thick pile of calc-alkaline rocks (Foster, 1985; Foster et al., 1986).

The explanation for the tholeiitic–calc-alkaline dichotomy in Au metallogeny in Zimbabwe is unlikely to be related to the primary Au contents of the magmatic rocks, which tend to be similar irrespective of parentage and fractionation (Tilling et al., 1973). More probably, the lack of Au mineralisation can be explained by the higher crustal position, hence lower grade of metamorphism, occupied by the calc-alkaline rocks and by the lack of potentially auriferous inter-flow sediments (Foster, 1985). Alternatively, the low Fe content of the volcanic rocks relative to tholeiitic basalts may not have facilitated extensive Au precipitation in the hydrothermal conduits, although this is unlikely to account for the extreme paucity of Au mineralisation in some calc-alkaline suites.

In marked contrast to the antipathetic relationship between lode gold deposits and calc-alkaline volcanism in Zimbabwe, many Au lode deposits are spatially and temporally associated with TTG intrusions which were comagmatic with the calc-alkaline volcanic rocks (Colvine *et al.*, 1984, 1988; Foster and Wilson, 1984). These intrusive rocks are unlikely to have been particularly important sources of gold (Saager and Meyer, 1984). They probably provided an important thermal impetus to fluid-flow during deformation and may have been responsible for the generation of slightly oxidised fluids which would have been particularly effective in the transport and ultimately the precipitation of the gold (Foster, 1985; Cameron and Hattori, 1987).

8.4 Chromite mineralisation: major layered intrusions

More than 90% of the world's proven Cr reserves are found within the Bushveld Complex, South Africa and the Great Dyke, Zimbabwe, both major layered instrusions of the early Precambrian. In addition, the early Precambrian Stillwater Complex (USA), Fiskenaesset intrusion (West Greenland), Bird River Sill (Canada) and the Shurugwi complex (Zimbabwe) all contain notable occurrences of chromitite. Within each of these layered intrusions, chromite segregations comprise thin (1–10 cm thick) layers of massive chromitite (> 90 modal % chromite), which form an integral part of the layered sequence.

8.4.1 *Stratigraphy*

Schematic stratigraphic columns of the Bushveld, Stillwater and Great Dyke layered intrusions are outlined in Figure 8.6. All three intrusions contain chromitite layers which vary from 1 cm to 1 m thick, and which extend laterally for tens of kilometres. Examination of the enlarged chromite-bearing portion of each column reveals subtle but significant differences between the intrusions. The chromitite layers of the Great Dyke are restricted to cyclic units within the ultramafic part of the succession at or very near its base (Figure 8.6). The individual cyclic units show olivine-rich bases to olivine-poor tops and comprise successions of basal dunite to harzburgite, olivine-bronzitite and finally bronzitite (Wilson, 1982). In the northern part of the complex, eleven major chromitite seams have been recognised and several minor seams also occur in the succession (Worst, 1960). The lowermost seams (nos. 4 to 11) are hosted by dunite, whereas the uppermost seams (nos. 1 to 3) are hosted by harzburgitic lithologies (Prendergast, 1987).

The major chromitite layers of the Bushveld occur within the Critical Zone which is the second major subdivision above the base (Von Gruenewaldt *et al.*, 1985). The chromitite layers have been subdivided into the lower (LG1–LG7), middle (MG1–MG4) and upper (UG1–UG3) groups, the LG6 layer being the main chromite producer. Typical host lithologies to the chromitite layers include feldspathic bronzitites in the Lower Critical Zone, and norites and anorthosites in the Upper Critical Zone (Cameron, 1977). Thus, in contrast to the Great Dyke, the chromite-bearing rocks are found within the increasingly fractionated part of the succession.

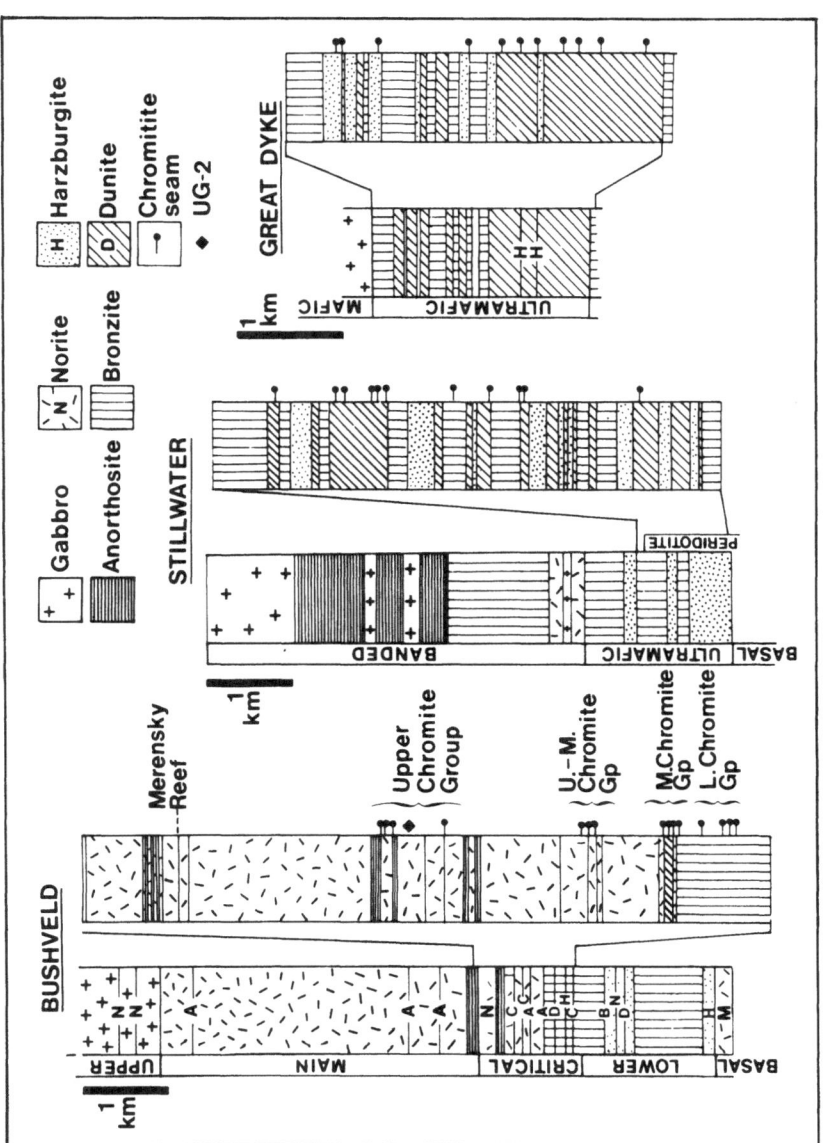

Figure 8.6   Schematic stratigraphic sections from the Bushveld, Stillwater and Great Dyke layered intrusions (see Hatton and Von Gruenewaldt, this volume). The part of the section containing the chromite deposits is enlarged in the right hand column in each case. Bushveld sections based on Von Gruenewaldt *et al.* (1985), Cameron (1980) and Gain (1985); Stillwater section after Jackson (1963), Irvine *et al.* (1983) and Raedeke and McCallum (1984); Great Dyke data after Wilson (1982) and Prendergast (1987).

Furthermore, although cyclic units are recognised within the Bushveld Critical Zone they are less well defined, and the chromite deposits do not occupy a regular stratigraphic position within the units, as they do in the Great Dyke.

The chromitite layers of the Stillwater Complex are restricted to cyclic units within the peridotite members of the ultramafic zone (Jackson, 1963; Irvine and Sharpe, 1986). Eleven chromite layers are recognised (denoted A to K), layer A being the lowermost seam and layers G and H showing the most economic potential. The chromite layers occur well above the base of cyclic units which typically comprise harzburgite and bronzitite. Thus, a brief consideration of the stratigraphic position of the chromitite layers within the various complexes reveals that the layers form integral parts of cyclic units within the cumulate pile. However, the position of the chromite layers within an individual cyclic unit and the nature of the host lithologies vary between each intrusion. For example, in contrast to the Bushveld and Stillwater complexes, the chromite cumulates in the 2.86 Ga Fiskenaesset complex of Southern West Greenland (Ashwal et al., 1989) occur in the upper, anorthosite units rather than in the ultramafic or gabbroic layers near to the base (Myers, 1985).

8.4.2   Textures

The chromitite layers of the Great Dyke complex comprise more than ninety-five modal % coarse-grained chromite, and vary between 2 and 12 cm thick. They have sharp bases and diffuse tops. Interlocking chromite grains 0.5–1.9 mm in diameter occur within the massive part, and olivine and pyroxene are the principal interstitial silicates in the more disseminated horizons of the lower and upper layers respectively (Wilson, 1982; Prendergast, 1987).

The Bushveld chromite layers also show sharp bases and diffuse tops, although within an individual layer the modal proportion of chromite varies from 5 to 100% (Eales, 1987). In the more disseminated horizons the spinel grains are small (0.05–0.1 mm), whereas spinel grains within the massive chromitite can exceed 2.5 mm in diameter. A variety of interstitial silicates are present including olivine (often replaced by pyroxene), orthopyroxene, clinopyroxene and plagioclase (Cameron, 1980). At Stillwater the chromitite layers, up to 30 cm thick, show sharp basal contacts of massive chromitite with disseminated tops. Spinel octahedra (1–4 mm diam.) are present within these layers and they often contain abundant silicate inclusions. Olivine is the main interstitial phase, in addition to bronzite and plagioclase (Jackson, 1961).

Each intrusion exhibits layers of massive chromitite with chromite grains commonly exceeding 2 mm in diameter. In contrast, the accessory spinels of the host silicate lithologies are rarely larger than 0.05 mm. The formation of these massive, coarse-grained chromitite layers remains problematic. Jackson (1961) demonstrated that following crystal settling and dense packing of spinel octahedra, layers of only 70% chromite could be achieved. He attributed the formation of the chromitite layers to the adcumulus growth of spinel grains via the inward diffusion of supernatant melt. More recently, doubt has been cast on the ability of interstitial melt to provide sufficient Cr ions to support such adcumulus growth, particularly given that conservative estimates of the partition coefficient between spinel and basaltic liquid are in excess of 100 (Hill and Roeder, 1974; Irving, 1978). Recent explanations of massive chromitite ore rely on postcumulus and subsolidus processes, involving compaction, grain annealing and textural re-equilibration (Hulbert and Von Gruenewaldt, 1985; Eales, 1987). The massive ore is the result of recrystallisation of the original chromite layer (*c.* 70% chromite) in the slowly cooling environment of the layered intrusion, resulting in the formation of coarse spinel grains with 120° triple junctions and the expulsion of interstitial melt.

8.4.3 *Geochemistry*

Cr-spinels from layered intrusions have straightforward compositions, simplified as $R^{2+}R^{3+}_2O_4$ (Irvine, 1967). The major divalent ions are Mg^{2+} and Fe^{2+} and the main trivalent species Cr^{3+}, Al^{3+} and Fe^{3+} with minor amounts of Ti, Ca, Zn and V. Spinel analyses from chromitite layers of the Bushveld, Great Dyke and Stillwater complexes (Figure 8.7) show X_{Cr} (Cr/(Cr + Al)) values of between 0.5 and 0.8, and X_{Mg} (Mg/(Mg + Fe)) of between 0.34 and 0.7. The X_{Mg} of silicate minerals is widely accepted as an index of fractionation. However, this ratio in spinels is particulary susceptible to sub-solidus re-equilibration processes (Irvine, 1967; Roeder *et al.*, 1979). For example, with falling temperature spinel grains exchange Mg for Fe, especially when an appreciable amount of co-precipitating silicates and/or intercumulus liquid are involved. The effects of equilibration processes are shown as vectors on Figure 8.7, and in order to overcome the problems of re-equilibration only chromite analyses from layers with > 95% modal chromite are illustrated.

By considering only the Great Dyke and Bushveld data (Figure 8.7), it is tempting to conclude that the more specific host lithologies and a lower position within the

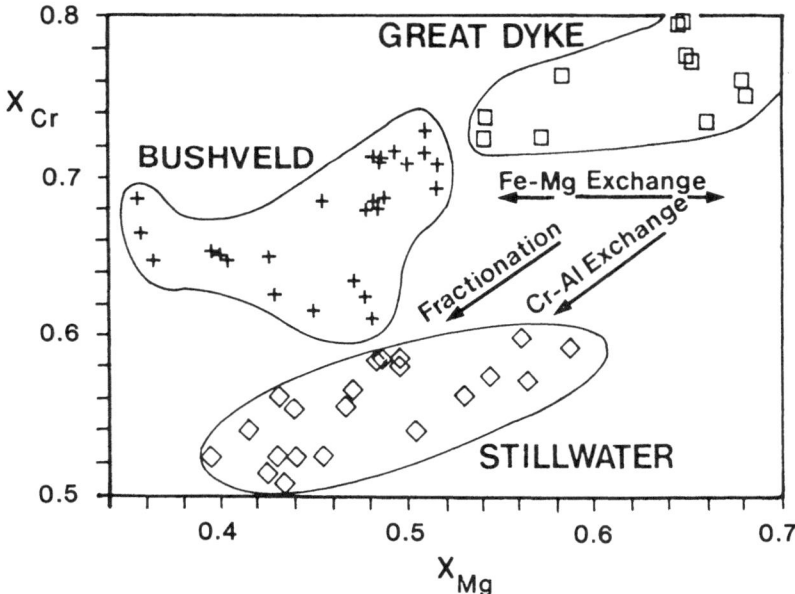

Figure 8.7 Plot of X_{Mg} versus X_{Cr} in chromites from massive chromitite layers of the Bushveld, Great Dyke and Stillwater complexes. Vectors outline predicted trends due to various cumulus, postcumulus and sub-solidus processes. Data sources: Jackson (1969), Cameron (1977) and Wilson (1982).

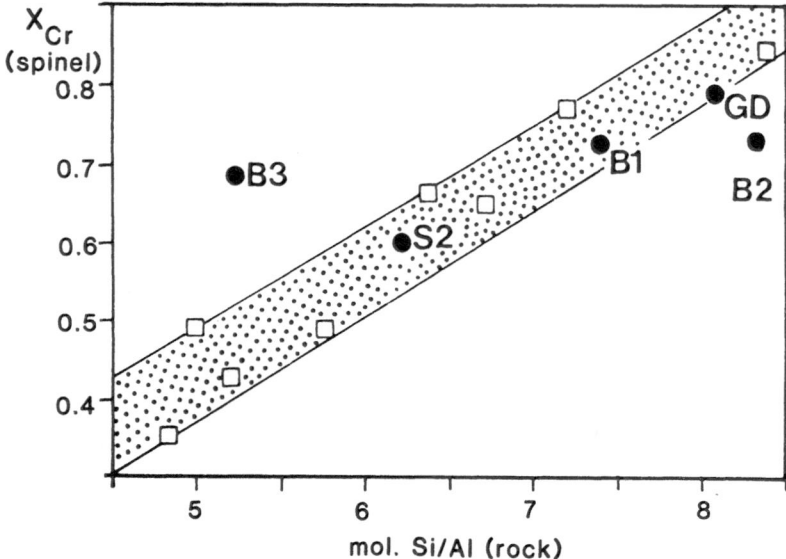

Figure 8.8 Plot of molecular Si/Al proportions in rock versus X_{Cr} in spinel. The shaded array represents the field of basalts and associated spinels (Dick and Bullen, 1984). Symbols: GD: Great Dyke parent (Wilson, 1982); B1: Bushveld parent of Davies and Tredoux (1985); B2 and B3: parent magmas of Lower and Upper critical zone respectively (Sharpe, 1982); S2: Stillwater parent magma of Helz (1985). Spinel data sources as for Figure 8.7.

stratigraphy are reflected by higher X_{Cr} values in spinels. However, the Stillwater chromite analyses cast doubt on this interpretation, given that their stratigraphic position is comparable to those from the Great Dyke. Figure 8.8 shows a good correlation between the Si/Al ratios of a series of basalts against the X_{Cr} value of the associated spinel, indicating that the X_{Cr} value of spinel is closely linked to the composition of the magma during crystallisation, increasing with increasing SiO_2 (Irvine, 1976; Dick and Bullen, 1984). Figure 8.8 also shows the parent magma compositions of the three complexes and the highest X_{Cr} value from their respective chromitite layers (to negate the effects of fractionation). Encouragingly, they fall on the basalt evolution trend. Furthermore, despite the 'ultramafic' setting of the Stillwater chromitites, the lower Si/Al ratios of the proposed parent magma are reflected in the analyses of the chromitite layers. Two contrasting parent magma types have been postulated for the upper and lower Critical zone of the Bushveld Complex (Sharpe, 1982). By plotting the compositions of these two parents against chromite analyses from the upper and lower zones, a reasonable correlation with the original array is observed. Nevertheless, the analyses do fall slightly above and below the array, implying that perhaps neither liquid is directly associated with spinel crystallisation in the chromite layers, although a mixture of the two parents could plot within the array. This may be an important observation in the light of magma mixing hypotheses.

Fe^{3+} versus Cr^{3+}, and Cr versus Ti for spinel grains from the various chromitite layers are plotted in Figures 8.9 and 8.10 respectively and outline the paths of spinel fractionation. The Great Dyke samples C1 to C12 are from increasing depths within the ultramafic sequence. It is notable that although a fractionation trend is observed in

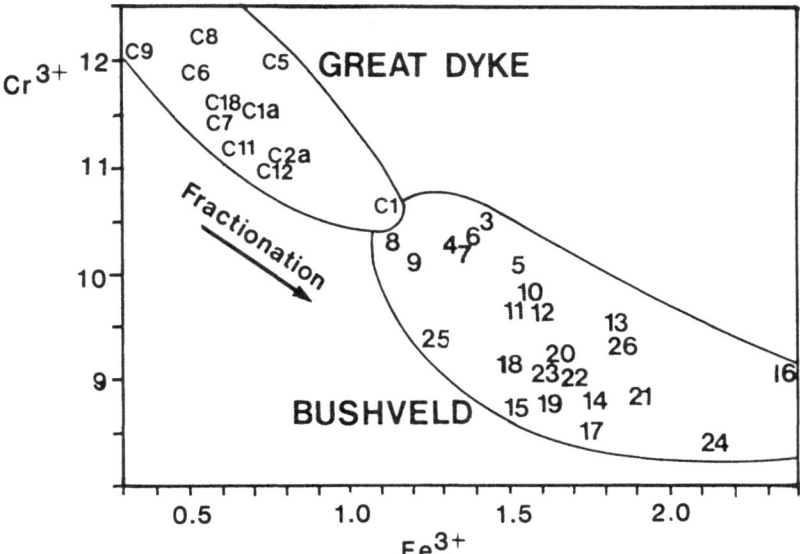

Figure 8.9 Fe^{3+} v Cr^{3+} cations for chromite grains from the Great Dyke and Bushveld (data sources as in Figures 8.7 and 8.8). Great Dyke analyses C1 to C12 represent increasing depth in the ultramafic succession. Bushveld analyses 26 to 1 represent increasing depth into the Upper and Lower Critical Zones.

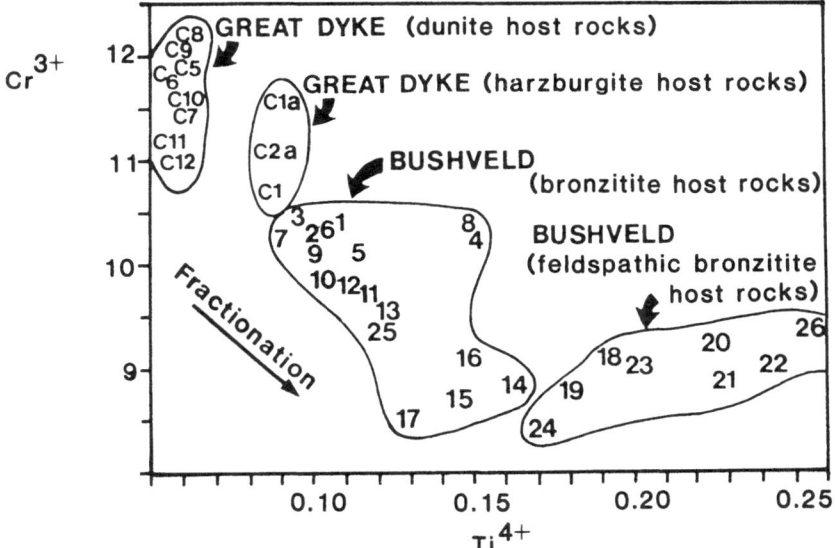

Figure 8.10   Variation of Cr^{3+} with respect to Ti^{4+} (cations) in chromites from contrasting host lithologies of the Bushveld Complex and Great Dyke. There is no simple fractionation trend with respect to height.

both plots, there is no systematic variation with respect to height. This is also the case for the Bushveld analyses, although the host lithologies and relative position of the chromitite layers are discernible. For example, the chromitite layers from the pyroxenite portion of the Great Dyke ultramafic sequence show higher Ti values compared to those from within the lower dunitic units. Similarly, the Bushveld data set distinguishes chromitite layers hosted by feldspathic bronzitites (higher Ti values) from the bronzitite-hosted chromitite layers of the lower critical zone.

8.4.4 *Petrogenesis*

Phase equilibria studies indicate that fractional crystallisation alone cannot yield a concentration of chromite once olivine, pyroxene or any other silicate minerals have begun to crystallise (Irvine and Smith, 1969). Therefore, the position of the chromite layers within cyclic units requires a mechanism to drive the magma into the spinel-only phase volume. In turn, the invoked mechanisms must be capable of producing the observed lateral extent of the chromite layers.

Cameron (1980) concluded that the only mechanism which could produce the extensive chromitite layers involved intermittent fluctuation in total pressure (P_T). In essence, magmas with liquidi close to the chromite boundary are likely to shift into the spinel-only phase volume given a sudden drop in P_T (Osborne, 1978). The fluctuations in pressure are considered to result from sporadic surface eruptions. A similar mechanism has been proposed to account for the chromite seams of the Stillwater complex. Mafic pegmatoids associated with the chromite layers are considered to be the result of a sudden P_T drop, which effected the precipitation of spinel and caused the vapour content of the magma to approach saturation, producing the associated pegmatites. Cameron and Desborough (1969) and Ulmer (1969) suggested that chromite precipitation could be controlled by fO_2. This argument gained impetus from experimental work on the effects of variable fO_2 on spinel precipitation from natural basalt starting compositions (Hill and Roeder, 1974), which showed that the Cr content of melts increases with decreasing fO_2. Consequently, sudden increases in fO_2 at constant temperature could lead to saturation of the magma in chromite. Recent studies have corroborated this argument (Murck and Campbell, 1986; Maurel and Maurel, 1982a, b). Wilson (1982) considered the chromite deposits of the Great Dyke to be the result of an increase in fO_2 due to a magma mixing event.

By virtue of the curved nature of the olivine–spinel cotectic in the system MgO–Cr_2O_3–SiO_2, Irvine (1975) demonstrated via a graphical solution that the contamination of a basaltic magma by sialic roof rocks would move the basalt liquidus into the spinel-only phase volume. A similar solution can be achieved by contaminating a basaltic magma resident within the magma chamber by a more primitive influx from below (Irvine, 1977). Two parental magmas can be recognised for both the Bushveld and Stillwater complexes (Irvine *et al.*, 1983; Todd *et al.*, 1982; Irvine and Sharpe, 1982). One parent ('U') was an early Mg-rich liquid crystallising harzburgite, bronzitite, norite and two-pyroxene gabbro. The second parent ('A') crystallised anorthosite, troctolite, olivine gabbro and two-pyroxene gabbro. Melting experiments by Irvine and Sharpe (1982) showed that whilst neither liquid 'U' nor 'A' have chromite on the liquidus, mixtures of the two lead to chromite as the only liquidus phase (Figure 8.11). Whilst magma mixing hypotheses currently receive the greater amount of support, some problems still remain with these models. Although the mixed hybrid magmas reside in

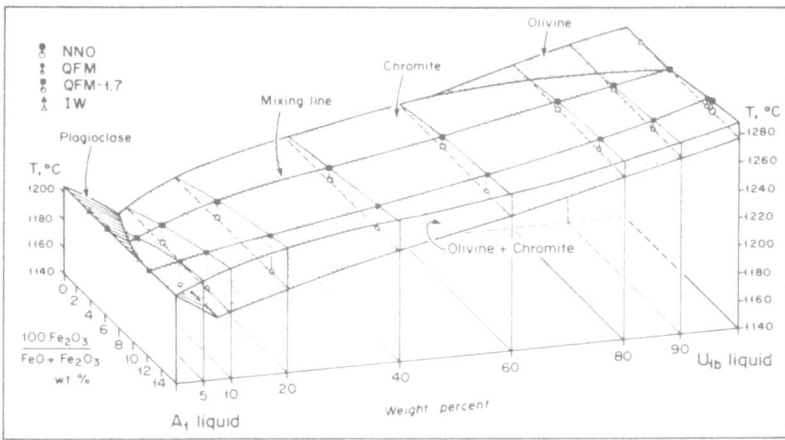

Figure 8.11 Experimentally established phase relations of 'A' and 'U' magma mixtures (from Irvine and Sharpe, 1986). Note that chromite is the liquidus phase for intermediate compositions, even when plagioclase and olivine are on the liquidi of the two end members.

the spinel-only phase volume, they rapidly fractionate to a cotectic boundary, thus yielding only thin chromite layers (Irvine and Sharpe, 1986). Lipin and Loferski (1983) considered that the magma-mixing theory outlined above is unable to account for the distinctive chemistry of the chromite layer 'B' of the Stillwater Complex. Mixing of the two postulated end-member compositions should produce an intermediate composition, and they found this not to be the case.

8.5 Platinum group elements (PGE): layered complexes

Outside the USSR, production of the platinum group elements (PGE) is focused on a series of early Precambrian layered intrusions, and in particular the Bushveld Complex of South Africa. Significant reserves are reported for the Stillwater and Lac des Iles complexes (North America) and the Great Dyke (Zimbabwe). Other early Precambrian reserves of PGE worthy of note, but not discussed in this section, include those associated with komatiite-related Ni-S deposits, and in particular those of the Kambalda region (Western Australia). The present account focuses on the PGE mineralisation of layered intrusions, notably the Bushveld and Stillwater complexes.

8.5.1 *Reef characteristics*

In each of the layered intrusive suites the PGE are associated with sulphide-rich intervals referred to as reefs, which are laterally continuous across the complex. The best described examples include the Merensky and UG-2 reefs of the Bushveld Complex and the J-M reef of the Stillwater Complex. The reefs occur above the ultramafic portion of the succession, often at the 'interface' between the ultramafic and more fractionated rocks, and in each case the sulphide assemblage is found in close association with pegmatoidal cumulates.

In the Bushveld Complex, the Merensky reef forms part of the Merensky cyclic unit

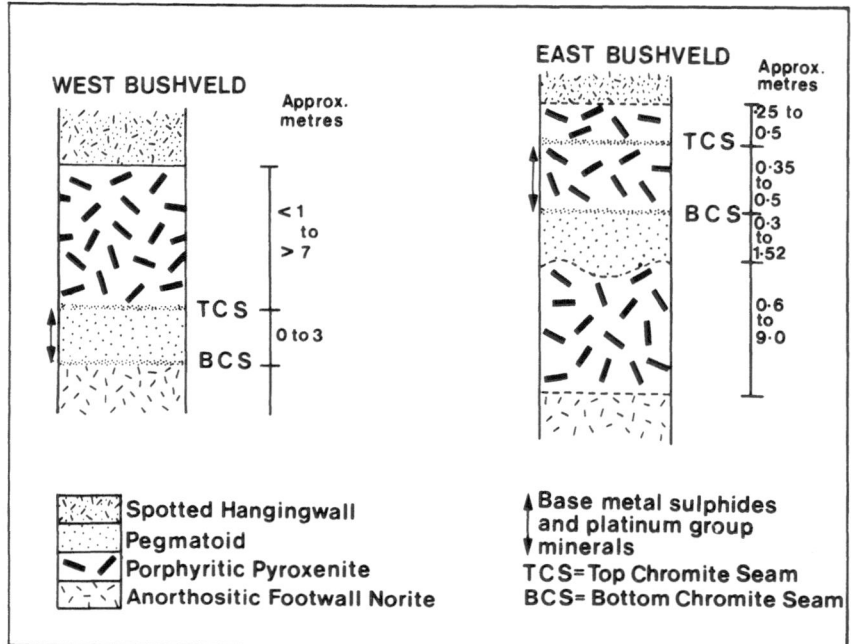

Figure 8.12   PGE reef characteristics from the Bushveld complex (after Hiemstra, 1979).

which shows the typical lithostratigraphic succession of chromitite, pyroxenite, norite and anorthosite. The reef shows considerable lateral variation. In the east of the complex it rests on and comprises in part a pegmatitic pyroxenite (up to 0.5 m thick) with the PGE mineralisation occurring in close proximity to two enclosed chromitite seams (Figure 8.12). To the west, the reef reaches as much as 4 m thick and starts with a lower chromite seam in sharp contact with anorthosite. It consists of porphyritic melanorite and minor pegmatite, with the PGE mineralisation concentrated in the vicinity of the upper chromite seam (Vermaak and Hendricks, 1976; Heimstra, 1979). Much attention has been paid recently to sites within the Merensky Reef where it cuts down through the footwall units for up to tens of metres. These transgression sites have a cyclindrical form and are referred to as potholes (Buntin *et al.*, 1985; Viljoen *et al.*, 1986). Individual potholes may coalesce to produce large irregular structures with dimensions in excess of 1 km which are filled with reef equivalent material. Buntin *et al.* (1985) reported pegmatitic dykes, an increase in the presence of graphite and low 'redox' halos specifically associated with these potholes.

The UG-2 layer of the Bushveld Complex is one of a series of chromitite layers associated with the cyclic units beneath the Merensky Reef (Figure 8.6). This chromitite seam contains a major PGE reserve associated with base metal sulphides within the layer (Hiemstra, 1979; Mclaren and De Villiers, 1982). The UG-2 reef varies from 0.15 to 2.55 m thick and comprises between 60 and 90% chromite. It is underlain by a pegmatoid zone consisting of coarse bronzitite with irregular layers of harzburgite and chromitite, and is situated from 15 to 370 m below the Merensky Reef.

The J-M reef of the Stillwater Complex occurs within the banded zone (Figure 8.6) where alternations of troctolite and anorthosite predominate. The reef is generally between 1 and 3 cm thick and contains small but visible concentrations of sulphides, which comprise between 0.5 and 1 vol% of the rock. Although spinel grains are scattered throughout the olivine-rich lithologies, no chromite seams are associated with the reef as they are in the Bushveld Complex (Page *et al.*, 1985; Todd *et al.*, 1982). Within the various reefs the sulphides are interstitial to the silicate minerals and/or chromite grains. Pyrrhotite, pentlandite and chalcopyrite, with sperrylite, braggite and cooperite are the main Pt-bearing minerals, and significant amounts of Pd partition into pentlandite.

8.5.2 Geochemistry

The present setting of the PGE deposits is readily appreciated by considering their geochemical affinities. The PGE form part of Group 8a of the periodic table along with Fe, Co and Ni, and are considered as noble metals, although not to the same extent as Au. All the PGE are siderophile in nature. High concentrations (ppm levels) occur in Fe-meteorites and Fe-Ni chondrites. They are also chalcophile, with high concentrations observed in Ni-Cu sulphides from mafic and ultramafic rocks. For example, Makovicky *et al.* (1986) produced Fe-sulphide charges at 900°C and reported that pyrrhotite dissolved up to 11 wt% Pd, 2.2 wt% Pt, 3.6 wt% Ru and 4.4 wt% Rh, and that pentlandite, while not dissolving detectable amounts of Pt, dissolved 12.5 wt% Pd, 12.4 wt% Rh and 11.4 to 12.9 wt% Ru. PGE affinities for oxides and silicates are less well understood. Chromite is presently the principal oxide of interest. PGE analyses of chromitites, regardless of their tectonic setting, cover a spectrum from low (less than chondritic values) to enriched 'Merensky reef-type' contents. The extent to which the PGE reside in solid solution or as sulphide inclusions within the chromite grains is debatable. Silicates are considered to be poor collectors of the PGE compared to the sulphides and oxides. However, there is limited evidence that silicates can fractionate PGE, particularly Ir and to a lesser extent Ru and Os.

The Bushveld Complex currently represents the largest exploited repository of PGE. Data for the marginal rocks of the complex, considered the best approximation to parental liquids for the complex (Sharpe, 1982), show a depleted PGE pattern with a positive slope from Ir to Au (Figure 8.13). The marginal rocks are enriched compared to MORB but show similar noble metal abundances to those of both komatiitic liquids and low-Ti basalts. That the Bushveld marginal liquids are not unique in terms of their noble metal content led Davies and Tredoux (1985) to the view that the huge noble metal reserves of the Bushveld-layered suite must be related to the crystallisation processes operating within the layered series.

The Bushveld cumulates show a wide range of PGE values, some quite enriched, distributed around the marginal rock values. Notably, all the analysed olivine-cumulates are relatively enriched in Ir and Ru compared to the potential source, evidence that the cumulates can fractionate the PGE and especially Ir and Ru, which apparently partition into olivine (Ross and Keays, 1979). The highest PGE values for the Bushveld occur in the Merensky reef and UG-2 chromitite layer. These horizons show a positive slope from Ir and Pd comparable, though relatively enriched, to the marginal rock pattern. This is consistent with the view that $K_{D(PGE)}^{sulphide} \gg K_{D(PGE)}^{silicate\ liquid}$ (where $K_{D(PGE)}^{sulphide}$ is the PGE partition coefficient for sulphides). Thus, the noble metal pattern of the original liquid is preserved unless the sulphide assemblage itself

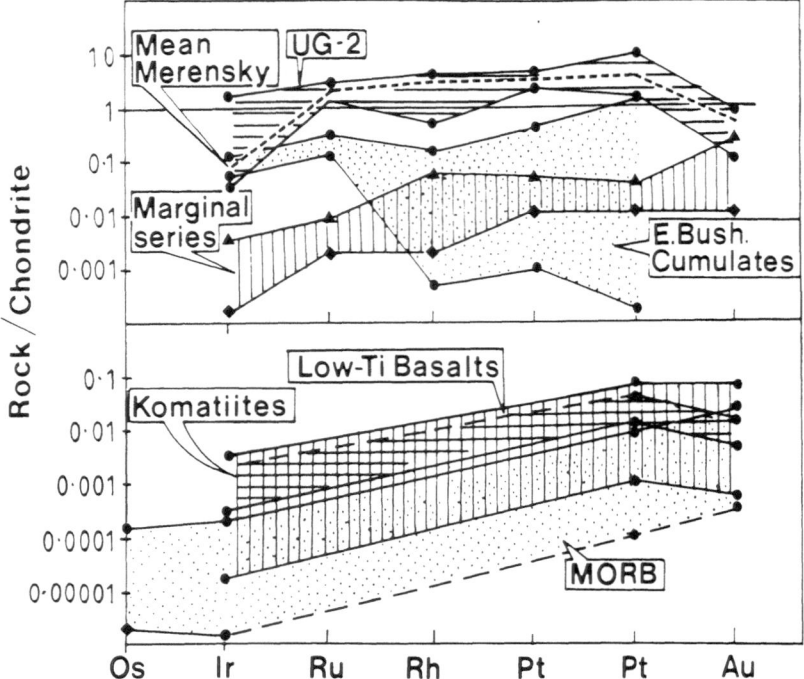

Figure 8.13 Chondrite-normalised PGE plots for the various lithologies of the Bushveld complex, and basic and ultrabasic magmas. Data sources: Bushveld: Gain (1985), Mclaren and De Villiers (1982), Sharpe (1982), Davies and Tredoux (1985); MORB: Morgan *et al.* (1981), Crocket and Teruta (1977), Hertogen *et al.* (1980); Komatiites: Crocket and MacRae (1986), Green and Naldrett (1981); Low-Ti Basalts: Hamylyn *et al.* (1985).

fractionates the PGE (e.g. the experiments of Makovicky *et al.* (1986) suggest that pentlandite would decouple Pt and Pd).

PGE data from the J-M reef, all chromite horizons and olivine-cumulates from the Stillwater Complex, are presented in Figure 8.14. Their most striking feature is the greatly enriched Pt and Pd values of the J-M reef (Pd up to 1000 × chondrite). The J-M reef shows a positive sloping pattern from Ir to Pd. The Stillwater chromitites show a range of PGE values, some of which compare to those of the Bushveld UG-2 chromite seam. However, no simple stratigraphic relationship exists as seam 'A' is the lowest in the sequence, whereas the second most enriched seam, 'J', is one of the highest. A limited data set is available for Stillwater olivine-cumulates, and these show strongly depleted patterns.

8.5.3 *Petrogenesis*

Broadly speaking there are two schools of thought concerning the petrogenesis of PGE deposits in layered intrusions. The first invokes a magma mixing model while the second emphasises the importance of late stage magmatic fluids. PGE reefs form an integral part of layered intrusions, occurring at or near the base of cyclic units. This fact together with their great lateral extent leads many workers toward a model which, in

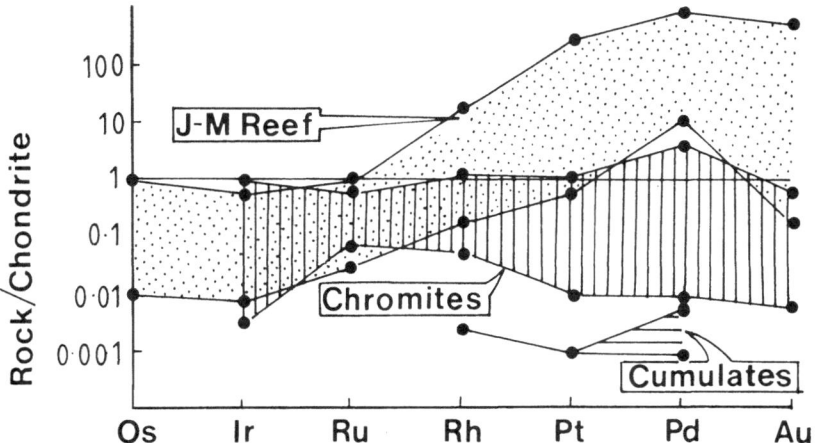

Figure 8.14   Stillwater noble metals data (from Barnes and Naldrett, 1985; Talkington and Lippin, 1986; Zientek *et al.*, 1986; Page *et al.*, 1985).

simple terms, is an extension of the mixing model postulated for chromitite formation. In this case the PGE (in particular Pt and Pd) are derived from the 'U' liquid which yields the ultramafic rocks and the sulphur from the 'A' liquid, the parent to the anorthositic lithologies (Todd *et al.*, 1982). In contrast to the chromite model in which mixing results in a hybrid magma with chromite the only liquidus phase, for PGE formation the mixing event ultimately results in S saturation by virtue of the reduced crystallisation temperatures in the hybrid melt and the consequent reduction in S solubility. The chalcophile nature of the PGE ensures almost complete partitioning of the PGE into the sulphide phase, and enhanced PGE values. Within the confines of this model the pothole structures are interpreted as resulting from the resorption of the footwall cumulates by the undersaturated A/U hybrids (Irvine *et al.*, 1983).

In parallel with the magma mixing hypothesis, an alternative model has emerged from the recognition of hydrous phases and reducing environments in the vicinity of PGE mineralisation (Boudreau *et al.*, 1986; Ballhaus and Stumpfl, 1986; Buntin *et al.*, 1985). In particular, the typically coarse-grained pegmatitic reefs contain primary magmatic phlogopite and hornblende which show unusually high Cl/F ratios, indicating reaction with Cl-rich magmatic fluids (Boudreau *et al.*, 1986). Resultant models invoke the exsolution of a Cl-rich fluid from the interstitial melt which dissolves and transports metals as it migrates upward through the cumulate pile and becomes trapped at the reef horizon resulting in PGE mineralisation.

8.6   Closing remarks

From the information presented in this chapter it is evident that basic and ultrabasic magmatism contributed directly or indirectly to the spectacular enrichment of metals in the Earth's early crust. The unique chemical and tectonic framework of the evolving early Precambrian crust obviously played an important role in the metallogeny (e.g. Anhaeusser, 1981; Groves and Batt, 1984). Some of the more important aspects highlighted in this chapter are summarised below.

In the case of Ni sulphides the key lies in the nature of komatiitic magmas. Their abundance in the Archaean and virtual absence in younger rocks (certainly post 2.0 Ga) points to a different thermal regime within the mantle in which 1600°C melts were generated and erupted. These melts were characterised by their high temperatures, high Mg and Ni contents, and viscosities which were much lower than basaltic melts. All of these characteristics imply either very high degrees of partial melting of a primitive mantle or lower degrees of partial melting of a residual refractory mantle. This latter alternative requires even higher melting temperatures for the komatiite. However, if the S is magmatic and of mantle origin, it is difficult to envisage why this refractory mantle should contain anomalous quantities of sulphur. Perhaps the key lies within Archaean plate-tectonic models. If we accept that the Archaean oceanic crust contained a high komatiitic component (Arndt, 1983), and speculate that it contained anomalously high S, it is then possible that this S was released during subduction and created sulphide-rich zones in the overlying wedge. Komatiite magmas from deeper parts of the mantle would be capable of assimilating these sulphide zones, and this would generate an immiscible sulphide/komatiite magma. This mixture would erupt through the overlying continental crust and consequently would be preserved within Archaean cratons (Figure 8.15).

The role of basic magmatism in gold metallogeny is less obvious, but the scavenging of the precious metal from high-Mg basalts by an immiscible sulphide phase probably played an important, albeit intermediary, role in the gold budget of some greenstone belts. The gold would have been liberated readily from the relatively unstable sulphide

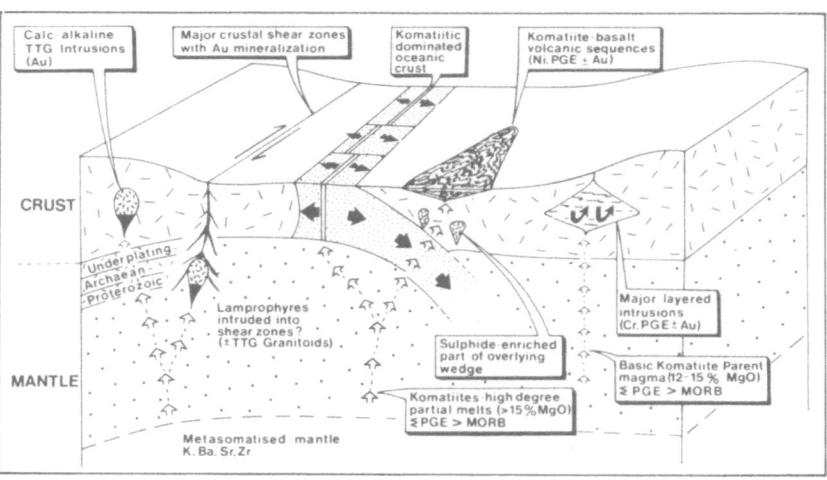

Figure 8.15 Schematic diagram outlining the tectonic setting of early Precambrian mineralisation associated with basic and ultrabasic magmatism. (a) Ni mineralisation occurs in the basal member(s) of the komatiitic volcanic succession. The S could be derived from either the mantle or crust (by assimilation of sulphidic metasediments). Alternatively, the komatiites may assimilate sulphides from a 'mantle wedge' which had been enriched by fluids derived from an older subducted oceanic crust. (b) Au mineralisation associated with crustal lineaments and both komatiitic and basaltic volcanic suites. (c) Chromite and PGE mineralisation restricted to cyclic units within layered basic intrusions, the PGE associated particularly with sulphide assemblages.

phases during any subsequent period of hydrothermal activity. Furthermore, the extrusion to burial history and consequent chemical behaviour of the basic and ultrabasic components of the Archaean sequences was somewhat distinct from that of the felsic and clastic components. Sea-floor hydration of the basic rocks generated enormous water reservoirs in most greenstone belts which in turn supplied the huge hydrothermal systems which affected the early crust during subsequent de-watering events. As a corollary, the more reactive mafic components of these basic crustal rocks provided the most suitable chemical traps for the precipitation of gold and other metals from the high temperature fluids.

Of fundamental importance, however, was the presence of major crustal dislocations which probably tapped the reservoir of fluids, metals and ligands as deep as the lower crust and perhaps even the upper mantle. It was these major fault zones which facilitated the de-watering of the crust during prograde metamorphism and which also may have permitted devolatilisation of the upper mantle and the transfer of calc-alkaline magmas into the upper crust. The geological record demonstrates that a worldwide Archaean hydrothermal Au event occurred at approximately 2.7–2.6 Ga, immediately prior to, and perhaps even as a consequence of, underplating and final cratonisation of the Earth's crust. The subsequent Proterozoic and Phanerozoic record reveals that major Au-bearing hydrothermal events were confined, temporally and geographically, to convergent plate margins e.g. the epithermal deposits of the Pacific 'rim of fire' (Nesbitt et al., 1986). Such a constraint emphasises the importance of deep crustal fault zones to the entrainment of major hydrothermal systems and the consequent transport and precipitation of gold. Thus, the marked abundance of lode Au deposits in Archaean terranes reflects the complex interplay of: (i) development of major fault zones which transected the complete profile of the crust prior to cratonisation; (ii) the thermal and tectonic framework which gave rise to the basic magmas which in turn were both important source rocks and important sinks for the various hydrothermal components; (iii) the transfer of gold into the upper crust as a minor but accessible component of sulphide phases in basic magmas.

The major layered basic intrusions of the early Precambrian are unique in terms of their size and chromite mineralisation. Proterozoic and Phanerozoic intrusions demonstrate similar igneous phenomena such as the development of cyclic units, cryptic variation and chromite mineralisation, but rarely on the scale of those of the early Precambrian. Enriched parental melts may be one factor to account for the enhanced mineralisation of the intrusions, but the processes responsible for the development of such large intrusions may be more important. In this case the simplest argument to account for the size of the intrusions and the contained mineralisation is that the higher thermal regime in the early Precambrian mantle enabled the generation of great volumes of melt (by more recent standards) and their subsequent emplacement into a relatively stable craton. For the PGE mineralisation of the early Precambrian, the observation that komatiites and the parental melts of the layered intrusions are both enriched in PGE compared to present day MORBs indicates a fundamental change in the nature of the melts or the melting regime during early Precambrian times. However, it is still an open debate as to whether this reflects enriched PGE abundance of the source regions or the development of S-undersaturated melts, which were unable to fractionate PGE until S saturation was achieved in large open-system magma chambers.

In contrasting Archaean and Proterozoic sequences the most striking and genetically significant factor regarding early Precambrian Ni, Au, Cr and PGE mineralisation is the higher thermal regime of the Archaean. In particular, komatiites provide compelling evidence of higher mantle temperatures, producing magmas with eruption temperatures of $c.$ 1600°C. This enhanced thermal regime may have facilitated the development of major transcrustal dislocations in the pre-2.5 Ga crust, the upward access of immiscible magma–sulphide mixtures into subvolcanic and submarine environments and the subsequent accumulation of Kambalda-type Ni mineralisation. Similar dislocations permitted de-watering of the lower crust, possibly even de-volatilisation of the upper mantle, and the focusing of fluids and ultimately gold into upper crustal shear zones. The same mantle de-gassing event may even have contributed to the widespread cratonisation of the Archaean crust at $c.$ 2.5 Ga.

The interplay between a high thermal gradient and the generation and, perhaps more important, the preservation of major layered intrusions may be subtle but important. Richter (1988) has demonstrated a crude correlation between the decline in the thermal regime and the preservation of a greater mass of crust, due to cessation of melting at the base of the crust. The metallogenic implications of this model are that large early Archaean basic intrusions were consumed as the early crust was recycled and thus the only evidence of Cr or PGE mineralisation are remnants of layered intrusions such as the $c.$ 3.5 Ga chromitite pods of Shurugwe, Zimbabwe. Thus, the cratonisation and stabilisation of the late Archaean crust at $c.$ 2.5 Ga were of considerable significance as they provided a unique thermal and tectonic window in which the thermal regime of the mantle remained capable of generating the source magmas and the crust had obtained a sufficient stability to preserve Bushveld-type layered complexes. However, the continuing decay of the mantle's thermal regime into the Proterozoic dictated that, although the development of layered intrusive bodies with or without Cr and PGE mineralisation remained possible, the genesis of such bodies became rare localised events with thermal and tectonic regimes incapable of generating major layered complexes.

Acknowledgements

The authors wish to express their thanks to R.N. Taylor for his constructive review of the manuscript, and Anthea Dunkley and Barry Marsh for preparation of the figures. S.R. was supported by an EEC research fellowship.

Part II
Regional Syntheses

9 Early Precambrian basic rocks of the USA

G.L. SNYDER, R.P. HALL, D.J. HUGHES and K.R. LUDWIG

9.1 Introduction

Early Precambrian basic rocks occur throughout a large part of the USA (Peterman, 1979; Harrison and Peterman, 1984). Archaean complexes outcrop in two major provinces, one forming the southwestern tip of the Superior Province in northern Minnesota and Wisconsin, and the other comprising the Wyoming craton, outcropping in the mountain ranges of Wyoming and its neighbouring states (Figure 9.1). Both of these provinces have significant plutonic, hypabyssal and volcanic basic components, ranging in age from early to late Archaean. In addition, early to mid-Proterozoic hypabyssal basic intrusions occur in virtually all of the Archaean cratonic regions, while early to mid-Proterozoic metavolcanic suites occur throughout western central USA (Condie, 1986).

This chapter aims to catalogue and review the most important basic units and to highlight their lithological and geochemical characteristics. This description is by no means comprehensive in either aspect, but in drawing together a lexicon of the varied and widespread mafic intrusive and extrusive units which have been given a more rigorous treatment elsewhere, we hope to draw attention to the petrological differences and similarities between the various units, and to summarise some of the tectonic interpretations which can be drawn from them.

9.2 The Minnesota–Wisconsin region

The early Precambrian basic rocks of the Minnesota and Wisconsin region comprise three different major groups. The oldest group is early Archaean and forms part of the 3.6 to 3.3 Ga Minnesota River Valley gneiss terrane in southwestern Minnesota (Figure 9.2). The second group is late Archaean (c. 2.7 Ga) and is part of a major belt of supracrustal rocks particularly in the Vermilion district of northern Minnesota (Morey, 1980; Schulz, 1980). The youngest rocks under consideration are early Proterozoic basaltic metavolcanics and gabbroic intrusive bodies in northern Wisconsin and central Minnesota (Greenberg and Brown, 1983; Horan et al., 1987).

9.2.1 Early Archaean suites

The early Archaean basic rocks occur mainly as amphibolitic layers within the high-grade (amphibolite to granulite facies) quartzo-feldspathic gneisses of southwestern Minnesota (Figure 9.2). Field relationships indicate that some of these layers originated as basic dykes or sills while others probably pre-date the mainly tonalitic

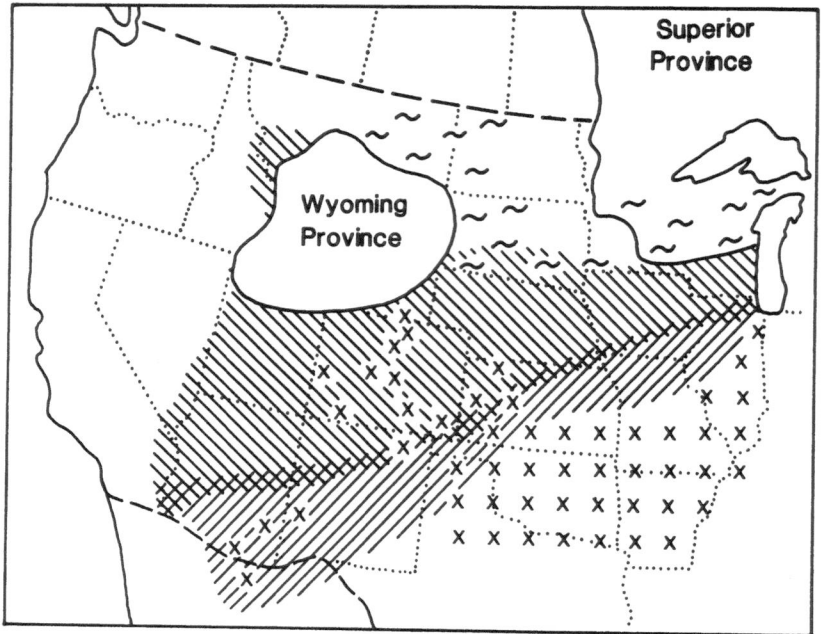

Figure 9.1 Sketch map of western and central USA showing the Wyoming and Superior Archaean provinces and accreted Proterozoic volcanic terranes: NW–SE line shading: 1.8–1.7 Ga; NE–SW line shading: 1.7–1.6 Ga; crosses: 1.5–1.4 Ga. Rejuvenated Archaean rocks indicated by the squiggle symbol. Dotted lines depict American state and Canadian provincial boundaries. (Adapted from Karlstrom and Houston, 1984.)

gneisses (Goldich *et al.*, 1980), which have been dated at between 3.6 and 3.3 Ga (Michard-Vitrac *et al.*, 1977; Goldich and Wooden, 1980; Peterman *et al.*, 1980). The basic rocks are considered to be of approximately the same absolute age as the quartzo–feldspathic gneisses (Schulz, 1982).

There are two main types of early Archaean basic rocks. The older type pre-dates the tonalitic gneisses and is plagioclase- and Al-poor but rich in Mg and Fe. The second, younger type carries abundant plagioclase and is tholeiitic in composition (Wooden *et al.*, 1980). The magnesian amphibolites (MgO = 7–15%) have FeO* contents between 12 and 17%, Cr and Ni up to 1300 and 530 ppm respectively, and Al_2O_3 contents ranging from *c.* 6–10%. Their chemical compositions are characteristic of komatiitic basalts, although they have mildly fractionated REE abundances, with Ce_N and Ce_N/Yb_N values of *c.* 20 and 2.5 respectively (Schulz, 1982). This suite of rocks is considered to have been derived by partial melting at a considerable depth (pressures greater than 30 kb) of primordial garnet–lherzolitic mantle with garnet as a residual phase.

9.2.2   *Late Archaean basic metavolcanics*

The Vermilion district of northern Minnesota forms the western extension of the major Wawa volcanic–plutonic belt of the Superior Province of the Canadian Shield. The geology of the Vermilion Greenstone Belt has been described by Morey (1980) and

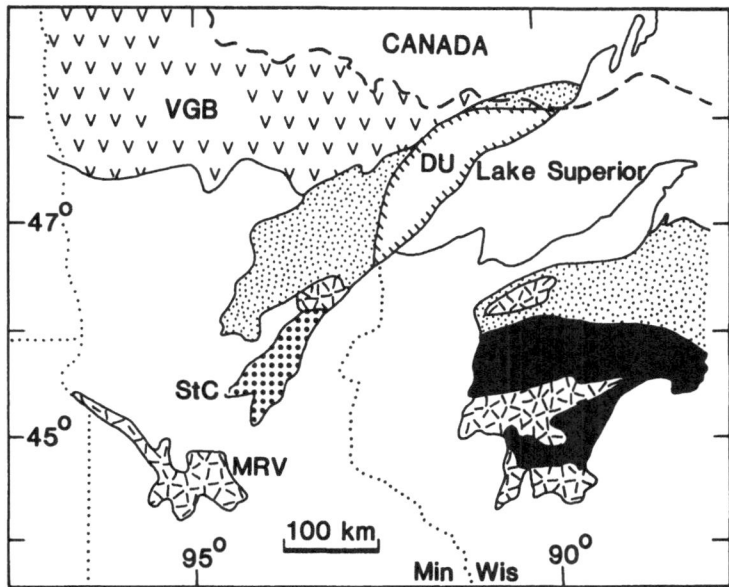

Figure 9.2 Sketch map of Minnesota and Wisconsin showing the locality of the early Proterozoic (Penokean; *c.* 1.8 Ga) metavolcanics (black), meta-sediments (light stipple) and plutonic rocks (heavy stipple), with respect to Archaean gneisses (cross-hatch) and late Archaean (*c.* 2.75 Ga) Vermilion greenstone belt (VGB). MRV: Minnesota River Valley gneiss terrane; StC: St Cloud region; DU: 1.15 Duluth complex. (Adapted from Horan *et al.*, 1987.)

Schulz (1980). It comprises a major volcano-sedimentary succession, younging to the north in a linear WSW-trending, NW-facing belt approximately 20 km wide and 160 km long. The supracrustal belt and neighbouring intrusive granitic (*s.l.*) rocks are around 2.7 Ga old and their low initial $^{87}Sr/^{86}Sr$ (Sr_i) values indicate that these rocks were not derived from the reworking of significantly older crust (Catanzaro and Hanson, 1971; Jahn and Murthy, 1975).

The lower, middle and upper units of the supracrustal belt comprise the Ely Greenstone (oldest), overlain by the Knife Lake Group in the central part of the region and the Lake Vermilion Formation in the western part, and by the (youngest) Newton Lake Formation. The Ely Greenstone consists predominantly of pillowed, massive and amygdaloidal basaltic rocks and associated hypabyssal intrusions which commonly retain doleritic textures. The lower portion of this unit has a basic assemblage compositionally similar to island arc tholeiites and calc-alkaline basalts while the upper rocks have geochemical affinities with ocean floor and island arc tholeiites (Sims, 1976; Morey, 1980; Schulz, 1980). In contrast, the overlying Lake Vermilion and Knife Lake Groups are dominated by various dacitic volcaniclastics. The uppermost unit of the Vermilion greenstone belt, the Newton Lake Formation, comprises a bimodal suite of intercalated basic and calc-alkaline intermediate and felsic metavolcanics (Sims, 1972). The mafic members consist of massive and pillow-structured meta-basaltic flows which have the chemical characteristics of basaltic komatiites and high-Fe tholeiites (Green and Schulz, 1977, 1978; Schulz, 1982). Intrusive bodies associated with the basaltic komatiitic metavolcanics show clear evidence of differentiated layering, and comprise peridotitic, pyroxenitic and gabbroic layers. Pyroxene spinifex textures similar to those found in equivalent rocks in Ontario (Arndt et al., 1977) are locally preserved and, together with the close spatial association with the komatiitic lavas, suggest a komatiitic parentage for these intrusives. The Vermilion district supracrustal assemblage as a whole is dominated by dacitic calc-alkaline rocks (Schulz, 1980), and may be consistent with its formation in a marginal basin or island arc environment (Jahn et al., 1974; Green and Schulz, 1977; Morey, 1980).

9.2.3 Early Proterozoic rocks

Early Proterozoic basic rocks of the Wisconsin–Minnesota region comprise two distinct units (Figure 9.2). Penokean supracrustals (1.9–1.8 Ga) are abundant to the south of Lake Superior, in northern Wisconsin (Sims et al., 1980, 1985; Van Schmus and Bickford, 1981), while post-orogenic gabbroic intrusives and NE-trending basaltic dykes penecontemporaneous with granitic intrusions dated at 1.8 Ga occur in the St Cloud region of central Minnesota (Horan et al., 1987). Few geochemical data are available for the Penokean metavolcanic rocks of northern Wisconsin. They constitute an essentially bimodal suite of island arc-type basalts and rhyolites. Basic metavolcanics of the same age in northern Michigan are compositionally similar to continental rift-type basalts (Sims et al., 1985). Structural and sedimentological studies on the associated and volumetrically dominant metasedimentary units suggest that these supracrustal rocks formed in continental marginal basins and that the Penokean orogeny involved a convergent plate boundary (Larue and Sloss, 1980; Van Schmus and Bickford, 1981; Holst, 1984; Sims et al., 1985).

The gabbroic rocks of central Minnesota include the Watab amphibolite and the St Wendel and Little Falls metagabbros (Horan et al., 1987). These metagabbros are geochemically similar to the basaltic dykes. They all have hypersthene-normative, tholeiitic basaltic compositions with MgO contents between 4 and 9%. However, they are distinctive in that they are relatively rich in LIL elements (Ba, Sr, K) and LREE (Ce_N = 40 to 170) and have strongly fractionated REE distributions (Ce_N/Yb_N = 3 to 12). Low concentrations of these elements in the nearby Mora anorthositic gabbro argue against their remobilisation and enrichment during metamorphism. Furthermore, combined REE and Nd and Pb isotope data render the possibility of crustal contamination of the parental magma to the gabbros highly unlikely. Instead, these data suggest the contamination of the mantle source at least 500 Ma prior to its partial melting and the formation of the gabbroic protoliths (i.e. at c. 2.3 Ga). The mantle metasomatism involved a LILE- and LREE-enriched fluid or melt, either from mantle-derived melts or from subducted sedimentary material (Horan et al., 1987). By contrast, the Mora anorthositic gabbro has unfractionated REE abundances ($Ce_N = c.$ 3; Ce_N/Yb_N = 1 to 1.5) with a slight positive Eu anomaly, and may have derived from a melt analogous to modern ocean ridge or marginal basin magmas (Horan et al., 1987).

9.2.4 Tectonic implications

Little can be deduced about the tectonic setting of the early Archaean basic rocks of central Minnesota because of their subsequent severe tectono-metamorphic history. They occur now mainly as fragmented layers within migmatitic gneisses. However,

Schulz (1982) has pointed out the close similarities between their compositions and those of the Fe-rich komatiitic basaltic volcanics in the Newton Lake Group of the 2.7 Ga Vermilion Greenstone Belt. This late Archaean belt has oceanic and island arc-type tholeiitic, komatiitic and predominantly calc-alkaline andesitic to dacitic components, and is considered to have developed in an environment most closely akin to a modern island arc. After cratonisation, this late Archaean greenstone terrane acted as a northern plate, and the 3.6 Ga Minnesota River Valley gneiss terrane as a southern block, between which the 1.8 Ga supracrustal rocks were deposited and accreted (Sims *et al.*, 1980). Island arc-type early Proterozoic volcanics occur to the south. These rocks were formed during south-dipping subduction beneath the early Archaean Minnesota River Valley gneiss continental margin (Larue, 1983; Greenberg and Brown, 1983). Further isotopic evidence for episodic north to south crustal formation and accretion has been presented by Nelson and DePaolo (1985).

9.3   Wyoming Province: Archaean metavolcanic rocks

The Wyoming Province incorporates rocks exposed in the abundant mountain ranges of Wyoming and those nearby in northern Utah, northern Colorado, eastern Idaho, southern Montana, western South Dakota and northern Nevada (Figure 9.3). Archaean basic metavolcanic rocks are generally only sparsely and fragmentally preserved throughout the province. They occur most commonly as high-grade metamorphosed remnants within quartzo-feldspathic gneisses. In the Wind River range, for example, fragments of early Archaean basic metavolcanic rocks associated with paragneisses are distinguishable from metavolcanic amphibolites which form around 20% of the late Archaean Medina Mountain supracrustal sequence and an intervening generation of basic dykes (the Victor dykes) (Koesterer *et al.*, 1987). Very few primary igneous structures are preserved in the Archaean metavolcanics, and these are most common in the late Archaean supracrustal belt of the central Laramie range. Otherwise metamorphism and tectonic disruption has left the metavolcanic rocks mainly as enclaves or discontinuous minor horizons which, to a large extent, appear to have been transformed isochemically. Igneous structures are well preserved in some late Archaean intrusive complexes, most notably the Stillwater Complex of southern Montana (Czamanske and Zientek, 1985).

9.3.1   Early Archaean

Early Archaean basic metavolcanics occur as disseminated high-grade relics of what were clearly complex supracrustal units within younger (*c.* 2.8 Ga) gneisses in the Beartooth Mountains (Henry *et al.*, 1982; Mueller *et al.*, 1985). Rb–Sr, Sm–Nd and Pb–Pb isotopic data suggest that these units are around 3.4 Ga old (Mueller *et al.*, 1982, 1985). The metavolcanics occur as pyroxene-amphibolites and include both mafic and ultramafic types. Geochemically, these rocks broadly correspond to MORB-like tholeiitic basalts ($MgO = 7\%$; $La_N/Lu_N = 0.8$) and komatiitic composi-tions ($MgO = 17\%$; $CaO/Al_2O_3 = 1.7$; $K_2O = 0.17\%$; $Na_2O = 0.89\%$; $La_N = 4$; $La_N/Lu_N = 1$) (Mueller *et al.*, 1985). Other samples have more strongly fractionated REE distributions ($La_N = 20$; $La_N/Lu_N = 2$). The metasedimentary assemblage with which they are associated is more strongly reminiscent of a shelf sequence rather than that of typical greenstone belts.

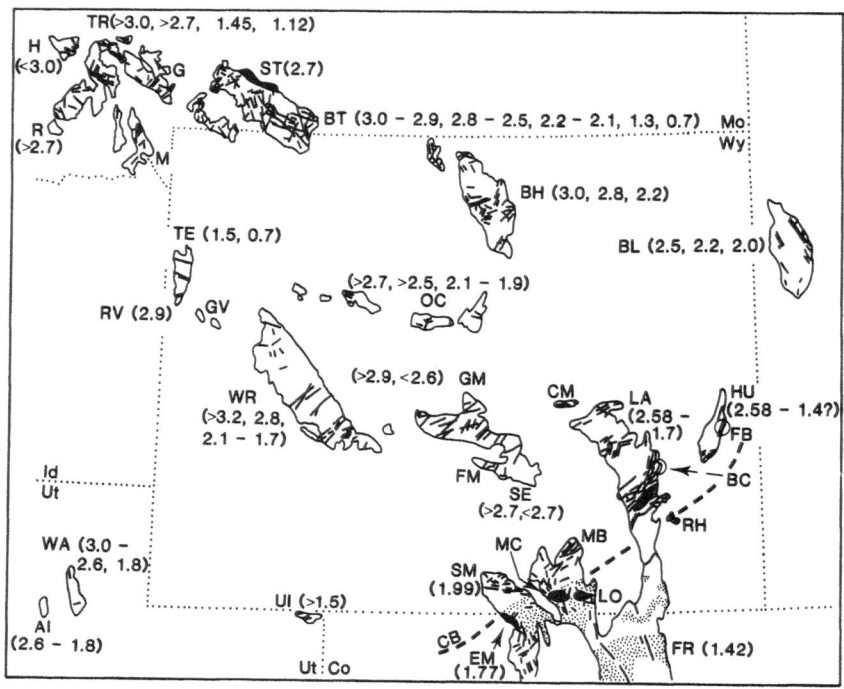

Figure 9.3 Map of the Wyoming Province (dotted lines: state boundaries) showing the distribution of Archaean gneiss outcrops and the minor basic intrusives within them, from data sources referred to in the text (see reference compilation by Snyder *et al.*, 1989). Numbers refer to approximate ages (Ga) of basic dykes.

Mountain ranges (in alphabetical order): AI: Antelope Island; BH: Bighorn; BL: Black Hills; BT: Beartooth; CM: Casper; FR: Front; G: Gallatin; GM: Granite; GV: Gros Ventre; H: Highland; HU: Hartville uplift; LA: Laramie; M: Madison; MB: Medicine Bow; OC: Owl Creek; R: Ruby; SE: Seminoe; SM: Sierra Madre; TE: Teton; TR: Tobacco Root; UI: Uinta; WA: Wasatch: WC: Weber Canyon.

Igneous complexes: BC: Boy Scout Camp; EM: Elkhorn Mountain; LO: Lake Owen; MC: Mullen Creek; FB: Flattop Butte; RV: Rendezvous; ST: Stillwater; stippled ornament (Front, Sierra Madre, Medicine Bow ranges): early Proterozoic metavolcanic rocks; CB (dashed line): Cheyenne Belt.

9.3.2   Late Archaean

The younger Archaean basic metavolcanic rocks of the Beartooth and Bighorn ranges comprise disseminated biotite-amphibolite rafts which vary from a metre to a kilometre in length. Rb–Sr and Nd–Sm isotope systematics suggest that the basic metavolcanics are all marginally older than the enveloping granitic (*s.l.*) gneisses, their interpreted ages being 2.9 Ga compared to 2.8 Ga respectively (Mueller *et al.*, 1983).

The late Archaean basic metavolcanics of these ranges differ from the early Archaean ones in that they are overwhelmingly andesitic in composition, although they do vary continuously from tholeiitic to calc-alkaline types (Armbrustmacher and Simons, 1977; Mueller *et al.*, 1983). The tholeiitic members tend to have lower Mg, Cr, Ni and Al and higher Ti and Fe than the calc-alkaline ones. The major element geochemistry of these andesitic rocks is not unusual compared to that of modern or other Archaean andesites (Gill, 1981; Condie, 1982). However, the metavolcanic suite of the Beartooth Mountains is apparently unique in that (a) the andesitic rocks have unusually large ranges of generally high incompatible element abundances (for example, Zr contents range from around 50 to 450 ppm, and Sr from 200 to 1000 ppm), and (b) the suite overwhelmingly comprises these meta-andesitic lithologies, with only rare basaltic and no acidic or alkalic components. One of their most distinctive chemical characteristics is that they all show marked LREE-enrichment (La_N = 200 to 500; $La_N/Lu_N = c.$ 25) (Mueller et al., 1983), approximately twice the LREE-enrichment in average Archaean andesites (Condie, 1976). As LREE and Zr contents are higher in these rocks than in the host granitic (s.l.) gneisses, a petrogenetic model to explain their geochemistry involving crustal contamination (Armbrustmacher and Simons, 1977) is considered unlikely (Mueller et al., 1983), and instead a model is invoked involving variable metasomatism by an incompatible element-rich fluid phase during varying degrees of partial melting of a plagioclase-poor mantle source, rather than a crustally contaminated melt (Wooden and Mueller, 1988).

The North Snowy Block of the northeastern Beartooth Mountains appears to be a severely tectonised late Archaean zone between two fundamentally different terranes (Mogk et al., 1988). The Beartooth and Bighorn ranges predominantly comprise quartzo-feldspathic gneisses in which amphibolites occur as included remnants, while the Tobacco Root and Ruby ranges consist mainly of supracrustal suites (quartzites, pelites, BIF and metavolcanic amphibolites). This tectonic zone represents a collisional orogeny that occurred along the edge of an older (3.4–3.2 Ga) continent at around 2.7–2.6 Ga, and it is possible that the dominantly andesitic compositions of the 2.9 Ga metavolcanics reflect the early stages of this collisional event. The tectonic evolution of the eastern Beartooth range is discussed by Mueller et al. (1985).

Metavolcanic rocks form a major part of the extensive late Archaean Elmers Rock greenstone belt in the central Laramie range (Figure 9.4), associated with various metasediments (Graff et al., 1982; Snyder; 1984). The abundant metasediments in these supracrustal belts include conglomerates which were deposited under transgressive conditions at the rifted margin of the Archaean continent (Snyder, 1984; Karlstrom and Houston, 1984). A minimum age of 2.6 Ga for the belt is given by the age of intrusive granites (Snyder, 1984). Similar supracrustal rocks occur, probably as a continuation of the belt, to the north and in the Hartville uplift (Figures 9.4, 9.5; Snyder, 1980b; Snyder and Peterman, 1982). The metavolcanics include a wide variety of compositional types, but comprise predominantly a bimodal basic–rhyodacitic suite. The basic members range from low-K tholeiitic basaltic to more Mg-rich types (Table 9.1). Two distinct types of basaltic assemblages are discernible geochemically in both regions. Most samples have flat, unfractionated REE distributions (type I), some having slight, MORB-like, LREE-depletion, while those of the second group (type II) show marked LILE and LREE-enrichment (Figure 9.6). The MORB-normalised element values shown in Figure 9.7 clearly show the flat, MORB-like patterns of the type I volcanics and the relative enrichment of the type II rocks from both the Elmers Rock greenstone belt and the Hartville uplift, but also illustrate the low overall element abundances of the un-enriched Archaean rocks compared to modern MORB (c. 0.6:1). The most magnesian samples (MgO up to 15%) have the chemical characteristics of komatiitic basaltic rocks ($CaO/Al_2O_3 > 0.9$; low alkalis and TiO_2; high MgO/Al_2O_3 and

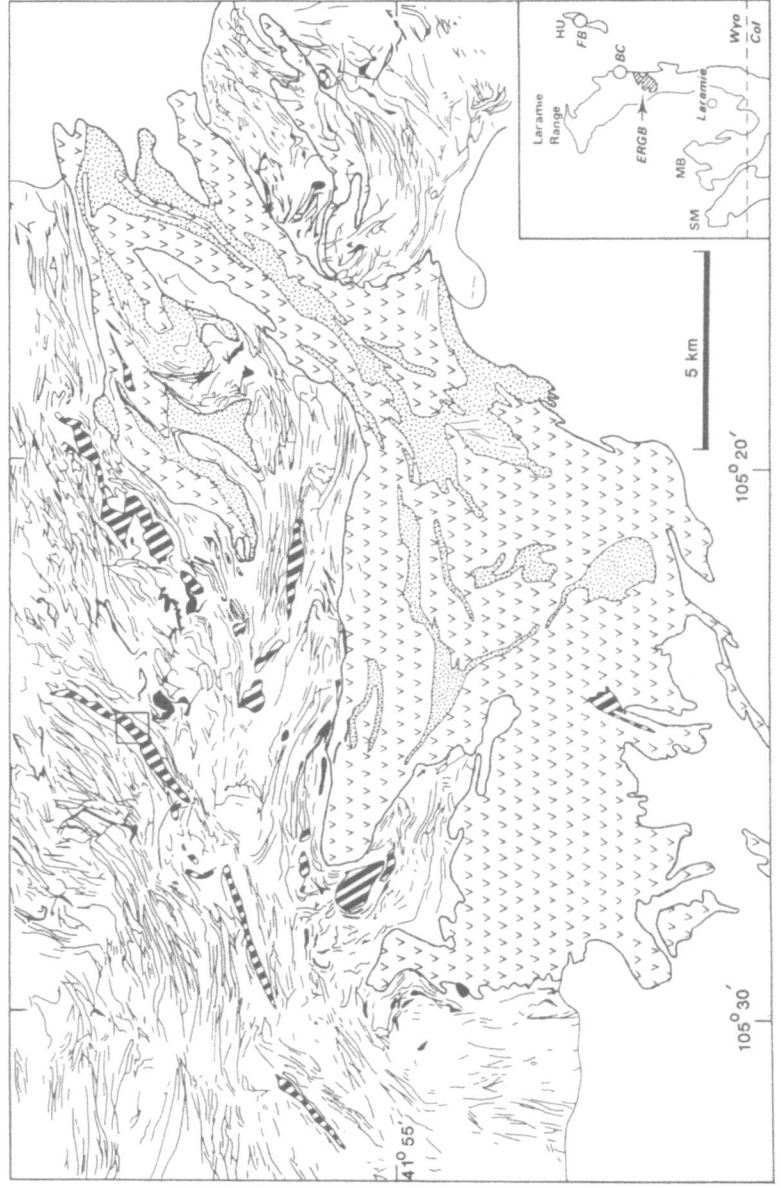

Figure 9.4 Geological sketch map showing the abundance of basic dykes (black lines; NW–SE line shading; harzburgitic intrusions) in the gneisses neighbouring the Elmers Rock greenstone belt (ERGB on inset map) in the central Laramie Mountains. 'V' shading: metavolcanic rocks; stipple: meta-sediments; Laramie Anorthosite Complex and cover rocks not shaded. Box shows approximate locality of intrusive harzburgite contact shown in Figure 9.8. Also shown on inset map are the Sierra Madre (SM), Medicine Bow (MB) and Hartville uplift (HU) mountains and the localities of Boy Scout Camp (BC) and Flattop Butte (FB) granitoid rocks shown in Figure 9.5. Adapted from Snyder (1984).

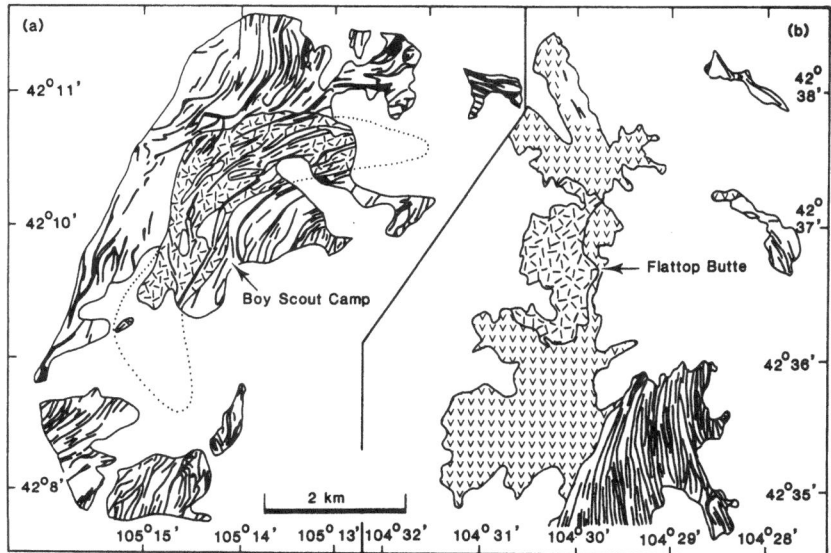

Figure 9.5   Sketch maps showing the contrast in the relationships between amphibolitic basic dykes (black lines) and (a) the 2.05 Ga granodiorite of Boy Scout Camp, central Laramie Mountains and (b) the 1.98 Ga granite of Flattop Butte in the northern part of the Hartville uplift (cross-hatch shading). Despite their apparent similarity, the dyke swarms of the two areas cannot be simply correlated. Some at least of the Laramie range dykes must post-date those in the Hartville uplift. 'V' shading depicts Archaean supracrustal rocks (including unmapped amphibolitic dykes in areas of mafic metavolcanic rocks). Archaean basement gneisses cut by basic dykes and cover rocks left unshaded. Geology from Snyder (1980b; 1986).

MgO/FeO* values) (cf. Jensen, 1976; Arndt *et al.*, 1977; Arndt and Nisbet, 1982a). The juxtaposition of the originally komatiitic basalts, two types of tholeiitic basalts and calc-alkaline acid metavolcanics suggests the involvement of a variety of different magmas and the virtually simultaneous eruption of these different magmas. It is possible that they formed in different provinces and that their present intercalation reflects a significant tectonic influence.

Both the supracrustal rocks and the gneisses marginal to the Elmers Rock and Hartville uplift greenstone belts contain abundant deformed amphibolitic dykes (Figure 9.4), although few dykes have been mapped within the supracrustal belt. However, no dykes of komatiitic composition have been identified within the quartzo-feldspathic gneisses while the numerous noritic and harzburgitic minor intrusives do not appear to have an extrusive, siliceous high-Mg basaltic (SHMB) or boninitic counterpart in the metavolcanic sequence. Most dykes cannot, therefore, be feeders to the volcanic pile. These supracrustals are the most southeasterly Archaean rocks of the Wyoming Province, which is truncated by a major SW-trending shear zone, the Cheyenne Belt (Figure 9.4; Karlstrom and Houston, 1984) (or Wyoming shear zone of Condie, 1982, 1987). This zone is interpreted as an early Proterozoic suture separating the Archaean craton from successive accretionary Proterozoic island arc sequences to the south (Karlstrom and Houston, 1984; Condie, 1986; Reed *et al.*, 1987).

Table 9.1 Representative analyses of early Precambrian metavolcanic rocks, USA

	1 WY52	2 4003	3 WY48	4 4005	5 2507	6 19MSC1	7 B-8	8 A-6	9 GM57	10 Av6
%										
SiO_2	46.1	50.0	49.3	54.5	53.9	50.2	48.1	50.4	50.24	47.2
TiO_2	0.44	1.06	0.76	0.70	0.93	1.40	1.16	0.96	0.72	1.90
Al_2O_3	10.82	14.5	10.87	14.3	14.4	12.3	14.9	14.0	17.7	15.6
Fe_2O_3	14.56	2.37	12.46	1.38	1.91	1.49	4.85	4.97	9.33	14.5
FeO	—	9.75	—	8.39	7.83	10.9	8.27	6.47	—	—
MgO	14.15	6.34	12.23	6.66	7.07	8.96	7.13	8.21	5.79	6.9
CaO	11.93	11.6	12.29	9.50	8.99	10.6	10.6	10.4	11.60	8.7
Na_2O	1.43	2.05	1.53	2.12	2.43	1.73	2.70	2.85	3.33	1.93
K_2O	0.22	0.24	0.32	0.42	0.13	0.50	0.61	0.20	0.66	0.42
MnO	0.24	0.23	0.20	0.17	0.14	0.20	—	—	0.26	0.21
P_2O_5	0.12	0.09	0.16	0.07	0.08	0.15	—	—	0.18	0.26
ppm										
Sc	42	39	40	37	47	32	48	45	40	42
Cr	1700	234	667	124	195	556	148	67	130	204
Ni	391	130	225	150	50	276	48	45	28	101
Rb	10	10	14	10	<2.5	8	26	nd	10	18
Sr	52	94	117	124	102	134	181	315	654	214
Y	10	22	14	16	15	21	—	—	17	44
Zr	19	62	47	63	78	109	98	102	35	178
Nb	—	—	3	8	5	6	—	1	5	—
Hf	1.1	1.73	—	2.0	1.25	2.86	—	2	1.3	2.9
Ta	1.3	0.18	0.5	2.29	0.13	0.47	—	0.2	0.1	0.4
Ba	17	41	28	89	36	350	172	148	197	110
La	2.00	3.55	4.20	8.73	2.73	16.9	7	20	17	14
Ce	3.81	7.73	9.28	19.1	7.76	37.4	19	37	31	35
Sm	0.99	2.57	2.00	2.48	1.97	4.75	4.0	4.2	3.3	6.2
Eu	0.35	0.93	0.75	0.82	0.69	1.36	1.6	1.9	0.6	1.9
Gd	1.34	3.40	2.54	3.08	2.70	4.95	—	—	3.3	—
Yb	1.16	2.64	1.43	2.00	2.20	1.94	4.4	3.9	1.5	3.8
Lu	0.23	0.40	0.23	0.31	0.33	0.30	0.8	0.64	0.2	0.7

nd: not detectable; —: no data available.

Nos. 1 to 6 from authors' unpublished data; 1 to 4: Elmers Rock greenstone belt, central Laramie Mountains; 5,6: Hartville uplift metavolcanics; 7,8: groups I and II early Proterozoic tholeiites, Dubois greenstone succession, western Colorado (Condie and Nuter, 1981); 9: early Proterozoic Green Mountain formation calc-alkali basalt, SE Wyoming (Condie and Shadel, 1984); 10: average of six early Proterozoic rift basalts, Dos Cabezas mountains, Arizona (Condie et al., 1985).

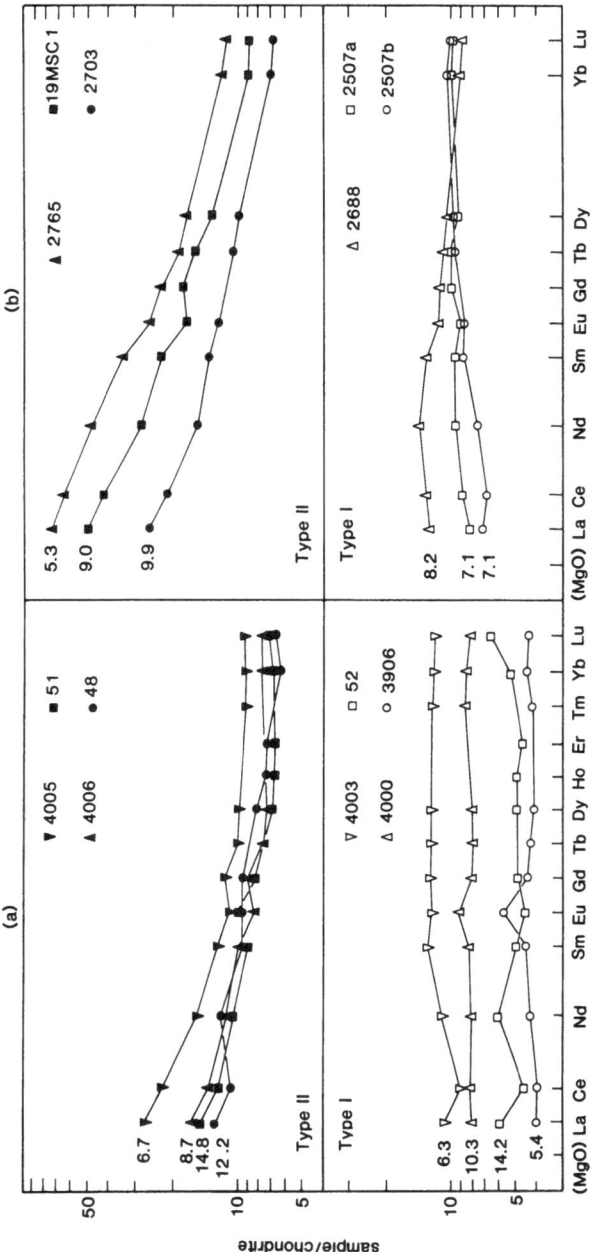

Figure 9.6 Unfractionated (type I) and LREE-enriched (type II) REE abundances in basic metavolcanics from (a) the Elmers Rock greenstone belt and (b) the Hartville uplift. The two types of REE profiles are not simply related to MgO content. (i.e. fractionation).

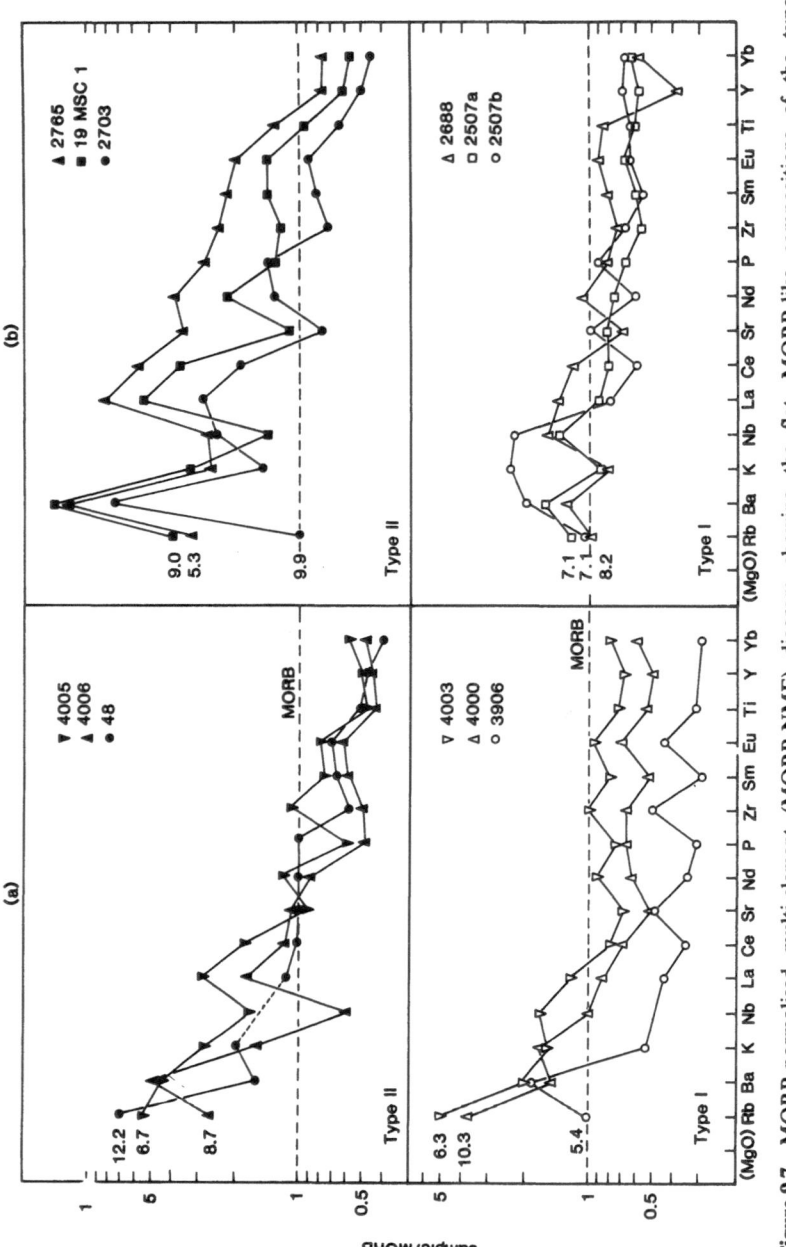

Figure 9.7 MORB-normalised multi-element (MORB-NME) diagram showing the flat, MORB-like compositions of the type-I metavolcanics and the more strongly fractionated patterns of the type-II rocks from (a) the Elmers Rock greenstone belt and (b) the Hartville uplift (authors' unpublished data).

9.4   Wyoming Province: intrusive rocks

This section describes the Archaean and early Proterozoic mafic and ultramafic intrusive rocks of the Wyoming Province. They occur as large-scale layered complexes, as abundant swarms of dykes variably constituting between 1 and 30% of the exposure, and as widespread disrupted lenses, boudins and irregular pods within the quartzo-feldspathic gneisses. A little additional information comes from boreholes through the Phanerozoic cover (Peterman, 1981; Peterman and Futa, 1987; Tweto, 1987) and from xenoliths in the Laramide intrusives (Brozdowski, 1985; Dudas and Eggler, 1986; Eggler, 1987).

Precambrian mafic intrusive rocks are common throughout North America. They range in age from c. 3.6 to 3.3 Ga in the early Archaean gneiss terrane of northern Minnesota (Schulz, 1980), to late Proterozoic (c. 1.4–1.3 and 0.8–0.7 Ga) sills and dykes in the Belt basin supracrustal rocks of northern Idaho and Montana (Obradovich et al., 1984; Zartman et al., 1982) and the Beartooth Mountains of southern Montana and northern Wyoming (Wooden and Mueller, 1979; Mueller et al., 1982). The early Archaean rocks occur as disrupted amphibolitic layers within high-grade gneisses and are mainly plagioclase-rich tholeiitic rocks, distinct from the komatiitic to tholeiitic metavolcanic suite of the region (Schulz, 1982). The late Proterozoic sills and subordinate dykes of the Belt basin (the Purcell sills) are correlated with the Moyie sills in Canada. Late Proterozoic intrusives are also associated with high-Ti within-plate basalts that vary from transitional to alkaline rocks in the Bannock range of southern Idaho (Harper and Link, 1986).

Mid-Proterozoic basic minor intrusions include a set of 1.4 Ga old lithologically diverse NW-trending dykes found on both sides of the Cheyenne Belt, the sheared southern margin of the Archaean craton (Karlstrom and Houston, 1984). The 'Iron Dike' of this set is a dolerite as much as 45 m thick and traceable for 130 km, which has 8.5 modal % titaniferous magnetite (Wahlstrom, 1956). A near-vertical flow lineation indicates that the '...magma rose vertically throughout the length of the dike and thus that the deep source was widespread...' (Tweto, 1987). In the northern Park range of Colorado, dykes of this general age include NW- and NE-trending porphyry dykes that respectively cut and are cut by a phase of the 1.47 Ga Mount Ethel pluton (Snyder and Hedge, 1978). The following account incorporates only the mid-Archaean to early Proterozoic basic intrusive rocks of the Wyoming Province (Figure 9.3). A more comprehensive account of the distribution of mafic dykes in the Wyoming Province and Belt Basin has been compiled by Snyder et al. (1989).

9.4.1   Southern Idaho, northern Utah and Nevada

The exposures of Archaean rocks in northern Utah contain only sparse amphibolitised mafic intrusives. Those in the Weber Canyon have been dated at 2.74 Ga. Two generations of basic dykes in the Wasatch range have ages of 1.8 Ga and between 3.0 and 2.6 Ga respectively, and those on Antelope Island are between 2.6 and 1.85 Ga old (Hedge et al., 1983; Bryant, 1988).

WNW-, E- and ENE-trending sills, dykes and irregular bodies up to 2 km long of amphibolite, hornblendite and epidiorite occur in the northeastern Uinta Mountains. They were intruded into strongly deformed metasediments and were themselves deformed and metamorphosed prior to 1.5 Ga (Hansen, 1965). Mafic intrusive bodies are not a major component of the Archaean terrane in the Albion range or nearby mountains on the Utah–Idaho border (Armstrong and Hills, 1967; Miller, 1980).

It has recently been recognised that the Wyoming Province extends some 160 km further southwest from the Albion range, into the northern East Humboldt range in northeastern Nevada (not shown in Figure 9.3). The recently dated 2.5 Ga gneisses here contain numerous amphibolitic bodies interpreted as the remnants of mafic intrusions (Lush *et al.*, 1988).

9.4.2   Black Hills, South Dakota

Three generations of early Proterozoic differentiated metagabbroic sills cut the supracrustal rocks on the north and east sides of the Black Hills uplift of South Dakota (Dewitt *et al.*, 1986). Archaean ultramafic rocks probably also form part of the basement in this region. Detrital chromite in early Proterozoic meta-conglomerates indicates that the ultramafic rocks were exposed in the Archaean terrane at this time, although none presently outcrops (Redden *et al.*, 1988).

The oldest Proterozoic mafic intrusive unit is the 2.2 Ga old, 900 m thick Blue Draw metagabbro, near Nemo (Redden, 1981). This metamorphosed intrusion displays relict gravity-stratified layering and consists, from its base upwards, of serpentinite, hornblendite and amphibolite, with granodiorite developed locally at the top. Similar but more strongly differentiated younger layered metagabbroic sills up to 300 m thick near Bogus Jim Creek and Prairie Creek have recently been dated at 2.0 Ga and 1.9 Ga respectively (Z.E. Peterman, 1988, pers. comm.) and are similar in age to the overlying metavolcanic rocks. These sills and metavolcanics are tholeiitic in composition with flat REE_N distribution patterns, and were deformed, amphibolitised and intruded by granite at about 1.7 Ga (Dewitt *et al.*, 1986). The granites were derived from the partial melting of mid-Archaean crust, but the isotopic composition of the 2.2–1.9 Ga tholeiitic metavolcanics and associated metasediments indicates the accretion of mantle-derived material during the very early Proterozoic (Walker *et al.*, 1986).

9.4.3   Southern Wyoming and northern Colorado

At least three generations of Proterozoic mafic intrusions (2.1–2.0, 1.77 and 1.4 Ga respectively) in southern Wyoming and northern Colorado penetrate early Proterozoic supracrustal rocks which unconformably overlie Archaean granitic basement containing an earlier set of Archaean mafic intrusives (> 2.5 Ga). The two oldest groups can be distinguished radiometrically but not lithologically, and the two youngest groups also contain lithologically similar types but can be distinguished by their structural history and lithological associations.

High-grade metamorphosed intrusives in the Medicine Bow and Sierra Madre Mountains (Figure 9.3) are deformed along with their host quartzo-feldspathic gneisses into NNW-trending folds. These intrusives, together with other garnet amphibolites and peridotites cutting the Archaean basement gneisses, are considered to be Archaean (Houston *et al.*, 1978; Divis, 1976; Karlstrom *et al.*, 1981). The ages of the early Proterozoic mafic and ultramafic intrusives in both ranges are relatively well constrained. For example, a 2.0 Ga Sm–Nd mineral and whole-rock isochron is reported from the 2 km long Spring Creek Lake olivine norite body that cuts Archaean granite in the central Sierra Madre (Karlstrom *et al.*, 1981; Shaw *et al.*, 1986), although the Rb–Sr systematics are imprecise. Four zircon fractions from a pegmatitic phase of the 1.3 km long Cushman Creek metagabbro plug have yielded a U–Pb age of 2.1 Ga (Premo and Van Schmus, 1988), and the altered Gaps intrusive in the Medicine Bow Mountains is also about 2.0 Ga (Karlstrom et al., 1981). Thus, different types of intrusions which have been described variously as tholeiitic and boninitic were clearly active virtually simultaneously at this time.

Dykes in the Hartville uplift include quartz amphibolites that are isotopically co-linear on a 2.58 Ga granite Rb–Sr isochron. The 1.98 Ga granite of Flattop Butte (Figure 9.5b; Snyder and Peterman, 1982) is younger than most of the amphibolite dykes that form such a dense swarm in the neighbouring Archaean granites. However, younger NW-trending amphibolitic dykes cut the 1.74 Ga Haystack Range granite and olivine basalts intrude folded metasediments northeast of Guernsey, Wyoming.

Several types of mainly NE-trending mafic and ultramafic dykes in the central Laramie Mountains and Hartville uplift comprise voluminous swarms (as much as 20% of outcrop) (Figures 9.4, 9.5). They range from older plagioclase-phyric meta-basalt to altered aphyric tholeiitic doleritic and noritic types and rare clinopyroxene-cumulate dykes with basaltic komatiitic compositions (Snyder, 1984). Some of the largest scale intrusions in the Laramie Range (up to 7 km long and 500 m thick) are ultramafic, and were derived dominantly from markedly discordant orthopyroxene-rich (harzburgitic) protoliths (Figure 9.8; Table 9.2). One differentiated body (Sellers Mountain) has a peridotitic base and a basaltic top. They are commonly associated with, and thought to be related to, smaller-scale noritic dykes. The tholeiitic, harzburgitic and noritic intrusions which occur together in the central Laramie Mountains are not well dated internally (Davis et al., 1977; Snyder et al., 1985). Recent zircon isotope data (K.R. Ludwig, unpublished data) show that garnetiferous amphibolite dykes are younger than a 2.05 Ga grandodiorite in the vicinity of Boy Scout Camp (Figure 9.5a); while some similar dykes pre-date a 1.74 Ga pyroxene granite in the Richeau Hills, 30 km to the southeast (Snyder, 1984, Part L). Most of the dykes probably fall within this 2.05–1.74 Ga age bracket.

The various dykes in the Laramie range appear to be derived from very different magma types. The metadolerites retain ophitic textures and mainly have normal tholeiitic compositions with mildly fractionated REE distributions ($La_N = 20$; $La_N/Lu_N = 2$) (Figure 9.9). The minor noritic intrusions are characterised by high modal orthopyroxene and the late crystallisation of plagioclase, and their relatively high normative hy:di ratios and lower normative feldspar contents than the neighbouring doleritic rocks (Table 9.2). They show an equal or slightly higher degree of LREE enrichment ($La_N/Lu_N = 3$) considering their higher MgO contents (c. 15%), and are slightly more siliceous than the doleritic dykes. The larger, harzburgitic sheets are predominantly cumulus-textured, comprising olivine enclosed by orthopyroxene oikocrysts, with minor interstitial clinopyroxene and plagioclase. These plagioclase harzburgites have dish-shaped chondrite-normalised REE patterns (Figure 9.9), with middle order REE slightly lower than both the light and heavy REE ($La_N/Gd_N = c.$ 2; $Gd_N/Lu_N = c.$ 0.6), and with overall REE abundances around 2 to 7 times those of average chondrite (Table 9.2). It is significant that this REE distribution is similar to those in some types of modern boninite (Hickey and Frey, 1982). Similar ranges of noritic and doleritic intrusives with contrasting REE geochemistry have been found in the eastern extension of the central Laramie Mountains, in the Hartville uplift (Table 9.2). The norites from this area also show marked LREE-enrichment, in contrast

Figure 9.8   Obliquely discordant intrusive contact between harzburgitic intrusion ('Tony Ridge peridotite'; dark grey, left of photograph) and host banded grey gneisses, North Fork Cherry Creek, central Laramie Range (see Figure 9.4). The banding in the gneisses slopes from upper left to lower right. Tholeiitic dykes between 1 and 20 m thick also cut the grey gneisses but are here truncated by the ultramafic intrusion. Thin (< 1 m) olivine norite dykes cut the ultramafic body at localities indicated by the asterisks. Harzburgitic intrusion at this locality is c. 150 m thick.

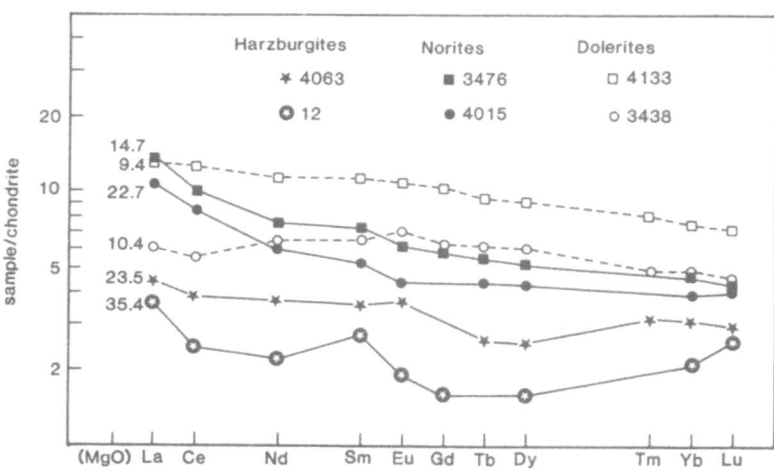

Figure 9.9   Chondrite-normalised REE abundances in harzburgitic, noritic and doleritic intrusive rocks, central Laramie range (authors' data). The dolerites have the highest levels of only mildly fractionated REE ($La_N/Lu_N = 1.5$). The harzburgites have slightly 'dish'-shaped REE patterns, the middle order REE being lower than both the LREE and HREE. The magnesian norites have the most strongly fractionated REE abundance ($La_N/Lu_N = 3$).

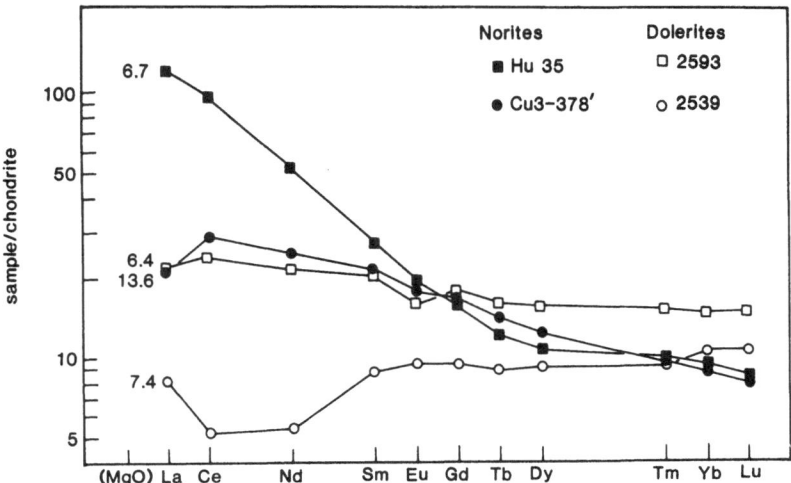

Figure 9.10   Contrasting REE abundances in noritic and doleritic rocks from the Hartville uplift.

to the unfractionated or mildly LREE-enriched distributions found in the associated tholeiitic dolerites (Figure 9.10).

The coupling of high Mg, Si and LREE in the noritic (and harzburgitic) intrusions means that these rocks cannot be derived directly from, or related by any sensible fractionation model to, the tholeiitic parental magma of the doleritic dykes, nor can these contrasting basic rocks simply reflect different degrees of partial melting of the same source. The increased partial melting required to give the higher MgO contents of the noritic rocks clearly would not produce the higher LREE contents (Figures 9.9, 9.10). While the crustal contamination of ascending basic magma cannot be discounted in explaining the slightly unusual geochemistry of the norites, it could only have-been partially responsible as many of the LIL elements are no more enriched in the Laramie norites and harzburgites than they are in the associated dolerites. The abundance and compositions of the mafic and ultramafic dykes suggest extensive post-2.6 Ga mantle partial melting (Holden and Snyder, 1983). The REE and Sr and Nd isotope geochemistry of the different types of dykes suggest two distinct mantle sources (Snyder *et al.*, 1985; Hall *et al.*, 1987b). ε_{Nd} values range from chondritic to positive and Sr_i ratios range from 0.7015 to 0.7045 (Snyder *et al.*, 1985). Most of the (meta-) dolerites appear to have been fed by a tholeiitic basic magma derived from a fertile source, while most ultramafic rocks appear to have originated from a depleted, but variably metasomatised harzburgitic mantle.

Three plutonic layered basic complexes and numerous smaller intrusions cut the early Proterozoic rocks just to the southeast of the Archaean craton in southernmost Wyoming and northern Colorado. These are (from NE to SW) the Lake Owens and Mullen Creek complexes and the gabbro of Elkhorn Mountain (Figure 9.3), which become increasingly amphibolitised towards the southwest (Houston *et al.*, 1978; Snyder, 1980a; Edwards, 1981; Karlstrom *et al.*, 1981; Myers *et al.*, 1987; Snyder *et al.*, 1989). The freshest intrusion is the Lake Owens complex which is a strongly compositionally layered body of gabbronoritic and troctolitic rocks totalling about 5.5 km in thickness and covering around 155 km^2. It is thought to have been derived

Table 9.2 Representative analyses of early Precambrian mafic intrusives, Wyoming, USA

	1 WY12	2 4063	3 3476	4 3438	5 HU35	6 2539	7 111	8 105A	9 104	10 93A
%										
SiO_2	47.69	49.4	52.2	50.9	48.3	49.7	47.60	51.27	51.15	48.65
TiO_2	0.16	0.28	0.54	0.60	0.84	0.76	0.84	0.63	0.63	0.88
Al_2O_3	3.32	7.59	10.3	12.9	16.4	14.4	12.36	14.88	12.91	15.54
Fe_2O_3	1.50	1.13	1.33	1.52	3.18	2.82	2.35	10.94	1.98	3.47
FeO	8.46	9.06	7.98	7.72	8.66	8.80	9.58	—	8.23	8.83
MgO	35.36	23.5	14.7	10.4	6.73	7.45	15.08	8.45	11.22	7.45
CaO	3.24	6.45	9.92	13.3	9.47	11.70	8.32	10.53	10.77	11.29
Na_2O	0.04	0.85	1.20	1.41	2.55	2.38	1.72	1.66	1.49	1.86
K_2O	0.03	0.05	0.28	0.11	0.74	0.78	0.65	1.29	0.39	0.66
MnO	0.17	0.18	0.17	0.18	0.18	0.21	0.19	0.19	0.19	0.21
P_2O_5	0.02	<0.05	<0.05	<0.05	0.25	0.05	0.22	0.16	0.14	0.18
ppm										
Sc	18	30	33	38	29	51	23	37	37	39
Cr	3537	2470	1400	893	143	141	951	215	830	161
Ni	1625	—	300	150	345	306	824	111	247	116
Rb	nd	<2.5	9	10	16	22	10	38	12	17
Sr	18	<75	80	123	524	103	296	143	120	133
Y	3	—	9	11	18	13	14	16	13	20
Zr	6	47	34	29	74	33	78	50	41	62
Ba	5	<35	66	98	709	102	320	226	128	133
La	1.21	1.47	4.31	2.03	39.3	2.69	11.6	6.49	4.96	6.51
Ce	2.15	3.27	8.64	4.90	83.4	5.30	22.3	13.20	9.21	13.42
Sm	0.56	0.74	1.48	1.34	5.49	1.81	2.93	2.16	1.94	2.64
Eu	0.14	0.29	0.47	0.54	1.46	0.73	1.06	0.72	0.67	0.91
Gd	0.44	—	1.60	1.70	4.13	2.50	2.95	2.53	1.87	3.24
Yb	0.47	0.70	1.07	1.03	2.03	2.27	1.44	1.86	1.51	2.36
Lu	0.09	0.10	0.15	0.16	0.28	0.36	0.21	0.29	0.25	0.36

norm %	1	2	3	4	5	6	7	8	9	10
Q	—	—	1.1	0.9	—	—	—	—	1.2	—
or	0.2	0.3	1.7	0.7	4.5	4.7	3.9	7.7	2.3	3.9
ab	0.3	7.3	10.3	12.0	22.0	20.3	14.7	14.2	12.7	15.9
an	8.8	16.9	22.2	28.8	31.7	26.6	24.3	29.7	27.6	32.4
di	5.7	12.0	22.0	30.3	6.9	25.9	12.8	17.9	20.5	18.7
hy	39.9	39.8	39.7	23.9	25.4	10.7	18.2	26.3	31.2	21.4
ol	42.7	20.8	—	—	0.5	6.1	20.4	1.0	—	0.5

Analyses from authors' unpublished data. 1.2: Tony Ridge peridotite, central Laramie Mountains; 3. 4: noritic and doleritic dykes, central Laramie Range; 5. 6: noritic and doleritic dykes, Hartville uplift; 7. 8: noritic and doleritic dykes, Beartooth Mountains; 9. 10: noritic and doleritic dykes, Bighorn Mountains. Normative composition of analysis 8 based on $FeO:Fe_2O_3$ ratio of 4:1

from a siliceous, noritic magma that was replenished in the magma chamber with more basic liquids (Patchen and Myers, 1987; R.R. Loucks, 1987, pers. comm.). In terms of its mineralogy, layering and geochemistry, this intrusion resembles the better known Stillwater Complex of southern Montana. Both the Mullen Creek Complex and the gabbro of Elkhorn Mountain have been precisely dated, zircon ages from dioritic phases being 1.77–1.78 Ga (Snyder, 1980a; Reed et al., 1987; Loucks et al., 1988). A set of NE-trending basaltic dykes cuts the gabbro of Elkhorn Mountain (Snyder, 1980a; Snyder et al., 1988).

9.4.4 Stillwater Complex and minor intrusions, Beartooth Mountains

Abundant mafic intrusions cut the Archaean basement of the Beartooth Mountains of northern Wyoming and southern Montana, including the Stillwater Complex (Czamanske and Zientek, 1985) and numerous variously oriented (mainly NW-trending) mafic dykes (Figure 9.3), that span a wide range of compositions and ages. They have been extensively studied, and relationships between different generations have been described by, among others, Ecklemann and Poldervaart (1957), Prinz (1964), Fraser et al. (1969), Reid et al. (1975), Wedow et al. (1975), Simons et al. (1979), Casella et al. (1982) and Elliot et al. (1983). Numerous attempts have been made to apply radiometric dating techniques to the mafic intrusions (e.g. Condie et al., 1969b; Baadsgaard and Mueller, 1973; Mueller and Rogers, 1973; Wooden, 1975; Wooden et al., 1982).

There are at least six different generations of intrusive basic rocks (Mueller et al., 1982) comprising: (i) 2.9–3.0 Ga old dioritic amphibolites, (ii) the 2.7 Ga old Stillwater Complex and associated minor sills, (iii) a range of 2.8–2.5 Ga old mafic dykes of various compositions, (iv) 2.2–2.1 Ga mafic dykes, (v) 1.3 Ga olivine-normative dykes, and (vi) 0.7 Ga quartz-tholeiites. However, despite these groupings, there is no simple relationship between intrusive age, orientation and composition (Condie et al., 1960a, b; Wooden, 1975), in contrast to the dyke swarms of the Canadian Shield (Fahrig and West, 1986) or southern Greenland (Hall et al., 1985). Wooden (1975) and Wooden and Mueller (1979) identified five generations and seventeen chemical groups of dykes, and a limited consistency in the orientation of the Proterozoic dykes, compared to a diversity of trends of the Archaean ones. The 2.1–2.0 Ga dykes are generally oriented N–S, the 1.3 Ga group is consistently WNW-trending, and the c. 0.7 Ga group trends NW. Most groups are severely limited in their geographic extent (Wooden and Mueller, 1979); only the 1.3 Ga dolerites occur throughout the Beartooth range.

Simons et al. (1979) also recognised six compositional varieties of basic intrusive rocks in the eastern Beartooth range, including quartz-dolerites, alkali olivine-dolerites, metadolerites, ultramafic rocks, amphibolites and hornblende gneisses. The length-weighted rose diagram (Figure 9.11) derived from the map of Simons et al. (1979) shows that these different types do appear to have fairly consistent orientations. Some are clearly equivalent to the groups identified by Wooden and Mueller (1979), but others are not easy to correlate. Without reliable age data for each dyke, the relationships between composition and orientation are difficult to assess in a region of such varied basic magmatism.

Some pre-2.8 Ga ultramafic pods up to 140 metres long in the eastern Beartooth Mountains were incorporated tectonically into the host granitic gneisses (Skinner,

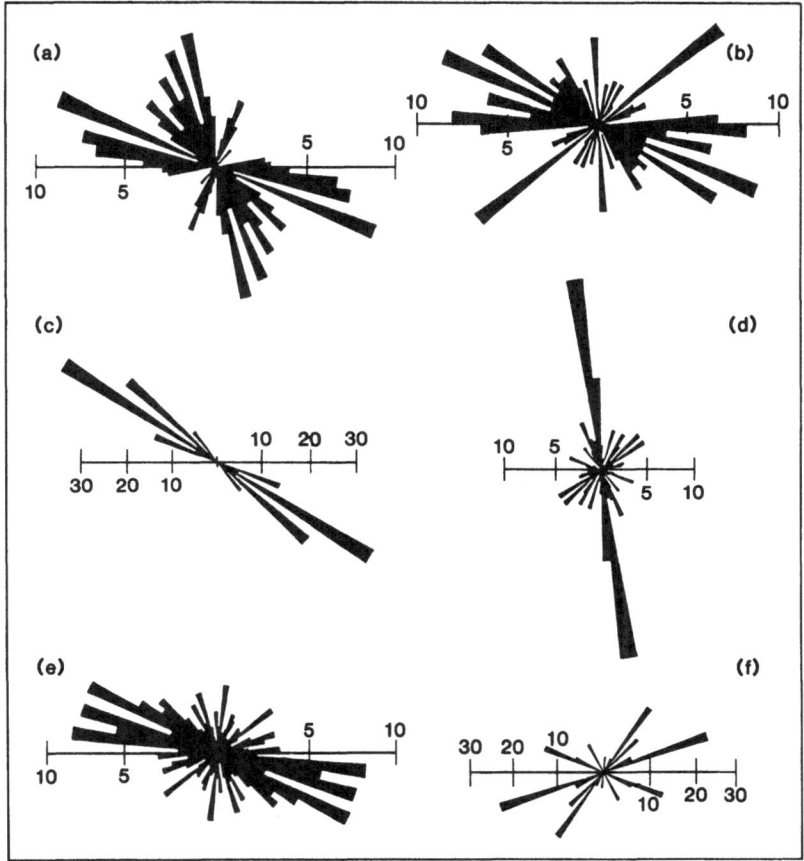

Figure 9.11   Length-weighted rose diagrams for Precambrian mafic dykes of the eastern Beartooth Mountains derived from the map of Simons *et al.* (1979). (a) quartz dolerites (n = 1311); (b) alkali olivine dolerites (n = 96); (c) metadolerites (n = 1393); (d) unassigned dykes (n = 318); (e) amphibolites and hornblende gneisses (n = 390); (f) ultramafic rocks (n = 29). Percentage scale lines oriented E–W in each diagram. Note different scales.

1969), and other chromite-bearing ultramafic bodies are the disrupted fragments of a large stratiform complex (Loferski, 1986). The wide variety of 2.8–2.5 Ga chemical groups and the variously orientated fractures that they occupy suggest that they represent the diachronous intrusion of a number of diverse magmas emplaced into a rapidly changing tectonic stress field. For example, fine-grained noritic dykes occur together with doleritic ones. The noritic dykes are characterised by high Mg, Ni and Cr (Table 9.2), but also relatively high Si and LREE contents ($La_N/Lu_N = 5$) (Figure 9.12). They have relatively low Ca contents and high normative orthopyroxene, although their modal pyroxene assemblages span a wide range of Mg-rich, low-Ca types (Figure 9.13), similar to those in modern boninites (Cameron *et al.*, 1979; C.P. Wood, 1980). Neighbouring doleritic dykes also possess a wide range of pyroxenes, but these are more Fe-rich augites (Figure 9.13). These dykes have normal continental tholeiitic compositions (Table 9.2), with slightly less strongly fractionated REE

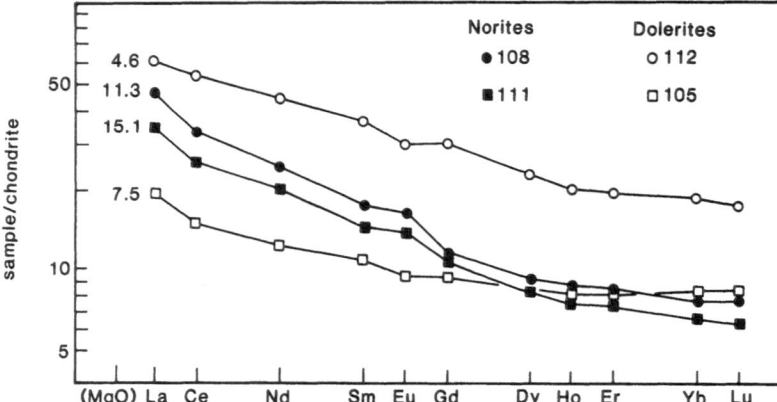

Figure 9.12 Contrasting chondrite-normalised REE abundances in representative doleritic (mildly LREE-enriched) and fine-grained Mg-rich noritic dykes (strongly fractionated) from the southern Beartooth Mountains (authors' data).

($La_N/Lu_N = c.$ 2.5) (Figure 9.12). The 2.2–2.1 Ga event includes two or three magma types, and the younger two intrusive events both represent single pulses of magma. The extensive intrusion of the 1.3 Ga dykes was probably related to the formation of the Belt basin, active between 1.4 and 0.9 Ga, the main axis of which is parallel to the dyke swarm (Wooden and Mueller, 1979).

By far the largest basic intrusion of the Beartooth range is, of course, the late Archaean Stillwater Complex (Czamanske and Zientek, 1985). This complex is a strongly differentiated, coarse-grained, post-orogenic mafic–ultramafic body up to 5.5 km thick in the northern Beartooth Mountains. It was emplaced at 2.7 Ga (DePaolo and Wasserburg, 1979; Nunes and Tilton, 1971; Lambert *et al.*, 1985) and comprises cumulus peridotites, bronzitites, norites, gabbros and anorthosites (Zientek *et al.*, 1985). The varied and frequently intricate rhythmic layering of these rocks appears to reflect both the complex interactive differentiation mechanisms operative in most large-scale basic intrusions (McBirney, 1985; Irvine, 1980; Parsons, 1987) and the repeated influxes of different types of basic magma into a relatively quiescent pool of differentiating magma (McCallum *et al.*, 1980; Bow *et al.*, 1982; Todd *et al.*, 1982; Irvine *et al.*, 1983; Ryder and Spettel, 1985). Component primary magmas with komatiitic, picritic, tholeiitic, anorthositic and boninitic compositions have been proposed.

The profusion of layered cumulates making up the Stillwater Complex severely impedes the direct recognition of parental magma compositions. Recent attempts to do so have thus concentrated on the surrounding dykes in the Beartooth Mountains, and on chilled margins of dykes and sills closely associated with the complex (Longhi *et al.*, 1983; Helz, 1985, 1987). One dyke which intrudes and is chilled against one of the eastern Stillwater cumulate zones has given a tentative (2-point) Sm–Nd mineral isochron indicating an age of 2.65 Ga. Of six geochemically distinct groups of Precambrian high-Mg dykes in the Beartooth Range, only two have been found to contain sufficiently magnesian olivine and orthopyroxene to be possible candidates as parental magma compositions for the Stillwater Complex, but none of these groups

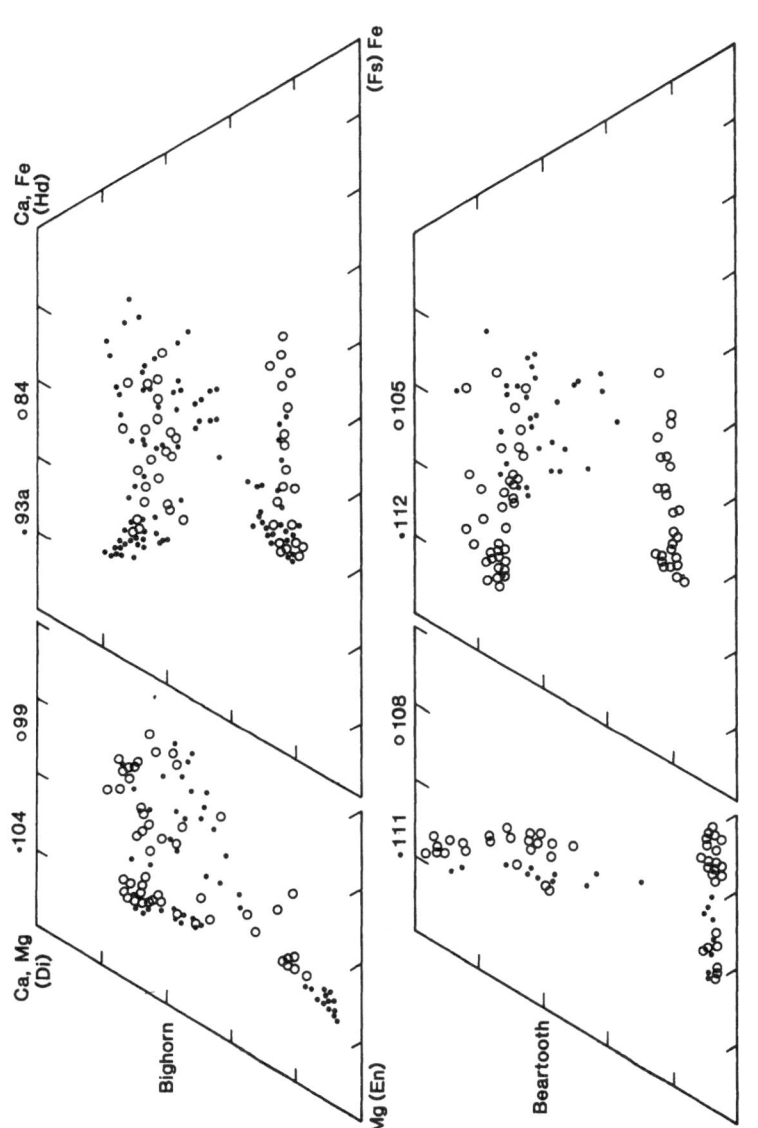

Figure 9.13 Contrasting pyroxene assemblages in noritic and doleritic dykes (left and right of diagrams respectively) from the southern Beartooth and northern Bighorn Mountains (authors' data).

is considered to have sufficiently calcic plagioclase (Longhi et al., 1983). These high-Mg dykes appear to have been derived principally from two distinct magmas, one tholeiitic and the other rich in normative orthopyroxene (noritic). The rocks derived from these magma types have around 11 and 13% MgO respectively, but also have relatively high K_2O (up to 0.8%) and LREE contents. Longhi et al. (1983) proposed that these magmas were derived from variously olivine-fractionated komatiitic parents, which underwent crustal contamination during their ascent. It is also possible that they were derived from metasomatised mantle sources, in comparison with the model proposed for the late Archaean andesitic amphibolites of the eastern Beartooth Mountains (Wooden and Mueller, 1988).

Six different groups of chilled-margin basic dykes and sills have also been recognised at the base of the Stillwater Complex (Helz, 1985, 1987). These vary up to 0.5 km in width and range in orientation from E-W in the west to NE-SW in the east. Many were emplaced along faults and some are sheared. The early minor intrusions which slightly pre-date the Stillwater Complex are the most appropriate rocks for investigating the Stillwater parental magma compositions and are different from the post-complex dykes which comprise late Archaean and Proterozoic tholeiitic dolerites, metadolerites and plagioclase-phyric 'leopard rock' basalts (Baadsgaard and Mueller, 1973). One early basal norite dyke has recently yielded 2.71 Ga zircons (Premo et al., 1986). Two groups, comprising mafic norite and high-Mg gabbronorite dykes with 9.6–13% MgO respectively, are interpreted as representing the main parental Stillwater liquids, while a liquid with the composition of the olivine gabbro sills (6.4–7.6% MgO) formed an additional minor component (Helz, 1985). The mixing of these different magmas obviously requires their simultaneous generation and intrusion, which might also apply to the associated discrete dykes in the Beartooth range, and to the dykes further to the east and south in the Bighorn and Laramie Mountains.

9.4.5   *Tobacco Root and neighbouring mountains, southwestern Montana*

Three major episodes of mafic intrusives have been recognised in the Tobacco Root and neighbouring mountain ranges west of the Beartooth Range in southwestern Montana (Figure 9.3) (James and Hedge, 1980). First, widespread emplacement of mafic dykes and sills occurred before c. 3.0 Ga. These intrusions now occur as metamorphosed and deformed sheets, lenticular bodies and boudins of amphibolite and hornblende gneiss. Subsequently, mafic sheets were locally emplaced after two episodes of folding, but between two major metamorphic events (granulite and amphibolite facies respectively), the younger of which culminated at 2.75 Ga. Lastly, dolerite dykes were intruded during regional uplift at about 1.45 Ga and again at 1.12 Ga (Wooden et al., 1978).

The 1.45 Ga dyke swarm in the Tobacco Root and Ruby Range Mountains comprises primitive tholeiite dykes with 0.2–0.8% K_2O, while other dykes are more highly differentiated or contaminated quartz-normative tholeiites with K_2O values up to 1.6% (Koehler 1976; Wooden et al., 1978). The 1.12 Ga dykes were derived from two different magmas, having FeO* values of around 9% and 15% respectively. The isotopic systems of some of the dykes were disturbed when subsequent basic dykes were intruded beside them. The dykes in the Tobacco Root Mountains were emplaced progressively northeastwards, into WNW-trending fractures. In addition to these unmetamorphosed basic dykes, an equal volume of older, metamorphosed mafic and ultramafic sheets and pods occurs in the Tobacco Root Mountains (McCulloch and Cummings, 1987). These are oriented NNE, at right angles to and cut by the later dyke swarms (Vitaliano *et al.*, 1979).

Similar pre- and post-metamorphic mafic intrusives occur in the other ranges of southwest Montana. In the Ruby Range, texturally and mineralogically zoned amphibolite facies ultramafic rocks are cut by NW-trending basic dykes (Garihan, 1979; Desmrais, 1981; Karasevich *et al.*, 1981). In the southern Madison Range and Centennial Mountains further south, metamorphosed Archaean volcanics contain bronzititic pods that may represent the remnants of ultramafic flows and sills. These are cut by NNW- to NNE-trending doleritic and gabbroic quartz-tholeiite dykes up to 5 km long (e.g. Erslev, 1983).

The northern Madison and Gallatin Ranges also contain lenses of wehrlitic intrusives and three generations of NW- to NE-trending mafic sheets. The oldest of these are foliated, folded and boudinaged amphibolites, the middle generation comprises unfolded but foliated amphibolites and the youngest dykes are unaltered doleritic and basaltic rocks (Spencer and Kozak, 1975). Two generations of basic dykes have been recognised in the Highland Mountains, trending E to NE, and E to SE respectively (O'Neill *et al.*, 1988). Dyke orientation analysis for the Tobacco Root, Ruby, Highland and Madison Ranges has been summarised by Schmidt and Garihan (1986).

9.4.6   *Bighorn Mountains*

Three generations of mafic and ultramafic intrusives are present in the Bighorn Mountains. There is no simple correlation between dyke age and orientation (e.g. Kiilsgaard *et al.*, 1972). The first two groups (*c.* 3.0 and 2.8 Ga respectively) were emplaced late in separate plutonic–metamorphic events, while the third swarms (*c.* 2.2 Ga) were intruded after the cessation of significant tectono-thermal activity (Stueber *et al.*, 1976; Arth *et al.*, 1980). The oldest group comprises up to 5% of the 3.0 Ga gneisses in the southern Bighorn Mountains, but their (K–Ar) age is poorly constrained (Heimlich and Banks, 1968). The 2.8 Ga dykes comprise metadolerites, dolerites and micronorites. The 2.2 Ga dykes occur as E–W, NE- and NW-trending dykes up to 20 km long and smaller bodies of dolerites and dolerite porphyry (Condie *et al.*, 1969b; Heimlich *et al.*, 1973; Stueber *et al.*, 1976; Armbrustmacher, 1977). The dolerite porphry ('leopard rock') dykes have plagioclase-rich flow-differentiated interiors and aphyric margins (Heimlich and Manzer, 1973), which have yielded the Rb–Sr isochron age 2.2 Ga (Stueber *et al.*, 1976).

Many of the dykes have amphibolitised margins in which the plagioclase becomes less calcic and the pyroxene less Fe-rich due to the formation of hornblende (Heimlich *et al.*, 1974). However, many of the unaltered dolerites contain complex, disequilibrium pyroxene assemblages (Figure 9.13) which record sequential low pressure crystallisation (Hall *et al.*, 1986, 1987b). The *c.* 2.8 Ga generation dykes include tholeiitic and ultramafic (pyroxenitic and hornblenditic) types whose trends are highly variable (Ross and Heimlich, 1972; Heimlich *et al.*, 1973, 1974; Manzer and Heimlich, 1974; Stueber *et al.*, 1976; Armbrustmacher, 1977).

Partially quenched varieties of high-Mg dykes have mineral assemblages and whole rock compositions similar to those of noritic and boninitic dykes in the Beartooth

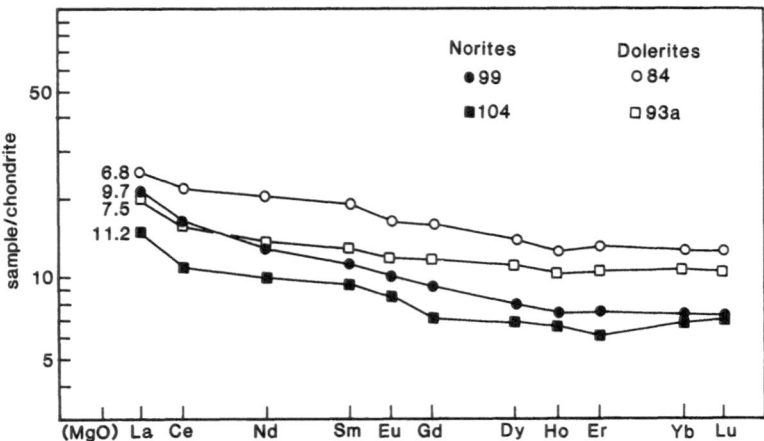

Figure 9.14   Chondrite-normalised REE abundances in doleritic and fine-grained noritic dykes from the northern Bighorn Mountains (authors' data).

Range and southern West Greenland (Longhi *et al.*, 1983; Hall and Hughes, 1987). Pyroxene assemblages span a range of enstatite, pigeonite and Ca-poor and calcic augite with X_{Mg} values of 0.8 to 0.9 (Figure 9.13). Plagioclase in these dykes is in the range An_{70}–An_{60}. They may be related to magmas which fed larger scale norite-rich intrusions such as the Stillwater Complex. The compositions of these noritic dykes contrast with those of the neighbouring tholeiitic dolerites in that they are more magnesian (MgO = 10–13% compared to 6–9%) and correspondingly rich in Ni (*c.* 250 ppm compared to 100 ppm) and Cr (600 ppm compared to 150 ppm), but have even more strongly fractionated REE distributions than the dolerites ($La_N/Lu_N = 2.5$ compared to 1.5) (Figure 9.14). These compositions strongly suggest the intrusion of two different types of basic magma during the late Archaean, as also seems to be the case in the Laramie and Beartooth ranges (Hall *et al.*, 1987b).

9.4.7   Teton Mountains

One of the largest basic intrusions in the Teton range is the Rendezvous metagabbro, which is an amphibolitic plutonic body occurring in the southern part of the range and measuring at least 4 × 8 km (Figure 9.3). It has a Rb–Sr whole rock age of 2.9 Ga, and was probably derived from the mantle only shortly before its metamorphism at this time (Reed and Zartman, 1973). The youngest Precambrian basic rocks in the region are a series of WNW-trending dykes up to 45 m wide and 16 km long. The chilled margin of one such dyke has given a K–Ar age of 775 Ma, but isotopic data from wall-rock biotites suggest that the dyke may have been intruded some time between 1.36 and 2.57 Ga, and perhaps just prior to the 1.3–1.5 Ga thermal event (Reed and Zartman, 1973; Houston *et al.*, 1988).

9.4.8   Wind River Range

At least three major generations of mafic intrusive events are recorded in the Wind River Range, at > 3.2 Ga, 2.8 Ga, and one (or more) between 2.1 and 1.7 Ga, between which there were two episodes of granulite facies metamorphism (Aleinikoff et al., 1987). All three events gave rise to tholeiitic and ultramafic rocks. The oldest (> 3.2 Ga) is represented by an E-trending, 5 km belt of meta-harzburgites and pyroxenites (Anderson, 1985). Other possible mid-Archaean basic rocks occur as irregular dyke-like bodies of clinopyroxene-bearing metagabbro and metadiorite, and other large and small granulite facies ultramafic bodies (e.g. Stuckless et al., 1985).

The late Archaean mafic-ultramafic event is exemplified by the greenstone belt rocks in the south of the range which pre-date the 2.63 Ga Louis Lake batholith. Part of this belt, consisting of meta-turbidites and calc-alkaline andesitic rocks, has yielded a Rb–Sr whole-rock age of 2.8 Ga (Stuckless et al., 1985). These rocks are sutured to a belt of ultramafic and tholeiitic gabbroic, doleritic rocks and associated pillow-structured lavas, which may represent ophiolitic oceanic crust (Condie, 1972; Bayley et al., 1973). Metagabbros cut by basic dykes elsewhere in the range could also be part of the 2.8 Ga intrusive sequence (e.g. Granger et al., 1971).

The early Proterozoic events consist of fresh or mildly metamorphosed doleritic and plagioclase peridotitic NE-trending dykes up to 32 km long (Granger et al., 1971; Pearson et al., 1971). K–Ar dating of these rocks falls in the wide range of 2.2–1.4 Ga (Condie et al., 1969b; Spall, 1971), and whether there are one or more intrusive events remains unclear. However, in the northern Wind River Mountains at least three generations of metadoleritic intrusives have been recognised (Worl, 1969, 1972; Granger et al., 1971).

9.4.9 Owl Creek Mountains

The oldest basic intrusive rocks in the Owl Creek Mountains are > 2.7 Ga old metagabbroic (amphibolite) sills in supracrustal units (Granath, 1975; Hausel et al., 1985). One younger set of Archaean dykes and sills trends ENE, parallel to the supracrustal unit, while a second discordant set trends almost at right angles to it. Both sets are amphibolitised and cut by late Archaean granite, but are not dated absolutely (Hausel et al., 1985). A swarm of dark WNW- to W-trending amphibolitic dykes has yielded K–Ar ages of 2.1–1.9 Ga (Condie et al., 1969b), and a set of fresh basaltic dykes is thought to be of approximately the same age. Deformed amphibolite and fresh discordant dolerite dykes also occur in the eastern Owl Creek Mountains.

9.4.10 Granite, Seminoe, Ferris, Casper and northern Laramie Mountains

Two generations of late Archaean tholeiitic intrusives occur in the Granite Mountains. The older set of discordant amphibolites and serpentinites pre-date the 2.9 Ga metamorphic event, while the younger set is constrained only in that it post-dates a 2.6 Ga granite (Peterman and Hildreth, 1978; Stuckless et al., 1981). The younger dykes are NE-trending and continuous for up 5 km. Some are deuterically altered, but most are fresh and have chilled margins.

Several sets of mafic dykes up to 2 km long occur in the Archaean gneisses of the Seminoe and Ferris Mountains. Pre-2.7 Ga metagabbroic sills and serpentinites in the Seminoe Mountains are cut by a NE- to E-trending swarm of doleritic and plagioclase-phyric ('leopard rock') dykes (Dixon, 1982; Klein, 1982). NNW- and NNE-trending dykes in the Ferris Mountains comprise an early set of fine-grained basaltic rocks and later, generally altered gabbros and hornblende diorites.

Three generations of mafic intrusives have been recognised on Casper Mountain and in the northern Laramie range (Eller and Friberg, 1982; Gable *et al.*, 1988). The oldest on Casper Mountain are pre-2.8 Ga deformed and metamorphosed (amphibolite facies) dykes and sills. A set of undeformed ultramafic and younger 'leopard rock' dykes pre-dates a 2.4–2.6 Ga granite, and the youngest set comprises fresh clinopyroxene-bearing E- to NNE-trending dykes up to 4 km long. In the northern Laramie Mountains, an older generation of altered ultramafic and doleritic dykes (some with garnet) of various orientations are cut by NE- to E-trending, locally composite, vertical dolerite dykes around 3 km long (Johnson and Hills, 1976).

9.4.11 *Summary of basic intrusive rocks, Wyoming Province*

Mafic and ultramafic magmas were intruded in varying quantities into the Archaean core of western America throughout the early Precambrian. By far the majority of these magmas were of tholeiitic basaltic composition, and are represented by various types of dolerites. Similar magmas were apparently intruded throughout the entire Precambrian. Primitive tholeiites are represented in nearly all regions, while more strongly differentiated quartz dolerites, alkali olivine dolerites, and suites containing magmas as siliceous as quartz porphyry rhyolite are restricted to the middle and late Proterozoic, mainly in NW-trending swarms at the northwestern and southeastern margins of the Archaean craton. Plagioclase-phyric, 'leopard rock' dykes are present in many mountain ranges. They are somewhat older than accompanying aphyric dykes in the Wind River and central Laramie ranges. Despite the longevity of tholeiitic magma intrusion throughout the early Precambrian, the penecontemporaneous injection of diverse types of basic magmas is also apparent in individual areas, and exemplified most notably by the virtually simultaneous intrusion of contrasting gabbroic and noritic rocks in the Beartooth, Bighorn and Laramie ranges.

Ultramafic rocks form only 5% or less of basic intrusive suites, and are mainly restricted to older, Archaean rocks. They may have formed in a variety of ways: (i) as oceanic or ophiolitic basement such as, perhaps, in the Wind River Range; (ii) as liquid intrusives (all ranges); (iii) as semi-solid intrusives (locally in Wind River and Tobacco Root Mountains); (iv) as differentiates of noritic intrusives (locally in the Black Hills, central Laramie, Park, and Beartooth ranges). Most mafic intrusives form sheet-like bodies that, where the swarms are voluminous, indicate crustal dilation of the country rocks by as much as a third. Large-scale plutonic basic complexes in the Wyoming Province include principally the Stillwater, Lake Owens and Mullen Creek layered complexes and the Elkhorn Mountain gabbro. These have been postulated to have formed by the simultaneous intrusion and mixing of a variety of magmatic liquids including tholeiitic and SHMB, or boninitic, types and the presence of fine-grained noritic dykes in several parts of the craton which have boninite-like compositions and mineral assemblages supports this interpretation.

Unravelling the tectonic, metamorphic and magmatic histories of the intrusive basic rocks is constantly under review. Most dyke swarms with consistent orientations would be considered by many to be *prima facie* evidence for a tensional environment. However, this might be disputed where plutons and dyke swarms are considered to underpin accreted former volcanic arcs (cf. Fahrig, 1987) as, for example, along the Cheyenne belt at the southeastern margin of the craton (Figure 9.3). Further difficulty is introduced where trend patterns are more complex, as in the northwestern part

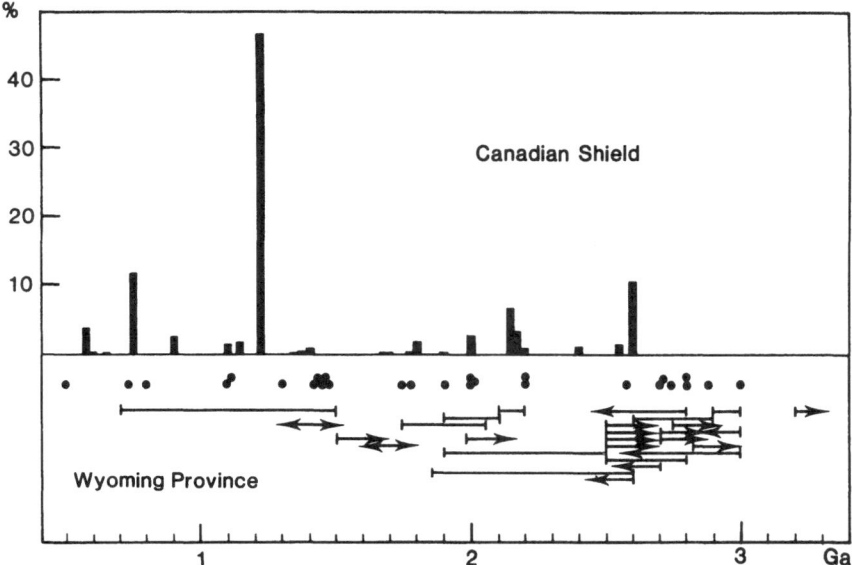

Figure 9.15   Comparison of dated mafic dykes in the Canadian Shield and Wyoming Province. For the Canadian Shield dykes, height of bars represents proportion of mapped dykes, as derived from the map of Fahrig and West (1986). The data are insufficient to plot volumetric proportions for the Wyoming dykes. Well constrained ages for Wyoming dykes are shown as dots; less well constrained age ranges are shown as bars, which are unconstrained at arrow heads.

of the craton. Furthermore, the episodes of basaltic dyke intrusion are interspersed with repeated episodes of metamorphism, deformation and the emplacement of synorogenic granitoid plutons, all of which are normally interpreted as being associated with compression or shearing. Much work remains to be done on the precise radiometric geochronology of the Wyoming Province, and it is obvious that additional knowledge is needed to understand its complex tectonic history.

The dyke intrusion events of the Wyoming Province do not correlate well with those in the Canadian Shield (Figure 9.15). The Canadian Metachewan (2.6 Ga), Priessac (2.14 Ga), MacKenzie (1.22 Ga), Franklin (0.75 Ga) and Grenville (0.58 Ga) dyke swarms which together account for 80% of the Canadian dykes (Fahrig and West, 1986) are absent or only poorly represented in the Wyoming Province, while major dyke events near 1.45 Ga or older than 2.65 Ga are important in Wyoming but apparently not in Canada.

9.5   Early to mid-Proterozoic metavolcanic belts of western central USA

Early to mid-Proterozoic supracrustal belts occur throughout western central North America. For detailed accounts of the geology of these belts the reader is referred to Bickford and Boardman (1984), Condie (1982, 1986), Karlstrom and Houston (1984), Karlstrom *et al.* (1983) and Bickford (1988). Broadly speaking, the belts become younger southwards away from the Wyoming Archaean province (Figure 9.1), but they comprise various terranes across the USA (Van Schmus and Bickford, 1981; Nelson and DePaolo, 1985). The oldest Proterozoic belts are around 1.8 Ga old.

Most fall in the range 1.8–1.6 Ga, and these pass southeastwards to mid and late Proterozoic belts (1.4–1.1 Ga) (Condie, 1982, 1987; Nelson and DePaolo, 1985).

The 1.8–1.6 Ga metavolcanic belts share some compositional features, primarily in that they comprise bimodal basic–acid suites. The basic members are most commonly tholeiitic basalts and the acid rocks are calc-alkaline dacites and rhyolites (Condie, 1986). The basaltic rocks tend to have moderate MgO (c. 7%), low alkali contents and show Fe-enrichment. They have REE$_N$ distributions which vary from MORB-like, LREE-depleted patterns to flat and slightly LREE-enriched types. In some cases, such as the 1.78–1.7 Ga Dubois, Cochetopa and Black Canyon green-stone successions in western Colorado (Nelson and DePaolo, 1984), geochemical criteria have been used to distinguish groups of metavolcanic amphibolites with these different REE concentrations (Condie and Nuter, 1981) which variously correspond to basalts from immature island arcs or N-MORBs, evolved island arcs and back-arc basins, and continental margin arc systems (Knoper and Condie, 1988).

Other basic metavolcanic suites, such as the Green Mountain formation in the Sierra Madre of southeastern Wyoming, have transitional tholeiitic to calc–alkaline basaltic affinities (Condie and Shadel, 1984). These rocks have relatively high incompatible element abundances (Rb, Ba, Th, K) and negative Ta, Nb, Hf, Zr and Ti anomalies similar to those in modern convergent margin basalts. Their Hf, Th and Ta proportions are also similar to those of calc-alkaline basalts (Saunders et al., 1980; D.A. Wood, 1980). However, they are transitional in that they show mild Fe-enrichment and have high Zr/Ti values (Table 9.1).

The geochemistry of the 1.7 Ga metavolcanics in the Dos Cabezas Mountains in central Arizona (Table 9.1) suggest that they formed in a continental back-arc setting (Condie et al., 1985; Condie, 1986). Isotopic data indicate that the continental basement involved in the back arc basin is only marginally older than the metavolcanics themselves, corresponding to that found in central Arizona and New Mexico (DePaolo, 1981; Conway and Silver, 1984).

These various types of early Proterozoic bimodal volcanic suites differ from the late Archaean bimodal units north of the Cheyenne belt (Karlstrom and Houston, 1984) only in the proportions of their different basic components. The late Archaean Elmers Rock greenstone belt includes Fe-rich komatiitic basaltic types and, mainly, tholeiitic basaltic rocks (Table 9.1), compared to the low-K tholeiitic and, in places, calc-alkaline basic volcanism of the early Proterozoic terranes. The temporal and geochemical evolution of these late Archaean and early and mid-Proterozoic supracrustal belts signifies sequential continental marginal rifting and crustal accretion with the progressive complex development of island arc series away from the Archaean cratonic nuclei (Condie, 1982, 1986). Indeed, it seems that all of the basic metavolcanic suites of northwestern USA reflect the sequential island arc-type accretion at the margin of the expanding early Precambrian continent, although the 1.7–1.6 Ga rocks of Arizona suggest that crustal growth involved the assembly of diverse tectono-stratigraphic terranes, rather than simple progressive southward accretion (Karlstrom and Bowring, 1988; Bickford, 1988).

Acknowledgements

We are grateful to W. Hamilton, W. Johnson, M. MacLachlan and J. Wooden for their comments on a previous version of this chapter.

10 Early Precambrian basic rocks of the Canadian Shield

P.C. THURSTON

10.1 Introduction

The Precambrian Shield of Canada is subdivided into five major Archaean provinces (Figure 10.1). This chapter concentrates on the Superior Province which covers much of Quebec, Ontario and parts of northern central USA, and is separated from the Rae and Hearne Provinces (formerly the Churchill Province; Hoffman, 1988), by the 1.8 Ga Hudsonian orogen. The Rae and Hearne Provinces are separated from the Slave Province to the northwest by the Great Slave Lake shear zone. The Nain Province comprises the western part of the North Atlantic craton (see Chapter 11), along the coastal strip of Labrador.

The Archaean is considered to terminate world-wide with a large-scale orogenic cratonising event at about 2.5 Ga (Plumb and James, 1985). However, this event occurred at about 2.68 Ga in the southern part of the Superior Province and slightly earlier in northern Superior Province (Stott et al., 1987). The major orogeny occurred at 2.63–2.58 Ga in Slave Province (Henderson, 1985). Therefore, consideration of mafic magmatism through to 2.5 Ga in the Canadian Shield must include post-cratonisation dyke events (cf. Fahrig, 1987). However, the emphasis in this review is on the volcanology and geochemistry of the Canadian Archaean greenstone belts. The principal difficulty in characterising Archaean mafic magmatism and comparing it with that in modern 'mobilist' tectonic regimes has been the lack of recognition of variations in tectonic environment that is discernible through stratigraphic analysis of even the least deformed Archaean supracrustal sequences. Preliminary evidence suggests that several lithostratigraphic assemblages comparable to those in a variety of Phanerozoic tectonic environments may be present in the Canadian belts (Thurston and Chivers, 1989). The mafic metavolcanics of the early Precambrian greenstone belts of the Superior Province include komatiites, Mg- and Fe-tholeiites and calc-alkaline basalts. Parental liquids of mafic intrusions vary from high-Al basalt to ultramafic compositions.

In modern tectonic settings, most variation in mafic magmatism is due to the mechanism and extent of partial melting, residence time in the crust, extent and depth of fractionation and the mechanics of eruptive processes. All of these processes are readily relatable to the tectonic environment of the magmatism, from oceanic floor and island to magmatic arc and continental flood basalts. Archaean greenstone belts have been considered to represent virtually every tectonic environment, including amalgamated island arcs (Langford and Morin, 1976), continental rifts (Goodwin, 1981) or

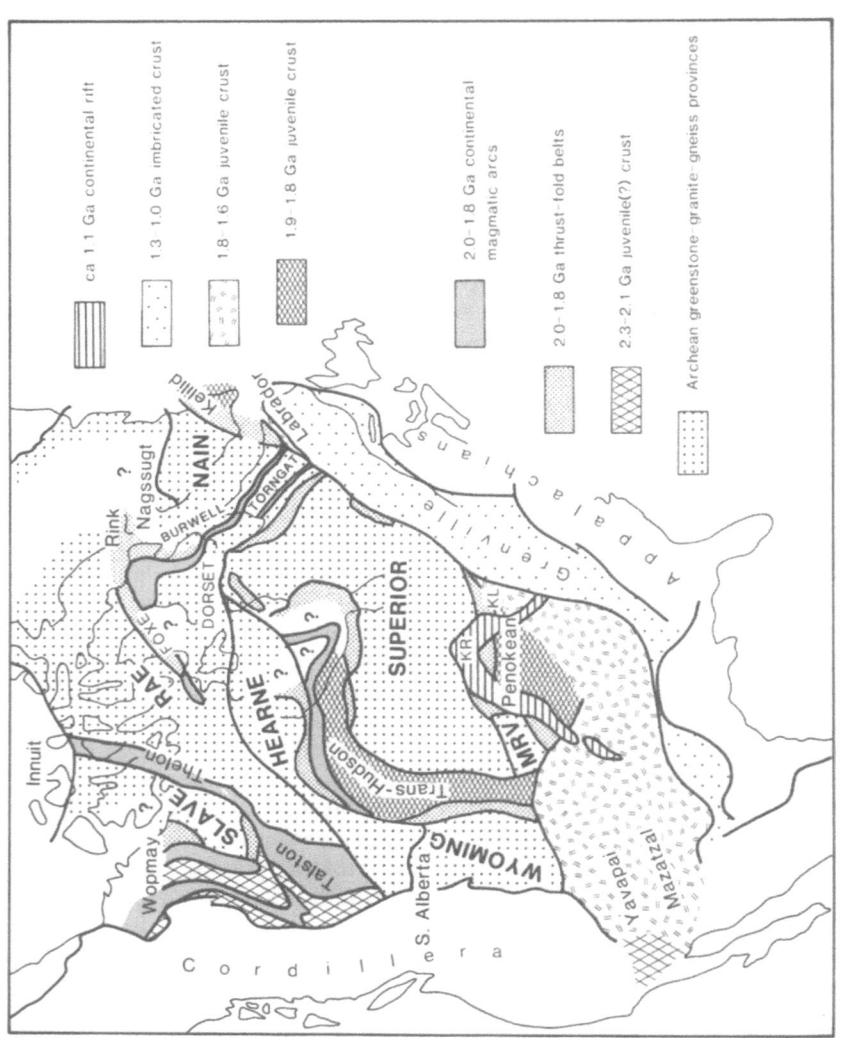

Figure 10.1   Precambrian tectonic elements of Laurentia (after Hoffman, 1988) with emphasis on Archaean (> 2.5 Ga) elements.   Relationship of the Superior Province to the Slave, Rae and Hearne Province is shown. Each province is generally bounded by major shear systems.

back-arc basins (Condie, 1986). However, a consensus of opinion has emerged that the components of individual greenstone belts probably represent a variety of tectonic settings (de Wit and Ashwal, 1986). Mafic magmatism in the Archaean greenstone belts of the Canadian Shield includes examples of oceanic tholeiites (Dimroth et al., 1982; Helmstaedt et al., 1986; Hall et al., 1987a), arc volcanism (Langford and Morin, 1976; Dimroth et al., 1982; Thurston and Chivers, 1989), and stratigraphically high continental tholeiites (Trowell et al., 1982). Late mafic magmatism expressed as dykes, mafic–ultramafic layered intrusions and volcanics include boninitic rocks (Thurston and Fryer, 1983; Hall and Hughes, 1987) and high-Al tholeiites (Simmons, 1980; Sutcliffe, 1984; Sutcliffe and Sweeney, 1985).

Recent work has shown that Archaean greenstones contain several distinct and consistent stratigraphic associations indicative of different environments, namely: (1) shallow marine platform sequences comprising quartz-arenite, BIF and komatiite associations; (2) thick massive basaltic flows and associated deep water pelagic sediments representing an oceanic volcanic or mafic plain environment; (3) cyclic mafic to felsic volcanics with or without an upper sedimentary unit indicative of arc volcanism; and (4) shallow water to fluviatile sediments and alkalic to calc-alkalic volcanics formed in pull-apart basins (Thurston and Chivers, 1989). These distinct associations form an ordered series in the Superior Province from (1) (oldest) to (4) (youngest). The series is preceded and followed by orogenic events. Petrological variation in the mafic volcanism of the various sequences has not previously been related to these different tectonic environments, nor has any systematic variation in mafic plutonism been documented. This work attempts to characterise mafic volcanism in the Superior Province and relate it to the above litho-stratigraphic associations. This examination of mafic magmatism concentrates on the Superior Province but will include briefly some Slave and Hearne Province examples.

Redman and Keays (1985) recognised three basic differentiation series in greenstone belts of the Yilgarn Block of Australia. These are: (1) high-Mg basalts (HMB) which include komatiites, komatiitic basalt, and high magnesian basalts of other authors, characterised by CaO/Al_2O_3 ratios of c. 1 and MgO from 40 to 5%; (2) siliceous high-Mg basalts (SHMB) characterised by $SiO_2 \geqslant 52\%$, and MgO from 15 to 5% with variable enrichment in LILE and $CaO/Al_2O_3 < 1$; and (3) low magnesian, tholeiitic basalts. This nomenclature fits well with observations in the Canadian Shield and is used here. The variation of mobile elements such as Rb, Sr, U and Th can clearly only be used with caution in attempting to characterise Archaean mafic magmas. However, work on the geochemistry of Superior and Slave Province metavolcanics (e.g. Thurston and Fryer, 1983; Jenner et al., 1981) has shown that, in general, REE and other incompatible trace elements are not mobile under the metamorphic and hydrothermal conditions prevalent in the Canadian Shield except within a few metres of very intense hydrothermal alteration such as that associated with volcanogenic massive sulphide deposits (Campbell et al., 1984). Analytical data from this and other Ontario Geological Survey (OGS) projects are obtainable from the PETROCH database of the Survey.

10.2 Superior Province

The major plutonic terranes (subprovinces) within the Superior Province are the Berens and Winnipeg River Subprovinces (Figure 10.2). They are characterised by a

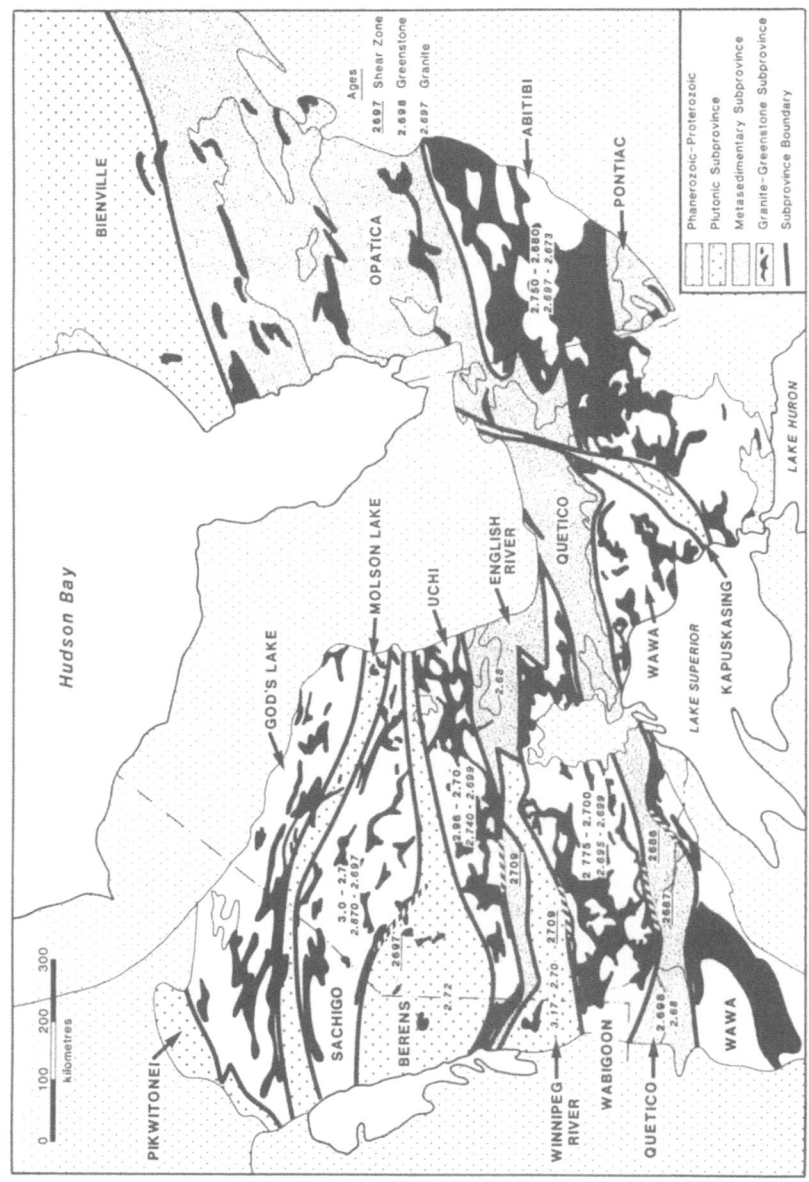

Figure 10.2 Subprovinces of the Superior Province of the Canadian Shield (after Card and Ciesielski, 1986, and modifications by Thurston *et al.*, 1987).

scarcity of supracrustal rocks and dominated by an older (3.17–2.76 Ga) suite of gneissic sodic dioritic to tonalitic rocks cut by younger (2.7 Ga) syn- to post-tectonic monzonitic to granitic batholiths (Corfu, 1988). The sodic suite includes metre to kilometre scale high-grade metamorphosed extrusive and intrusive mafic enclaves varying from tholeiitic basalt (Breaks *et al.*, 1978) to peridotitic sills of komatiitic affinity (G. Stott, pers. comm., 1988) in the northern Berens Subprovince. Little is known of the geochemistry of the mafic rocks in these terranes. Field relations suggest that the volcanics within the Winnipeg River Subprovince represent the oldest mafic volcanism in the Superior Province.

The metasedimentary subprovinces in the Superior Province such as the English River and Quetico units comprise linear belts up to several hundred kilometres long of wacke and pelite with subordinate coarse clastic rocks and high-grade schist, paragneiss and migmatite (Figure 10.2). These terranes have greenschist facies margins, with local nodes of low pressure granulites in the interior. Percival and Williams (1989) described the eastern part of the Quetico belt immediately south of the granite–greenstone terrane of the Wabigoon Subprovince as a series of thrust-bounded panels of wacke and tholeiite. From sedimentological, structural and geochemical criteria, the Quetico belt is considered to be equivalent to a modern accretionary prism. Preliminary data show that the tholeiites intercalated with the metasediments of the Quetico Subprovince exhibit depletion in the LREE, suggestive of an oceanic tholeiite geochemistry (H.R. Williams, pers. comm., 1989). The application of similar models to the other sedimentary Subprovinces of the Canadian Shield remains to be tested.

The Pikwitonei and Kapuskasing Subprovinces (Figure 10.2) are characterised by high pressure upper amphibolite to granulite facies gneisses of plutonic and supracrustal origin cut by layered gabbro-anorthosite bodies, and tonalitic, granodioritic and syenitic plutons. Percival and Card (1985) described the Kapuskasing Sub-province as a section through > 25 km of Archaean crust. From east to west the terrane comprises linear belts of vertical melanocratic meta-igneous gneisses cut by basement-type anorthosites (Windley, 1971) with 0.6–0.8 GPa metamorphic assemblages overlain to the west by 10–15 km of tonalitic gneisses forming shallow dipping domes and basins representing middle crust, all superseded by 5–10 km of upper crustal granite–greenstone suites with 0.3–0.5 GPa mineral assemblages. The Pikwitonei Sub-province consists of granulite facies granites and greenstones adjacent to the Proterozoic Trans-Hudson orogen (Green *et al.*, 1985) succeeded eastward by lower grade assemblages. This terrane represents the uplift of lower crustal levels brought about by a continent–continent collision at about 1.8 Ga (Hoffman, 1988). The Ashuanipi (Percival and Girard, 1988) and the Minto Subprovinces both comprise granulite and upper amphibolite facies gneissic rocks and supracrustal sequences. These subprovinces represent granite–greenstones and sedimentary subprovinces at deep crustal levels.

10.3 Granite–greenstone terranes

10.3.1 *Platform sequences*

Platform greenstones occur throughout the Sachigo Subprovince and the older part of the Wabigoon Subprovince (Davis and Jackson 1988; Thurston *et al.*, 1987). U–Pb zircon geochronology suggests that the platform greenstones are > 2.85 Ga and in many instances approach 3.0 Ga (see Thurston and Chivers, 1989). The platform sequences consist of basal quartz arenites and carbonate sediments overlain by banded iron formation (BIF), komatiites and tholeiites. A shallow water depositional environment is suggested by primary structures and biota (Hofmann *et al.*, 1985; De Kemp, 1987; Thurston *et al.*, 1987). The lithological association and areal extent of the units suggests widespread quartz-rich sedimentation indicative of extensive sialic areas and platforms reminiscent of modern passive margin sequences.

The volcanic part of the platform sequences of Superior Province consist of tholeiites with or without komatiites, no andesites and minor felsic airfall and ash-flow tuffs (Thurston *et al.*, 1987). Based on the structure of the mafic flows (using the criteria of Dimroth *et al.*, 1978) and the limited areal extent of komatiites, most occurrences represent a relatively proximal facies. Ultramafic flows display olivine spinifex and 'stringbeef' spinifex textures and are up to a few metres thick. Most occurrences have greenschist facies assemblages, but relict clinopyroxene occurs in some tholeiitic and komatiitic flows (Jensen, 1987).

In general, the platform tholeiites and komatiites display low Ti and Zr and high Mg,

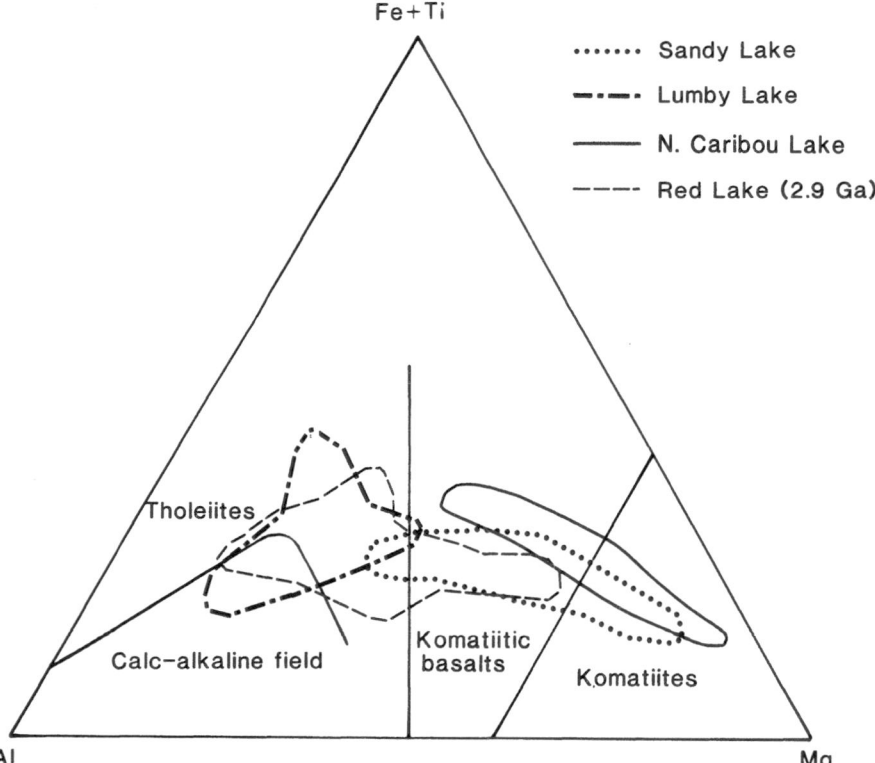

Figure 10.3 Al: (Fe + Ti): Mg (cation %; after Jensen, 1976) showing the affinities of a variety of platform sequences: Sandy Lake (Thurston *et al.*, 1987), Lumby Lake (Jackson, 1985), North Caribou Lake (Breaks *et al.*, 1985) and Red Lake (Andrews *et al.*, 1986). From unpublished data, available through the PETROCK database of the Ontario Geological Survey.

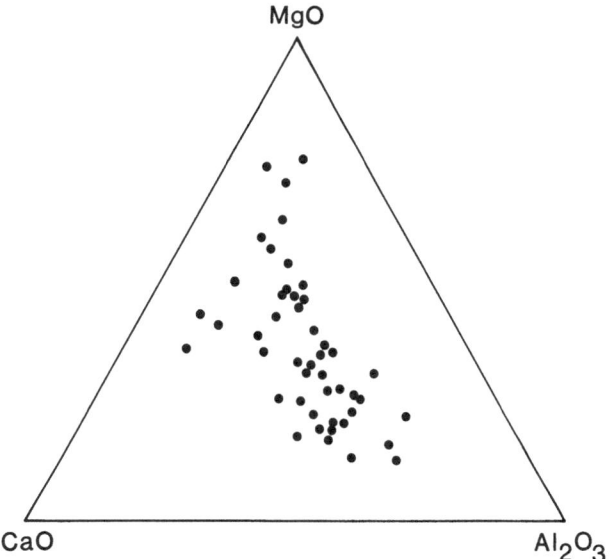

Figure 10.4 CaO:Al$_2$O$_3$:MgO(wt%) plot for platform sequence komatiitic metavolcanics plotted in Figure 10.3.

Ni and Cr contents (cf. Pearce and Norry, 1979; Arndt *et al.*, 1977; Sun *et al.*, 1979). In terms of their Al:(Fe + Ti):Mg proportions (Figure 10.3) the platform sequences of Agutua Arm (Breaks *et al.*, 1988), Sandy Lake (Thurston *et al.*, 1987), Red Lake (Corfu and Wallace, 1986) and Lumby Lake (Jackson, 1985) display tholeiitic Fe-enrichment trends and the komatiites have a range of MgO contents varying from *c.* 10 to 29%. The high-Mg samples have elevated Cr (1190 ppm) and Ni (2360 ppm) and low Ti and Zr contents. They also have low Al$_2$O$_3$ (6–12%), K$_2$O, and SiO$_2$ (36–54%), high CaO/Al$_2$O$_3$ ratios (Figure 10.4) and only slightly variable incompatible element ratios (Figure 10.5). These characteristics are similar to those of other komatiite suites (Arndt and Nisbet, 1982a).

Large decreases in Cr (2000 to 10 ppm) and Ni (300 to 10 ppm) relative to the increase in Zr (20 to 200 ppm) suggest fractionation of olivine, spinel and/or clinopyroxene. Zr/Y increases due to the crystallisation of a phase which accepts Y in preference to Zr (and Yb) such as clinopyroxene (Pearce and Norry, 1979) although the range of variation of Zr/Y also reflects processes such as differing degrees of partial melting, different source regions and contamination. Stringbeef clinopyroxene spinifex is observed in platform sequences at Caribou Lake (Breaks *et al.*, 1984), Lumby Lake (Jackson, 1985), Sandy Lake and Heaven Lake (Thurston *et al.*, 1987). Relict clinopyroxene phenocrysts occur at Lumby Lake (Jackson, 1985) and Horseshoe Lake (Jensen, 1987). Overall, Zr increase reflects simple incompatible element enrichment in residual liquids through low pressure fractionation of phases such as olivine and plagioclase. The platform sequences demonstrate appreciable fractionation relative to most arc sequences.

In the simplified normative basalt system the basaltic rocks cluster on the clinopyroxene- and olivine-poor side of the komatiites. If these suites were co-genetic,

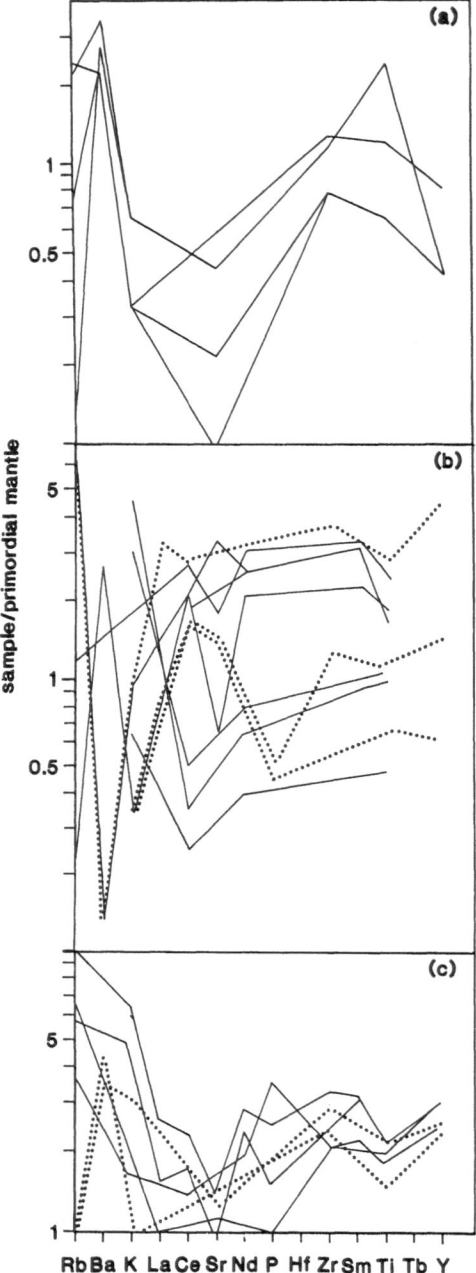

Figure 10.5 Mantle–normalised incompatible element abundances for komatiites of various litho-stratigraphic associations: (a) Prince Albert Group platform sequences (Fryer and Jenner, 1978), (b) Munro (solid lines) and Lamotte (dotted lines) mafic plain komatiites (Arndt *et al.*, 1977; Ludden and Gelinas, 1982), (c) Kambalda (dotted lines) and Swayze (solid lines) arc sequences (Arndt and Jenner, 1986; Cattell and Arndt 1987).

either olivine and clinopyroxene were both significant fractionating phases, or the komatiitic members originated by higher degrees (20–25%) of partial melting of mantle than the tholeiitic rocks (10–20%) (cf. Arth and Hanson, 1975; Elthon, 1986). Few of the platform sequence komatiite suites have been described and comprehensive trace element data are not yet available. Limited geochemical data for platform sequence komatiites in the Prince Albert Group of Rae Province are described later. The trace element geochemistry of komatiites and tholeiites from this litho-stratigraphic association are compared to the same rock types from the arc association in section 10.3.3.

10.3.2 Mafic plain volcanism

Mafic plain and central volcanic complex volcanism (Greeley, 1982; Dimroth *et al.*, 1985) is typified by high eruption rates of low viscosity magmas giving rise to thick massive tholeiitic and/or ultramafic flows often with relatively thin pillowed zones at the flow tops, and with low primary dips. This pattern has been recognised in the Canadian Shield in the form of volcanic cycles consisting of tholeiites and komatiites intercalated with deep water sulphidic shales (sulphide facies BIF). Archaean mafic plain volcanism occurs in all granite–greenstone subprovinces of the Superior Province. Archaean mafic plain sequences were recognised by Dimroth *et al.* (1985) within the Abitibi Subprovince, and have also been identified in the Wabigoon Subprovince (Thurston and Chivers, 1989). These c. 2.7 Ga sequences are of two types, namely predominantly massive tholeiite and/or komatiite flows with interflow argillites and airfall tuffs forming units of relatively constant thickness with a lateral extent of a few kilometres, and secondly, sequences with more pillowed flows throughout the extent of the sequence with shallow water features such as hyalotuff, hyaloclastite, vesicles and compound pillow rims becoming common high in the stratigraphy. Typical examples include those in the Bethune, Manitou and Sandy Beach Lakes areas of Wabigoon Subprovince (Edwards, 1983; Blackburn, 1982; Berger, 1987) and the Kinojevis Group of the Abitibi Subprovince (Dimroth *et al.*, 1985). In general, these sequences are found towards the base of the greenstone belts and are slightly older than most of the arc volcanic sequences (see Thurston and Chivers, 1989). Many Archaean komatiite-bearing mafic/ultramafic sequences have similar stratigraphies and primary structures. However, the geochemical characterisation of these sequences is very sketchy. In general, there is a greater proportion of dense, fractionated material in the arc sequences (Thurston and Sutcliffe, 1986).

The mafic plain sequences consist largely of HMB, tholeiites and SHMB in ascending stratigraphic order (Thurston, 1986; OGS/MERQ, 1983) often capped by ocellar, variolitic or plagioclase-phyric SHMB, chemical sediments or felsic tuffs. This type of unit tends to occur at the base of the greenstone belts. Examples occur in the Red Lake gold area within the older units of the Uchi Subprovince (Andrews *et al.*, 1986; Wallace *et al.*, 1986). Mixed 2.9 Ga komatiite–tholeiite successions with minor felsic metavolcanics form the bulk of this greenstone belt. In the external zone of the Abitibi Subprovince, magnesian units form the Stoughton–Roquemaure Group, which includes the Munro Township komatiites (Pyke *et al.*, 1973; Arndt *et al.*, 1977), the lower formation of the Deloro and Tisdale Groups, the Larder Lake Group, the Wabewawa Group, and the Lamotte–Vassan Group (terminology after OGS/MERQ, 1983). The exposed komatiite-bearing units are up to 15 km wide.

A typical example is the 6 km wide upper part of the Stoughton–Roquemaure Group (Jensen and Langford, 1983) which consists of fourteen major units. The units are 300–1400 m thick with basal komatiites overlain by magnesian basalts, and capped by Fe-tholeiite units. The komatiitic units decrease in thickness upward. Komatiitic flows are as much as a few hundred metres thick and consist of high-Mg flows (up to 40% MgO) 15 cm to 5 m thick overlain by thinner, less magnesian flows. The flows are separated by sedimentary graphitic and sulphide-rich horizons. Similar units within the Tisdale Group to the south of Timmins are described by Robinson and Hutchinson (1982). The flows here range from thick massive units with cumulus textures (Johnston and Trowell, 1987) to thinner units with basal cumulates grading upward through the standard vertical sequence of primary structures described in the type example in Munro Township (Pyke et al., 1973). In the Lamotte sequence in Quebec (Ludden and Gelinas, 1982), komatiitic pyroclastic rocks range from tuff breccia to tuff (Imreh, 1979). Associated syngenetic Fe-Ni-Cu ores occur in the basal flows of these units (e.g. Green, 1978).

Some of the komatiites of Munro Township in the Abitibi Subprovince contain unaltered cumulus olivine and clinopyroxene in a matrix of acicular pyroxene and sparse chromite (Pyke et al., 1973; Arndt et al., 1977). However, they are generally altered to tremolite- and chlorite-rich assemblages locally with serpentine, talc and carbonate. The overlying magnesian basalts range from 9% to 16% MgO with relict olivine phenocrysts with skeletal overgrowths, discrete prismatic grains and thin skeletal plates in a matrix of dendritic or plumose clinopyroxene and sparse chromite and relict glass (Arndt et al., 1977).

In general, komatiites are characterised by high CaO/Al_2O_3 ratios ($\geqslant 0.8$) (Figure, 5.2; Figure 10.4), $TiO_2 \leqslant 1.0\%$, high Cr and Ni, and chondritic ratios of incompatible trace elements. They show variable degrees of enrichment or depletion in Rb, Ba, Th, K and LREE relative to HREE, P, Zr, and Y (Figure 10.5). In the Abitibi Subprovince komatiites (Table 10.1), significant and similar decreases in both Ni and Cr with decreasing MgO suggest that Ni removal was probably related to an olivine fractionation control and not the generation of immiscible sulphide melts. Similar relationships are seen in the Stoughton–Roquemaure (Arndt et al., 1977; Pyke et al., 1973), Wabewawa (Jensen, 1985) and Larder Lake groups (Jensen, 1985). TiO_2/Al_2O_3 and TiO_2/CaO relationships show that during differentiation of the magnesian basalts, as measured by decrease in MgO, Ti was concentrated in the residual liquid while Ca and smaller amounts of Al were removed from the melt. These relationships suggest some clinopyroxene fractionation. REE data for the tholeiites within the largely komatiitic Wabewawa Group are shown in Figure 10.6. The unit includes a basal SHMB overlain by HMB flows generally with flat REE patterns at or under ten times chondrite. The SHMB have elevated LREE abundances at a MgO content of c. 10%. These patterns reflect variable olivine and plagioclase fractionation. The high-Fe tholeiites forming the upper part of komatiitic volcanic cycles such as the Stoughton–Roquemaure Group contain up to 14.5% Fe_2O_3* (cf. Jensen and Langford, 1983), but Y and Zr levels in these tholeiites suggest they were derived from separate batches of magma.

The high-Fe tholeiites are overlain by andesitic to rhyolitic pyroclastics with flat REE patterns ($REE_N = 100$). Modelling of the REE data, major elements (Thurston, 1981) and Y and Zr contents suggests that these pyroclastics were derived by fractionation of olivine and plagioclase from tholeiitic basaltic parent magmas.

Table 10.1 Representative analyses of komatiitic–tholeiitic metavolcanic rocks from Munro Township (1–4), Wabewawa (5–10) and Kinojevis (11–14) Sub-provinces of the Abitibi Belt, Superior Province

Analysis	1	2	3	4	5	6	7	8	9	10	11	12	13	14
SiO_2	48.2	52.2	50.9	49.5	45.0	48.9	50.9	44.0	48.4	47.1	50.15	50.22	53.07	61.31
TiO_2	0.49	0.59	1.58	0.45	0.70	0.57	0.58	1.14	1.14	0.93	0.72	3.09	2.62	1.42
Al_2O_3	8.89	12.30	12.60	12.5	11.40	14.40	15.20	16.80	14.20	14.80	14.74	12.62	13.13	11.84
Fe_2O_3	2.63	1.97	2.93	1.88	11.20	10.80	9.83	10.40	13.00	12.90	2.37	6.74	3.71	3.31
FeO	8.70	9.80	12.60	10.00	—	—	—	—	—	—	7.54	10.10	11.47	8.43
MnO	0.20	0.18	0.22	0.21	0.21	0.19	0.16	0.15	0.19	0.19	0.13	0.25	0.20	0.19
MgO	21.80	10.60	4.89	13.20	10.30	6.58	6.59	9.68	7.10	8.17	8.46	3.78	3.46	1.55
CaO	8.60	11.50	9.84	10.40	9.58	12.30	11.30	11.60	10.60	11.60	10.68	7.96	6.13	4.92
Na_2O	0.49	0.58	4.81	2.01	2.64	0.60	1.19	1.00	0.83	1.13	1.87	2.61	3.29	3.85
K_2O	0.03	0.29	0.05	0.01	1.38	—	—	—	—	0.14	0.51	0.19	0.07	0.02
P_2O_5	—	—	—	—	0.43	—	—	—	—	—	0.04	0.36	0.43	0.64
Sc	29	44	21	35	20	35	35	18	25	30	—	—	—	—
V	—	—	—	—	239	262	252	205	324	286	—	—	—	—
Cr	2009	143	309	—	675	204	300	280	276	264	343	38	6	—
Ni	986	93	99	—	134	41	50	230	132	183	186	5	1	—
Rb	5	8	1	0.2	25	1	—	—	2	1	12	—	<1	—
Sr	15	97	67	75	93	14	20	74	76	78	230	—	150	—
Y	—	—	—	—	25	21	20	17	29	27	24	—	52	—
Zr	—	—	—	—	92	26	29	37	60	50	43	—	153	—
Nb	—	—	—	—	3	3	1	2	1	2	—	—	—	—
Ba	<1	56	20	20	360	50	40	50	50	80	—	—	—	—
La	3.9	3.6	—	—	38	0.5	1.3	2.1	4.3	3.8	2.3	7.2	7.7	10.7
Ce	2.6	3.2	20.3	4.1	90.5	3.27	3.76	5.67	12.1	8.9	5.8	18.9	21.6	27.7
Nd	2.6	3.2	14.0	3.2	45.0	2.69	2.81	5.18	9.03	6.88	—	—	—	—
Sm	0.31	0.46	3.99	1.16	8.18	0.91	1.68	1.56	3.51	1.96	2.0	5.3	5.9	8.0
Eu	—	—	1.41	0.43	0.66	2.21	0.98	1.12	0.99	0.87	0.7	1.4	2.2	2.7
Yb	0.93	1.45	2.51	1.36	1.04	2.41	2.72	0.88	2.55	2.86	2.1	4.4	6.5	7.5

1–4: Samples P9-130 (STPK); P9-108 (komatiitic basalt); P9-195 (tholeiitic basalt); P9-185 (pyroxenitic komatiite), Munro Township, Abitibi Belt (Arndt et al., 1977).
5–10: Samples 126, 127, 128, 129, 130, 132 of pillowed tholeiites from the Wabewawa Group komatiite–tholeiite suite, Kirkland Lake, Catherine Township, Abitibi Belt (Jensen, 1985).
11–14: Samples 21, 2, 55 and 29a, Kinojevis Group high-Fe tholeiites (Jackson, 1980).

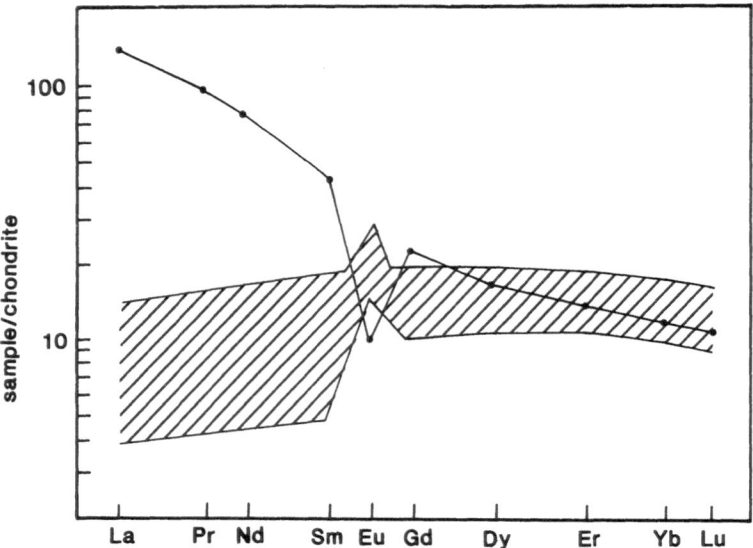

Figure 10.6 Chondrite–normalised REE abundances in tholeiites (shaded field) and a sample of
SHMB (dots) of the Wabewawa Group (OGS/MERQ 1983), Abitibi Subprovince.

In summary, within komatiitic units in Superior Province mafic plain sequences, the
proportion of tholeiite increases upward and komatiitic units are capped by high-Fe
tholeiites and minor volumes of tholeiitic rhyolite. SHMB increase in abundance
upward in the stratigraphy and can be interpreted in terms of mantle melting, as was
proposed to explain contrasting types of early Proterozoic mafic intrusives (Hall *et al.*,
1987b), or the contamination of a komatiitic melt by sialic crust (Arndt and Jenner,
1986; Barley, 1986; see Chapter 5). The latter origin is preferred for the Archaean
SHMB metavolcanics in Canada on stratigraphic grounds which suggest that these
magmas must have travelled through extensive older sialic basement (cf. Gariepy *et al.*,
1984). However, the geochemical data are equivocal and confirmation of this origin
must await isotopic evidence. Komatiites associated with the mafic plain association
are relatively depleted in LREE (La, Ce and Nd) and LILE (Ba, Rb and K) (cf. Wood *et
al.*, 1979) and perhaps depleted in Ti (Figure 10.5). If the SHMB were produced by the
contamination of komatiitic liquid by granitoids (Arndt and Jenner, 1986) their
occurrence high in the stratigraphy is explained by komatiitic liquid being slowed on
ascent and becoming progressively contaminated by surrounding sialic crust. Komatiitic magmas emplaced at *c.* 1600°C (Arndt *et al.*, 1977; Elthon, 1986) are capable of
substantial thermal erosion (Huppert *et al.*, 1984; Huppert and Sparks, 1985) and
therefore komatiitic liquids are capable of assimilating large volumes of surrounding
rocks, especially low melting temperature siliceous types such as granitoids. Therefore,
crustal contamination as well as olivine, clinopyroxene and plagioclase fractionation
probably played a role in the variation seen in the komatiite-bearing volcanic units.

10.3.3 *Arc volcanism*

Arc volcanism in the Superior Province has resulted in the development of major
stratigraphic units in the order of kilometres thick, which generally consist of a

Table 10.2   Cyclic volcanism in Canadian Archaean greenstone sequences

---

1. Peridotitic komatiite → dacite
2. Peridotitic komatiite → tholeiitic basalt → rhyolite → calc–alkalic basalt → rhyolite → alkalic basalt
3. Tholeiitic basalt → andesite → calc–alkalic dacite → rhyolite
4. Tholeiitic basalt → andesite → calc–alkalic basalt → rhyolite → alkalic basalt
5. Calc–alkalic basalt → rhyolite
6. Tholeiitic basalt → calc–alkalic dacite → rhyolite → tholeiitic basalt

---

volcanic base often overlain by a sedimentary upper unit. While there is abundant evidence in some Archaean terranes for the tectonic disruption of the stratigraphy (de Wit *et al.*, 1987), a relatively intact stratigraphy is postulated for some sections in the Superior Province greenstone belts (Stott, 1986) and corroborated by numerous U–Pb zircon geochronological studies (see Davis *et al.*, 1986). The stratigraphic analysis of volcanic geochemistry in Canadian Archaean greenstones has revealed six major types of cyclic volcanism (Table 10.2). Most display a mafic base giving way to a felsic upper part. The mafic base generally represents deep water shield volcanism (Thurston *et al.*, 1986) which gives way upward to shallow water volcanism, with the products of Plinian eruptions and ash-flows predominating in the upper parts of volcanic cycles. The nature of volcanic processes clearly has direct application to deciphering the rate of supply of mafic magma and its potential for interaction with sialic crust and hence the overall nature of mafic magmatism in the arc sequences.

Canadian arc volcanic sequences are 2.77–2.70 Ga (Davis *et al.*, 1986) and generally lie disconformably above mafic plain and platform sequences where these are present (Thurston and Chivers, 1989) but are themselves unconformably overlain by pull-apart basin assemblages (Ayres and Thurston, 1985). Arc volcanics occur at the southern and eastern margins of the Uchi Subprovince, and form the upper units in two Sachigo Subprovince greenstone belts (Corfu and Wood, 1986; Corfu and Ayres, 1987). The Wabigoon Subprovince consists almost entirely of 2.75–2.71 Ga arc volcanics (Thurston and Chivers, 1989) with *c.* 3 Ga platform sequences within the older central granitic area (Davis and Jackson, 1988). The external zone of the Abitibi Subprovince (Dimroth *et al.*, 1982) forms the southern and eastern margin of the Subprovince and consists of arc lithologies ranging from 2.75 to 2.7 Ga.

Tholeiitic units within arc volcanic sequences as defined in earlier geochemical and stratigraphic studies (Gelinas *et al.*, 1977; Goodwin, 1977) include low- and high-Fe tholeiitic basalts. Representative units include the middle formations of the Deloro and Tisdale Groups in the Timmins area, the Kinojevis Group and the Catherine Group (cf. OGS/MERQ, 1983) and the tholeiitic base of 2.7 Ga volcanic cycles in the Uchi (Thurston and Fryer, 1983) and Wabigoon (Trowell *et al.*, 1980) Subprovinces.

When examined in stratigraphic order (Figure 10.7), the Catherine Group tholeiites in the Abitibi Subprovince display three evolutionary cycles. One defines an irregular Fe-enrichment cycle similar to other Archaean tholeiite successions (cf. Thurston and Fryer, 1983) although some samples display increased LREE abundances and high MgO, best explained by sialic contamination (cf. Arndt and Jenner, 1986). The evolution of this cycle is interpreted to have been controlled by olivine and plagioclase fractionation, substantiated by Cr and Ni relations (Table 10.1). Basalts in the next

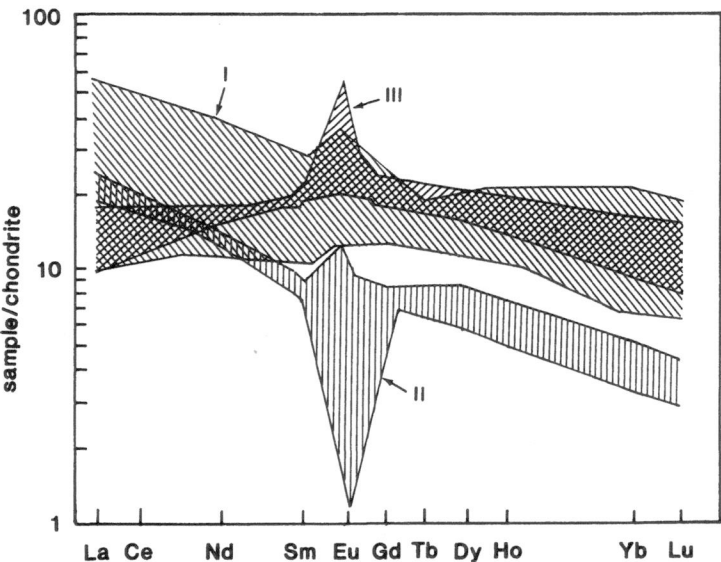

Figure 10.7   Chondrite–normalised REE abundances in cycles I, II and III tholeiitic basalts of the Catherine Group, Abitibi Subprovince.

stratigraphic unit upward (Figure 10.7) have high LREE abundances and are best explained by the fractionation of clinopyroxene, olivine and plagioclase. The easternmost cycle (III) contains HREE-depleted samples, LREE-depleted samples and those with a flat REE pattern. These groups of samples are respectively explained in terms of basalts derived from a garnetiferous source (cf. Cattell and Arndt, 1987; Arth and Hanson, 1975), mantle derived basalts with no crustal contamination component (Redman and Keays, 1985; Arndt and Jenner, 1986) and contaminated basalts (Arndt and Jenner, 1986).

Jackson (1980) examined a representative 3 km section of Kinojevis Group basalts and described low- and high-Fe tholeiites forming alternating units of 10–100 m thick massive and pillowed flows. The low-Fe flows have higher MgO, CaO and Al_2O_3 for given SiO_2, while the high-Fe flows show a steady increase in TiO_2 and P with increasing Fe (Table 10.1). The low-Fe flows have high Ni and Cr, and with decreasing mg' there is a decrease in Cr and Ni and an increase in Y and Zr. Groundmass pyroxenes range from $Wo_{36}En_{50}Fs_{12}$ to $Wo_{42}En_{38}Fs_{24}$ and their Fs content increases with decreasing whole-rock mg'. REE concentrations vary from about eight to forty times chondrite values (Figure 10.8). Petrogenetic modelling suggests that the variations are accounted for by the fractionation of plagioclase, clinopyroxene and olivine in the proportions 5:3:1, from an initial low-Fe tholeiite, until about 50–60% of the liquid was utilised. The upper part of the Kinojevis Group includes minor volumes of tholeiitic andesites, dacites and rhyolites. Thurston (1981) modelled the genesis of the tholeiitic rhyolites by extensive olivine, clinopyroxene and plagioclase fractionation from a tholeiitic basaltic liquid. This genetic model is in accord with the limited volume of these siliceous derivatives relative to the basalts.

Mafic volcanic rocks in the Wabigoon (Trowell *et al.*, 1980) and the upper part of the

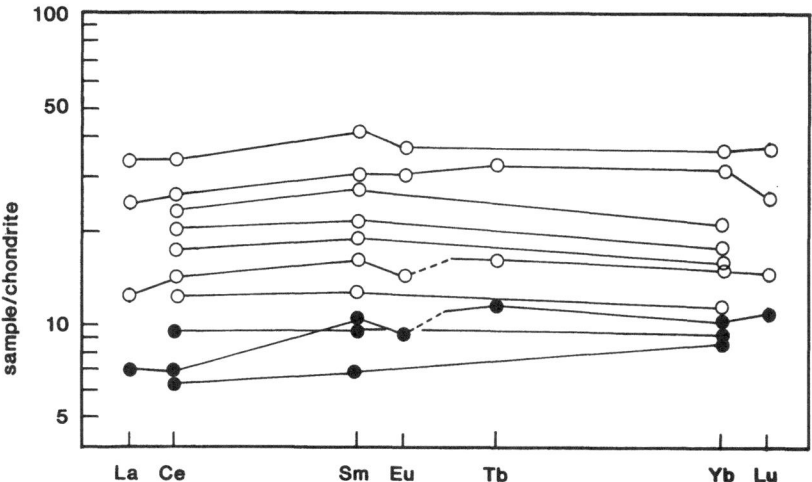

Figure 10.8 Chondrite–normalised REE abundances of tholeiitic flows of a large scale Fe tholeiite unit within the Kinojevis Group in the Abitibi Subprovince (Jackson, 1980).

Abitibi Subprovinces (the Blake River Group; Gelinas *et al.*, 1977; Ludden *et al.*, 1982) display consistent stratigraphic variations. In both belts tholeiites are stratigraphically low, forming the mafic part of a locally bimodal volcanic assemblage (Thurston *et al.*, 1985). These are succeeded by basaltic to andesitic units transitional in major element chemical characteristics between tholeiites and calc-alkaline basalts, which are overlain by high-Fe tholeiites.

In detail there are important differences between the two subprovinces, but the styles of mafic volcanism make it logical to consider them together. The basal tholeiites (termed Mg tholeiites by Trowell *et al.*, 1980) occur throughout the Wabigoon Subprovince (Northern Volcanic belt; Jutten volcanics; and Wapageisi volcanics) and as local units within the Blake River Group (Pelletier Series of Gelinas *et al.*, 1977). These units comprise both pillowed and massive magnesian tholeiites which display slight LREE depletion brought about by their derivation from depleted mantle (Ludden *et al.*, 1982). They contain olivine pseudomorphs (Gelinas *et al.*, 1977) and their parallel REE patterns are explained by olivine and plagioclase fractionation. Mg-tholeiites (*c.* 12% MgO) in the Wabigoon Subprovince (Morrice, 1988) have flat REE patterns with REE values six to nine times those of chondrites.

The basal Katimiagamak tholeiites (Edwards, 1985) have Fe_2O_3* varying from 12 to 18% and TiO_2 from 0.8 to 3.9%. REE data (Figure 10.9) show that the sequence evolved by fractionation of olivine, clinopyroxene and plagioclase. The sequence consists of two sub-units which may represent either separate magma batches involving the fusion of relatively normal and depleted mantle, or separate magma systems (Edwards, 1985). In terms of their HFSE and LILE chemistry, these basal basalts appear to represent an oceanic environment in that they have E-MORB and N-MORB affinities.

A transitional unit (TCFP of Trowell *et al.*, 1980) containing tholeiitic and calc-alkaline sub-units occurs in Wabigoon Subprovince, and alternating tholeiitic and calc-alkaline units also form a major part of the Blake River Group. The units range

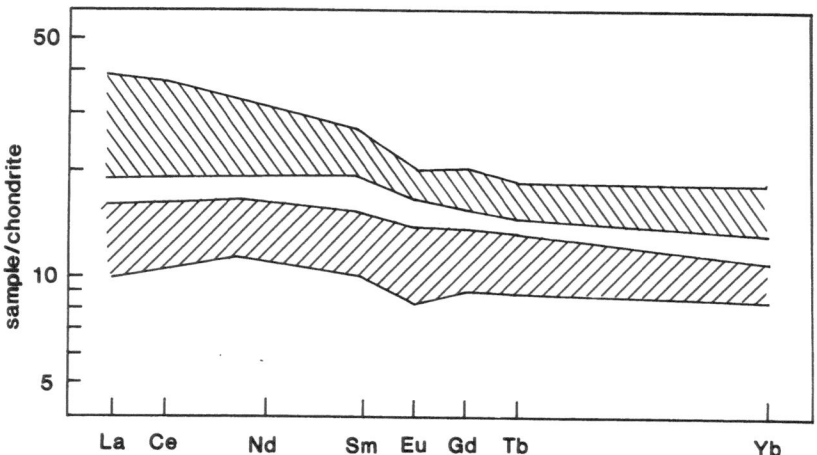

Figure 10.9   Chondrite–normalised REE abundances in two groups of Katimiagamak basalts in southwestern Wabigoon Subprovince. Upper group is comparable to E–MORBs, lower group to N–MORBs (after Edwards, 1985).

in major element composition from basalt to rhyolite of tholeiitic and calc-alkaline affinity. With increasing stratigraphic height, basaltic units become richer in Fe, pillowed flows and andesites become more abundant and more siliceous pyroclastic units occur. The mafic part of the sequence has typical tholeiitic flat REE patterns with REE_N between 10 and 20 (Morrice, 1988). The upper mafic units vary from basaltic to andesitic with up to 20–30% clinopyroxene phenocrysts and fractionated REE patterns, suggesting later clinopyroxene fractionation (Figure 10.10).

The transitional unit of the Wabigoon Subprovince is volcanologically and stratigraphically similar to the Blake River Group in the Abitibi Subprovince (OGS/MERQ, 1983), of which the Duprat–Monbray complex is the most extensively analysed example. This complex consists of four mafic to felsic cycles each with an andesitic base and a rhyolitic top. Stratigraphically higher andesites are successively

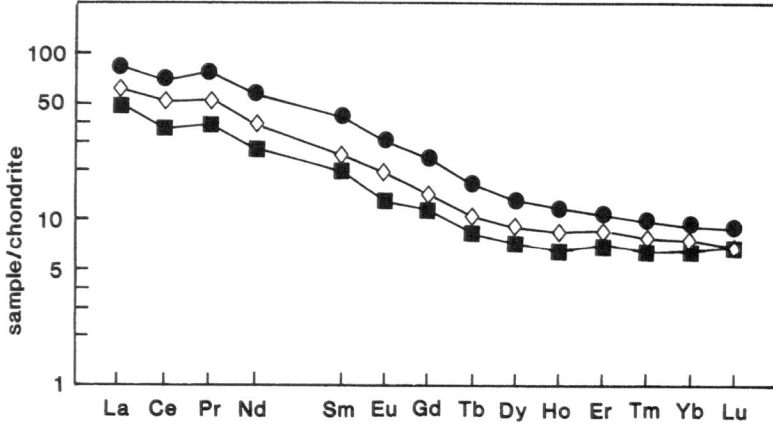

Figure 10.10   Chondrite-normalised REE abundances of stratigraphically high, clinopyroxene-phyric mafic flows in the southwestern part of the Wabigoon Subprovince in the Lake of the Woods area (Morrice, 1988).

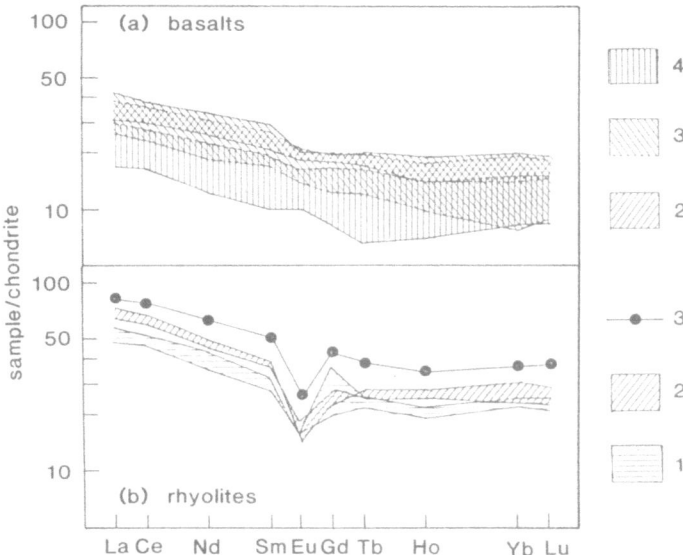

Figure 10.11   Chondrite–normalised REE abundances of (a) basalts and (b) rhyolites of cycles 1 to 4 of the Duprat–Monbray complex of the Blake River Group, Abitibi Subprovince (Verpaelst, 1985).

poorer in incompatible elements, whereas in younger cycles the rhyolites are enriched in incompatible elements (Figure 10.11). The petrogenesis of these rocks probably involved successive mantle-derived batches of basaltic to andesitic magma representing progressively depleted source materials, whereas the rhyolites represent the products of fractionation in a compositionally zoned magma chamber, involving the varying input of surrounding sialic material. Ta–Th–Hf systematics (Figure 10.12) show that parts of the Blake River Group resemble Phanerozoic arc volcanics (Ujike and Goodwin, 1987).

In the Oxford Lake, Uchi–Confederation Lake, Yellowknife and Rice Lake greenstone belts and the Kinojevis Group in the Abitibi Subprovince, there is a progression within basaltic units from basal low-Al tholeiites upward to high-Al tholeiites, capped by a feldspar glomeroporphyritic or variolitic unit (Green, 1975; Thurston and Fryer, 1983). The lower mafic part of the uppermost volcanic cycle at Confederation Lake (Thurston and Fryer, 1983) consists of basaltic to andesitic rocks. Below a variolitic marker horizon, the tholeiitic basalts exhibit decreasing Cr and Ni contents with increasing Ti, Fe, Zr and Y. The geochemistry of the system evolves smoothly upward to an extremely fractionated variolitic basalt horizon with relatively flat REE patterns ($La_N = 100$) and a pronounced negative Eu anomaly (Figure 10.13). On the basis of bulk chemistry, partitioning of trace elements, petrography, and the presence of large varioles, the variolitic unit appears to have originated by liquid immiscibility. Above this, high-Al tholeiitic basalts fractionated olivine, plagioclase and minor clinopyroxene to produce andesitic rocks also with flat REE patterns. The unit comprises a typical Fe-enrichment cycle ($Fe_2O_3^*$ from 9.7 to 17.6%) in which Zr varies from 28 to 119 ppm over a range of SiO_2 from 48 to 55%. Incompatible

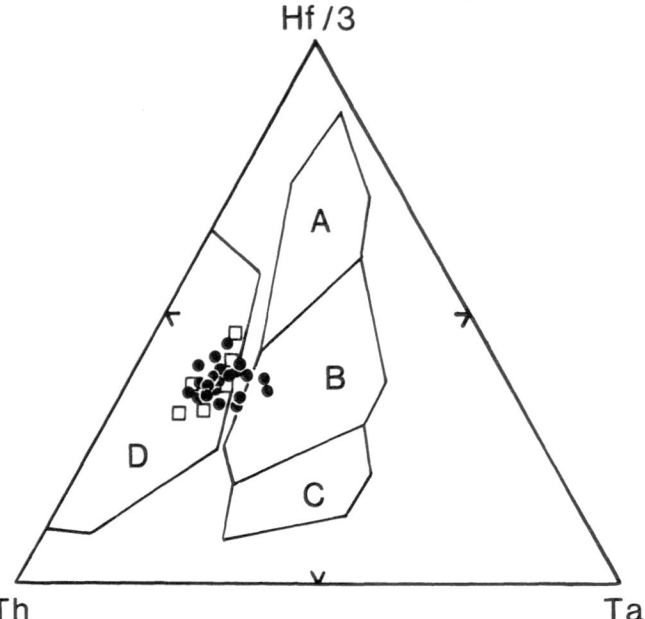

Figure 10.12   Th-Hf-Ta diagram for upper Noranda Subgroup andesites, dacites and rhyolites of the Blake River Group, Abitibi Subprovince. Data from Capdevila *et al.* (1982) (squares) and Ujike and Goodwin (1987) (dots). Compositional fields (from Wood *et al.*, 1979) are: A: N–type MORB; B: E–type MORB; C: within plate basalts; D: magmas at destructive plate margins.

element abundances suggest eight-fold fractionation which should have brought about the existence of a felsic liquid. Thurston and Fryer (1983) suggested that open-system fractionation (cf. O'Hara, 1977) must have been operative leading to immiscibility in some high-Fe tholeiite systems. The glomeroporphyritic units represent expulsion of evolved tholeiitic magmas (Phinney *et al.*, 1988) toward the later stages of evolution of tholeiitic volcanism.

The high-Fe tholeiites in the Wabigoon Subprovince generally overlie tholeiitic to calc-alkaline units. Within the Abitibi Subprovince, Fe-tholeiites are represented by units such as the Dufresnoy and Destor Series within the Blake River Group (Ludden *et al.*, 1982) which are slightly LREE-enriched. These units occur close to a peripheral fracture and represent leakage along a border fault during ring complex volcanism dominated by the bimodal andesite–rhyolite volcanism described above. Given their chemistry and distribution, these units represent fractionated tholeiitic liquids from a crustal chamber released during the late stages of volcanism.

Komatiites may occur within felsic sequences high in the arc-related lithostratigraphic associations. Examples include the Deloro Group in the Abitibi Subprovince south of Timmins (OGS/MERQ, 1983) and an occurrence in the southern part of the Abitibi Subprovince termed the Swayze greenstone belt (Cattell and Arndt, 1987). A komatiite unit forms the uppermost part of the Swayze belt overlying felsic metavolcanics. Three komatiites and three komatiitic basalts form the upper part of the belt. They are subdivided on the basis of Al_2O_3/TiO_2 ratios into six units characterised by variable enrichments and depletions of the REE. The lower komatiites were

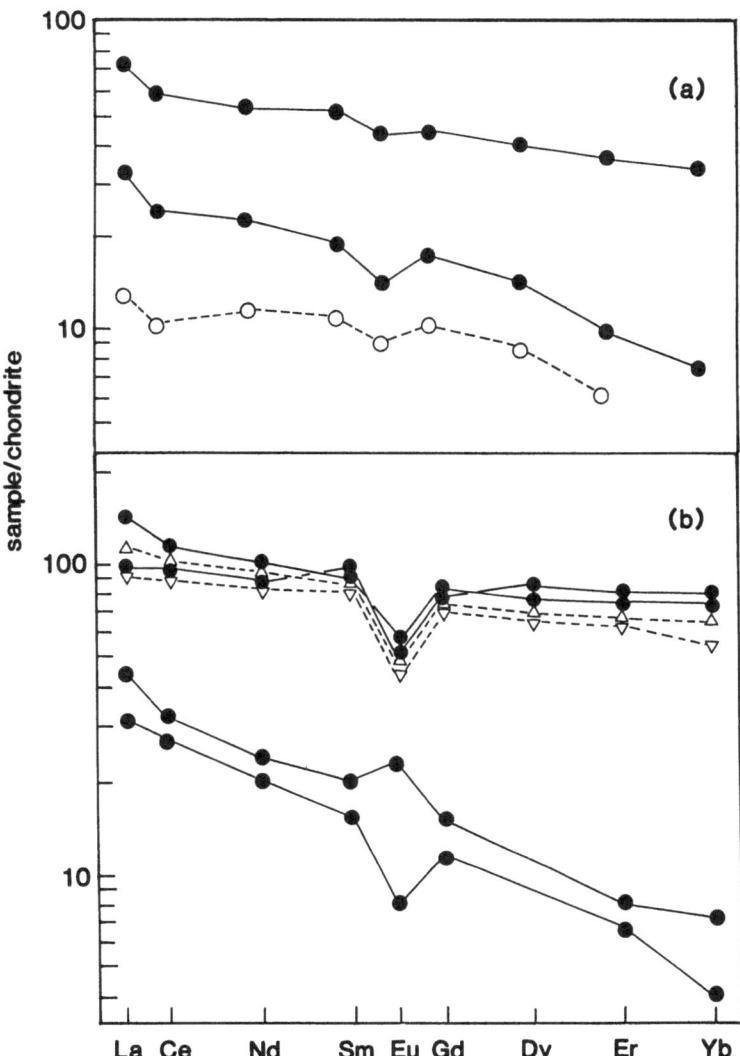

Figure 10.13 Chondrite–normalised REE abundances of (a) basalts above variolitic flows and (b) intermediate variolitic metavolcanics mid-way through the third cycle of volcanism in the 2.74 Ga Uchi Subprovince (Thurston and Fryer, 1983). Open circles in (a): average of 5 basalts ($TiO_2 = 0.72$–0.9%). In (b), solid symbols: mafic matrix; open symbols: felsic spherultic textured varioles.

apparently influenced by garnet fractionation. The middle komatiitic basalts were modified by assimilation of upper crustal material into the magma chamber which produced a layered magma chamber, while the middle komatiites are the product of magma mixing. The upper komatiitic basalts and komatiites are also the result of magma mixing, and the latest komatiites represent a relatively pristine mantle melt.

10.3.4 *Pull-apart basins*

Areally restricted fault-bounded sequences characterised by alkalic to calc-alkalic metavolcanics and fluviatile sediments form the youngest units in several Superior Province greenstone belts. Named 'Timiskaming' sequences after the type locality in the Abitibi Subprovince, they can be compared to pull-apart basins developed during late deformation of greenstone belts during the process of the juxtaposition of one Subprovince against another (Thurston and Chivers, 1989). They are characterised by either calc-alkalic to alkalic mafic trachytes and leucite-bearing flows and minor mafic to felsic pyroclastic rocks (Cooke and Moorhouse, 1969), or shoshonitic volcanism, typical of 'Timiskaming' sequences in the Sachigo Subprovince (Smith and Longstaffe 1974; Brooks *et al.*, 1982), the Uchi Subprovince (Wallace, 1985) and the Abitibi Subprovince (Shegelski, 1980).

In the best documented example, Brooks *et al.* (1982) described the Oxford Lake sequence as high-K andesite–dacite–rhyolite to shoshonite in geochemical affinity. The sequence is marked by high abundances of Sr, Rb, K and La, with high La_N/Yb_N ratios and unfractionated HREE (Figure 10.14). This geochemical signature is considered to represent genesis within a thickened crust in an Archaean arc of possible continental affinity. The age of the pull-apart basins and the fact that they are uniformly developed after the first major deformation and prior to late deformation concentrated at major Subprovince boundaries (Stott *et al.*, 1987) suggests that the pull-apart basins are a late stage in the accretion of arc terranes around an older sialic nucleus which forms the central part of the Superior Province.

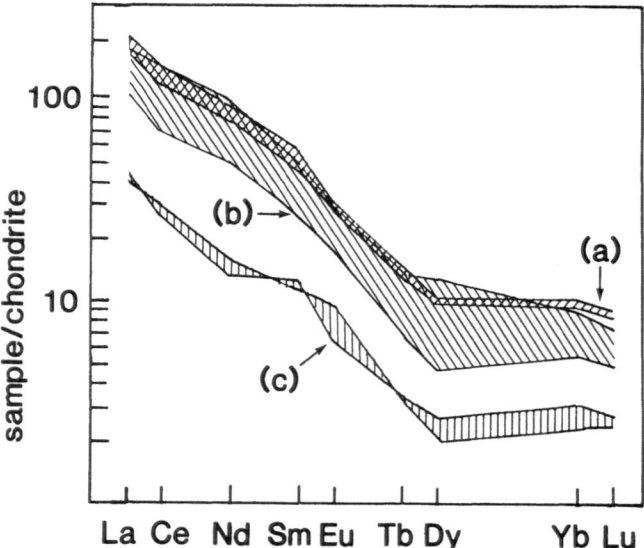

Figure 10.14   Chondrite–normalised REE abundances of shoshonitic metavolcanic rocks of the Oxford Lake Greenstone belt, Sachigo Subprovince, with $SiO_2 = 53$–55% (a), 56–66% (b), and 66–75% (c).

10.3.5 Late-stage mafic magmatism

Shoshonitic lamprophyre dykes cut all lithologies of the granite–greenstone sub-provinces of the Superior Province and are preferentially concentrated near the major shear zones defining subprovince boundaries associated spatially with the rocks of the pull-apart basins (Wyman and Kerrich, 1989). The dykes contain phlogopite or hornblende phenocrysts with feldspar restricted to the groundmass. They have rounded segregations of K-feldspar and calcite, panidiomorphic textures and rare crustal xenoliths. Unaltered dykes have $SiO_2 = 47–51\%$, $TiO_2 = 0.55–0.60\%$, $P_2O_5 = 0.36–1.58\%$, $K_2O/Na_2O = 1.5–2.3$, and $mg' = 0.8$. The dykes are characterised by extreme enrichment of LILE and LREE and high abundances of Cr, Co, Ni and Sc. MORB-normalised values show enrichment of LILE (Rb, K, and Ba) relative to Sr and LREE, and troughs at Ta, Nb and Ti. The late lamprophyre dykes are similar in age to the pull-apart basins, and were emplaced diachronously from north (2.7 Ga) to south (2.67 Ga) across the Superior Province. Wyman and Kerrich (1989) related the genesis of this suite to their derivation from a depleted mantle source and emplacement during localised extension during successive accretion of individual granite–greenstone terranes.

10.4 Mafic–ultramafic intrusions

10.4.1 Platform terranes

In the Sachigo Subprovince (Figure 10.2), minor mafic–ultramafic complexes composed of pyroxenite, gabbro and amphibolite (Breaks et al., 1986) cut the platform metasediments. Field relations suggest that the complexes are similar in age to the early, possibly syn-volcanic granitoid units. No geochemical data are available for these mafic–ultramafic intrusions.

In the older part of the Wabigoon Subprovince, four sets of mafic dykes are associated with the platform lithostratigraphic association (Wilks and Nisbet, 1988). Early mafic dykes with up to 50% pseudomorphs after plagioclase in a chloritic matrix cut the early c. 3.0 Ga granitoids (Davis and Jackson, 1988) but not the unconformably overlying supracrustal units. These dykes are olivine-normative tholeiites (Wilks, 1986). High-Mg mafic dykes cut the older granitoids and the overlying supracrustal units. Their field relations, petrography and geochemistry suggest that these dykes were the feeders for the komatiitic volcanics of the platform sequence (Wilks and Nisbet, 1988). Two types of late mafic dykes cut platform sequence supracrustal rocks and were intruded parallel to the regional foliation in the granitoids. These are 10 m wide olivine-metadolerites which comprise 15% pseudomorphs after olivine and 10% opaques in a matrix of actinolite, chlorite and serpentine, and metadolerites with a relict ophitic texture. High-Fe tholeiitic dykes cut > 3 Ga tonalitic rocks of the Caribou Lake area (Davis et al., 1988) and the older part of the Wabigoon Subprovince, suggesting that there is a great deal of variation of dyke types associated with the platform sequences.

10.4.2 Arc terranes

In the arc sequences, there are three types of mafic–ultramafic intrusions. One type is dominated by anorthosite and is closely related to the mafic volcanism. The second comprises the mafic components of syn-volcanic granitoid complexes, closely related to felsic volcanism, and the third occurs as complexes which are in part related to syn- to post-tectonic granitoid intrusions.

Major gabbro to anorthosite bodies in the Canadian Shield can be regarded as part of the mafic volcanism of arc volcanic systems. These bodies form plagioclase-megacrystic anorthosite intrusions and post-tectonic dykes. Examples include the Dore Lake complex (Allard, 1970) and Shawmere complex (Simmons et al., 1980) in the Abitibi Subprovince, Bad Vermilion Lake in the Wabigoon Subprovince (Ashwal et al., 1983), and the Bird River Sill (Scoates, 1983) in the Winnipeg River Subprovince. They contain plagioclase (An_{80-90}) megacrysts up to 30 cm across in a basaltic matrix. Phinney et al. (1988) surveyed these greenstone belt anorthosite units and noted that the megacrysts are similar to those found in plagioclase-phyric basaltic flows in greenstone belts (Green, 1975) and other dyke and sill units. The REE trends and fractionation patterns of the anorthosite bodies are similar to the tholeiitic trends observed in arc mafic volcanic sequences (Figure 10.15) which suggests a tholeiitic parental liquid to the anorthosite complexes. Some megacrysts have thin, more sodic rims indicating that the megacrysts formed under deeper crustal conditions than the matrix material. Intrusions related to mafic volcanism are numerous in the Abitibi Subprovince (Naldrett and Mason, 1968; MacRae, 1969) where sill-like bodies up to 1 km thick with basal peridotite overlain by clinopyroxenite and gabbro occur amongst arc volcanics. These intrusions consist of complex sills with cyclic repetition of layers, simple sills with one repetition of the above units and unlayered gabbroic intrusions. Their parental magma was tholeiitic, and petrogenetic modelling suggests that fractionation involved olivine, clinopyroxene and plagioclase. Jolly (1977) suggested

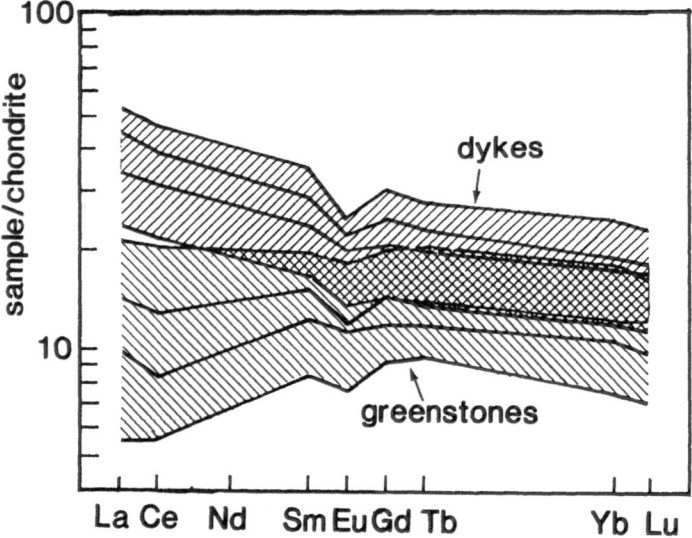

Figure 10.15 Chondrite-normalised REE abundances of basaltic matrices of megacrystic anorthosite units associated with tholeiitic flows of Superior Province greenstones compared with matrices of post-tectonic plagioclase megacrystic dolerite dykes (Phinney et al., 1989).

that the mafic–ultramafic intrusions and tholeiitic volcanics in the Abitibi greenstone belt represent extrusive and intrusive parts of the same magma system. This has been verified by subsequent high precision geochronology and trace element geochemistry.

Late tectonic magmatism within the Superior Province generally ranges from tonalite to quartz monzonite with relatively rare granite (Ayres and Cerny, 1982). However, the late granitoids include mafic to ultramafic intrusions at Lac des Iles and Entwine Lake (Sutcliffe, 1988) in the Wabigoon granite–greenstone Sub-province, and the Abitibi Batholith (Smith and Sutcliffe, 1988). At Lac des Iles, a concentrically zoned series of mafic to ultramafic intrusions form a complex 30 km in diameter with wehrlite, lherzolite, clinopyroxenite and websterite with marginal zones of hornblendite, pyroxene-hornblendite and hornblende-gabbro. The intrusions are bounded by tonalitic gneiss cut by hornblende-bearing tonalite and invaded by numerous mafic to intermediate dykes. Evidence for the contemporaneous mafic and felsic magmatism is provided by the gradational contacts between the ultramafic and felsic units, the development of skeletal amphibole in the hornblende tonalite, the margins of mafic intrusions and small intrusions and dykes, and net-vein textures where tonalite back-veins mafic rocks along intrusive contacts. The tonalitic rocks are seen as juvenile mantle-derived units similar to sinukitoids (Shirey and Hanson, 1984) based on their high mg' values and high Cr and Ni contents as well as trace element arguments (R. Sutcliffe, pers. comm., 1989). The parental magma for the Lac des Iles complex was a high-Al, mantle-derived basaltic liquid. The relationships suggest emplacement from a zoned magma chamber with a mafic base and a tonalitic top, a scenario envisaged for many Archaean felsic volcanic systems (Ayres and Thurston, 1985). This zoned magma chamber in turn suggests the possibility of large-scale mafic underplating during the final stages of the Kenoran orogeny.

10.5 Slave Province

The Slave Province (Figure 10.16) has recently been subdivided into four major terranes (Kusky, 1989). The Anton terrane of 3.5–3.2 Ga tonalitic rocks represents an Archaean microcontinent cut by mafic dykes and unconformably overlain by quartzites and komatiites of an early platform sequence (Covello et al., 1988). The Sleepy Dragon terrane comprises 2.8–2.7 Ga mafic quartzo-feldspathic gneisses with isolated dioritic and gabbroic inclusions unconformably overlain by quartz-rich metasediments and mafic–ultramafic volcanics representing a shallow water platform (Covello et al., 1988). The Contwoyto terrane is comprised of extensive greywacke–mudstone turbidites with locally developed black shale and BIF at the base. The sediments are unconformable on allochthonous greenstone belts which have an ophiolite-like stratigraphy (Helmstaedt et al., 1986). The few available geochemical data (Easton, 1985; Jenner et al., 1981) show the basalts to be typical Archaean low-K tholeiites of oceanic affinity (Figure 10.17). The fourth terrane, the Hackett River Arc, consists of 2.67 Ga NW-trending predominantly andesitic–rhyolitic volcanic piles and syn-volcanic granitoids. The metavolcanics exhibit cauldron subsidence structures, rhyolitic ring intrusions, domes and textures suggestive of subaerial deposition. These features and their association with contemporaneous greywackes suggest that this terrane represents arc volcanism. However, no detailed trace element data are available.

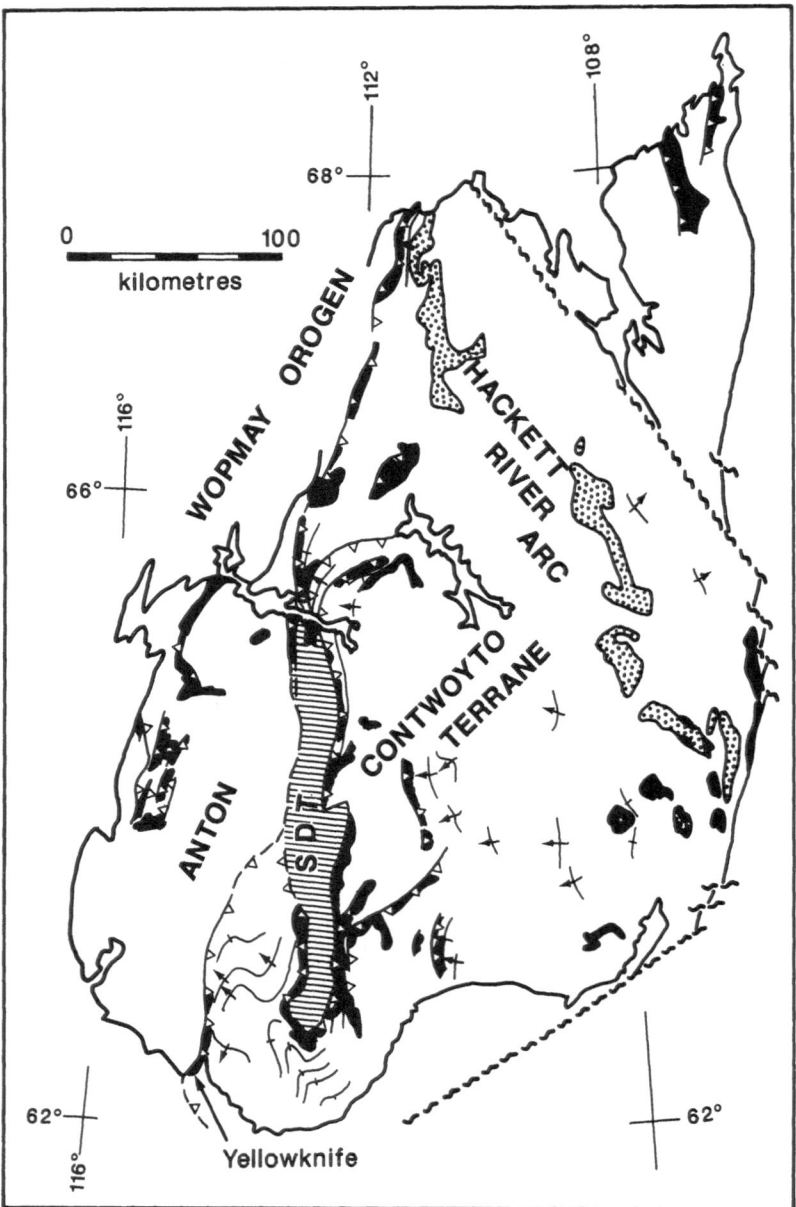

Figure 10.16   Terrane Map of Slave Province showing locations of the Anton, Sleepy Dragon
(SDT) and Contwoyto terranes and Hackett River volcanic arc (after Kusky, 1989), and major
occurrences of mafic (black) and intermediate (stippled) metavolcanic rocks. Triangles indicate
major thrust contacts; arrows and lines indicate structural orientation.

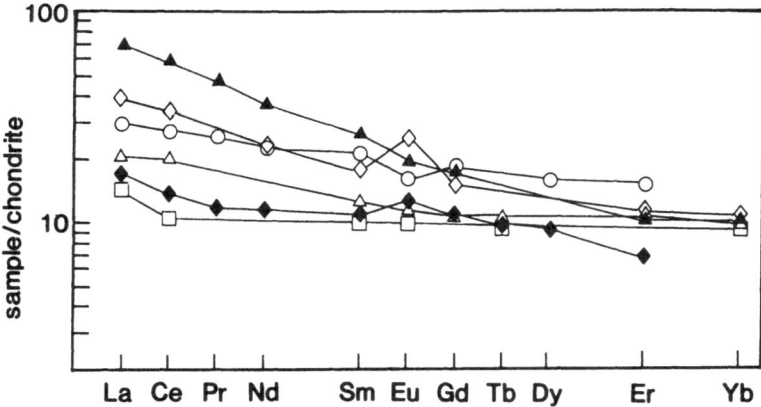

Figure 10.17   Chondrite–normalised REE abundances of representative Yellowknife Supergroup basalts, Contwoyto terrane, Slave Province (Jenner *et al.*, 1981).

10.6   Rae Province

The Rae Province is the northwestern part of what was formerly known as the Churchill Province (Hoffman, 1988). Lithological assemblages are similar to those in the Superior Province (Lewry *et al.*, 1985). The Prince Albert Group is a platform sequence found in the NE extremity of the Province and is the only platform sequence for which trace element geochemical data are available, The lower part of the stratigraphy consists of tholeiites and komatiites, spatially associated with cross-bedded quartzites and sedimentary carbonates (Schau, 1977). Fryer and Jenner (1978) provided trace element geochemical data for the mafic and ultramafic metavolcanic units. These komatiites are unusual in having relatively elevated LILE abundances (Rb, Ba, etc.) and anomalously high Ti contents.

10.7   Post-tectonic dykes

Early Precambrian mafic dykes of the southern Superior Province include the 2.65 Ga N- to NW-trending Matachewan swarm, which extends from the north shore of Lake Huron to north of Lake Nipigon (Halls and Fahrig, 1987). Individual dykes are generally *c.* 10 m wide, with exceptional members up to 250 m. The dykes cut all Archaean rock types including the late shear zone-related lode gold mineralisation. The dykes are not found within the granulite facies gneisses of the major Kapuskasing Structural Zone (Percival and Card, 1985), suggesting that the swarm was emplaced during late Archaean events related to the uplift of the zone. The dykes are all quartz-tholeiites with altered calcic plagioclase (An_{80}), pyroxene, amphibole, quartz, iron oxides and sulphides. The plagioclase typically forms glomeroporphyritic megacrysts up to 10 cm across. Geochemically, the dykes are characterised by relatively enriched LILE and a negative Ti anomaly with respect to primordial mantle relative abundances (Condie *et al.*, 1987).

10.8   Summary and conclusions

It has been demonstrated in this review that it is possible to subdivide the Archaean metavolcanic sequences of the Canadian Shield into lithostratigraphic associations comparable to those formed in Phanerozoic platform, mafic plain, arc, and pull-apart basin environments. The geochemistry of the mafic–ultramafic volcanic rocks is different in each of the above lithostratigraphic associations.

1. Mafic–ultramafic rocks of the platform sequences are relatively enriched in Zr and Y, similar to modern flood basalts, and in contrast to MORB. Platform sequence komatiites are enriched in LILE and Ti compared to mafic plain and arc komatiites.
2. Mafic plain sequences demonstrate physical volcanological characteristics typical of high volumes of magma per unit time and extensive fractionation relative to platform sequences. The depletion in highly incompatible elements (Figure 10.5) is similar to that in basalts of modern oceanic environments (cf. Wood *et al.*, 1979). There is a marked increase in dense fractionated material in the Wabigoon and Abitibi Subprovince mafic plain basalts.
3. The Archaean arc volcanics show geochemical characteristics of younger arc sequences in terms of REE patterns (Figure 10.12), normalised multi-element diagrams and Hf-Th-Ta systematics (Figure 10.13) (Ujike and Goodwin, 1987). Archaean arc-related basalts differ from younger sequences in terms of their generally higher Fe and Ti contents, and the basaltic systems display extreme Fe enrichment probably through open-system crystal fractionation (Thurston and Fryer, 1983). Some Archaean arc-related basaltic systems evolved to fractionate clinopyroxene in the late stages (Morrice, 1988). Others developed varioles due to liquid immiscibility (Thurston and Fryer, 1983), and some systems develop plagioclase megacrystic units toward the top. These variations are probably a function of variations in fO_2 and depth of fractionation.
4. Mafic arc and mafic plain sequences contain a small proportion of material, generally low in the stratigraphy, which displays elevated LREE together with relatively high MgO contents. These units are succeeded upward by rocks with conventional basaltic flat REE patterns. Limited evidence suggests that these units may be developed through sialic contamination of mafic magmas.
5. Mafic arc metavolcanic sequences show evidence of deep-level fractionation, producing calcic plagioclase megacrysts and then removal to shallow magma chambers where extensive fractionation has produced high-Fe basaltic liquids and variolites through liquid immiscibility.
6. The predominance of ash-flow volcanism in the felsic part of arc-type lithostratigraphic associations is significant in several respects in terms of mafic volcanism. In most instances the variation developed within compositionally zoned magma chambers (Smith, 1979), which had a silicic upper part and a more mafic lower part. The silicic upper part develops compositional zonation over extended time periods (*c.* 10^7 years), being maintained in the liquid state by continued input of fresh mantle-derived mafic liquid. However, ash-flow volcanism is the product of a delicate physico-chemical balance (Shaw, 1985). If the flow of mafic magma is excessive, mafic volcanism is the result, if the input is too little, the felsic magma system crystallises at depth rather than erupting. The dominance of ash-flow

volcanism in Archaean arc volcanism (Thurston *et al.*, 1985) suggests the existence of a continued supply of mafic magma during the late stages of greenstone belt volcanism, similar to that of modern ash-flow dominated volcanic systems and, therefore, the rate of mafic magma supply was similar to that in modern magma systems, implying that arguments about, for example, a thinner crust during the Archaean must take into account the minimal thickness necessary to sustain compositionally zoned magma chambers as large as 10^4 km^3.

7. Syn-volcanic plutonism developed during both mafic and felsic volcanism and produced minor mafic–ultramafic plutons spatially associated both with the greenstone belts and associated granitoids. The most significant aspect of these plutons, as well as the granitoid-related mafic/ultramafic intrusions such as Lac des Iles, is that they represent the simultaneous intrusion of mafic and felsic magmas. The concentric zoning of ultramfic–mafic units at Lac des Iles (Sutcliffe, 1988), the juvenile mantle-derived nature of the magma and magma mixing textures point to the simultaneous existence of the two magmas. The tectonic interleaving of tonalite sheets and gabbroic bodies at deep crustal levels in Greenland (Myers, 1981) and the Kapuskasing Structural Zone (Percival and Card, 1985) suggest that large-scale mafic and felsic magmatic underplating occurred during the late stages of arc lithostratigraphic associations in greenstone belts.

8. Mafic volcanism in the pull-apart basin environment is represented by alkalic to shoshonitic volcanism and plutonism marked by extensive hornblende fractionation and enrichment in LREE similar to the late stages of arc volcanism in younger orogens.

9. The geochemical signature of late lamprophyric dykes cutting all greenstone belt lithologies are similar to those developed in the late stages of modern arc-related magmatism.

10. The overall development of basic magmatism in the early Precambrian Canadian Shield can be equated with that in various modern plate-tectonic related volcanic provinces.

11 Early Precambrian basic rocks of Greenland and Scotland

R.P. HALL, D.J. HUGHES and J. TARNEY

11.1 Introduction

Early Precambrian basic rocks in Greenland and Scotland comprise suites with a wide range of ages, compositions and metamorphic grade, and include plutonic, hypabyssal and volcanic types. The two provinces are conventionally considered to have been originally part of the same craton, and contiguous with the Nain Province of eastern Labrador before the opening of the North Atlantic (Bridgwater et al., 1973, 1978; Korstgård et al., 1987). The Archaean Craton from southern West Greenland to the Lewisian of Scotland comprises a high-grade gneiss complex (Bridgwater et al., 1976; Park and Tarney, 1987a). The development of the best preserved and most extensive part of the craton in West Greenland has been described by several authors (e.g. Bridgwater et al., 1976; Myers, 1976; Kalsbeek, 1981; McGregor et al., 1986). The craton is bounded to the north and south by the Proterozoic Nagssugtoqidian and Ketilidian mobile belts respectively (Figure 11.1) (Allaart, 1976; Escher et al., 1976; Korstgård, 1979a). The composite lithostratigraphic units of these early Precambrian terranes include a wide spectrum of basic lithologies ranging from early Archaean to early Proterozoic. The Archaean Complex of East Greenland, on the other hand, contains relatively few basic units and the craton is dissected by the Ammassalik Proterozoic mobile belt (Chadwick et al., 1989). Abundant high-grade metasediment-dominated supracrustal belts in the centre of the mobile belt (Hall et al., 1989) are probably also early Proterozoic. Reworked Precambrian gneisses also occur sporadically throughout the rest of Greenland (Figure 11.1) and often include supracrustal and intrusive basic units (Andersen and Pulvertaft, 1986; Kalsbeek, 1981; Kalsbeek and Taylor, 1985a; Kalsbeek et al., 1984; Mengel, 1983a, b; Hansen et al., 1987). However, they have been less extensively studied than those of southern West Greenland.

The Lewisian gneiss terrane of northwest Scotland, although smaller than the West Greenland region (Figures 11.1, 11.2) has been studied intensively. It is a high-grade Archaean terrane (Park and Tarney, 1987a) with a minor low-grade volcano-sedimentary succession, but has been dissected by major Proterozoic shear zones which have juxtaposed different granulite and amphibolite facies crustal segments. It has also been intruded by the northwest-trending Scourie dyke swarm. Recent work on the geology of this region has been collated by Park and Tarney (1987b).

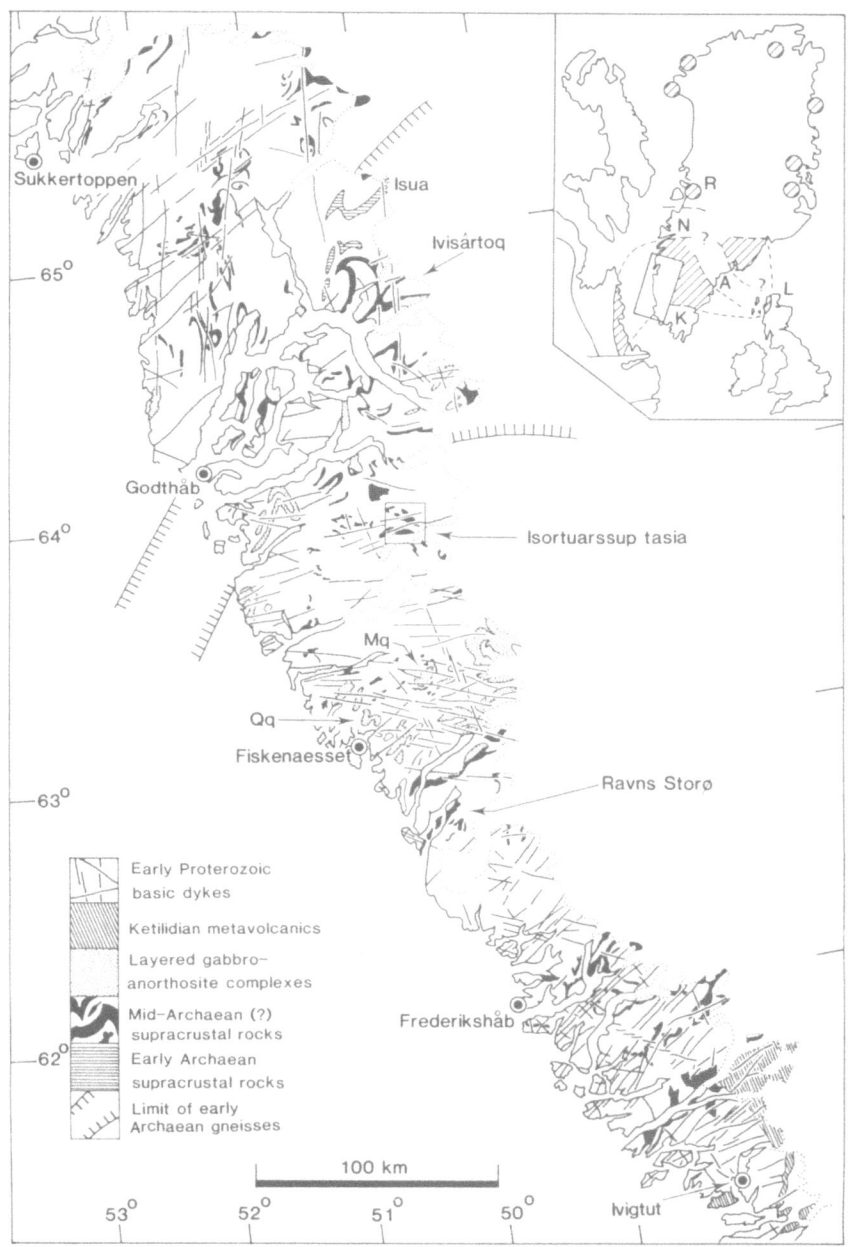

Figure 11.1 Simplified map showing the distribution of Archaean and early Proterozoic basic rocks in southern West Greenland and the limit of early Archaean (*c.* 3.7 Ga) Amîtsoq gneisses in the Godthåb region in which Ameralik basic dykes are abundant. Mq and Qq are metagabbro-anorthosite complexes at Majorqap qâva and Qeqertarssuatsiaq respectively. The inset map depicts the extent of the map in relation to the Archaean craton (line shading) and the Proterozoic mobile belts (Ketilidian: K; Nagssugtoqidian: N; Rinkian: R; and Ammassalikian: A). Occurrences of Archaean rocks in northern Greenland are adapted from Kalsbeek (1981, 1986). Correlation with the Lewisian terrane of northwest Scotland (L) is not certain. (Compiled from sources mentioned in the text.)

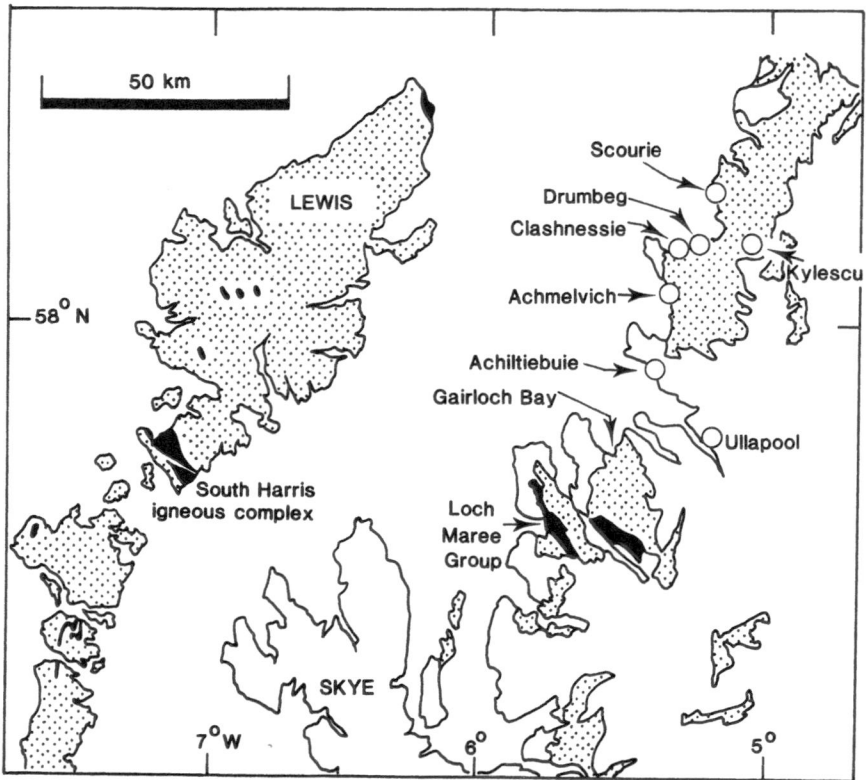

Figure 11.2 Sketch map of northwest Scotland showing the limit of outcrop of the Lewisian complex (shaded) and the main basic rock localities mentioned in the text. (Modified after Park and Tarney, 1987a.)

11.2 Age and distribution of the basic rocks

The sequence of Archaean and early Proterozoic basic magmatism in southern West Greenland and northwest Scotland is summarised in Table 11.1 and the distribution of the larger units is shown in Figures 11.1 and 11.2. The best preserved basic rocks occur within and marginal to the craton exposed for some 600 km in southern West Greenland, and collectively constitute around 15% of that terrane (Figure 11.1). There is generally only indirect evidence for the absolute ages of the basic units.

11.2.1 *Archaean suites*

The oldest basic rocks of the North Atlantic Craton are the early Archaean metavolcanics (3.8 Ga) (Hamilton *et al.*, 1983; Baadsgaard *et al.*, 1984) which form part of the Isua supracrustal belt in West Greenland (Figure 11.1). They are interleaved with various metasedimentary schists, cherts and various banded iron-formation (BIF) lithologies (Nutman, 1986), and local discordances show that the surrounding 3.7–3.4 Ga dioritic to granitic Amîtsoq gneisses (McGregor, 1973; Baadsgaard

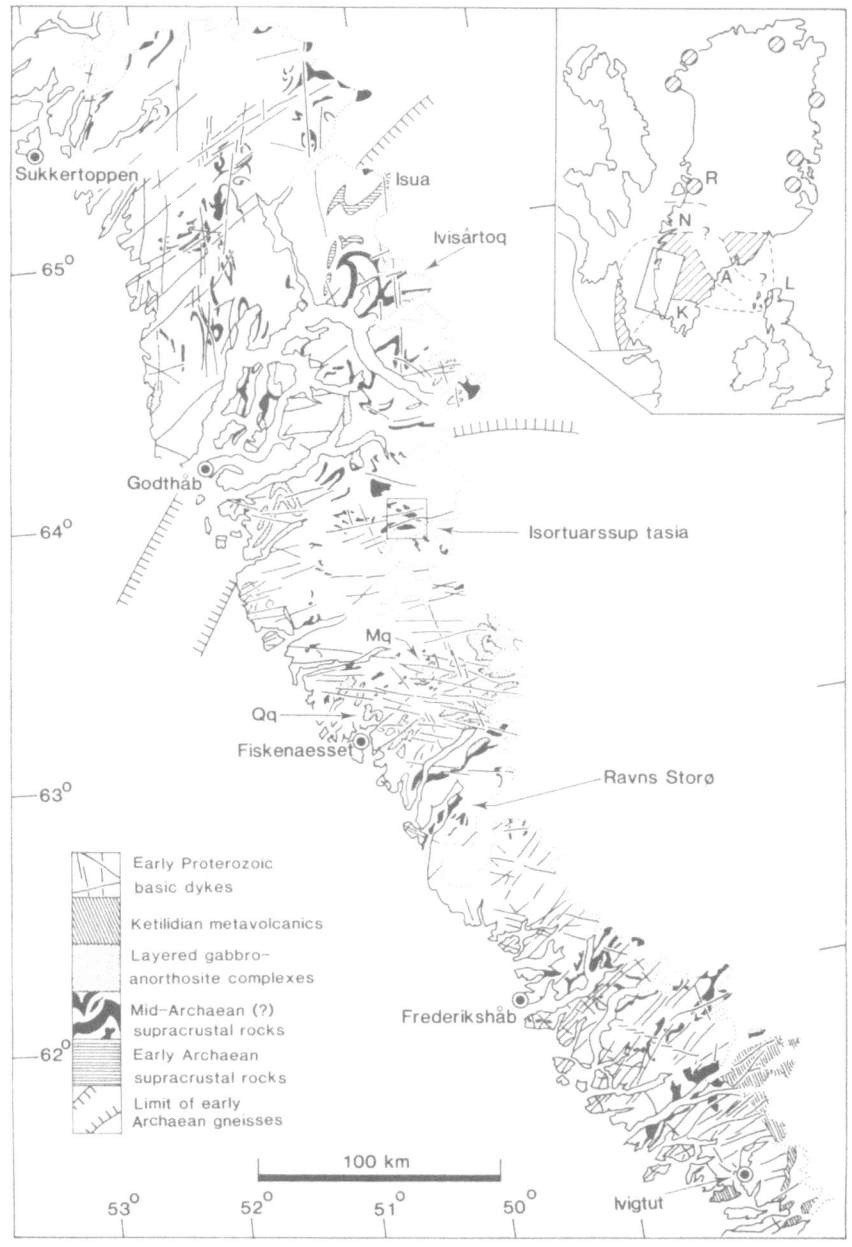

Figure 11.1 Simplified map showing the distribution of Archaean and early Proterozoic basic rocks in southern West Greenland and the limit of early Archaean (*c.* 3.7 Ga) Amîtsoq gneisses in the Godthåb region in which Ameralik basic dykes are abundant. Mq and Qq are metagabbro–anorthosite complexes at Majorqap qâva and Qeqertarssuatsiaq respectively. The inset map depicts the extent of the map in relation to the Archaean craton (line shading) and the Proterozoic mobile belts (Ketilidian: K; Nagssugtoqidian: N; Rinkian: R; and Ammassalikian: A). Occurrences of Archaean rocks in northern Greenland are adapted from Kalsbeek (1981, 1986). Correlation with the Lewisian terrane of northwest Scotland (L) is not certain. (Compiled from sources mentioned in the text.)

that these rocks are around 3.2 Ga old (Stecher *et al.*, 1986). Most now comprise homogeneous or banded, fine-grained amphibolites. Large outcrops of layered metagabbro–anorthosite complexes are also abundant, particularly in the classic Fiskenaesset region (Figure 11.1). Sm–Nd isotope data have yielded a crystallisation age of 2.86 ± 0.05 Ga for the Fiskenaesset Complex (Ashwal *et al.*, 1989). The various layered gabbroic and anorthositic units become deformed into coarser-grained banded amphibolitic rocks which usually retain textural evidence of their layered gabbroic origin (Myers, 1981, 1985). Despite the lack of isotopic data, the age of the Ameralik Dykes is constrained to the extent that they post-date the 3.6 Ga granulite facies metamorphism and the youngest (3.4 Ga) Amîtsoq pegmatitic gneisses, but are older than the 3.1–2.8 Ga Nûk gneisses. They are broadly equivalent to the Saglek Dykes of Labrador (Bridgwater *et al.*, 1978). The Ameralik Dykes are conventionally tabulated as the second phase of Archaean basic magmatism since they do not cut the Malene supracrustal rocks. This could mean that (a) the dykes are older than the Malene metavolcanics, (b) the volcanics were fed by the dykes and are simply their extrusive equivalents, or (c) the volcanics and the dykes and their enveloping gneissic host rocks were juxtaposed tectonically. The Fiskenaesset-type layered gabbro–anorthosite complexes have been shown locally to be the youngest of the three basic suites by their rare discordant contacts with pillow-structured amphibolites in the type region (Escher and Myers, 1975).

The Tartoq supracrustal group (Bridgwater *et al.*, 1976; Allaart, 1976; Appel and Secher, 1984) occurs in the northern border zone of the Ketilidian mobile belt at the southern margin of the Archaean craton, in the Ivigtut region (Figure 11.1). This group comprises greenschist to epidote–amphibolite facies metasediments (mainly psammitic), basic metavolcanics (locally pillow-structured), and gabbroic and ultrabasic sills (Higgins, 1968; Berthelsen and Henriksen, 1975). They are affected by metamorphic and intrusive granitic activity of at least 2.6 Ga, but are considered to be younger than the neighbouring 2.8 Ga gneisses (Berthelsen and Henriksen, 1975; Allaart, 1976; Taylor *et al.*, 1980). However, it is also possible that they are an allochthonous sequence equivalent to the mid-Archaean supracrustal units further north.

The Archaean Complex of northwest Scotland (Figure 11.2) is characterised by the presence of a vast number of ultramafic and mafic pods and lenses, particularly within the granulite facies gneisses, which may represent severely dissected portions of larger bodies such as those found in Greenland. The later, early Proterozoic shear zones are often concentrated where the ultramafic components are most abundant. Mid-Archaean basic units include metavolcanic horizons associated with paragneisses (Bowes, 1976; Okeke *et al.*, 1983; Cartwright *et al.*, 1985) and fragments of layered peridotitic, gabbroic and anorthositic complexes which occur at Achmelvich Bay, Achiltibuie, Drumbeg and Scouriemore (Tarney, 1978; Sills *et al.*, 1982; Fettes and Mendum, 1987; Rollinson and Fowler, 1987). No early Archaean rocks have been identified in northwest Scotland (Chapman and Moorbath, 1977).

11.2.2 *Proterozoic units*

The most significant early Proterozoic basic rocks occur as large swarms of dykes and as less extensive metavolcanic horizons within supracrustal units in the mobile belts (Figure 11.1). Many of the dykes are completely fresh and have widths of between 10 and 50 m and are continuous for several tens of kilometres, but they become deformed and metamorphosed in the marginal Proterozoic mobile belts. In southern West Greenland they were originally known collectively as the MD (meta-dolerite) dykes (Bridgwater et al., 1976), but it has since been recognised that there are at least three separate swarms: the MD, the BN (boninitic–noritic) and the PP (plagioclase-phyric) dykes (Hall and Hughes, 1986, 1987; Hall et al., 1987b, c). The BN and MD dykes of the Frederikshåb and Godthåb regions have given isotopic ages of around 2.1 Ga (Bridgwater et al., 1985; Kalsbeek and Taylor, 1985c). The abundant dykes in the southern part of the Nagssugtoqidian mobile belt (the Kangâmiut dykes) (Escher et al., 1975, 1976) are around 1.95 Ga old (Kalsbeek et al., 1978).

The dense early Proterozoic Scourie dyke swarm in northwest Scotland (Weaver and Tarney, 1981a) trends between E–W and NW–SE (Tarney and Weaver, 1987b; Park and Tarney, 1987b) and comprises a variety of different types which were emplaced into a transtensional regime at c. 2.4–2.2 Ga (Evans and Tarney, 1964; Park et al., 1987; Chapman, 1979; Heaman and Tarney, 1989). The dykes cutting the central zone granulites of the mainland Lewisian were emplaced into country rocks which were at amphibolite facies temperatures and depths of 15–20 km (Tarney and Weaver, 1987b, c; Sills and Rollinson, 1987). Some are fresh but many others are metamorphosed. Dykes in the flanking Laxfordian amphibolite facies gneiss zones are generally much thinner and usually metamorphosed to foliated amphibolites. In this respect the Scottish dykes are similar to those within and north of the Ammassalik Proterozoic mobile belt of East Greenland, rather than the pristine, apparently higher level MD and BN dykes in the centre of the Archaean gneiss terrane of West and East Greenland, which retain complex igneous mineral assemblages (Hall et al., 1986, 1988).

Early Proterozoic supracrustal belts in West Greenland occur in the Nagssugto-qidian and Rinkian mobile belts (Escher and Pulvertaft, 1976; Escher et al., 1976) and in the northern border zone of the Ketilidian belt (Allaart, 1976). Basic metavolcanic rocks form a minor part of these belts and supracrustal belts in southeast Greenland are also dominated by metasediments (Hall et al., 1989). In northwest Scotland, the Loch Maree Group in the Gairloch district (Figure 11.2) comprises two supracrustal belts around 4 km and 2 km wide respectively of amphibolite facies metabasitic rocks and subordinate semipelitic metasediments, thin marbles and BIF (Johnson et al., 1987). This group has given a Sm–Nd model age of c. 2.0 Ga (O'Nions et al., 1983).

11.3 Early Archaean volcanism: the Isua–Akilia association

The basic rocks of the 3.8 Ga Isua supracrustal belt (Nutman, 1986) fall into two groups: (i) homogeneous plagioclase-hornblende-amphibolites together with banded rocks with varying amounts of hornblende, Ca-poor amphibole, clinopyroxene, garnet, feldspar and biotite, and (ii) sills of garbenschiefer-textured amphibolitic and chloritic schists. The homogeneous amphibolites were derived from basic volcanics with chemical characteristics typical of Archaean magnesian low-K tholeiites (Gill, 1979; Gill et al., 1981). Their mg' numbers ($100Mg/(Mg + Fe^*)$) vary from 73 to 42, and dfferentiation was controlled by olivine and subsequent plagioclase and subordinate clinopyroxene fractionation. No komatiitic rocks have been identified in the suite. However, Nutman (1986) and Nutman et al. (1984) interpreted the metavolcanics as a high-Fe tholeiite suite of which some layers represent '...clinopyroxene-rich and residual liquid parts respectively of layered picritic–tholeiitic sills and flows'. The Akilia association enclaves of the Godthåb–Isua region also include clinopyroxene-rich rocks which may represent komatiitic metavolcanics or clinopyroxene cumulates (McGregor and Mason, 1977). Clinopyroxene fractionation is invoked for these rocks. The two petrogenetic models for these early Archaean metavolcanics are clearly not compatible and perhaps simply reflect the original complexity of the basic magmas.

According to Gill *et al.* (1981) the garbenschiefer-textured amphibolites (mg' = 80 to 65) originated from a highly aluminous basaltic magma (Al_2O_3 = 18%; MgO = 10%) in which the progressive accumulation of olivine (*c.* Fo_{87}) gave rise to the common Mg-rich rocks (MgO up to 18%; Table 11.2). A model for these amphibolites involving the fractionation rather than accumulation of olivine would require a parental magma with around 18% MgO and 15% Al_2O_3 which, as Gill *et al.* (1981) pointed out, would have been unlikely. Nutman *et al.* (1983) suggested that the aluminous garbenschiefer amphibolites originally formed bodies with 'leucogabbroic' affinities, while it has also been proposed that these amphibolites originated from a 'primitive liquid' derived from a '...clinopyroxene-depleted mantle that had already undergone partial melting to supply basic magma, possibly the parental liquids of the tholeiitic rocks...of the Isua supracrustal belt...', and that the petrogenesis of the sills was controlled by olivine followed by orthopyroxene fractionation (Nutman, 1986). Once again this interpretation differs significantly from that of Gill *et al.* (1981).

11.4   Mid-Archaean basic magmatism

11.4.1   *Ameralik Dykes*

The significance of the field and geochemical characteristics of the Ameralik dykes have been discussed by many authors (McGregor, 1973; Gill and Bridgwater, 1979; Chadwick, 1981; Hall, 1982; Nutman *et al.*, 1983; Hall *et al.*, 1987a). Several types have been recognised by their mineralogy, internal structures and geochemistry (Chadwick, 1981; Gill and Bridgwater, 1979). As a whole they form a suite with mainly low-K tholeiitic affinities and geochemically they are indistinguishable from many of the Isua metavolcanics (Gill *et al.*, 1981). Most samples have MgO values between 9 and 5% (Table 11.2). They have low Cr, Al, P, Zr and Ti and high Y and Fe/Mg values compared to modern oceanic basalts. As in many other Archaean tholeiite suites (Gill, 1979), they have somewhat higher Ni contents than modern MORB at equivalent mg' values. They also have characteristically low Zr/Y and Ti/Y ratios (Figure 11.3). The least evolved dykes have REE contents around ten times chondritic values with $La_N/Lu_N = 1$, but more commonly they show LREE enrichment ($La_N = 50$; $La_N/Lu_N = 5$) (Figure 11.4). Their parental magma was derived by moderate degrees of partial melting of undepleted mantle, the Mg-poor dykes having evolved by the fractionation of plagioclase, olivine and clinopyroxene. Despite the apparent influence of early plagioclase precipitation, very few have Eu anomalies.

Rare high-Mg dykes (MgO from 9 to 18%) (Table 11.2) that occur both at Isua and near to Godthåb are remarkable in that they appear to be the only rocks of this age in which primary igneous minerals and textures remain intact (Gill and Bridgwater, 1979; Nutman, 1986). These dykes are petrologically distinctive in that they are orthopyroxene-rich, whereas fractional crystallisation models for the Ameralik Dykes do not involve orthopyroxene (Gill and Bridgwater, 1979). These high-Mg noritic dykes are texturally and chemically similar to early Proterozoic ones, abundant in the

Table 11.2 Representative analyses of Archaean basic rocks from Greenland and Scotland

	1 292258	2 171756	3 Av(17)	4 225964	5 200887	6 207616	7 129978	8 129970	9 177148	10 207037	11 MA(11)	12 71657	13 Fi (37)	14 92571	15 200401	16 146	17 109
SiO_2	50.86	45.58	48.55	51.93	44.52	46.61	47.26	49.20	45.70	47.50	47.17	46.98	47.80	47.50	44.66	48.24	50.88
TiO_2	1.10	0.22	1.11	0.34	0.38	0.57	0.32	0.84	0.51	0.72	0.65	0.54	1.08	0.16	0.12	1.21	0.55
Al_2O_3	13.70	15.33	13.59	9.47	8.65	10.28	6.63	10.78	3.62	12.67	15.83	13.67	14.80	24.60	7.28	10.37	11.53
Fe_2O_3	—	2.01	15.89	—	12.87	14.35	11.91	13.73	3.97	1.61	12.25	4.98	12.59	4.21	9.11	—	—
FeO	12.83	7.31	—	9.82	—	—	—	—	9.37	9.86	—	6.64	—	—	—	12.97	10.91
MnO	0.31	0.18	0.25	0.17	0.24	0.29	0.23	0.24	0.25	0.18	0.22	0.20	0.21	0.08	0.14	0.24	0.24
MgO	5.43	16.30	6.86	17.44	21.78	11.93	23.95	11.46	24.67	11.68	7.82	12.61	7.29	5.36	36.82	8.99	8.53
CaO	9.23	7.83	10.39	4.21	10.49	13.74	9.19	10.92	7.98	12.88	13.77	9.05	12.05	14.53	1.69	11.58	11.10
Na_2O	2.61	0.78	2.19	1.10	0.77	2.00	0.47	2.45	0.47	1.20	1.26	1.69	1.82	1.89	0.12	1.68	2.57
K_2O	1.01	0.07	0.95	0.97	0.19	0.23	0.01	0.29	0.09	0.13	0.35	1.06	0.31	0.37	0.02	0.68	0.56
P_2O_5	0.12	0.05	—	0.04	0.10	0.01	0.03	0.09	0.05	0.06	—	0.22	0.08	0.01	0.05	0.01	0.03
Sc	—	—	—	—	36	42	20	39	—	—	—	—	—	—	19	—	—
V	140	—	260	129	191	223	138	260	—	—	255	—	239	—	113	285	563
Cr	119	1090	198	2669	2467	811	2633	310	1804	909	410	153	132	813	8800	110	146
Ni	53	540	123	599	723	322	1312	110	1649	208	187	155	—	81	1700	5	5
Rb	34	nd	18	40	9	14	nd	nd	3	2	5	—	7	5	nd	—	—
Sr	106	55	105	51	13	56	11	114	36	89	67	—	81	102	nd	123	110
Y	25	10	34	12	7	8	8	17	10	15	28	—	23	3	3	21	12
Zr	109	18	71	55	24	33	19	56	26	33	28	—	63	3	nd	19	29
Nb	—	—	—	—	nd	nd	1	nd	nd	2	—	—	3	10	nd	nd	1
Ba	133	21	156	58	8	20	nd	62	13	19	52	158	36	4	nd	204	196
Ce	—	—	—	—	2	6	4	14	—	—	—	40	11	28	2	4	33

nd = not detectable; — = no data available.

1: c. 3.8 Ga metavolcanic amphibolite, Isua (Nutman, 1986).
2: Garbenschiefer amphibolite Isua (Nutman, 1988).
3: Average of 17 D-type Ameralik dykes, Godthåb (Chadwick, 1981).
4: High-Mg Tarssartoq dyke, Isua (Nutman, 1986).
5, 6: High- and low-Mg komatiites, Ivisârtoq (Hall et al., 1987a).
7, 8: High- and low-Mg komatiites, Ravns Storø (Friend et al., 1981).
9, 10: High- and low-Mg komatiites, Isortuarssup tasia (Stecher, 1981).

11: Average of 11 metavolcanic amphibolites, Godthåb (Chadwick, 1981).
12: Metavolcanic amphibolite, Neria (Ivigtut) (Kalsbeek and Leake, 1970).
13: Average of 37 metavolcanic amphibolites, Fiskenaesset (Weaver et al., 1982).
14: Leucogabbro, Fiskenaesset Complex (Weaver et al., 1981).
15: Harzburgite, Fiskenaesset Complex (R.P. Hall, unpubl. data).
16: Archaean amphibolites, northwest Scotland (Rollinson, 1987).
All Greenlandic sample numbers should carry the prefix GGU

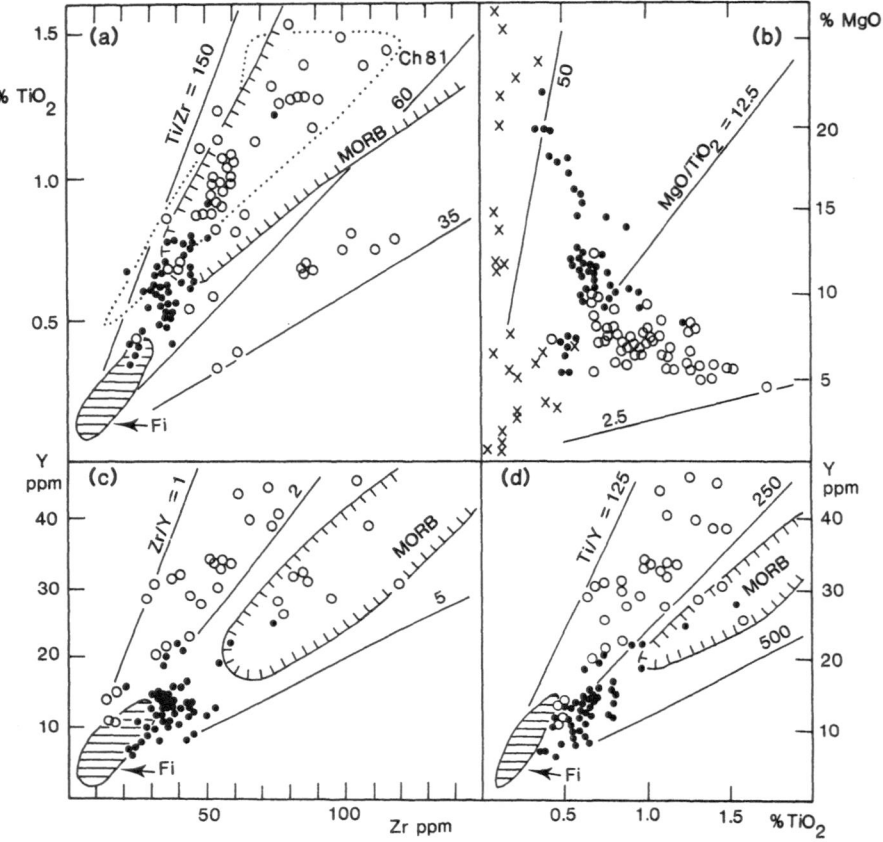

Figure 11.3 Relative Mg, Ti, Zr and Y contents of Malene komatiitic–tholeiitic metavolcanics from Ivisârtoq (dots), Ameralik dykes (open circles), and Fiskenaesset complex gabbroic and anorthositic rocks (crosses and 'Fi' fields). The dotted line in (a) depicts the range of Ameralik dyke compositions described by Chadwick (1981). Other data from sources mentioned in the text. Field of modern MORB shown for comparison in (a), (c) and (d).

northern part of the craton (Hall and Hughes, 1987). There are no komatiitic dykes amongst the Ameralik Suite, and in general the range of magmas they represent is 'more restricted than in any major greenstone belt' (Gill and Bridgwater, 1979).

11.4.2 Metavolcanic suites

Work on the metavolcanic amphibolites in the southern part of the Archaean Craton of southern West Greenland was summarised by Rivalenti (1976). They originated as basic lavas mainly with oceanic basaltic affinities, but include a few alkali basaltic and komatiitic as well as tholeiitic types. The supposed alkali basalts are mildly nepheline-normative but otherwise they do not have the trace element compositions of modern alkali basalts. For instance, their average Rb and Ba contents are only

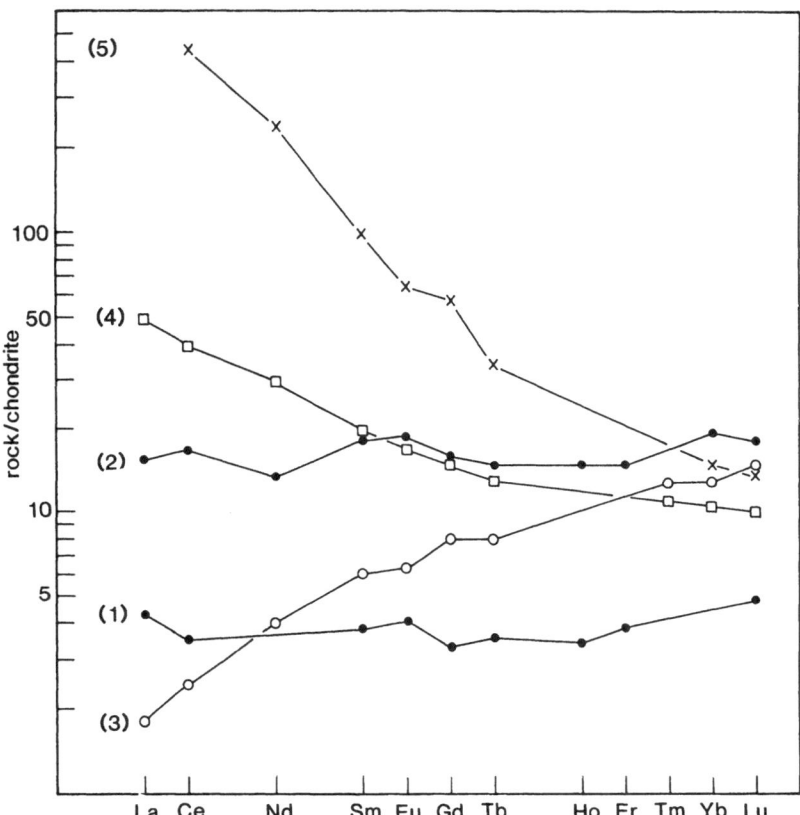

Figure 11.4   Flat chondrite–normalised REE abundance patterns of (1) high-Mg and (2) low-Mg (MgO = 19.65% and 5.7% respectively) members of the komatiitic–tholeiitic metavolcanic suite compared to those of (3) the MORB-type REE abundances of a Malene metavolcanic amphibolite and (4) a typical Ameralik dyke both from the southern Godthåb region (Chadwick, 1981), and (5) a LREE-rich Neriunaq 'intra-Nûk' dyke (Compton, 1978).

11 and 15 ppm respectively. The metavolcanic amphibolites of the northern Ivigtut region (Kalsbeek and Leake, 1970) are notable in that they have some chemical characteristics similar to those of continental tholeiites. Of fifty-four analysed samples, over 80% are hypersthene-normative but average K_2O, Ba and Ce values are 1.03%, 140 ppm and 35 ppm respectively (Table 11.2). Only nine have K_2O/TiO_2 values less than 0.5, compared to the range for modern MORB of between 0.5 and 0.05 (Pearce et al., 1975; Sun et al., 1979; Tatsumi and Ishizaka, 1982). Although the concentration of these elements can be affected by metasomatism and metamorphism (e.g. Gill and Bridgwater, 1979; Zeck et al., 1983), the basic, low-K tholeiitic metavolcanics slightly further to the north have the same metamorphic grade but do not have this geochemistry (Rivalenti and Rossi, 1972, 1975; Friend et al., 1981; Hall et al., 1987a). Thus, it appears that the Ivigtut basic metavolcanics represent a quite different suite.

The rocks described by Rivalenti (1976) as having komatiitic affinities are pyroxenitic and hornblenditic bands within quartzo-feldspathic gneisses, but well preserved pillow-structured metavolcanic rocks in large supracrustal belts at Ravns Storø, Isortuarssup tasia and Ivisârtoq (Figure 11.1) all comprise suites ranging in chemical affinity from oceanic tholeiites to basaltic komatiites in roughly equal proportions (Hall, 1980; Friend et al., 1981; Stecher, 1981; Hall et al., 1987a). Fresh komatiites are normally recognised in part by their quenched, spinifex textures (Arndt and Nisbet, 1982) which are not preserved in these amphibolite facies rocks. They do, however, meet all the geochemical criteria used to define komatiites and relict igneous structures clearly show that the Mg-rich rocks are not simply olivine cumulates. The metavolcanics form a geochemical continuum, komatiitic rocks being distinguished by their high MgO (up to 22% in the Ivisârtoq Belt) and correspondingly high Cr and Ni contents (up to 5000 and 1500 ppm respectively) and high CaO/Al_2O_3 values (> 0.9). They have low Al, Ti, Zr, Y and LILE contents and flat, unfractionated REE distributions, between two and ten times chondritic levels (Hall et al., 1987a). The tholeiitic members of the suite have similar but slightly higher (up to eighteen times chondrite) REE abundances (Figure 11.4). Petrogenetic modelling suggests that the komatiites and tholeiites in these belts were derived from a Mg-rich komatiitic parent, either by the fractionation of olivine and clinopyroxene (liquidus phases in many komatiitic suites) or by varying degrees of partial melting of the same relatively clinopyroxene-rich lherzolitic mantle (Hall et al., 1987a).

Although the Ivisârtoq Belt is within the heart of the c. 3.7–3.4 Ga Amîtsoq gneiss terrane while the Ravns Storø Belt is remote (150 km away) from these older gneisses, there are only minor geochemical differences between the basic metavolcanic units within them, and they provide no evidence for invoking different igneous provinces (Friend et al., 1981). The Ivisârtoq metavolcanic suite is clearly not genetically related to the Ameralik Dykes, which are less magnesian and have no komatiitic affinities (Gill and Bridgwater, 1979; Hall, 1982; Hall et al., 1987a). The lack of komatiitic Ameralik Dykes and the chemical similarity between the Ivisârtoq metavolcanics and those of the Isortuarssup tasia and Ravns Storø Belts (Friend et al., 1981; Stecher, 1981), which are the oldest rocks in those areas, suggest that they were all erupted in an oceanic environment and that the Ivisârtoq Belt and surrounding Amîtsoq sialic crust were juxtaposed tectonically, as has been suggested on structural grounds (Bridgwater et al., 1974; Chadwick and Nutman, 1979; Hall and Friend, 1979).

Elsewhere the situation is slightly different. For example, to the south of Godthâb some of the Malene volcanics and Ameralik Dykes are compositionally similar (Chadwick, 1981), although the Malene rocks tend to have modern MORB-like REE abundances whereas the Ameralik Dykes are predominantly LREE-enriched (Figure 11.4). However, some of the metavolcanics in this area may have been extruded through and onto an Amîtsoq sialic basement and fed by Ameralik Dykes (Chadwick, 1981), as suggested by local contact relationships (Chadwick and Nutman, 1979; Nutman and Bridgwater, 1983). Other basic metavolcanics near Fiskenaesset (Figure 11.1) also seem to be different to the komatiitic–tholeiitic association. Weaver et al. (1981, 1982) showed that the petrogenesis of these volcanics was governed by plagioclase-dominated fractional crystallisation of an aluminous tholeiitic basaltic liquid (Table 11.2). They are slightly LREE-depleted and are thought to be the volcanic equivalents of the abundant layered gabbro–anorthosite complexes of the region, derived from a relatively refractory, garnetiferous mantle source (Henderson et al., 1976; Weaver et al., 1981).

11.4.3 *Fiskenaesset-type gabbro–anorthosite complexes*

Remnants of large-scale layered gabbro–anorthosite complexes occur throughout the Archaean Craton of southern West Greenland, and are typified by the Fiskenaesset anorthosite complex (Figure 11.1). Cumulus textures and gravity-stratified (?) layering are frequently preserved within these anorthosite complexes, despite their being highly deformed and metamorphosed to amphibolite or granulite facies (Windley, 1973; Myers, 1981, 1985). Plagioclase accumulation is a characteristic feature and many of the metagabbroic rocks are very coarse-grained with cumulus plagioclase crystals commonly reaching up to 100 mm in diameter.

The Fiskenaesset intrusion was originally around 500 m thick and comprises roughly equal proportions of gabbroic and anorthositic rocks. The best preserved and stratigraphically most complete outcrops of the complex occur at Majorqap qâva and Qeqertarssuatsiaq (Figure 11.1). They show progressively evolved cryptic mineralogical and petrochemical variation with respect to height in the intrusion (Windley *et al.*, 1973; Windley and Smith, 1974; Steele *et al.*, 1977; Henderson *et al.*, 1976; Myers, 1975; Myers and Platt, 1977). Plagioclase compositions range from around An_{97} to An_{76} and whole rock mg' values vary from 0.65 to 0.29. Chromite cumulate horizons occur in the upper, anorthositic units, in contrast to those which occur near to the base of intrusions such as the Stillwater and Bushveld Complexes (see Chapter 8). This indicates the hydrous and highly aluminous nature of the Fiskenaesset Complex parental magma (Steele *et al.*, 1977). Myers (1985) concluded that the Middle Gabbro unit and the higher leucogabbroic and anorthositic lithologies of the Majorqap qâva outcrop (Figure 11.1) were derived from a second batch of magma, on the basis of the different mineral chemistry of this unit (Myers and Platt, 1977). Steele *et al.* (1977) and Weaver *et al.* (1981) calculated that the parental magma of the Fiskenaesset gabbro–anorthosite complex approximated to an aluminous tholeiitic basalt equivalent to the composition of the enveloping fine-grained metavolcanic amphibolites (Table 11.1). Henderson *et al.* (1976) suggested that these magmas were derived from a garnet-bearing source, on the grounds of REE studies. Recent Sm–Nd isotope studies have revealed an emplacement age of 2.86 Ga and an ε_{Nd} of 2.9 ± 0.4 (Ashwal *et al.*, 1989). This means that the intrusion of the anorthosite complex and the formation of the enveloping tonalitic gneisses must have occurred within a period of around only 70 Ma. The positive ε_{Nd} value indicates a long-term LREE-depleted mantle source.

There are sporadic occurrences of highly deformed and fragmented gabbroic and anorthositic lithologies similar to those of the Fiskenaesset Complex in the northern part of the Archaean Craton (Andersen and Pulvertaft, 1986) and in East Greenland (Andrews *et al.*, 1973; RPH unpublished data). Chaotic leucogabbroic and anorthositic units occur as a large deformed sheet and as disseminated blocks within the reworked Archaean gneisses of the Nagssugtoqidian mobile belt (Andersen and Pulvertaft, 1986). The anorthositic units which occur in North Greenland are probably of Proterozoic age (Nutman, 1984).

11.4.4 *'Intra-Nûk' dykes*

Thin basic dykes are preserved sporadically throughout southern West Greenland as deformed but discordant homogeneous amphibolites within the *c.* 3.0–2.8 Ga quartzo-feldspathic gneisses (the Nûk gneisses of the Godthåbsfjord region; McGregor, 1973). These gneisses are younger than the Malene-type metavolcanic and Fiskenaesset-type metagabbroic rocks. The wispy basic dykes comprise earlier (Neriunaq) and later (Qáqatsiaq) sets, but are usually referred to loosely as 'intra-Nûk' dykes (Chadwick and Coe, 1984). They are typically less than a metre wide and not continuous for significant distances. The limited data available for these dykes show that their geochemistry is quite different to that of the other basic units in that they have some chemical characteristics akin to those of evolved alkali basalts (Compton, 1978; Chadwick, 1981). They have low Mg, Cr, Ni and Ca and high Fe, alkali, Ti and light REE contents. They have distinctive, strongly fractionated REE patterns with La_N/Lu_N values up to 30, and La and Lu contents of up to 500 and 15 times respectively those of average chondrite (Figure 11.4).

11.4.5   Archaean basic rocks of Scotland

Most studies of the Lewisian Complex of northwest Scotland have concentrated on its tectonic and metamorphic evolution (Park and Tarney, 1987b). Bowes et al. (1971) compiled analyses of the various meta-igneous units of the 'Kylesku group' (Figure 11.2), with MgO contents ranging from 9 to 0.7%, and of the metaperidotites to anorthosites derived from layered complexes, with MgO values from 35 to 0.7%. The peridotitic rocks are dominated by lherzolitic and harzburgitic compositions, and minor dunitic and pyroxenitic rocks, interpreted as tectonically disrupted fragments of larger layered bodies.

Most of the amphibolite facies basic rocks from the Gruinard Bay area have geochemical affinities with olivine-normative T-type MORB (Rollinson, 1987; Rollinson and Fowler, 1987). They have MgO contents ranging from around 6 to 12% and high CaO/Al_2O_3 values (> 0.9). Ti, Zr and Y abundances and ratios are very similar to those of their Greenlandic counterparts (Figure 11.3). Their REE abundances vary from slightly LREE-depleted ($La_N = 4$, $La_N/Lu_N = 0.6$) to LREE-enriched ($La_N = 40$, $La_N/Lu_N = 8$). The granulite facies basic rocks of Drumbeg typically have moderately fractionated REE distributions with La_N/Lu_N values of around 2 to 4, and La_N abundances varying from around 15 to 50 (Weaver and Tarney, 1980, 1981b). Tarney and Weaver (1987a) also concluded that the basic rocks represent low pressure fractionated oceanic tholeiites derived from undepleted mantle (Sills et al., 1982; Rollinson, 1987) and that the Laxford amphibolite facies mafic bodies appear to have a similar origin. Associated ultramafic gneisses have 'undepleted' REE abundances. The layered gabbroic complexes were derived from tholeiitic magmas with MgO contents of around 15% (Sills et al., 1982) and may represent cumulates derived from the same parental liquid as the neighbouring amphibolites (Rollinson, 1987). Fractionation and the settling of olivine, orthopyroxene and clinopyroxene produced ultramafic rock types with MgO contents of between 25 and 35% and also depleted the magma in Ni and Cr. The complexes show a marked Fe-enrichment trend, and TiO_2 contents range from around 0.1 to 0.7%. These features, and the progressive increase in V contents, reflect the delayed crystallisation of ilmenite and magnetite, in comparison with the evolution of the Fiskenaesset Complex of Greenland (Weaver et al., 1981). However, they differ from the anorthosite-dominated Fiskenaesset Complex in being predominantly ultramafic.

Thin discordant late Archaean basic dykes with epidote-amphibolite facies mineral assemblages are preserved within the Scourian Gneisses at Clashnessie in northwest Scotland (Figure 11.2), and have been tentatively dated at between 2.7 and 2.4 Ga (Sills and Windley, 1982). These dykes appear to be stratigraphically equivalent to the 'intra-Nûk' dykes of West Greenland, but compositionally they are very different in that the Clashnessie Dykes have chemical affinities with evolved low-K tholeiites (Table 11.3). They have fairly high Si and Fe, and low Al, Ca, alkali, Cr and Ni contents. Their tholeiitic affinity is suggested by their Fe-enrichment ($MgO/Fe_2O_3^* = 0.35$). They are LREE-enriched, with La_N and Y_N values around 50 and 15 respectively.

11.5 Early Proterozoic basic magmatism

11.5.1 Ketilidian, Nagssugtoqidian and Rinkian volcanics, Greenland

Relatively little geochemical work has been done on the early Proterozoic metavolcanics of Greenland. Archaean and early Proterozoic basic volcanic rocks are juxtaposed in the northern border zone of the Ketilidian mobile belt of South Greenland (Allaart, 1976). The Proterozoic rocks occur mainly as part of the upper, Sortis Group of the Ketilidian supracrustals, the lower, Vallen Group being dominated by sediments which are unconformable with the pre-Ketilidian gneisses and Tartoq supracrustals. The Ketilidian supracrustals in this border zone are metamorphosed only to low greenschist facies. Further to the south they are increasingly deformed, metamorphosed to granulite facies and intruded by Ketilidian granites of around 1.8 to 1.7 Ga (Patchett and Bridgwater, 1984; Kalsbeek and Taylor, 1985b). However, they post-date the *c.* 2.1 Ga basic dykes (Allaart, 1976; Kalsbeek and Taylor, 1985c).

The Sortis group metavolcanics are dominated by lavas, which are often pillow structured, and variously sized sills (Bondesen, 1970; Higgins, 1970; Berthelsen and Henriksen, 1975). They approximate to (just-)saturated olivine tholeiites with MgO values between 6 and 8.5% and SiO_2 between 47 and 51%. They have low K_2O contents (< 0.25%) and TiO_2 between 0.75 and 1.6%. Nd isotope analysis of samples from the centre of the mobile belt suggests that the lower lavas at least were derived from a depleted mantle (Patchett and Bridgwater, 1984). The lavas and sills have broadly the same compositional ranges, but Henriksen (1969) demonstrated that they have slightly different compositions to those of the doleritic dykes which cut the neighbouring Archaean gneisses. The Ketilidian belt represents mainly new 1.8 to 1.7 Ga crust, rather than reworked Archaean material (Patchett and Bridgwater, 1984; Kalsbeek and Taylor, 1985a).

Supracrustal rocks in the Nagssugtoqidian mobile belt (Figure 11.1) contrast with the Archaean ones further south in that they are dominated by quartzitic, carbonate and graphitic metasediments rather than basic metavolcanics, and in this respect they resemble the Rinkian supracrustals (Escher and Pulvertaft, 1976; Escher *et al.*, 1976). Both the Rinkian and some of the Nagssugtoqidian supracrustal rocks were probably deposited onto an Archaean gneissic basement and subsequently metamorphosed at between 1.9 and 1.8 Ga (Kalsbeek, 1981; Kalsbeek *et al.*, 1984). Locally, deformed pillow-structured metavolcanics are intimately associated with garnet amphibolite and pyroxenitic intrusive units (Talbot, 1979). Mengel (1983b) demonstrated the overall tholeiitic affinities of the basic rocks in the Nordre Stromfjord region (although they comprise a rather incoherent suite). However, nearby metamorphosed andesitic to rhyolitic pyroclastic rocks have calc-alkaline affinities (Rehkopff, 1984) and the

Table 11.3 Representative analyses of early Proterozoic basic rocks Greenland and Scotland

	1 SV(5)	2 SS(4)	3 207645	4 149208	5 179092	6 120526	7 KD(18)	8 N1	9 919	10 MM5A	11 LGA1
SiO_2	48.38	50.39	52.78	46.39	49.48	52.05	50.5	51.6	51.8	51.33	50.00
TiO_2	0.96	0.91	0.48	1.04	2.10	2.61	1.40	0.61	1.76	0.88	2.32
Al_2O_3	13.72	14.08	11.49	16.23	13.56	14.59	13.9	10.1	14.2	14.80	13.76
Fe_2O_3	1.62	1.88	2.00	3.34	3.39	2.49	3.6	10.60	14.87	12.33	19.86
FeO	11.32	8.15	7.78	9.99	12.85	10.5	9.6	—	—	—	—
MnO	0.19	0.33	0.18	0.22	0.25	0.19	0.21	0.21	0.21	0.19	0.23
MgO	7.23	7.21	17.34	8.45	5.54	4.43	6.2	15.27	4.50	7.88	6.50
CaO	10.62	10.20	6.66	11.89	9.11	7.75	10.0	7.53	9.32	10.49	7.49
Na_2O	2.71	3.38	1.37	1.96	2.62	3.51	2.6	1.87	3.10	2.34	3.33
K_2O	0.22	0.40	0.56	0.22	0.63	1.26	0.61	1.00	0.67	0.14	0.15
P_2O_5	0.15	0.10	0.16	0.17	0.24	0.62	0.10	0.08	0.18	0.05	0.21
Cr	—	—	2345	151	25	38	—	1435	102	289	54
Ni	—	—	687	111	41	35	87	563	53	127	42
Rb	—	—	13	15	17	16	17	22	16	2	nd
Sr	—	—	199	141	151	442	212	171	161	133	103
Y	—	—	6	25	30	27	28	12	30	24	44
Zr	—	—	37	53	109	199	103	80	118	55	148
Nb	—	—	nd	2	6	13	8	3	6	4	9
Ba	—	—	175	56	149	809	200	274	220	48	67
Ce	—	—	22	9	24	82	—	29	34	8	25

1, 2: Average of five Sortis Group volcanics and four sills respectively (Higgins, 1970). 3: BN dyke, Ivisârtoq (Hall and Hughes, 1987). 4, 5: MD dykes, Fiskenaesset (Hall et al., 1987c). 6: PP dyke, Fiskenaesset (Hall et al., 1985). 7: Average of 18 Kangâmiut dykes (Zeck and Kalsbeek, 1981). 8, 9: Scourie norite and dolerite dykes (Weaver and Tarney, 1981). 10, 11: Loch Maree Group metavolcanics (A. Cattell, pers. comm., 1988)

occurrence of these rocks supports Pb and Rb–Sr isotopic data which suggest a possible Proterozoic suture between two Archaean continental plates, of which remnants occur as the Archaean Gneisses to the north and south (Kalsbeek and Taylor, 1985a).

11.5.2   Early Proterozoic basic dyke swarms of Greenland

Three types of fresh igneous basic dykes occur in the Archaean Craton of southern West Greenland. The c. 2.1 Ga old MD dykes predominate in the southern part of the craton (Kalsbeek and Taylor, 1985c; Hall and Hughes, 1987) and are mainly sub-ophitic dolerites. The oldest (MD1) of these dykes are the most primitive and the youngest (MD3) the most evolved (Table 11.3) (Rivalenti, 1975; Hall et al., 1985). They show a clear Fe-enrichment trend and many other chemical characteristics common to continental tholeiitic dyke swarms (Figures 11.5, 11.6). Some show considerable variation both laterally and vertically. Many contain highly complex mineral assemblages, some including a wide range of orthopyroxenes, pigeonites and subcalcic and Ca-poor augites, with X_{Mg}^{px} ($Mg/(Mg + Fe)^{pyroxene}$) values from 0.83 to 0.53, together with olivines ranging from Fo_{70} to Fo_{33} (Hall et al., 1986). These assemblages demonstrate the complex sequential crystallisation of the dykes under relatively shallow, low pressure conditions. These dolerites have only moderately high LILE contents and unfractionated to mildly LREE-enriched REE abundances (Figure 11.7) (Hall and Hughes, 1987), similar to continental tholeiitic dykes of many parts of the world. The earliest MD1 and MD2 dykes show the least evidence of a crustal contamination component. Two MD dykes in the Frederikshåb area have given Sr_i values of 0.70155 and 0.70277. Kalsbeek and Taylor (1985a, c) considered that the latter value indicates contamination at depth of their parental magma by crustal material with the geochemical characteristics of the 3.7 Ga Amîtsoq gneisses. This indirect evidence gives the only indication of the possible presence of early Archaean crust in this part of the craton.

A set of Mg-poor and Ti-rich, generally fine-grained plagioclase-phyric dykes (the PP dykes) occur in the Frederikshåb–Fiskenaesset region and were considered to be the youngest and most evolved of the MD dykes ('MD3b' dykes). However, differences in the proportions of their higher Ti, LILE and LREE, and lower Fe, Cu, Sc and V contents (Figures 11.6, 11.7) show that they had a petrogenetic evolution different to that of the MD dykes (Hall et al., 1987c). The absence of Eu anomalies (Figure 11.7) precludes simple plagioclase extraction or accumulation as a petrogenetic link between the PP and MD dyke magmas. The PP dykes appear to have been derived from a separate magma which precipitated Fe-Ti oxide far earlier than did that which fed the MD dolerites.

The basic dykes of the Sukkertoppen region in the north of the Archaean Craton of West Greenland are predominantly noritic (the BN dykes) (Hall and Hughes, 1986, 1987). Rb–Sr and Pb isotope studies suggest that these dykes are around 2.1 Ga old and derived from Mg-rich magmas with high Sr_i values (0.7036 to 0.7045)(Bridgwater et al., 1985). They have a completely different geochemical character to that of the MD dolerites (Table 11.3, Figure 11.5), most closely resembling modern boninitic compositions (Hickey and Frey, 1982). Most have high MgO, Cr and Ni contents (c. 17%, 2000 ppm and 750 ppm respectively) coupled with relatively high Si, LILE and LREE concentrations. The similar composition of a partially quenched dyke suggests that these chemical characteristics reflect those of the parental magma. This dyke consists of microphenocrysts of olivine, orthopyroxene and, most abundantly, Mg-rich pigeonite

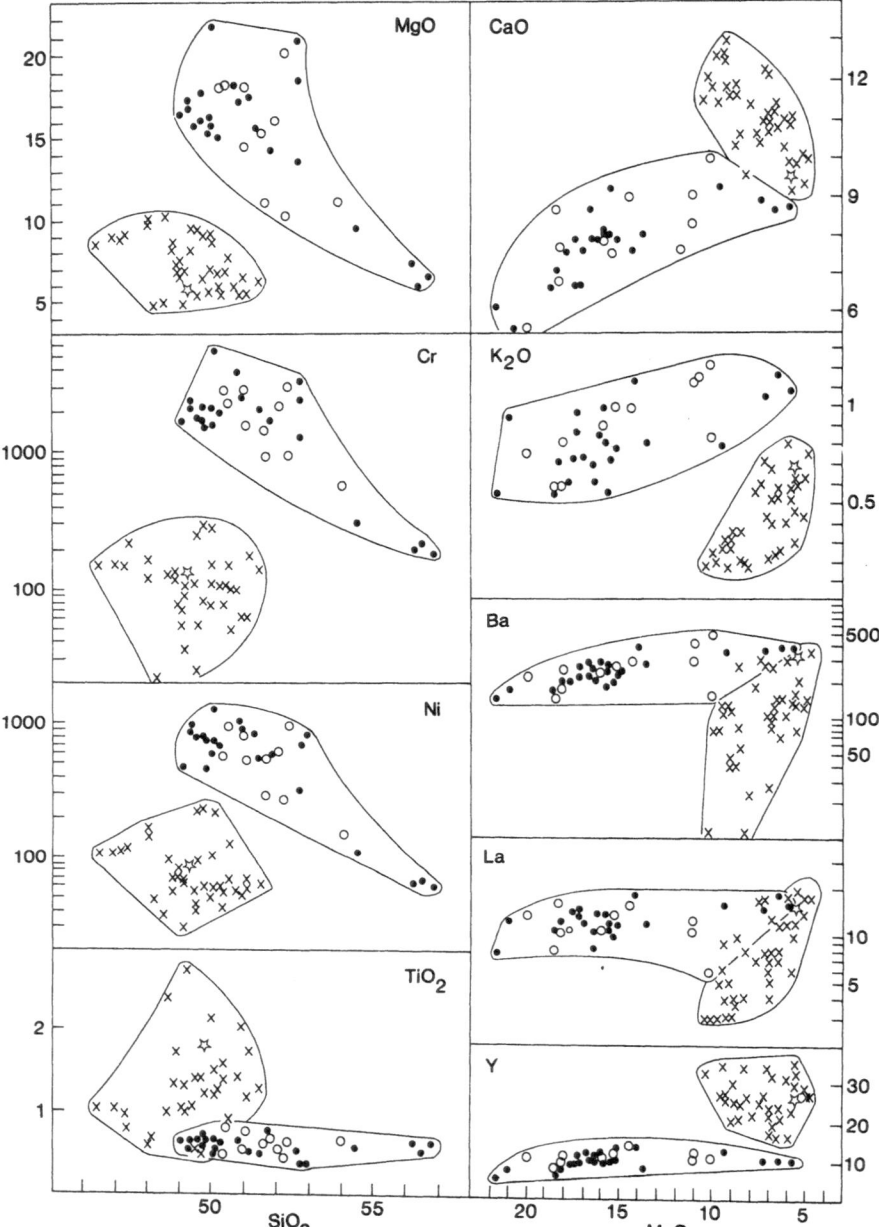

Figure 11.5 Chemical variation diagrams showing the contrasting geochemical characteristics of the early Proterozoic BN norite (dots) and MD dolerite (crosses) basic dyke swarms of West Greenland (Hall and Hughes, 1987; Hall *et al.*, 1987b). Compositions of Scourie norites/picrites (open circles) and average of twenty-three Scourie dolerites (asterisk) from northwest Scotland shown for comparison (data from Tarney, 1973; Weaver and Tarney, 1981).

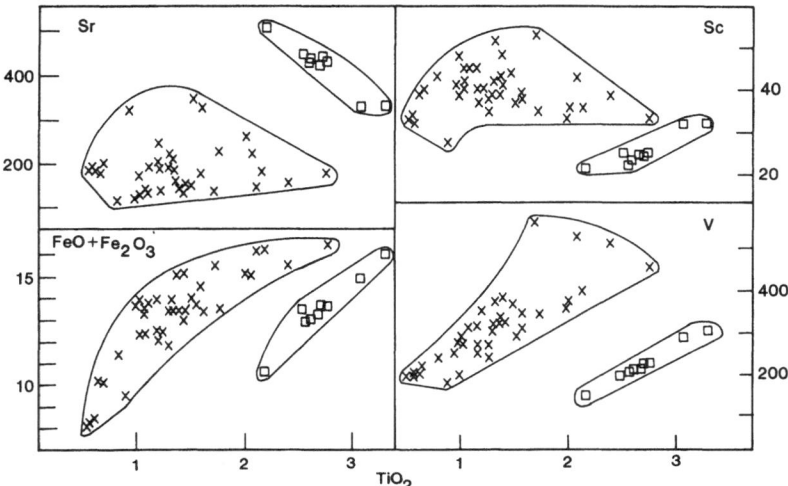

Figure 11.6 Different geochemical trends of the early Proterozoic MD dolerite (crosses) and PP (plagioclase-phyric) basic dykes (squares) of southern West Greenland, showing that the PP dykes are not simply related to the MD dykes by fractionation or plagioclase accumulation (Hall *et al.*, 1987c).

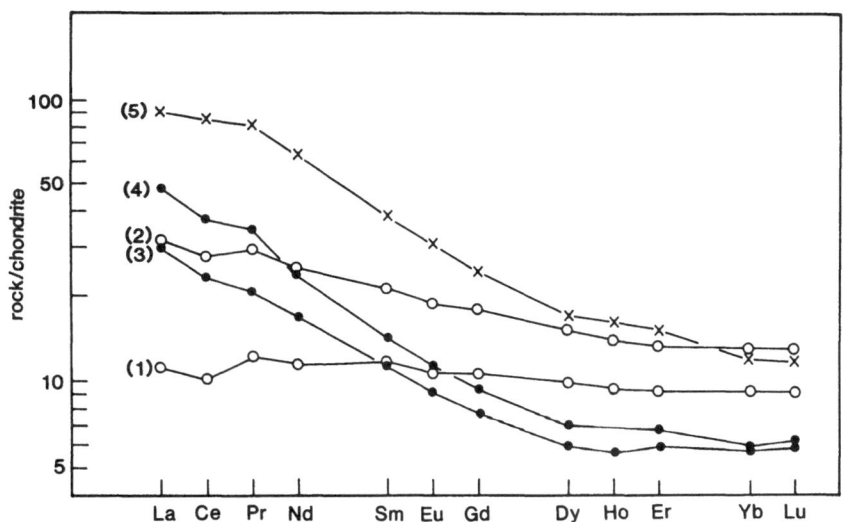

Figure 11.7 REE contents of typical MD (1, 2; open circles), BN (3, 4; dots) and PP (5; crosses) dykes showing the contrast between relatively unfractionated REE distributions in MD dolerites compared to LREE-enriched BN and PP dykes. Sample numbers (and MgO values) are: 1: MD2 149208 (8.45); 2: MD3 179092 (5.54); 3: BN 290511 (16.18); 4: BN 290506 (5.72); 5: PP 120528 (5.36). (All sample numbers should carry the Geological Survey of Greenland prefix GGU.)

mantled by augite, within a fine-grained groundmass of assorted pyroxenes and dendritic plagioclase (Hall and Hughes, 1987; see Chapter 5). The close similarity between the composition of this dyke and the more abundant medium-grained olivine-augite-norites suggests that their composition has not been significantly affected by crystal accumulation. The marked geochemical contrast between the tholeiitic MD dolerites and these noritic BN dykes (Figure 11.5) shows that the two swarms cannot have derived from the same source by differing degrees of melting, or be related by fractionation or crystal accumulation, but that they represent fundamentally different magmas.

The Kangâmiut Basic Dykes occur as a dense swarm in the southern part of the Nagssugtoqidian Mobile Belt, north of Sukkertoppen (Escher et al., 1975, 1976; Myers, 1984). These syntectonic dykes vary from fresh igneous to amphibolite and garnet–granulite facies rocks (Korstgård, 1979b; Glassley and Sørensen, 1980; Zeck and Kalsbeek, 1981). Most are coarse-grained and they have a tholeiitic geochemistry (Zeck and Kalsbeek, 1981). They define a clear Fe-enrichment trend and despite the contrast in their tectonic setting, they have broadly the same geochemistry as the younger and more evolved Fe-rich MD dolerites, but do not include magnesian compositions equivalent to the MD1 and MD2 dykes (MgO of 8.5 to 10.5%). The fresh and metamorphosed dykes are geochemically similar apart from, in particular, their K and Rb contents which are strongly affected by dyke–host rock ion exchange, and several other minor though significant elemental variations (Zeck et al., 1983). An early phase of Mg-rich norite dykes (Zeck and Kalsbeek, 1981) is probably related to the BN norite dyke swarm.

Deformed basic dykes are abundant in the vicinity of the Ammassalik Mobile Belt of East Greenland. They include BN-type noritic ones and, in the south of the region, dolerites with complex pyroxene assemblages similar to those of the MD dykes (Hall et al., 1988). Those in the northern part of the region comprise high-grade metamorphic assemblages similar to those of the Kangâmiut Dykes in the central Nagssugtoqidian belt at the west coast, and the Scourie Dykes of northwest Scotland (Tarney and Weaver, 1987b), indicating their intrusion at deeper crustal levels than the MD dykes which comprise unmodified low-pressure igneous assemblages (Hall et al., 1986, 1988). It is possible that dykes equivalent to the Kangâmiut, MD, BN and PP swarms of the west coast are all present in this region.

11.5.3 Loch Maree Group metavolcanics, northwest Scotland

Basic amphibolitic horizons up to 1 km thick occur in the region around Loch Maree in northwest Scotland (Figure 11.2). Most are interpreted as metamorphosed lavas, but some homogeneous horizons locally retain ophitic textures and may represent sills. The contacts between these Loch Maree Group (LMG) basic rocks and the surrounding Archaean Gneisses are tectonic and interpretations of their original disposition are based on structural and geochemical evidence (Johnson et al., 1987; Park et al., 1987).

The geochemistry of the LMG metavolcanic amphibolites corresponds broadly to that of oceanic tholeiitic basalts (Table 11.3). They show an Fe-enrichment trend with $Fe_2O_3^*$ contents from around 12 to 20%, matched by a three-fold increase in incompatible element abundances ($Zr = 50–150$ ppm; $TiO_2 = 0.5–2\%$). Most have flat, unfractionated REE distributions at around ten times chondritic levels, although

Johnson et al. (1987) recognised a sub-group (group B) which are LREE-enriched (fifty times chondrite) and have higher LILE, Si, Ti, and Zr contents. These group B amphibolites represent late-stage sills possibly derived from a different, contaminated mantle source to that which bore the majority of the (group A) metavolcanics and having suffered crustal contamination during ponding of the parental magma. The volcanics were erupted as the result of continued dextral transtensional rifting of a highly attenuated continental crust (Park et al., 1987) and ascended rapidly and directly from the mantle. The group B sills were derived from basic magma reservoirs held in, and contaminated by, granulite facies gneisses of the lower crust, and were intruded after the cessation of rifting (Johnson et al., 1987). Alternatively, it has been suggested that the LMG comprises an almost continuous differentiation series linked by the fractionation of olivine, clinopyroxene and plagioclase, joined by Fe–Ti oxide at the latest recorded stage of differentiation (A. Cattell, pers. comm., 1988).

11.5.4   Early Proterozoic intrusive basic rocks of Scotland

The Lewisian Complex of northwest Scotland is dissected by abundant, closely spaced, WNW-trending basic dykes, collectively called the Scourie Dykes. An account of their geochemistry and petrogenesis has been presented by Weaver and Tarney (1981a) and Tarney and Weaver (1987b, c). The swarm exhibits a wider range of dyke compositions than individual Proterozoic dyke swarms elsewhere and four main different types of dykes have been recognised. Most are Fe-rich metadolerites, but olivine-gabbros, bronzite-picrites and norites also occur.

The Mg-rich varieties show significant transverse compositional variation from their centres to margins, interpreted to result from turbulent flow and crystallisation during intrusion, and subsequent crystal settling closely following intrusion in inclined dykes. These dykes were emplaced with most of their early-crystallising olivines in suspension and concentrated toward the centres of the up-flowing dyke channels, permitting the second liquidus phase (orthopyroxene in the case of the picrites and clinopyroxene in the olivine-gabbros) to crystallise on the dyke walls, forming medium- to coarse-grained pyroxenitic margins. The tholeiitic dolerite dykes have chilled margins with few plagioclase microphenocrysts and show little transverse compositional variation. None of the dykes shows any field evidence of thermal interaction with the country rocks.

Despite their wide range of compositions, the Scourie Dykes are mostly hypersthene-, and often quartz-normative. They have broad chemical similarities with modern continental flood basalts (Weaver and Tarney, 1983). Their mantle-normalised trace element patterns show moderate to strong enrichment in the more incompatible elements including the LREE (Figure 11.8). A strongly negative Nb anomaly is characteristic of all of the dykes and the Fe-rich tholeiites display negative Sr and (in some cases) Eu anomalies, although evidence for crystal differentiation in these dykes is lacking.

Although these trace element characteristics would be generally consistent with appreciable interaction of the basic magmas with continental crust, the composition of the granulitic country rocks in northwest Scotland (Weaver and Tarney, 1981b) is so distinctive that this can be ruled out. Detailed consideration of their geochemistry suggests the derivation of the picrites from a more refractory source at different depths, whereas the olivine-gabbro and dolerite dykes were derived from melting at various

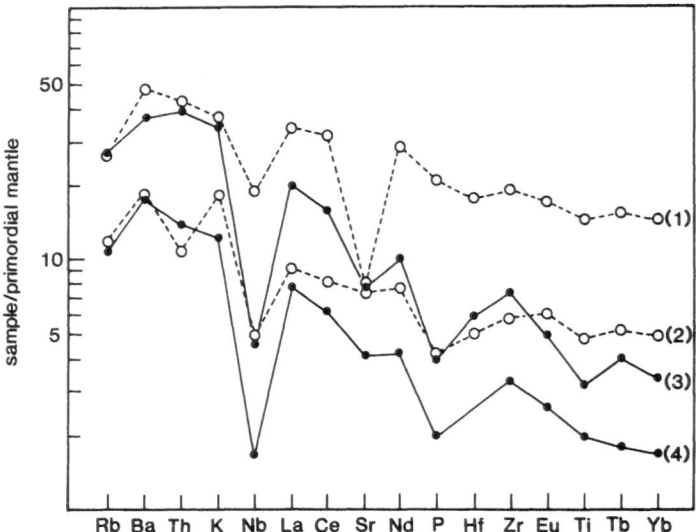

Figure 11.8 Mantle-normalised multi-element diagram showing relative incompatible element abundances in representative dolerites (1,2) and a norite (3) and average picrite (4) Scourie dykes, northwest Scotland (redrawn from Tarney and Weaver, 1987c). The noritic and picritic dykes have notably high LILE and LREE concentrations in view of their high Mg contents, and generally steeper patterns with more pronounced negative Nb anomalies than the dolerites. Sample numbers and wt% MgO values: L636, 5.1%; 2: L613, 6.1%; 3: N1, 15.3%, 4: average 5 picrites, 32%.

depths of a more normal lherzolitic mantle (Weaver and Tarney, 1981a). Both sources are inferred to have inherited their 'continental' trace element characteristics at the time of generation of the Lewisian gneisses.

The noritic Scourie Dykes and the BN norite dykes of West Greenland have the same geochemical characteristics, including their high Mg, Ni, Cr, LILE, LREE and alkali contents (Figure 11.8) (Weaver and Tarney, 1981a; Tarney and Weaver, 1987b; Hall et al., 1987b). The juxtaposition of the various types of dykes in northwest Scotland means that the Scourie Dykes probably constitute several different penecontemporaneous swarms intruded either syntectonically or partly brought into parallelism (NW–SE) during subsequent tectonism, as are the Kangâmiut Dykes in West Greenland and the East Greenland dyke swarms (Dawes et al., 1989; Escher et al., 1989). The Scourie Dykes were injected into a dextral strike-slip regime mainly at around 2.4 Ga, with some emplaced possibly as late as 2.0 Ga (Evans and Tarney, 1964; Chapman, 1979; Heaman and Tarney, 1989).

There are clear geochemical distinctions between the Scourie Dykes and the Loch Maree Group (LMG) metavolcanic rocks (Johnson et al., 1987). This would be consistent with the Scourie Dykes being earlier than the extensional regime which produced the LMG lava pile at around 2.0 Ga. On the other hand, the LMG may have originated through the melting of the rising asthenosphere during extension, whereas the Scourie Dykes were derived from the sub-continental lithosphere (Weaver and Tarney, 1981a).

The South Harris igneous complex (Figure 11.2) is a NW-trending belt 12 km long and 3 km wide bounded by major dextral shear zones (Dearnley, 1963; Graham, 1980). It comprises an eastern unit of deformed gabbroic and anorthositic rocks, and adjacent tonalites to the west. The anorthositic rocks have been dated at about 2.2 Ga, slightly younger than the emplacement age of the Scourie Dykes and the tonalites are slightly younger still (< 2.06 Ga)(Cliff *et al.*, 1983). Both have been subject to garnet-granulite facies metamorphism (Wood, 1975). The few available geochemical data indicate that the gabbroic and anorthositic rocks appear to have developed from an alkali-poor, aluminous tholeiitic basic magma, as evidenced by their Fe-enrichment trend. The tonalites do not show Fe-enrichment but are calc-alkalic, and are probably not co-genetic with the basic, metagabbro–anorthosite complex, which is interpreted to have originated from a Fiskenaesset-type layered intrusion, severely disrupted by shearing into tectonic pods (Graham, 1980). Nearby ultrabasic pods have been interpreted as the remnants of alpine-type peridotites (Livingstone, 1976).

11.6   Basic magmatic evolution and crustal growth in the North Atlantic craton

There is little evidence for the existence of sialic crust before 3.8 Ga and it is generally accepted that the Earth's earliest crust must have been dominated by basaltic material, very little of which is preserved. The progressive formation of Precambrian sialic continental crust occurred in cycles comprising four stages: (i) the generation of basaltic oceanic crust, (ii) the subduction and melting of this basaltic crust to form the dominant tonalitic–trondhjemitic–granodioritic (TTG) gneiss component of early continental nuclei, (iii) tectonism, variable reworking of the TTG suite and granitic plutonism during the final stages of cratonisation, and (iv) the intrusion of basic dyke swarms associated with the rifting of the new cratons, or the initiation of a new cycle. The North Atlantic Craton displays a wide range of early Archaean to Proterozoic mafic lithologies which provide information on the nature and variety of early basic magmas and mantle components during the development of the Earth's early crust.

Archaean crustal development in Greenland clearly represents two cycles. The Isua/Akilia supracrustals, Amîtsoq TTG–granite gneiss suite and the Ameralik dykes represent the first, early Archaean crustal growth stage, and the Malene and Fiskenaesset-type basic rocks, Nûk-type TTG and slightly younger (2.6–2.5 Ga) granitic rocks, and various early Proterozoic basic dykes comprise the second cycle. Their distribution has been severely modified by late Archaean tectonism (Friend *et al.*, 1988). The development of the Lewisian of northwest Scotland can be broadly correlated with the second tectono-magmatic cycle in Greenland (Park and Tarney, 1987b). Because of the variety of basic complexes, two fundamental aspects of early Precambrian basic magmatism can be addressed, namely the compositions of pene-contemporaneous magmas formed in different tectonic settings, and those formed in similar tectonic settings but at different times during the Precambrian.

Figure 11.1 illustrates the enormous volume of basic material that was incorporated into the West Greenland Craton. Field relationships demonstrate that the disrupted Malene-type metavolcanic and Fiskenaesset-type layered gabbro–anorthosite complex rocks are the oldest recognisable crustal component through much of the craton and could, therefore, indicate the nature of early Precambrian oceanic crust. Simple physical considerations argue against the dispersed metagabbroic and anorthositic outcrops belonging to a single major disrupted intrusion; they more likely constitute a series of lithologically similar complexes. Similar considerations apply to the abundant basic metavolcanic horizons.

Geochemical results have confirmed the primitive character of many of the pillow-structured metavolcanic units, but have also demonstrated the considerable diversity of the basic magmatism (e.g. Hall *et al.*, 1987a). The extent to which these metavolcanics can be correlated with basaltic rocks of modern tectonic environments on the basis of their geochemistry remains uncertain. Distinguishing between island arc and MORB-type basaltic compositions, for example, in Archaean suites is rather difficult, particularly in high-grade terranes. One important finding is that there are no significant andesitic rock types within the supracrustal suites that might correspond with the calc-alkaline andesites which are predominant in modern continental margin arc volcanic suites. Instead, the metavolcanics are dominated by komatiitic or tholeiitic basaltic compositions. In this regard a comparison with Archaean greenstone belts is interesting. Early Archaean greenstone belts have an abundance of tholeiitic and komatiitic lavas typically in the lower parts of their successions, but the appearance of andesitic and dacitic lavas is not uncommon in the upper parts. Late Archaean greenstone belts typically have a higher proportion of these more silicic 'calc-alkaline' lavas. Strictly, these are not the same as modern calc-alkaline andesites, in that the late Archaean ones normally show much greater HREE depletion. However, the lack of lavas with intermediate and silicic compositions within the Archaean metavolcanic suites of the North Atlantic Craton implies that their provenance differs from that of greenstone belts. It is also highly significant that most Archaean greenstone belts appear to have formed on a sialic crust, whereas the West Greenlandic high-grade basic units are the oldest recognisable rocks. The simplest and most logical model is that they represent tectonically emplaced slivers of oceanic crust incorporated into the gneiss complex at the time of the late Archaean TTG suite generation. In other cratons, such as in Australia and the Baltic Shield (see Chapters 12 and 14), bimodal basic–acid volcanism is observed which is more typical of extensional environments in modern arc terranes.

The tholeiitic magmas which gave rise to the earliest Archaean volcanics at Isua were formed again some 500 million years later to feed the Ameralik Dykes (Gill *et al.*, 1981). Some Malene metavolcanics in the Godthåb region (Figure 11.1) have broadly the same geochemistry (Chadwick, 1981), suggesting that the Ameralik Dykes may have acted as feeders to the Malene lavas in this area, which could, therefore, have been erupted through and onto the margin of an Amîtsoq continent. This raises the question of whether the same magma type can be erupted repeatedly in time. In detail the correspondence is not so close. The larger supracrustal belts have a significant komatiitic component which is not found in the Ameralik Dykes. Furthermore, the petrogenesis of the Ameralik Dykes was controlled by plagioclase-dominated fraction-ation, whereas olivine and clinopyroxene were the principal phases controlling the compositional evolution of the komatiitic volcanic suites. The trace element data also indicate that these extensive metavolcanic belts cannot have come through the Amîtsoq continent as we understand it. The basic magmas which fed the large Fiskenaesset-type gabbro–anorthosite complexes also seem to have been significantly different to the Malene komatiitic magmas. Both petrogenetic modelling and abundant field evidence demonstrate the overwhelming importance of plagioclase fractionation in the layered complex rocks (Myers, 1985). The parental magma was clearly aluminous (Weaver *et al.*, 1981) and its source garnetiferous (Henderson *et al.*, 1976). In contrast, the Malene komatiitic lavas have high Ca/Al ratios, and may have been derived from a relatively clinopyroxene-rich source (Hall et al., 1987a). These data, together with the wide range of observed trace element patterns, indicate not only different depths of magma generation and varying degrees of melting for the early Precambrian basic magmatism, but that there may have been significant variation in mantle source compositions too.

The abundant ultramafic lenses preserved within the gneiss complexes of Greenland and Scotland possibly provide a direct means of assessing mantle compositions. Some are as much as 2 km long. They are commonly associated with the supracrustal belts, but elsewhere occur as discrete pods enclosed within the gneisses. They most probably represent dismembered tectonic slices of the uppermost mantle, incorporated with slivers of oceanic crust and the crust-generating phase during the widespread early, sub-horizontal, thrust-style tectonics (Bridgwater et al., 1974; Hall and Friend, 1979). Although a few retain primary textures (Friend and Hughes, 1977, 1978), most are completely recrystallised and are frequently altered (Sharpe, 1980). They range from dunites to lherzolites and clinopyroxenites, but it is significant that in both Greenland and Scotland a large number of these lenses are harzburgitic (Friend and Hughes, 1978; Crewe, 1984)(Table 11.2), as are similar lenses in Labrador (Collerson et al., 1976b). Whereas it has been suggested that they represent flows or sills associated with the metavolcanics (Collerson et al., 1976b; Chadwick, 1981, 1986), their orthopyroxene-rich character clearly contrasts with the composition of the komatiitic metavolcanics, and indeed komatiites elsewhere (Arndt and Nisbet, 1982). It is more likely that they are tectonically emplaced relics of harzburgitic upper mantle. The most obvious explanation is that they represent residual mantle from which basaltic and komatiitic magmas have already been extracted, and which forms the buoyant sub-continental lithosphere (Jordan, 1988). However, not all of these ultramafic fragments have the depleted trace element characteristics expected of the residues resulting from basalt extraction (cf. Weaver and Tarney, 1980; Sills et al., 1982). An additional factor may be that the sub-continental mantle is contaminated by the silicic magmas resulting from the crust-generating phase, which would have the effect of transforming a lherzolite into a harzburgite (cf. Weaver and Tarney, 1981a).

Further evidence for the widespread presence of harzburgitic mantle is provided indirectly by the abundant noritic (orthopyroxene-rich) magmas which appear in the early Proterozoic (the BN dykes of West Greenland and noritic and bronzite-picrite Scourie dykes in northwest Scotland). These rocks are distinctive in being rich in Mg, Ni and Cr, but also relatively rich in Si, LILE and LREE (Weaver and Tarney, 1981a; Hall and Hughes, 1987). There are problems in modelling these compositional characteristics by magma contamination through crustal assimilation because the available country rocks do not have the appropriate composition (Weaver and Tarney, 1981a). Hence, alternative models invoking the derivation of the noritic dyke suite from harzburgitic upper mantle which had been earlier contaminated with crustal components (as outlined above) is preferred (Weaver and Tarney, 1983; Tarney and Weaver, 1987b; Hall and Hughes, 1987). Such mantle may include a hydrous phase (phlogopite or hornblende), the breakdown of which would have aided magma formation during subsequent thermal events. It is perhaps significant that the BN noritic dyke swarm of West Greenland is most dense in the Sukkertoppen region, just south of the Nagssugtoqidian mobile belt, potentially an old continental margin.

The dominant dolerite dykes in both Scotland and Greenland were derived from a more normal lherzolitic mantle (Weaver and Tarney, 1981a; Hall and Hughes, 1987), though most share (though less markedly) some of the 'continental' trace element characteristics of the noritic suite – characteristics that are typical of many Phanerozoic continental flood basalt provinces (Weaver and Tarney, 1983). The penecontemporaneous emplacement of noritic and doleritic swarms during the early Proterozoic indicates the availability of both harzburgitic and lherzolitic mantle components in the sub-continental lithosphere. The stretching of the lithosphere to permit dyke emplacement would have allowed the rising hot asthenospheric mantle (the MORB reservoir) to interact with the sub-continental lithosphere, generating noritic and quartz-tholeiitic magmas from that lithosphere. Equally diverse magma types have been produced more recently, where rising mantle plumes ('hotspots') interact with sub-continental lithosphere, as on Kerguelen in the Indian Ocean (Storey et al., 1988).

There is a clear contrast in the North Atlantic Craton between Archaean and Proterozoic high-Mg magmas. The Archaean high-Mg magmas are mostly komatiitic whereas the early Proterozoic types are dominantly noritic. The former type probably reflects moderately high degrees of melting of rising mantle diapirs at a time when, or in a situation where, the lithosphere was thin and imperfectly developed. Conversely, by the Proterozoic the sub-continental lithosphere had matured and thermal activity induced the melting of the hydrous (?) harzburgitic mantle component. However, in terms of the tholeiitic magmas, it is the consistency of compositions generated throughout an extended time period which is remarkable. For instance, the mid-Archaean Ameralik Dykes which intrude the early Archaean Gneisses of the Godthåb region are similar to some of the Malene basaltic metavolcanic rocks of virtually the same age. Moreover, they are not significantly different to low-K tholeiitic volcanic rocks from many other parts of the world (Gill and Bridgwater, 1979). They are also geochemically similar to the early Proterozoic dolerites which cut the Archaean craton (the MD dykes and the Scourie quartz-dolerites), to Proterozoic doleritic dyke swarms worldwide, and to Phanerozoic continental flood basalts. Nor is there any difference between the MD dykes intruded into a stable cratonic region of Greenland and the syntectonic Kangâmiut Dykes emplaced in the bordering active mobile belt. This common tholeiitic basic magma type does not seem to change significantly with time.

Acknowledgement

The Greenland data are published with the kind permission of the Director of the Geological Survey of Greenland.

12 Early Precambrian basic rocks of the Baltic Shield

T.S. BREWER and T.C. PHARAOH

12.1 Introduction

The Baltic Shield comprises lithologies ranging in age from 3.5 to 0.9 Ga in Finland, Norway, Sweden and the USSR (Figure 12.1). The shield is zoned chronologically from NE (Archaean) to SW (mid to late Proterozoic). It was formed by crustal accretion onto the Saamian (3.1–2.9 Ga) nucleus during the Lopian (2.9–2.6 Ga), Svecofennian (1.9–1.8 Ga) and Gothian (1.7–1.5 Ga) orogenies. The Sveconorwegian orogeny was mainly a period of late Proterozoic (1.25–0.9 Ga) crustal reworking, and is beyond the scope of this chapter. Reviews of the geological evolution of the Baltic Shield are presented by Gaál and Gorbatschev (1987) and Starmer (1987). The nomenclature for geological domains, provinces and tectonic phases used in the following account follows the scheme of Gaál and Gorbatschev (1987). Basic intrusive and extrusive rocks occur throughout the shield and are important markers in its evolution (Table 12.1). Few of these magmatic suites retain much of their primary mineralogy, the layered basic–ultrabasic complexes of northern Finland being a notable exception, because of recrystallisation during greenschist or amphibolite facies metamorphism. As a result, the isotopic and whole-rock geochemical data provide more petrogenetic information than do the mineralogical data. In this text, the early Proterozoic tholeiitic meta-volcanics which occur throughout much of the northern Baltic Shield (Kiruna greenstones and equivalents) are correlated with the Jatulian stratigraphy of Finland (Table 12.1). However, an alternative scheme (Geol. Survs. Finland, Norway and Sweden, 1987, 1988) places these volcanics at the top of the Lapponian Supergroup and reserves the term Jatulian for a thin interval at the top of the early Proterozoic sequence.

12.2 The Archaean Domain

The Archaean Domain can be subdivided into three provinces (Figure 12.1b). The Archaean basement of Soviet Karelia and adjacent East Finland (the Karelian Province) is a typical granite–greenstone terrane, while the Belomorian and Kola Peninsula Provinces are high-grade gneiss terranes. The Kola Province is separated from the others by the early Proterozoic Lapland Granulite Complex. The oldest lithologies are 3.1 Ga tonalitic, trondhjemitic and granodioritic gneisses (TTG), which represent basement to Archaean greenstones in the Karelian Province (Kröner et al., 1981; Jahn et al., 1984; Gaál and Gorbatschev, 1987). Many of these TTG plutonic rocks had crustal residence ages of between 200 and 500 Ma and were formed in two stages during the 3.1 to 2.9 Ma Saamian orogeny (Martin, 1987). Partial melting of the

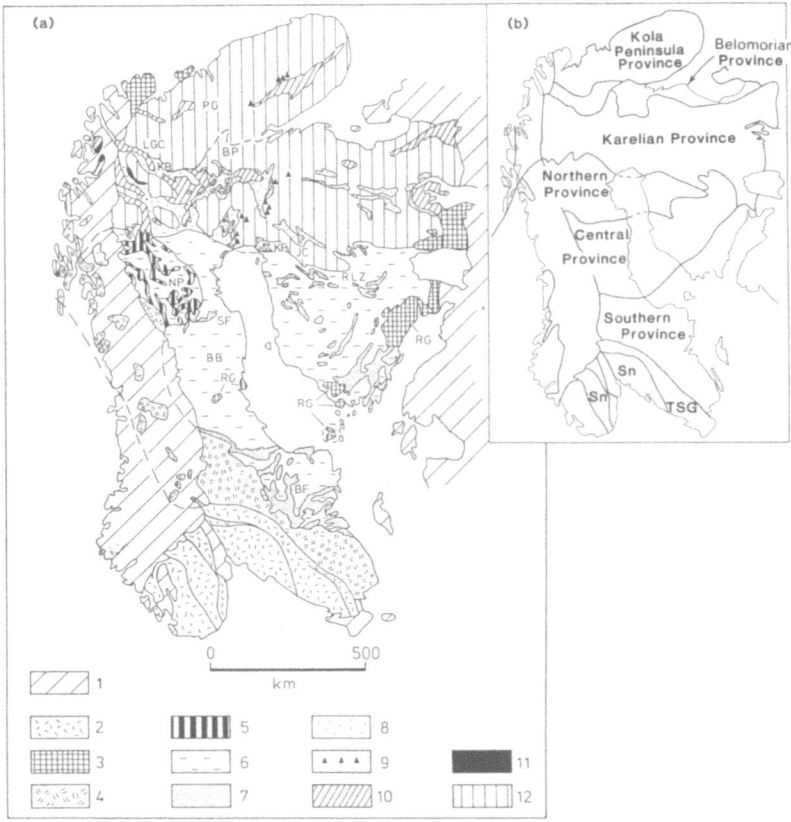

Figure 12.1   (a) Geological map of the Baltic Shield, modified from Gaál and Gorbatschev (1987) and Pharaoh *et al.* (1987). 1: Caledonides (west) and platform cover (east); 2: Sveconorwegian orogen; 3: post-orogenic intrusions and sediments; 4: trans-Scandinavian granite–porphyry belt, Svecofennian Domain; 5: porphyry volcanics; 6: metasediments and plutonic complexes; 7: Svecofennian and Kalevian metavolcanic suites, early Proterozoic cratonic cover; 8: Sariolian–Jatulian sediments; 9: layered gabbro–ultrabasic complexes; 10: Jatulian volcanics; 11: Lapponian supracrustals, Archaean Domain; 12: granitic gneiss and greenstone belts. BB: Bergslagen Field; JC: Jormua complex; KB: Karasjok Belt; NP: Norrbotten porphyries; O: Ostrobothnian Schist Belt; PG: Pechenga Group; RG: Rapakivi granite suite; RLZ: Raahe–Ladoga Zone.

(b) Major tectonic domains of the Baltic Shield. The Svecofennian domain is divided into northern, central and southern provinces. LGB: Lapland Granulite Belt; Sn: Sveconorwegian Domain; TSG: trans-Scandinavian granite porphyry belt.

mantle gave rise to a tholeiitic oceanic crust. This underwent subduction and was transformed mainly into garnet amphibolites, the partial melting of which produced the TTG suite. One point of speculation concerns the extent and coherence of the Saamian crust and whether it was welded together during the 2.9–2.6 Ga Lopian orogeny (Gorbunov *et al.*, 1985; Gaál and Gorbatschev, 1987).

The oldest basic lithologies in the Baltic Shield are the greenstone belts of the

Table 12.1 Correlation of early Proterozoic supracrustal suites of the Baltic Shield

Ga	Division	A. Eastern Finnmark Kola Peninsula	B. Western Finnmark	C. Northern Norrbotten Northern Finland	D. North central Finland	E. Soviet Karelia	F. Southern Norrbotten Northern Västerbotten	G. Southern Finland Central and Southern Sweden
1.5	Post-Svecofennian							W — E Rapakivi Suite / Trans-Scandinavian igneous belt
1.84	Svecokarelian orogeny	Svecokarelian orogeny	Raudfjell Suite and Synorogenic Intrusive Suites	Younger Plutonic series; Lina granite; Hauki, Mattavaara quartzite etc.; Kiruna Porphyry Group; Haparanda Suite etc.	Post-orogenic granites		Sorsele granitoids; Dobblon volcanics; Revsund granitoids; Vargfors Group; Arvidsjaur/Jörn Suite porphyry GP; Skellefte Group ~ Argillites	Laterogenic intrusions; Earlyorogenic intrusions; Leptite volcanics ~ Slate FMS
2.0	Kalevian	Vespan Group South Pechenga Group and Tominga Group			Outokumpu alloch Jormua ophiolite Kiiminki belt	Vespian Group		
	Jatulian	? Kdanov Suite	Upper Raipas and Caravarri Groups	Pahakurkio and Kilavaara Groups	Synorogenic granitoids	Suisaarian Ljudian — Upper / Middle / Lower Jatulian		
		Petsamo Group and Pechenga Group. Karasjok Belt?	Nussir Group, Kvenvik greenstone, Turelv FM Suolovuobmi FM and Cas'kejas Group	Kiruna and Vittangi Greenstone Groups Kemi greenstones				Vastervik quartzite Zircon source?
2.2	Sariolian	Kuetsjärvi Suite Ahmalahti Suite	Saltvatn and Masi Groups	Tjärro quartzite	Sariolian present locally	Sariolian; Vermas; Sumian Paanajärvi Ozhijärvi Horizon Tunguda Suite Okun Suite		
0.40	Upper Lapponian	? U. Strelna Group?	U. Holmvatn Group	U. Lapponian				
0.45	Intrusive complexes	Monckegorsk Federova etc.	? ? ?	Koitelainen Kemi etc.				
	L. Lapponian	Bjornevatn Group, Strelna Group and Tundra series	L. Holmvatn, Kviby GPS and Gal'denvarri FM	L. Lapponian				
0.50	Archaean	Kola Complex	Granite–Gneiss Complex	Granite–Gneiss Complex	Granite–Gneiss Complex	Granite–Gneiss Complex	Archaean basement not recognised	Archaean basement not recognised

Left margin stratigraphic brackets: Svecofennian; Karelian; Post-Svecofennian; Kalevian; Jatulian; Sariolian; Upper Lapponian; Intrusive complexes; L. Lapponian; Archaean.

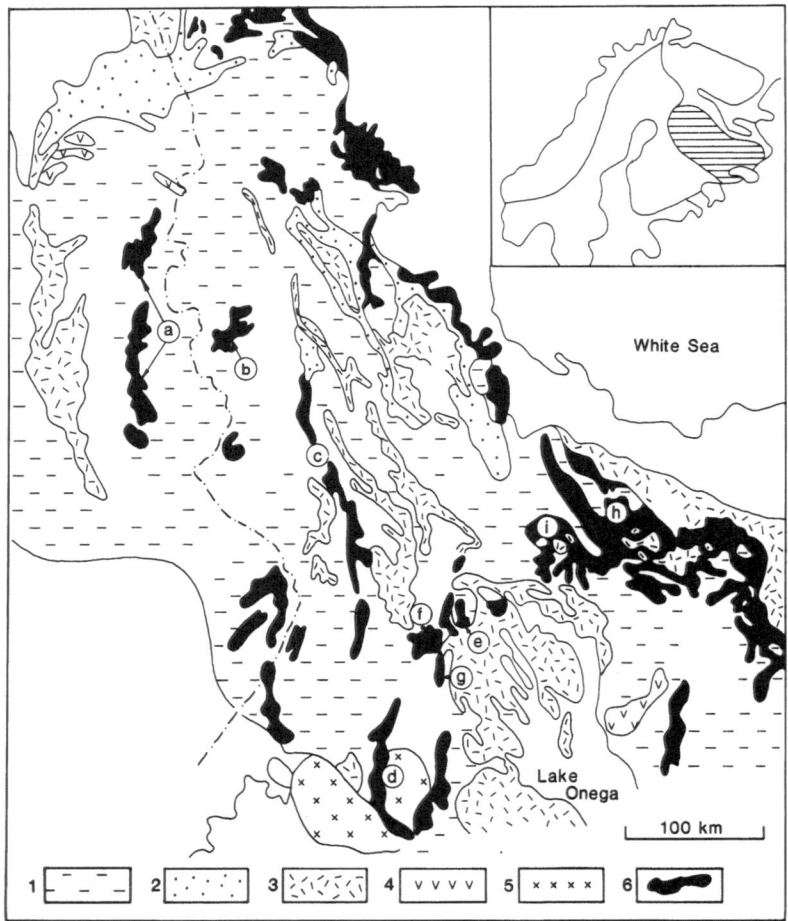

Figure 12.2 Simplified geology of the Karelian Province, modified from Lobach-Zhuchenko *et al.* (1986). Key to shading symbols: 1: Archaean granites and gneisses; 2: Sumian and Sariolian Groups; 3: Jatulian Group; 4: layered gabbro complexes; 5: post-orogenic granites; 6: Lopian greenstone belts: (a) Suomussalmi–Kuhmo–Tipasjärvi, (b) Kostomuksa, (c) Bolschozero, (d) Hautavaara, (e–f) Palaja Lamba, Semch and Oster, (g) Koikary, (h) Kamennozero, (i) Schilos.

Karelian Province. The N to NW-trending belts (Figure 12.2) have undergone low to medium grade metamorphism and several episodes of deformation (Gaál *et al.*, 1978; Lobach-Zhuchenko *et al.*, 1986; Barbey and Martin, 1987). The stratigraphic sequences of the greenstone belts vary considerably. Radiometric age determinations indicate that they formed between 3.0 and 2.65 Ga ago and were deposited upon the 3.1 Ga crust (Vidal *et al.*, 1980; Kröner *et al.*, 1981; Hupponen, 1983; Martin, 1983; Martin and Quérré, 1984; Bibikova, 1984). The greenstone belts of western and eastern Karelia (Figure 12.2) were formed at 3.0–2.9 Ga and are characterised by bimodal volcanism (Lobach-Zhuchenko *et al.*, 1986). In contrast, the greenstone belts of eastern Finland and central Karelia were formed between 2.8 and 2.65 Ga and comprise a highly varied volcanic suite. Lobach-Zhuchenko *et al.* (1986) and Gaál and Gorbatschev (1987) state that the lower volcanic units in the Suomussalmi–Kuhmo–Tipasjärvi Belt of eastern Finland are calc-alkaline, whereas Blais *et al.* (1978) indicate that they contain komatiitic and tholeiitic series. Despite this controversy, it is clear that the Archaean greenstone belts are lithologically diverse and that there is a chronological variation from east (oldest) to west.

The Suomussalmi–Kuhmo–Tipasjärvi greenstones form a composite N-trending belt approximately 20 km wide and 200 km long (Figure 12.2), the most complete (3–5 km thick) lithostratigraphic sequence being in the northern part of the Suomussalmi belt (Blais *et al.*, 1978; Jahn *et al.*, 1980; Vidal *et al.*, 1980; Barbey and Martin, 1987). The belt preserves upper greenschist to garnet-amphibolite facies assemblages and can be divided into three units, namely an early volcanic cycle, a middle sedimentary sequence and an upper volcanic cycle, affected by three phases of deformation and two episodes (2.5 and 2.4 Ga) of granitic intrusion (Blais *et al.*, 1978; Jahn *et al.*, 1980; Martin *et al.*, 1983; Martin and Quérré, 1984).

The lower volcanic sequence contains a komatiitic–tholeiitic series, whereas the upper volcanic series contains calc-alkaline andesitic to rhyolitic lithologies. The relationship between the komatiitic and tholeiitic series is unclear. In some instances, the hydration of ultrabasic lithologies now containing $> 20\%$ volatiles (Blais *et al.*, 1978) has had a significant effect on their geochemical composition. The two distinct evolutionary trends of the lower and upper volcanic sequences are reflected in an AFM plot (Figure 12.3a), the lower komatiite–tholeiite sequence defining an Fe-enrichment trend and the upper sequence a calc-alkaline trend towards the alkalis apex (Blais *et al.*, 1978). All of the basaltic rocks are compositionally similar to modern oceanic low-K tholeiitic basaltic types. The lower volcanic sequence can also be subdivided into a komatiitic and tholeiitic series on the basis of TiO_2 content (Figure 12.3b). The komatiitic series includes rocks with compositions ranging from peridotitic to basaltic, with high Ca/Al ratios and low alkali and Ti contents.

The crystallisation trends within this series are complex. In general, lavas with 45–18% MgO were controlled mainly by olivine fractionation, while for compositions with between 18 and 10% MgO, both olivine and clinopyroxene were major fractionating phases. Below 11% MgO the general trend splits into three, controlled by variations in clinopyroxene + plagioclase fractionation (Blais *et al.*, 1978). The tholeiitic series is predominantly composed of basaltic rocks with rare clinopyroxene and olivine cumulates. This series is relatively depleted in Ca, Mg and K, and enriched in Ti and Fe compared to the komatiitic series (Figure 12.3). With the exception of the cumulate rocks, the various tholeiitic series rocks define linear differentiation trends controlled by the fractionation of clinopyroxene, plagioclase, olivine and Fe-opaques. The transition elements confirm the importance of fractional crystallisation in the petrogenesis of the lower volcanic series and that these rocks were not derived from a single parental magma (Blais *et al.*, 1978; Jahn *et al.*, 1980).

REE data have been particularly important in elucidating the petrogenesis of the various groups (Figure 12.4). In the Suomussalmi belt, the komatiitic series rocks have essentially flat REE profiles between four and twenty-four times chondritic values. Within this series, there are two types of profiles, those which are slightly LREE-enriched ($La_N/Sm_N = 1.0–1.5$) and those which are slightly LREE-depleted ($La_N/Sm_N = 0.8–0.9$). In the Kuhmo belt, all of the komatiitic series have flat (although somewhat altered) REE profiles. Three types of REE profiles are found in the tholeiitic

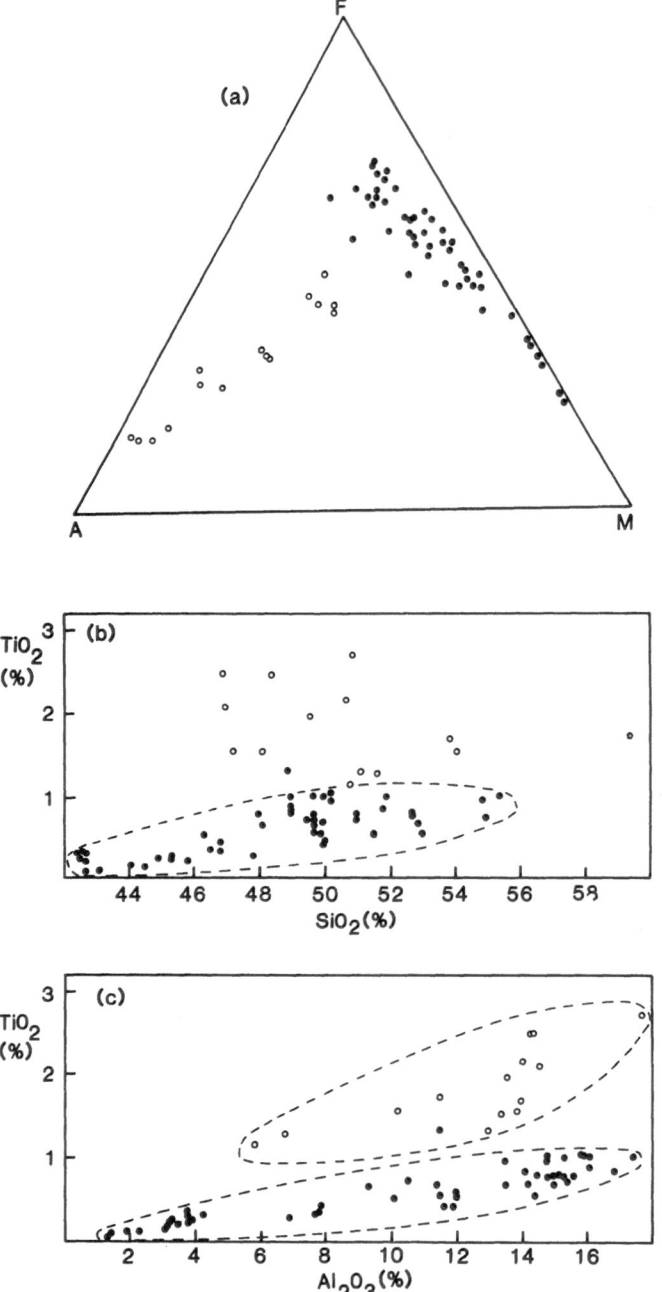

Figure 12.3 Geochemical characteristics of the Suomussalmi–Kuhmo–Tipasjärvi greenstone belts (after Jahn *et al.*, 1980). (a) AFM plot of Upper calc-alkaline sequence (open circles) and Lower komatiitic–tholeiitic series (dots). (b) and (c) the Lower volcanic series subdivided into komatiitic (dots) and tholeiitic (open circles) members.

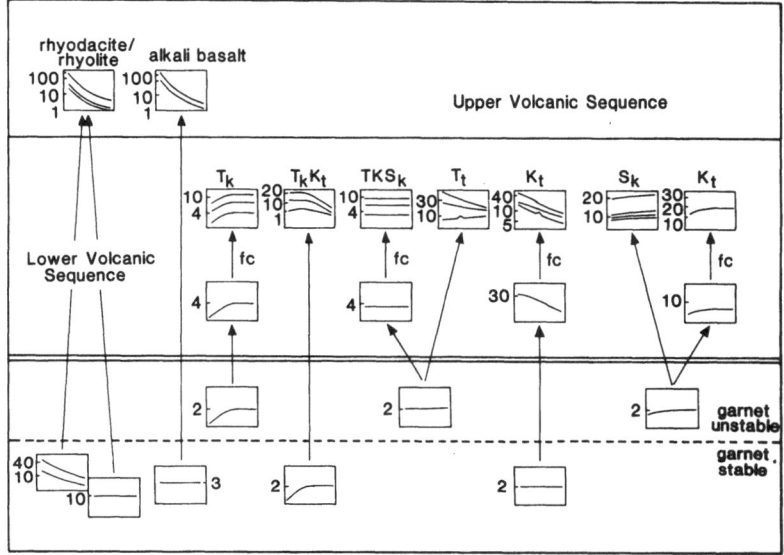

Figure 12.4   Schematic magmatic evolution of the Suomussalmi (S)–Kuhmo (K)–Tipasjärvi (T) greenstone belts in terms of their REE geochemistry. t: tholeiites; k: komatiites; fc: fractional crystallisation. Numbers next to REE profiles refer to enrichment factor relative to chondritic values.

series, (i) those with flat HREE and LREE-depleted, (ii) those enriched in LREE ($La_N/Sm_N = 1.5-1.9$) and fractionated HREE ($Gd_N/Yb_N = 1.5-2.2$), and (iii) those with upward convex LREE patterns and fractionated HREE ($Gd_N/Yb_N = 2.2$). Types (ii) and (iii) constitute the majority of the series and have parallel HREE profiles.

In the Tipasjärvi belt the komatiitic series contains three groups considered to represent distinct source regions and processes of magma genesis: (i) strongly LREE-depleted ($La_N/Sm_N = 0.15-0.3$) with uniform HREE abundances approximately three to twelve times chondritic values, (ii) slight to moderate LREE-depletion ($La_N/Sm_N = 0.5-0.9$) with fractionated HREE ($Gd_N/Yb_N = 1.2-2.0$) and (iii) overall uniform chondrite-normalised abundances, with slight LREE-depletion ($La_N/Sm_N = 0.75-0.9$). In general, the komatiitic series have $La_N/Sm_N < 1.0$ and consistent, near flat HREE, suggesting that garnet was not an important phase in their genesis. In contrast, the tholeiites have more variable but generally higher La_N/Sm_N and Gd_N/Yb_N values, possibly reflecting the petrogenetic importance of garnet. The upper volcanic series rocks are characterised by strongly fractionated REE profiles, some of which are extremely HREE-depleted (Figure 12.4) and could be interpreted as alkali basalts (Jahn *et al.*, 1980).

From the geochemical data, Jahn *et al.* (1980) suggested that three different mantle source regions were involved in the petrogenesis of the various volcanics and that the degrees of partial melting and fractional crystallisation were also variable. Environments of formation proposed for the Archaean greenstone belts have included (a) proto-oceanic rifts (Blais *et al.*, 1978), (b) fault-bounded rifts (Gaál *et al.*, 1978), (c) island arc or back-arc environments (Gaál, 1986; Sokolov and Heiskanen, 1985; Lobach-Zhuchenko *et al.*, 1986; Gaál and Gorbatschev, 1987), or (d) a subduction-related arc

(Barbey and Martin, 1987). The principal depositional environment of the greenstones was onto rifted Saamian crust and hence a plate tectonic model has been developed to explain the juxtaposition of the Karelian, Belomorian and Kola Peninsula Provinces (Gaál, 1986; Gaál and Gorbatschev, 1987). In this model the Belomorian represents a mobile belt region and the Karelian Province the foreland complex of the Lopian orogeny, under which the Belomorian was subducted westwards beneath the Saamian craton. The position of the Kola Peninsula Province suggests the possibility of a continent–continent collisional system (Gaál and Gorbatschev, 1987).

12.3 Lapponian Supergroup (earliest Proterozoic)

Following cratonisation during the Lopian orogeny, the Archaean Domain underwent an episode of uplift and erosion prior to the deposition of the Lapponian Supergroup supracrustal suites in NW- to NNW-trending rifts (Laajoki, 1983; Saverikko, 1983; Gorbunov et al., 1985; Gaal and Gorbatschev, 1987). This supergroup is dominated by metasediments and mafic metavolcanics, age estimates for which range from 2.7 to 2.3 Ga (Silvennoinen, 1985; Sjöstrand et al., 1986; Gaál and Gorbatschev, 1987). The lower Lapponian contains ultramafic, mafic, intermediate and felsic volcanics interbedded with clastic sediments (Gorbunov et al., 1985; Gaál and Gorbatschev, 1987). These are cross-cut by 2.44 Ga gabbros which form part of an extensive E to NE-trending suite in northern Finland and the Kola Peninsula. The upper Lapponian group contains basaltic volcanics and various types of sediments. The Lapponian sequences are overlain, usually unconformably, by sedimentary and volcanic sequences of the Karelian Supergroup (Table 12.1).

In Soviet Karelia (Figure 12.2) the Sumian Suite, which has been dated at 2.5 Ga (Gorbunov et al., 1985; Ryabchikov, 1988), rests unconformably on Archaean basement and is composed mainly of volcanic rocks. Volumetrically, basalts are progressively replaced upwards by andesite and felsic lavas. The Strel'na Group occupies a similar stratigraphic position in the Pechenga–Varzuga zone of the Kola Peninsula (Table 12.1). This group is dominated by basic lavas and intercalated clastic sediments which are cross-cut by 2.44 Ga gabbroic sills. Geochemical criteria suggest that Strel'na komatiitic basalts and the 2.44 Ga old layered gabbroic sills are comagmatic. In northern Finland the Lapponian Supergroup forms an extensive NW-trending belt which rests on Archaean basement (Silvennoinen, 1985; Laajoki, 1986). The lower part comprises basic and intermediate metavolcanics capped by arkosic metasediments. The upper part is mostly pillowed basaltic lavas (Silvennoinen, 1985) and metasediments of which some have been interpreted as glacial deposits (Marmo and Ojakangas, 1984).

In the Repparfjord window of Finnmark, the Holmvatn Group (Table 12.1) is probably equivalent to the Lapponian Supergroup. The lower part of the Holmvatn Group comprises basic, intermediate and felsic metavolcanics interbedded with clastic meta-sediments of mixed volcanogenic and continental provenance, while the upper part is composed of metamorphosed pillow lavas and minor mafic volcaniclastics (Pharaoh et al., 1983). Comprehensive geochemical data sets are only available for the Holmvatn Group (Pharaoh, 1980; Pharaoh et al., 1987). The lower Holmvatn Group contains basic, intermediate and felsic calc-alkaline volcanics (Figure 12.5) which appear to have been generated in a volcanic arc (Pharaoh and Pearce, 1984; Pharaoh et al., 1987). In contrast, the upper part of the group (the Magerfjell Formation)

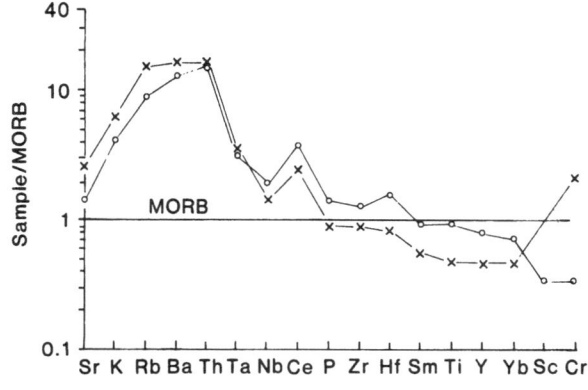

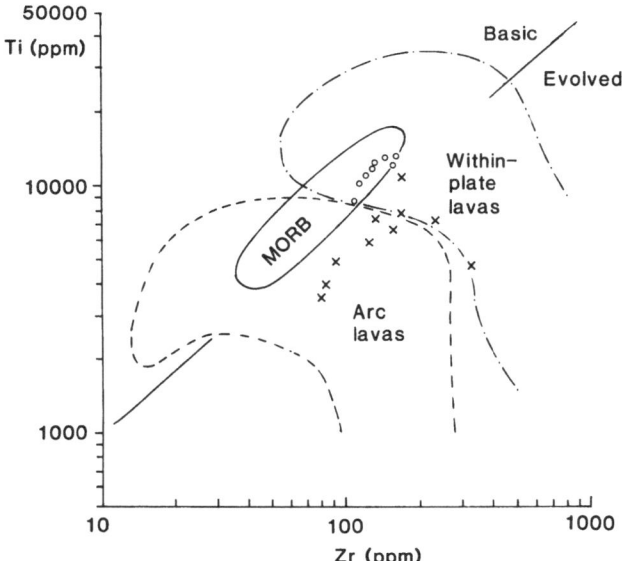

Figure 12.5 Geochemical characteristics of the lower (crosses) and upper (circles) Holmvatn Group (from Pharaoh and Pearce, 1984).

contains tholeiitic pillow lavas showing marked Fe-enrichment and geochemical signatures consistent with their formation in a within-plate setting (Pharaoh, 1980; Pharaoh and Pearce, 1984).

12.4   Early Proterozoic layered gabbroic complexes

Layered gabbroic complexes follow E- to NE-trending faults in a 300 km long belt in northern Finland, e.g. at Kemi, Koillismaa and Mustavaara and the Kola Peninsula (Figure 12.1) (Alapieti, 1982; Lahtinen, 1985; Gorbunov *et al.*, 1985). These intrusions were emplaced at about 2.44 Ga into both Archaean basement and lower Lapponian

Supergroup rocks, and locally contain significant amounts of Cr, Ni, Cu and V (Lahtinen, 1985).

The Koillismaa Complex of northeastern Finland has a volume greater than $2000 \, km^3$ and outcrops over $240 \, km^2$ (Alapieti, 1982; Lahtinen, 1985). It comprises a western complex and an eastern unit (the Narankavaara intrusion) connected by a hidden feeder dyke identified from a major gravity anomaly (Alapieti, 1982). The complex can be subdivided into three principal units, the marginal series, the layered series and the uppermost granophyre. The marginal series is usually discordant to the layered units (Alapieti, 1982; Lahtinen, 1985), grading inwards from gabbros at the contact with the thermally metamorphosed and brecciated gneisses, to pyroxenites and peridotites. No chilled contact has been recognised. The layered series forms the major part of the complex. The scale of the layering varies from less than a centimetre to hundreds of metres and can be further subdivided on the basis of cryptic layering. Interstitial granophyric quartz–alkali feldspar intergrowths occur in the middle and upper zones of the layered series. The granophyric cap to the complex is locally a kilometre thick and shows no layering or lamination.

The mineralogy of the complex essentially comprises olivine, Ca-poor pyroxene, calcic pyroxene, plagioclase and spinel. Accessory minerals include garnet, apatite and zircon. Olivine compositions range from Fo_{88} to Fo_{68}. The olivine in the marginal series is extensively serpentinised. In the layered series it is restricted to certain ultramafic and gabbronoritic zones. Ca-poor pyroxenes occur throughout the complex and vary from hypersthene ($c. \, En_{70}$) in the marginal gabbros to bronzite (En_{88}) in cumulate harzburgitic and gabbronoritic layers. The En values of the orthopyroxenes decrease systematically with height in the intrusion and hypersthene is superseded by inverted pigeonite in the uppermost layers. Calcic pyroxene also occurs throughout the complex. Its modal proportion increases and its mol. En content decreases from the ultramafic to the gabbroic rocks (Cr-rich endiopside occurs in the ultramafic rocks). The clinopyroxene in the gabbroic rocks is extensively uralitised. Zoned plagioclase is also ubiquitous, the An content varying sympathetically with Mg content of the ferromagnesian phases, decreasing both upwards in the intrusions and outwards in the marginal series.

The whole-rock geochemistry reflects the cryptic mineralogical variations, except in some of the marginal series rocks where the assimilation of country rocks produces unusual compositions. The fine-grained marginal facies of the Koillismaa Complex is characterised by high Mg and low Ti and K values (see Chapter 4). Generally, there is an increase in Fe and alkalis and decrease in Mg from the bottom to the top, and Cr contents decrease and Ti increases progressively upwards in the intrusions. The 2.44 Ga gabbroic complexes are probably contemporaneous and comagmatic with upper Lapponian volcano-sedimentary complexes – for example, the upper Holmvatn Group of the Repparfjord window (Pharaoh and Brewer, in prep.) – and this episode of combined plutonism and volcanism appears to be part of the continued rifting of the Archaean crust (Alapieti, 1982; Gorbunov et al., 1985; Lahtinen, 1985; Gaál and Gorbatschev, 1987) and to reflect the initiation of a new Wilson Cycle.

12.5 Supracrustal belts marginal to the Lapland Granulite Complex

The Lapland Granulite Complex (LGC) is a 50 km wide belt in the northern Baltic Shield (Figure 12.1b) composed of sillimanite–garnet–cordierite gneisses, garnet gneisses and hypersthene–plagioclase gneisses, for which medium pressure granulite facies metamorphic conditions ($P = 0.7 - 0.8$ GPa, $T = 800°C$) have been inferred (Barbey *et al.*, 1982). Around 80% of the complex may have been derived from flysch-type sedimentary protoliths (khondalitic suite) Barbey *et al.*, 1980, 1982) or calc-alkaline volcanics (Hormann *et al.*, 1980). The mafic components of the complex appear to have been derived from an upper mantle source immediately prior to 1.9 Ga (Bernard-Griffiths *et al.*, 1984) and show no evidence for an earlier crustal history.

The contiguous Karasjok and Kittilä Belts of northern Norway and Finland have a combined length of more than 200 km. They abut against, and are overthrust by, the western side of the LGC (Figure 12.6) (Krill, 1985). The Karasjok Belt is itself allochthonous with respect to the Archaean substrate of the Karelian Domain

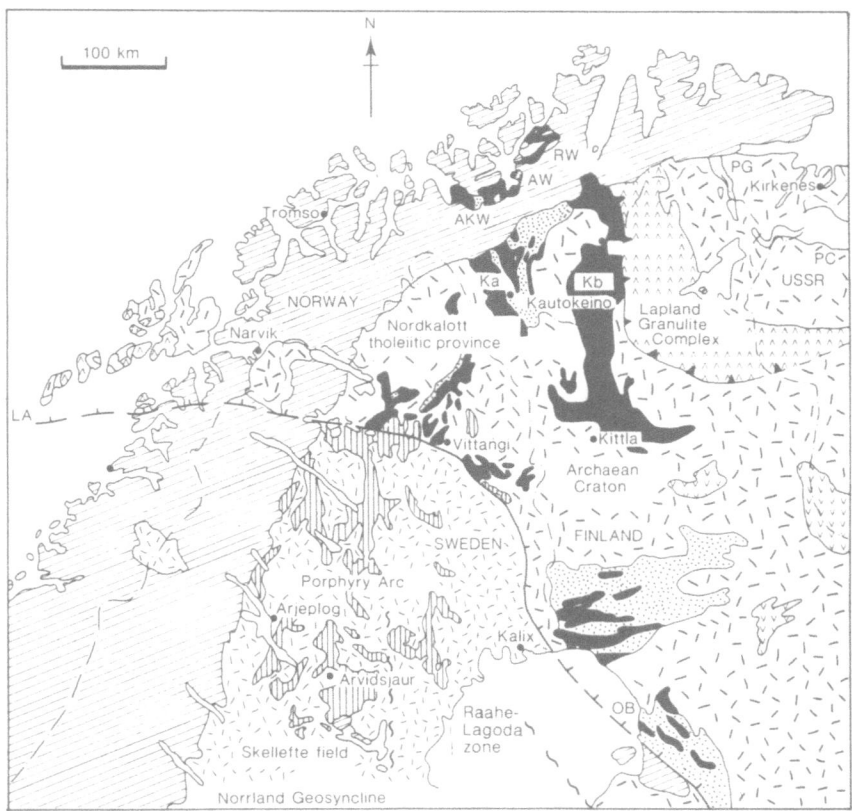

Figure 12.6   Geology of the northern Baltic Shield emphasising the distribution of the early Proterozoic supracrustals (after Pharaoh *et al.*, 1987). Coarse hatch shading: Archaean granites and gneisses; fine hatch shading: Svecofennian granites, gneisses and metasediments; ' ∧ '-shading: Archaean greenstones; black: Jatulian tholeiite metavolcanic suites; stipple: Jatulian–Sariolian metasediments; N–S line shading: Svecofennian arc volcanics. AKW: Alta–Kvaenangen window; RW: Repparfjord window; AW: Altenes window; Ka: Kautokeino Belt; Kb: Karasjok Belt; LA: southern limit of isotopically defined Archaean crust; OB: Ostrobothnian Schist Belt; PC: Pechenga Complex; PG: Petsamo Group.

(Pharaoh, 1984). Recent research does not support the Archaean age previously inferred from the presence of komatiitic lavas (Gaál, 1978; Pharaoh, 1985). Geological constraints suggest a younger, possibly Lapponian age for the Kittilä Belt (Gaál and Gorbatschev, 1987) while the Karasjok komatiites have yielded an imprecise errorchron age of 2085 ± 85 Ma (Krill *et al.*, 1985) which could represent either a Jatulian eruptive age or metamorphic homogenisation of an older, perhaps Lapponian, sequence. The Karasjok–Kittilä Belt is lithologically and geochemically distinct from, and not easily correlated with, either the Lapponian or Jatulian supracrustal sequences in Finnmark.

The northern part of the Karasjok Belt is composed of schistose amphibolites of the Bakkilvarri formation (Siedlecka *et al.*, 1985), which in places contain pillow structures and chlorite–amphibole schists interpreted as metakomatiites (Henriksen, 1983). The age and origin of the belt is of crucial importance in understanding the uplift history of the LGC, and the crustal evolution of this region (see later discussion). The Bakkilvarri amphibolites are considered to represent a metavolcanic suite comprising basaltic komatiites, high-Mg tholeiites and high-Fe tholeiites (Pharaoh *et al.*, 1987). On the basis of Ti, HFS element and REE contents, the basaltic komatiites have been subdivided into a depleted and an enriched series. Those of the depleted series have MORB-like geochemical characteristics, whereas the remaining more enriched geochemical patterns are comparable to those of basaltic rocks erupted in a within plate setting (Figure 12.7). These differences were originally attributed to source region heterogeneity (Pharaoh *et al.*, 1987) but derivation of the enriched series by the crustal contamination of depleted series-type liquids, perhaps by thermal erosion (Huppert and Sparks, 1985) of Fe- and Ti-rich tholeiitic basalt, cannot be ruled out.

12.6 Karelian Supergroup (early Proterozoic)

In Finland and the Soviet Union the Lapponian Supergroup is unconformably overlain by the Karelian Supergroup. In Soviet Karelia the lower Jatulian contains a diachronous basal sedimentary sequence overlain by basaltic lavas. The middle Jatulian has a basal conglomeratic unit overlain by a 350 m thick basaltic lava sequence, while the upper Jatulian is dominated by mixed sediments, lava flows and tuffs. In the Kola Peninsula the Pechenga Group is restricted to thrust-bounded basins (Gorbunov *et al.*, 1985). The Kola superdeep borehole penetrated thick sequences of metamorphosed basaltic and andesitic lavas (the Luostarin Series) and basalts, picrites and tuffs (the Nikel Series) (Kozlovsky, 1984; Kremenetsky and Ovchinnikov, 1986). In Finland, clastic sediments of the Sariolian Group are succeeded by Jatulian sandstones with intercalations of basaltic lavas (Silvennoinen, 1985; Perttunen, 1985; Laajoki, 1986). The uppermost Jatulian (Marine Jatuli) is composed of black schists, dolomites and basaltic lavas (Perttunen, 1985; Laajoki, 1986).

In northern Norway (western Finnmark) and adjacent parts of northern Sweden and Finland (Figure 12.6), the Jatulian succession is dominated by 1–2 km thick sequences of basaltic pillow lavas and aquagene tuffs, with subordinate amounts of epiclastic sediments. The huge volumes of lava erupted in this region, and the good lithostratigraphic correlation between individual supracrustal belts (Pharaoh and Pearce, 1984) led Pharaoh (1985) to propose the existence of a tholeiitic magmatic province in this region, distinct from Jatulian sequences further southeast in Finland and the USSR where volcanic rocks are subordinate. Examples of these sequences include the

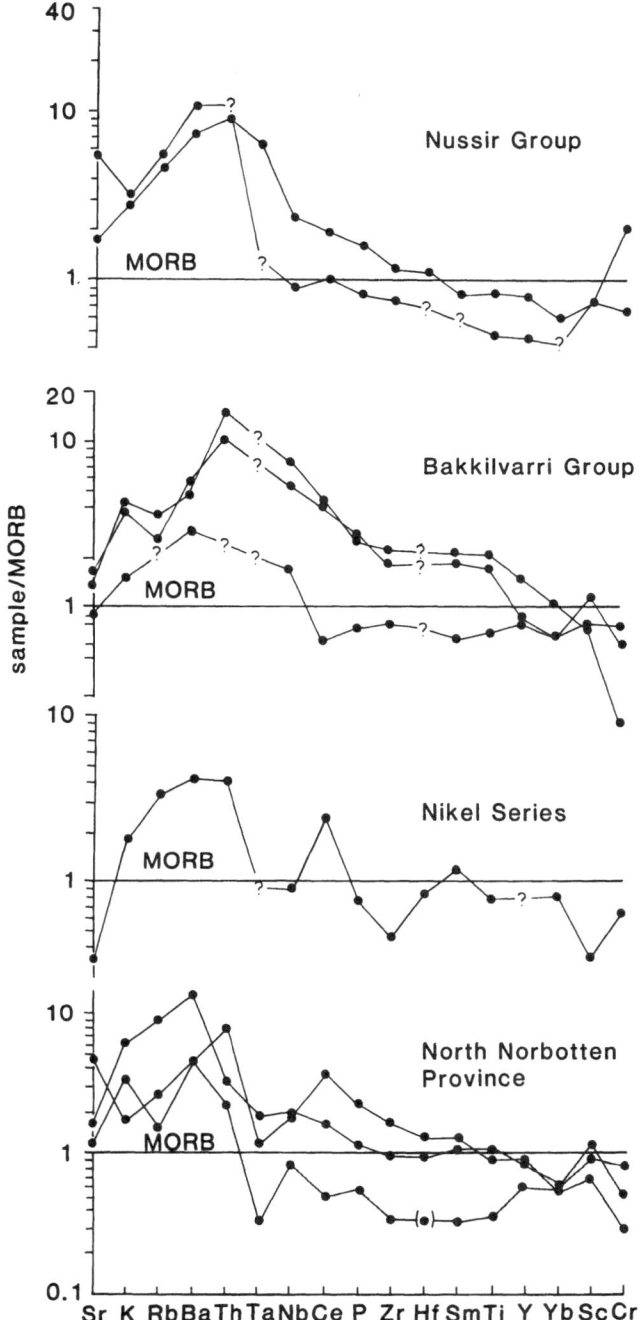

Figure 12.7 MORB–normalised multi-element diagrams for early Proterozoic basaltic lithologies from the northern Baltic Shield (data from Pharaoh *et al.*, 1987).

Cas'kejas, Raipas and Nussir Groups of northern Norway (Pharaoh, 1985; Siedlecka *et al.*, 1985), the Kiruna and Vittangi Groups of northern Sweden (Witschard, 1984) and the Joutiappa Volcanic Formation of northern Finland, which has yielded a Nd isochron age of 2.09 Ga (Huhma, 1984). NW-trending diabasic dykes yielding U–Pb ages of 2.25–2.0 Ga (Sakko, 1971; Pekkarinen, 1979) are also part of this magmatic episode, which marks a period of extension of the Archaean cratonic lithosphere.

The Jatulian metabasalts invariably show strong Fe-enrichment (Pharaoh and Pearce, 1984; Pharaoh *et al.*, 1987). Metabasalts in Finnmark have Nb/Y values of 0.25 to 0.75, indicative of a sub-alkaline magma series. Metabasalts of the Nussir Group have MORB-normalised patterns with a strong within-plate component (Pharaoh, 1985), while lavas from elsewhere in Finnmark and northern Sweden have a more depleted trace element composition comparable to that of MORB or LKT (Pharaoh and Pearce, 1984). It is clear from their sedimentary associations (e.g. quartzites and stromatolitic dolomites) that basalts throughout the Nordkalott Province were erupted within rifted Archaean crust (Torske, 1978; Pharaoh, 1985). These compositional differences are believed to reflect heterogeneity of the mantle source region. Lavas from the Pechenga Complex exhibit much stronger enrichment in Th and LREE with respect to HFS elements such as Nb in their MORB-normalised patterns (Figure 12.7), from which Pharaoh *et al.* (1987) inferred a strong arc-type magmatic component.

12.7 Marginal zone of the Svecofennian Domain

In north central Finland, the Kalevian Group unconformably overlies the Jatulian sequences described above and is predominantly composed of metaturbidites with tholeiitic volcanics. Equivalents in the USSR include the Vespian Group of Karelia, which comprise conglomerates and sandstones with few basaltic lavas, and the South Pechenga and Tominga Groups of the Kola Peninsula, comprising metamorphosed basaltic, andesitic and felsic volcanics, tuffs and sediments. The upper part of the Kalevian Group in Finland is dominated by distal turbidites (Honkamo, 1985; Perttunen, 1985; Laajoki, 1986) and tectonic slices of ophiolitic material, dated at about 1.96 Ga (Kontinen, 1987; Ward, 1987; Gaál and Gorbatschev, 1987; Park, 1988). This group marks the onset of the Svecofennian orogenic cycle and was preceded by intense faulting and the intrusion of 2.25–2.0 Ga NW-trending diabasic dykes (Gaál and Gorbatschev, 1987). A 150 km long NW-trending zone of early Proterozoic metasediments and metavolcanics forms the Karelidic Schist Belt between the eastern margin of the Karelian and Svecofennian Provinces in north Finland (Figure 12.1). In central Finland, the marginal zone has been severely modified by the development of the Raahe–Ladoga Zone, a major dextral transpressional structure (Halden, 1982), during the Svecofennian orogeny.

12.7.1 *Karelidic Schist Belt*

The lithologies and tectonic setting of the Karelidic Schist Belt are exemplified by the Ostrobothnian Schist Belt of northern Finland (Figure 12.6), which is composed of Kalevian (2.1–1.9 Ga) metasediments and metavolcanics (Honkamo, 1985, 1987). Archaean basement occurs to the north and east of this belt, and younger Proterozoic migmatites and granites occur to the south and west. The Ostrobothnian Schist Belt

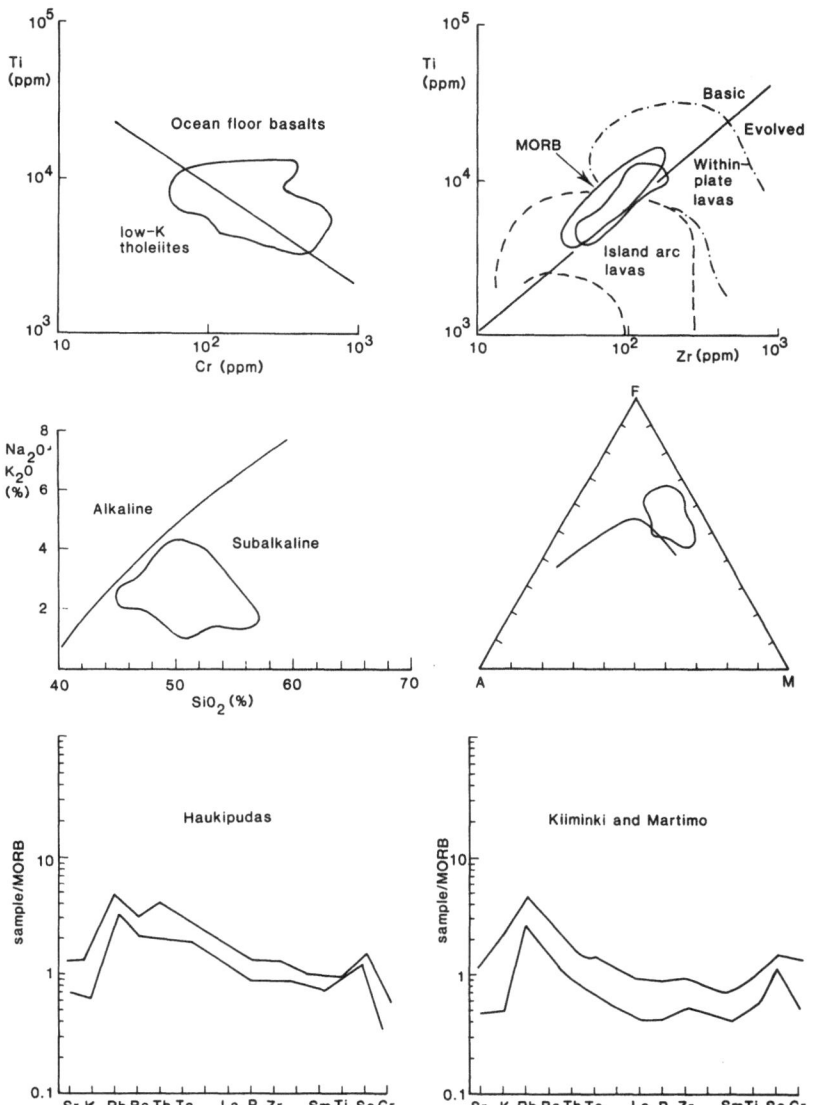

Figure 12.8 Geochemical characteristics of metavolcanics from the northern Ostrobothnia Schist Belt (shown as fields for clarity; from Honkamo, 1987).

comprises a series of metamorphosed (amphibolite facies) greywackes and volcanics up to 8 km thick which lie unconformably upon Archaean basement rocks. This area is interpreted as a remnant of an early Proterozoic intra-continental or marginal basin (Honkamo, 1985, 1987). The associated lavas are dominated by quartz- or olivine-normative tholeiitic metabasalts. Trace element data suggest that the basalts are comparable to low-K tholeiites, back-arc tholeiites or MORB (Figure 12.8).

12.7.2   *Early Proterozoic ophiolitic (?) complexes*

The 1.96 Ga old Jormua Complex occurs in the Kainuu Schist Belt of northeast Finland
(Kontinen, 1987). The main outcrop is 20 km long and up to 5 km wide and comprises
an intensely faulted body of serpentinite and mafic extrusives and intrusives. It contains
four main lithological units, (i) serpentinite with cross-cutting basaltic and gabbroic
dykes, (ii) metagabbroic bodies cross-cut by basic dykes, (iii) a sheeted dyke complex,

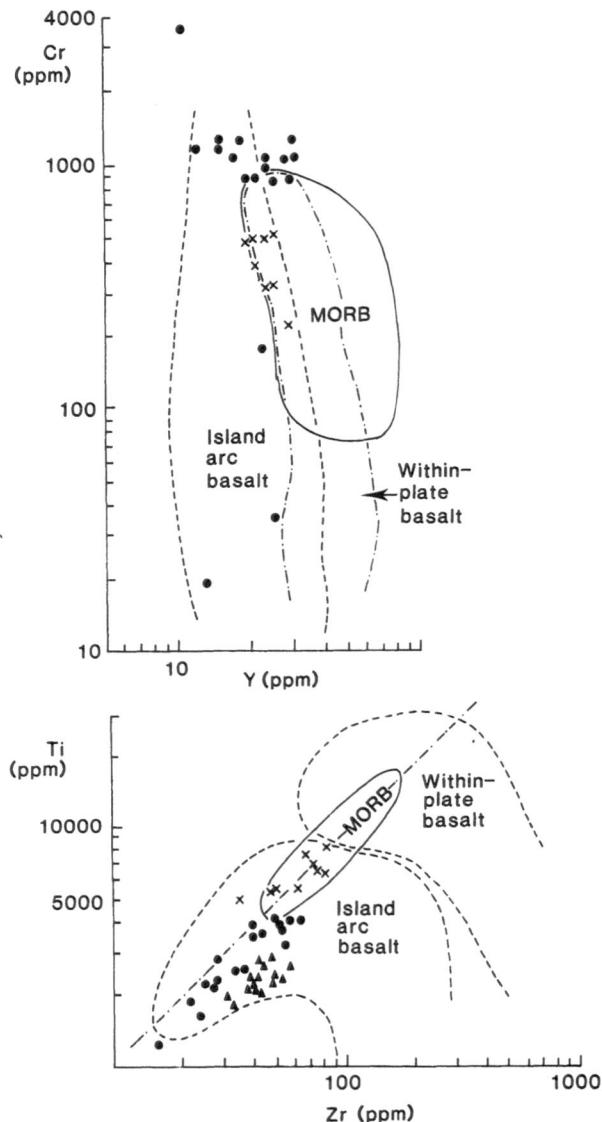

Figure 12.9   Cr/Y and Ti/Zr relationships in the Jormua ophiolite (crosses) and the Outokumpu
association (circles and triangles) (after Park, 1988).

and (iv) metabasaltic pillow lavas and breccias. The lavas and dykes are considered to be cogenetic and comprise sub-alkaline MORB-type tholeiites (Figure 12.9) (Kontinen, 1987). Their REE profiles are akin to those of transitional (T-)MORB. The metagabbros show an Fe-enrichment trend and can be subdivided into tholeiitic magnesian and ferrogabbros. They are slightly LREE-depleted with positive Eu anomalies, and have lower overall REE concentrations than the lavas and dykes (Figure 12.7). They are similar to gabbros dredged from mid-ocean ridges and are considered to be cumulates derived from the same source as the lavas. The origin of the serpentinites is more problematical because of the effects of low temperature alteration, but they are probably derived from harzburgitic parents (Kontinen, 1987).

The Outokumpu Complex consists of various metasediments (carbonaceous shale, quartzite, dolomite, talc and fuchsitic schists), intrusive saxonites, pyroxenites, serpentinised dunites and wehrlitic gabbros, and basic metavolcanics (Park, 1984, 1988). Well preserved contacts in the Kaavi district demonstrate that the ultramafic units were intruded as sills. Associated ore deposits are interpreted as the products of sea floor hydrothermal systems (black smokers) which preceded the deposition of the Kalevian (Koistinen, 1981; Park, 1983, 1988). It has been proposed that these volcanic rocks were derived from a source more depleted than that which gave rise to the Jormua Complex (Figure 12.9), possibly in an arc or back-arc setting (Park, 1988).

12.8   Basic rocks of the Svecofennian Domain

Volcanism in the Svecofennian Domain between 1.9 and 1.87 Ga was distinctly provincial in nature (Pharaoh and Pearce, 1984; Gaál and Gorbatschev, 1987). The northern and southern provinces contain calc-alkaline volcanic suites which rest on thick basal greywacke sequences. Metamorphosed felsic lavas and tuffs (volcanic leptites) dominate the lower part of the sequence in each region, whereas tholeiitic metabasalts are common in the upper part (Simonen, 1980; Pharaoh and Pearce, 1984; Vivallo and Rickard, 1984; Vivallo and Claesson, 1987; Lundstrom, 1987; Gaál and Gorbatschev, 1987). The predominant lithologies in the central province are meta-turbidites with subordinate metabasalts, deposited in the Bothnian Basin (Hietanen, 1975).

12.8.1   Northern Province

The Skellefte Group of northern Sweden is a NNW-trending volcano-sedimentary sequence between 1.91 and 1.86 Ga old (Lundberg, 1980; Skiöld and Cliff, 1984; Claesson, 1985; Skiöld, 1987; Vivallo and Claesson, 1987). This group is essentially bimodal, composed of felsic and basic volcanics, and it can be divided into three units based on the proportions of felsic and mafic volcanics and sedimentary material. The lower volcanic unit contains calc-alkaline felsic volcanics and sub-alkaline meta-basalts, which are tholeiitic in the east and mainly calc alkaline in the west (Vivallo and Claesson, 1987). REE data and Ti–Zr variations demonstrate that the basic and felsic units cannot easily be related by simple fractionation processes. The metabasalts of the middle unit are extremely primitive and show geochemical similarities with basaltic komatiites and boninites (Vivallo and Claesson, 1987).

Tectonic reconstructions for the Skellefte district are based on the interpretation that the Skellefte Group formed in an island arc setting (Rickard, 1979; Pharaoh and Pearce,

1984; Claesson, 1985; Vivallo and Claesson, 1987). However, Claesson and Vivallo (1987) have argued that after an initial stage of felsic volcanism related to an island arc system, there followed a period of rifting during which the tholeiitic basalts were erupted.

12.8.2 Southern Province

A 200 km long NNE to E curving belt of 1.9–1.84 Ga volcanics, sediments and associated plutonic rocks and Fe-Mn and massive sulphide deposits occurs in the Bergslagen Province of central Sweden (Figure 12.1) (Vivallo and Rickard, 1984; Lundstrom, 1987; Parr and Rickard, 1987; Welin, 1987). The supracrustal sequence is dominated by leptites but exhibits considerable lithological variation laterally, metavolcanics predominating in the west and metasediments in the east (Lundstrom, 1987). In the Garpenberg region in the centre of the province, the sequence can be divided into lower and upper units, each dominated by pyroclastics and sediments, and a middle unit which is composed of submarine basalts. Marbles, sediments and mafic flows are more abundant upward in the succession, whereas mafic sills and dykes are more common in the lower part (Vivallo and Rickard, 1984; Oen, 1987; Parr and Rickard, 1987). However, felsic volcanics (70–75% SiO_2) are by far the most common type and, as intermediate volcanics are rare, the assemblage is essentially bimodal. The felsic volcanics are calc-alkaline, whereas the basic rocks are tholeiitic and the two populations appear to be genetically unrelated (Vivallo and Rickard, 1984). The felsic suite is characterised by 'I-type' calc-alkaline magmatism related to Andean-type subduction processes (Vivallo and Rickard, 1984), but the presence of tholeiitic basalts indicates areas of rifting within a predominantly destructive plate margin.

12.8.3 Central Province

The Central Province of the Svecofennian Domain is dominated by metamorphosed and migmatised greywackes and pelites which contain minor intercalations of quartzo-feldspathic schists (leptites) and amphibolitic metabasalts (Pharaoh and Pearce, 1984; Gaál and Gorbatschev, 1987; Lundqvist, 1987). Geochemical data for the metabasaltic rocks are limited, but suggest that these rocks have affinities with MORB-type tholeiites. However, there are still major uncertainties concerning the provenance of the associated sediments, the age of the pre-volcanic basement (if any) to this region, and the extent and distribution of such basement to the south and west of the region.

12.9 Syn- and post-orogenic intrusions

The 'Wilson Cycle' in which the Jatulian, Kalevian and Svecofennian sequences developed was brought to a close during the Svecofennian orogeny (0.90–1.84 Ga), during which event synorogenic plutonic suites were emplaced throughout the orogen (Table 12.1). Granites and granitoid rocks form the greatest volume of these plutonic suites, particularly in central and southern Scandinavia. However, in many areas the synorogenic intrusive suites also contain basic components. In Finnmark, for example, the Raudfjell Suite comprises a number of differentiated basic–ultrabasic pipe-like intrusions, emplaced after early orogenic deformation but prior to the regional metamorphic peak (Pharaoh, 1980; Pharaoh et al., 1983). The 'albite-diabases' of the Finnmarksvidda and the gabbroic, dioritic and syenitic Haparanda Series of northern Sweden (Witschard, 1984) are further examples in northern Scandinavia. The U-Pb zircon ages of these two suites are 1815 ± 8 Ma (Krill et al., 1985) and 1880 ± 28 Ma (Skiöld, 1981) respectively.

The Raudfjell Suite displays strong enrichment in Fe, Ti and V, and represents a typical differentiated tholeiitic suite, emplaced in a synorogenic setting. The Haparanda Series exhibits a similar geochemical range and is believed to be comagmatic with the Kiruna Porphyry lavas (Öhlander, 1986; Pharaoh and Brewer, unpublished data), which host the famous Kiruna apatite–magnetite ore deposits. The suite exhibits enrichment of Th and LREE with respect to the HFS elements, particularly Nb, giving a distinctive arc-type signature to MORB-normalised geochemical profiles. The Haparanda Suite is believed to mark a return to subduction-related magmatism, with underflow directed eastward beneath the recently accreted early Proterozoic crust of the Svecofennian Domain.

The phase of renewed subduction recognised above led to the formation of the trans-Scandinavian granite–porphyry belt and came to a close during the 1.75–1.5 Ga Gothian orogeny (Gorbatschev, 1980). Beyond the confines of the Gothian orogen, a post-Svecofennide rapakivi granite suite is associated with gabbros and anorthosites (Vaasjoki, 1977).

12.10 Concluding remarks

In the broadest of terms, the Baltic Shield developed in the early Proterozoic by sequential accretion to the margins of Archaean continental nuclei. The Archaean terranes were cratonised during the Saamian (3.1–2.9 Ga) and Lopian (2.9–2.6 Ga) orogenies, subsequent accretion and reworking occurring during the Svecofennian (2.0–1.75 Ga), Gothian (1.75–1.5 Ga) and Sveconorwegian (1.25–0.9 Ga) orogenies. In recent years the Proterozoic tectonic evolution of the shield has been explained in terms of modern plate tectonic processes (e.g. Hietanen, 1975; Berthelsen, 1980; Pharaoh and Pearce, 1984; Barbey et al., 1984; Gaál and Gorbatschev, 1987). Such models have become even more robust with the recent discovery of ophiolitic relics (Kontinen, 1987). There are many strong parallels between the Proterozoic evolution of the Baltic and the Laurentian Shields (Pharaoh and Brewer, in prep.)

A 'consensus' plate model is presented in Figure 12.10. According to this model, the majority of the Lapponian and Jatulian supracrustals formed in ensialic rift basins (Torske, 1978; Pharaoh, 1985; Pharaoh et al., 1987), whereas the Svecofennian development mainly involved subduction processes. Several major problems arise from the model, including (i) the spatial disposition of the Archaean cratons during the early Proterozoic, (ii) the interpretation of arc-type basaltic rocks in the lower Lapponian group on the basis of geochemical signatures, (iii) the age of the Karasjok–Kittilä Belt and significance of the Lapland Granulite Complex, (iv) the degree of tectono-magmatic reprocessing involved in the generation of Svecofennian arc systems and (v) whether or not pre-volcanic Archaean or early Proterozoic crust was involved in the latter.

The arc-type geochemical signatures of the lower Lapponian lavas are problematical. Do these reflect a real subduction-related magmatic episode during the earliest Proterozoic, or is the signature inherited from earlier, perhaps Archaean arc volcanism? If the latter is true, an ensialic rift environment would be an equally plausible model. Equally problematical is the correlation of the distinctive, metakomatiite-bearing Karasjok–Kittilä sequences with the Lapponian and Jatulian. Do these belts mark a major crustal suture between Archaean cratons, as has been

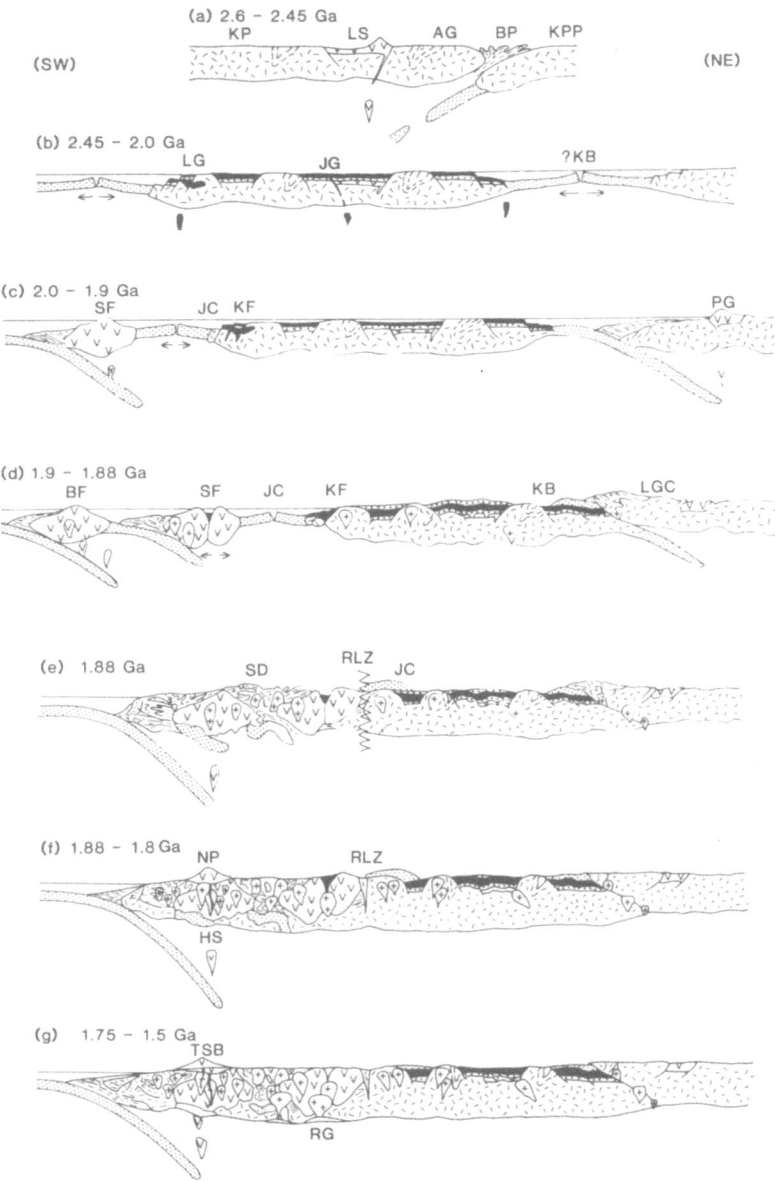

Figure 12.10 Tectonic synthesis of the early Proterozoic development of the Baltic Shield (modified after Barbey *et al.*, 1984; Krill, 1985; Gaál and Gorbatschev, 1987; Pharaoh *et al.*, 1987). Ages are approximate, based on the best available radiometric data. AG: Archaean greenstone belts; BB: Bothnian basin; BF: Bergslagen Fjeld; BP: Belomorian Province; HS: Haparanda Series; JC: Jormua Complex; JG: Jatulian Group; KB:Karasjok Belt; KF: Kalevian flysch and ophiolite complexes; KP: Karelian Province; KPP: Kola Peninsula Province; LG: layered

suggested by several authors (Barbey et al., 1982; Krill, 1985; Pharaoh et al., 1987)? Or is the LGC simply a mid-continental orogenic 'pop-up', analogous to the Kapuskasing and Wind River uplifts of North America and the Arunta Block of central Australia (which, coincidentally, is associated with some unusual supracrustal sequences)? Until the supracrustal sequences are dated with a precision comparable to that achieved for the plutonic suites, it is difficult to constrain these uncertainties.

During the Svecofennian orogenic stage, the volume and predominance of felsic volcanism and the presence of within-plate tholeiitic basalts in the Bergslagen arc are not consistent with simple models for a subduction-related arc. Lithological criteria suggest a need for the tectono-magmatic reprocessing of primitive, possibly oceanic arc-type material along a continental margin, and if such a margin can be recognised, the nature and extent of continental lithosphere needs to be defined.

Thus, although the concept of Proterozoic crustal accretion onto northern Archaean continental nuclei by processes akin to plate tectonics is now widely accepted, many details of the processes by which this was accomplished are poorly understood at present.

Acknowledgements

The contribution by T.C.P. is published with the approval of the Director of the British Geological Survey.

---

gabbroic complexes; LGC: Lapland Granulite Complex; LS: Lapponian Supergroup; NP: Norrbotten porphyries; PG: Pechenga Group; RG: Rapakivi granite suite; RLZ: Raahe–Ladoga zone; SD: Svecofennian domain; SF: Skellefte Field; TSB: trans-Scandinavian granite-porphyry belt. Key to shading: black: continental rift tholeiitic suites; V: calc-alkaline suites; stipple: oceanic and marginal basin crust; dots: cratonic sediments; hatched shading: granitic gneisses and Archaean greenstones; crosses: granitoid plutonic series.

(a) Early Lapponian arc volcanism and sedimentation (2.6–2.45 Ga). Eruption of lower Holmvatn Group, Kviby Group and other calc-alkaline suites, post-orogenic with respect to the Lopian (late-Archaean) orogeny.

(b) Late Lapponian–Jatulian rift volcanism and sedimentation (2.45–2.0 Ga). Initiation of new Wilson Cycle. Emplacement of layered basic complexes and dyke suites, and eruption of tholeiitic basalts within rifted Archaean crust. Possible development of oceanic crust (e.g. Karasjok Belt).

(c) Oceanic closure (2.0–1.9 Ga). Northeastward subduction beneath Kola continent marked by emplacement of plutonic rocks into Lapland Granulite Complex and eruption of calc-alkaline Pechenga Group lavas. Formation of marginal basin lithosphere (Jormua, Outokumpu?) and sedimentary basins (Kalevian) behind immature volcanic arcs (Skellefte Field, Tampere) and associated forearc accretionary prisms (Bothnian basin).

(d) Early Svecofennian orogeny (1.89 Ga). Collision (?) of the Kola and Karelian continents, uplift of the LGC to upper crustal levels accompanied by overthrusting and metamorphism in the Karasjok Belt. Main phase of Svecofennian calc-alkaline arc volcanism, magmatic reprocessing of primitive arcs and continued intra-arc sedimentation.

(e) Svecofennian orogeny (1.88 Ga). Collision of Svecofennian arc complexes with Karelian continent, obduction of ophiolites, folding and metamorphism of Svecofennian accretionary complex followed by dextral transpression on the Raahe–Ladoga zone and possible warping of the arc complexes.

(f) Late Svecofennian orogeny (1.88–1.80 Ga). Eastward directed subduction, eruption of subaerial synorogenic volcanics (Kiruna and Arvidsjaur porphyries) and emplacement of the Haparanda Series. Extensive anatexis in the accretionary complex and 'S'-type granitic plutonism.

(g) Post-orogenic intrusive suites (including Rapakivi granites) and Gothian orogeny (1.75–1.5 Ga). Continued eastward subduction beneath Andino-type margin leads to the formation of the trans-Scandinavian granite-porphyry belt. High heat flow in back-arc region maintained and many Svecofennian isotope systems remain open.

13 Early Precambrian basic rocks of China

B.-M. JAHN

13.1 Introduction

Basic volcanism has been a fundamental process in the evolution of the Earth as well as other terrestrial planets of the solar system. It was particularly important in the early stages of planetary evolution. From the results of experimental and geochemical studies, it is clear that basaltic magmas can be regarded as first-order differentiation products of mantle melting, hence they (and komatiitic magmas) play an important role as probes of chemical composition, chemical evolution and physical conditions of planetary mantles. By contrast, granitic magmas may be considered as second or higher-order differentiation products and they play key roles in the study of continental development and accretion, crustal recycling, particular trace element concentration processes and mineralisation.

Early Precambrian basic rocks (including komatiites) occur most abundantly in greenstone belts and locally include rocks which preserve fine-grained chilled margins, pillowed structures and spinifex textures. In general, greenschist facies mineral assemblages predominate (Condie, 1981). However, basic rocks are also found in high grade terranes, metamorphosed to amphibolite or granulite facies while yet others have undergone partial melting and migmatisation. In the Precambrian terranes of China, supracrustal rocks and associated plutonic granitoids have been subjected to strong deformation and commonly metamorphosed to amphibolite and granulite facies, followed by subsequent retrograde recrystallisation. Lower greenschist facies supracrustals also occur, but are less important volumetrically and are generally restricted to the early Proterozoic formations (Sun et al., 1984).

13.2 Distribution, metamorphism and age

The Archaean and early Proterozoic rocks of China occur mainly in the Sino-Korean Craton (or paraplatform; Huang, 1978) in the northern and northeastern parts of China (Figure 13.1). The principal outcrops are found in six broad regions: (1) the eastern Hebei Province, (2) the east and west sides of the gigantic Tan-Lu Fracture Zone in the Shandong Province, (3) the Liaoning and Jilin Provinces of northeastern China (Manchuria), (4) the Inner Mongolia autonomous region (Neimongol), spanning about 800 km east to west between 40° and 42° N, (5) the Taihang–Wutai mountains of the Shanxi Province and (6) the Dengfeng–Songshan region of the Henan Province. The stratigraphic sequences, probable age ranges, metamorphic grades and protolithic

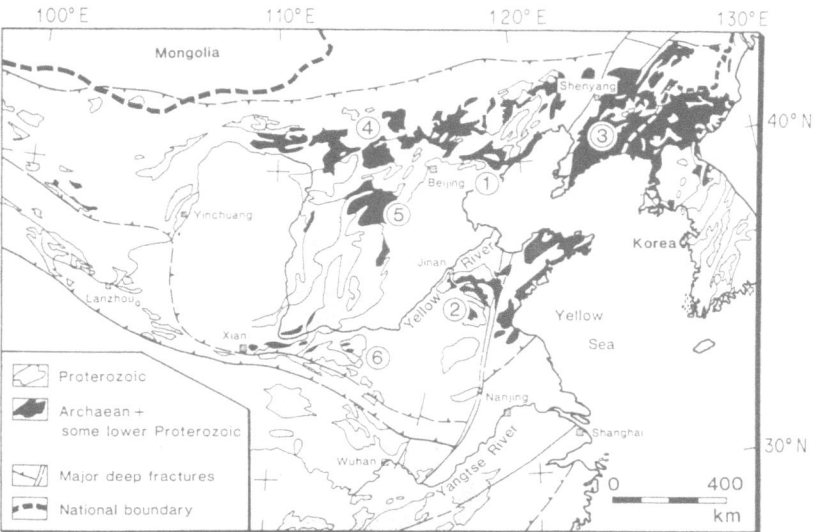

Figure 13.1 Distribution of Archaean (black) and Proterozoic (stippled) rocks in the Sino-Korean Craton. The craton is roughly bounded by major deep fractures. Numbered regions are 1: eastern Hebei Province; 2: Shandong Province; 3: NE China; 4: Inner Mongolia; 5: Shanxi Province; 6: Henan Province (see section 13.2 and Table 13.1).

characters of the Archaean and early Proterozoic formations in individual regions are summarised in Table 13.1.

13.2.1 *Eastern Hebei Province: the Qianxi Group*

More than 90% of the Archaean outcrops in this region are characterised by high-grade metamorphic rocks (granulite and upper amphibolite facies), and the remainder by lower amphibolite and greenschist facies assemblages (Sun *et al.*, 1984; Qian *et al.*, 1985). Most basic rocks have been metamorphosed to two-pyroxene granulites and amphibolites. In the Caozhuang area, amphibolites are common and generally occur as discrete bodies. Such enclaves are typical of high-grade Archaean gneiss terranes (Bridgwater *et al.*, 1974, 1976; Windley, 1984; Chadwick, 1981, 1985; Hunter *et al.*, 1984). These enclaves or blocks are embedded in grey gneisses which have been further intruded and migmatised by late Archaean K-rich granitic magmas (U–Pb zircon age = 2.6 Ga). The entire assemblage (amphibolite, grey gneiss and granite) was then cut by early Proterozoic pegmatites (U–Pb zircon age of 2.2 Ga; Liu *et al.*, 1986). The amphibolite–grey gneiss assemblage has been interpreted as a bimodal magmatic suite engulfed in younger granitic masses (Jahn *et al.*, 1987). Moreover, in a few localities amphibolites occur as layers intercalated with garnet-and sillimanite-bearing metasediments. These basic units are believed to be of volcanoclastic origin, probably representing volcanic tuff–pelite alternations. The amphibolites are also associated with minor banded iron-formation (BIF) and other metasedimentary rocks including fuchsite-bearing quartzite, cordierite–biotite–sillimanite gneiss and garnet–ortho-

Table 13.1 Stratigraphy of early Precambrian basic and associated rocks of China

Region	Stratigraphic unit	Age	Metamorphic facies	Protolith
1. Eastern Hebei	Qianxi Group: (Caozhuang, Zunha amphibolites, Qianxi granulites) (Table 13.2)	3.5–2.4 Ga	Amphibolite–granulite	Basic, intermediate and acid volcanics, volcano-sedimentary deposits, BIF, tonalite–granodiorite
	Shuanshanzi Group	Early Proterozoic	Upper greenschist–low amphibolite	Basic to acid volcanics, pelites and BIF
	Qinglonghe Group	Early Proterozoic	Greenschist	Conglomerates, pelites and BIF
2. Shandong	Taishan Group (west) (Table 13.2)	2.75–2.45 Ga	Amphibolite	Basaltic–komatiitic flows, basic tonalite, granodiorite, greywacke
	Jiaodong Group (east) (Table 13.2)	Middle-late Archaean 3.0 Ga–early Proterozoic	Amphibolite	Basic volcanics, clastic sediments, greywacke, limestone.
3. NE China	Anshan Group (and Longgang Group)	3.2–2.5 Ga	Amphibolite, some greenschist (upper part) and granulite	Basaltic–komatiitic volcanics, tonalite–granodiorite, pelitic sediments, BIF
4. Inner Mongolia	Jining Group (east)	2.6–2.45 Ga	Mainly granulite, some amphibolite	Basic and acid volcanics, pelitic and calcareous sediments
	Wulashan Group (west)		Mainly amphibolite, some greenschist	Basic volcanics and pyroclastics, calcareous and pelitic sediments
5. Shanxi	Fuping Group	2.8–2.6 Ga	Amphibolite, some granulite	Calcareous and semi-pelitic sediments
	Wutai Group	2.6–2.5 Ga	Greenschist–amphibolite	Basic, intermediate and acid volcanics, ultrabasic rocks, pelitic sediments, BIF
6. Henan	Dengfeng Group (and Taihua Group)	2.6–2.5 Ga	Mainly amphibolite, locally granulite	Basic volcanics, pelitic sediments, tonalite–granodiorite
	Songshan Group	Early Proterozoic (> 1.8 Ga)	Greenschist	Pelitic and calcareous sediments

Compiled from references mentioned in the text. Regions 1 to 6 shown in Figure 13.1

pyroxene–plagioclase–quartz gneiss. However, their field relationships are not clear. Using several geothermometers and geobarometers, Sills *et al.* (1987) determined the metamorphic conditions for the garnet-bearing metasediments as 600–650°C and 0.45–0.6 GPa, whereas an amphibolite enclave retrogressed from earlier granulite facies assemblage gave a temperature of 835 ± 25°C and pressure of 0.7–0.9 GPa (Jahn *et al.*, 1987).

The amphibolite enclaves of the Caozhuang area have been dated by the Sm–Nd method. A nine-point data set yielded an age of 3.47 ± 0.11 (2 sigma) Ga and $\varepsilon_{Nd}(t) = +2.7 \pm 0.6$ (Jahn *et al.*, 1987). Similar results have also been obtained by Huang *et al.* (1986), who reported an age of 3.5 ± 0.08 Ga and $\varepsilon_{Nd}(t) = +3.3 \pm 0.3$ from a six-point data set. The combined fifteen-point data set gives a new age of 3.5 ± 0.08 Ga and $\varepsilon_{Nd}(t) = +3.0 \pm 0.4$ (Figure 13.2). If the amphibolite enclaves are remnants of ancient mafic crust, this age of *c.* 3.5 Ga records the earliest crustal formation event in the Sino-Korean Craton and perhaps in the whole of eastern Asia.

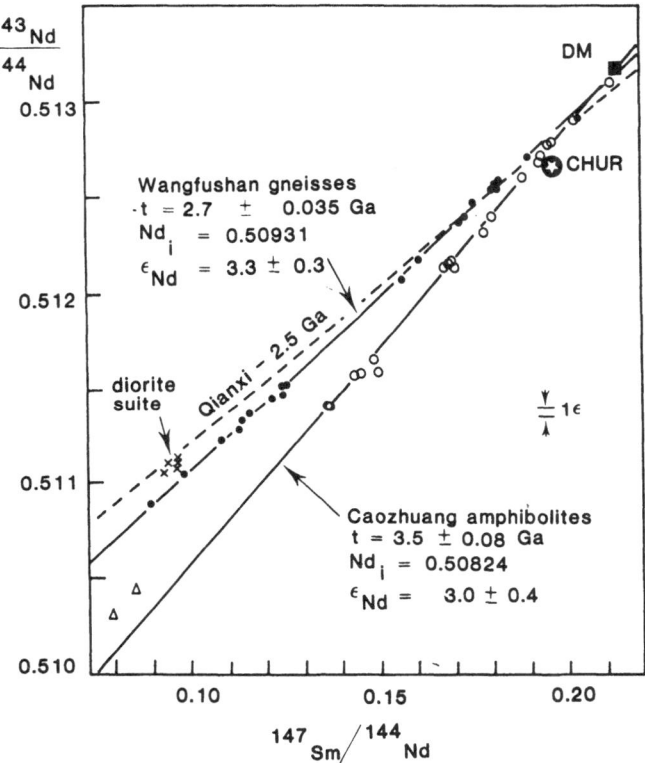

Figure 13.2   Sm–Nd isochron diagram for the Caozhuang amphibolites of the Qianxi Group (open circles), and the Wangfushan Gneisses (mostly metabasic rocks; dots) of the Taishan Group (data from Huang *et al.*, 1986; Jahn *et al.*, 1987, 1988). Two data points (triangles) representing grey gneisses with model ages of about 3.3 Ga and the 2.5 Ga Qianxi granulite isochron are shown for comparison. The intrusive diorite suite in the Wangfushan gneisses (crosses) has a U–Pb zircon age of 2.56 Ga and a distinct Nd isotopic composition corresponding to $\varepsilon_{Nd} = 3.1$ to 4.7, suggesting their derivation from a long-term depleted mantle source.

About 40 km to the north of the Caozhuang area, basic rocks occur as two-pyroxene granulites in the type locality (Taipingzhai–Guojiago) of the Qianxi Group. These basic granulites have long been regarded as the oldest rocks in China based on structural interpretations and metamorphic facies implications. However, repeated radiochronometric studies using various methods (Rb–Sr, Sm–Nd, U–Pb) have consistently produced ages of about 2.5 Ga (Pidgeon, 1980; Compston et al., 1983; Jahn and Zhang, 1984; Liu et al., 1985). The most likely interpretation is that the protolithic basaltic magmas were emplaced about 2.5 Ga ago and immediately followed by granulite facies metamorphism (Jahn and Zhang, 1984). The metamorphic temperature and pressure conditions have been determined at 700°C and 0.7–0.8 GPa respectively (Sills et al., 1987).

The western part of the Qianxi Group (near Zunha) is characterised by a high-grade belt of amphibolites and grey gneisses with intercalations of layered ultrabasic–gabbro–anorthosite complexes (Wang, R.M. et al., 1985). The metabasic and associated rocks have not been dated systematically, but a few unpublished results give Sm–Nd depleted mantle model ages greater than 2.8 Ga.

13.2.2   Shandong Province: the Taishan Complex

The Taishan Complex broadly consists of two geological entities: the Wangfushan basement gneisses and a late Archaean intrusive plutonic association (Ying, 1980; Cheng, 1986; Jahn et al., 1988). The basement gneisses are made up mainly of grey gneisses with tonalitic, trondhjemitic and granodioritic (TTG) compositions and amphibolites in various proportions. They are strongly folded, metamorphosed to amphibolite facies assemblages, and have locally undergone migmatisation.

The genetic and temporal relationships between the amphibolites and grey gneisses are not clear. In the Tai-An area, abundant amphibolites and minor hornblendites occur as mappable units or as small lenses or boudins several metres long enclosed within grey gneisses. Amphibolites and accompanying ultrabasic rocks are volumetrically important in the vicinity of Yanlingguan (about 40 km to the southeast), where metavolcanic rocks of a greenstone belt association, including rocks of komatiitic affinity, have been reported (Cheng et al., 1982a). The amphibolites and TTG gneisses of the Taishan Group are well dated by both the Rb–Sr and Sm–Nd isochron methods. A whole-rock Rb–Sr isochron gave an age of 2.69 ± 0.08 Ga with $Sr_i = 0.7006 \pm 0.0004$, and a whole-rock Sm–Nd isochron defined an age of 2.70 ± 0.03 Ga with $\varepsilon_{Nd}(t) = +3.3 \pm 0.3$ (Jahn et al., 1988; see Figure 13.2). Both the low Sr_i and high positive ε_{Nd} values suggest that the amphibolites and grey gneisses were derived from long-term depleted sources, regardless of their mutual genetic relationship. These isochron ages of around 2.7 Ga for the basement gneisses and the amphibolites are consistent with their stratigraphic relationships with the plutonic association (2.56 Ga for the diorite–granodiorite association and 2.45 Ga for the anatectic granites; Jahn et al., 1988). In conclusion, basaltic volcanism appears to have been the earliest magmatic phase, or at least was contemporaneous with the emplacement of TTG magmas in the Taishan Complex. No older rocks have been identified so far.

Metabasic rocks also occur in the Jiaodong Group which is distributed to the east of the Tan-Lu Fracture Zone in the Shandong Peninsula (Figure 13.1). Archaean ages of 2.6–2.8 Ga have been confirmed by recent zircon U–Pb isotopic analyses on one metabasic and five associated granitic gneisses (unpublished data; Liu, D.Y., pers. comm.).

13.2.3   Northeastern China (Manchuria): the Anshan Group

Archaean rocks are extensively exposed in eastern Liaoning and southeastern Jilin Provinces, and extend east into North Korea (Figure 13.1). They are collectively termed the Anshan Group, the major rock types being grey gneisses, amphibolite, abundant BIF and granulitic rocks of both igneous and sedimentary protoliths.

In the Qingyuan region, about 200 km east of Shenyang (Figure 13.1), a well-developed greenstone belt is surrounded by gneisses of TTG and monzonitic compositions (Zhai et al., 1985). The volcano-sedimentary piles are composed of a lower basic volcanic formation (Shipenzi Formation) including komatiitic rocks, a middle calc-alkaline volcanic formation with interbedded thin sedimentary rocks (Hongtoushan Formation), and an upper sedimentary formation of greywackes, pelites, volcanic tuffs, BIF, carbonates and quartzites (Nantianmen Formation). The greenstone belt is believed by Chinese geologists to be underlain by granulitic gneiss basement. The entire supracrustal sequence was metamorphosed to amphibolite facies and was invaded by tonalitic plutons and monzonitic granites.

The age of the greenstones is quite well bracketed by recent geochronological studies. An amphibolite from the lower Shipenzi Formation has been dated by K–Ar and $^{40}Ar/^{39}Ar$ methods. The results obtained for three aliquots of hornblende separates gave an age of 2.98 ± 0.07 Ga by the conventional K–Ar measurements and two excellent $^{40}Ar/^{39}Ar$ plateau ages of 2.99 Ga (Wang, S.S. et al., 1987; Figure 13.3a). Thus, the hornblende crystallisation during amphibolite facies metamorphism took place about 2.99 Ga ago, which sets the minimum age for the primary basaltic volcanism of the Qingyuan greenstone belt. On the other hand, a tonalitic pluton which intrudes the Hongtoushan Formation has been dated by the U–Pb zircon method (Peucat et al., 1986), and the nearly concordant zircon data yield a very precise age of 2.511 ± 0.001 Ga. An $^{40}Ar/^{39}Ar$ study on biotite from a nearby but separate tonalitic intrusion gave an age of 2.58 Ga (Wang, S.S. et al., 1986). In addition, a tonalitic intrusion at Huadian, in Jilin Province, has been dated at 2.51 ± 0.02 Ga by the U–Pb zircon method (Zhu et al., 1986). New Sm–Nd isotopic analyses on six metabasic rocks from other parts of the Anshan Group give an isochron age of 2.66 ± 0.075 Ga with $\varepsilon_{Nd}(t) = +4.4 \pm 0.5$ (Jahn and Ernst, 1989; Figure 13.3b). It appears that the formation of the Anshan Group greenstones took place in more than a single period. The deposition of Qingyuan greenstone belt ended at least 2.5 Ga ago and its initial volcanism is likely to have predated 3 Ga.

In the Anshan–Benxi region the Tiejiashan Complex is believed to be the oldest Archaean unit, consisting of tonalitic gneisses and various metamorphic rocks of supracrustal origin such as quartzites, talc + mica schists, amphibolites, BIF and retrogressed chlorite schists, although there are some granulite facies relics which have recrystallised at upper amphibolite facies (Ernst et al., 1988). The superjacent lower Proterozoic series (the Liaohe Group) rests with an angular unconformity on the Archaean rocks. It has been subjected to greenschist facies metamorphism, and now comprises chlorite + sericite phyllites, metaconglomerates, marbles, quartzites and metavolcanic rocks. Basic volcanism in this region is not as important as in the

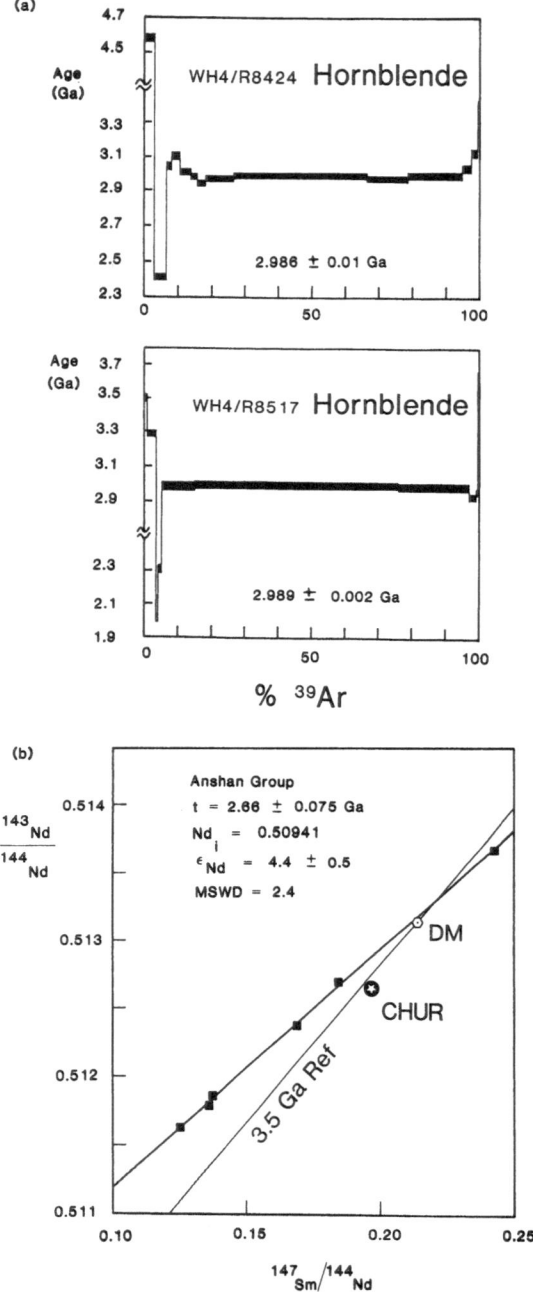

Figure 13.3 (a) $^{40}Ar/^{39}Ar$ plateau ages for hornblende separates from amphibolite of the Shipenzi Formation of the Anshan Group, northeastern China (data from Wang, S.S. *et al.*, 1987). (b) Sm–Nd isochron diagram for metabasic rocks from the Anshan Group (data from Jahn and Ernst, 1989), compared to a 3.5 Ga isochron reference line.

Qingyuan greenstone belt, and often occurs in close association with the voluminous BIFs.

Archaean basic volcanics are also preserved in other parts of northeastern China (Ernst *et al.*, 1988) in similar associations: mafic amphibolites as enclaves in or interlayered with voluminous tonalitic gneisses and BIFs. The temporal relationship between the tonalitic gneisses and amphibolites and BIFs is debatable. A systematic geochronological study is currently under way in Rennes and Beijing (Chinese Academy of Geological Sciences). Presently available data show that the tonalitic gneisses of the Tiejiashan type locality were intruded about 2.83–2.86 Ga ago, based on whole-rock Rb–Sr and Pb–Pb analyses (Zhong, 1984). However, the near concordant ages of about 2.98 Ga obtained by ion microprobe analyses on individual zircon grains suggest that these gneisses were derived from protoliths about 3 Ga old (Zhong, 1984).

The mineral assemblages of the Anshan Group suggest late Archaean lower granulite facies metamorphism (600–800°C; 0.6–0.7 GPa) and progressively higher crustal levels for early and mid-Proterozoic amphibolite facies (400–600°C; 0.4–0.6 GPa) and greenschist facies (350–450°C; 0.2–0.3 GPa) events (Ernst *et al.*, 1988).

13.2.4 Inner Mongolia (Yingshan region): the Jining and Wulashan Groups

The Yingshan region of Inner Mongolia has extensive exposures of Archaean rocks (the Jining and Wulashan Groups) which constitute the northern border of the Sino-Korean Craton (Figure 13.1). Mineral paragenesis data indicate a considerable variation in metamorphic grade, ranging from lower amphibolite to granulite facies (Dong *et al.*, 1986). According to Cheng (1986) and Shen *et al.* (1987), the Archaean sequence shows the following general upward succession: (1) basic pyroxene granulites, pyroxenites and charnockitic gneisses, (2) amphibolites, (3) garnet–biotite–sillimanite gneisses, (4) leptynites and tonalitic gneisses, and (5) marbles and graphitic beds. Their protoliths include basic to acid volcanic rocks and pelitic, semipelitic and carbonate sedimentary rocks. Judging from the resemblance in lithological characters and mineral parageneses with the rocks of the Qianxi Group in eastern Hebei, the basic volcanism in the Yingshan region might also have been initiated as early as 3.5 Ga. Unfortunately, the few available radiometric data (U–Pb zircon and Rb–Sr whole-rock isochron) on granulite facies rocks from the type locality near Jining have not produced any age greater than 2.5 Ga (Shen *et al.*, 1987).

13.2.5 Shanxi (Taihang–Wutai–Luliang region): the Fuping and Wutai Groups

The Archaean rocks of this region, southwest of Beijing (Figure 13.1), comprise a variety of amphibolite facies gneisses with variable proportions of dolomitic marbles, amphibolites and leptynites. Two-pyroxene basic granulites and hypersthene-bearing BIF also occur in the lower parts of the Fuping Group. Amphibolites and mafic granulites are interpreted as metavolcanic sequences and in places exhibit well-preserved amygdales and pillow structures. The unconformably overlying Wutai Group is thought to be the base of the Proterozoic succession and consists of amphibolite facies mica schists, carbonates and amphibolites in its lower part, and greenschist facies clastic sediments and volcanics in the upper part (Ma *et al.*, 1957; Liu *et al.*, 1985). Metamorphic grades of the various units in the Fuping Group are closely related to stratigraphic positions. They range from granulite facies in the lower part to amphibolite facies in the middle and upper parts. The stratigraphically higher Wutai Group is characterised by upper amphibolite facies assemblages in the lower part and greenschist facies parageneses in the upper part.

A recent U–Pb geochronological study has conclusively demonstrated that the Fuping and Wutai Groups were formed during the late Archaean (2.8–2.5 Ga; Liu *et al.*, 1985). A minimum age for the Fuping Group and maximum age for the Wutai Group have been derived from zircon measurements on the Lanzhishan granite (2.56 Ga) which intrudes the Fuping Group and is unconformably overlain by the Wutai Group. Zircons from a low-grade metamorphosed keratophyric lava in the upper part of the Wutai Group suggest an eruption age of 2.52 ± 0.017 Ga. Thus, the Wutai Group was deposited in a time interval of about 40 Ma (2.56–2.52 Ga).

13.2.6 *Henan Province: the Dengfeng and Taihua Groups*

The Archaean rocks of the Dengfeng Group have the essential features of a granite-greenstone terrane (Zhang *et al.*, 1985). The general succession of the greenstone belt can be divided into three lithological units, comprising (1) a lowermost ultramafic complex, including some rocks of komatiitic affinity; (2) a bimodal suite of tholeiitic and felsic metavolcanics, in which pillow structures are locally preserved and andesitic volcanics are conspicuously lacking; and (3) an upper metasedimentary suite of molasse and turbidite formations, with minor mafic volcanic intercalations. The associated gneisses are mainly of TTG composition with late K-rich granitic intrusions. Both the supracrustal and plutonic rocks are intensely and penetratively deformed and have been metamorphosed to amphibolite facies (600°C, 0.6–0.7 GPa; Zhang *et al.*, 1985).

In the south, the rocks of the Dengfeng Group are enclosed by the Taihua Group rocks, which consist primarily of tonalitic gneisses and tectonically interbedded supracrustals including metatholeiites, metapelites and lenses of komatiitic(?) amphibolites. The Taihua Group was metamorphosed to upper amphibolite to granulite facies. The peak metamorphic conditions have been estimated at 750–850°C and 0.8–1.0 GPa, suggesting that the Taihua Group corresponds to a higher grade terrane than the Dengfeng Group (Zhang *et al.*, 1985).

Li *et al.* (1987) performed Sm–Nd analyses on six amphibolites and two acid metavolcanics from the Dengfeng greenstone belt and they obtained an isochron age of 2.51 ± 0.03 Ga with $\varepsilon_{Nd}(t) = 2.2 \pm 0.8$. Kröner *et al.* (1988) carried out ion microprobe analyses on single zircons extracted from a Dengfeng Group rhyodacite and obtained a concordant U–Pb age also of 2.51 ± 0.02 Ga. Both the Sm–Nd and U–Pb zircon ages have been interpreted as the crystallisation ages of the original greenstone volcanics. Moreover, a monzonite from the Shipaihe pluton (Dengfeng County) which intrudes the Dengfeng Group has been dated at 2.52 Ga by the conventional U–Pb zircon method (Wang, Z.J. *et al.*, 1987). Kröner *et al.* (1988) dated zircons from a tonalitic gneiss of the Taihua high-grade gneiss terrane using the whole-grain evaporation technique of Kober (1986) and obtained two $^{207}Pb/^{206}Pb$ ages of 2.84 and 2.81 Ga, clearly older than the Dengfeng greenstone sequence. Thus, it appears that the Dengfeng Group greenstone belt was deposited on an older continental crust probably represented by the Taihua Group.

13.3 Geochemistry and petrogenesis

In this section a broad geochemical characterisation of the Archaean basic rocks of China will be attempted and general petrogenetic implications will be discussed. No detailed petrogenetic modelling will be given for the volcanic sequences or rock associations because of the limitation of space. More comprehensive geochemical treatment on the Qianxi and Taishan Groups are presented elsewhere (Jahn and Zhang, 1984; Jahn et al., 1987, 1988).

13.3.1 Bulk-rock chemistry and trace element differentiation

Because metamorphism is so widespread and intense in Archaean amphibolites of China, classification schemes based on present mineral assemblages are evidently impractical. Moreover, because of the probable redistribution of some 'mobile' elements (such as K, Rb, Ba, U and Th) during metamorphism, schemes based on normative mineralogy and alkali contents, and plots such as the commonly used AFM diagram, must be treated with caution. This is particularly true for granulite facies metamorphic rocks.

The bulk-rock chemistry of metabasic rocks from five selected assemblages – the 3.5 Ga old amphibolites from the Caozhuang area, the Qianxi basic granulites and the Zunha amphibolites of the Qianxi Group, the amphibolites of the Taishan Complex and the basaltic to komatiitic amphibolites of northeastern China – is presented in Table 13.2 and represented on an Al:Fe + Ti:Mg cation plot (Jensen, 1976) in Figure 13.4. High-Mg and high-Fe basalts seem to constitute the majority of analysed samples. The widespread BIF may be consequential of the abundant high-Fe basalts in China. Some of the samples appear to be basaltic komatiites on the basis of their bulk-rock major element proportions, but their precise identification requires a complete set of chemical data, particularly some critical immobile elemental ratios.

Figure 13.5a shows that the TiO_2 contents (0.2–1.8%) do not vary systematically with MgO (18–3.5%) in any given rock assemblage or for any given region. Clearly, complex petrogenetic processes, such as the involvement of heterogeneous mantle source regions, fractional crystallisation, and possible metamorphic/metasomatic effects must have contributed to the wide scatter of data points. In the Taishan Group two distinct types of amphibolites have been recognised in terms of their REE patterns (Figure 13.6) and other element contents (Jahn et al., 1988). Consequently, more than one liquid line of descent, or at least two trends of Ti–Mg evolution might be expected.

The Ti-Zr plot (Figure 13.5b) shows that the majority of data points deviate significantly from the Ti/Zr ratio of chondrites (= 110) or the primitive Earth's mantle as established from the compositions of komatiites (Nesbitt and Sun, 1976). Ti depletion relative to Zr appears to be a general characteristic of the Archaean basalts of China. The present Ti–Zr data are comparable with those of modern island arc basalts (Pearce, 1982).

The data for most regions show good correlation of CaO and Al_2O_3 with respect to TiO_2 (Figure 13.7). Ratios of CaO/TiO_2 and Al_2O_3/TiO_2 higher than chondritic values (c. 18 and 22, respectively) are rather common and they resemble those in certain low-Ti basic assemblages formed in island arc tectonic settings (Sun and Nesbitt, 1978). Such high CaO/TiO_2 and Al_2O_3/TiO_2 ratios in many samples (Figure 13.7) and the low Ti/Zr ratios argue for an arc-type signature and suggest the involvement of

Table 13.2   Average compositions of Chinese Archaean metabasic rocks

Age (Ga): Region:	3.5 Caozhuang amphibolite		2.5–2.7 Qianxi basic granulite		2.5–2.7 Zunha amphibolite		2.7–2.75 Taishan amphibolite		2.7–3.0 NE China amphibolite		Total	
No. of samples*:	43		52		21		38		14		168	
	Mean	σ	Mean	σ	Mean	σ	Mean	σ	Mean	σ	Mean	σ
%												
SiO$_2$	49.89	1.82	49.50	2.42	48.38	1.88	48.45	2.97	48.33	2.33	49.01	2.48
Al$_2$O$_3$	13.58	1.89	14.11	3.36	13.67	1.87	13.32	2.76	8.95	3.41	13.31	3.04
Fe$_2$O$_3$*	12.76	2.21	12.66	2.65	14.01	1.99	12.58	2.59	12.18	2.76	12.80	2.54
MnO	0.22	0.04	0.18	0.05	0.18	0.05	0.20	0.05	0.19	0.03	0.19	0.05
MgO	7.38	1.66	8.09	3.28	8.43	3.56	8.61	3.61	17.61	6.47	8.79	4.33
CaO	10.37	0.90	9.87	2.07	9.74	1.84	10.21	1.38	9.40	2.70	9.96	1.68
Na$_2$O	2.31	0.89	2.82	1.00	2.72	0.77	2.36	0.69	1.12	0.77	2.44	0.97
K$_2$O	0.92	0.30	1.01	0.69	0.90	0.42	1.10	0.55	0.69	1.06	0.97	0.60
TiO$_2$	0.83	0.28	0.83	0.38	0.86	0.32	1.04	0.40	0.43	0.30	0.85	0.38
P$_2$O$_5$	0.13	0.13	0.17	0.14	0.14	0.11	0.17	0.13	0.26	0.39	0.16	0.17
LOI	0.97	0.51	1.02	0.62	1.63	0.94	1.99	1.13	1.37	0.23	1.39	0.92
ppm												
V	237	97	199	90	—	—	289	77	—	—	232	94
Cr	335	301	340	373	—	—	356	392	1084	651	484	508
Co	54	50	75	111	—	—	50	11	76	34	62	66
Ni	165	86	134	116	—	—	131	92	594	525	229	295
Rb	25	16	17	21	—	—	36	34	27	14	29	40
Sr	149	60	355	310	—	—	260	189	196	162	233	214
Y	23	7	23	11	—	—	31	13	—	—	23	9
Zr	76	28	66	27	—	—	75	27	72	22	73	27
Nb	5.5	2.3	4.4	2.0	—	—	7.7	2.9	—	—	5.0	2.2
Ba	134	69	242	145	—	—	223	155	—	—	196	133

*Due to the lack of availability of data, the mean values for trace elements are usually based on fewer samples than the major element statistics. All Fe presented as Fe$_2$O$_3$.

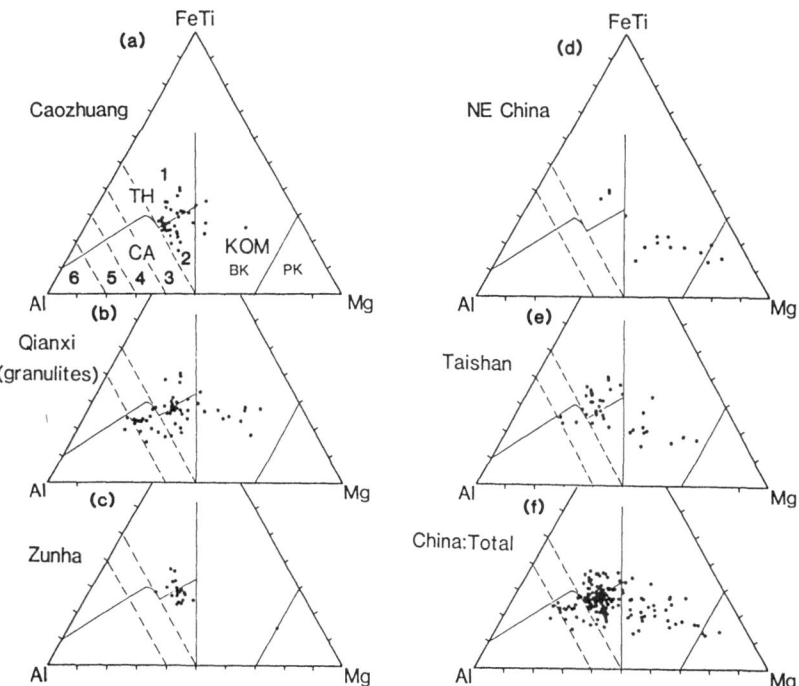

Figure 13.4 Al:(Fe, Ti):Mg cation plot and classification scheme (Jensen, 1976) for Archaean metabasic rocks of China. (a) 3.5 Ga Caozhuang amphibolites; (b) 2.5 Ga Qianxi basic granulites; (c) 2.7–2.5 Ga Zunha amphibolites (a to c from the Qianxi Group); (d) 3.0–2.6 Ga Anshan Group amphibolites of komatiitic to basaltic compositions; (e) 2.7 Ga Taishan Group amphibolites; (f) all data points (168) from the above groups. Data from Jahn and Zhang (1984), Sun, D.Z. *et al.* (1984), Wang, K.Y. *et al.* (1985, 1988), Wang, R.M. *et al.* (1985), Zhai *et al.* (1985), Qian *et al.* (1986), Jahn *et al.* (1987, 1988), Ernst *et al.* (1988) and author's unpublished data. Fields of komatiitic (KOM; BK = basaltic komatiite; PK = peridotitic komatiite), tholeiitic (TH) and calc-alkaline (CA) suites are shown in (a). Subdivisions in the TH and CA fields are: 1, high-Fe basalts; 2, high-Mg basalts; 3, basalts; 4, andesites; 5, dacites; and 6, rhyolites.

subduction zone related processes in the genesis of many Chinese Archaean basalts.

Ni and Cr variations with MgO are shown in Figure 13.8. Such sympathetic variations are typical of Archaean greenstone belts and of metabasic rocks in high-grade terranes (Nesbitt and Sun, 1976; Arndt *et al.*, 1977; Hawkesworth and O'Nions, 1977; Nisbet *et al.*, 1977; Jahn *et al.*, 1980; Hall, 1982). Ni–Mg variation is dominantly controlled by olivine fractionation and Cr–Mg by clinopyroxene and chromite fractionation. Amphibolites of the Taishan Group have the largest variations in both Ni and Cr, and some of the very low values may favour their derivation from mantle sources in an arc-type tectonic setting.

Many basic granulites of the Qianxi Group show very high K/Rb values (1000–2000) following the well-known depleted granulite trend, and very low Rb/Sr ratios (< 0.03) similar to those found in N-MORB (Figure 13.9). These features, however, probably do not reflect the original compositions, but indicate alkali depletion and preferential Rb loss (coupled with U and Th losses) during granulite facies metamorphism. In this respect the Qianxi granulites are comparable with the granulites of the Lewisian

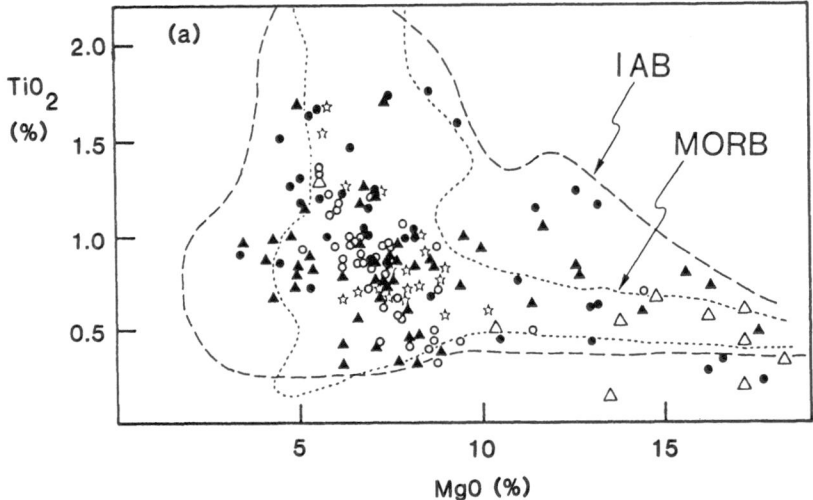

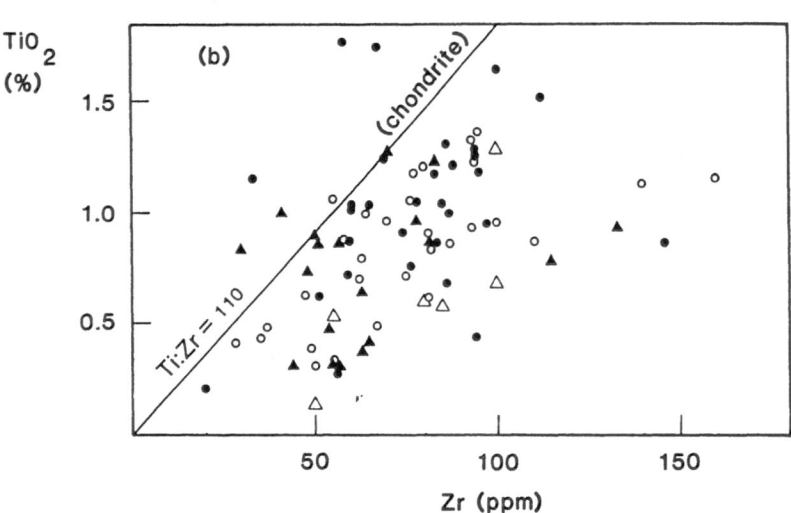

Figure 13.5 Plots of $TiO_2(\%)$ versus $MgO(\%)$ (a) and $Zr(ppm)$ (b) for Archaean metabasic rocks from China. Open circles: Caozhuang amphibolites; filled circles: Taishan amphibolites; open triangles: Anshan Group amphibolites; filled triangles: Qianxi basic granulites; asterisks: Zunha amphibolites (no Zr data). Note that nearly all data points fall within the IAB envelope in (a) and have lower Ti/Zr values than that of chondrite or mantle (= 110; Nesbitt and Sun, 1976), clearly suggesting a depletion of Ti relative to Zr.

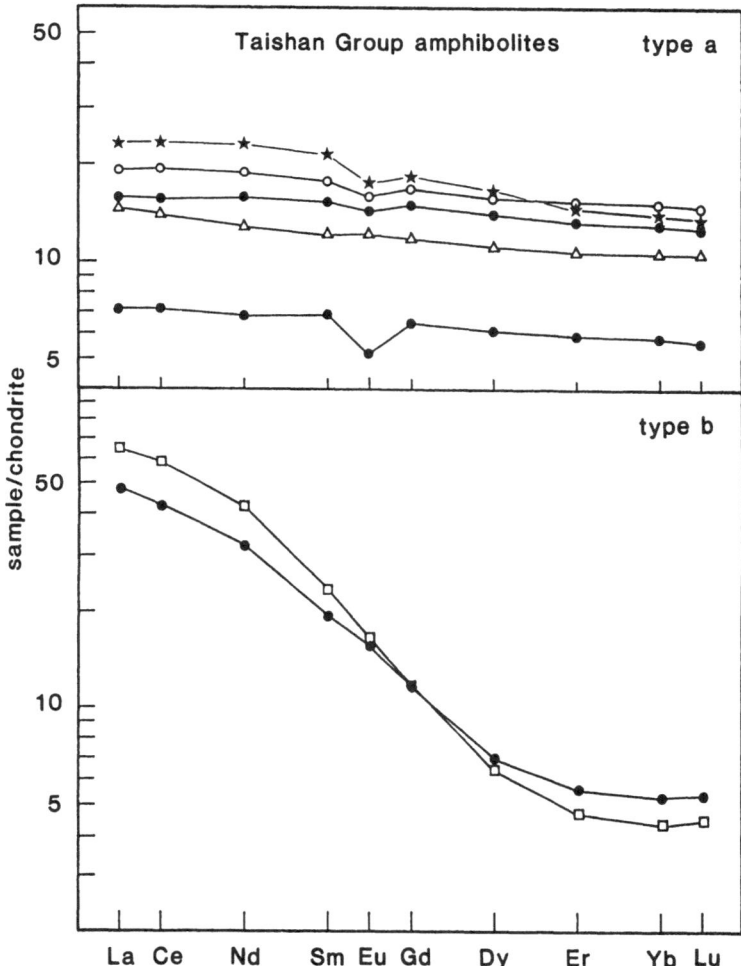

Figure 13.6   Contrasting chondrite-normalised REE patterns for the two types of amphibolites of the Taishan Group showing their derivation from distinct sources (data from Jahn *et al.*, 1988).

complex of northwestern Scotland (Tarney and Windley, 1977; Weaver and Tarney, 1980, 1981; Jahn and Zhang, 1984). Non-depleted granulites also occur in the Qianxi Group. The early Archaean amphibolites of the Qianxi Group from the Caozhuang area and the late Archaean Taishan Group amphibolites show comparable ranges in Rb contents (6–100 ppm) and K/Rb (150–800) and Rb/Sr ratios (0.03–0.3).

13.3.2   *REE geochemistry*

A substantial number of precise isotope dilution REE analyses on Chinese Archaean rocks have been published recently (Jahn and Zhang, 1984; Jahn *et al.*, 1987, 1988; Jahn and Ernst, 1989), and REE analyses by other methods are also available in recent

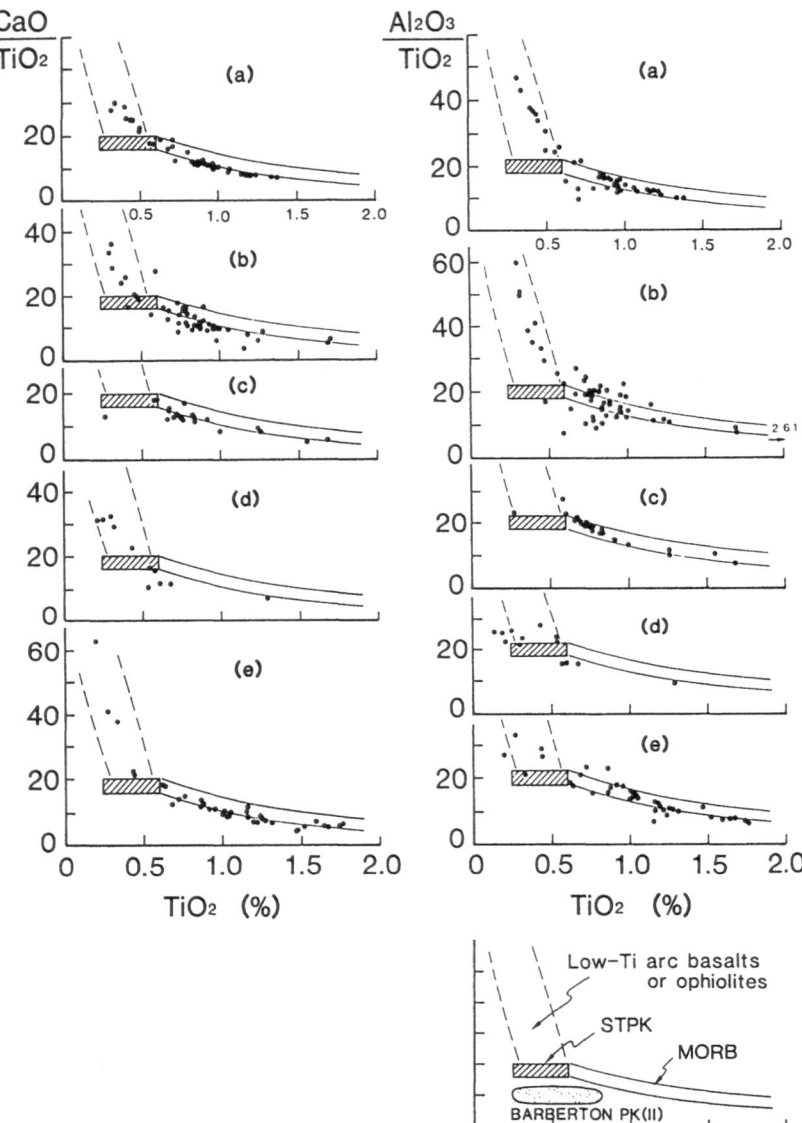

Figure 13.7 CaO/TiO_2 and Al_2O_3/TiO_2 versus TiO_2 plots for Archaean metabasic rocks of China. (a) Caozhuang amphibolites; (b) Qianxi basic granulites; (c) Zunha amphibolites; (d) Anshan amphibolites; (e) Taishan amphibolites. Reference fields shown in bottom right schematic diagram are from Sun and Nesbitt (1978). Stippled field represents Barberton-type Al-depleted group II komatiites. Note that (i) early and late Archaean amphibolites (a to e respectively) do not differ significantly, (ii) the majority of samples follow the MORB trend, (iii) few samples show PK-type values, and (iv) high CaO/TiO_2 and Al_2O_3/TiO_2 values from low TiO_2 abundances suggest magma genesis in a subduction zone environment. Note that the negative correlation designated as MORB trend may be interpreted as a consequence of fractional crystallisation of pyroxene and plagioclase in basaltic magma, but not necessarily representing a series of partial melting involving sequential entry of pyroxene and garnet in the melt as suggested by Sun and Nesbitt (1978).

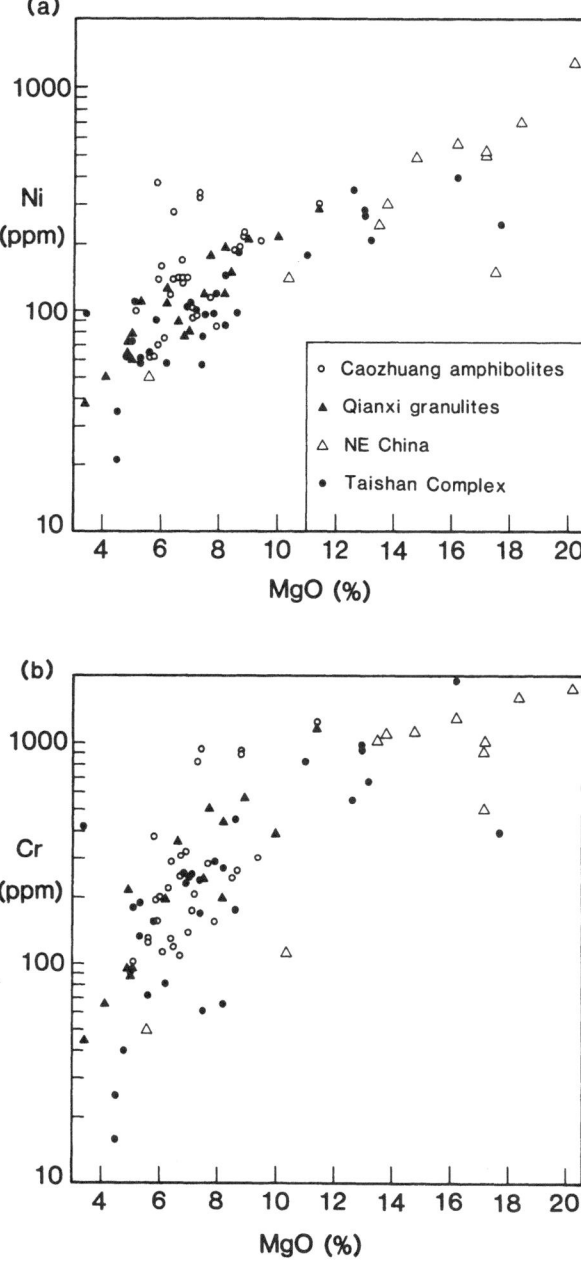

Figure 13.8   Plot of (a) Ni and (b) Cr versus MgO (%) for Archaean metabasic rocks of China (symbols as in Figure 13.5).

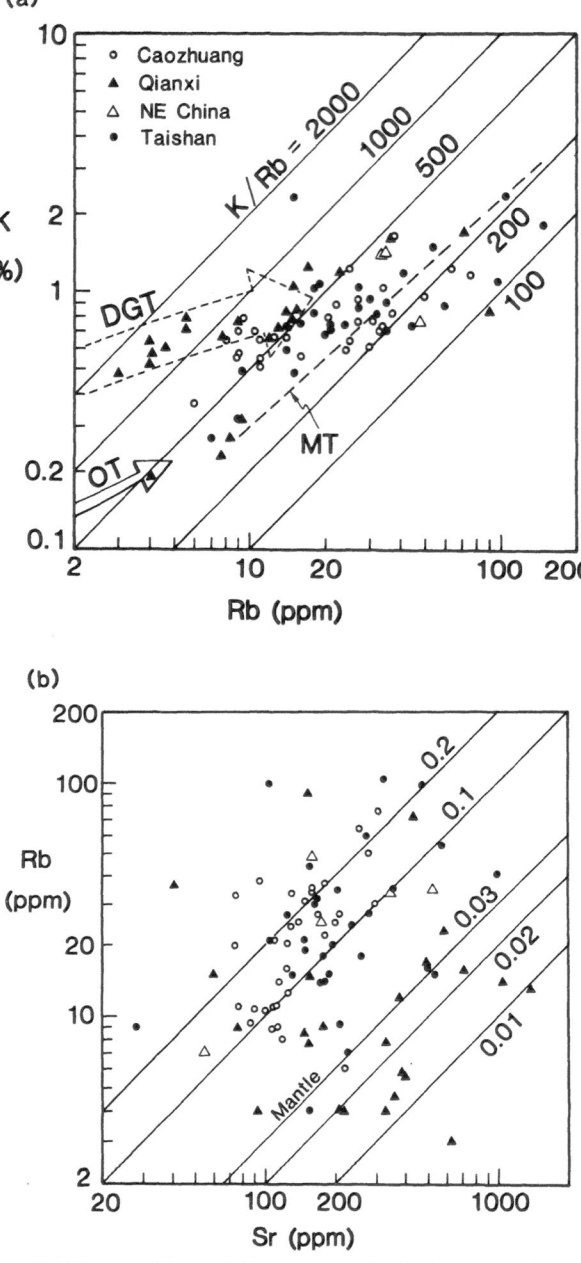

Figure 13.9 Plot of (a) K versus Rb, and (b) Rb versus Sr for Archaean metabasic rocks of China (symbols as in Figure 13.5). The bulk of the data do not follow trend of oceanic basalts (OT). However, the Qianxi metabasic rocks which are depleted in alkalis are shown to follow the depleted granulite trend (here designated as DGT) and extend into the field for non-depleted metabasic rocks. MT is the main trend of continental sialic rocks. Early and late Archaean amphibolites are indistinguishable with respect to these elements.

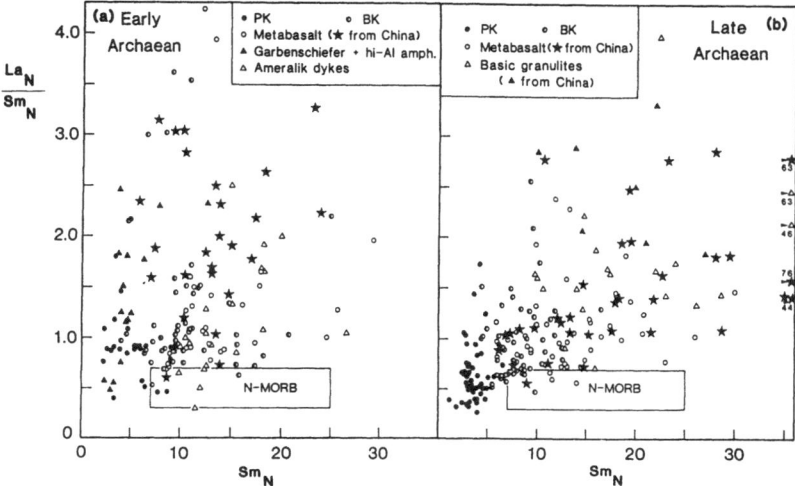

Figure 13.10   Plot of La_N/Sm_N versus Sm_N for (a) early Archaean and (b) late Archaean basic and ultrabasic rocks from China and elsewhere (data sources mentioned in the text). Field of N-MORB (characteristically LREE-depleted) is also shown. La_N/Sm_N represents degree of LREE fractionation and Sm_N indicates the general level of REE abundances. $La_N/Sm_N < 1.0$ suggests derivation from depleted mantle source. $La_N/Sm_N > 1.0$ indicates an enriched source, crustal contamination of the magma, or advanced fractional crystallisation. Most Caozhuang amphibolites and Qianxi granulites show LREE enrichment, which together with their high La/Nb ratios (Figure 13.11) indicates the involvement of crustal contamination or a crustal component in the source via subduction processes. Nd isotopic constraints require that crustal contaminants should have isotopic compositions not much different from that of the contemporaneous mantle, effectively limiting its age to around 3.5 Ga. The relatively low La_N/Sm_N values of many PK and BK and metabasaltic suites indicate the largely depleted nature of the late Archaean mantle. Differentiation of uncontaminated PK magmas derived from a depleted source would displace their data points to the right, into the field of MORB.

publications (Wang, R.M. *et al.*, 1985; Wang, K.Y. *et al.*, 1985, 1988; Zhai *et al.*, 1985; Zhang *et al.*, 1985; Ernst, 1989). Except for a few patterns which show LREE depletion, the majority of Chinese Archaean rocks are characterised by LREE enrichment ($La_N/Sm_N > 1$). Figure 13.10 illustrates some important geochemical consequences of the REE distributions of these and other Archaean basic rocks.

The La_N/Sm_N ratios are less than unity for many, and perhaps most Archaean komatiites and tholeiitic basalts (Figure 13.10). Some peridotitic komatiites (PK) and basaltic komatiites (BK) are as depleted as modern MORB, suggesting that Archaean mantle sources were depleted in nature. This evidence agrees with that deduced from Nd isotope studies (see section 13.4). However, a substantial number of both early and late Archaean basic and ultrabasic rocks have enriched REE patterns with La_N/Sm_N values of between 1 and 4. This is particularly true for the amphibolites and basic granulites of China. Most data for BK and tholeiites shown in Figure 13.10 are comparable with those of modern island arc basaltic rocks, being less depleted than MORB and many even enriched in LREE. This implies that many Archaean basic volcanic rocks formed by subduction zone related processes. However, this does not necessarily suggest a modern plate tectonic regime in the Archaean. Instead, it indicates that a pre-melting enrichment process was important in the generation of many

Archaean basic volcanic rocks. Such an enrichment process could have been achieved by some sort of mantle metasomatism, including recycling of continental crustal material through subduction processes.

13.3.3 *Nb anomalies*

Continental and oceanic island alkali basalts and associated rocks (nephelinites, lamprophyres, carbonatites, etc.) often possess positive Nb and Ta anomalies which can be clearly shown on a so-called 'spidergram' (MORB-normalised multi-element diagram). In contrast, many rock types of continental affinity (granitoids and sediments), island arc and continental margin volcanics, and some continental basalts show markedly negative Nb and Ta anomalies (Sun, 1980; Tarney *et al.*, 1982; Weaver and Tarney, 1981, 1983; Thompson *et al.*, 1983; Pearce, 1983; Sun and McDonough, 1989). As continental rocks have high La/Nb ratios and very low Nb/Nb* ratios (where Nb* is an interpolated value based on mantle elemental ratios), the involvement of continental material in magma generation processes would probably result in negative Nb anomalies or high La/Nb ratios in the basic rocks so formed. Except for a few from the Taishan Group, most Chinese Archaean amphibolites show high to very high La/Nb ratios (Figure 13.11), suggesting that their basaltic (and in some cases komatiitic) protoliths were produced by melting of mantle sources containing contributions from older continental material (source contamination hypothesis), or their parental magmas were contaminated by continental rocks en route to the surface (melt contamination hypothesis).

13.4 Implications for early continental evolution

13.4.1 *Nd-Sr isotopic constraints*

The available Nd isotopic data indicate that the Archaean basaltic magmas in China had positive $\varepsilon_{Nd}(t)$ values (+ 2.5 to + 4.4), suggesting their derivation from long-term depleted mantle sources (Figure 13.12). It has been shown earlier that the majority of the Archaean amphibolites and basic granulites are LREE-enriched (Figure 13.10). For such long-term depleted isotopic signatures to be preserved, the enrichment processes must have taken place in each case shortly before the melting event. Such a scenario seems to be true not only for the Chinese metabasalts, but also for similar Archaean basic volcanic rocks in other parts of the world. Moreover, Figure 13.12 demonstrates that the majority of Archaean igneous rocks appear to have been derived from heterogeneously depleted mantle sources, perhaps also involving various degrees of crustal contribution in their magma generation processes.

The oldest known rocks in China are the Qianxi Group amphibolites from the Caozhuang area, which are *c*. 3.5 Ga old and have $\varepsilon_{Nd}(t)$ value of + 3.0. Similarly, early Archaean rocks from the Isua supracrustal belt and the enveloping Amîtsoq gneisses of West Greenland, the Warrawoona Group in the Pilbara Block of Western Australia, and some Onverwacht Group volcanics (South Africa) also show $\varepsilon_{Nd}(t)$ values lying around the depleted mantle path (DM, Figure 13.12). Mantle depletion is believed to be related to the extraction of LILE-enriched continental material. If these early Archaean rocks were also derived from source regions characterised by strong LREE depletion such as that found in the modern N-MORB sources, those depletion events must have

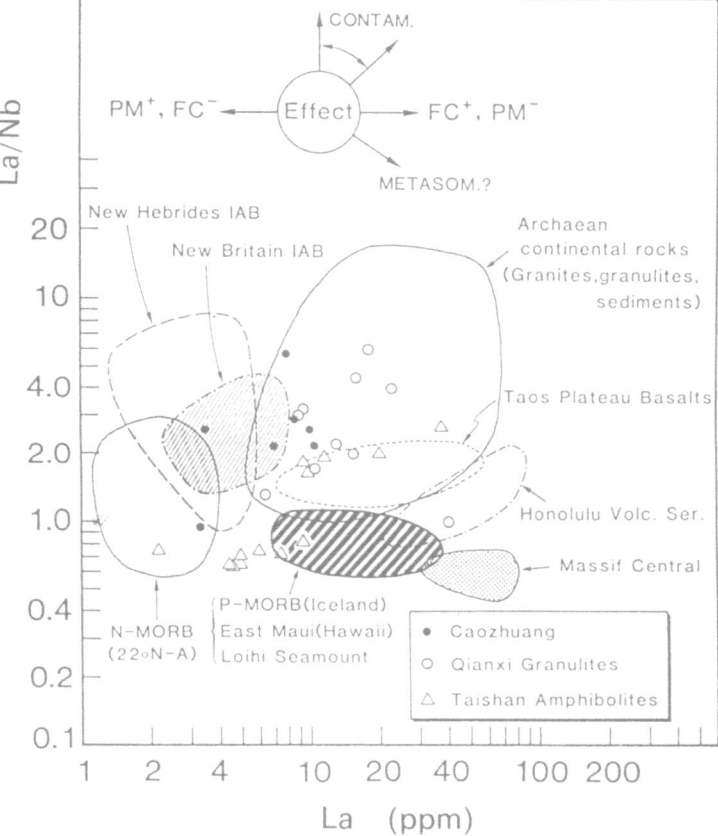

Figure 13.11   La/Nb versus La diagram for metabasic rocks of China compared to various other groups of basic rocks. The effects of mantle metasomatism, crustal contamination, increased or decreased partial melting (PM$^+$, PM$^-$) and higher or lower degrees of fractional crystallisation (FC$^+$, FC$^-$) are shown as vectors for comparison. OIB and alkali basalts from Massif Central do not appear to have been influenced by a continental crustal component, in contrast to IAB and the Taos plateau basalts. The data suggest variable but significant continental components in the generation of the Chinese Archaean basic rocks.

taken place around 4 Ga ago. That this type of depleted source may not have been uncommon in the early Archaean is exemplified by the geochemical nature of some komatiites with strong LREE depletion (Figure 13.10). However, if the mantle source was intrinsically less depleted or was possibly refertilised through the reinjection of continental material into a highly depleted source, then the LREE depletion could have occurred earlier. In any case, the Nd isotopic data suggest that sialic material must have formed in substantial volume prior to 4 Ga ago and possibly as early as 4.5 Ga ago, even though its survival has been minimal or non-existent because of the high rate of crust–mantle turnover.

Reliable Sr isotopic data are relatively meagre. However, the low initial ^{87}Sr/^{86}Sr

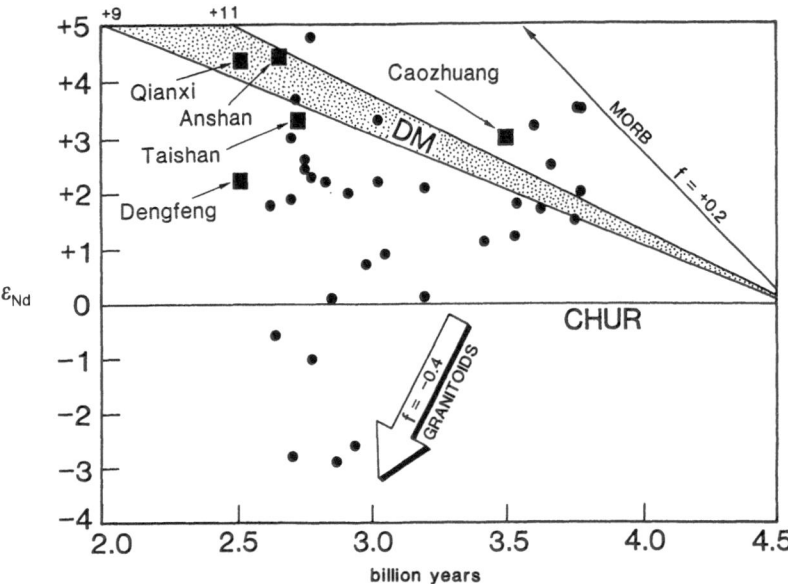

Figure 13.12 Nd isotopic evolution diagram for representative Chinese (squares) and other world Archaean rocks (dots; for details see Jahn *et al.*, 1987). Isotopic evolution trends in the depleted mantle (DM; present day $\varepsilon_{Nd} = +9$ to $+11$) and chondritic mantle (CHUR) and rates of increase in ε_{Nd} value (or Nd isotopic ratio) in the MORB-type sources ($f = +0.2$) and decrease in granitoids or continental crust ($f = -0.4$) are shown for comparison. Parameter f is an index of Sm/Nd fractionation and is defined as $((Sm/Nd)_{Sample}/(Sm/Nd)_{CHUR}) - 1$. The general spread of the data points strongly suggests that recycling of continental crust in depleted mantle has been significant since early Archaean times, in agreement with the conclusions reached from trace element geochemical arguments.

ratios for the Wangfushan Gneisses (0.7004–0.7006) and associated diorite group (0.7008) of the Taishan Complex provide supplementary evidence for continental growth by the addition of juvenile material ultimately separated from long-term depleted mantle sources (Jahn *et al.*, 1988)

13.4.2   *Amphibolite–TTG gneiss relationships*

In Archaean terranes, metabasic rocks are invariably found in close association with orthogneisses of TTG composition. Metabasic rocks may occur as weakly metamorphosed greenstone belt assemblages with original depositional structures (e.g. pillows) and they may also occur as enclaves or mappable units within high grade TTG gneisses. However, the temporal and genetic relationship between basic amphibolites and TTG gneisses has always been one of the most controversial issues in discussions about the early development of continental crust. It is well understood that gneiss complexes are largely made up of early intrusive TTG granitoids and late K-rich granites. Their emplacement and attendant metamorphism suggest crystallisation and recrystallisation at considerable crustal depths which imply the existence of an earlier cover series, perhaps volcanic or volcano-sedimentary sequence, while in many cases the deposition of the cover series would have required a pre-existing basement.

The temporal and stratigraphic relationships of the Chinese Archaean basic rocks with respect to their surrounding gneisses have not been clearly established in many cases. In the Caozhuang area, the 3.5 Ga old amphibolite enclaves of the Qianxi Group and the nearly contemporaneous grey gneisses (depleted mantle model age = 3.3 Ga; Sm–Nd isochron age = 3.5 Ga; Wang, K.Y. *et al*., 1988) have been interpreted as an early Archaean bimodal suite, which has been regarded as a series of giant enclaves engulfed in late Archaean granitoids (Jahn *et al*., 1987). Amphibolites from the Qingyuan greenstone belt in northeastern China have been dated at 2.99 Ga (Ar–Ar), whereas a granodiorite gneiss (Tiejiashan gneisses) gives a slightly younger age of 2.86 Ga (Rb–Sr and Pb–Pb), but also contains an inherited 2.97 Ga zircon component (Zhong, 1984). Within the Taishan Group, the oldest Wangfushan gneisses are made up of amphibolites and grey gneisses of about 2.7–2.75 Ga. With the exception of the Taihua Group high-grade gneiss terrane in Henan where the TTG gneisses appear to have formed a basement (*c*. 2.82 Ga) to the Dengfeng and Taihua Group supracrustal rocks (*c*. 2.5 Ga; Kröner *et al*., 1988), basic rocks generally appear to have developed either nearly contemporaneously with or prior to the emplacement of grey gneisses, judging from their enclave–host relationships. It is understood, however, that some enclave–host relationships are ambiguous. Basic enclaves may be derived from either boudined dykes (younger) or disrupted fragments of older mafic crust. Nevertheless, geochemical arguments strongly suggest that the TTG gneisses are probably derived by the remelting of pre-existing mafic rocks or greywackes (Arth and Hanson, 1975; Condie and Hunter, 1976; Glikson, 1979; Jahn *et al*., 1981, 1984; Martin *et al*., 1983; Glikson and Jahn, 1985). This by no means suggests that the observed bimodal suites must have had a direct genetic relationship. While the amphibolites of a region may not have been the source of the igneous precursors of the surrounding grey gneisses, the existence of the latter requires a pre-existing mafic crust, irrespective of the observed age relationship between exposed amphibolites and grey gneisses. In general, continental evolution starts with basic magmatism and formation of mafic crust, and is followed by subduction and partial melting leading to the production of TTG magmas. Commonly, the assemblage of amphibolite and grey gneisses is intruded by late K-rich granites, as witnessed in all Archaean terranes. Geochemical and isotopic arguments indicate that the K-rich granites were often produced by the remelting of earlier basement TTG gneisses. Excellent examples can be found in Finland (Martin and Quérré, 1981; Martin *et al*., 1984; see Chapter 12) and in the Taishan Group in China (Jahn *et al*., 1988).

13.5 Conclusions

1. Archaean basic magmatism in China started about 3.5 Ga ago and seemed to have culminated in the late Archaean (2.7–2.5 Ga).
2. Most basic and associated supracrustal rocks in China have undergone amphibolite or granulite facies metamorphism.
3. Nd isotopic data suggest that most early Precambrian basic magmas of both China and elsewhere were derived from long-term depleted mantle sources.
4. In view of the geochemical diversity of modern volcanic rocks formed in a single tectonic setting, for example at oceanic spreading centres, as a result of the interplay of asthenospheric and deep mantle plume components, precise assignment of Archaean palaeotectonic regimes based on geochemical characteristics is

considered to be tenuous (see Chapter 3). Consequently, only general implications are presented. The overall chemical data argue for the derivation of many of the Chinese Archaean basic magmas in the equivalent of a subduction zone environment. The involvement of sialic components either in the source regions or as contaminants of mantle-derived liquids is evident from trace element considerations.

Acknowledgements

Critical comments by G. Ernst, P. Hall and S.S. Sun are most appreciated. The research at Rennes on Archaean crustal evolution in China has been supported by the INSU-CNRS of France through the ATP and DBT projects. This is a contribution to IGCP Project 280: The oldest rocks on Earth.

14 Early Precambrian basic rocks of Australia

A.C. PURVIS

14.1 Introduction

Australia has many geological similarities with the other fragments of the former supercontinent Gondwanaland (India, Africa, South America, Antarctica), and the Archaean of Australia is most similar to that of southern Africa (Chapter 16). Parts of the Yilgarn Block (Figure 14.1) contain greenstone belts unconformably overlying older gneisses, and to this extent are similar to Peninsular India (Chapter 15). The recent discovery of Archaean high-grade sapphirine + quartz assemblages in South Australia points to similarities with the Napier Complex in Antarctica (Oliver and Purvis, 1986; Harley and Black, 1987). Most of the early Proterozoic rocks of Australia occur in folded, deformed 'mobile belts' (Figure 14.1) (e.g. Etheridge *et al.*, 1987), which appear to be a world-wide phenomenon. Although some Archaean rocks also occur in mobile belts, most of the Archaean mafic rocks of Australia occur within granite-greenstone terranes, known in Australia as 'blocks'. A considerable thickness of Archaean basalt also occurs in the gently folded Hamersley Basin (Pidgeon, 1984; Blake and Groves, 1987). Post-tectonic suites of dykes and sills occur commonly both in the mobile belts and in the Archaean blocks, and extensive sills occur in the epicratonic Kimberley Basin (Bofinger, 1967).

14.2 Distribution and stratigraphy

14.2.1 Archaean rocks

The oldest known rocks in Australia are the 3.8 Ga gneisses at Mt Narryer in the northern part of the Western Gneiss Belt of the Yilgarn Block (Figure 14.2), which include fragments of a gabbro–anorthosite complex (the Manfred Complex) (Niewland and Compston, 1981; de Laeter *et al.*, 1981). However, the oldest well-studied mafic magmatic suites are those in the Pilbara Block (Figure 14.1), which consists of domal areas of granitoids and gneiss–migmatite–amphibolite complexes (e.g Bickle *et al.*, 1985) and intervening synclinal keels occupied by metavolcanic and metasedimentary rocks (Gee, 1979; Hickman, 1981, 1983).

Two major sequences in the Pilbara Block contain mafic magmatic rocks (Barley and Bickle, 1982; Sun *et al.*, 1989). One is the 3.55 Ga Warrawoona Group, which is similar in age and lithology to the greenstone belts in the Kaapvaal Craton of South Africa. The other is in the 2.85 Ga Negri volcanics (Barley, 1987) containing siliceous high-Mg basalts (SHMB) and post-dating all but the youngest granitoids in the area. The latter sequence represents a superposed greenstone sequence. The Warrawoona Group has Al-depleted (Barberton-type) komatiites and tholeiites, and thick,

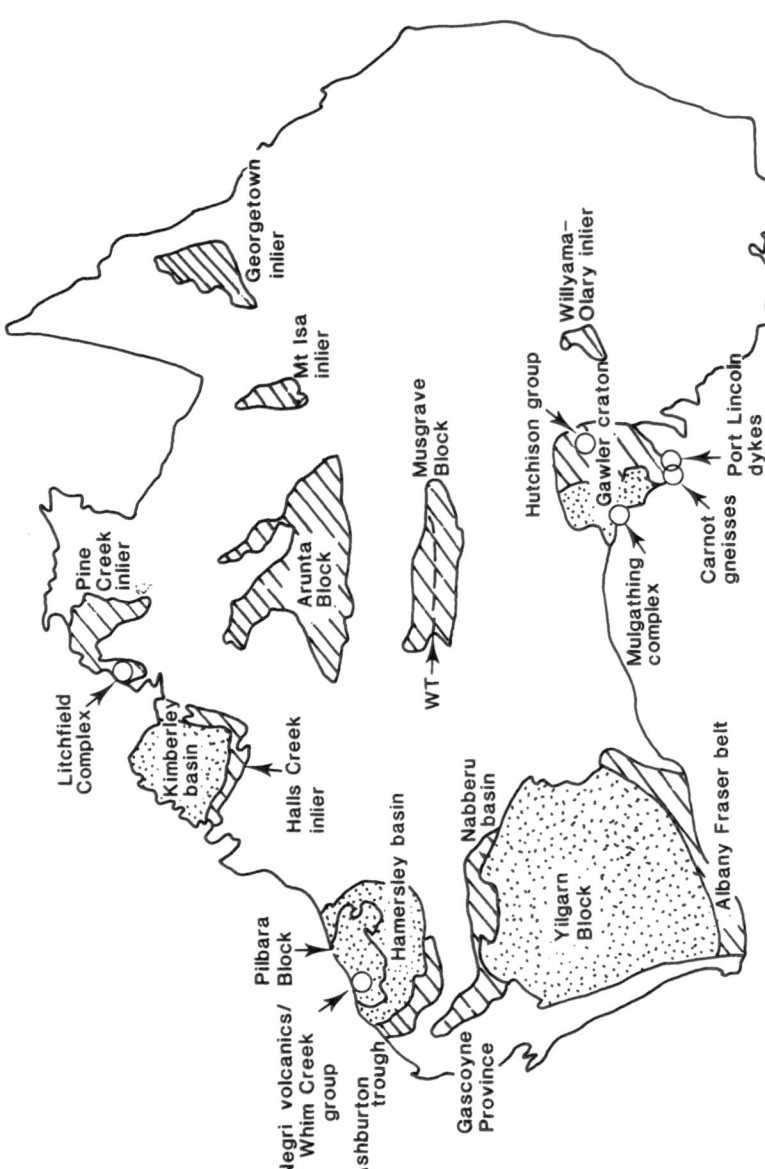

Figure 14.1 Sketch map of Australia showing the major Archaean (stippled) and early Proterozoic (line shading) terranes. WT: Woodward thrust collision zone.

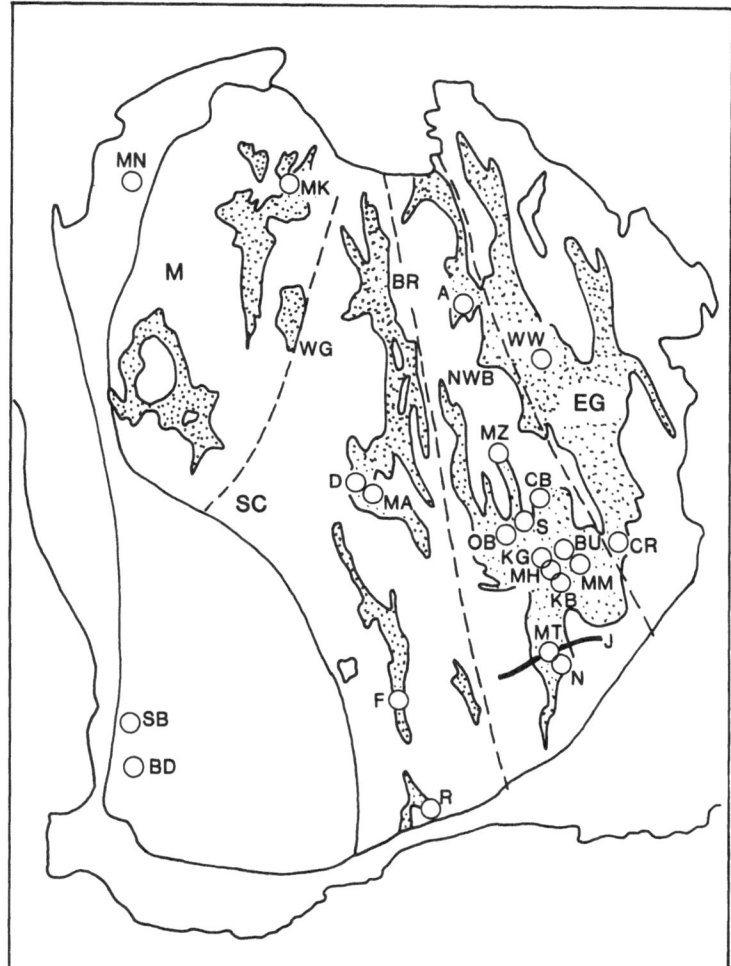

Figure 14.2   Sketch map of the Yilgarn Block of Western Australia showing the Murchison (M), Southern Cross (SC) and Eastern Goldfields (EG) provinces and the major localities and complexes mentioned in the text.

Key (in alphabetical order): A: Agnew; BD: Boddington; BR: Booylgoo Range; BU: Bulong; CB: Carr Boyd Rocks; CR: Cowarna Rocks; D: Diemals; F: Forrestania; J: Jimberlana intrusion; KB: Kambalda; KG: Kalgoorlie; MA: Marda complex; MH: Mt Hunt; MK: Meekatharra; MM: Mt Monger; MN: Mt Narryer; MT: Mt Thirsty; MZ: Menzies; N: Norseman; NWB: Norseman-Wiluna belt; OB: Ora Banda; R: Ravensthorpe; S: Scotia; SB: Saddleback; WG: Windimurra gabbro; WW: Welcome Well.

widespread banded iron-formation (BIF) (Hickman, 1983), indicating predominantly shallow water conditions (Barley, 1981). Flows and sills are abundant in both the Warrawoona Group and the Negri volcanics. Peridotitic komatiites have been described from the Warrawoona Group at Ruth Well (Nisbet and Chinner, 1981). The komatiitic sills commonly have chilled komatiitic basalt margins (Mathison and Marshall, 1981; Donaldson, 1974). Metamorphism reached amphibolite facies and is marked by rare occurrences of Cr-bearing kyanite (Purvis, 1983), and metasomatism, especially by CO_2-rich fluids, is widespread.

The Yilgarn Block comprises a number of distinct provinces with different ages, stratigraphies and magma types (Compston and Arriens, 1968; Gee, 1975; Gee et al., 1981; Fletcher et al., 1984), of which the oldest is the Western Gneiss Belt (Figure 14.2). The southwestern part of this province contains a 3.3–3.1 Ga gneiss–granulite–migmatite terrane (Niewland and Compston, 1981; McCulloch et al., 1983a, b) incorporating both mafic and ultramafic granulites (Wilson, 1978), metasedimentary belts and superposed younger (2.67–2.65 Ga), less strongly metamorphosed greenstone belts, notably the Saddleback Belt which contains the Boddington gold deposit (Wilde and Pidgeon, 1986; Groves et al., 1987a).

The Southern Cross and Murchison Provinces contain 3.0 Ga supracrustal sequences dominated by tholeiitic basalts (Fletcher et al., 1984), although komatiites are well developed in the Southern Cross Province (Nesbitt et al., 1979, 1982). A large layered tholeiitic body, the Windimurra gabbro (Ward, 1975; Ahmad and de Laeter, 1982; Parks and Hill, 1986), occurs close to the boundary between these two provinces (Figure 14.2) and has layers rich in vanadiferous magnetite. Thick and extensive BIF units are abundant, as in the Pilbara Block, and there are also felsic volcanics and associated detrital sediments. Komatiites at Forrestania and Ravensthorpe are of Barberton type (Al-depleted) (Nesbitt et al., 1979; Purvis, 1980), whereas those in the Diemals area have almost chondritic Al_2O_3/TiO_2 ratios (Nesbitt et al., 1982). Flows and sills of both komatiitic and tholeiitic parentage are common in these areas, but komatiitic basalts are rare. A single cycle, with a komatiite–tholeiite unit overlain by a felsic volcano-sedimentary unit, appears to be most common (Binns et al., 1982), but two cycles occur at Meekatharra (Hallberg et al., 1976b). The greenstones at Diemals give a Nd–Sm age of 3.05 Ga (Fletcher et al., 1984) but the samples used in calculating this age include calc-alkaline volcanics from the Marda area, which give a separate Rb–Sr age of 2.63 Ga (Hallberg et al., 1976a).

The Eastern Goldfields Province has younger (2.7–2.8 Ga; Chauvel et al., 1985), larger and more complex supracrustal belts than those in the Southern Cross and Murchison Provinces (Hallberg and Williams, 1972; Gemuts and Theron, 1975; Naldrett and Turner, 1976; Gresham and Loftus-Hills, 1981; Hallberg, 1986). These include repeated cycles of basal tholeiite–komatiite–SHMB units overlain by felsic volcanics and sediments. BIF occurs only in the southwestern and northeastern extremities of the Province, but cherty to dolomitic sediments, commonly rich in iron sulphides and/or graphite, are abundant within the tholeiite–komatiite–SHMB units, particularly in the Norseman–Wiluna Belt (Groves et al., 1987b). Giles and Hallberg (1982) and Hallberg and Giles (1986) have mapped calc-alkaline volcanics, including some LILE-enriched basalts and basaltic andesites.

The youngest supracrustal sequences occur in narrow grabens and are mainly polymict conglomerates which post-date the bulk of the granites in the area (Durney, 1972; Marston and Travis, 1976). As in Pilbara and Western Yilgarn, flows and sills are well developed, but pyroclastic rocks of all types (komatiitic, tholeiitic, high-Mg basalt) are more common than in the older sequences (Stolz and Nesbitt, 1981). Metamorphism ranges from prehnite–pumpellyite to lower granulite facies (Binns et al., 1976). Rare occurrences of Cr-bearing kyanite have been noted, but andalusite and sillimanite are the principal alumino-silicates (Purvis, 1983; Binns et al., 1976). CO_2-metasomatism is a common indication of gold mineralisation (Groves et al., 1987a; Cameron, 1988; see Chapter 8).

The basal Hamersley Basin units (Figure 14.1) are of Archaean age and represent the rift and sag stages of an extensive stable epicratonic basin (Hickman, 1983; Blake and

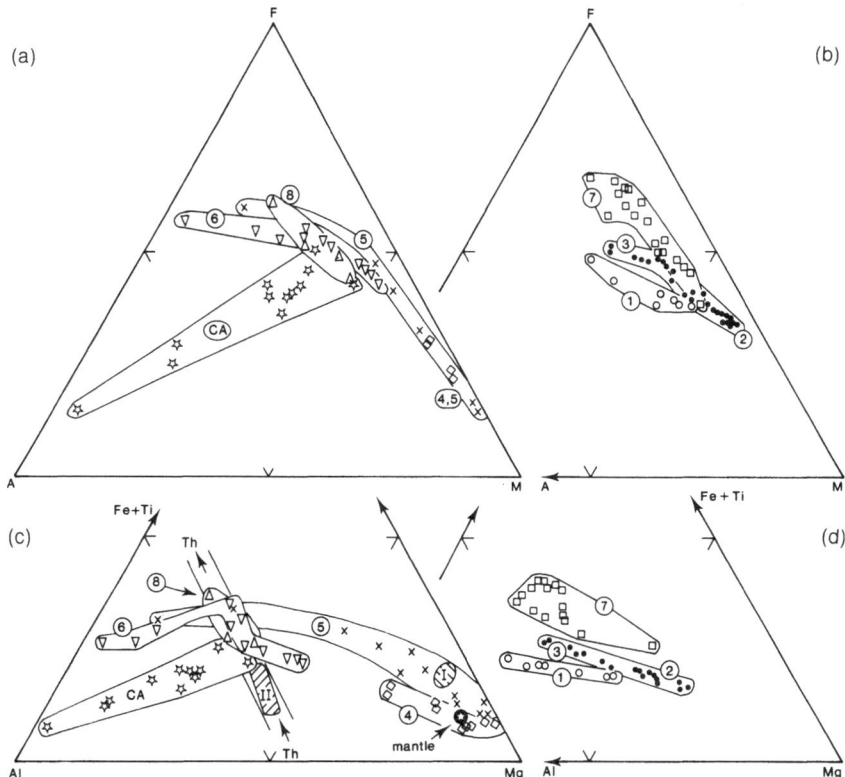

Figure 14.3   AFM (a and b) and Mg:Al: (Fe, Ti) (c and d) plots for: 1: Cowarna Rocks dykes, Yilgarn Block (2.4 Ga); 2: picritic dykes, Yilgarn Block (2.4 Ga); 3: SHMB lavas and dykes, Gawler, Yilgarn and Pilbara Blocks and Antarctica (2.85–1.6 Ga); 4: Munro-type komatiites, Yilgarn Block (2.75 Ga); 5: Barberton-type komatiites, Pilbara and Yilgarn Blocks (3.55 and 3.05 Ga); 6: Archaean tholeiites, Yilgarn Block (3.05–2.7 Ga); 7: Proterozoic tholeiites, Yilgarn Block and Gawler craton (2.4–1.9 Ga); 8: Mt Isa tholeiites (1.75 Ga). Data from sources mentioned in the text. Plots include both cumulate and quenched (approximate liquid composition) rocks. Cumulate rock compositions plotted in (c) include plagioclase-free (I) and plagioclase-rich (II) cumulates from Carr Boyd Rocks and Forrestania (Purvis *et al.*, 1972; Purvis, 1980). The positions of the Mg-rich cumulates show that a komatiitic trend (Jensen, 1976) can be effected by crystallising plagioclase-free cumulates from both tholeiitic and komatiitic parents, and the tholeiitic trend (Th) is governed simply by cumulus, probably cotectic-controlled, gabbronorites. The crystallisation of plagioclase-free cumulates, in SHMB sills (e.g. Williams and Hallberg, 1973) also gives rise to a 'komatiitic' fractionation trend (cf. Kuehner, 1989).

Groves, 1987), like the Ventersdorp Supergroup of South Africa (McIver *et al.*, 1982). Flows and sills of tholeiitic basalt and rare komatiite are present (Blake and Groves, 1987). The Fortescue Group is about 2.7 Ga (Pidgeon, 1984) whereas overlying BIF is about 2.49 Ga (Trendall, 1983). The Hamersley Basin grades southwestwards into the more tectonically active Ashburton Trough and into the Gascoyne Mobile Belt (Gee, 1979). The metamorphic grade ranges from zeolite facies to lower greenschist facies in the Hamersley Basin (Smith *et al.*,1982), to greenschist and lower amphibolite facies in the Ashburton Trough, and reaches granulite facies in the Gascoyne Mobile Belt (Gee, 1979).

Medium to high-grade mafic and ultramafic meta-igneous rocks occur throughout the remaining Archaean blocks of Australia (Figure 14.1). For example, mafic granulites occur in the Carnot gneisses and the Mulgathing complex in the Gawler craton (Cooper *et al.*, 1976; Daly, 1986; Fanning *et al.*, 1986) and in the Litchfield complex in the Pine Creek inlier of the Northern Territory (Hammond *et al.*, 1984). Many stratigraphic relationships are as yet not clearly defined because of structural and metamorphic complexities.

14.2.2 *Proterozoic mobile belts and dyke swarms*

Most Early Proterozoic mafic rocks of Australia occur in deformed and metamorphosed mobile belts, namely the Albany–Fraser, Gascoyne, Halls Creek, King Leopold, Litchfield, Arunta, Willyama–Olary and Georgetown belts and blocks, the Gawler Craton, and the Mt Isa inlier (Etheridge *et al.*, 1987; Wyborn, 1987). Reliable ages are available for some of the mafic rocks, but in some areas only felsic volcanics or granitoids have been dated. Reliable stratigraphies are avaliable for the Gawler (Parker and Lemon, 1981), Mt Isa (Derrick, 1982), Arunta (Stewart *et al.*, 1984) and Willyama–Olary areas. Geochemical data indicate that the volcanic activity was bimodal (Sun and McCulloch, 1985; Wyborn *et al.*, 1987b). Similarities between the units in the younger (1.8–1.6 Ga) eastern Australian Proterozoic inliers have been described by Laing and Beardsmore (1986). Post-tectonic dykes and sills occur in some of these areas and include tholeiitic and rare SHMB compositions. A major collision zone and associated gabbroic to tonalitic intrusive suites occur in the Arunta Block (Lawrence *et al.*, 1987; Foden *et al.*, 1988; James and Ding, 1988).

Dyke swarms of Proterozoic age occur throughout the Yilgarn Block (Parker *et al.*, 1987), the most abundant belonging to the 2.4 Ga Widgiemooltha dyke swarm (Hallberg, 1986). This swarm contains individual dykes up to 585 km long and 1.5 km wide. The Jimberlana dyke (Keays and Campbell, 1981; Fletcher *et al.*, 1987) has complex cumulus layering and a structure similar to that of the Great Dyke of Zimbabwe (see Chapter 4). There are four types of post-cratonic dykes in the Yilgarn Block (Hallberg, 1987). Two show a tholeiitic trend with respect to their AFM components but not on a $Mg:Al:(Fe + Ti)$ cation plot (Jensen, 1976) (Figure 14.3). Both have weak LILE enrichment and Ti/Zr ratios of 80 to 110, but the two suites differ in their Ti and Zr contents for a given mg' value, and are low- and high-Ti tholeiites respectively. Other dykes show less Fe-enrichment, and have calc-alkaline affinities (Giles and Hallberg, 1982). These dykes show LILE enrichment and negative Ti anomalies. The Cowarna Rocks dykes are SHMB which also have negative Nb, Sr and P anomalies compared to MORB (Figure 14.4c) (Purvis and Moeskops, 1981; Sun *et al.*, 1989). They appear to be variants of a more abundant picrite–olivine gabbro suite (Hallberg, 1987). Older dykes of the Black Range dyke suite (*c.* 2.9 Ga; Lewis *et al.*, 1975) are present in the Pilbara Block and may be feeders to lavas in the Negri volcanics (Parker *et al.*, 1987; Sun *et al.*, 1989).

A major suite of continental tholeiitic dolerites, the Zamu dolerites (> 1.8 Ga), is present in the Pine Creek fold belt in the Northern Territory (Ferguson and Needham, 1978; Allen, pers. comm., 1987). Metabasalts and dolerite sills are also abundant in the Halls Creek–King Leopold mobile belts and are referred to as the Woodward dolerites. These bodies are about 2.0 Ga (Page, 1985) and were metamorphosed in a greenschist to amphibolite facies event at 1.92 Ga. The McIntosh gabbro and Alice Downs ultramafic body (Plumb *et al.*, 1981) are syn-metamorphic sills, while the younger

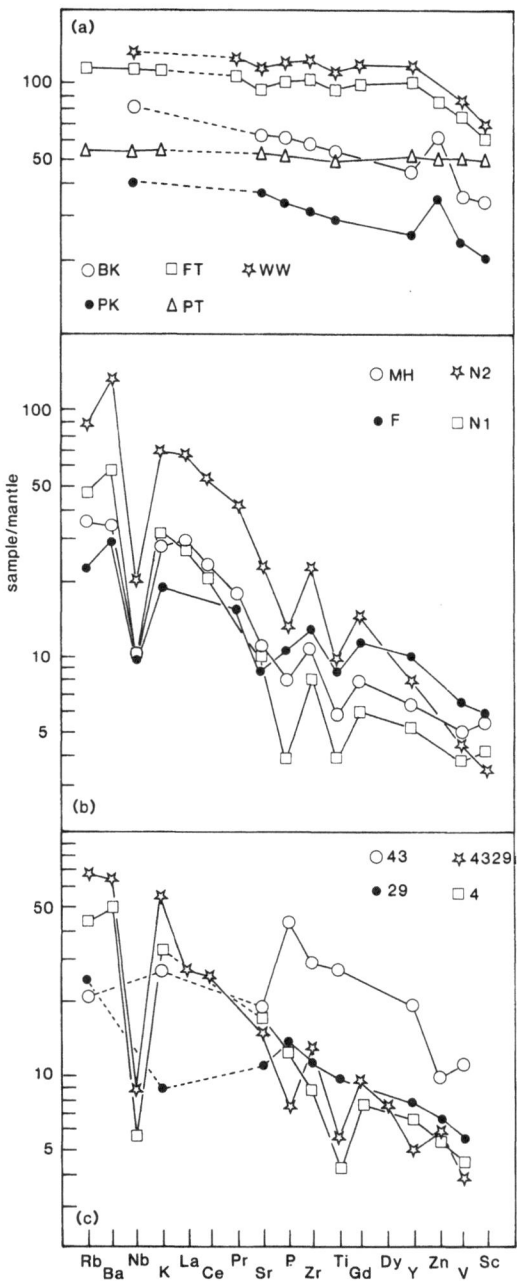

Figure 14.4 Mantle–normalised multi-element (mantle-NME) diagrams (normalised to mantle values of Sun, 1982) for (a) Archaean komatiites and tholeiites from Forrestania (Purvis, 1980) and Welcome Well tholeiite (WW; Giles and Hallberg, 1982). PK: peridotitic komatiite; BK: basaltic komatiite; PT: primitive tholeiite; FT: fractionated tholeiite, (b) Archaean SHMB from Forrestania (F) (c. 3.0 Ga; Purvis, 1980), Mt Hunt (MH) (c. 2.75 Ga) and the Negri volcanics (N1, N2) (c. 2.85 Ga; Sun et al., 1989). (c) Post-cratonisation dykes (c. 2.4 Ga) of the Yilgarn Block. 43: high-Ti tholeiite, Forrestania; 29: low-Ti tholeiite, Southern Cross Province; 4: picrite, Widgiemooltha (all from Hallberg, 1987); 4329: clinobronzite basalt, Cowarna Rocks (Sun et al., 1989).

Kimberley Basin (1.8 Ga) is intruded extensively by Hart dolerite sills and dykes (Bofinger, 1967). The 1.6 Ga dykes intruding the Gawler craton (Mortimer et al., 1988) range from SHMB-like norites to more typically tholeiitic dolerites (cf. Hall and Hughes, 1987). Dykes in the Mount Isa area (< 1.8 Ga) are typical continental tholeiites (Ellis and Wyborn, 1984). The volcanic sequences in this area are 1.68–1.81 Ga old, overlying a 1.87 Ga basement, and include continental tholeiites as well as acid volcanics (Bultitude and Wyborn, 1982). The oldest of these basalts (the Magna Lynn basalt) is similar to the Cobbold metadolerites (and associated Dead Horse metabalts) in the Einasleigh metamorphics of the Georgetown Inlier (Withnall et al., 1980) of uncertain lower Proterozoic age.

14.3 Diversity of basic magmatism

Models for the evolution of basic magmatism in Australia during the early Precambrian are based on four types of sequences:

1. older Archaean greenstone belts (> 3.0 Ga);
2. granulite–gneiss terranes;
3. younger greenstone belts other than superposed belts (3.0–2.5 Ga); and
4. superposed greenstone belts, epicratonic basin and mobile belt sequences generally lacking or poor in komatiites (2.7–1.8 Ga).

Komatiites occur in the older and younger Archaean greenstone belts but are rare elsewhere. The older Archaean greenstone belts have Barberton-type komatiites whereas the younger belts have Munro-type komatiites (see below). Tholeiites occur in all terranes and SHMB occur in the younger Archaean greenstone belts and to a lesser extent in the early Proterozoic sequences. Calc-alkaline volcanics with mafic components occur sparsely in the Yilgarn Block and mafic calc-alkaline intrusions occur in the Arunta Block. The broad geochemical characteristics of the various suites are presented in Tables 14.1 and 14.2 and in Figures 14.3 to 14.6.

14.3.1 Komatiites

Komatiites (Viljoen and Viljoen, 1969; Arndt and Nisbet, 1982a, b) are derived from high degrees of partial melting of the mantle and have liquidus temperatures up to 1600°C (Green, 1975). At these high temperatures, interactions between turbulent magma and cold supracrustal rock may be expected, and thermal erosion channels have been described from the basal parts of komatiitic volcanic sequences where they overlie easily fusible felsic sediments (Groves et al., 1987b; Barnes et al., 1988). Barnes et al. (1988) interpreted that lava rivers and lakes are present in the Agnew area of the Yilgarn Block (Figure 14.2), where olivine-rich cumulates overlie sediments, and occur along strike from 'overbank' lava flows. A similar situation appears to exist in the Forrestania area, where olivine-rich cumulates also overlie sediments and occur along strike from lava flows (Purvis, 1980; Porter and McKay, 1981). The incorporation of tholeiitic and sedimentary material has affected the ε_{Nd} values (Chauvel et al., 1985) and the Sm/Nd ratios of komatiitic flows at Kambalda (McCulloch and Compston, 1981; Arndt and Jenner, 1986), and the addition of sediment-derived lead has resulted in uniform, sedimentary, lead isotope characteristics in these rocks. Crustal contamination has resulted in the formation of SHMB in many areas (e.g. Arndt and Jenner, 1986; Redman and Keays, 1985).

Table 14.1 Geochemical analyses of representative Archaean basic rocks from Australia

	1	2	3	4	5	6	7	8	9	10	11	12	13	14	15	16
%																
SiO_2	43.4	49.97	48.73	51.18	50.62	67.34	54.00	51.84	51.66	54.67	51.00	45.85	44.95	48.56	50.29	50.03
TiO_2	0.25	0.35	0.57	0.63	1.25	1.01	0.97	0.39	0.59	0.55	0.55	0.24	0.28	0.66	1.01	1.47
Al_2O_3	4.5	3.92	6.00	11.70	14.31	13.21	14.61	9.93	11.94	11.18	11.01	5.24	5.84	12.22	13.69	15.78
Fe_2O_3	n.d.	1.17	1.15	1.39	2.48	1.68	1.78	10.63	11.49	9.60	—	11.58	10.72	12.31	10.99	12.14
FeO	7.0	11.86	10.13	9.57	10.25	4.57	6.73				10.66					
MnO	n.d.	0.17	0.20	0.19	0.21	0.20	0.18	0.23	0.17	0.14	0.19	0.21	0.18	0.24	0.18	0.17
MgO	39.0	24.27	21.47	12.87	6.73	0.53	6.18	16.50	13.22	11.04	15.61	30.74	31.94	12.24	11.37	6.82
CaO	3.1	6.42	9.68	10.61	10.86	6.71	11.68	9.06	7.91	7.63	8.72	5.85	5.68	11.54	8.98	10.03
Na_2O	0.4	0.75	1.05	1.70	3.05	3.54	2.59	0.98	1.73	2.12	1.34	0.40	0.52	1.70	1.86	2.50
K_2O	0.2	0.08	0.13	0.09	0.19	0.47	0.28	0.34	0.94	0.76	0.28	0.01	0.06	0.17	0.97	1.00
P_2O_5	n.d.	0.04	0.07	0.06	0.12	0.29	0.11	0.04	0.07	0.07	0.04	0.02	0.02	0.07	0.37	0.25
ppm																
Sc	—	18	30	44	51	24	42	28	39	—	32	—	—	—	28	26
V	—	118	171	244	362	38	311	147	193	157	202	112	117	—	197	209
Cr	—	2263	1912	1260	—	—	—	2535	1460	1150	2019	2870	2890	—	800	331
Ni	—	920	440	277	96	nd	75	471	341	286	468	1628	1562	—	317	203
Rb	—	2	3	4	4	10	8	8	41	23	6	1	1	—	24	22
Sr	—	42	82	72	114	—	100	30	92	154	54	19	24	—	421	262
Y	—	7	13	14	30	53	25	11	15	12	15	6	7	—	23	28
Zr	—	20	38	34	75	174	77	41	57	74	57	13	15	—	111	125
Nb	—	1.6	3.1	2.1	4.6	10.8	3.5	2.8	3.5	3.0	3	1.8	16	—	6	7
Ba	—	—	—	—	—	—	114	135	145	235	131	—	—	—	270	299

1: basal margin of Bulong Complex (2.75 Ga) (Moeskops, 1977).
2 to 7: 3.05 Ga Forrestania complex (Purvis, 1980).
2: peridotitic komatiite.
3: chilled margin of Purple Haze sill.
4: magnesian tholeiite.
5: fractionated tholeiite.
6: granophyre.
7: high-Si tholeiite.
8: SHMB, Negri Volcanics (2.85 Ga; Sun et al., 1988).

9: SHMB, Mount Hunt (2.75 Ga; ibid).
10: Clinobronzite bearing dyke, Cowarna Rocks (2.41 Ga; ibid).
11 to 14: c. 2.75 Ga.
11: SHMB Kambalda (Arndt and Jenner, 1986).
12: Komatiite, Mount Burges (Nesbitt and Sun, 1976).
13: Komatiite, Scotia (ibid).
14: Magnesian tholeiite, Scotia (Sun and Nesbitt, 1978).
15, 16: Calc-alkali basalts, Wellcome Well, Yilgarn Block (> 2.6 Ga; Giles and Hallberg, 1982).

Table 14.2 Geochemical analyses of representative early Proterozoic basic rocks from Australia

	1	2	3	4	5	6	7	8	9	10	11	12	13	14	15
%															
SiO_2	48.59	49.30	49.90	50.49	48.25	47.21	47.76	48.99	52.43	47.95	52.21	51.43	49.99	51.83	47.54
TiO_2	0.81	1.04	1.40	1.30	2.77	2.97	1.39	0.56	0.67	2.59	1.20	1.46	0.62	0.88	2.50
Al_2O_3	15.13	15.05	15.10	14.47	13.28	12.25	12.93	9.69	12.85	12.39	13.60	13.32	11.05	14.92	13.43
Fe_2O_3	1.96	1.46	3.81	4.99	5.22	17.83	14.81	1.84	2.17	3.65	2.49	3.50	11.35	10.69	17.40
FeO	8.32	11.11	8.08	6.71	9.66	—	—	9.01	8.44	12.41	10.24	11.27	—	—	—
MnO	0.26	0.20	0.35	0.19	0.23	0.25	0.22	0.18	0.18	0.25	0.29	0.41	0.17	0.17	0.23
MgO	10.71	9.42	7.04	6.62	4.75	6.65	7.23	18.62	9.10	5.51	6.70	6.21	15.86	7.66	6.06
CaO	11.65	9.81	7.97	9.51	7.50	10.33	12.90	7.58	11.40	10.14	9.92	9.08	9.04	10.31	9.39
Na_2O	1.20	2.13	3.12	2.70	2.44	2.16	0.83	1.60	2.10	2.22	1.49	2.01	1.29	2.28	2.32
K_2O	1.30	0.50	1.52	1.04	1.63	0.27	0.98	0.43	0.27	0.36	0.95	1.03	0.53	0.84	0.91
P_2O_5	0.03	0.08	0.20	0.13	0.45	0.24	0.07	0.13	0.05	0.41	0.12	0.14	0.08	0.12	0.30
ppm															
Sc	15	18	—	—	—	38	59	—	—	—	55	55	37	35	35
V	240	270	—	—	—	383	476	171	235	509	—	—	231	256	405
Cr	970	105	—	—	—	279	91	2215	380	46	168	101	2401	568	170
Ni	345	170	70	30	64	87	54	640	119	72	41	31	526	160	79
Rb	89	19	100	181	160	2	9	14	8	7	30	28	19	30	35
Sr	183	132	32	26	51	270	377	180	121	188	115	111	112	182	177
Y	18	19	180	120	290	43	26	16	22	48	33	31	16	25	44
Zr	40	54	18	6	18	139	55	49	62	157	75	85	74	116	150
Nb	—	—	—	—	—	9.1	2.3	—	—	—	3	4	5	6	12
Ba	506	139	380	330	443	159	158	183	130	115	277	595	151	273	266

1, 2: dolerite dykes, Mount Isa inlier (1.7 Ga; Ellis and Wyborn, 1984).
3: Magna Lynn Basalt, Mount Isa inlier (1.8 Ga; Bultitude and Wyborn, 1982).
4: Eastern Creek Volcanics, basalt, Mount Isa inlier (1.75 Ga; ibid.).
5: Cromwell Basalt, Mount Isa inlier (1.75 Ga; ibid.).
6, 7: Basic amphibolites, Entia Gneiss Complex, Harts Range, Central Australia
(>1.74 Ga; Sivell and Foden, 1988).
8 to 10: 2.4 Ga (Hallberg, 1987).

8: Olivine picrite dyke, Widgiemooltha.
9: Low-Ti tholeiite dyke, Yilgarn Block.
10: High-Ti tholeiite dyke, Forrestania.
11, 12: Amphibolites Middleback Range (2.0 Ga; A.J. Parker, pers. comm., 1988)
13, 14, 15: Norite (SHMB) and gabbronorite dykes, Port Lincoln (1.6 Ga; Mortimer et al., 1988)

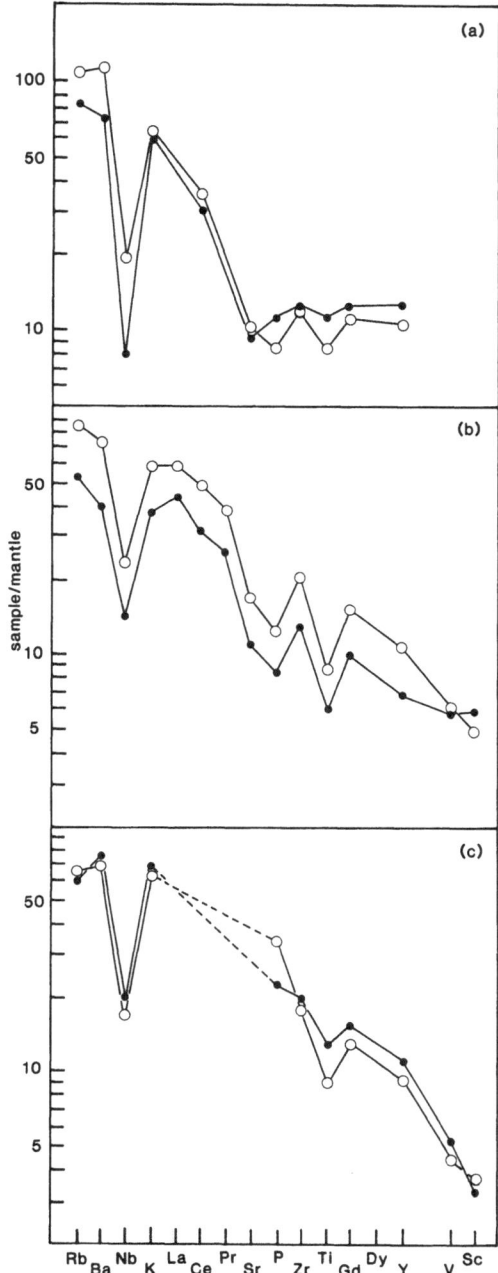

Figure 14.5 Mantle-NME diagrams (normalised to mantle values of Sun, 1982) for (a) Amphibolites, Middleback Range (*c.* 1.9 Ga; SADME unpublished; A.J. Parker, pers. comm., 1988). (b) Norite dykes, Port Lincoln, Gawler craton (*c.* 1.6 Ga) (Mortimer *et al.*, 1988). (c) Archaean calc-alkaline basalts, Welcome Well (Giles and Hallberg, 1982).

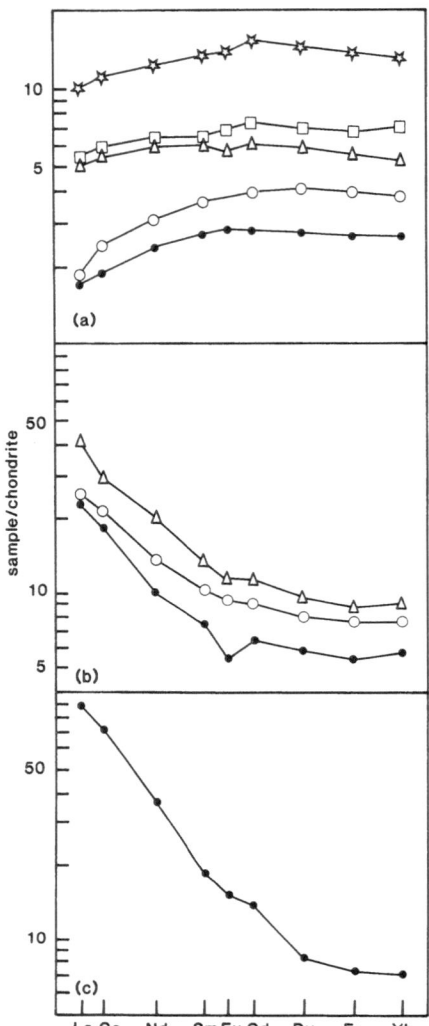

Figure 14.6 Chondrite-normalised REE abundances in representative Archaean basic rocks of Australia. (a) Munro-type komatiites (29–24% MgO) and tholeiites (13–7% MgO) of the Norseman–Wiluna belt (*c.* 2.7–2.8 Ga) (Sun and Nesbitt, 1978). dots: Mt Burges komatiite; open circles: Agnew komatiite; open squares: Agnew tholeiite; triangles: Scotia tholeiite; stars: Mt White tholeiite (Barnes *et al.*, 1988). (b) SHMB from the Negri volcanics (dots), Kambalda (open circles) and Port Lincoln (triangles) (Mortimer *et al.*, 1988; Sun *et al.*, 1989). (c) Basaltic andesite of the *c.* 2.63 Ga Marda complex (Taylor and Hallberg, 1977).

In contrast to the thermal erosion channels seen at Agnew and on a smaller scale at Kambalda (Figure 14.2) (Groves *et al.*, 1987b; Barnes *et al.*, 1988), sheet-like flows extending for up to 100 km (Hill *et al.*, 1987) and smaller flows with spinifex-textured or harrisitic zones (Stolz and Nesbitt, 1981) are also preserved. These flows commonly have a basaltic substrate, less amenable to erosion than the felsic volcanics and sediments at Agnew and Forrestania. The larger ones appear to have had a very high magma flow rate compared to the smaller flows, which apparently cooled quickly and are possibly totally uncontaminated. Large volumes of these flows occur in the Norseman–Wiluna belt especially in the southern part where there are only small scale nickel sulphide deposits (Nesbitt and Sun, 1976; Sun and Nesbitt, 1978; Stolz and Nesbitt, 1981).

Komatiitic pyroclastic rocks are common in some areas (e.g. Scotia; Stolz and Nesbitt, 1981). They occur as lenses up to 100 m thick interlayed with small bodies of spatter material, pillowed and spinifex-textured horizons, and may be flow-top breccias derived from fire-fountaining. These pyroclastics reflect the low viscosity of komatiitic magma.

Barberton-type komatiites Barberton-type komatiites are those which have chondrite-normalised La to Sm ratios (La_N/Sm_N) equal to or greater than 1 (e.g. Sun and Nesbitt, 1978). Mafic and ultramafic rocks from the Barberon (3.55 Ga) greenstone belt of South Africa and from the Isua belt (3.8 Ga) in southern West Greenland have ε_{Nd} values of about zero (Zindler 1982). Jahn *et al.* (1982) give a value of $+2(\pm 4.6)$ for Barberton Mountainland komatiites consistent with their having formed from primitive mantle. They are of two types: Al-depleted and Al-undepleted. Al-depleted types are the most common and have Al_2O_3/TiO_2 ratios of about 11. Undepleted types occur at the base of the sequence in the Barberton Mountainland and have chondritic Al_2O_3/TiO_2 ratios (*c.* 20). CaO/TiO_2 ratios are commonly higher in the Al-depleted types than in the undepleted types, but as Ca is a mobile element in many of these rocks, interpretation of the CaO/TiO_2 ratios is equivocal.

In Australia, Barberton-type Al-depleted komatiites occur in the Warrawoona Group of the Pilbara Block and in the Southern Cross and Murchison Provinces of the Yilgarn Block (Nesbitt *et al.*, 1979) together with some Al-undepleted types (Glikson and Hickman, 1981). Peridotitic and basaltic komatiites are well developed at Forrestania in the Southern Cross Province (Nesbitt *et al.*, 1979), but komatiitic basalts are well developed as flows and in the chilled margins of sills in the Pilbara Block (Mathison and Marshall, 1981). The komatiites in the Forrestania area occur as:

1. Flows of peridotitic komatiite, with chilled margins locally displaying spinifex textures, containing about 25% MgO. These flows are 5–40 m thick and most of them contain zones rich in cumulus olivine.
2. Flows of basaltic komatiite, with chilled margins containing about 18% MgO. These flows are 1–10 m thick, with accumulative zones of olivine, olivine + clinopyroxene, and olivine + clinopyroxene + chromite.
3. Dunite–peridotite lenses up to 300 m thick, containing 45–95% accumulated olivine, 5–45% liquid of basaltic komatiite composition (22% MgO), and 1–15% (rarely 90%) trapped sulphide liquid.
4. In the Purple Haze Sill, as a layered, 120–200 m thick sill with a chilled margin containing 22% MgO, and layers of olivine, olivine + chromite, and olivine + clinopyroxene cumulates. An actinolite schist layer appears to have originated as an orthopyroxene + clinopyroxene-bearing komatiitic basalt with about 13% MgO, and a small offshoot of komatiitic ferrodiorite with 2.9% MgO has pyroxene-cumulate zones.

Mantle-normalised multi-element diagrams for these rocks are presented in

Figure 14.4. The komatiites show a regular depletion of elements from at least La to Sc. The amount of liquid capable of having been produced from primitive mantle has been estimated using La or Nb as a totally incompatible element (K_D values of about 0 for all processes involved), and deviations from primitive mantle ratios have been attributed to garnet removal (Green, 1975). A primitive mantle composition similar to that determined by Nesbitt and Sun (1976) can be derived from:

(a) 45% peridotitic komatiite + 11% garnet + 44% olivine;
(b) 24% basaltic komatiite + 13% garnet + 63% olivine;
(c) 25% Purple Haze sill chilled margin + 12% garnet + 63% olivine.

This is based on an olivine of Fo_{92} and a chromiferous pyrope garnet with 6–7% CaO and 0.5% TiO_2. The Ca content is higher than that in garnets of granular and sheared lherzolite nodules (e.g. Cox et al., 1973), but the TiO_2 and Cr_2O_3 contents are comparable.

A possible model for the formation of these rocks (Green, 1975) firstly involves a thermal perturbation in deep-seated garnet–peridotite facies primitive mantle, creating sufficient porosity to have allowed the 11–13% garnet to settle out. Secondly, the resulting mantle, having lost its most dense component (garnet), was considerably less dense than its surroundings, and rose adiabatically, resulting in further melting. Thirdly, olivine separated from the melt, which became komatiitic. The depth of magma separation was probably about 100 km (cf. Thompson, 1987) so that the komatiite liquid would be less dense than the mantle. It is not envisaged that the liquid present at the first stage would have been komatiitic, or that garnet and komatiitic liquid were ever in equilibrium.

In the Pilbara Block, the Munni Munni complex (Donaldson, 1974) contains ultramafic cumulates with Al_2O_3/TiO_2 ranging from 8 to 12, suggesting that an Al-depleted Barberton-type komatiitic basalt was the parental magma. The lack of cumulus orthopyroxene from the Purple Haze sill suggests a similar parental magma for this complex. A basal zone of interlayered peridotites and pyroxenites is overlain by gabbros and more fractionated rocks with as little as 1–5% MgO, which possibly represent the extreme fractionation of a komatiite liquid.

Munro-type komatiites Munro-type komatiites are those which have low La_N/Sm_N ratios (Figure 14.6a), generally c. 0.5–0.75 (Sun and Nesbitt, 1978; Arndt and Nisbet, 1982). Most are not depleted in Al and have Al_2O_3/TiO_2 values of between 19 and 26 (Nesbitt et al., 1979), but Al-depleted types with Al_2O_3/TiO_2 of 14 to 15 and high Gd_N/Yb_N ratios have been described from Canada (Cattel and Arndt, 1987). Munro-type komatiites occur in the Eastern Goldfields Province and are exemplified by those occurring in flows at Scotia (Figure 14.2), where igneous textures and primary olivine are commonly well preserved (Stolz and Nesbitt, 1981). These flows are commonly 40 m thick and are either strongly zoned with well developed spinifex-textured and locally olivine-dominated layers, or weakly zoned, with dendritic 'harrisitic' olivine and amygdaloidal horizons. In the vicinity of the Scotia Ni–Fe sulphide ore body, there is abundant pyroclastic material, possibly resulting from fire-fountaining, and small thin lenses of locally pillowed flow units (seen only in drill cores).

The mineralogy of Munro-type komatiites is dominated by forsteritic olivine which is locally rich in Ca and/or Cr and is commonly quenched (Nesbitt, 1971; Stolz and Nesbitt, 1981), together with quenched Al-rich Cr-diopside or endiopside (Arndt and Fleet, 1979) in altered glass. Pyroxene and chromite occur in areas of segregated quenched liquid adjacent to vesicles (segregation vesicles) (Stolz and Nesbitt, 1981), but plagioclase and orthopyroxene are known only from layered flows and sills. For example, the Bulong Complex (Moeskops, 1977) consists of five sills apparently formed from a komatiitic magma. The sills are 200–500 m thick, with the sequence dunite–(chromitite)–bronzitite–websterite–gabbro in which the order of crystallisation was olivine–chromite–orthopyroxene–clinopyroxene–plagioclase (cf. Arndt, 1976). The 'chilled margin' composition (Table 14.1) is approximately that of Archaean mantle (Nesbitt and Sun, 1976; Moeskops, 1977), which suggests that it contains accumulated olivine.

14.3.2 Tholeiites

A clear distinction can be made between early Precambrian komatiitic and tholeiitic rocks in Australia. For example, in the Forrestania area in the Southern Cross Province of the Yilgarn Block (Figure 14.2; Purvis, 1980), the basaltic lavas contain from 4 to 14% MgO and 0.6 to 1.4% TiO_2, and those elements which retain total incompatibility at any given stage of fractionation remain proportional to primitive mantle values. The parental basalts have around 13% MgO, chondritic CaO/TiO_2 ratios, but slightly low Al_2O_3/TiO_2 ratios (c. 19) which, even so, are higher than those of the interlayered komatiites (c. 11). They have totally flat mantle-NME profiles (Figure 14.4). Normative calculations (Thompson, 1987) suggest magma separation at about 10 kb, within the spinel peridotite field, and the level of incompatible element enrichment suggests 30–35% melting of primitive mantle with about twice chondritic levels of REE, leaving a dunite residue of 98% olivine and 2% Cr–Al spinel ($Al_{70}Cr_{30}$). No isotope data are available for these rocks. Samples from the nearby Diemals greenstone belt indicate an ε_{Nd} value of 0.9 (± 0.7) (Fletcher et al., 1984), which is significantly lower than that for the Kambalda komatiites (Chauvel et al., 1985), but is derived from mafic and felsic as well as komatiitic volcanics.

The Forrestania parental tholeiitic magmas show three distinct fractionation trends. The first is dominated by moderate pressure fractionation controlled by either orthopyroxene or clinopyroxene and plagioclase (evidenced by outcropping bronzitite and gabbro cumulates), leading to fractionated basalts with lower Si and Mg contents than those of the parental basalt, but higher Fe and Ti contents. The second type is low pressure, olivine-controlled fractionation leading to high-Si, low-Mg basalts only slightly richer in Ti than the parental basalt, while the third trend produces low-Si, high-Al, low-Mg basalts, possibly by high pressure granulite or eclogite facies olivine pyroxenite separation.

Outcrops in the Forrestania area suggest the presence of individual flows about 40 m thick, some with vesicular zones. The tholeiitic lavas locally preserve intergranular–ophitic textures. Small gabbro–dolerite sills are also present, and a large layered sill at the western margin of the greenstone belt, informally termed the 'White Nellie Complex', covers an area of about 15 × 3 km, and has bronzitites, gabbros and granophyres (MgO from 31 to 0.5%). Minor basaltic types at Forrestania include basalts with around 8% MgO, derived from mildly depleted mantle (Zr/Nb about 22), and siliceous basalts with 4–6.5% MgO and irregular trace element distributions, similar only to those reported for the distinctive 'Theo's Flow' of the Munro Township area in Canada (Arndt and Nesbitt, 1982). These may reflect crustal or subcontinental lithospheric contamination (cf. Sun et al., 1989), as they are similar to, but more subdued than, typical SHMB patterns (Figure 14.4b).

Basalts with c. 11–15% MgO are widespread in the Pilbara and Yilgarn Blocks

(Hallberg, 1972; Stolz and Nesbitt, 1981; Gresham and Loftus-Hills, 1981; Giles and Hallberg, 1982). Those in the Eastern Goldfields Province lack the strong LREE-depletion of the interlayered komatiites and are therefore not simple fractionation products derived from the Munro-type komatiites of this area. These basalts have CaO/TiO_2 and Al_2O_3/TiO_2 ratios which are more nearly chondritic than those of the interlayered komatiites (Sun and Nesbitt, 1978), and they were possibly derived from near-primitive mantle. They have more variable trace element patterns than those of the Forrestania basalts, suggesting the possibility of magmatic interaction with crustal or subcrustal lithosphere (cf. Arndt and Jenner, 1986), and show a range of TiO_2 contents for a given MgO content, possibly reflecting a fractionation pathway between trends (1) and (2) as defined above for the Forrestania tholeiites.

The Carr Boyd Rocks Complex in the Yilgarn Block (Figure 14.2) (Purvis *et al.*, 1972) comprises tholeiitic cumulates, with three cumulate cycles reflecting three different orders of crystallisation:

1. dunite–harzburgite–olivine-bronzitite–bronzitite cycles (olivine–orthopyroxene–[clinopyroxene–plagioclase]);
2. dunite–harzburgite–olivine-norite–norite–gabbro cycles (olivine–orthopyroxene–plagioclase–clinopyroxene);
3. dunite–feldspathic dunite–troctolite–anorthosite cycles (olivine–plagioclase–[orthopyroxene–clinopyroxene]).

Despite these multiple cycles, the complex does not show progressive cryptic variation, suggesting the repeated influx and extraction of magma in an open system chamber. Tholeiitic cumulates similar to those in cycle (3) are overlain by fractionated non-cumulus gabbro, ferrogabbro and granophyre in the Kilkenny Sill (Jaques, 1976), the upper chilled margin of which is a tholeiitic basalt with about 16% Al_2O_3.

Ti-rich ferrobasalts are a minor constituent of some greenstone belts, notably the Lawlers Belt, in the Eastern Goldfields Province (Nesbitt and Sun, 1976; Naldrett and Turner, 1976). These lavas are all highly fractionated and are LREE-enriched. Tholeiitic basalts, dolerites and gabbro–anorthosite complexes are common in the Lower Proterozoic of Australia (Ferguson and Needham, 1978; Wilson, 1978; Thornett, 1981; Hammond *et al.*, 1984; Sivell *et al.*, 1985; Sivell and Foden, 1988; Mortimer *et al.*, 1988). Most of these rocks appear to be related to continental tholeiites, although LREE-depleted, MORB-like rocks are also present (Sun and McCulloch, 1985). Some of these suites, such as the Zamu dolerite (Ferguson and Needham, 1978), contain essentially unmetamorphosed dolerites, whereas others such as the Entire anorthosite (Sivell *et al.*, 1985) are extremely deformed, with the ultramafic layers reduced to boudins within a meta-anorthosite host. Continental tholeiites occur in the 1.9 Ga Hutchison Group and as dykes of around 1.6 Ga old (Mortimer *et al.*, 1988) in the Gawler craton (A.J. Parker, pers. comm., 1988).

14.3.3 *Siliceous high-Mg basalts (SHMB) and high-Mg basalts (HMB)*

Magnesian lavas and co-magmatic sills characterised by pyroxene-dominated quenched and cumulus assemblages occur in the Yilgarn and Pilbara Blocks, and post-cratonisation dykes with similar textures and compositions occur in the Yilgarn Block and the Gawler craton (Purvis and Moeskops, 1981; Chauvel *et al.*, 1985; Arndt and Jenner, 1986; Mortimer *et al.*, 1988; Sun *et al.*, 1989). Similar dykes also occur in

Antarctica (Kuehner, 1989; Sun et al., 1989). The terminology for these rocks is far from established. They have some similarities with Archaean komatiites and others with modern boninites (Arndt and Nisbet, 1982a; Cameron and Nisbet, 1982; Crawford, 1989). They are referred to here as siliceous high-Mg basalts (SHMB) and high-Mg basalts (HMB) (after Hallberg et al., 1976b; Sun et al., 1989).

Lavas of the SHMB suite generally have fine- to medium-grained pyroxene-dominated quench textures (Nesbitt et al., 1982; Redman and Keays, 1985). The dykes are generally glassy to variolitic, with skeletal phenocrysts of zoned pyroxene (Al- and Cr-rich diopside and bronzite rimmed by pigeonite or augite) and/or olivine. The Si-rich groundmass contains pyroxenes which range from magnesian pigeonite to augite and ferroaugite and microphenocrysts of chromite. Clinobronzite occurs in dykes at Cowarna Rocks (Purvis and Moeskops, 1981). Pyroxene compositions in these dykes indicate crystallisation temperatures of 1350–1050°C (author's unpublished data; Lindsley, 1983). These dykes are similar to early Proterozoic noritic dykes of Greenland in their textures, mineralogy and inferred crytallisation temperatures (Hall and Hughes, 1987). Sills of SHMB and HMB parentage have been described from the Eastern Goldfields Province (Williams, 1972; Williams and Hallberg, 1973). Recent work has shown that they are characterised by abundant bronzitite and norite, minor granophyre and locally minor peridotite (R.E.T. Hill, pers. comm., 1988). The best known of these sills are those at Norseman Mission, Mt Thirsty, Ora Banda and Mt Monger (Figure 14.2). Mineral separates from the Ora Banda sill yielded an age of 2.75 Ga and a negative ε_{Nd} (Chauvel et al., 1985).

The SHMB and HMB lavas and dykes are similar texturally but different compositionally (Hallberg et al., 1976b). SHMB rocks have between 20 and 7% MgO, 51 to 57% SiO_2 and low TiO_2 contents. They have variable Al_2O_3/TiO_2 values of between 10 and 30. The very high ratios characteristic of boninites are not seen (cf. Bloomer and Hawkins, 1987). The 2.9–1.6 Ga SHMB from Australia and Antarctica have uniform REE and trace element distributions (Figures 14.4 to 14.6) and similar AFM and Al:Mg:(Fe + Ti) proportions (Figure 14.3b, d), although the data given by Kuehner (1989) suggest less uniformity than those presented by Sun et al. (1989). Most SHMB have primitive mantle-type K/La ratios (c. 380; Figure 14.7), the main exception being the Cowarna Rocks dyke which has a higher value (c. 800). This dyke forms part of a low-Fe suite (Purvis and Moeskops, 1981), with more extreme depletion of Nb, Sr and P (Figure 14.4c). HMB rocks from Meekatharra (Figure 14.2) (Hallberg et al., 1976b) have less SiO_2 (46–53%) than SHMB at equivalent MgO contents, and are relatively depleted in LIL elements. They possess negative P and Ti anomalies, but these are less strong in the HMB than the SHMB.

SHMB in Australia and Antarctica occur in three tectonic settings. Those occurring in 'initial' greenstone belts, such as those at Kambalda, Mt Hunt, Mt Monger of the Yilgarn Block, which do not have a recognisable basement and may therefore represent oceanic crust, have Ti/Zr ratios of 57–60 (Sun et al., 1989). Evidence of older crust is provided only by zircon xenocrysts (up to 3.5 Ga old) in the lavas (Compston et al., 1986). SHMB in 'superposed' greenstone belts (those unconformably overlying older greenstones), such as the Negri volcanics of the Pilbara Block, have Ti/Zr in the range 47–57 (Sun et al., 1989). The third type of SHMB occurs as post-cratonisation dykes, such as those of the Gawler Craton and at Cowarna Rocks which have Ti/Zr values of 43–50. SHMB in Antarctica have Ti/Zr values as low as 38 (Mortimer et al., 1988; Kuehner, 1989; Sun et al., 1989).

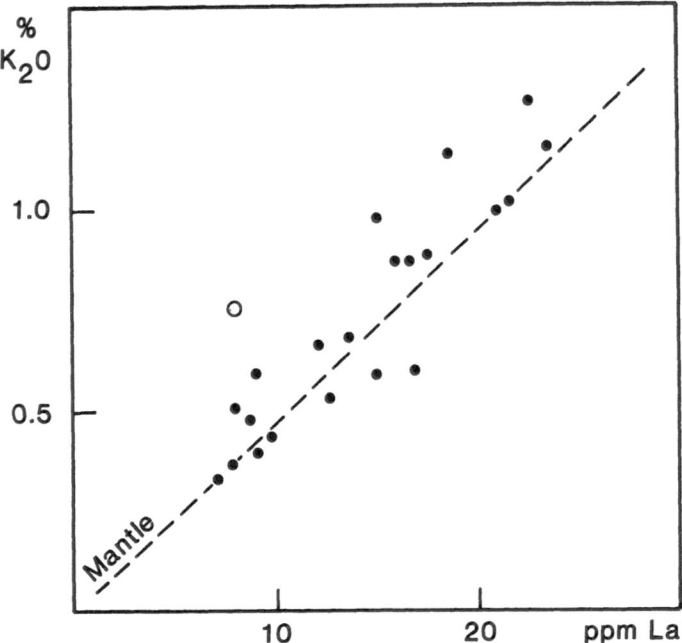

Figure 14.7 K_2O vs. La plot for SHMB from Australia and Antarctica compared to the primitive mantle ratio (K/La = 380; Sun, 1982). Open circle: Cowarns Rocks dyke. Data from Mortimer *et al.* (1988), Kuehner (1989) and Sun *et al.* (1989).

There is some suggestion that the Ti/Zr ratios of SHMB reflect the thickness of the crust or lithosphere through which the magma passed (cf. Redman and Keays, 1985) rather than the nature of any supposed crustal contaminant (Arndt and Jenner, 1986; Sun *et al.*, 1989). The Meekatharra HMB have higher Ti/Zr than SHMB, mostly in the range 70–90 (Hallberg *et al.*, 1976b), which may reflect an even thinner crust and lithosphere than that in the Kambalda–Mt Hunt–Mt Monger area. It is still not clear whether there is a continuum from unmodified mantle melts (Ti/Zr about 110) through HMB to SHMB in its various settings, reflected in decreasing Ti/Zr values. The uniform negative ε_{Nd} values (cf. DePaulo and Wasserberg, 1979) within any given suite of SHMB (Sun *et al.*, 1988) suggest that fractionation was superimposed on an already modified magma, probably with an MgO content of around 17%. The constancy of K/La and ε_{Nd} values in so many SHMB suites and their negative ε_{Nd} values argue strongly against upper crustal contamination. Granites and sediments in the Kambalda–Kalgoorlie region, for example, have $\varepsilon_{Nd} = 0$ (Redman and Keays, 1985; Arndt and Jenner, 1986; Sun *et al.*, 1989). The difference in age ranges between Australian komatiites (3.6–2.6 Ga) and SHMB (2.9–1.6 Ga) also suggests that SHMB is generally not simply modified komatiite, although such komatiite alteration has been demonstrated at Kambalda (Arndt and Jenner, 1986).

14.3.4 *Calc-alkaline mafic rocks*

Early Precambrian calc-alkaline suites with mafic components are sparse in Australia compared to komatiite, tholeiite and SHMB suites. An Archaean example is provided by the Welcome Well complex in the Yilgarn Block (Giles and Hallberg, 1982), in which lavas have a continuous range from 50 to 70% SiO_2 and from 11 to 1% MgO. Trace element plots for the most basic rocks (50% SiO_2) are typical of calc-alkaline volcanics, with depletion in Nb and Ti and LIL-enrichment (Figure 14.5c). They are richer in Ni and Cr than recent calc-alkaline basalts and andesites (Giles and Hallberg, 1982) and appear to have formed from parental magmas representing 20–25% melting under hydrous conditions of LIL-enriched mantle. Basaltic andesites in the Marda complex (Taylor and Hallberg, 1977) have typical calc-alkaline REE distribution patterns (Figure 14.6c).

Proterozoic calc-alkaline suites are not well known in Australia. The 1.74 Ga old Huckitta tonalite gneiss suite (Arunta Block) varies from gabbro to tonalite and was intruded into a major collision zone (Foden et al., 1988; James and Ding, 1988). Amphibolites in this area formed partly from cumulate rocks and may be related to the nearby Huckitta tonalite gneiss suite (Foden et al., 1988).

14.4 Magmatic and thermal evolution

Geological evidence concerning the thermal history of Australia during the early Precambrian is summarised below:

1. The oldest rocks in Australia are 3.8 Ga old gneisses with detrital zircons as old as 4.2 Ga.
2. Greenstone belts with komatiites, tholeiites and SHMB formed at 3.6–3.5 Ga, 3.1–3.0 Ga and 2.9–2.7 Ga.
3. Superposed greenstone belts and basalt-dominated cratonic basins (the Hamersley Basin) developed initially at about 2.85 Ga on an older metamorphosed basement.
4. Early Proterozoic basic magmatism is represented by tholeiitic and SHMB dykes at about 2.4 Ga, by the basalts of bimodal suites in mobile belts, and by post-tectonic suites of dykes and sills.
5. Trace element evidence of LREE depletion is seen in komatiites younger than 3.0 Ga and in tholeiites of early Proterozoic age. LREE-enrichment is seen in SHMB as old as 2.85 Ga, although in the Barberton Mountainland there appear to be similar rocks as old as 3.5 Ga.
6. According to Zindler (1982) and Jahn et al. (1982) the Isua and Barberton mafic and ultramafic rocks have ε_{Nd} statistically indistinguishable from zero. However, younger komatiites have positive ε_{Nd} values of $+2$ to $+4$, and SHMB have ε_{Nd} values of -1 to -3 (Fletcher et al., 1987; Sun et al., 1989). Re–Os isotope data from the LREE-depleted Munro Township komatiites suggest that the source had primitive mantle Re/Os ratios and was not modified by continental crust removal (Walker et al., 1988). Hf–Lu isotope data on Barberton komatiites do not support the long-term existence of garnet-depleted mantle (S.-S. Sun, 1988, written comm.) and are thus consistent with the model for the development of these rocks as outlined above, that is without either a dynamically stable layered mantle or a widespread magma ocean (cf. Sun, 1987).

The data presented above can be incorporated into a thermal history similar to that presented by Lambert (1980), with heat flux at a minimum at around 3.3 Ga, corresponding to the gap in greenstone belt ages from 3.5 to 3.1 Ga (see above), and with smaller and larger peaks in heat flux at 3.0 and 2.7 Ga respectively, corresponding to the older and younger parts of the Yilgarn Block, the Whim Creek Group in the

Pilbara Block and the Fortescue Group basalts. The trace element data suggest that primitive mantle was available for both komatiites and tholeiites up to about 3.0 Ga, but that long term depleted mantle was involved, initially in the formation of Munro-type komatiites, and later in the formation of tholeiites. The isotope data are not unequivocal as to the nature of the mantle source of older komatiites and tholeiites (e.g. Jahn *et al.*, 1982; Fletcher *et al.*, 1984).

14.4.1 *Earliest Archaean (4.2–3.8 Ga)*

The oldest known rocks in Australia are the 3.8 Ga old gneisses at Mt Narryer in the Western gneiss belt (Figure 14.2), and some water-lain metasediments in this area have detrital zircons as old as 4.2 Ga. The bulk of the Earth's surface at this time may have been occupied by oceans, and the crust underlying these oceans must have been as self-destructive as that underlying modern oceans, and has not been preserved. In this sense, Archaean granite–greenstone terranes cannot represent typical Archaean oceans, because they have been preserved perhaps by having been made buoyant by extensive early granite formation.

14.4.2 *Early to mid-Archaean (3.6–2.9 Ga)*

The komatiites in the Pilbara Block are Al-depleted (Barberton-type). The formation of these komatiites suggests considerably more sluggish convection in the deeper parts of the mantle than was the case during the formation of the younger Munro-type komatiites. It was only after considerable buoyancy had been given to the source diapirs of these rocks by the removal of garnet that they were able to rise sufficiently to form komatiitic magmas. Convection appears to have been better established at shallow levels in the mantle, as evidenced by the abundant tholeiites.

The gap in greenstone belt ages from 3.5 to 3.1 Ga may reflect a heat flux minimum at this time (Lambert, 1980). The 3.1–3.0 Ga komatiites are of the same type as those formed at 3.5–3.6 Ga, indicating the re-establishment of similar conditions. The tholeiites at Forrestania were formed from a mantle from which spinel was removed, either before or during partial melting, and it is possible that this is the response of spinel–peridotite facies mantle to the same conditions which required garnet to separate out from the source of the komatiites.

The trace element patterns of the older Archaean lavas are all consistent with their having been derived from primitive mantle, although the isotope data do not particularly clarify this point. Nd isotope data suggest that depleted mantle became significant at *c.* 3.0–2.5 Ga (Zindler, 1982). Lu–Hf isotope data from rocks from the southern Superior Province in Canada also suggest depleted mantle separation in this area at *c.* 2.9 Ga (Smith *et al.*, 1987), and the observed flat mantle-NME patterns for the Forrestania parental tholeiites (6.0 ± 0.1 times primitive mantle; Figure 14.4) and for K, Rb, Nb, P and Zr are also consistent with this possibility.

14.4.3 *Late Archaean (2.9–2.5 Ga)*

Late Archaean komatiites are all of Munro type and both trace element patterns and isotope data are consistent with their having a depleted mantle origin. The trace element patterns of associated tholeiites are more varied, but isotope data are at present insufficient to determine to what extent they were formed from depleted or primitive mantle, or mantle metasomatised by a crustal or lithospheric component (cf. Barley,

1986). Those SHMB with LREE enrichment and negative ε_{Nd} values have clearly been affected by interaction with a sialic crustal or lithospheric contaminant, a crustal influence being favoured by Sun (1988, pers. comm.) and Sun et al. (1989).

14.4.4   Early Proterozoic (2.7–1.8 Ga)

The early Proterozoic was a period of progressive mantle cooling after the late Archaean thermal peak (Lambert, 1980) and saw the end of komatiite formation in Australia. Thick continental crust was clearly present, but no oceanic crust of this age appears to have been preserved. Late Archaean to early Proterozoic komatiite-free or komatiite-poor greenstone belts occur in the southwest of the Yilgarn Block and the intra-cratonic Hamersley Basin (Blake and Groves, 1987). Extensive tholeiitic and SHMB dyke formation was initiated in the Yilgarn Block at about 2.4 Ga (Hallberg, 1986; Parker et al., 1987).

Early Proterozoic lavas and sill and dyke swarms in the mobile belts are generally of two geochemical types, namely suites with depleted to flat REE patterns and positive ε_{Nd} values (e.g. the Arunta Block, Einasleigh metamorphics, Woodward dolerites, Halls Creek area), and suites which are LREE-enriched with negative ε_{Nd} values, including both tholeiite and SHMB (Eastern Creek volcanics, Mt Isa; Hart dolerite, Northern Territory; Lamboo complex, Halls Creek) (Sun and McCulloch, 1985; Sun, 1987; Sun et al., 1989). Negative Nb anomalies are common in these otherwise enriched suites, as they are in modern continental tholeiites. However, similar anomalies are also observed in modern island arc tholeiites (Pearce, 1983; Thompson et al., 1983).

14.5   Petrogenetic synthesis

The early Precambrian of Australia contains various types of terranes:

1. 3.8 and 3.3–3.1 Ga granulite–gneiss terranes with supracrustal sediments.
2. 3.5–3.6, 3.0–3.1 and 2.7–2.9 Ga granite–greenstone belts without outcropping basement.
3. 2.7–2.5 Ga superposed greenstone belts and intracratonic basins with few or no komatiites. Intracratonic basins such as the Hamersley Basin persisted until the early Proterozoic (2.3 Ga).
4. Proterozoic mobile belts containing amphibolite to granulite facies metabasites and post-tectonic dykes and sills.
5. Proterozoic dyke suites cutting stabilised Archaean granite–greenstone and granulite–gneiss terranes.

The evidence presented in this chapter lends itself to a model involving granitoid production as a necessary adjunct to high rates of mafic magma formation. Such areas are preserved mainly as granite–greenstone terranes. Those areas with low mafic magma production rates formed oceanic crust which has not been preserved, at least in Australia. In areas of extremely rapid mafic magma production, the basic rocks would have formed a thick and dense lithosphere, which could have partially melted at its base to form abundant granitoids. The abundance of these granitoids would be sufficient to ensure that most later mafic magmas would underplate the granite and would to a large extent also melt to form further granitoids. Such areas became granulite–gneiss terranes (cf. Wilks, 1988).

Areas of rather less rapid mafic magma production would also produce abundant granitoids, but if such areas were subjected to extension sufficient to cause considerable attenuation of this light crustal material, further mafic magmatic material would have broken through, rather than underplated it. However, sufficient underplated material must have been produced in order to account for the extensive high-level granitoid plutons. It should be noted that the average mafic magmatic material formed during the Archaean would have, for example, about 2 ppm Rb (Nesbitt and Sun, 1976), whereas the average Archaean granitoid has 40–80 ppm Rb, indicating that a very large volume of underplated material must have been involved in the formation of Archaean granitoids, unless there was a considerable amount of LILE-enriched volatile flux produced during the formation of the granitoids. The residue from granitoid formation would be relatively dense and could sink into the deeper mantle, mixing with primitive mantle to form the depleted source of the Munro-type komatiites, although this interpretation is not clearly supported by Re–Os isotope data (Walker et al., 1988). Areas of this type became granite–greenstone terranes and the influence of (at least attenuated) crust and an enriched lithosphere is indicated by the occurrence of xenocrystic older zircons (Compston et al., 1986) and perhaps by the isotopic composition of some SHMB lavas (Chauvel et al., 1985; Sun et al., 1989). The return of the underplated material to the mantle would account for the lack of evidence for its existence in seismic profiles of granite–greenstone areas (Wyborn et al., 1987a). It is likely that both the granulite–gneiss terranes and the granite–greenstone terranes developed deep lithospheric keels, during and after their formation, as suggested by the Archaean age of some diamonds (Richardson et al., 1984).

Oceanic areas in which mafic magma production was low could not have produced sufficient granitoid material to become buoyant, and became self-destructive. Basins and mobile belts developed in the late Archaean and early Proterozoic on and between areas of older crust (Sun and McCulloch, 1985; Sivell and Foden, 1988). These are characterised by MORB-type tholeiites (depleted to flat REE patterns and positive ε_{Nd} values) and plume or rift-related continental tholeiites or, less probably, island arc tholeiites with LREE-enrichment, negative ε_{Nd} values and negative Nb anomalies. However, mafic magmatic rocks in a collision zone setting have been reported from the eastern Arunta inlier (Foden et al., 1988).

Acknowledgements

I wish to thank S.-S. Sun (BMR, Canberra), J.D. Foden (Univ. Adelaide), R.E.T. Hill (CSIRO, Perth), A.J. Parker (SADME, Adelaide), J.A. Hallberg (Hallberg and Assoc., Western Australia) and M. Sandiford (Univ. Adelaide) for unpublished manuscripts, data and comments. Prof. Krumm is thanked for access to the library at the J.G. Goethe University, Frankfurt, and Prof. R.W. Nesbitt for the use of facilities at the University of Southampton.

15 Early Precambrian basic rocks of India

B.L. WEAVER

15.1 Introduction

Precambrian rocks are widely exposed in the Indian Shield, which forms the majority of peninsular India (Figure 15.1). Although radiometric dating of these rocks is not very comprehensive, they range from crystalline basement rocks as old as 3.8 Ga in the Singhbhum craton to widespread supracrustal sequences, and dykes intruding crystalline basement which are only 0.9 Ga old. Archaean and early Proterozoic igneous and meta-igneous rocks comprise a large proportion of the exposed Precambrian Indian Shield basement. Basic igneous and meta-igneous rocks occur in all of the three major cratonic nuclei in the Indian Shield. Detailed geochemical data are rather sparse for many of these occurrences, and this review concentrates on those regions (primarily the Dharwar craton; Figure 15.1) for which a reasonable body of petrological and chemical data are available.

Naqvi *et al.* (1974) identified the three major separate Archaean cratonic nuclei, namely the Dharwar, Singhbhum and Aravalli cratons (Table 15.1), although Naqvi and Rogers (1987) subdivided these, recognising seven different nuclei. The shield appears to have formed by the accretion of these individual cratonic nuclei, and may have been intact either by the late Archaean (Radhakrishna and Naqvi, 1986), or at least by 1.5 Ga (Rogers, 1986).

15.2 The Dharwar craton

The Dharwar craton as discussed here (Figure 15.1) is equivalent to the Dharwar protocontinent of Naqvi *et al.* (1974), and therefore includes the Eastern and Western Dharwar cratons and the Southern Granulite craton of Naqvi and Rogers (1987). Rogers (1986) used the term 'Dravidian Shield' for this craton, and considered it to have been a coherent crustal block since *c.* 3.0 Ga. The geology and boundaries of the craton are reviewed by Naqvi and Rogers (1987). The craton grades southwards from greenschist facies greenstone–supracrustal successions (the Dharwar supracrustals) through an amphibolite facies gneissic terrane to a granulite facies gneissic terrane, representing a more or less vertical section through the structurally complex craton. Pressure–temperature $(P-T)$ estimates for the metamorphism range from 0.2 to 0.4 GPa and 300–400°C in the north of the craton (Harris and Jayaram, 1982; Sivaprakash, 1983), through *c.* 0.6 GPa and 750°C for the amphibolite–granulite transition zone, to maximum metamorphic conditions of about 1 GPa and 700–800°C in the granulite facies terrane (Raith *et al.*, 1983; Raase *et al.*, 1986). Raase *et al.* (1986) traced north to south $P-T$ variations in basic gneisses of the Dharwar craton from 0.4

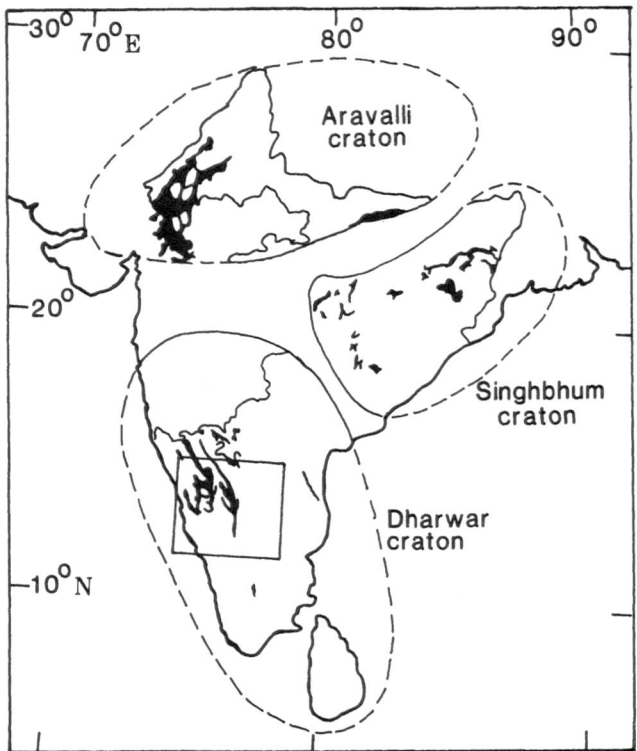

Figure 15.1 Sketch map of India showing the distribution of the early Precambrian Dharwar, Singhbhum and Aravalli cratons and the major metabasic rocks (black). The box depicts the area of the map shown in Figure 15.2.

to 0.5 GPa and c. 500°C in greenschist facies metavolcanics through 0.5–0.7 GPa and c. 600°C in amphibolite facies basic gneisses to 0.7–0.8 GPa and 700–750°C in granulite facies basic gneisses, although maximum P–T conditions in the granulite facies terrane are somewhat higher. Large and irregular variations in P–T conditions are reflected in the granulite facies terrane (Raith $et~al.$, 1983). Radiometric ages for the greenschist and amphibolite facies rocks of the craton generally fall between 3.4 and 3.0 Ga. Granulite facies metamorphism seems to have occurred at c. 2.6–2.5 Ga, according to limited Rb–Sr and U–Pb data (for summary see Naqvi and Rogers, 1987).

The Dharwar Craton is dominated by tonalitic to trondhjemitic gneisses (the Peninsular Gneisses), although in much of the granulite facies terrane the gneisses are more potassic. A number of basic rock types occur in different associations within the craton, ranging from low-grade greenstone belt metavolcanics (the Dharwars), through conformable and cross-cutting meta-igneous lenses and layered complexes in the high-grade (amphibolite and granulite facies) gneiss terranes (the Sargurs), to late basic dykes (Naqvi and Rogers, 1987). In describing the higher grade portions of the Dharwar Craton (as defined here), Naqvi and Rogers (1987) distinguished coherent amphibolite facies 'greenstone' belts, ultrabasic–basic enclaves in the amphibolite to granulite facies transition zone (the Sargurs proper), and basic gneiss suites in the

Table 15.1   Major early Precambrian basic components of the Indian Shield

	Age (Ga)
Dharwar craton	
Mafic dykes; various generations of abundant tholeiitic dolerites	2.4–0.9
Kabbaldurga tholeiitic–andesitic amphibolites	
Pallavaram basic granulites	2.76–2.36
Sittampundi layered gabbro–anorthosite complexes	
Kolar schist belt komatiitic–tholeiitic volcanics	2.9
Bababudan, Chitradurga, Javanahalli, Kudremukh, Shigegudda, and Shimoga greenstone belts	< 3.0
Sargur supracrustal belt tholeiitic metabasalts and associated layered ultrabasic–gabbro–anorthosite complexes	⩾ 3.0
Holenarasipur ultramafic–anorthosite complex	3.1
Dharwar supracrustal metavolcanics	3.4–3.0
Sighbhum craton	
Dolerite dykes	1.5–1.0
Nanogaon Group olivine- and quartz-tholeiites (and rhyolites)	2.3–2.2
Amgaon Group metabasalts	Archaean?
Ultramafic–basaltic flows associated with Singhbhum granite	2.9
Older Metamorphic Group low-K tholeiites	3.8–3.2
Aravalli craton	
Scarce dolerite dykes	2.4–1.7
Bijawar Group metabasalts	2.6–2.4
Bundelkhand complex metavolcanics, gabbros and dykes	> 2.6
Aravalli Group minor K-rich tholeiites and andesites	
Banded Gneiss Complex low-K tholeiitic amphibolites	3.5

granulite facies terrane. These Archaean suites (> 3.0 Ga) are all older than the low-grade greenstone belts (2.9–2.5 Ga), although there are many similarities between the basic rocks of the three suites. The main differences are only the relative proportions of the various rock types (Naqvi and Rogers, 1987). However, there are significant differences in the metasedimentary assemblages associated with these suites (Naqvi and Rogers, 1987). For simplicity, the higher grade Archaean ultrabasic–basic layered complexes and basic lenses will be considered as a single suite, although anorthositic complexes comprise slightly different assemblages. All of the Archaean complexes are cut by abundant early to late Proterozoic basic dykes.

15.2.1   Greenstone belts

The low-grade (greenschist to amphibolite facies) greenstone belts and associated supracrustal sequences of the Dharwar Craton (Figure 15.2) range in age from 2.9 to 2.5 Ga, and are collectively termed the Dharwar Group, or Dharwars (Naqvi and Rogers, 1987). These belts (e.g. Bababudan, Chitradurga, Shimoga, Javanahalli, Kolar and other minor belts) typically comprise greenschist facies metavolcanic and clastic metasedimentary rocks, with a high proportion of sediments relative to volcanics. The

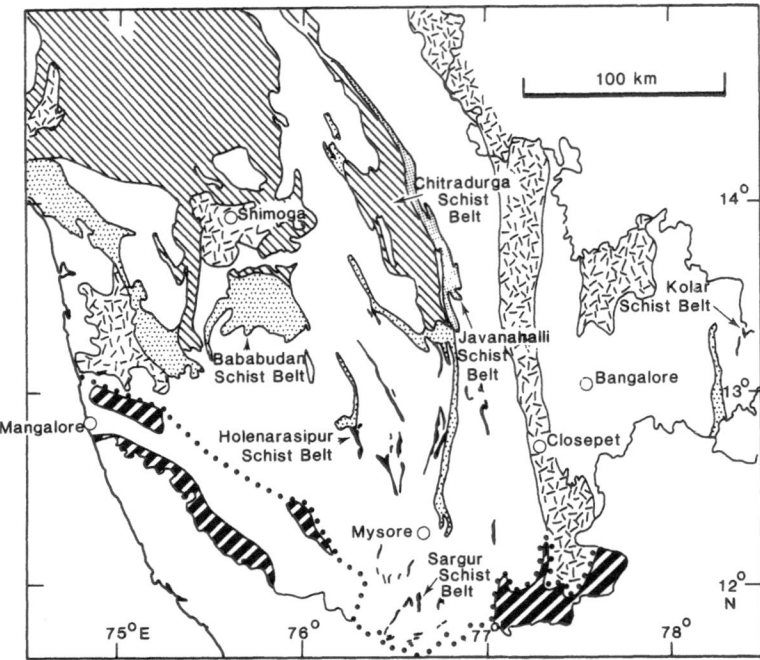

Figure 15.2 Sketch map showing the distribution of the major greenstone and high-grade metabasic suites in the central part of the Dharwar craton (Figure 15.1). NW–SE line shading: Chitradurga Group; dense stipple: Javanahalli Group; light stipple: Bababudan Group; black: Holenarasipur Group; bold NE–SW line shading: charnockites; cross hatching: younger granites; dotted line: boundary between low- and high-grade terranes. Redrawn from Naqvi et al. (1981).

compositions of the metavolcanic rocks range from basalts through andesite to rhyolite. Ultrabasic volcanic rocks (komatiites *sensu stricto*) are very rare or absent. The belts generally display multiple deformation, and often lie unconformably on the *c.* 3 Ga Peninsular Gneiss basement. Basic metavolcanics usually constitute only 10–20% of these belts, with a typical metamorphic mineralogy of amphibole + plagioclase ± quartz ± garnet ± epidote (Naqvi and Rogers, 1987). A reasonable body of petrological and geochemical data is available for these belts, although many belts have only been studied on a reconnaissance basis.

Drury (1983) studied the ultrabasic to acidic metavolcanics from the greenschist to amphibolite facies Bababudan, Kudremukh, Shigegudda (Dharwar to the west of the Clospet granite), and Kolar (Dharwar to the east of the Clospet granite) greenstone belts (Figure 15.2). The overall petrogenetic evolution of the metavolcanics in the different belts was apparently very similar. Few of the basic metavolcanics represent primary liquids, most having undergone substantial crystal fractionation of mafic phases from primary liquids. Both light rare-earth element (LREE)-depleted and enriched metavolcanics occur in all of the belts, and Drury (1983) suggested that this may be a function of partial melting of a single mantle source, with vapour phase transfer of LREE occurring in ascending mantle diapirs. Both the ultrabasic and basic metavolcanics are depleted in the high-field strength elements (HFSE) Nb and Ta relative to the LREE (La) and immobile large-ion lithophile elements (LILE) such as Th. They have La/Ta and Th/Ta ratios higher than those of modern MORB, but similar to ratios in modern subduction-related magmas. From this geochemical evidence, and considering that the belts are unconformable on slightly older crust, Drury (1983) concluded that the greenstone belts were formed in an ensialic back-arc environment, with magma production being associated with a shallowly dipping subduction zone.

Rajamani *et al.* (1985) described the geochemistry of metavolcanics from the Kolar schist belt, to the east of the Clospet granite. This belt is dominated by metavolcanic amphibolites, some of which are komatiitic (14–21% MgO), while the majority are meta tholeiites with 5.5–10% MgO, similar to the Archaean THI and THII tholeiite types defined by Condie (1981). Pillow structures are not uncommon. LILE abundances vary erratically and have apparently been affected by alteration. Both komatiitic and tholeiitic amphibolites from the eastern part of the Kolar schist belt have LREE-enriched chondrite-normalised REE patterns. Chondrite-normalised Ce/Yb (Ce_N/Yb_N) ratios range from 1.4 to 3.0, although most have Ce_N/Yb_N ratios greater than 2.2. In the central and western parts of the belt, komatiitic amphibolites vary from LREE-enriched ($Ce_N/Yb_N = 1.8$ to 2.1, but with $Ce_N < Nd_N$) to LREE-depleted ($Ce_N/Yb_N = 0.7$–0.8), while tholeiitic amphibolites with 6–7% MgO are LREE-depleted or have unfractionated REE patterns (Ce_N/Yb_N ratios from 0.7 to 1.1). These REE geochemical distinctions seem to represent a fundamental difference in mantle source compositions, the amphibolites from the eastern part of the belt having been generated from LREE-enriched mantle, those from the western and central parts of the belt having been generated from LREE-depleted mantle. Rajamani *et al.* (1985) considered that the tholeiites and komatiitic amphibolites were derived by similar degrees of melting (10–25%) but at different depths, the tholeiites having been derived from mantle at a depth of less than 80 km, and the komatiitic rocks from depths greater than 80 km. None of the metavolcanics appears to represent primary mantle melts. All have undergone extensive crystal fractionation. The komatiitic amphibolites, for example, may have undergone up to 20% olivine separation. All komatiitic amphibolites from the Kolar schist belt define a 2.9 Ga Sm–Nd isochron, and have ε_{Nd} values ranging from +2 to +8 (Balakrishnan *et al.*, 1986), implying long term depletion of Nd relative to Sm in the mantle source for all of these basic volcanics. LREE enrichment of the mantle source for the eastern Kolar amphibolites must therefore have occurred shortly prior to, or contemporaneously with, magma generation.

15.2.2 High-grade ultrabasic–basic complexes

Deformed and metamorphosed intrusive igneous complexes dominated by ultrabasic lithologies occur widely in the amphibolite to granulite facies regions of the Dharwar craton (Naqvi and Rogers, 1987). These complexes occur both in association with belts of supracrustal rocks, and as isolated enclaves in the Peninsular Gneisses. Some of these complexes display unequivocal relict igneous layering, while others have been interpreted as being of extrusive origin, largely based on tenuous textural interpretations such as relict pillow structures and spinifex textures which may equally be misidentified deformational textures, and ambiguous geochemical arguments (Naqvi and Rogers, 1987). However, most of these complexes have the characteristics of stratiform layered bodies (Janardhan *et al.*, 1978; Srikantappa *et al.*, 1984), and they are all considered as such here. These layered complexes typically show only minor development of plagioclase-rich cumulates.

Some of the most comprehensively described layered ultrabasic–basic complexes in the Dharwar craton are those occurring in the Sargur supracrustal belt in Karnataka (Srikantappa *et al.*, 1984). These layered complexes are intimately associated with metasediments. The complexes are either older than or synchronous with the enclosing Peninsular Gneisses. The $P-T$ conditions estimated for granulite facies metamorphism of the layered complexes suggest metamorphic pressures of 0.9 GPa and temperatures of 650–750°C (Srikantappa *et al.*, 1984).

The rock types present in the Sargur layered complexes range from dunite, harzburgite, lherzolite and bronzitite to gabbro, gabbroic anorthosite, anorthositic gabbro and minor anorthosite (Srikantappa *et al.*, 1984). The ultrabasic layers locally display cumulus textures and compositional layering, including the development of chromitite layers. The gabbroic components of the complexes are often gradational with the ultrabasic units, and also grade into the anorthositic gabbro and minor anorthositic rocks, which contain calcic plagioclase (An_{80}).

The ultrabasic rocks contain up to 44% MgO and have very high Cr and Ni contents (6000–2500 ppm Cr, 2200–1200 ppm Ni), while the gabbros have *c.* 13% MgO and 16% Al_2O_3, and gabbroic anorthosites and anorthosites 20–23% Al_2O_3 (Srikantappa *et al.*, 1984). Incompatible trace element abundances are uniformly low (e.g. Zr < 80 ppm) in all rock types. REE abundances in the gabbros and anorthosites are low ($Ce_N = 3$–12; $Yb_N = 2$–7), and these rocks vary from slightly LREE-depleted to slightly LREE-enriched types ($Ce_N/Yb_N = 0.7$ to 1.6). Small positive Eu anomalies occur in the anorthositic gabbros and anorthosites, and negligible or slight negative Eu anomalies occur in the gabbros. The chemical characteristics of all the units in the complexes are typical of cumulates. Srikantappa *et al.* (1984) considered the layered complexes to have been emplaced into the Sargur supracrustals, which include tholeiitic metabasalts which have some chemical affinities with modern ocean floor basalts (Janardhan *et al.*, 1978).

The metamorphosed ultrabasic to tholeiitic rocks of the amphibolite facies Holenarasipur supracrustal belt metavolcanics (Drury, 1983) have chemical characteristics very similar to those of the Sargur layered complexes described above, and it is likely that these are, in fact, fragments of layered complexes. In the Holenarasipur area, very fine-grained anorthosite lenses and sheets up to 20 m thick with a mineralogy of plagioclase (An_{92-95}) + hornblende ± garnet ± quartz, are concordant with enclosing ultramafics and were initially interpreted as anorthositic flows (Drury *et al.*, 1978; Naqvi and Hussain, 1979). These anorthosites have low REE abundances ($La_N = 7$–12; $Yb_N = 2$–5; $Ce_N/Yb_N = 1.6$–2.6) and moderate to large positive Eu anomalies ($Eu/Eu^* = 1.2$–9.0). Kutty *et al.* (1984) obtained a 3.1 Ga Rb–Sr isochron age for the Holenarasipur anorthosites and a low initial $^{87}Sr/^{86}Sr$ ratio (Sr_i) of 0.7016, and re-interpreted these anorthosites as minor differentiates of ultrabasic–basic layered complexes, based on the identification of relict cumulus textures in the less deformed bodies, and their chemical characteristics which are typical of plagioclase cumulates. The associated ultramafics are olivine + pyroxene cumulates, while the anorthosites are plagioclase + clinopyroxene cumulates (Kutty *et al.*, 1984). The recrystallisation of the anorthosites and destruction of primary igneous textures occurred during deformation.

15.2.3 Anorthosite complexes

In contrast to the ultrabasic–basic complexes described above, some high-grade layered complexes in the Dharwar craton show a major development of plagioclase cumulates (e.g. Windley and Selvan, 1975). The occurrence of this type of complex appears to be restricted to the amphibolite–granulite transition zone and the granulite facies terrane. The best documented of these layered ultrabasic–gabbro–anorthosite complexes is the Sittampundi complex (Subramaniam, 1956; Janardhan and Leake, 1975; Ramadurai et al., 1975), although this is only one of a large number of such complexes within the Dharwar craton.

Chromitite layers are commonly associated with the anorthosites in these complexes, in contrast with the chromitites which occur in the ultramafic units of the ultrabasic–basic complexes. The Sittampundi complex is approximately 1 km thick, and consists of three distinct cumulate zones (Ramadurai et al., 1975). A lower zone of gabbros with pyroxenite inclusions is overlain by a middle zone of chromite- and hornblende-bearing anorthosites displaying internal layering, which is in turn overlain by an upper zone of clinozoisite-bearing anorthosite. The gabbros comprise an assemblage of garnet + clinopyroxene + orthopyroxene + plagioclase + amphibole. There is so little plagioclase remaining in the Sittampundi gabbros that they have been considered to be eclogitic (e.g. Chappell and White, 1970). The anorthositic rocks comprise an assemblage of plagioclase + edenitic hornblende + anthophyllite + corundum + clinozoisite + garnet. The major element compositions of the Sittampundi gabbros and anorthosites are overwhelmingly controlled by the cumulus phases (plagioclase in anorthosites; plagioclase + mafic minerals in gabbros), while the very low abundances of incompatible trace elements reflect the small proportion of trapped intercumulus liquid relative to cumulus phases (Weaver, unpublished data).

As is the case with the layered ultrabasic–basic complexes, these anorthositic bodies are often associated with metasediments and metavolcanics. In both their occurrence and chemistry the gabbro–anorthosite complexes are comparable to similar complexes which occur widely in other Archaean terranes (e.g. Windley et al., 1981). There are numerous similarities between the Sittampundi complex and, for example, the well documented Fiskenaesset complex of southern West Greenland (e.g. Windley et al., 1981; Weaver et al., 1981; see Chapter 11), with the exception that there are significant mineralogical differences due to the higher grade of metamorphism in southern India. The Fiskenaesset complex and the enclosing metavolcanic amphibolites have been considered to represent pieces of Archaean oceanic crust on the basis of their chemical characteristics (Weaver et al., 1981, 1982; Hall et al., 1987a), and similar interpretations might be applicable to the Sittampundi and other similar southern Indian layered complexes.

15.2.4 High-grade basic lenses

Lenses of basic meta-igneous rocks are ubiquitous throughout the amphibolite and granulite facies gneiss terranes of the Dharwar craton. The amphibolite facies basic gneisses have a mineralogy of plagioclase + hornblende $\pm$ biotite, while the basic granulite facies gneisses comprise plagioclase + clinopyroxene + orthopyroxene + hornblende $\pm$ garnet $\pm$ biotite. The best studied basic lenses are granulite facies rocks in the type charnockite area of Pallavaram, Madras (Howie, 1955). Metamorphic conditions in this area are estimated at 750–850°C and 0.65–0.75 GPa (Bhattacharya and Sen, 1986). The basic granulites appear to pre-date the enclosing acid granulites (tonalitic to granitic charnockites) in this area (Sen *et al.*, 1970). A number of distinct types of basic granulite lenses have been identified (Weaver *et al.*, 1978: Weaver, 1980; Chakraborty and Sen, 1983). Most comprise pyroxene, hornblende–pyroxene, or metasomatised biotite granulites (Sen and Ray, 1971; Sen and Sun, 1980). Two compositionally different groups of basic granulites (groups I and II) are distinguished in that the group I granulites are more Fe-rich and have higher abundances of the incompatible trace elements Zr, Nb, Y and REE (Weaver *et al.*, 1978). Both groups are LREE-enriched ($Ce_N/Yb_N = 1.4$–3.9), but group I granulites are somewhat more so (Weaver, 1980). These two groups represent differentiated liquids (group I) and cumulate rocks (group II) respectively, although Chakraborty and Sen (1983) considered that the two groups evolved from different parental magmas and followed different low-pressure fractionation paths.

Sm–Nd isotope data for the Pallavaram ultrabasic, basic and acid granulites define an isochron age of 2.55 ± 0.14 Ga and an ε_{Nd} value of *c.* + 1.0 (Bernard-Griffiths *et al.*, 1987). T_{CHUR} ages for the ultrabasic and basic granulites range from 2.76 to 2.36 Ga. The isochron age could represent either the age of the protoliths (and imply derivation from a slightly depleted mantle source at this time), or a metamorphic age, particularly in view of the evidence for pervasive metasomatism of the Pallavaram granulites (Weaver, 1980), implying major REE mobility during metasomatism (Bernard-Griffiths *et al.*, 1987). The Sm–Nd systematics could, of course, date both protolith formation and metamorphism, if these occurred within a short time (*c.* 50 Ma) of each other.

Basic lenses from the amphibolite to granulite facies transition zone (e.g. Kabbaldurga quarry; Stahle *et al.*, 1987) appear to be compositionally similar to the lenses from the even deeper parts of the craton. The tholeiitic to andesitic plagioclase-bearing amphibolites in Kabbaldurga quarry are mildly LREE-enriched ($Ce_N/Yb_N = 1.7$). Their chondrite-normalised REE patterns are slightly concave downwards ($La_N/Sm_N > Gd_N/Yb_N$) and they have small negative Eu anomalies. Metasomatism of the amphibolites in association with the incipient granulite facies metamorphism has apparently resulted in increases in the LILE (K, Rb, Sr, Ba), and Zr and REE abundances (Stahle *et al.*, 1987).

15.2.5 *Dykes*

Basic dykes are relatively common in the gneiss terranes of the Dharwar craton. They predominantly trend E to ENE (Murthy, 1987). Radiometric ages for dolerite dykes in the Dharwar craton vary from *c.* 2.4 to 0.9 Ga, but most are older than 1.6 Ga (Murthy, 1987; Venkatesh *et al.*, 1987). The early Proterozoic tholeiitic dolerite dykes are distinct from younger, alkaline dykes (Murthy, 1987; Naqvi and Rogers, 1987; Venkatesh *et al.*, 1987).

In a palaeomagnetic and geochemical study of the dolerite dykes intruding the granulite facies terrane of Tamil Nadu, Venkatesh *et al.* (1987) identified four groups of dykes based mainly on differences in direction of remanent magnetisation. Pole positions for the dykes are consistent with an age of 2.0–1.6 Ga. All of these dykes are quartz-normative tholeiites, but two chemically distinct groups were recognised. Most of the dykes are relatively evolved, with low MgO, Ni and Cr abundances (Venkatesh et al., 1987).

Although comprehensive geochemical (and particularly trace element) data are lacking for these dykes, they appear to be mineralogically and compositionally similar to early Proterozoic dolerite dyke suites which intrude many other Archaean terranes (e.g. Tarney and Weaver, 1987a, b). These dykes have the chemical characteristics of continental tholeiites, such as LREE enrichment and low HFSE (Nb and Ta) contents, consistent with their derivation from an enriched sub-continental lithosphere formed during the generation of the overlying continental crust (e.g. Tarney and Weaver, 1987a, b).

15.3 The Singhbhum craton

The Singhbhum craton (Figure 15.1) incorporates the Bhandara and Singhbhum cratons described by Naqvi and Rogers (1987), and is equivalent to the Singhbhum protocontinent as defined by Naqvi et al. (1974). The southwestern part of the Singhbhum craton is dominated by post-Archaean gneisses and granites. These gneisses form the basement to a number of supracrustal sequences, although some granites/gneisses appear to intrude these supracrustals (Naqvi and Rogers, 1987). The only possibly Archaean supracrustal rocks are those of the Amgaon Group of the Dongargarh Supergroup (Sarkar et al., 1981), which includes metabasalts interbedded with metasediments and granitic gneisses. Little information is available on the petrology or chemistry of these metavolcanics. The younger Nandgaon Group of the Dongargarh Supergroup (c. 2.3–2.2 Ga; Sarkar et al., 1981) comprises olivine- and quartz-normative tholeiites and rhyolites, but few petrological data are also available for these rocks. Many other supracrustal sequences whose ages are highly uncertain also contain variable amounts of metavolcanics.

In the northeastern part of the Singhbhum craton, the nucleus of Archaean gneisses, granites and supracrustals is separated from younger rocks to the north (possibly with a late Archaean component) by the Singhbhum thrust (Naqvi and Rogers, 1987). Radiometric ages obtained for the oldest components of the Singhbhum nucleus vary considerably. A Sm–Nd age of c. 3.78 Ga was obtained for the Older Metamorphic Group (OMG) from a poorly constrained isochron by Basu et al. (1981). A possibly more reliable Rb–Sr isochron age of c. 3.2 Ga (Sarkar et al., 1979) suggests that the Singhbhum craton may not have the antiquity claimed by Basu et al. (1981). The Singhbhum Granite intrudes the OMG and has yielded a Rb–Sr isochron age of c. 2.9 Ga (Sarkar et al., 1979). These and younger (undated) units of the craton are intruded by 1.5–1.0 Ga old dolerite dykes (Sarkar and Saha, 1983). The OMG contains low-K tholeiitic metavolcanic amphibolites (Saha and Ray, 1984), although their low K_2O contents may also be a metamorphic feature. The REE patterns of these metavolcanics are mildly LREE-enriched and display small negative Eu anomalies (Saha and Ray, 1984).

A number of belts of supracrustals which contain metavolcanic rocks also occur around the Singhbhum granite (Sarkar and Saha, 1983) and although they are not radiometrically dated, it has been suggested that they are slightly older than, or possibly synchronous with, the Singhbhum granite (2.9 Ga). Basaltic and ultramafic pillowed flows and metavolcanic amphibolites and tuffs are common in these belts. Their major element chemistry suggests that some of these rocks represent island arc basalts, some erupted as flows on continental platforms, with compositions between those of modern continental flood basalts and oceanic tholeiites, and some were intrusive into the supracrustals (Banerjee, 1982). Olivine- and quartz-normative dolerite dykes which are quite common in the Singhbhum nucleus have yielded radiometric ages which range from c. 1.6 to 1.0 Ga (Naqvi and Rogers, 1987).

15.4 The Aravalli craton

Most of the rocks forming the Aravalli craton (Figure 15.1) are less than 2.0 Ga old. The main Archaean and early Proterozoic units are the Banded Gneiss Complex (BGC), Bhilwara Group and Aravalli Group in the western part of the craton, and the Bundelkhand Complex and Bijawar Group in the eastern part of the craton (Naqvi and Rogers, 1987). The BGC comprises tonalitic to adamellitic gneisses, granites and metasediments, and includes a component of amphibolites which have a mineralogy of amphibole + plagioclase + quartz ± clinopyroxene ± garnet ± biotite. The metamorphic grade of the craton ranges from amphibolite to granulite facies. Only a limited amount of chemical data is available for the amphibolites within the BGC (McDougall *et al.*, 1983). The amphibolites are low-K (0.24–0.5% K_2O) metabasaltic rocks, with slight to moderate LREE enrichment. These amphibolites and enclosing grey gneisses have yielded a Sm–Nd isochron age of 3.5 Ga with an ε_{Nd} value of +3.5 (McDougall *et al.*, 1983), and both were derived from similarly depleted sources. Granulite facies basic rocks also occur in the BGC (e.g. Naqvi and Rogers, 1987).

The Bhilwara Group is entirely metasedimentary, while the Aravalli Group is predominantly composed of greenschist facies metasedimentary rocks with minor amounts of metavolcanics. The metasedimentary associations and chemistry of the metavolcanics (medium- to high-K tholeiitic andesites and andesites) suggest that the Aravalli Group developed in an active continental margin setting (McDougall *et al.*, 1985). The Bundelkhand Complex comprises volcano–sedimentary schists and granitic gneisses, with an apparent age of > 2.6 Ga (Naqvi and Rogers, 1987). Metavolcanics, gabbroic intrusives and basic dykes are all present in the complex (Banerjee, 1982; Sharma, 1982, 1983), but there is no detailed information about any of these units. The Bijawar Group consists predominantly of a sedimentary sequence with some pillow-structured basalts towards the base of the succession (Naqvi and Rogers, 1987) which have been dated at 2.6–2.4 Ga (Crawford and Compston, 1970). Petrographic and chemical data are lacking. Mafic dykes are relatively scarce in the Aravalli craton, but appear to range in age from c. 2.4 to 1.7 Ga (Murthy, 1987). Very little petrological data are yet available for these dykes.

15.5 Discussion

15.5.1 *Geochemical variation of basic rocks with metamorphic grade*

Certain trace element characteristics of basic igneous rocks may be used to model their petrogenesis and, by comparison with modern basic suites, perhaps to assign a formative tectonic setting. However, in the comparison of rocks of different metamorphic grade, account must be taken of the potential mobility of trace elements during metamorphism. In greenschist to amphibolite facies greenstone belts, LILE abundances vary erratically compared to other less mobile trace elements (e.g. REE, Zr, Cr,

Y etc.), and are of little use in elucidating the petrogenesis of the metavolcanics (Drury, 1983; Rajamani *et al.*, 1985). However, the REE, Zr, Nb, Ta, Y, Cr, Ni have been successfully used in petrogenetic modelling (Drury, 1983; Rajamani *et al.*, 1985).

The characteristic depletion of radioactive heat-producing elements (Rb, Th, U) in granulite facies relative to lower metamorphic grade gneisses is well known (e.g. Tarney and Windley, 1977). This depletion may be the result of either the removal of a melt phase from the lower crust during granulite facies metamorphism, or of the flushing of the lower crust by a mantle-derived CO_2-rich fluid during granulite facies metamorphism (carbonic metamorphism). In both cases radioactive heat-producing elements are transferred from refractory granulite facies lower crust to higher crustal levels via migration of a melt or fluid phase. Metasomatism of intermediate level (amphibolite facies) crust may occur in response to the introduction of lower crustally derived melts/fluids. The role of carbonic metamorphism is stressed in the production of the granulite facies gneisses in the Dharwar craton (Janardhan *et al.*, 1979, 1982; Condie *et al.*, 1982; Condie and Allen, 1984; Hansen *et al.*, 1987).

It is generally considered that the transition from amphibolite to granulite facies metamorphic assemblages is isochemical for all rock compositions, with the exception of depletion in the radioactive heat-producing elements (Tarney and Windley, 1977; Weaver and Tarney, 1981a; Condie and Allen, 1984). In the amphibolite–granulite transition zone and lower pressure granulite regions of the Dharwar craton, the situation is complicated by the occurrence of metasomatism and migmatisation shortly prior to granulite facies metamorphism (e.g. Weaver, 1980; Stahle *et al.*, 1987), and LILE depletion does not necessarily correlate with metamorphic grade (Condie and Allen, 1984; Stahle *et al.*, 1987). Metasomatism of ultrabasic and basic gneisses is certainly evident in the granulites of Pallavaram area (Bernard-Griffiths *et al.*, 1987) and the incipient granulites at Kabbaldurga (Stahle *et al.*, 1987), and can result in significant REE and LILE enrichment which survives granulite facies metamorphism. Therefore the interpretation of the trace element characteristics of metasomatised high-grade basic gneisses must be made with caution.

15.5.2 *Early Precambrian basic magmatism and evolution of the Indian Shield*

The Indian subcontinent clearly comprises a number of early Precambrian nuclei which are separated from each other by major rifts or lineaments (Naqvi and Rogers, 1987). There may be as few as three or as many as seven such individual nuclei (Naqvi *et al.*, 1974; Naqvi and Rogers, 1987). However, it is not clear if the rifts are intra-continental features developed within a single large craton, or whether they mark suture lines between smaller, originally independent cratonic nuclei which have been assembled into the Indian Shield (Naqvi and Rogers, 1987). Substantial differences in the geology of the exposed cratons of the subcontinent perhaps favour the interpretaion that accretion of separate, small nuclei occurred. In this case, the precise timing of the assembly of these nuclei to form the Indian Shield is uncertain, but the accretion of the shield was certainly complete by 1.5 Ga ago (Naqvi and Rogers, 1987).

If the accretion of small continental nuclei to a growing Indian Shield occurred via the operation of some form of plate tectonics, then ancient convergent plate boundaries should be marked by ophiolite suites and arc-related volcanics (although these may be neither well preserved nor exposed). Unfortunately, it is extremely difficult to fit the early Precambrian basic rocks of the various cratonic nuclei of the Indian shield into any overall tectonic scheme. The geochemical data on which such interpretations rely are lacking for many of the basic suites, and the chronological relationships are also not well known. This is especially true of the Aravalli and Singhbhum cratons. The chronology of events within the Dharwar craton is somewhat better known, and a reasonable body of geochemical data exists for both basic and acid components of this craton. The following general sequence of cratonic development and associated basic igneous activity can be suggested.

Layered basic complexes and associated metasediments and metavolcanics are arguably the oldest components of the craton, or at least they are synchronous with the protoliths of the Peninsular Gneisses. The distinction between the layered ultrabasic-basic complexes and anorthositic complexes seems to be simply the extent of development of plagioclase-rich cumulates and their crystallisation sequences, with chromite crystallising earlier in the ultrabasic–basic complexes. The geochemical differences between the two types of layered complexes are small. These complexes (and the associated metavolcanics) may represent pieces of Archaean oceanic crust emplaced into the deeper levels of an accreting Archaean continent, as suggested for comparable complexes elsewhere (Weaver et al., 1982; Hall et al., 1987a).

The amphibolite and granulite facies Peninsular Gneisses are similar to the tonalitic–trondhjemitic–granodioritic (TTG) grey gneisses which dominate most Archaean terranes (Condie et al., 1982; Condie and Allen, 1984). The chemical characteristics of these gneisses are comparable to those of modern moderate to deep crustal intrusions in active continental margins, and a widely advocated model for the origin of Archaean TTG gneisses is that they were formed from magmas generated by partial melting of (perhaps garnet-bearing) amphibolite chemically similar to Archaean THII tholeiites (e.g. Condie et al., 1982; Condie and Allen, 1984), and underplated to the base of accreting crust above a subduction zone. In this environment slices of subducted oceanic crust could be emplaced into the deeper levels of the accreting continental crust (Weaver et al., 1982; Tarney and Weaver, 1987a). This could account for the chemical, chronological and structural relationships between the layered ultrabasic–basic and anorthositic complexes and the enclosing grey gneisses.

The concordant basic lenses in the gneisses are chemically distinct from the layered complexes and associated metavolcanics in that they have more fractionated REE patterns than the metavolcanics, and also tend to have higher La/Nb ratios which are more closely comparable to ratios in subduction-related magmas (see Chapter 3). These basic lenses most probably represent highly disrupted, deformed dykes which were intruded synchronously with the emplacement of the TTG precursors to the gneisses.

Interpreting the origin of the ultrabasic and basic metavolcanics within the greenstone belts involves the more general problems concerning the origin of komatiitic magmas and tectonic models for greenstone belt formation. An ensialic back-arc basin setting is one model proposed for greenstone belt formation (Condie, 1981; see Chapter 3), and certainly some of the characteristics of the greenstone belts of the Dharwar craton are consistent with this model (Drury, 1983). However, the precise origin of the greenstone belt metavolcanics is still somewhat enigmatic.

The dominantly early Proterozoic dolerite dykes intruded into the Dharwar, Aravalli and Singhbhum cratons (Murthy, 1987) appear to be comparable both mineralogically and chemically to the ubiquitous quartz-normative tholeiitic dolerite dyke suites intruded into many Archaean cratons, apparently during final craton stabilisation (Halls and Fahrig, 1987). The general chemical characteristics of these dykes are consistent with the derivation of the dyke magmas from an enriched sub-continental lithosphere source which was perhaps formed synchronously with the new overlying crust (Tarney and Weaver, 1987a, b). The subduction-related chemical signature (HFSE depletion relative to LILE and LREE) of these dykes is then a primary feature, imparted to the accreting sub-continental lithosphere by subduction zone hydrous fluids, and does not reflect crustal contamination.

Some of the elements of such a tectono-magmatic model are still highly tentative, especially applied to the evolution of the Indian Shield where in general the structural, chronological and geochemical relationships between the various components of the shield are not so well established as in other Archaean terranes. However, the Indian Shield, and especially the crustal section displayed by the Dharwar craton, provide excellent terranes in which with further geochemical data such crustal evolution models can be tested in the future.

16 Early Precambrian basic rocks of Africa

H.S. SMITH

16.1 Introduction

African geology has an important place in the study of the history of our planet because it contains rocks of all ages from the early Archaean to the Quaternary. Unfortunately, '...the details of the geology of Africa are still not quite clear...' (Kogbe, 1983) or at least they are heterogeneously known. Africa contains large tracts of early Precambrian rocks. Whereas a great deal of detailed information is available for the southern African terranes, less is known about the Precambrian terranes of West Africa and only sparse information is available from the other old cratons. A large proportion of this information is not available internationally as it is contained in unpublished company reports, local survey publications and unpublished theses. This review of mafic–ultramafic volcanism in the early Precambrian of Africa is based on information contained, as far as possible, in the international literature. This is not to underrate the efforts of all those geologists who have puzzled over the problems of the Archaean in Africa, but is done to enable the reader to access the available information.

The early Precambrian rocks of Africa mainly consist of gneisses, migmatites, granites (*sensu lato*), and remnants of supracrustal greenstone belts. The greenstone remnants are preserved on all scales from belts covering thousands of square kilometres to small xenoliths in gneissic rocks. The metamorphic grade of the basic metavolcanic rocks varies from very rare, virtually unaltered lava flows (Nisbet *et al.*, 1987) to more common lower greenschist to granulite facies. In this chapter the geological features of early Precambrian basic rocks in Africa will be discussed mainly with respect to those preserved within the larger greenstone belts. In this respect the Barberton greenstone belt is one of the classic exposures, as it was here that Viljoen and Viljoen (1969a, d, e) first recognised and defined the new rock types komatiite and basaltic komatiite (now termed komatiitic basalt; Arndt and Nisbet, 1982b) and that detailed geological work by Anhaeusser *et al.* (1969) stimulated renewed interest in greenstone belts worldwide.

It should be noted that in Africa the boundary between the Archaean and Proterozoic is taken as the age of the Great Dyke (2.46 Ga) in Zimbabwe (Nisbet, 1982). This definition is not without its problems, particularly in the Kaapvaal Craton, but is useful for most of the early Precambrian rocks in Africa. The definition of komatiites and komatiitic basalts from Viljoen *et al.* (1983) is used in this chapter. Volcanic rocks with $> 24\%$ MgO (volatile free) are considered to be komatiites and associated lavas with $8–24\%$ MgO (volatile free) are considered to be komatiitic basalts.

16.2 Distribution of Archaean rocks in Africa

The areas of outcrop of Archaean rocks and extrapolated continuations under younger cover are shown in Figure 16.1. In places these cratonic blocks are bounded by rocks originally of Archaean age which have been reworked during younger tectono-thermal events. Most of the central and eastern portions of Malagasy (Madagascar) consist of this type of material (see Figure 24.2 in Cahen *et al.*, 1984). The cratonic nuclei themselves contain Archaean mobile belts such as the Kasila Belt in West Africa and the Limpopo Belt in southern Africa. In fact, some of the oldest crustal material identified on the African continent has been postulated to exist in one of these belts as Barton and Ryan (1977) have published an Rb–Sr isochron age of *c.* 3.86 Ga for the Sand River gneisses of the central Limpopo Belt. Recent U–Pb determinations on zircons from a quartzite from the central Limpopo Belt indicate that at least some components of the quartzite were derived from a very ancient (> 3.8 Ga) provenance area (Armstrong *et al.* unpubl. data).

The question of basement rocks to supracrustal greenstone belt remnants has long been a controversial topic in most Archaean terranes. The oldest ages reported for Archaean cratons adjacent to the Limpopo Belt are from gneisses and not greenstone rocks. For one greenstone belt at least, the basement/cover relationships are clear. For instance, the younger greenstones of the Ngezi Group, part of the Belingwe (Mberengwa) greenstone belt in Zimbabwe, were deposited on older tonalitic crust (Bickle *et al.*, 1975).

Figure 16.1   Distribution of Archaean rocks in Africa (after Cahen *et al.*, 1984).

The oldest ages reported for the Kaapvaal Craton are also from zircons in tonalitic gneiss from the Ancient Gneiss Complex. Compston and Kroner (1988) have reported U–Pb ages from these zircons in the range 3.64–2.87 Ga. They interpret the oldest age as the time of magmatic precipitation of zircon and consequently demonstrate that at least part (some zircon crystals) of the Ancient Gneiss Complex is older than anything yet identified within the adjacent Barberton greenstone belt. However, it remains to be demonstrated what the precise geological relationships were between the rock units that encased these old zircon crystals and the Barberton greenstone belt at the time of volcanic activity. Clearly, the recognition of basement to the greenstone belts is critical to the development of tectonic models. Altered and/or metamorphosed greenstone belt rocks have proven extremely difficult to date absolutely, and consequently the best age constraints that can be obtained for the belts are often derived by dating intrusives and basement granite-gneisses. Only in a few instances have reliable ages been obtained directly from the greenstone belt mafic and ultramafic volcanic rocks.

16.3 North and West Africa

Three Archaean cratonic blocks have been identified in North and West Africa (Figure 16.1), namely the Reguibat Rise on the northern side and the Guinea Rise on the southern side of the West African Craton, and the Uweinat Inlier in central North Africa. The most detailed information comes from the Guinea Rise, while comparatively little is known about the other areas. Recently, Klemm et al. (1984) have proposed that metavolcanic–sedimentary sequences in southwestern Nigeria are similar to typical Archaean greenstone belt assemblages and are probably of Archaean age. However, isotopic age determinations have not yet been obtained.

16.3.1 Uweinat Inlier

The basement rocks of the Uweinat Inlier consist of gneisses and migmatitic gneisses in the central and southeast areas, while in the south a group of pyroxene granulites, charnockitic gneisses and quartzites have undergone retrogressive metamorphism at amphibolite facies (Klerkx and Deutsch, 1977; in Cahen et al., 1984; Klerkx, 1980). Cahen et al. (1984) have recalculated the Rb–Sr ages and re-evaluated some of the isotope data obtained by Klerkx and Deutsch (1977) and suggest that the granulite facies metamorphism occurred at 2.9 Ga and retrogressive metamorphism at 2.63 Ga. Model Nd ages (Harris et al., 1984) indicate that the charnockites at Karkur Murr are 3.2–3.0 Ga.

16.3.2 Reguibat Rise

Bessoles (1977) has suggested that the Reguibat Rise should be subdivided into a southwestern province consisting predominantly of Archaean basement gneisses, and an eastern province consisting of Proterozoic granites and associated rhyolites, metamorphosed volcanic and sedimentary rocks and small enclaves of Archaean basement gneisses (see Figure 18.1 in Cahen et al., 1984). In the southwestern province, the Amsaga Group is made up of migmatites, granites and pegmatites and the Saouda metamorphic rocks (Barrere, 1967). The rocks from the Saouda Group consist of charnockitic rocks, granulites, pyroxene amphibolites, sillimanite and biotite gneisses, marbles and ferruginous quartzites. This group is intruded by metagabbros, amphibolites, anorthosites, serpentinites and granitoid rocks (Barrere, 1967). The Amsaga assemblage is about 10 km thick (Cahen *et al.*, 1984), and Bessoles (1977) suggested that the stratigraphy established for this assemblage is applicable to the Ghallaman assemblage further to the east. Cahen *et al.* (1984) recalculated the available Sr isotope data from the Amsaga granulite facies gneisses to obtain an isochron age of 2.7 Ga. They suggested that this age dates an episode of migmatisation with the formation of granites and pegmatites. Older events are identifiable in the Ghallaman assemblage gneisses.

16.3.3 *Guinea Rise*

The Guinea Rise forms the southern part of the West African craton and Cahen *et al.* (1984) considered the area as two domains. The western region, the Kenema–Man domain, consists mainly of rocks older than 2.7 Ga while the eastern region, the Baoule–Mossi domain, comprises rocks mostly of 2.1 Ga or younger. The Archaean of the Kenema–Man domain covers Sierra Leone, Liberia, the bordering parts of Guinea and the Ivory Coast (Figure 16.2; after Rollinson, 1978) and consists of typical Archaean greenstone belts surrounded by granite-gneisses and intrusive granitoids. The granite–greenstone terrane on the southwestern margin of Sierra Leone is bordered by a linear belt of highly metamorphosed (granulite facies) supracrustal

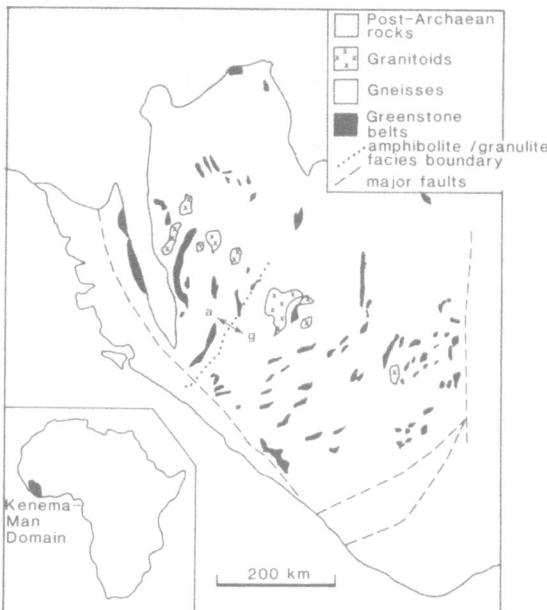

Figure 16.2 Kenema-Man granite–greenstone terrane (western Guinea Rise) of Sierra Leone, Guinea, Liberia and Ivory Coast showing the distribution of major greenstone belts and the boundary between principally amphibolite (a) and granulite (g) facies rocks (after Rollinson, 1978).

rocks, the Kasila Group mobile belt (Williams and Williams, 1976; Williams, 1978, 1988).

Two tectono-thermal events and two ages of greenstone belt formation have been identified in Sierra Leone (Beckinsale et al., 1980; MacFarlane et al., 1981). The older greenstones, the Loko Group, consist of amphibolites, serpentinites and meta-sediments which have been affected by the Leonian tectono-thermal event. This was followed by the formation of the Kambui Supergroup greenstone belt assemblages which were in turn affected by the Liberian tectono-thermal event. There is some confusion in the literature as to what constitutes the Kambui Supergroup. Williams (1988) considered that it incorporates all Archaean supracrustals in Sierra Leone and several authors consider that there was only one greenstone-forming event (Williams, 1978, 1988; Rollinson, 1978; Rollinson and Cliff, 1982; Umeji, 1983). Clearly, this problem is fundamental in interpreting the crustal evolution of the region.

The ages of the two tectono-thermal events are fairly well established. The older Leonian event is considered to have occurred at about 2.97 Ga based on a Rb–Sr isochron age of c. 2.98 Ga of whole-rock granite–gneiss samples from southeastern Sierra Leone and a Pb–Pb whole-rock isochron age of 2.96 Ga for granite–gneisses west of the Sula Mountains. These latter rocks have a Rb–Sr whole-rock isochron age of 2.75 Ga which is discordant with the Pb–Pb age. Beckinsale et al. (1980) interpreted the slightly younger age as being definitive for the younger Liberian tectono–thermal event. Cahen et al. (1984) have reviewed other isotopic data available for the Liberian event and support the age of 2.75 Ga. The time of formation of the greenstone belts is not established but minimum ages can obviously be ascertained from the ages of the tectono-thermal events. The Loko Group rocks described by Beckinsale et al. (1980) and MacFarlane et al. (1981) must be older than 2.97 Ga and the Kambui Supergroup must be between this age and c. 2.75 Ga.

Rollinson and Cliff (1982) established Rb–Sr isochron ages for the older tonalitic gneisses at 2.79–2.77 Ga, which is interpreted as either the time of formation of the original tonalites or their metamorphism. They also reported three new Rb–Sr age determinations for the younger granites of virtually the same age: 2.78 ± 0.14 Ga, 2.78 ± 0.079 Ga and 2.77 ± 0.05 Ga. These ages are all within error of each other, and clearly suggest a major event at 2.78 Ga, in agreement with the Cahen et al. (1984) estimate. If the interpretation of Rollinson and Cliff (1982) is correct that the Rb–Sr age of the old gneisses represents their protolith age, then the greenstone belts must have formed at c. 2.78 Ga.

16.3.4 Greenstone belts of Sierra Leone

The stratigraphy of the Sierra Leone greenstone belts has been described by Williams (1978), Rollinson (1978), MacFarlane et al. (1981) and Umeji (1983). As noted previously, some workers have interpreted two greenstone belt formation events and others only one. Rollinson (1978) presented a comparative stratigraphy of the supracrustal rocks in the West African Archaean craton (Figure 16.2). They fall into two groups based on size, thickness of the lithologies, metamorphic grade and geographical distribution. One group occurring in western Sierra Leone consists of greenstone belts up to 130 km long, with successions up to 6.5 km thick and metamorphosed to amphibolite facies (565–595°C; 0.49–0.55 GPa; Rollinson, 1982). BIF constitutes a relatively minor component. The second group occurs as belts in southeast Sierra Leone, Liberia and Ivory Coast and are much smaller in extent (< 40 km long), comprise much thinner stratigraphic sections (usually less than 1 km thick), contain BIF as a dominant component and are metamorphosed to granulite facies (770°C and 0.75 GPa). The proposed division between dominantly amphibolite facies and granulite facies belts is shown in Figure 16.2. Rollinson (1978, 1982) considered that the two groups are coeval.

The older group of greenstones lies on older sialic basement and mainly comprises metavolcanic amphibolites overlain by subordinate quartzites and BIF (MacFarlane et al., 1980). These rocks belong to the Loko Group and are generally poorly exposed and largely confined to the Kamakwie area (Figure 16.2). A suite of mafic intrusions is contemporaneous with the Loko Group. The structural relationship between the older gneisses and the Loko Group rocks are not documented in detail. These rocks were all affected by the Leonian (2.97 Ga) tectono-metamorphic event.

Between the Leonian and Liberian (2.75 Ga) events, typical Archaean greenstone belts were formed which belong to the Kombui Supergroup. There are four major greenstone belts in Sierra Leone, namely those of the Sula–Kangari belt, the Kambui Hills, the Nimini Hills and the Gori Hills (Figure 16.2). The most important member of the Kambui Supergroup is the Sula Group (Williams, 1978; Beckinsale et al., 1980; MacFarlane et al., 1980) which is divided into the lower predominantly metavolcanic Sonfon Formation and the overlying metasedimentary Tonkalili Formation.

The Sonfon Formation volcanics consist of massive lavas with pillowed tops, the earliest phase consisting of ultramafic schists. The ultramafic rocks can be as much as 1 km thick (Williams, 1978) and are overlain by amphibolites which reach 2.5 km thick. MacFarlane et al. (1980) suggested that the ultramafic rocks may be komatiitic in composition and Rollinson (1978) suggested that some may have been extrusive. Chromite is common in the ultramafic rocks of the Kambui and Gori greenstone belts, often forming chromite-rich bands (Umeji, 1983). Olivine (Fo_{92}) is also preserved in these ultramafic rocks, but is replaced by serpentine in places.

The mafic rocks consist of a wide variety of originally gabbroic or basaltic units which occur now as massive amphibolites. Pillows from a few tens of centimetres to 1.5 m wide are present with vesicles concentrated towards their upper parts. Umeji (1983) attempted to fit the greenstone belt assemblages into the Barberton-type stratigraphic model of Anhaeusser et al. (1969). While there are obvious similarities between the Sierra Leone greenstone belt stratigraphy and the Barberton rocks, there are some striking differences and the similarities may be superficial and should be treated with caution until detailed structural information is available for the belts. For example, no komatiitic basalts have been reported and the sediments contain significant amounts of BIF which is relatively rare in the Barberton greenstone belt volcanic successions.

16.4 Equatorial Africa

The Archaean cratons in equatorial Africa are poorly defined but cover a huge area in Cameroun, Gabon, Angola, Zaire, Uganda, Kenya and Tanzania (Figures 16.1 and 16.3). This Archaean province is surrounded and, in part, reworked by younger Proterozoic mobile belts (Kröner, 1977), suggesting that the province may once have been much larger. The best preserved supracrustal rocks are in the eastern and northeastern areas (Figure 16.3) where the metamorphic grade is low. Remnants of

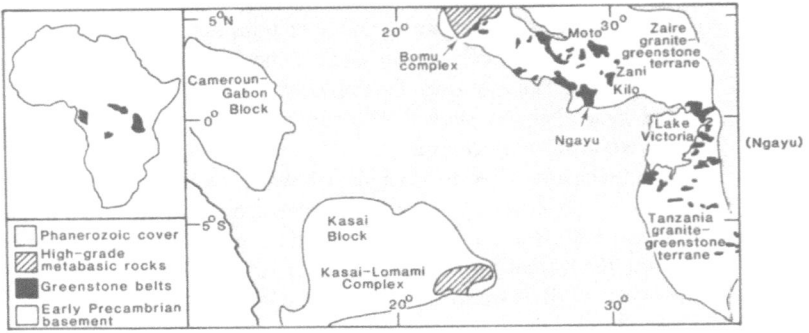

Figure 16.3 Archaean terranes of Equatorial Africa.

amphibolite and granulite facies supracrustal rocks occur in the West Nile Complex, the Dodoman System and in the Bomu Complex. Prian *et al.* (1988) described metamorphosed BIF and an interesting occurrence of early Proterozoic komatiitic rocks associated with andesitic material, pelites and basal conglomerates, resting on Archaean granite–gneiss basement in central Gabon in the Cameroun–Gabon block. Caen-Vachette *et al.* (1988) have provided age constraints on at least two greenstone belt formation events in Gabon.

The Archaean rocks of the Kasai block (Figure 16.3) comprise two major units (Cahen *et al.*, 1976). The Dibaya Complex consists of granitic to tonalitic migmatites and anatectic alkaline granites. Sphene extracted from one of the migmatites gave concordant U–Pb ages of 2.68 Ga (Delhal *et al.*, 1975) and an Rb–Sr isochron age defined by nine granite samples is 2.59 ± 0.09 Ga. South of this complex, the Kasai–Lomami Complex is comprised of gabbros, norites, enderbites, metasedimentary granulites, alaskitic rocks and metadolerites. The basic rocks may originally have been intrusive or volcanic. All the rocks from this complex have undergone granulite facies metamorphism at about 2.82 Ga (Cahen *et al.*, 1984). Locally, the rocks underwent retrograde metamorphism at or before 2.42 Ga.

Two major units have been recognised in the Cameroun–Gabon Archaean block (Cahen *et al.*, 1976). The 'Complexe calcomagnesien' contains enderbites, charnockites and metadolerites whereas the second unit consists of a heterogeneous suite of granitoid rocks with an older grey granitoid suite and a younger potassic granite suite. Recent geochronological work and a more detailed breakdown of the Archaean geological history of Gabon are summarised by Caen-Vachette *et al.* (1988) (Table 16.1).

The Archaean block in northeast equatorial Africa covers a vast area from 5°N to 5°S and 24° to 35°E (see Figure 16.3). The Archaean block in northern Zaire and the Central African Republic is separated from the Archaean block in Tanzania by later geological events in the vicinity of the Western Rift (Cahen *et al.*, 1984). For simplicity the northwestern Archaean block which covers the southern Central African Republic, southern Sudan, western Uganda and northeastern Zaire will be referred to as the Zaire granite–greenstone terrane. The second area covering southern Uganda, southern Kenya and most of Tanzania will be termed the Tanzania granite–greenstone terrane.

Table 16.1 Summary of the Archaean geological history of Gabon

Age range (Ga)	Geological events
2.5–2.4	Lower Proterozoic events
2.7–2.6	Epizonal metamorphism in Chaillu and magmatism in northern Gabon and Chaillu/
2.85–2.7	High-grade metamorphism of western Makokou gneisses and main magmatic phase in northern Gabon and Chaillu
3.0–2.85	High-grade metamorphism of eastern Makokou area and Chaillu gneisses and emplacement of the Nounah–Ivindo greenstone belt/
3.15–3.0	High-grade metamorphism of bimodal gneisses in Northern Gabon and emplacement of the Afoumadzo greenstone belt
3.9–3.2	Formation of protoliths of a basic sequence followed by the proto-liths of the enderbitic sequence/

Data from Caen-Vachette *et al.*, 1988/

16.4.1 *Zaire granite–greenstone terrane*

Granitic–gneissic rocks of the Zaire granite–greenstone terrane have been divided into three groups. The oldest generation consists mainly of tonalites, diorites and granodiorites. A second group is made up of quartz monzonites which are intrusive into the first group (Cahen *et al.*, 1984), and a third group of potassic granites has also been recognised, some of which may not be Archaean. The areal extent of each group is poorly known although the older tonalites have been reported only in close association with greenstone belts. The oldest tonalites from the region have yielded ages of 2.89–2.72 Ga (Lavreau, 1980, in Cahen *et al.*, 1984). Monzonites from the second generation granitoids gave ages of 2.51 ± 0.06 Ga and 2.46 ± 0.026 Ga. The third group of granitoids has not yet been dated.

The greenstone belts have not been dated directly, but minimum ages of 2.9–2.7 Ga can be inferred from the intrusive granitic rocks. In the Moto and neighbouring Zani greenstone belts two lithostratigraphic units can be distinguished: a lower unit of metavolcanic rocks from mafic to intermediate composition with associated BIF, and an upper, probably younger unit of volcanic agglomerates, meta-andesites, quartzites and BIF. The lower unit at Moto was intruded by the 2.89 Ga old tonalites while the upper unit was intruded by the 2.51 Ga old monzonites. Cahen *et al.* (1984) reviewed the geochronological data from the area and suggested that the Moto, Zani, Kilo and Ngayu greenstone belts (or at least the older portions of the belts) are older than *c.* 2.84 Ga. The three greenstone belts in the west of the Upper Zaire Granitoid Massif are referred to collectively as the Ganguan belts, for which a minimum age of 2.98 Ga has been established (Cahen *et al.*, 1984). The Bomu amphibole–pyroxene gneisses in this area have been dated at 2.96 ± 0.068 Ga, which is believed to be an age of reworking of pre-existing rocks. The northern part of the Ganguan belts rests unconformably on the Bomu mafic gneisses.

Two big greenstone belts occur in the Central African Republic (Poidevin *et al.*, 1981). The Bandas belt is about 250 km long and the Dekoa belt about 150 km long. They are believed to be a continuation of the greenstone belts in the Zaire granite–

greenstone terrane. At present, the geochronological constraints are not sufficiently good to establish if there was more than one greenstone formation event in this Archaean block.

16.4.2 *Tanzania granite–greenstone terrane*

The central plateau region of Tanzania is essentially composed of Archaean rocks and is entirely surrounded by *c.* 2.0 Ga mobile belts. Clifford (1970) divided the Tanzanian Craton into a central region of granites, gneisses and migmatites associated with metamorphosed supracrustals of the Dodoman System, and a northern portion embracing northern Tanzania, western Kenya and southern Uganda where the Nyanzian System (ultrabasic rocks, basic, intermediate and acid volcanics and sediments) is overlain unconformably by the Kavirondian System, consisting mainly of sediments and minor volcanics. The status of the Dodoma schist belt in the Dodoman System is uncertain (Bell and Dodson, 1981). It is not seen in contact with the Nyanzian or Kavirondian but may be equivalent to the former.

There are no age determinations for the supracrustal rocks. Rb–Sr ages for granitoid rocks across 800 km of the craton fall in the range 2.6–2.4 Ga (Bell and Dodson, 1981). Dodson *et al.* (1975) obtained ages for two granitic bodies. The post-Kavirondian Mumias–Buteba granite yielded an age of *c.* 2.47 Ga (as computed by Cahen *et al.*, 1984), and the Migori granite which intrudes the Nyanzian supracrustal rocks yielded in isochron age of *c.* 2.74 Ga, indicating a minimum age for the Nyanzian rocks. The 3.0 Ga age reported for the Masaba granite (Old and Rex, 1971) was revised by Dodson *et al.* (1975) to be 2.45 Ga. Bell and Dodson (1981) considered that at least two granitoid-forming events can be identified at *c.* 2.74 Ga and 2.54 Ga in the Tanzanian Craton and that there is little evidence to indicate any reworking of more ancient continental crust. At least two ages of greenstone belt formation have been identified in this crustal segment. The Nyanzian System is constrained at > 2.74 Ga and the younger Kavirondian System pre-dates 2.47 Ga. The age of the Dodoma schist belt is unknown. It may represent either a third greenstone formation event or a part of the Nyanzian System.

16.4.3 *Greenstone belts of Equatorial Africa*

Although greenstone belts make up a significant portion of the Zaire and Tanzania granite–greenstone terranes, there is only limited information on the geology of these belts. The stratigraphic units of two sequences will be briefly described. The Bandas belt in the Central African Republic is considered to be the northerly extension of the greenstone belts in Zaire, while the Nyanzian System occurs to the east of Lake Victoria.

Poidevin *et al.* (1981) described the Bandas greenstone belt and divided it into two lithostratigraphic units. The lower unit is composed predominantly of volcanic rocks and the upper unit (*c.* 500 m thick) mainly of metasediments (greywackes) intercalated with acid volcanic tuffs. The first 500 m of the lower unit consists of quartzo-feldspathic schists intercalated with garnet amphibolites and itabirites (quartz and iron oxides). These rocks are overlain by a sequence of alternating basalts and itabirites *c.* 2.6 km thick. This unit is in turn overlain by 1 km of itabirites, basalts and andesites. The metavolcanics in the lower unit occur as massive flows but in places contain pillow lavas, with pillows up to 1.5 m in diameter. The rocks have been recrystallised to low-grade amphibolites and primary minerals such as pyroxene and plagioclase are preserved only rarely. The relative proportions of basalts:andesites:acid volcanics are 80:19:1 (Poidevin *et al.*, 1981). No komatiites or komatiitic basalts have been reported from this greenstone belt.

The early descriptions of the Nyanzian System remain the definitive work on the area (Pulfrey, 1938; Shackleton, 1946). Four groups have been defined in the succession (Cahen *et al.*, 1984), namely a basal unit of mainly pillowed basic lavas with local BIF horizons, a group of intermediate to acid volcanics intercalated with tuffs and agglomerates, a group of greywackes with andesitic tuffs, and a top-most unit consisting of andesitic rocks with ferruginous shale and BIF near its base. The thickness of the complete succession is *c.* 7.5 km. The metamorphic grade is predominantly greenschist facies, the degree of alteration of the volcanics increasing towards the granite contacts (Davis and Condie, 1977). The Dodoma schist belt also contains metamorphosed ultrabasic rocks.

16.5 Southern Africa

The Archaean cratons in southern Africa comprise the Zimbabwe (Rhodesian) Craton in the north (Stagman, 1978) and the Kaapvaal Craton in the south, separated by the Archaean Limpopo Belt (SACS, 1980). Clifford (1970) has termed the whole region the Rhodesia–Transvaal Craton. Because of the large amount of information available from these two cratonic areas they will be discussed separately. The metamorphic rocks of the Limpopo Belt contain both mafic and ultramafic components but due to their higher grade of metamorphism and deformation compared to similar rock types on the adjacent cratons they will not be discussed in detail.

16.5.1 *Zimbabwe*

The Zimbabwe (Rhodesian) Craton (Figure 16.4) covers most of Zimbabwe, the eastern portion of Botswana and western Mozambique. The craton is 650 km long (NE–SW) and 350 km wide (Stowe, 1971). The characteristic assemblage of a 'basement complex' intruded by granitoids is well established (Macgregor, 1951, 1953; Bliss and Stidolph, 1969; Wilson, 1979; Wilson *et al.*, 1978). The craton is covered by younger rocks in the west and is bounded by Proterozoic mobile belts in the north and east. To the south, the craton passes into the Limpopo Belt and the southern and southwestern margin is arbitrarily delineated at the northern limit of granulite facies development (Wilson, 1979). Wilson *et al.* (1978) noted that there are possibly three different ages of granite–greenstone terrane and have tentatively subdivided the basement complex into 3.5 Ga, 2.9 Ga and 2.7 Ga terranes (but note that revision of the Rb–Sr ages due to a refinement of the Rb decay constant requires that these ages of the three terranes should be lowered by *c.* 100 Ma).

The *c.* 3.5 Ga terrane occurs in the southern and central portions of the craton (Figure 16.4) in and around the crustal segment at Selukwe (now Shurugwe), Fort Victoria (Nyanda) and Shabani. The basement consists mainly of highly deformed tonalitic gneisses with infolded greenstone belt remnants. The greenstones in the Shurugwe area consist largely of basaltic to peridotitic metavolcanics, and the succession includes a chromite-bearing ultramafic intrusion. Granite of the Mont d'Or

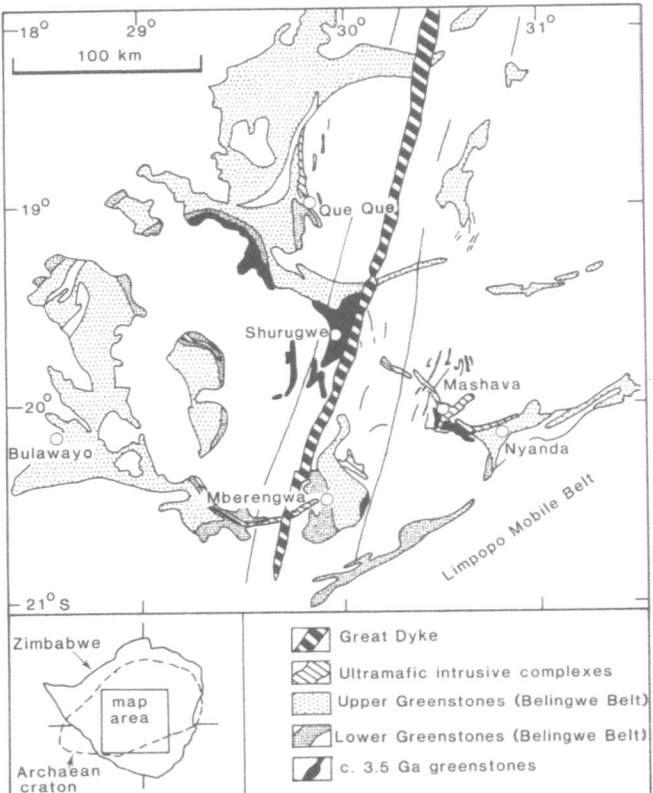

Figure 16.4 Simplified geological map of the central region of the Zimbabwe craton showing the distribution of the major metabasic rocks (after Wilson, 1981; reproduced by Nisbet, 1987).

formation dated (Rb–Sr) at *c*. 3.3 Ga (Moorbath *et al.*, 1976) intrudes these greenstones and near the *c*. 3.37 Ga Mushandike granite (Hickman, 1974) intrudes the tonalitic gneisses, providing a minimum age for the oldest known greenstone belts in this area. No radiometric ages have been obtained on the greenstone belt material, although Hawkesworth *et al.* (1975) have obtained a poorly defined age of *c*. 3.29 Ga (as recalculated by Cahen *et al.*, 1984) from the highly banded gneisses in the Mashaba area. The occurrence of a granitic basement to at least part of the 3.5 Ga greenstone belts is indicated by the presence of boulders of tonalitic gneiss, granodiorite, monzogranite and pegmatite in the metasedimentary Wanderer Formation, which occurs about mid-way up the greenstone belt succession (Wilson *et al.*, 1978).

The oldest, *c*. 3.5 Ga greenstone belts in the Shurugwe–Mashaba area consist of two groups of rocks separated by an unconformity (Stowe, 1968). The lower group has the sedimentary Mont d'Or Formation as the lowest member and is overlain by a lower greenstone formation which is made up of mafic lavas (now chlorite and actinolite schists) interbedded with jaspilites, quartzites and tuffs. This unit is overlain by altered ultramafic rocks consisting of talc–carbonate schists, serpentinites and chromite-rich rocks. An unconformity separates the ultramafic rocks from the sedimentary Wanderer Formation which in turn is overlain by an upper greenstone formation of altered basaltic lavas interbedded with BIF and chloritic argillites.

The 2.9 Ga terrane is defined by Rb–Sr whole rock ages on the 2.86 Ga Mashaba tonalite (Hawkesworth *et al.*, 1979) and, further southwest, the 2.82 Ga Chingezi gneiss (Hawkesworth *et al.*, 1979). Numerous biotite and hornblende K–Ar ages from granitic rocks also indicate an event at 2.85 Ga, but are unreliable indicators of the age of intrusion (Wilson *et al.*, 1978). The major Belingwe (now Mberengwa) greenstone belt in this terrane is subdivided into two main units, the Lower Greenstones and Upper Greenstones (Wilson, 1979; Wilson *et al.* 1978), also known as the Mtshingwe or Belingwean and Ngezi or Bulawayan Groups respectively (Nisbet, 1987). Wilson (1979) correlated the main stratigraphic elements across the greenstone belts of the central cratonic area. The base of the Lower Greenstones is not well exposed and their relationship with the basement granites and gneisses remains obscure. The Upper Greenstones, on the other hand, are perhaps the best example know of greenstone belts laid down entirely on old, established continental crust (Nisbet, 1987).

Direct dating of the Lower Greenstones has been attempted on the Hokonui Formation tuffs (2.52 ± 0.03 Ga; Hawkesworth *et al.*, 1979) from the Belingwe belt and from the Nyanda area (2.74 ± 0.28 Ga, Hawkesworth *et al.*, 1975). These ages are difficult to interpret. It seems likely that the Rb–Sr systems were disturbed by tectono-thermal effects associated with the emplacement of the late granites (Cahen *et al.*, 1984).

The age of the Upper Greenstones is well established and constitutes the 2.7 Ga terrane. Nisbet *et al.* (1987) reported a Pb–Pb isochron age of c. 2.69 Ga (obtained by Chauvel *et al.*, unpubl. data) for komatiitic basalts from the Reliance Formation. Recalculated Rb–Sr isochron ages (Cahen *et al.*, 1984) obtained by Hawkesworth *et al.* (1975, 1979) and Jahn and Condie (1976) for the greenstone belts of Zimbabwe indicate an age of 2.68 ± 0.07 Ga for the Maliyami Formation of the QueQue area (Hawkesworth *et al.*, 1975), in good agreement with the Pb–Pb age for the Upper Greenstones of the Belingwe belt. Hamilton *et al.* (1977) obtained a Sm–Nd age of 2.65 ± 0.05 Ga for mafic and ultramafic samples from a number of greenstone belts. Despite the possibility that small differences in initial $^{143}Nd/^{144}Nd$ ratio may have existed, this age probably reflects the time of volcanic activity and is consistent with the Pb–Pb age for the Upper Greenstones. Other age determinations range from 2.38 to 2.63 Ga but generally have poor regression statistics and large errors. The pooling of the isotope data for samples from wide geographic and stratigraphic positions, as attempted by Jahn and Condie (1976), produces ages of no geological significance (Wilson *et al.*, 1978).

The Lower Greenstones of the Belingwe greenstone belt (Bickle *et al.*, 1975; Nisbet *et al.*, 1977; Wilson *et al.*, 1978) show variations in lithology from the southeast to the western sides of the Belingwe belt. The lowest unit in the west consists of poorly exposed amphibolites and the relationship of these rocks to the overlying Hokonui Formation felsic flows and pyroclastic material is obscure. The top unit (the Bend Formation) consists of alternating pillowed mafic lavas, spinifex-textured peridotites (komatiites) and intercalated BIF. This formation is capped by a thick conglomerate with an interbedded unit of felsic agglomerate (Wilson *et al.*, 1978).

The basal Manjeri Formation of the Upper Greenstones in the Belingwe belt (Bickle *et al.*, 1975; Nisbet *et al.*, 1977; Nisbet *et al.*, 1987) contains up to 300 m of steeply dipping shallow-water sedimentary rocks (clastic sediments, limestones and sulphide facies BIF) which lie unconformably on the Lower Greenstones. Well developed stromatolites have been identified in the carbonate units (Bickle *et al.*, 1975) and from elsewhere in the Archaean greenstone belts (Martin *et al.*, 1980; Orpen and Wilson, 1981; Abell *et al.*, 1985). In places, well exposed outcrops show the Manjeri Formation lying unconformably on the *c.* 3.5 Ga Shabani gneisses (Oldham, 1968; Wilson, 1973; Bickle *et al.*, 1975). This is one of the few places in Africa where unambiguous cover relationships between greenstone belts and basement gneisses are exposed. The unconformity can be traced around the whole greenstone belt except where younger intrusive granites disrupt the contact in the north and south. The basal sediments are overlain successively by the Reliance Formation komatiite and komatiitic basalt flows (0.5–1.0 km thick), and the Zeederbergs Formation (~ 6 km thick), also predominantly comprised of volcanic rocks.

Ultramafic rocks form over 40% of the Reliance Formation, occurring as well-developed pillow-structured horizons alternating with massive flows (Bickle *et al.*, 1975), in which spinifex-textured komatiitic basalts are common. The Zeederbergs Formation consists mainly of tholeiitic basalts and basaltic andesites as pillow lavas and massive flows. Pyroclastic material including lapilli tuffs, graded and cross-bedded tuffs and coarser agglomerates occur in the upper part of the Zeederbergs Formation (Nisbet *et al.*, 1977). The Cheshire Formation comprising shallow-water sediments and stromatolitic limestones lies above the Zeederbergs Formation.

Three deformation events have been recognised in the Upper Greenstones of the Belingwe belt. Because of the stratigraphic continuity and abundant and consistent 'way-up' evidence, it is unlikely that any major hidden tectonic discontinuity exists, and the entire succession was probably deposited above the basal unconformity of the Manjeri Formation (Nisbet *et al.*, 1977).

Late granites (*s.l.*) which post-date the Upper Greenstones and the uppermost Shamvaian Group (sediments and minor andesitic–dacitic volcanics) include the Sesombi tonalite (Hawkesworth *et al.*, 1975), the Somabula tonalite, (Moorbath *et al.*, 1977), the Chilimanzi batholith (Hickman, 1976), the Zimbabwe batholith (Hickman, 1976) and the Victoria porphyroblastic granite (Hickman, 1976). The age of all of these granitic rocks (as recalculated by Cahen *et al.*, 1984) is *c.* 2.6 Ga. On the basis of the recent Pb–Pb ages for the Belingwe Upper Greenstones it would seem that the contention that the Maliyami Formation calc-alkaline volcanic rocks and the Sesombi and Somabula tonalites can be regarded as a late plutonic expression of the same magmatic episode (Wilson *et al.*, 1978), needs to be re-evaluated.

16.5.2 *Kaapvaal Craton*

The Archaean history of the Kaapvaal Craton has been the focus of many detailed geological investigations. Despite the enormous amount of information that has been obtained, many unsolved problems and controversial issues remain, particularly with respect to the geological relationships between the older rock units. Mafic and ultramafic rocks form a significant component of the supracrustal assemblages in the Archaean craton. Vast volumes of these rocks occur in the ancient greenstone belts, the Pongola Supergroup, the Dominion Group and the Ventersdorp Supergroup. Relatively minor volumes of mafic–andesitic lavas also occur in the Witwatersrand Supergroup. Intrusive rocks such as the Rooiwater Igneous Complex, the Messina and Usushwana Intrusive Suites, and mafic dykes of various ages also occur through-

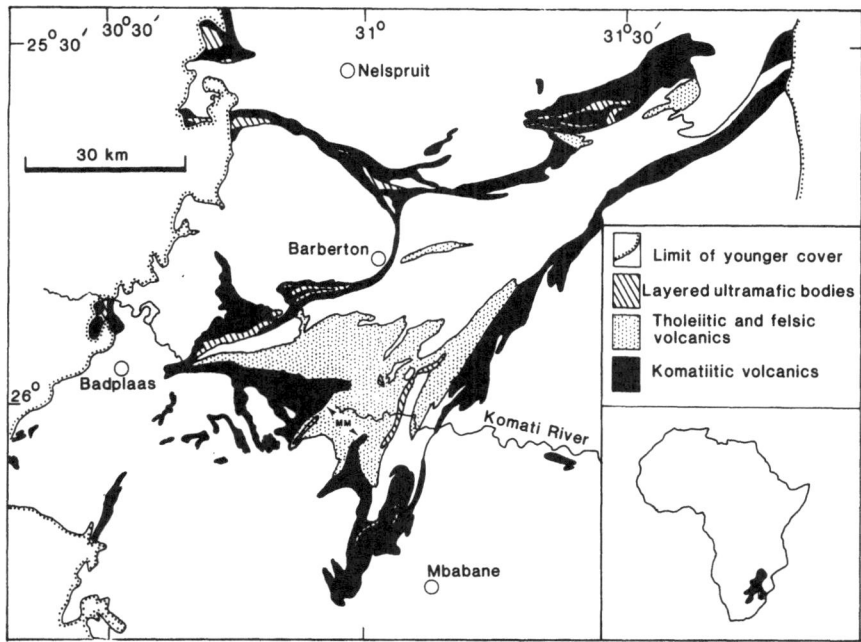

Figure 16.5 Simplified geological map of the Barberton greenstone belt showing the relative distribution of the tholeiitic and komatiitic volcanics and the layered ultramafic bodies. Sediments associated with the greenstone belt and granites and gneisses are not shaded. MM: Middle marker horizon. (After Viljoen *et al.*, 1982.)

out the craton. These are important both for their economic mineralisation and for the age constraints that can be obtained from these units. Some of the age relationships, as they are understood today, between these units and the surrounding granite-gneisses are shown in Tables 16.2 and 16.3.

Two types of Archaean terrane are known in the Kaapvaal Craton (Anhaeusser, 1975), those dominated by high-grade gneisses and migmatites such as the Ancient Gneiss Complex, and those consisting of granites and low-grade greenstones such as the Barberton Mountain Land (Figure 16.5). The geological relationship between the two has long been controversial. Anhaeusser (1973a, 1975) considered that the greenstone remnants represent the earliest decipherable events while Hunter (1970, 1974) and Jackson (1984) considered that the Ancient Gneiss Complex may have been a basement to the greenstones. Selected isotopic ages are given in Table 16.2. Sm–Nd and Rb–Sr isochron ages of whole rock samples generally have errors that are too large to establish the age of the oldest unit with certainty. The very precise U–Pb ages obtained from zircons by ion microprobe analysis (Compston and Kröner, 1988; de Wit *et al.*, 1987a) in both units indicate that the oldest zircons are from the Ancient Gneiss Complex. The oldest zircon age of 3.644 ± 0.004 Ga (Table 16.2) is interpreted by Compston and Kröner (1988) as the event in which the zircons were originally precipitated and younger ages are considered to represent periods of zircon recrystallis-

Table 16.2 Radiometric age determinations of the Barberton greenstone belt and Ancient Gneiss Complex

Method	Location and material	Age (Ma)	Reference
Barberton Greenstone Belt			
Rb–Sr internal isochron	Komati Formation, komatiitic basalt	3430 ± 200	Jahn and Shih (1974)
Sm–Nd whole rock	Onverwacht mafic and ultramafic rocks	3510 ± 60	Hamilton *et al.* (1979)
Sm–Nd whole rock	Onverwacht mafic rocks	3560 ± 240	Jahn *et al.* (1982)
^{40}Ar–^{39}Ar	Komati Formation, komatiitic rocks	3489 ± 34	Martinez *et al.* (1984)
Pb–Pb whole rock	Onverwacht basic rocks	3460 ± 70	Brevart *et al.* (1986)
U–Pb in zircons	Onverwacht felsic rocks	3445 ± 8	de Wit *et al.* (1987a)
Sm–Nd whole rock	Onverwacht lavas	3420 ± 70	Chauvel *et al.* (1987)
Sm–Nd whole rock	Hooggenoeg Formation, komatiitic flows	3440 ± 90	Gruau *et al.* (1988)
Ancient Gneiss Complex			
Rb–Sr whole rock	Northeast Swaziland, orthogneisses and amphibolites	3555 ± 111	Braton *et al.* (1980)
Sm–Nd whole rock	Bimodal Suite	3417 ± 34	Carlson *et al.* (1983)
Sm–Nd and Pb–Pb	Bimodal Suite	$3550 \pm ?$	Carlson *et al.* (1985)
U–Pb in zircons ion microprobe	Piggs Peak Tonalite	3644 ± 4 3504 ± 6 3433 ± 8 2986 ± 20 2867 ± 30	Compston and Kröner (1988)

ation or precipitation during tectono-metamorphic events and periods of intrusion of felsic veins. The present contacts between the greenstone belt and the Ancient Gneiss Complex in the Piggs Peak area are sheared (Compston and Kröner, 1988). Consequently, the original relationship between these units remains unknown.

All the isotopic dating techniques give ages for the Barberton greenstone belt which agree within the quoted errors (Table 16.2). This is rather surprising as certain techniques such as ^{40}Ar–^{39}Ar (Martinez *et al.*, 1984) and the Rb–Sr internal isochron of a komatiitic basalt (Jahn and Shih, 1974) have been interpreted as the age of low-grade metamorphism rather than the extrusion age of the lavas. The age data indicate that alteration or metamorphism occurred shortly after extrusion (Martinez *et al.*, 1984; Duchac and Hanor, 1987), or at least within a time frame that cannot be resolved by the present methods.

The felsic rocks of the Hooggenoeg Formation have, conventionally, been considered to be part of the greenstone belt stratigraphy (e.g. Viljoen and Viljoen, 1969d; Anhaeusser, 1973a). However, de Wit *et al.* (1987a) have recently re-mapped the area and considered the upper felsic unit in the Hooggenoeg Formation to have been emplaced along thrust zones and to postdate the Onverwacht Group by some

Table 16.3 Radiometric age determinations of greenstone belt remnants, intrusive suites and basins, Kaapvaal Craton

Area and material	Method	Age (Ma)	Reference
Greenstone belts and related rocks			
Rooiwater Igneous Complex	Inferred	> 2650	Vearncombe *et al.* (1987)
Murchison greenstone belt Antimony line sulphides	Pb–Pb	3020	Saager and Koppel (1976)
Murchison greenstone belt Rubbervale Formation	Rb–Sr whole rock	2837 ± 25	Vearncombe *et al.* (1987)
Murchison greenstone belt Rubbervale Formation	Pb–Pb whole rock	2953 ± 38	Vearncombe *et al.* (1987)
Rooiwater Igneous Complex hornblende diorite	U–Pb in zircon	2961 ± 150	Burger and Walraven (1979)
Pietersburg greenstone belt	Inferred	> 2600	SACS (1980)
Pietersburg greenstone belt	Stated	> 3450	Barton *et al.* (1986a)
Marydale Group metabasalts	Pb–Pb whole rock	2915 ± 130	Barton *et al.* (1986b)
Marydale Group metabasalts	Sm–Nd model t_{CHUR} age	2900	Reid (1981)
Basins and intrusives			
Pongola Supergroup Nsuze lavas	Pb–Pb zircon conventional	3090 ± 90	Burger and Coertze (1973)
Pongola Supergroup Nsuze rhyolite	U–Pb zircons	2940 ± 22	Hegner *et al.* (1984)
Pongola Supergroup rhyolite and basalts	Sm–Nd whole rock	2934 ± 114 (errorchron)	Hegner *et al.* (1984)
Pongola Supergroup rhyolite and basalts	Rb–Sr whole rock	2883 ± 69 (errorchron)	Hegner *et al.* (1984)
Usushwana Intrusive Suite	Rb–Sr whole rock	2870 ± 38	Davis *et al.* (1970)
Usushwana Intrusive Suite pyroxenite	Sm–Nd internal isochron	2871 ± 30	Hegner *et al.* (1984)
Usushwana Intrusive Suite gabbros	Sm–Nd whole rock	3124 ± 123	Hegner *et al.* (1984)
Dominion Group sulphide concentrate	Pb–Pb leached solution	2800 ± 60	Van Niekerk and Burger (1969)
Ventersdorp Supergroup Makwassi Formation, quartz porphyry	U–Pb zircons	2300 ± 100	Van Niekerk and Burger (1964)
Ventersdorp Supergroup Plantation porphyry	Rb–Sr whole rock	2154 ± 120	Harding *et al.* (1974)
Ventersdorp Supergroup Klerksdorp lavas	U–Pb zircons	2643 ± 80	Van Niekerk and Burger (1978)
Ventersdorp Supergroup Makwassi Formation quartz porphyry	U–Pb zircons	2693 ± 60	Retief (unpubl. data reported in R.A. Armstrong *et al.*, 1986)
Ventersdorp Supergroup Makwassi Formation quartz porphyry	U–Pb zircons ion microprobe	2699 ± 16	R.A. Armstrong *et al.* (1986)
Ventersdorp Supergroup Makwassi Formation quartz porphyry	Pb–Pb whole rock	2350 ± 180	R.A. Armstrong *et al.* (1986)
Ventersdorp Supergroup Makwassi Formation quartz porphyry	Rb–Sr whole rock	2140 ± 260	R.A. Armstrong *et al.* (1986)
Ventersdorp Supergroup Klipriviersberg Group lava flow	Pb–Pb whole rock	2370 ± 70	R.A. Armstrong *et al.* (1986)

40–50 Ma. The precise zircon age of 3.445 ± 0.008 Ga reported by de Wit et al. (1987a) for this unit cannot, therefore, be used to constrain the age of the mafic and ultramafic volcanicity of the Onverwacht Group. An age of 3.49 Ga from an igneous zircon in an Onverwacht metagabbro (Armstrong et al., in prep.; reported in de Wit et al., 1987a) is consistent with the other ages reported in Table 16.2. The average of all the ages for the Onverwacht Group mafic and ultramafic rocks given in Table 16.2 is 3.47 Ga. Eliminating two values with large errors reduces this slightly to 3.46 Ga, currently considered to be the best age estimate of the Onverwacht Group. The ages of the surrounding granitic (s.l.) plutons are summarised by Barton et al. (1983), Barton (1983) and de Wit et al. (1987a).

While the Barberton greenstone belt has received the most attention, there are remnants of numerous other greenstone belts within the Kaapvaal Craton (Figure 16.5) for which some age information is available (Table 16.3). The larger belts, namely the Murchison, Pietersburg and Sutherland belts, occur in the northern part of the craton. Numerous smaller belts include the Dwalile, Assegaai, De Kraalen, Commondale and Nondweni belts in Swaziland and northern Natal (Hunter and Wilson, 1988), the Kraaipan Group in the western Transvaal (SACS, 1980), the Amalia belt (Vearncombe, 1986), the Muldersdrif Ultramafic Complex and the Roodekrans Greenstone Complex northwest of Johannesburg (Anhaeusser, 1973b, 1974, 1977), the Rhenosterkoppies belt north of Pietersburg and the Marydale Group on the southwestern edge of the craton (Barton et al., 1986b). Although many of these smaller remnants contain areas of excellent exposure and well preserved volcanic textures, in many cases the age constraints are poor.

The Archaean Pongola Supergroup developed on the stabilised basement of the Kaapvaal Craton and is exposed in Swaziland, southeastern Transvaal and northern Natal (Tankard et al., 1982; N.V. Armstrong et al., 1982, 1986). The lower Nsuze Group has a lower sedimentary component (c. 800 m thick), a middle volcanic unit (c. 7500 m thick) and an upper volcaniclastic sedimentary unit (c. 600 m thick). Overlying the Nsuze Group is the 1800 m thick Mozaan Group consisting predominantly of sandstone and shale formations. The style of volcanism and sedimentation in the Pongola Supergroup is more typical of Proterozoic basins than of Archaean environments (Armstrong et al., 1982) and the volcanics show a continuous spectrum of compositions from basalt to rhyolite with no ultramafic lavas being present (N.V. Armstrong et al., 1986). The age of the Pongola Supergroup is constrained by the age of the basement granites (c. 3.1 Ga) on which it rests and the c. 2.87 Ga intrusive Usushwana Suite (Davis et al., 1970; age recalculated by Hegner et al., 1984). Hegner et al. (1984) obtained a precise U–Pb age of 2.94 ± 0.022 Ga from zircons in the felsic volcanics. This age is considered to be the best estimate of the Pongola Supergroup and Sm–Nd and Rb–Sr whole rock errorchron ages (Table 16.3) are consistent with the zircon age. Rb–Sr whole rock ages of 2.35 and 2.73 Ga (Hegner et al., 1981) are now considered to represent Rb and/or Sr redistribution during subsequent metamorphic events.

The Usushwana Intrusive Suite consists of basic to ultrabasic rocks in layered dyke-like structures in northeastern Swaziland and a sill at or near the base of the lower Pongola Nsuze Group in southern Swaziland and the adjoining portion of South Africa (Figure 16.5). This unit intrudes and is clearly younger than the Pongola rocks. The age of the Usushwana suite is c. 2.87 Ga (Table 16.3) and there is good agreement between the Rb–Sr age of Davis et al. (1970; as recomputed by Hegner et al., 1984) and the Sm–Nd mineral isochron obtained by Hegner *et al.* (1984) from a pyroxenite sample. However, the latter authors have also obtained an Sm–Nd whole rock errorchron of 3.12 ± 0.12 Ga for gabbro samples from this unit and internal isochrons from two of the gabbro samples of 2.77 ± 0.08 and 2.56 ± 0.35 Ga. They attribute the young ages obtained from the gabbro minerals to partly disturbed mineral isotopic systems and the old age of the gabbro whole rock samples to mixing processes.

The Witwatersrand Triad comprises the basal Dominion Group, the Witwatersrand Supergroup and the Ventersdorp Supergroup, and constitutes a volcano-sedimentary sequence that is famous for its fossil gold placers. The Dominion Group contains some 1100 m of basaltic andesite lavas and tuffs in the Renosterhoek Formation and a further 1550 m thickness of acid lavas, tuffs, andesitic lavas and quartz–feldspar porphyry in the overlying Syferfontein Formation (Tankard *et al.*, 1982). Sediments and lavas of this group rest unconformably on Archaean basement granites. The Pb–Pb age of *c.* 2.8 Ga obtained for sulphide concentrate samples from lavas at the Dominion Reef Mine (Van Niekerk and Burger, 1969) is widely accepted as the age of this volcanic succession.

No reliable age information has been published for the lavas within the Witwatersrand Supergroup. U–Pb and Pb–Pb isotope studies carried out on various components of the sediments (e.g. Burger *et al.*, 1962; Nicolaysen *et al.*, 1962; Allsopp *et al.*, 1986) have provided an indication of the original crystallisation ages of the minerals or ages of rejuvenation events rather than the age of sedimentation. The period when these auriferous sediments were deposited is constrained by the age of the older Dominion Group lavas (*c.* 2.8 Ga) and the overlying Ventersdorp Supergroup lavas, although the age of the Ventersdorp Supergroup has not been clearly established. The available isotopic age data (see Table 16.3) place it as late Archaean or early Proterozoic (2.7–2.64 Ga). Part of the difficulty in dating this huge outpouring of lava, which covers some 200 000 km^2 and is *c.* 7.9 km thick, is that the lavas have been metamorphosed to lower greenschist facies (Tankard *et al.*, 1982), and R.A. Armstrong *et al.* (1986) suggested that the Rb–Sr and Pb–Pb whole rock ages have been reset. No. 2.3 Ga ages were obtained from the zircons of the Ventersdorp Supergroup (R.A. Armstrong *et al.*, 1986), and it is assumed that the approximate age of the Ventersdorp is 2.7 Ga (Table 16.3). If the ages for the Dominion Group and Ventersdorp Supergroup are correct (2.8 Ga and 2.7 Ga respectively) it means that the Witwatersrand sediments were deposited in a time span of *c.* 100 Ma or less.

The geological character of the Pongola, Witwatersrand and Ventersdorp Supergroups is more typical of Proterozoic cover sequences. They have been preserved with a minimum of alteration and deformation, in marked contrast to the usually highly deformed and metamorphosed Archaean greenstone belts seen in the older portions of the Kaapvaal Craton and elsewhere in Africa. The Kaapvaal Craton appears to have achieved local crustal stability long before the end of the Archaean (Tankard *et al.*, 1982).

16.5.3 *Barberton greenstone belt*

The geology of the Barberton greenstone belt has become well known since Viljoen and Viljoen (1969a–g) mapped the Onverwacht Group in detail and documented a suite of previously unknown Mg-rich lavas for which they coined the name komatiite after the Komati River flowing through the Barberton Mountain Land. There is a long history of geological research in the Barberton Mountain Land (see SACS, 1980; Viljoen and

Viljoen, 1982). A recent review of the geology and structure of the Onverwacht Group has been carried out by Watkeys (1988). Some aspects of the 'classical' interpretation of the geology and geochemistry of the belt by Viljoen and Viljoen (1969a–g) have recently been challenged (e.g. Williams and Furnell, 1979a; de Wit et al., 1987a,b).

The stratigraphy of the Barberton greenstone belt (Viljoen and Viljoen, 1969d) modified to account for the new nomenclature (SACS, 1980) is given by Anhaeusser (1978) and the distribution of the major units is shown in Figure 16.5. The belt is 130 km long and between 50 and 8 km wide, and is relatively small compared to some Canadian and Australian greenstone belts. The type sections collectively indicate a volcano-sedimentary pile more than 20 km thick (Anhaeusser, 1975), and while this is an unrealistic estimate of the true thickness at any one point, it does illustrate the considerable volumes of mafic and ultramafic rocks within the belt. The predominantly volcanic Onverwacht Group is made up of the Tjakastad Subgroup (the Lower Ultramafic Unit of Viljoen and Viljoen, 1969a,d) consisting mainly of komatiites, komatiitic basalts and tholeiites, and the upper Geluk Subgroup (Upper Mafic to Felsic Unit of Viljoen and Viljoen, 1969b,d) made up of mafic to felsic rocks and cherts. The two subgroups are separated by the Middle Marker band of chert–carbonate rocks together with mafic tuffs up to 9 m thick which can be traced for about 72 km around the belt. The Onverwacht Group is overlain by the dominantly sedimentary units of the (largely argillaceous) Fig Tree Group and the (largely arenaceous) Moodies Group.

Each of the two subgroups of the Onverwacht Group has been divided into three formations. The lower Tjakastad Subgroup consists of the Sandspruit, Theespruit and Komati Formations. The Sandspruit Formation is mainly preserved as large xenoliths within the granite–gneiss terrane around the southwestern margin of the belt and consists of deformed mafic and ultramafic schists with intercalated metasedimentary bands. The ultramafic rocks contain crystalline quenched (spinifex) textures consisting of interlocking olivine blades (now altered to antigorite and magnetite) (Viljoen and Viljoen, 1969a). Some of the xenoliths, such as the Schapenburg remnant (Anhaeusser, 1983), contain well preserved komatiite flow structures and spinifex textures although they have been metamorphosed at 500–700°C and 0.25–0.35 GPa.

The Theespruit Formation is widespread around the margins of the belt and is characterised by the presence of water-lain tuffs interlayered with mafic and ultramafic horizons as well as a variety of talc, chlorite and carbonate schists, and bands and lenses of black chert containing abundant carbonaceous material in which Engel et al. (1968) identified possible evidence of primitive life forms. The Theespruit Formation is metamorphosed to greenschist facies except within thin zones on the immediate granite contacts where the grade is locally amphibolite or granulite facies (Viljoen and Viljoen, 1969d).

The Komati Formation is c. 35 km thick and is separated from the Theespruit Formation in the type area by the Komati Fault. Mafic and ultramafic volcanic rocks are essentially the only rock types present although numerous intrusive bodies of feldspar and quartz porphyry also occur. The ultramafic rocks occur as dykes or sills and spectacular komatiite flows. The flows vary from 33 cm to many tens of metres thick (Williams and Furnell, 1979a; Smith et al., 1980; Viljoen et al., 1983; Smith et al., 1988) and have chilled margins, spinifex textures and peridotite units similar to those documented by Pyke et al. (1973) in the flows in Munro Township, Canada. In some places more than twenty 1–3 m thick komatiite flows have been identified (Smith et al.,

1988). The contention (de Wit and Stern, 1980) that these units represent sheeted dykes has been refuted by Viljoen *et al.* (1983) who provided textural and geochemical evidence to support their origin as flows. However, there does appear to be a paucity of komatiitic dykes that could have fed lava to the flows. Williams and Furnell (1979a) and Smith and Erlank (1984) have identified a series of thin intrusive units of komatiite composition which strike parallel to the flows and may, therefore, represent sills rather than dykes. Massive and pillowed flows of komatiitic basalts and tholeiites are also common and become more abundant towards the top of the formation. Greenschist facies metamorphism and alteration have affected all the rocks of this formation, but in many localities original igneous textures are well preserved, and in some instances relict primary minerals occur.

Ultramafic bodies which occur as large sill or pod-like features are common in the Komati Formation (Viljoen and Viljoen, 1969c, 1970) and have been divided into various types based on their mineralogy. They consist of layers of dunite (now serpentinite), peridotite, pyroxenite and some have gabbro and anorthosite horizons. These bodies often contain economic deposits of secondary minerals such as chrysotile asbestos, talc and magnesite. Geochemical studies suggest that the bodies represent cumulates from the differentiation of komatiite magmas (Anhaeusser, 1985; Viljoen and Viljoen, 1970). Tholeiitic basalts, andesites, dacites and rhyodacites predominate above the Middle Marker (Viljoen and Viljoen 1969d, f). Consequently, the three upper formations of the Onverwacht Group have been defined as a separate subgroup, the Geluk Subgroup, consisting of the Hooggenoeg, Kromberg and Swartkoppie Formations. The volcanics are arranged in cycles with the mafic components at the base grading up to acid lavas at the top. Frequently, each cycle is capped by a substantial and persistent chert unit. Anhaeusser (1971a) further developed the concept of cyclic volcanicity and sedimentation in the evolution of Barberton and other greenstone belts. The origin of the felsic volcanics is 'somewhat problematical' (Viljoen and Viljoen, 1969f). They consist mainly of massive feldspar porphyries. However, in places fragmental rocks are present consisting of agglomerates and felsic tuffs often showing graded bedding and well preserved flame structures. Ultramafic bodies up to 11 km long are also present in the Geluk Subgroup. Some of the bodies, such as the Rosentuin body, have undergone magmatic segregation with dunite/peridotite at the base and pyroxenites at the top while others are simply massive, often sheared serpentinite pods (Viljoen and Viljoen, 1969g).

The 35 km long Jamestown schist belt (Figure 16.5) on the northwest flank of the Barberton greenstone belt is the largest of several narrow arcuate schist belt remnants marginal to the main body of the greenstone belt. It consists of mainly low greenschist facies mafic and ultramafic rocks (Anhaeusser, 1972). These rocks have been correlated with the Tjakastad Subgroup of the main Barberton belt, although the Jamestown schist belt appears to be made up of intrusive rhythmically layered successions of dunite, peridotite, pyroxenite and gabbro rather than extrusive rocks (Anhaeusser, 1972).

The geology of the Barberton greenstone belt and the surrounding granitic terrain has been used to model the evolution of granite–greenstone terrains in other parts of the world (Anhaeusser *et al.*, 1968, 1969; Anhaeusser, 1971b; Viljoen and Viljoen, 1969c). However, it has become apparent that many details of the Barberton model are not applicable in general to other greenstone belts including those in the Kaapvaal craton (e.g. see Vearncombe *et al.*, 1987).

Williams and Furnell (1979a) remapped the areas of the Komati and Hooggenoeg Formations and noted a more complex layering of komatiites and komatiitic basalts than documented previously (Viljoen and Viljoen, 1969a, f) and a tendency for extrusive komatiites to be concentrated in the centre of the Komati Formation rather than at the base. They also mapped cross-cutting dykes and thin (1–5 m thick) but persistent layers of intrusive peridotite generally conformable to the strike of the extrusive units. Within the Hooggenoeg Formation a 1 km thick unit of komatiitic basalts is traceable over a strike length of 15 km, indicating that substantial volumes of komatiitic magmas occurred above the Middle Marker and that the differences in compositions of the lavas between the Tjakastad and Geluk Subgroups may not be as striking as previously suggested. The komatiitic basalts are identical in texture and composition to those found in the Komati Formation. Williams and Furnell (1979a) also suggested that deformation is more widespread and complex in the Theespruit Formation than in the Komati Formation and that the Komati Fault separating the two formations may be the line of an unconformity. From these observations, there may be two distinct sequences of greenstone belt and not a geological continuum across the fault. However, these different interpretations are by no means resolved (Viljoen and Viljoen, 1979; Williams and Furnell, 1979b).

Structural complexities exist in both the Tjakastad and Geluk Subgroups (de Wit, 1982; de Wit et al., 1983) and the simple 'layer cake' stratigraphy originally suggested by Viljoen and Viljoen (1969a) thus needs to be re-evaluated. De Wit (1982) documented the existence of nappes, olistostromes, regionally inverted stratigraphy and the juxtaposition of deep and shallow water sediments in the upper Onverwacht and Fig Tree Groups. In an interesting development of the greenstone–basement problem, de Wit et al. (1983) mapped the Theespruit Formation along the granite contacts in detail and identified small outcrops of granitoid gneisses within the greenstone sequence. This gneiss is in contact with a diamictite which, along strike, contains large gneiss clasts which are similar in appearance to the small gneiss outcrops. This gneiss possibly represents a pre-greenstone belt sialic basement equivalent to the Ancient Gneiss Complex. De Wit et al. (1983) indicated that large-scale upright folds and structural dislocations within the rocks of the Theespruit Formation invalidate the existing stratigraphy in this part of the belt.

The question of cyclic volcanicity in the Hooggenoeg Formation which has been proposed as an essential part of the 'Barberton model' (Viljoen and Viljoen, 1969f; Anhaeusser, 1971a) has been severely criticised by a number of workers. Williams and Furnell (1979a) noted that a rock described by Viljoen and Viljoen (1969f) as a pillowed intermediate lava is really an olivine basalt in which serpentine pseudomorphs after olivine have been replaced by quartz, thus increasing the SiO_2 content of the rock. The silicification of mafic and ultramafic rocks has now been widely recognised in the Barberton greenstone belt (de Wit, 1982; Paris et al., 1985; Duchac and Hanor, 1987; Smith, 1988). The increase in SiO_2 often occurs without the destruction of igneous features such as pillow structures and spinifex textures, and in places spinifex textures (originally in komatiites) have been observed in rocks mapped as chert units (de Wit, 1982; Smith, 1988). A further problem with the cyclic volcanicity model for the Hooggenoeg Formation is that field and geochemical evidence indicate that the thick felsic unit in the upper Hooggenoeg Formation is a high level equivalent of the surrounding granitic plutons (de Wit et al., 1987a) and may represent shallow-level intrusions and subsurface felsic domes associated with minor volcanics that were emplaced into the greenstone belt stratigraphy along thrust zones. The precise age data obtained from zircons by ion microprobe analysis indicate that these units may postdate the mafic and ultramafic assemblages by as much as 40–50 Ma. Clearly, the concept of volcanic cycles is in need of some re-evaluation.

Recently, de Wit *et al.* (1987b) have proposed that the Barberton greenstone belt be renamed the Jamestown Ophiolite Complex as the stratigraphy of the mafic and ultramafic rocks is comparable in many respects to that of Phanerozoic ophiolites. The essential characteristics are metamorphosed and deformed peridotites overlain by intrusive and extrusive rocks ranging from komatiites to tholeiites which are in turn capped by a chert–shale sequence. De Wit *et al.* (1987b) consider that the whole sequence is less than 3 km thick and that the 3.5 Ga oceanic crust was also relatively thin. Recent geophysical studies on the structure of southern African greenstone belts (De Beer *et al.*, 1984; Kleywegt *et al.*, 1987; Stettler *et al.*, 1988; De Beer and Stettler, 1988) indicated that the maximum depth to which the belts extend is in the order of only 2–5 km. Initial interpretation of geophysical results from the Barberton greenstone belt (De Beer and Stettler, 1986, 1988) suggests that this belt is also in the order of 3–6 km. Such estimates are clearly at variance with the concept of a *c.* 20 km thick volcano-sedimentary pile (Viljoen and Viljoen, 1969d) and more consistent with the de Wit *et al.* (1987b) interpretation. On the other hand, Lowe *et al.* (1985) have mapped sections of the Geluk Subgroup in detail, and while they confirm that some sections of the stratigraphy have been affected by complex thrusting and folding, there are other areas with unrepeated stratigraphic sections at least 8 km thick. There are still problems in unravelling the structural history of critical parts of the greenstone belt, and, until a consensus is reached regarding the stratigraphy of the belt, models for its evolution.

16.5.4 *Pongola Supergroup*

The Pongola Supergroup (*c.* 2.94 Ga) has many of the characteristics of Proterozoic belts, in that it developed on the stabilised basement and unconformably overlies older Archaean granites (Tankard *et al.*, 1982). This sequence is exposed in Swaziland, southeastern Transvaal and northern Natal (Figure 16.5) and consists of two groups. The base of the Nsuze Group usually has a thin sedimentary unit (including lavas and volcanogenic sediments) overlain by a 2–7.5 km thick sequence of basaltic, andesitic and acidic lavas. The upper Mozaan Group consists entirely of quartzites, shales and thin BIFs (SACS, 1980; Armstrong *et al.*, 1982).

The lavas in the Nsuze Group are predominantly low greenschist facies inter-mediate rocks but display a complete spectrum of compositions from basalt to rhyolite (N.V. Armstrong *et al.*, 1986). They were extruded subaerially and flows of different composition are complexly interfingered. No ultramafic lavas have been identified. The Nsuze lavas define a tholeiitic trend on an AFM diagram (N.V. Armstrong *et al.*, 1986) and detailed geochemical modelling suggests that the basaltic magmas have been contaminated by continental crust (Hegner *et al.*, 1984). The Nsuze Group magmatism is somewhat different to the volcanic activity occurring elsewhere during this period. Their subaerial nature, the widespread lack of deformation and the abundance of intermediate volcanics strongly contrast with typical greenstone assemblages such as the Barberton belt.

16.5.5   *Ventersdorp Supergroup*

Of the mafic and ultramafic volcanics in the Witwatersrand Triad the best documented group is the Ventersdorp Supergroup, relatively little being known about the Dominion Group and Witwatersrand Supergroup lavas. Recent discussions of the tectonic models and geochemistry of the lavas in the Dominion Group are given by Crow and Condie (1987) and Bowen *et al.* (1986), and for the Witwatersrand Supergroup by Tankard *et al.* (1982) and Winter (1987).

The Ventersdorp Supergroup occupies a large elliptical basin that exceeds 200 000 km^2 in area (Figure 16.5). Like the underlying Witwatersrand Supergroup, it has not been tectonically disturbed, although it has been metamorphosed to lower greenschist facies (Tankard *et al.*, 1982). The stratigraphy of the Ventersdorp Supergroup has taken a long time to unravel because of such factors as poor exposure, low relief, and alteration. Boreholes drilled through the succession by gold mining groups have assisted in resolving the stratigraphic relationships between the various units (Winter, 1976). Stratigraphic columns obtained by Winter (1976), Tyler (1979) and Buck (1980) are summarised by Tankard *et al.* (1982). The succession has been divided into three groups, namely the predominantly volcanic basal Klipriviersberg Group, overlain by the Platberg Group consisting of sediments and volcanics which, in turn, is overlain by the Pneiel Sequence that contains the Allanridge andesite as the uppermost unit. The thicknesses of the volcanic and sedimentary rocks in the type area near the town of Bothaville (Winter, 1976) are 5.1 km and 2.9 km respectively although there are considerable lateral variations at any one location (Bowen *et al.*, 1986).

Lavas at the base of the Klipriviersberg Group fill topographic irregularities on the underlying Witwatersrand Supergroup and overlie older rocks outside the Witwatersrand basin. The volcanic assemblages at the type locality comprise four compositionally distinct units interstratified with sediments. The Klipriviersberg Group consists largely of basic lavas (1.5–1.7 km thick), for the most part in a monotonous sequence of low-Mg tholeiitic basalts flows (Wyatt, 1976). In places komatiitic basalts (14–17% MgO) alternate with more felsic horizons, constituting the Meredale member (Wyatt, 1976; McIver *et al.*, 1982). The Makwassi Formation consists of acid lavas and quartz porphyries which grade into intermediate to basic lavas of the Rietgat Formation. The volcano-sedimentary pile is capped by the Allanridge Formation tholeiitic andesites (Bowen *et al.*, 1986). A large geochemical data base has now been obtained on the Ventersdorp lavas (Grobler *et al.*, 1982; De Bruiyn *et al.*, 1984; Bowen *et al.*, 1986), and Bowen *et al.* (1986) considered that a distinct geochemical stratigraphy is present within the volcanics.

The Ventersdorp Supergroup volcanics are considered to represent Precambrian flood basalts (Wyatt, 1976) that were extruded along faults induced by subsidence under the mass of the Witwatersrand sedimentary succession (Tankard *et al.*, 1982). The occurrence of komatiitic basalts towards the base of the succession is of major interest. Grobler *et al.* (1986) also documented the occurrence of komatiitic lava types in the upper Allanridge Formation lavas.

16.6   Tectonic settings

Much of the geological work on greenstone belts has been geared towards the understanding of their tectonic development and evolution. However, it is clear from the work carried out on the Archaean of Africa that very few areas are sufficiently well understood to enable realistic tectonic models to be developed for these complex terranes. Nevertheless, many tectonic models have been proposed for their origin (Table 16.4) and have been discussed in detail by Windley (1977), Condie (1981), Hunter (1981), Kröner (1981, 1985) and Tankard et al. (1982). Not surprisingly, the Barberton greenstone belt has featured strongly in many of the models.

The models have become increasingly complex as more detailed geological data are incorporated, and almost every conceivable environment has been proposed for the generation of greenstone belts (Table 16.4). However, paramount to the tectonic processes during the Archaean is the thermal regime in the crust and mantle. Bickle (1978) made the point that if 'modern' plate tectonics did not take place in the Archaean, then some other very vigorous process of plate creation (and destruction) must have occurred to account for the metamorphic record (see Chapter 6). The occurrence of diamonds within the Witwatersrand Supergroup sediments and the

Table 16.4 Tectonic models proposed for the development of Archaean greenstone belts in Africa

Model	Author
Density inversion model where mafic rocks overlie sialic crust	Macgregor (1951) and Condie (1976)
Onverwacht Group in Barberton equivalent to an ophiolite assemblage	Anhaeusser et al. (1968)
Greenstones developed in downwarps at the interface between continental and oceanic crust or they developed in parallel down warps or fault bounded troughs on thin sialic crust	Anhaeusser et al. (1969)
Progressive downsagging of volcano-sedimentary basin into thin unstable crust	Anhaeusser (1971b)
Greenstone belts are equivalents of lunar maria formed as a result of meteorite impact	Green (1972)
Abyssal tholeiites and peridotites form in oceanic crust and incipient island arc development produced island arc tholeiites calc–alkaline volcanic and plutonic rocks	Anhaeusser (1973a)
Continental rift model with greenstones forming at the rift valley and proto-oceanic rift stage	Windley (1973)
Greenstones formed at proto-oceanic ridges above an ascending plume	Condie and Hunter (1976)
Small-scale plate tectonics with greenstones regarded as arc trench type or rift structures founded on the ocean floor	Glikson (1976)
Evolving arc system from oceanic rise tholeiites in marginal basin to andesites in mature emerging arc	Condie and Harrison (1976)
Marginal back-arc basin 'Rocas Verdes'	Tarney et al. (1976)
Back arc basin environment for greenstone belts and main arc environment for high-grade gneissic complexes	Windley (1977)
Greenstone belts developed along linear 'rift-type' structural weaknesses over small scale secondary upwelling cells in the mantle	Williams (1977)
Complex model involving disruption of early ultramafic crust, generation of tholeiitic magmas during plume ascent at oceanic ridges, recycling of basalts in subduction zones, meteoritic impacts, and andesitic arc systems	Condie (1981)
Numerous models	Kröner (1981)

identification of Archaean diamond inclusions derived from the lithosphere by Cretaceous kimberlites (Richardson *et al.*, 1984) are significant constraints on the temperature of the lithosphere below the Kaapvaal Craton. The generation of high-temperature Mg-rich komatiitic magmas at proposed Archaean ocean proto-ridges (Table 16.4) would be consistent with a cool lithosphere below the Kaapvaal Craton.

The relatively cool and thick lithosphere indicated by the diamond data indicates that very high temperature komatiite magmas could not have existed as large laterally linked Archaean magma anomalies as envisaged by Nisbet and Walker (1982), at least, not below the Kaapvaal Craton *c.* 3.5–3.3 Ga. Between 3.5 Ga and 3.0 Ga there appears to have been a dramatic change in the behaviour of the Kaapvaal Craton as it became rigid and enabled 'platform-type' sedimentation of the Pongola Supergroup and Witwatersrand Triad. The occurrence of these old diamonds indicates that this change may have been in response to the development of a thick, cold sub-cratonic lithosphere. However, if the Barberton greenstone belt is allochthonous, the komatiite magmas need not necessarily have been derived from the mantle underlying the Kaapvaal Craton.

In Zimbabwe there is indisputable stratigraphic evidence for the development of at least one younger greenstone belt (the Belingwe belt) on a pre-existing granitic crust (Bickle *et al.*, 1975). However, this does not mean that the same relationship can be applied to all greenstone belts. The relationships in the older greenstones are not so clear (Nisbet *et al.*, 1981; Vearncombe *et al.*, 1987). There is no positive evidence of Archaean ophiolite complexes or ancient oceanic crust and these environments can be rejected as being responsible for the generation of the komatiite–tholeiite volcanic phases in the younger Zimbabwean greenstones. At present, there is insufficient evidence from the geological record either to support or refute modern-style plate tectonics as the major tectonic mechanism operative during the Archaean in Zimbabwe.

In a discussion of one of the most recent tectonic models for the evolution of the Barberton greenstone belt, it has been proposed that the belt should be renamed the Jamestown Ophiolite Complex (de Wit *et al.*, 1987b). Thus, tectonic models have now come in a complete circle, as Anhaeusser *et al.* (1968) also proposed an ophiolitic origin. De Wit *et al.* (1987b) considered that a reconstructed section through the greenstones of the Barberton belt would consist of a lower peridotitic tectonite zone (e.g. the Stolzburg Layered Complex) overlain by an array of vertical magma conduits intruding and covered by a substantial carapace of pillow-structured lavas and thin chert horizons. While this model is tantalisingly simple, it is hampered by some unexplained details:

(a) the asymmetry in the composition of what we hitherto called komatiite flows and are now referred to as vertical magma conduits (dykes);
(b) the occurrence of komatiitic basalt flows with well preserved flow top breccias intercalated between, and with the same strike and orientation as the komatiite units;
(c) thin sedimentary units within the komatiite and komatiitic basalt units showing excellent graded bedding and with the same orientation as the komatiitic units (Viljoen *et al.*, 1983) and a thicker sedimentary unit, the Middle Marker, which again has graded bedding features and the same orientation and strike as the so-called vertical magma conduits.

The problem of whether the spinifex-textured komatiite units represent flows or dykes has not been completely resolved.

Table 16.5 Summary of age estimates of units containing significant mafic and ultramafic rocks in Africa

Location	Unit	Age (Ga)
North and West Africa		
Uweinat Inlier	—	—
Reguibat Rise	—	—
Guinea Rise	Kambui Supergroup	> 2.75
	Loko Group	> 2.97
Equatorial Africa		
Cameroun–Gabon Block	Nounah-Ivindo greenstone belt	3.0–2.85
	Afoumadzo greenstone belt	3.15–3.0
Kasai Block	—	—
Zaire granite-greenstone terrane	Moto, Zani, Kilo and Ngayu Greenstone belts	> 2.84
	Ganguan schist and greenstone belts	> 2.98
Tanzania granite-greenstone terrane	Kavirondian System	> 2.47
	Nyanzian System (may include Dodoma System)	> 2.74
Southern Africa		
Zimbabwe Craton	Great Dyke	2.46
	Ngezi Group (Belingwe)	2.69
	Mishingwe Group	2.9
	Sebakwian	3.5
Kaapvaal Craton	Ventersdorp Supergroup	2.7
	Dominion Group	2.8
	Usushwana Intrusive Suite	2.87
	Pongola Supergroup	2.94
	Barberton greenstone belt	3.46

De Wit *et al.* (1987a) have also integrated the accumulating evidence of large-scale sub-horizontal displacements in the belt into a broader tectonic model in which the continuous stacking and subsidence of hydrated simatic thrust piles is proposed. When these imbricated simatic sheets were buried to depths greater than *c.* 20 km during overthrusting, partial melts of trondhjemite–tonalite–granodiorite (TTG) compositions were derived at varied *P*, *T* and H_2O conditions. This model satisfactorily accounts for problems such as the mechanism for generating the TTG magmas and their prolonged stoping into the greenstone belt, the regional granitoid–greenstone interference patterns, and a mechanism for sustaining high fluid pressures necessary for the continuous thrusting along the granitoid–greenstone interface (de Wit *et al.*, 1987a). As such, this model has a lot to commend it, and may provide many of the answers to the tectonic evolution of greenstone belts.

After the formation of the Barberton greenstone belt and probably of other major greenstone belts on the Kaapvaal Craton, a fundamental change in the stability of the craton occurred which was almost certainly intimately related to granitic magmatism (Tankard *et al.*, 1982). Sufficiently stable crustal conditions enabled the accumulation and preservation of vast thicknesses of volcanics and sediments in the Pongola Supergroup and Witwatersrand Triad. Burke *et al.* (1986) suggested that the Witwatersrand basin developed in response to the collision of the Kaapvaal and Zimbabwe cratons causing subsidence and deposition in a foreland basin. The Kaapvaal Craton provides a nearly complete spectrum of events spanning a billion years of Archaean history and is obviously a key area in which to attempt to unravel the early history of the Earth. It would appear that the cratonisation of this area occurred well before that of any other Archaean block in Africa.

16.7 Final comment

There is very little information that can be condensed out of the Archaean rocks in Africa to imply a uniform tectono-magmatic evolution. A summary of selected age data is given in Table 16.5 for African Archaean mafic and ultramafic metavolcanic rocks. Reliable ages have been established for only two greenstone belts, namely, the Belingwe (c. 2.7 Ga) and Barberton (c. 3.5 Ga) belts. Other well established ages include the Pongola Supergroup and Usushwana Intrusives in the Kaapvaal Craton and the Great Dyke in Zimbabwe. In most Archaean cratons in Africa at least two ages of greenstone belt formation have been identified, and three occur in the Zimbabwe Craton. In the Kaapvaal Craton the situation is not so clear. Until recently most greenstone fragments were considered in terms of the Barberton model and were thought to be of similar age to the Barberton greenstone belt. However, there may be two ages of greenstone development within the Barberton belt and it now appears that the Barberton 'model' is no longer applicable to other greenstone belt assemblages. The geochronological information (Table 16.3) for other greenstones in the Kaapvaal Craton is not sufficiently well constrained at present to establish any clear history of the early to mid Archaean evolution of this terrane. After the craton had stabilised at c. 3.0 Ga, the sequence of geological events is more clearly established to the end of the Archaean. It should be emphasised finally that while the Barberton granite–greenstone terrane has undoubtedly had the most work done on it, there is still little agreement between the various workers on almost any aspect of its geology and geochemistry. This reflects the complexity of these Archaean terranes and why they will continue to fascinate geologists for many years to come.

Acknowledgements

Financial support from the CSIR Foundation of Research and Development and the University of Cape Town is gratefully acknowledged. Chris Harris and Mike Watkeys are thanked for reading and critically commenting on earlier drafts of this chapter. J.R. Vail and R.P. Hall are thanked for their comments on the manuscript.

17 Early Precambrian basic rocks of South America

K.R. WIRTH, E.P. OLIVEIRA, J.H. SILVA SÁ and J. TARNEY

17.1 Introduction

Precambrian mafic–ultramafic igneous rocks of South America are diverse in composition, age and origin. They include greenstone belts, layered complexes, and anorogenic volcanics and dyke swarms, exposed on the three main cratonic regions: the Amazonian Craton, the São Francisco Craton and the Goiás Massif (Figure 17.1). The geology of these regions has been reviewed in Almeida et al. (1981), Gibbs and Barron (1983), Almeida and Hasui (1984), Hasui and Almeida (1985) and Cordani et al. (1988). The geology has been interpreted in terms of a number of orogenic cycles, notably the Guriense (> 3 Ga), Jequié (2.7 Ga), Transamazonian (2.0 Ga), Uruaçuano–Espinhaço (1.5–1.1 Ga) and Brasiliano (1.1–0.5 Ga) Cycles (see Wernick, 1981), during which the Precambrian crust progressively accreted. It has been estimated that 80% of the Precambrian of South America had formed by the end of the Transamazonian (Cordani et al., 1988).

The Amazonian Craton is exposed over more than 3 million km^2 in Bolivia, Brazil, Colombia, French Guiana, Guyana, Suriname and Venezuela. The craton is divided by the Amazonas Basin into the Guiana Shield (north) and the Guaporé Shield (south) (Figure 17.1). Several basement terranes, each with different geological histories, are recognised within the Amazonian craton; these include the Archaean Imataca Complex (Venezuela), Archaean metavolcanic and metasedimentary belts (central Brazil) and early Proterozoic granite–greenstone terranes (Venezuela, Guyana, Suriname, French Guiana, northern Brazil). All of these terranes were affected by early to middle Proterozoic magmatism following the Transamazonian orogeny.

The São Francisco Craton is one of the best studied geotectonic units in Brazil, and achieved its cratonic status following the Transamazonian Cycle. It is bounded by late Precambrian fold belts related to the Brasiliano Cycle. In the Archaean, both high-grade granulite–gneiss and low-grade granite–greenstone terranes are represented; the early Proterozoic displays a great variety of sedimentary, volcanic and plutonic rocks, invariably metamorphosed during the Transamazonian Cycle. In total, more than a dozen volcano-sedimentary sequences are known from the Craton (Mascarenhas et al., 1984). Two of these belts, the Rio Itapicuru and the Rio das Velhas, have typical greenstone belt morphology. The Goiás Massif, in Central Brazil, comprises both Archaean and early Proterozoic terranes that are bordered by mid-Proterozoic (Uruaçú) and late-Proterozoic (Brasilia and Paraguai–Araguaia) fold belts. The

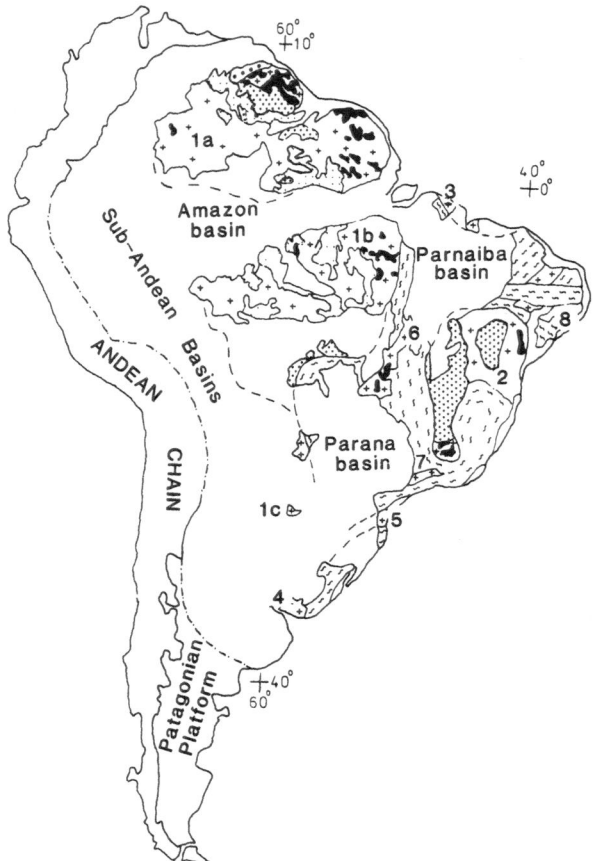

Figure 17.1   Generalised geology of South America (modified from Cordani and Brito Neves, 1982; Almeida *et al.*, 1981; Wernick, 1981; Gibbs and Wirth, 1985). Crosses: cratonic areas; 1: Amazonian craton; 1a: Guiana Shield; 1b: Guapore Shield; 1c: East Paraguay Massifs; 2: São Francisco craton; 3: São Luis craton; 4: Rio de la Plata craton; 5: Luis Alvez craton; 6: Goiás Massif; 7: Guaxupé Massif; 8: Pernambuco-Alagoas Massif.

Heavy stipple: Imataca Complex (northern Guiana Shield); black: greenstone belts; v-shading: early Proterozoic Uatumã Group (Amazonian craton); fine stipple: early to middle Proterozoic continental sediments; sigmoid shading: mid- to late-Proterozoic fold belts with re-worked basement; unshaded: Phanerozoic rocks; dashed lines: inferred boundaries of cratonic areas.

Archaean is represented by both greenstone and gneiss–granulite terranes, but a notable feature of the latter is the frequent occurrence of large mafic–ultramafic layered complexes. The Proterozoic also has a variety of metamorphosed volcanic and sedimentary sequences (Fuck *et al.*, 1982).

Because of the large number and diversity of Precambrian mafic and ultramafic rocks in South America, it is necessary to focus on a few selected examples, grouped according to (1) Archaean greenstone belts and other basic volcanic sequences, (2) early Proterozoic greenstone sequences (which are particularly well represented in South America), (3) layered mafic–ultramafic complexes and (4) Proterozoic mafic dyke swarms (Table 17.1). The localities of the more important units are shown on Figures 17.1 and 17.2.

17.2 Archaean basic volcanism

17.2.1 *Imataca Complex, Venezuela*

The Imataca Complex of Venezuela consists of amphibolite to granulite facies quartzo-feldspathic gneiss (Kalliokoski, 1965), amphibole-pyroxene gneiss, amphibolite, banded iron-formation (BIF), manganiferous metasediment, migmatite and marble. U–Pb isotope analyses of the gneisses and BIF indicate that the protoliths to the complex are Archaean (> 3.5 Ga) (Montgomery and Hurley, 1978). Many of the stratigraphic and geochemical features of the belt have been obscured by metamorphism and deformation that occurred at approximately 2.7 and 2.0 Ga (Hurley *et al.*, 1968).

The close association of amphibolite and amphibole-pyroxene gneiss with BIF (Kalliokoski, 1965), and the structural conformity of lithologies within the Imataca Series, suggest a volcanic, rather than an intrusive origin for the complex (Dougan, 1977). Major and trace element data from the amphibolites, amphibole-pyroxene gneisses and mafic granulites are most similar to continental tholeiitic volcanic rocks (Dougan, 1977) and the compositions of the felsic–intermediate gneisses and granulites are most similar to calc-alkaline continental volcanic rocks (Dougan, 1977).

17.2.2 *Bimodal volcanism, Serra dos Carajás*

The Serra dos Carajás region of Pará State in northern Brazil is currently one of the world's most active areas of mineral exploration. Prior to the discovery in 1967 of the large Carajás iron deposits (Tolbert *et al.*, 1971), the geology of the region was relatively poorly known. Since then, important deposits of Mn, Cu, Au, Ni and bauxite have also been found. The geology and ore deposits of the region have been reviewed by Tolbert *et al.* (1971), Santos (1981), Bernardelli *et al.* (1982), Hirata *et al.* (1982), Tassinari *et al.* (1982), Santos and Logercio (1984) and Olszewski *et al.* (1989).

The Grão Pará Group, which hosts the large iron deposits at Serra dos Carajás, is exposed in a broad synform and is considered to be unconformable on the surrounding Archaean (> 2.86 Ga; Machado *et al.*, 1988) gneisses, greenstone belts and granulite belts that are included in the Xingú Complex. From bottom to top, the Grão Pará Group includes a mafic metavolcanic sequence, a BIF unit (the Carajás Formation) and an upper metavolcanic and metasedimentary sequence. The lower metavolcanic unit consists of 4–6 km of mafic metavolcanic flows and breccias, and felsic flows and tuffaceous rocks (Gibbs *et al.*, 1986a; Olszewski *et al.*, 1989). The mafic flows are massive, porphyritic, and composed of plagioclase, clinopyroxene and low-grade metamorphic minerals. The felsic metavolcanic rocks are quartz-phyric and comprise 10–15% of the total volcanic section. The overlying Carajás Formation consists of 100–400 m of BIF (Tolbert *et al.*, 1971; Beisiegel, 1982), and is overlain by mafic metavolcanic flows, which are indistinguishable from metavolcanic rocks in the lower sequence, and by low-grade metamorphosed conglomerates, siltstones, cherts and schists. U–Pb isotope analyses of zircons from felsic rocks of the lower metavolcanic sequence indicate an age of 2.76 Ga (Wirth *et al.*, 1986; Machado *et al.*, 1988; Olszewski

Table 17.1 Major early Precambrian basic magmatic events in South America

Age	Orogenic Cycles	Amazonian Craton	São Francisco Craton	Goiás Massif
Late Proterozoic	Brasiliano Cycle (1.1–0.57 Ga)		Continental and marine sedimentation	Continental and marine sedimentation
(1.1 Ga) -------				
Middle Proterozoic	Espinhaço–Uruacuano Cycle (> 1.5–1.1 Ga)	Mafic dykes and sills (Avanavero, etc.) Continental sedimentation	Mafic dykes	Volcano-sedimentary sequences Granites
(1.9 Ga) -------				
Early Proterozoic	Transamazonian Cycle (2.2–1.8 Ga)	Mafic intrusions Felsic magmatism and sedimentation (Uatumã Group) Greenstone belts (Mazaruni Group, etc.)	Mafic dykes (Uauá Swarm) Mafic-ultramafic complexes (Caraíba, etc.) Greenstone belts (Rio Itapicuru)	Volcano-sedimentary sequence Layered complexes (Americano do Brasil, etc.)
(2.6 Ga) -------				
Archaean	Jequié Cycle (2.8–2.6 Ga)	Mature clastic sedimentation Bimodal volcanism and sedimentation (Carajás region)	Granites Greenstone Belts (Rio das Velhas, etc.) and other volcano-sedimentary sequences Granite-gneiss complex	Layered complexes (Niquelândia, etc.) Granites Mafic dykes (Goiás Swarm)
	Pre-Jequié or Guriense Cycle (> 3.0 Ga)	Gneisses, high-grade belts and greenstone Sedimentation and volcanism (Imataca Complex)		Granulite belts Greenstone belts (Crixás, etc.) Granite-gneiss complex

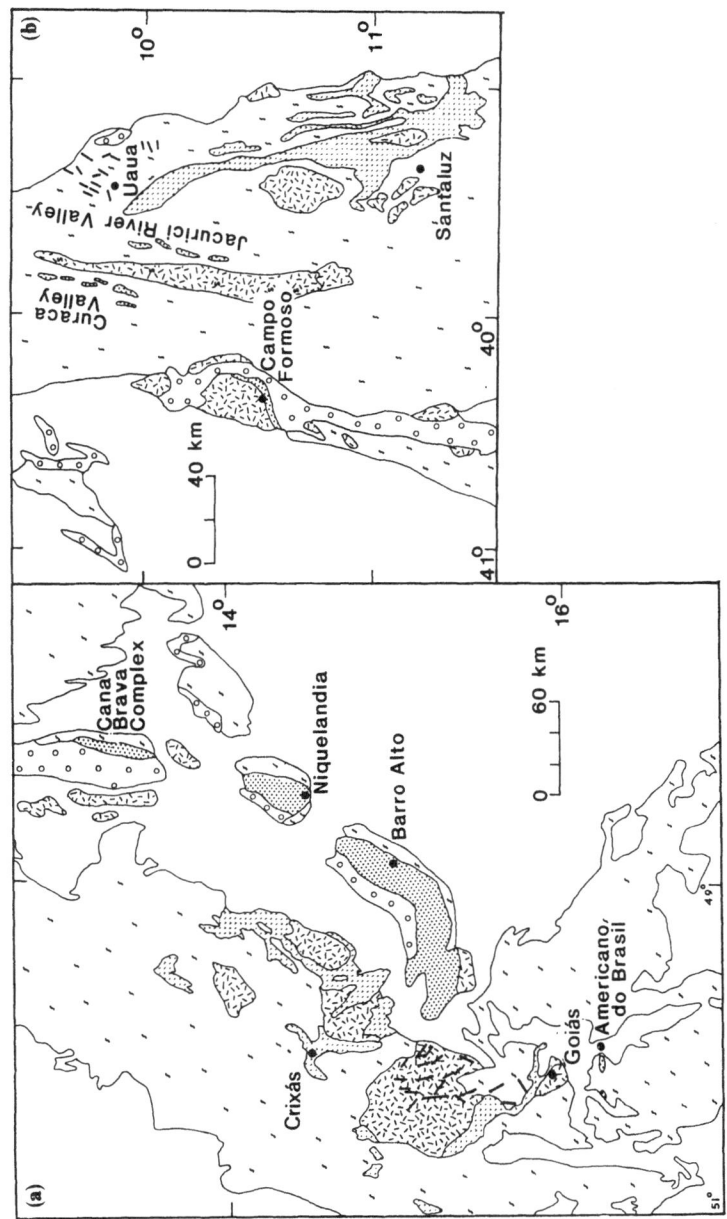

Figure 17.2 Simplified geology of part of the Goiás Median Massif (a) and of the northernmost part of the São Francisco craton (b). (After Marini *et al.*, 1984; Inda and Barbosa, 1978.) Symbols: sigmoids: high-grade gneisses: v-shading: greenstone belts; stipple: mafic–ultramafic complexes; circles: metamorphosed supracrustal sequences; cross-hatching: granite and syenite; black lines: mafic dyke swarms; unshaded: younger rock units.

et al., 1989). Rb–Sr whole-rock analyses of mafic flows from the lower metavolcanic sequence yield an isochron age of 2.69–0.06 Ga (Olszewski *et al.*, 1989).

The petrography and major-element chemistry of metavolcanic rocks of the Grão Pará Group indicate that the group is bimodal, consisting of basalts and basaltic andesites, and trachyandesites and rhyolites (Gibbs *et al.*, 1986a; Olszewski *et al.*, 1989). The good correlation between many of the more mobile alkali and trace elements indicates that the rocks have not been strongly altered. Most of the mafic metavolcanic rocks have enriched light rare-earth element (LREE) and flat HREE patterns (Figure 17.3a). The concentrations of the HREE correlate with mg' values (Mg/(Mg + Fe)). In contrast, the LREE do not correspond to varying mg', but vary proportionally with Si, K, Rb, and Ba concentrations. The Grão Pará Group mafic metavolcanics are enriched in K, Rb and Ba, relative to MORB, and are depleted in Th, Sr, Ta, Ti, and P. The ε_{Sr} and ε_{Nd} values of the various metavolcanic rocks range from 7.2 to 206 and -7.0 to $+4.6$ respectively.

The trace element and isotopic variations of the Grão Pará Group metavolcanic rocks cannot be explained by any single magmatic process. The co-variation of the HREE with mg' and some trace elements (Ta, Ti, P) can best be attributed to fractional crystallisation. However, neither the LREE nor the more incompatible trace elements (K, U, Th, Ba, and Rb) can be satisfactorily modelled by fractional crystallisation. The relatively high $^{87}Sr/^{86}Sr$ initial ratio ($Sr_i = 0.7057 \pm 0.001$) and negative ε_{Nd} values of some of the rocks require the assimilation of an old, radiogenic continental crust component. Assimilation of both upper and lower continental crust would have also enriched the Grão Pará magmas in the more incompatible elements, while simultaneously producing the observed relative depletions of Th, Ta, Sr, Ti and P (contained in more refractory phases) (Gibbs *et al.*, 1986a). The ε_{Nd} data also imply the existence of a significantly LREE-depleted mantle 2.76 Ga ago (Olszewski *et al.*, 1989).

The lithological, trace element and isotopic characteristics of the Grão Pará Group metabasalts are distinct from those of most greenstone belts (Condie, 1981; Sun, 1984). Many greenstone belts contain significant amounts of intermediate composition rocks, and the mafic metavolcanic rocks from most greenstone belts also generally lack strong enrichment of the LREE and LIL elements, and depletion of the HFS elements. In comparison with most modern oceanic basalts, the Grão Pará metabasalts are enriched in Rb, Ba, Th, U, K, and the LREE (Figure 17.3). The depletion of Th, Ta, Sr, P and Ti relative to the other trace elements is characteristic of magmas that have been contaminated with continental crust (Thompson *et al.*, 1983; Weaver and Tarney, 1983) either during the ascent of magmas through the crust, or by the subduction of continental material into mantle source areas. The bimodal composition of the Grão Pará Group metavolcanics, the negative ε_{Nd} values and the presence of mature sediments interlayered with the upper part of the volcanic sequence suggest that the group formed in a continental setting (Gibbs and Wirth, 1985; Dardenne *et al.*, 1988; Olszewski *et al.*, 1989). Many of the geochemical features of the Grão Pará Group basic rocks are similar to those of basalts in the British Tertiary Volcanic Province (Thompson *et al.*, 1982), the Paraná Basin (Fodor *et al.*, 1985; Mantovani *et al.*, 1985) and Mesozoic basalts of the eastern United States (Puffer and Philpotts, 1988; Whittington, 1988). The stratigraphy of the group is similar to the Mount Bruce Supergroup (Fortescue, Hamersley, and Turee Groups) in the Hamersley Basin, Western Australia (Trendall, 1983), which Blake and Groves (1987) have suggested formed on continental crust, adjacent to a rift.

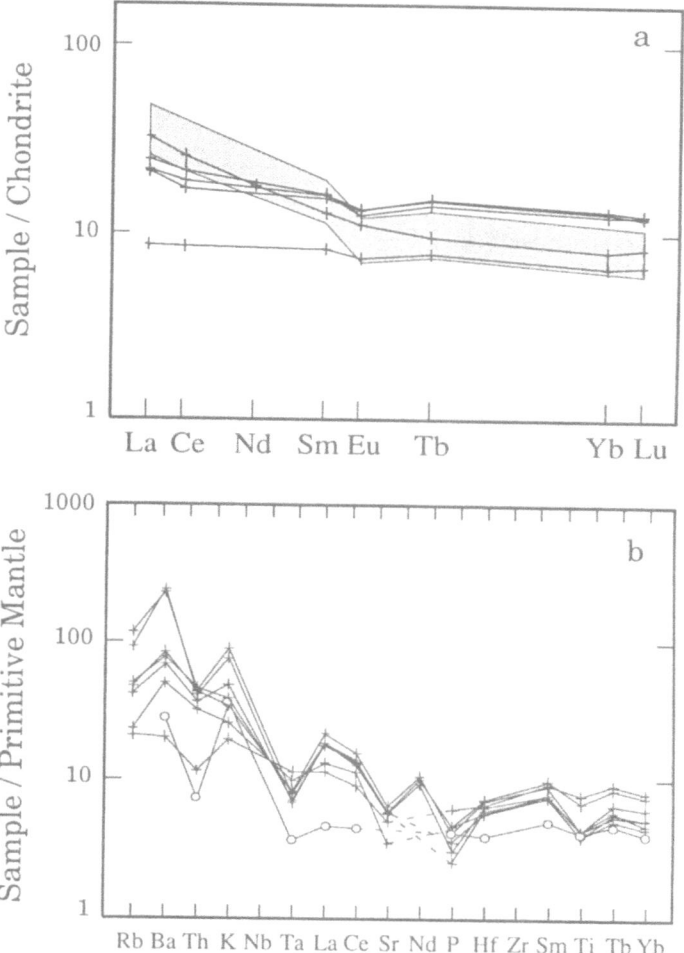

Figure 17.3   (a) Chondrite–normalised REE patterns of Grão Pará Group (Serra dos Carajás) basalt compared to those of basaltic andesite and trachyandesite (stipple), and (b) multi-element plots of Grão Pará Group basaltic andesites and trachyandesites (crosses) compared with a 'primitive' basalt (circles) (data from Olszewski et al., 1989). Normalising chondrite and primitive mantle values from Masuda et al. (1973) and Wood et al. (1979) respectively.

Several more highly metamorphosed volcanic–sedimentary belts occur in the Serra dos Carajás region (Hirata et al., 1982). The Bahia Group, west of the Carajás iron formations, hosts exhalative Cu sulphides and BIF (Ferreira Filho and Danni, 1985; Dardenne et al., 1988). The Bahia sequence consists of metamorphosed mafic flows, silicic pyroclastics and BIF that are overlain by interlayered siltstones, arenites and minor volcaniclastic sediments. A Rb–Sr reference isochron for pyroclastic rocks and an intrusive granophyre suggest that the Bahia metavolcanics formed at 2.6 Ga (Ferreira Filho et al., 1987).

Mafic volcanic rocks in the Bahia Group are relatively LILE- and LREE-enriched, and are depleted in Ti. Dardenne *et al.* (1987) concluded that the Bahia metavolcanics are similar to modern calc-alkaline volcanic rocks formed along continental margins during the early stages of subduction and suggested that the Grão Pará Group metavolcanics represent shoshonites formed during the rifting of a mature volcanic arc.

17.3 Archaean greenstone belts

17.3.1 *Crixás greenstone belt*

In central Goiás State (Figure 17.2b) extensive flows of basic and ultrabasic lavas make up the lower part of a volcanic–sedimentary sequence with many characteristics typical of greenstone belts (Sabóia, 1979) and have yielded an apparent Sm–Nd isochron age of 2.825 ± 0.1 Ga (Arndt *et al.*, 1989). The belt is surrounded by domes of tonalitic, granodioritic and granitic plutons. The basal sequence (600 m) comprises peridotitic lava flows with spinifex textures and polyhedral joints, which occur in association with serpentinites, magnesian schists and minor intercalations of pelitic and chemical sediments. Succeeding the ultramafic lavas are pillowed lava flows of komatiitic to tholeiitic composition, which are overlain by chemical and pelitic sediments (now graphite schists, chlorite schists, BIF and metacherts). The top of this lower unit is marked by the ubiquitous presence of talc-chlorite schists. The overlying unit is composed of metasedimentary schists and quartzites. The entire sequence has been metamorphosed to greenschist facies and isoclinally folded.

Available chemical data for the Crixás greenstone belt show that the meta-komatiites and basaltic komatiites have MgO contents of 31–38% and *c.* 24% respectively, high Ni (0.1–0.25%) and Cr (0.05–0.1%), low alkalis ($K_2O + Na_2O$ < 0.5%), low TiO_2 (0.1–0.2% and 0.4–0.6% respectively) and have TiO_2/P_2O_5 values of about 10 (Sabóia and Teixeira, 1980; Kuyumjian and Dardenne, 1982; Montalvão *et al.*, 1982). Figure 17.4 summarises the chemical data from the Crixás and other Brazilian greenstone belts. The abnormal behaviour of elements thought to be immobile (e.g. Al, Zr, REE) led Arndt *et al.* (1989) to speculate that the chemistry of the Crixás komatiites has been strongly modified during late-stage hydrothermal metamorphism.

Other volcanic–sedimentary greenstone sequences occur to the south of the Crixás belt, in the Goiás Velho and Rio do Côco regions (Danni *et al.*, 1982), and also comprise ultramafic rocks with pillow structures and minor intercalations of chemical and pelitic metasediments. MgO contents reach 42% in the serpentinised ultramafics and range from 15 to 22% in magnesian schists. The TiO_2 contents of these rocks are low (< 0.5 wt%).

17.3.2 *Rio das Velhas greenstone belt*

The Rio das Velhas greenstone belt (Schorscher, 1978) is situated in the south of the São Francisco craton, within the 'iron-quadrangle' of Minas Gerais State. This metamorphosed and deformed Archaean volcano-sedimentary sequence is in tectonic contact with the migmatitic gneissic basement, and both form a basement to the Minas Supergroup, an early Proterozoic supracrustal sequence which hosts large iron deposits.

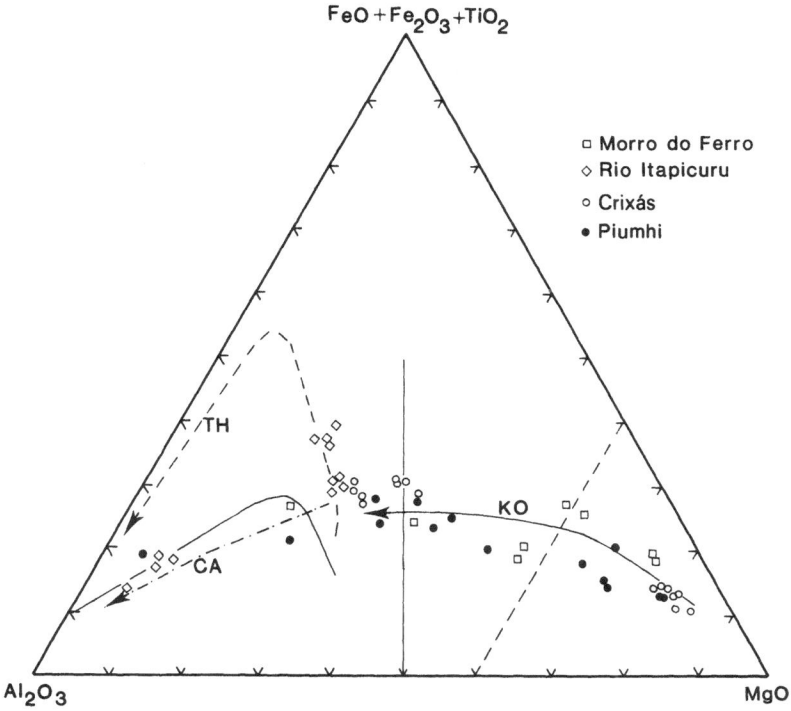

Figure 17.4 Al_2O_3:$(FeO + Fe_2O_3 + TiO_2)$:MgO (cation) plot (Jensen, 1976) for Brazilian Morro do Ferro (squares), Rio Itapicuru (diamonds), Crixás (open circles) and Piumhi (dots) greenstone belts. Data from Teixeira and Danni (1979), Jahn and Schrank (1983), Kishida and Riccio (1980) and Kuyumjian and Dardenne (1982). KO, TH and CA represent komatiitic, tholeiitic and calc–alkaline trends respectively.

The Rio das Velhas sequence comprises a lower mafic–ultramafic group, followed by a mafic group with interleaved sediments, and an upper group formed predominantly of clastic sediments. The 600 m thick lower group has massive flows of mafic and ultramafic lavas and breccias showing local development of spinifex textures, pillow and agglomerate structures. The rocks are altered to serpentine, chlorite, talc and tremolite, with Cr-spinels being the only relict primary minerals. The ultramafic rocks have high Cr and Ni contents (*c.* 2000 ppm) and may have been komatiitic (Schorscher *et al.*, 1982). Rare and narrow bands of metacherts and BIF are intercalated with the lavas. The group grades upwards into a 4000 m thick sequence of chlorite schists, metagreywackes, basic and intermediate metavolcanics intercalated with BIF, meta-cherts, carbonate rocks, graphitic phyllites, quartzites and minor coarse-grained detrital sediments. Important Au deposits are found in connection with the BIF and sulphide and carbonate facies rocks. The upper group (< 600 m thick) is un-conformable on the lower groups and comprises clastic sediments, pelites and minor lenses of conglomerate.

17.3.3   *Piumhi greenstone belt*

The Piumhi greenstone belt, near the southwestern edge of the São Francisco craton, comprises a volcano-sedimentary sequence which includes ultrabasic to intermediate volcanic flows, pyroclastics, immature volcanogenic and chemical sediments, together with late granitic intrusions and rhyolitic dykes and sills. The lavas include both komatiitic and tholeiitic types (Jahn and Schrank, 1983). Komatiitic basalts have TiO_2 contents in the range 0.5–0.66%, broadly chondritic Al_2O_3/TiO_2 ratios (23–17), and high Ni (79–236 ppm) and Cr (424–2700 ppm). The peridotitic komatiites have higher MgO contents (21–36%), Ni (707–1886 ppm) and Cr (948–7540 ppm) (Figure 17.4). Some peridotitic rocks display cumulus textures resulting from Cr-spinel accumulation.

The basaltic komatiites are mildly LREE-enriched and have relatively flat HREE distributions (Figure 17.5). Jahn and Schrank (1983) interpreted these basaltic komatiites as having been derived from a mantle source also enriched in LREE relative to chondritic abundances. Large negative Eu anomalies in two of the peridotitic komatiites are thought to be related to alteration; a third (Figure 17.5) has a pattern similar to that of the tholeiites, and may be related to them. Two tholeiitic basalts have very similar major element chemistries but rather different REE distributions (Figure 17.5; samples A38A and A25), and Jahn and Schrank (1983) concluded that these two

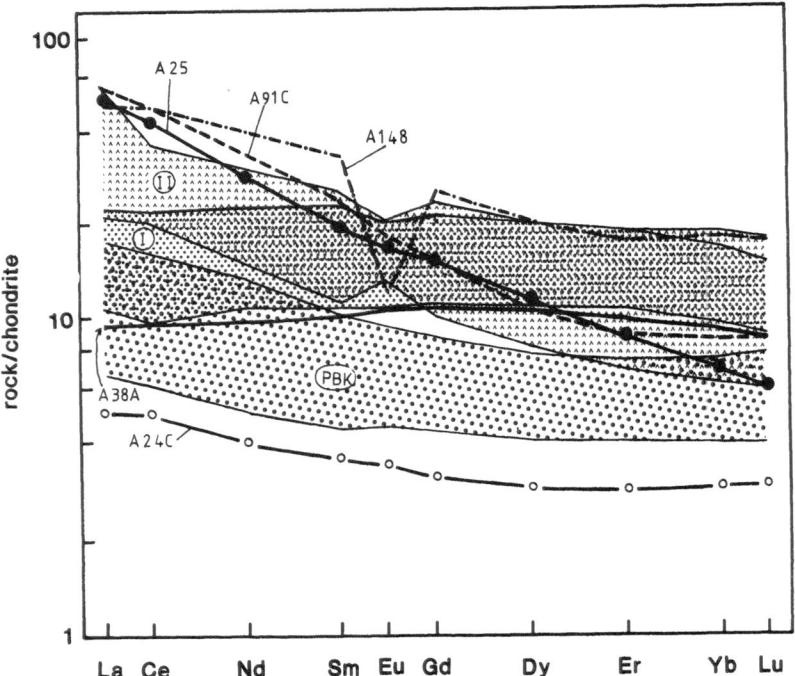

Figure 17.5   Chondrite-normalised REE patterns for rocks from the Rio Itapicuru and Piumhi greenstone belts showing fields of Rio Itapicuru Group I and Group II tholeiites and Piumhi basaltic komatiites (PBK) (after Silva, 1987: Jahn and Schrank, 1983).

basalts cannot be genetically related to each other, nor can they be related to the basaltic komatiites by simple fractionation mechanisms. A number of different mantle sources are inferred. The rhyolites have REE patterns with a strong negative Eu anomaly and less fractionated REE (Figure 17.5), similar to Archaean post-tectonic granites.

17.3.4 Morro do Ferro greenstone belt

The Morro do Ferro greenstone belt outcrops near the southwestern edge of the São Francisco craton. It occurs within a migmatitic gneiss terrane and is tightly folded and has greenschist facies metamorphic assemblages (Teixeira and Danni, 1979). The lower part is composed of locally spinifex-textured ultramafic lava flows, with interleaved BIF, chert and aluminous tuffs, overlain by altered massive basic rocks (tremolite–chlorite–albite–epidote schists), talc–schist lenses, acidic tuffs and ferruginous cherts, and an upper sedimentary group of muscovite–chlorite schists, quartz–chlorite schists, phyllites, marbles, BIF and minor reworked volcanic rocks. Using chemical data, Teixeira and Danni (1979) identified a fractional crystallisation trend ranging from komatiitic to tholeiitic (Figure 17.4), with a decrease in the MgO content at relatively constant CaO/Al_2O_3. Olivine was the first phase to crystallise followed by clinopyroxene and plagioclase.

17.3.5 Greenstone belts of the southeastern Guaporé Shield

Low-grade metavolcanic rocks of probable Archaean age are exposed in the southern part of the Amazonian craton (Amaral, 1984). Mafic metavolcanic rocks are associated with intermediate to felsic metavolcanics, tuffs, amphibolites, greenschists and metasediments in the Serra das Andorinhas, Serra do Inajá, Gradaús, Babaçu-Lagoa Seca, Pedra Preta and Sapucaia sequences. Pillow structures are preserved locally and minor amounts of pyroxenite and ultramafic schist are present in some of the belts. Metasedimentary and metavolcanic rocks of the Serra das Andorinhas greenstone belt are cross-cut by a granodiorite with a Rb–Sr whole rock isochron age of 2.56 Ga (Gastal et al., 1987). Basalts from the Serra das Andorinhas greenstone belt are relatively enriched in LREE ($La_N/Sm_N = c.$ 2.3), Rb, Ba, Th and K, and depleted in P and Ti (Wirth, unpublished data), similar to calc-alkaline basalts in other Archaean greenstone belts.

17.4 Proterozoic greenstone belts

Early Proterozoic volcanic and sedimentary rocks similar to those in Archaean greenstone belts are exposed across most of the northern Guiana Shield. From west to east, these rocks comprise the Pastora Supergroup of Venezuela (Menendez, 1972; Espejo, 1974), the Barama–Mazaruni Supergroup of Guyana (Gibbs and Barron, 1983), the Marowijne Group of Suriname (de Roever and Bosma, 1975; Bosma et al., 1983) and the Maroni Supergroup of French Guiana (Choubert, 1974; Marot et al., 1984). The greenstone belts are typically exposed in branching synforms surrounded by gneisses. The metamorphic grade ranges from amphibolite facies, around the peripheries of the belts, to greenschist facies near their centres. The greenstone belts and surrounding gneisses are in fault contact with the Archaean Imataca Complex. The nearly continuous exposure of greenstone belts in the northern part of the Guiana Shield is interrupted by the high-grade rocks of the Central Guyana granulite belt.

17.4.1 Mazaruni Group, Guyana

The lithostratigraphy of the Mazaruni Group in the Issineru River region (Gibbs, 1987a; Renner and Gibbs, 1987) is typical of the early Proterozoic greenstone belts of the Guiana Shield. The Mazaruni Group is divided into a lower metavolcanic unit (Issineru Formation) and an upper metasedimentary unit (Haimaraka Formation). The lower 3.5 km of the metavolcanic unit comprises massive and pillowed basalt flows, gabbros and minor interflow tuffs and cherts. The pillow structures, tuffaceous interflow sediments and spilitic alteration are clear evidence that the metavolcanic rocks were erupted in a subaqueous environment. The upper 3.5 km of the metavolcanic section consists of basalt flows interstratified with andesite, dacite and rhyolite flows with minor metavolcanic breccias and tuffs. The proportion of felsic volcanics and interlayered volcaniclastic sediments increases near the top of the metavolcanic section, which is transitional into the overlying metasedimentary section. Clasts in the metasediments indicate that they were derived entirely from volcanic rocks exposed within the belt (Gibbs et al., 1986b). U–Pb analyses of zircons indicate that the age of the volcanism is 2.25 Ga while zircons from the Batica gneisses surrounding the greenstone belt are 2.23 ± 0.39 Ga (Gibbs and Olszewski, 1982).

Both tholeiitic and calc-alkaline differentiation trends have been recognised in the Barama–Mazaruni metavolcanic rocks (Gibbs, 1987a; Renner and Gibbs, 1987). High-alumina tholeiitic rocks and calc-alkaline rocks are generally more abundant near the upper parts of the metavolcanic sections. Although all volcanic rock types (basalt to rhyolite) are represented in the Guyana greenstone belts, rocks with silica contents of 63–68% are not common, demonstrating that the volcanic suites are weakly bimodal (Gibbs, 1987a). Fragments of felsic volcanic rock comprise the bulk of the metasediments of the Haimaraka Formation.

Mafic to felsic Fe-rich metavolcanics form a geochemical continuum, suggesting that they developed by fractional crystallisation (Renner and Gibbs, 1987). REE patterns of the basalts are flat (Figure 17.6a) or a slightly LREE-depleted (Renner and Gibbs, 1987). Ni, Co and Cr contents in these rocks are as high as those in Archaean greenstone basalts (Condie, 1985). The calc-alkaline basalts of this group have higher Al and REE, and lower Cr and Ni abundances than the tholeiites. The basalts of both series have HFSE concentrations similar to MORB. However, they are relatively depleted in P and Ti, and are variably enriched in Th and LILE (Figure 17.6b). Trace element data from these rocks are generally similar to those of MORB or volcanic arc basalts (Gibbs, 1987a; Renner and Gibbs, 1987) and are similar to the basalts from the East Scotia Sea back-arc spreading centre (Tarney et al., 1981). Some of the calc-alkaline andesites are similar to continental or transitional arc volcanic rocks (Gibbs, 1987a). The Himaraka Formation sediments are similar to those in recent marginal basins and to those in many Archaean greenstone belts (Gibbs et al., 1986b).

17.4.2 Paramacca Series, French Guiana

The Paramacca Series in the Inini Belt of French Guiana (Choubert, 1974) consists of ultramafic, mafic, intermediate and felsic metavolcanic rocks overlain by metasedi-

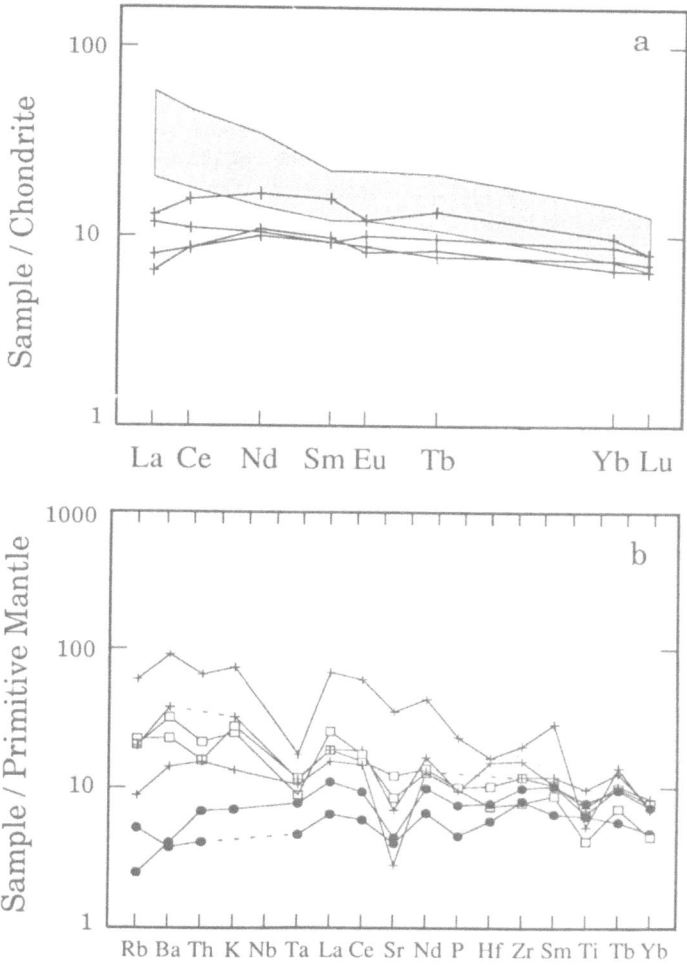

Figure 17.6   Trace element patterns of basaltic and andesitic volcanic rocks from Proterozoic greenstone belts of the Guiana Shield (data Gibbs, 1987a; Renner and Gibbs, 1987; Wirth, unpublished data). Symbols and normalising values as in Figure 17.3.

ments. Although spinifex textures have not been recognised, conformably interlayered talc- and tremolite-bearing schists and serpentinites are probably metamorphosed komatiites.

Sm–Nd whole-rock analyses of peridotitic meta komatiites, tholeiitic basalts and andesites from the Paramacca Series have yielded an isochron age of 2.11 ± 0.09 Ga (Gruau *et al.*, 1985), and an initial ^{143}Nd/^{144}Nd ratio of 0.51002 ($\varepsilon_{Nd} = +2.1$). Rb–Sr whole-rock analyses of orthogneisses indicate an age of 2.0 ± 0.07 Ga (Sr$_i = 0.7019$). The isotope data indicate that volcanism and plutonism occurred contemporaneously

(Gruau *et al.*, 1985). The low Sr_i of the gneisses is essentially that of undepleted upper mantle at 2.0 Ga, suggesting that they were derived either directly from the mantle, or from partially melted mafic crust. Similarly, the high ε_{Nd} of metavolcanic rocks from the Paramacca Series does not allow for significant crustal contamination of the Paramacca magmas by older Archaean crust. These isotope data suggest that the early Proterozoic greenstone–gneiss terranes of the Guiana Shield represent additions of new material to the continental crust reservoir and also imply that LREE-depleted source regions existed in the early Proterozoic mantle.

17.4.3   *Vila Nova and Serra do Navio belts, Amapá, Brazil*

In Amapá, northern Brazil, belts of metavolcanic and metasedimentary rocks trend roughly NW–SE (Montalvão *et al.*, 1975; Jorge João *et al.*, 1979), parallel to the trend of belts in the Carajás region. The stratigraphy of the Amapá belts has been largely obscured by medium-grade metamorphism and pervasive deformation. The Vila Nova Group consists of basal amphibolites (Jornal Group) overlain by quartzites, schists, manganiferous carbonates, BIF and gondites. The mafic to ultramafic Jornal Group amphibolites consist variably of talc, actinolite, tremolite, anthophyllite or cumming-tonite schists and local serpentinite. Some ultramafic amphibolites in the Vila Nova and Cupixi regions contain significant amounts of chromite. Chemical analyses of the amphibolites (Jorge João *et al.*, 1979; Wirth, unpublished data) indicate that they follow komatiitic and tholeiitic trends (e.g. on a Jensen (1976) Mg:(Fe + Ti):Al cation plot). Talc-, actinolite- and tremolite-bearing amphibolites from the Ipitinga area are high in MgO (9–35%), Cr (1000–5600 ppm) and Ni (50–250 ppm). Their REE patterns are slightly LREE depleted, with overall REE contents 5–10 times chondritic values (Wirth, unpublished data).

The pervasive deformation and metamorphism of the Amapá belts indicates that they formed prior to the Transamazonian orogeny (*c.* 2.0 Ga). The common association of ultramafic metavolcanic rocks and mature platform sediments in the Amapá region is not commonly observed in other Archaean greenstone belts either in the Amazonian craton or world-wide. It is possible that the volcanic and sedimentary rock sequences formed at different times or in different tectonic settings.

17.4.4   *Rio Itapicuru greenstone belt*

The Rio Itapicuru is possibly the only Proterozoic greenstone belt known outside the Guiana Shield. Mascarenhas (1976) first recognised the sequence in the São Francisco craton in east-central Bahia as a greenstone belt (Figure 17.2b), and this was supported by the geochemical studies of Kishida and Riccio (1980). The greenstone rocks crop out as N–S elongate strips, surrounding domes of granitic material and they are subdivided into a mafic volcanic domain (MVD) composed of tholeiitic lavas, a felsic volcanic domain (FVD) comprising calc-alkaline andesitic to dacitic lavas and pyroclastics and a sedimentary domain (SD) represented by chemical, pelitic and clastic sediments (Teixeira *et al.*, 1982; Silva, 1987). The MVD includes flows of metamorphosed massive, cumulophyric to porphyritic basaltic lavas, minor pillowed lavas, mafic tuffs, flow breccias and minor intercalations of sedimentary rocks such as charts, BIF and locally graphitic pelites. Only two thin ultramafic bodies have been recorded within the FVD pyroclastic rocks (Silva, 1987). They contain serpentinised euhedral olivines, laths and needles of pyroxene and post-cumulus clinopyroxene, and are mineralogically, texturally and chemically similar to komatiitic peridotites of South Africa (Kishida and Riccio, 1980; Silva, 1987). Small granitic bodies, gabbroic sills, tholeiitic dykes and felsic subvolcanic bodies intrude the central part of the supracrustal sequence (Silva, 1987).

Brito Neves *et al.* (1980) derived a Rb–Sr isochron of 2.1 Ga ($Sr_i = 0.7017$) for the volcanics and 2.1 Ga ($Sr_i = 0.704$) for the associated granitoids. Zircons and monazites from the FVD andesitic rocks have given ages of c. 2.18 Ga (Gaal, pers. comm.); monazites from the Poço Grande dome have given ages of c. 2.1 Ga, but zircons from the Ambrósio dome have an age of 2.94 Ga. However, according to Davidson *et al.* (1988) this Archaean age corresponds to a megaxenolith of magmatitic gneiss basement carried up during ascent of the granitoids. Sm–Nd model ages for the basaltic and ultramafic rocks lie in the range 2.51 ± 0.4 Ga with $\varepsilon_{Nd} = +3$, which led Silva (1987) to suggest a late Archaean to early Proterozoic evolution for the belt.

Geochemical studies by Silva (1987) have revealed three magmatic lineages: a komatiitic series, represented by peridotites and rare basalts; a series of tholeiitic gabbros, basalts, ferro-basalts and high-Mg basalts; and a series of calc-alkaline andesitic to dacitic lavas and pyroclastics. The tholeiitic group has fractionated REE patterns and Silva (1987) identified two sub-groups (Figure 17.5), one with flat REE patterns (type I) and a second, LREE-enriched type (type II). Petrogenetic modelling suggests that the two groups cannot be related by fractional crystallisation or by partial melting of the same source. However, magmas within each group are compatible with derivation through various degrees of partial melting of a similar source, followed by fractional crystallisation. The type II tholeiites ($La_N/Lu_N > 1$) probably originated from a mantle source with twice chondritic REE abundances at depths greater than 60 km (garnet zone), while the type I tholeiites ($La_N/Lu_N = 1$) originated at shallower depths. Whereas the tholeiites have some similarities with MORB, other trace element characteristics resemble those of low-K island-arc tholeiites. The andesitic and dacitic rocks have REE patterns which are HREE-depleted and much more highly fractionated, suggesting little direct link with the tholeiites. These volcanics and the syntectonic granitoids have been interpreted by Silva (1987) as partial melts of an eclogitic source. The komatiitic rocks show flat to slightly enriched LREE, but flat HREE patterns. Their ε_{Nd} value ($+3$) would be compatible with small degrees of crustal contamination (e.g. through thermal erosion), but there is no direct petrogenetic link between the komatiitic and tholeiitic rocks (Silva, 1987).

The Rio Itapicuru greenstone belt probably developed in an en-sialic marginal basin (back-arc) environment. The sequence underwent low-grade hydrothermal (seawater) metamorphism soon after, producing greenschist facies assemblages. A second metamorphic event related to the syn-tectonic emplacement of granitoids resulted in metamorphic zoning around the granitoids, producing amphibolite facies and ultimately hornfels (up to 650°C; 0.4 GPa) assemblages near the granite contacts.

17.5   Early Precambrian mafic–ultramafic plutonic complexes

A striking feature of the Goiás Massif is the occurrence of major mafic–ultramafic intrusive complexes lying along a 400 km long N- to NE-trending lineament. Figure 17.2a shows the location of the Barro Alto, Niquelândia and Cana Brava complexes, which are the most important of these bodies. These complexes are very similar, suggesting they are related to a common event. Radiometric dating suggests they are

Archaean in age (Cordani and Hasui, 1975; Girardi *et al.*, 1978). All three host important mineral deposits: Ni in Niquelândia and Barro Alto, and asbestos (chrysotile) in Cana Brava.

17.5.1 *Cana Brava complex*

The layered complex of Cana Brava (Figure 17.2a) is a 40 km long lenticular body which has faulted contacts with the country rocks to the south and east, but is in structural conformity with rocks of a volcanic-sedimentary fold belt to the west. Two minor satellite bodies are assumed to be related (Girardi and Kurat, 1982). Although the age of the Cana Brava complex is not established, an early Proterozoic or late Archaean age is probable on the basis of field relationships. Girardi and Kurat (1982) describe the complex as being composed mainly of layered metamorphosed gabbros, gabbronorites, norites, amphibolites, pyroxenites, serpentinites and minor magnesian schists and rodingites. The ultramafic members are concentrated toward the eastern part of the complex and are associated with amphibolites.

The layered rocks were recrystallised at granulite facies (900°C; 0.6–0.7 GPa) and partially re-equilibrated under amphibolite facies conditions. Later hydrothermal events resulted in the formation of serpentinites, rodingites, talc-schists and the economically important asbestos deposits. Girardi and Kurat (1982) interpreted the complex as a major intrusive basaltic magma body that underwent gravitative differentiation at depth and which was elevated to its present position partially by block faulting.

17.5.2 *Niquelândia complex*

The Niquelândia Complex is a lensoid layered body 40 km long and 20 km wide, bounded by faults, and composed of gabbronorites, pyroxenites, peridotites and anorthosite–gabbros (Figueiredo *et al.*, 1975; Rivalenti *et al.*, 1982; Girardi *et al.*, 1986). Many primary igneous structures and mineral assemblages are still preserved (Girardi *et al.*, 1986). According to Rivalenti *et al.* (1982) the complex has the following stratigraphy:

Upper Sequence	Upper amphibolites (UA)
	Upper gabbro–anorthosite zone (UGAZ)
Lower Sequence	Layered gabbro zone (LGZ)
	Layered ultramafic zone (LUZ)
	Basal peridotite zone (BPZ)
	Basal gabbro zone (BGZ)

Danni and Leonardos (in Rivalenti *et al.*, 1982) interpreted the two sequences as separate units: an older peridotite–pyroxenite–gabbro association metamorphosed under granulite facies conditions, and a younger metagabbro–anorthosite–amphibolite unit which grades upwards into a silicic to basic volcanic–sedimentary cover. The BGZ is made up of gabbros, websterite and peridotite, in faulted contact (via a mylonitic metagabbro) with country rocks (Rivalenti *et al.*, 1982; Girardi *et al.*, 1986). It grades upwards into harzburgites and dunites of the BPZ, the upper part of which is marked by a narrow chromitite horizon. The LUZ is characterised by alternate layers of websterite and peridotite, with two gabbro horizons at the base and a chromitite layer near the top. Pegmatitic bronzitite marks the contact with the overlying LGZ, which comprises layered melanocratic to leucocratic gabbros in which amphibole is the main mafic phase associated with the appearance of biotite and quartz. The roof of the Niquelândia complex is defined by layered amphibolites (UA) with occasional anorthosite intercalations in the lower portion. Hornblende, plagioclase and minor amounts of clinopyroxene, opaques and garnet are the main minerals of this zone. The complex was metamorphosed at greenschist to granulite facies conditions, but there has been no large-scale ductile deformation. On the basis of field observations, whole-rock chemistry and mineral data, Girardi *et al.* (1986) concluded that the Niquelândia complex represents a single magmatic body, the diverse rock types having derived through fractional crystallisation of a parental relatively aluminous, low-Ti picritic basaltic magma, which was emplaced in a rift-like environment. The Cana Brava and Barro Alto complexes may be linked to the same tectono-magmatic event.

17.5.3   Barro Alto complex

The Barro Alto complex has an arcuate shape and is more than 150 km long (Figure 17.2a). It is composed of a sequence of granulite facies to ultramafic rocks associated with a second sequence of metamorphosed anorthositic gabbros to the northwest (Figueiredo *et al.*, 1975). The relation between these two sequences is controversial. Some authors have argued that they represent two completely distinct magmatic units (Danni and Teixeira, 1981; Danni *et al.*, 1982; Fuck *et al.*, 1981). The mafic–ultramafic sequence comprises norites with pyroxenite lenses in tectonic contact with a strip of harzburgite and minor dunite. Despite the intense metamorphism, primary cumulate textures and rhythmic layering are still preserved in the noritic and gabbronoritic units (Nilson, 1984). The contact with the anorthositic gabbro sequence is tectonic, and the latter grades through fine-grained amphibolites into a volcano–sedimentary sequence (Nilson, 1984). In the upper anorthositic gabbros, orthopyroxene is subordinate to clinopyroxene and plagioclase. Although metamorphosed, primary igneous layering and textures and preserved. Olivine, plagioclase (An_{80-60}) and clinopyroxene are cumulus phases and hypersthene and hornblende post-cumulus phases (Danni and Teixeira, 1981). Mineral chemistry studies by Girardi *et al.* (1981) indicate metamorphic conditions at upper amphibolite to granulite facies (*c.* 800°C; 0.5 GPa). These authors suggested that the anorthositic unit and the upper gabbro are related, and were derived from a common source by fractional crystallisation. From the whole-rock and mineral chemistry data, they interpreted the parental magma of the Barro Alto Complex to be a 'normal' basalt having a MgO/FeO ratio somewhat lower than that deduced for the Niquelândia Complex (Nilson, 1984).

17.5.4   Americano do Brasil layered complex

Numerous small mafic–ultramafic bodies occur in the Archaean and lower Proterozoic terranes of the Goiás Massif. Of these, the Americano do Brasil complex is the best studied. It is an elongate body (9 × 2.5 km) partly bounded by faults and composed of a cumulate sequence of olivine–Cr spinel, olivine–clinopyroxene, olivine–clinopyroxene–plagioclase, and plagioclase–hypersthene–clinopyroxene rock types

(Nilson, 1984). The complex suffered amphibolite facies metamorphism but primary textures, igneous layering and cryptic variations are still preserved. MgO/FeO ratios in the rocks range from 2.8 in the lower part to 0.4 in the upper cumulates. Olivine compositions vary from Fo_{88} to Fo_{74}, orthopyroxene from En_{80} to En_{56}, and plagioclase from An_{87} to An_{45}. The cumulates in the lower-middle section host an important Ni-Cu ore deposit with average Ni and Cu contents of 0.62% and 0.65% respectively.

17.5.5   Caraiba complex

The Caraiba mafic–ultramafic complex of northern Bahia, São Francisco craton, hosts the second largest Brazilian copper deposit, having known reserves of 130 million tonnes of ore averaging 1% Cu. The mineralisation, which occurs mainly as chalcopyrite and bornite, is of disseminated type and is hosted by lensoid hypersthenitic and noritic bodies up to 100 m wide enclosed within high-grade gneisses of the Curaca Valley (Figure 17.2b). The hypersthenite–norite association of Caraiba is comparable to the Koperberg Suite in South Africa (Conradie and Schoch, 1986).

Jardim de Sá et al. (1982) and the Caraiba Mine staff (e.g. Hasui et al., 1982; D'El Rey Silva, 1985) have shown that the mafic and ultramafic rocks were emplaced into a supracrustal sequence represented by various banded, graphitic, and Al-rich gneisses, BIF, calc-silicate rocks and minor amphibolites and quartzites. The rocks have undergone three main phases of deformation accompanied by metamorphism, migmatisation and granitic intrusions. The present mushroom-like shape of the Caraiba deposit is the result of interference between F2 and F3 structures (D'El Rey Silva, 1985). The age of the Caraiba complex is not well constrained. Zircons from the G2 tonalitic orthogneiss have given an age of 2.3–2.25 Ga (Gaal, pers. comm.), which provides a minimum age for the mafic–ultramafic complex.

In spite of subsolidus recrystallisation, primary igneous features, such as cumulus, intergranular and poikilitic textures and flow structures are ubiquitous in the Caraiba mafic–ultramafic complex. It was first thought to be a differentiated tholeiitic sill (Townend et al., 1980; Lindenmayer, 1980) comprised of three cycles of hypersthenite with peridotitic 'enclaves' at the base, grading upwards to melanorite, norite and gabbro (Table 17.2; Mandetta, 1982, cited by D'El Rey Silva, 1985). However, the common presence of breccia-like structures, the multiple intrusion of norite into hypersthenite and vice-versa, off-shoots of hypersthenite in already deformed pyroxene-granulites, and the lack of either mineral or chemical layering in the hypersthenites and norites, suggest a different petrogenetic model (Oliveira, unpub. data). The complex consists of multiple injections of hypersthenites and norites emplaced during the final stage of the F1 deformation, and the peridotite 'enclaves' and some of the gabbros are probably xenoliths of mantle or lower crust. This interpretation is supported by the rounded shape of the peridotitic inclusions against which the hypersthenite is chilled, the contrasting mineral compositions of host and enclave material, and prograde metamorphic reactions found in gabbroic inclusions embedded in norites.

The inclusions comprise varied proportions of magnesian olivine (Fo_{87-73}), bronzite (En_{85-76}), pargasite, Fe-rich spinels and minor calcic pyroxene. Massive dunites, olivine–bronzitites and pargasite–olivine–bronzitites are the dominant rock types. The hypersthenites and norites have a calc-alkaline geochemical signature

Table 17.2 Representative rock analyses of the Caraiba mafic-ultramafic complex, and the Uauá and Goiás mafic dyke swarms

Analysis:	1	2	3	4	5	6	7	8	9	10	11	12	13	14	15
	2516 260.72	NO33 92.29	NO33 94.24	NO33 93.77	2516 288.33	NO33 89.08	1194 02	12/84 86.3	12/84 89	EB-51	EB-52	EB-53	EB-42	EB-49.2	EB-81.1
SiO_2	48.49	43.66	42.49	46.63	49.25	55.30	48.98	50.80	51.71	50.70	50.23	50.81	50.49	50.78	54.71
TiO_2	0.39	0.14	0.41	0.82	1.17	0.03	0.88	0.92	0.46	1.06	1.48	0.52	1.02	1.93	0.57
Al_2O_3	4.93	2.62	2.71	12.55	18.55	23.66	14.61	14.17	8.43	14.40	14.12	12.29	11.57	14.67	13.24
Fe_2O_3	21.31	19.56	25.75	17.54	10.99	2.86	12.58	12.76	11.25	11.92	13.70	9.69	12.90	13.51	10.40
MnO	0.35	0.43	0.44	0.29	0.16	0.06	0.18	0.19	0.18	0.18	0.20	0.16	0.19	0.19	0.16
MgO	20.01	17.56	19.80	12.10	6.94	2.00	8.64	7.84	20.03	8.23	6.91	12.65	11.21	5.59	9.36
CaO	0.87	7.77	0.46	5.76	7.74	8.87	11.68	11.43	6.60	11.51	10.76	11.17	10.41	9.38	8.85
Na_2O	0.88	0.22	0.00	1.45	2.18	5.53	2.09	1.95	0.91	1.91	2.10	1.12	1.55	2.83	2.03
K_2O	0.61	0.16	0.25	0.37	0.27	0.73	0.09	0.22	0.36	0.12	0.28	0.18	0.24	1.02	0.70
P_2O_5	0.00	5.29	0.06	0.36	0.61	1.11	0.06	0.09	0.08	0.09	0.14	0.04	0.08	0.54	0.11
V	298	115	335	270	257	22	275	280	170	277	351	202	286	159	166
Cr	1599	410	2632	826	298	41	408	305	2574	399	254	1025	883	110	737
Ni	912	348	928	420	157	33	157	132	581	150	95	289	231	30	238
Zn	414	373	561	279	153	66	83	93	89	86	108	70	97	123	81
Ga	22	16	20	26	30	27	17	18	12	18	23	13	17	24	16
Rb	23	6	6	10	7	6	2	6	10	2	4	5	5	15	21
Sr	51	76	5	219	485	725	101	120	78	104	112	97	98	336	166
Y	9	112	10	17	13	23	20	22	13	24	31	12	17	35	24
Zr	62	13	37	16	64	—	42	57	54	65	95	27	46	145	90
Nb	1	3	2	4	6		2	3	3	4	6	3	3	11	6
Ba	314	52	96	151	122	319	23	345	167	34	57	49	41	577	232
La	—	113	—	10	22	43	3	4	5	3	6	2	2	24	18
Ce	5	329	4	28	52	97	5	7	9	13	14	1	6	46	35
Nd	1	204	0	15	30	52	5	7	6	9	11	3	4	28	20
Th	17	36	4	4	6	—	4	2	4	3	3	5	4	7	6

Analyses 1–3: Caraiba hypersthenites; 4–6: Caraiba norites; 7–8: Uauá tholeiites; 9: Uauá norite; 10–13: Goiás group I tholeiites; 14–15: Goiás group II tholeiites. (Unpublished data of Oliveira.)

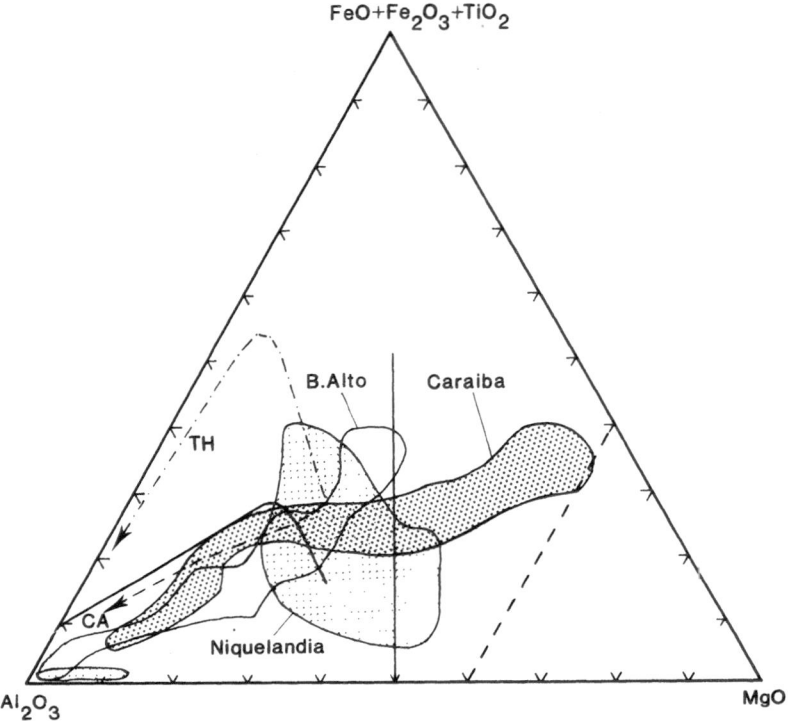

Figure 17.7 $Al_2O_3:(FeO + Fe_2O_3 + TiO_2):MgO$ (cation) plot (Jensen, 1976) for Brazilian plutonic mafic–ultramafic complexes. Data from Girardi *et al.* (1981, 1986) and E.P. Oliveira (unpublished). TH and CA represent tholeiitic and calc–alkaline trends respectively.

(Figure 17.7; Table 17.2) and are composed of varying proportions of hypersthene (En_{69-58}), plagioclase (An_{42-80}), phlogopite, magnetite, Cu sulphides, ilmenite and apatite. No calcic pyroxene has been recorded in these rocks. The hypersthenites and norites show fractionated REE patterns ($La_N/Yb_N = 2$–27) and high Ba/Nb, La/Y, Th/La, Ba/La and Rb/La ratios, indicating an enriched lithospheric source.

17.5.6   *Rio do Jacaré sill*

The Rio do Jacaré sill is a N–S trending layered complex around 40 km long and 1 km wide in the São Francisco craton (Galvão, 1980; Brito, 1984). It is sandwiched between the early Proterozoic volcanic rocks of the Contendas–Mirante volcanic–sedimentary belt to the west and the Archaean Jequié high-grade gneiss terrane to the east. The mafic–ultramafic body is exposed in the steep limb of a N–S trending antiform with an E-dipping axial plane.

The 400 m thick lower zone is composed mainly of gabbros with minor anorthosite. The upper gabbro zone comprises three members, each with intercalations of pyroxenites, seams of massive magnetite rock, and topped by tonalitic horizons. Brito (1984) suggested that these three units represented three distinct phases of magma influx. Despite metamorphism (most of the pyroxenes have been replaced by amphibole), primary magmatic features such as mineral layering and cumulus textures are preserved. A pipe-like body also occurs in the lower zone, composed of gabbro, pyroxenite and massive magnetitite. The latter has economic concentrations of vanadium-bearing magnetite. High V contents (0.8% V_2O_3) have been found in the magnetite seams of the upper zone. The whole exposed gabbro complex is notably Fe-rich. Apart from the pyroxenite layers, a major ultramafic component appears to be absent. In these respects it differs from the previously described complexes.

17.5.7 *Jacurici River Valley and Campo Formoso complexes*

Other mafic–ultramafic complexes found in the Jacurici River valley of the São Francisco craton (Figure 17.2b) host the only known economic deposits of chromite in Brazil. The bodies are distributed along a N–S zone some 70 km long. Tens of bodies are embedded in a migmatitic gneiss complex which is broadly similar to that which hosts the Caraiba complex.

The Medrado-Ipueira sill is as much as 300 m thick and hosts the chromite deposits of Medrado-Ipueira mines. It was emplaced into a supracrustal sequence believed to be the oldest rocks in the area (Marinho *et al.*, 1986). The sill occupies a synform, its upper part being in contact with paragneisses and the lower part conformable with underlying granulite facies granitic gneisses. No chilled margins have been observed. The sill is composed of olivine–orthopyroxene–spinel cumulates with chromitite layers in the lower part, giving way to orthopyroxene–olivine–spinel, orthopyroxene, and plagioclase–orthopyroxene cumulates successively towards the top (Deus and Viana, 1982). The chromitite seam is up to 7 m thick and chromite may reach 80% of the spinel-bearing assemblages. There is no evidence of multiple magma injection and a regular increase in Fe/Mg ratio occurs from bottom to top. However, the complex is pervasively altered to serpentinite.

The Campo Formoso complex is a similar mafic–ultramafic body around 40 km long and 100 m to 1.1 km wide. It is bounded by the 2.0 Ga Campo Formoso granite to the NNW and by the Jacobina Group metasediments to the ESE (Figure 17.2b). Its base lies on Archaean granulites (Deus *et al.*, 1982), but the upper contact is unconformably overlain by quartzites of the Jacobina Group which contain detrital chromite. The intrusion of the Campo Formoso granite resulted in widespread alteration of the complex, which is now pervasively replaced by serpentine, chlorite, talc and tremolite. Chromitite horizons are the only relics of original layering, and at least seven chromitite horizons have been located through drilling, the largest being *c.* 9 m thick. The Campo Formoso complex has been interpreted as part of a once larger layered complex which was tectonically thinned and eroded (Gonçalves *et al.*, 1972, quoted by Deus *et al.*, 1982).

17.5.8 *Mafic intrusive complexes of the Guiana Shield*

There are several small mafic and ultramafic complexes that were intruded into the granite–greenstone terranes of the Guiana Shield, including the De Goeje suite (Suriname), Tapuruquara suite (Brazil), Kaburi anorthosite, Appinitic suite and Itaki gabbro (Guyana), Tampoc gabbros (French Guiana) and San Juan de Manapiare complex (Venezuela). Only a few of these complexes have been dated, but their contact relationships, deformation and metamorphism indicate that they are pre- or syn-Transamazonian. Some of the intrusive complexes may be the deep-seated magma chambers to the younger, mid-Proterozoic dyke and sill complexes (Gibbs, 1987b).

The De Goeje suite of Suriname comprises numerous small, zoned and layered ultramafic to gabbroic, and locally granodioritic bodies that intrude the early Proterozoic basement. Isotope data suggest an emplacement age of $1.85 \pm 0.2\,Ga$ (Bosma *et al.*, 1983). Olivine, orthopyroxene and magnetite are common cumulus minerals, with intercumulus clinopyroxene, hornblende and biotite. The cumulus textures, compositional zoning and calc-alkaline geochemical trend of the gabbro suites suggest that they were derived by fractional crystallisation of a high-Al basalt (Bosma *et al.*, 1983).

17.6 Proterozoic dyke swarms and sills

17.6.1 *Avanavero Suite*

Thick sequences of felsic volcanic rocks (and related granitic intrusives) and continental sediments were deposited throughout much of the Amazonian Craton during the middle Proterozoic (Gibbs, 1987b; Sial *et al.*, 1987). The volcanic and sedimentary rocks are unconformable on early Proterozoic basement, and lack the tectonic fabric that is characteristic of the Transamazonian Orogeny. Mafic magmatic rocks of this age occur primarily as dykes, sills and differentiated intrusive complexes that intrude the continental sediments and felsic volcanics.

The felsic volcanics of the Uatumã Supergroup are the most extensive of the middle Proterozoic units in the Amazonian Craton, and consist mostly of andesitic to rhyolitic flows and tuffs. U–Pb, Rb–Sr and K–Ar analyses of felsic intrusive rocks related to the Uatumã volcanics have given ages of *c.* 1.8 Ga (Priem *et al.*, 1973; Gomes *et al.*, 1975; Tassinari *et al.*, 1982; Wirth *et al.*, 1986; Machado *et al.*, 1988). The Roraima Group sediments which unconformably overlie the Uatumã Supergroup volcanics in the Guiana Shield are cut by extensive dolerite dykes, sills and differentiated minor intrusive complexes collectively referred to as the Roraima intrusive complex or the Avanavero suite. Isotopic ages determined for this suite suggest two periods of intrusion, one from 1.80 to 1.84 Ga (Priem *et al.*, 1973; Onstott *et al.*, 1984) and another from 1.61 to 1.67 Ga (Priem *et al.*, 1973; Hebeda *et al.*, 1973). Isotopic ages determined from tuffs interlayered with sediments of the Roraima Group range from 1.66 to 1.73 Ga (Priem *et al.*, 1973; Gaudette and Olszewski, 1985).

The Avanavero suite sills range up to 400 m thick and occur within the Roraima Group sediments, along its basal unconformity and within the underlying basement rocks. Dykes occur only in areas of deeply eroded basement below the unconformity. Many of the intrusive bodies are compositionally zoned, consisting of early crystallised orthopyroxene dolerites that are succeeded by pigeonite-dolerites, ferrodolerites, hornblende-granophyres and leucogranophyres (Hawkes, 1966a). Inclined sills and dykes in the basement consist of pigeonite- and ferrodolerites, while the horizontal sills contain more Mg-rich lithologies near their bases and are differentiated to Fe-rich lithologies near their tops.

One of the largest intrusive complexes of the Avanavero suite is the zoned dyke and sill complex Tumatumari–Kopinang intrusion in Guyana. These rocks comprise layered dolerites, ferrodolerites, hornblende-dolerites and hornblende-granophyres

(Hawkes, 1966b). Variations in grain size and modal composition define rhythmic and stratiform layering. Very fine-grained dolerite along the chilled margins of the complex consists of labradorite, endiopside and bronzite. In layers that successively overlie the contact dolerite, the modal orthopyroxene content decreases, clinopyroxene becomes more Fe-rich, plagioclase becomes more sodic and hornblende replaces pyroxene as the major ferromagnesian phase. The upper zone of the Kopinang sill contains significant amounts of hornblende-granophyre. Ferrodolerite consisting of ferro-augite, andesine, fayalite, hornblende, biotite and opaque minerals occupies the core of the dyke feeding the Tumatumari intrusion. The contact dolerites are classified as basaltic andesites ($c. 52\% \; SiO_2$), and normatively they are quartz-tholeiites similar to many other continental dolerites (Hawkes, 1966b; Choudhuri, 1978).

Tholeiitic dolerite dykes and sills intruded into a thick sequence of mature continental platform sediments (Urupi Formation) in the northern part of Brazil are collectively referred to as the Quarentas Ilhas Formation. Although their ages are not established, stratigraphic and intrusive relationships constrain the age to $c.$ 1.69–1.84 Ga (Daoud and Fuck, 1987). The association of continental sediments and doleritic intrusives is similar to that between the Roraima Group sediments and Avanavero suite in the northern part of the Guiana Shield (Daoud and Fuck, 1987).

17.6.2   Mafic dyke swarms

The best preserved early Precambrian dyke swarms in Brazil occur in the northern and southern portions of the São Francisco craton and in the Goiás Massif. They have been named the Uauá, Pará de Minas and Goiás swarms (Table 17.2; Oliveira and Montes, 1984; Sial et al., 1987). Other minor dyke swarms are known, but only the Goiás and Uauá swarms will be dealt with here. They intrude Archaean to early Proterozoic granite–gneiss terranes (Figure 17.2a, b).

The Uauá dykes (Figure 17.2b) generally trend SW–NE and are up to 4 km long and a few metres wide. Winge and Danni (1980) described field and petrographic aspects of the Uauá swarm and its possible relationship with the Archaean volcano-sedimentary sequence of Rio Capim, which is not intersected by dykes. This feature suggests that the dykes could be feeders to the basal metavolcanic units of the Rio Capim sequence. However, the dykes post-date the third deformation phase which affected the Rio Capim sequence, and they are not much older than 2.2 Ga (Jardim de Sá et al., 1984). Moreover, the Rio Capim sequence has been thrust over the granite–gneiss terrane with the result that the dykes are intensely deformed along the shear zone. Similar relationships occur in the Goiás dyke swarm (Figure 17.2a), which does not cut the Guarinos, Goiás Velho, Pilar de Goiás and Crixás greenstone belts (Danni et al., 1982). Some of these dykes are several kilometres long and in the order of a hundred metres wide. They trend N–S, NW–SE, and less often NE–SW and have been affected by low to middle grade metamorphism.

The Goiás dykes have ophitic textures and mineral assemblages comprising plagioclase, calcic pyroxene (mostly uralitised), Fe-Ti oxides and interstitial quartz. One group of dykes from Uauá has ophitic and minor granoblastic textures, gabbroic to gabbronoritic compositions, and typically contains minor quartz and green amphibole. Another group displays cumulus textures and is noritic, with mineral assemblages comprising olivine, brown amphibole and biotite, plagioclase, orthopyroxene, minor clinopyroxene and Fe-Ti oxides. The gabbroic dykes of both Goiás and

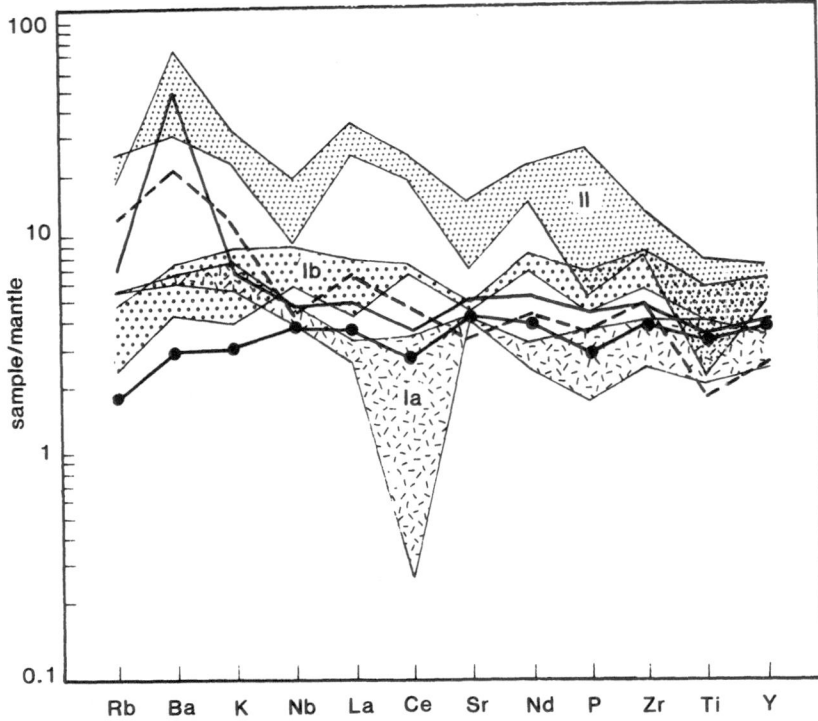

Figure 17.8   Mantle-normalised multi-element plot for Uauá and Goiás dyke swarms (E.P. Oliveira, unpublished data) normalised to primordial mantle values given by Wood *et al.* (1979), showing fields of Group I (a and b) and Group II dykes from Goiás. Dashed and continuous lines represent typical Uauá norites and tholeiites respectively (see text).

Uauá are all quartz- or olivine-normative tholeiites. Trace element distributions (Oliveira, unpublished data; Table 17.2) expressed as multi-element plots normalised to the primordial mantle of Wood *et al.* (1979) permit the recognition of two separate dyke groups in Goiás and at least three in Uauá (Figure 17.8). Group I dykes of Goiás have relatively flat patterns, slight LILE depletion and in some cases a slightly positive Nb anomaly. Group II dykes have more strongly fractionated patterns, higher LFSE (low field strength element) concentrations, and pronounced negative Nb and Sr anomalies. Group I samples appear to be related by fractional crystallisation according to their variation in MgO, mg', Zr, Cr, Ni and Ti/Zr values. Group Ia samples show a relative enrichment in Rb, Ba and K, and generally lower HFSE and LREE abundances than Group Ib. Geochemical modelling suggests that significant crustal contamination occurred either during magma ascent through the crust, or as the result of post-emplacement metasomatic fluids.

Although the MgO and total alkali contents of the group II dykes from Goiás can be explained by fractional crystallisation, variations in trace element contents suggest that the various magmas were produced by different degrees of melting of the same source rather than by crustal assimilation. The two Goiás dyke groups appear to be derived from different mantle sources, Group I from a source with less fractionated REE ($La_N/Yb_N = 1.3$) and Group II from a more fractionated source ($La_N/Yb_N = 5.4$)

The two tholeiitic groups from the Uauá swarm (Figure 17.8) have similar HFSE (Nb, Ti, P, Zr) distributions, perhaps reflecting their derivation from the same mantle source and the effects of fractional crystallisation. However, their respective LFSE (Rb, Ba, K) distributions are very different, requiring a much larger degree of crystal fractionation to explain the observed variation using the Rayleigh equation. The REE patterns of the two groups are similar for the middle REE and HREE, but completely different for the LREE. The LFSE-enriched tholeiites have a higher La_N/Yb_N ratio (1.4 compared to 0.85). Thus the Uauá tholeiites appear to have been derived from a LFSE- and LREE-depleted source followed by fractional crystallisation and crustal contamination and/or metasomatic alteration.

In contrast to the tholeiitic dykes, the noritic dykes have trace element patterns with negative Sr, Nb and Ti anomalies on mantle-normalised plots, and relatively high concentrations of Rb, Ba, K and LREE. The higher MgO (20%) and LFSE contents and more fractionated REE ($La_N/Yb_N = 4$) of the norites suggest they were not consanguineous with the tholeiites. Furthermore, the fresh nature of the norites (unmetamorphosed and no deuteric alteration) suggests that they are younger than the tholeiites. The Uauá norites have SiO_2, MgO, Zr/Y and Ti/Zr values similar to those of boninites and may form, together with the tholeiitic dykes, a bimodal mafic association similar to those described elsewhere (Weaver and Tarney, 1981; Hall and Hughes, 1987).

17.7 Conclusions

The early Precambrian of South America is characterised by a wide variety of mafic and ultramafic rocks that formed in diverse tectonic settings. The oldest rocks in the craton (Imataca Complex) are high-grade gneisses that include mafic metavolcanic rocks, and low-grade terranes are represented by a number of different types of volcano-sedimentary successions. Most have lithostratigraphic and geochemical compositions similar to those of greenstone belts elsewhere in the world. Those in the Serra dos Carajás region are composed of bimodal volcanic flows and mature clastic metasediments, and have a lithostratigraphy distinct from typical greenstone belts. The trace element compositions of these mafic volcanic rocks are similar to those of basaltic rocks occurring in modern continental dyke swarms, flood basalt provinces and rifts. Their high ε_{Sr} and low ε_{Nd} values are evidence for the involvement of continental crust at some stage in the petrogenesis of these rocks. The stratigraphic, trace element and isotope geochemical data indicate that the most likely tectonic setting was an extending continental crust.

More typical greenstone belts range in age from early Archaean to early Proterozoic. In general, the trace element and isotopic data for these greenstone mafic volcanics show little evidence of interaction with continental crust or sub-continental mantle. Early Proterozoic greenstone belts (2.25–2.1 Ga) in the northern part of the Amazonian craton comprise mafic to intermediate metavolcanic and volcano-sedimentary rocks. The compositions of the mafic volcanics are most similar to those of modern back-arc basins. Nd and Sr isotope data indicate that they were derived from a LREE-depleted mantle source. The early Proterozoic greenstone belts of the northern Guiana Shield represent the addition of new material to the continental crust reservoir, implying that a significant proportion of the present continental crust formed during the Proterozoic.

South America is also characterised by numerous mafic–ultramafic complexes of late Archaean–early Proterozoic age, some of which carry important mineral deposits. The petrogenesis of these important bodies still needs to be clarified. Their occurrence along lineaments suggests they may have been emplaced along major continental sutures. Study of these bodies may provide insights into the nature of the subcontinental mantle during the late Archaean.

The Archaean and early Proterozoic granite–greenstone belts and metavolcanic terranes were deformed and metamorphosed during the Transamazonian orogeny (c. 2.0 Ga). Following this major orogeny, the older metavolcanic terranes acted as the basement to further extensive suites of volcanic, intrusive and sedimentary material. The early to middle Proterozoic (1.85–1.6 Ga) mafic rocks are typical continental Fe-rich dolerites, emplaced principally as dykes, sills and differentiated intrusive complexes. Although the dykes are dominantly Fe-rich quartz tholeiites, noritic suites occur locally. Mid-Proterozoic mafic sills occur most abundantly in the Guiana Shield and northern Brazil. Most of these mafic suites have compositions typical of Phanerozoic continental flood basalts.

Acknowledgements

E.P. Oliveira and J.H. Silva Sá wish to thank Dr Glória Silva for making available unpublished data on the Rio Itapicuru greenstone belt, and the CVCP (UK) and CNPq (Brazil) for financial support. K.R.W. acknowledges support from NSF Grant EAR-8410379 to Allan Gibbs, and additional field support provided by DOCEGEO and ICOMI. Early drafts of this chapter were reviewed by C. Barron and D. Sarewitz.

This chapter is dedicated to the memory of Allan Gibbs, whose work has been instrumental to our understanding of the Precambrian evolution of the Amazonian Craton.

References

Abbott, D. and Ferguson, J. (1965) The Losberg intrusion, Fochville, Transvaal. *Trans. Geol. Soc. S. Afr.* **68**, 31–52.

Abell, P.I., McClory, J., Martin, A., Nisbet, E.G. and Kyser, T.K. (1985) Petrography and stable isotope ratios from Archaean stromatolites, Mushandike Formation, Zimbabwe. *Precamb. Res.* **27**, 385–398.

Agee, C.B. and Walker, D. (1988) Static compression and olivine flotion in ultrabasic silicate liquid. *J. Geophys. Res.* **93** (B), 3427–3449.

Agee, C.B. and Walker, D. (1989) Comments on 'Constraints on element partition coefficients between $MgSiO_3$ perovskite and liquid determined by direct measurements' by T. Kato, A.E. Ringwood and T. Irifune. *Earth Planet. Sci. Lett.* **94**, 160–161.

Ahmat, A. (1986) Petrology, structure, regional geology and age of the gabbroic Windimurra Complex, Western Australia. Unpubl. PhD thesis, Univ. W. Aust.

Ahmat, A. and de Laeter, J.R. (1982) Rb–Sr isotopic evidence for Archaean–Proterozoic crustal evolution of part of the central Yilgarn Block, Western Australia: constraints on the age and source of the anorthositic Windimurra gabbroid. *J. Geol. Soc. Aust.* **29**, 177–190.

Ahren, J.L. and Turcotte, D.L. (1979) Magma migration beneath an oceanic ridge. *Earth Planet. Sci. Lett.* **45**, 115–122.

Aitken, B.G. and Echeverria, L.M. (1984) Petrology and geochemistry of komatiites and tholeiites from Gorgona Island, Colombia. *Contrib. Mineral. Petrol.* **86**, 94–105.

Alapieti, T. (1982) The Koillismaa layered igneous complex, Finland – its structure, mineralogy and geochemistry, with emphasis on the distribution of chromium. *Bull. Geol. Surv. Finland* **319**, 116 pp.

Alapieti, T. and Lahtinen, J.J. (1986) Stratigraphy, petrology, and platinum-group element mineralization of the early Proterozoic Penikat layered intrusion, northern Finland. *Econ. Geol.* **81**, 1126–1136.

Alapieti, T. and Piirainen, T. (1984) Cu–Ni–PGE mineralisation in the marginal series of the early Proterozoic Koillismaa layered igneous complex, northeast Finland. In *Sulphide deposits in mafic and ultramafic rocks*, eds Buchanan, D.L. and Jones, M.J., Inst. Min. Metall. London, 123–131.

Alapieti, T., Hugg, R. and Piirainen, T. (1979) Structure, mineralogy and chemistry of the Syote section of the Early Proterozoic Koillismaa layered intrusion, northeastern Finland. *Bull. Geol. Surv. Finland* **299**, 33 pp.

Aleinikoff, J.N., Williams, I.S., Compston, W., Stuckless, J.S. and Worl, R.G. (1987) Conventional and ion microprobe U–Pb ages of Archaen rocks from the Wind River Range, Wyoming, *Geol. Soc. Amer. Abs. Prog.* **19**, 569.

Allaart, J.H. (1976) Ketilidian mobile belt in South Greenland. In *Geology of Greenland*, eds. Escher, A. and Watt, W.S., Geol. Surv. Greenland, Copenhagen, 121–151.

Allaart, J.H. (1982) (compiler) 1:500 000 scale geological map, Frederikshåbs Isblink–Søndre Strømfjord. *Geol. Surv. Greenland.*

Allard, G.O. (1970) The Dore Lake Complex, Chibougamau, Quebec – a metamorphosed Bushveld-type layered intrusion. *Geol. Soc. S. Afr. Spec. Publ.* **1**, 477–491.

Allard, G.O. (1976) Dore Lake Complex and its importance to Chibougamau geology and metallogeny. *Quebec Min. Rich. Nat. Doc.* **DP-368**, 446 pp.

Allègre, C.J. and Ben Othman, D. (1980) Nd–Sr isotopic relationship in granitoid rocks and continental crust development: a chemical approach to orogenesis. *Nature* **303**, 762–766.

Allègre, C.J., Dupré, B., Richard, P., Rousseau, and Brooks, C. (1982) Subcontinental versus suboceanic mantle, II. Nd–Sr–Pb isotopic comparison of continental tholeiites with mid-ocean ridge tholeiites, and the structure of the continental lithosphere. *Earth Planet. Sci. Lett.* **57**, 25–34.

Allen, P., Condie, K.C. and Narayana, B.L. (1985) The geochemistry of prograde and retrograde charnockite-gneiss reactions in southern India. *Geochim. Cosmochim. Acta* **49**, 323–336.

Allsopp, H.L., Evans, I.B., Giusti, L., Hallbauer, D.K., Jones, M.Q.W. and Welke, H.J. (1986) U–Pb dating and isotopic characterization of carbonaceous components of the Witwatersrand reefs. *Geocongress '86, Johannesburg, Ext. Abs.*, 85–88.

Almedia, F.F.M. and Hasui, Y. (eds.) (1984) Pre-Cambriano do Brasil. Editora Edgard Blucher, São Paulo, Brazil.

Almedia, F.F.M., Hasui, Y., Brito Neves, B.B and Fuck, R.A. (1981) Brazilian structural provinces: an introduction. *Earth Sci. Rev.* **17**, 1–29.

Andersen, M.C. and Pulvertaft, T.C.R. (1986) Occurrences of anorthositic rocks in the reworked basement in the Umanak area, central West Greenland. *Rapp. Grønlands Geol. Unders.* **129**, 18 pp.

Anderson, C.I. (1985) Petrography and geochemistry of Archaean pyroxenites, Horseshoe Lake quadrangle, Wind River Mtns., Wyoming. *Geol. Soc. Amer. Abs. Prog.* **17**, 206.

Anderson, A.T., Bunch, T.E., Cameron, E.N., Haggerty, S.E., Boyd, F.R., Finger, L.W., James, O.B., Keil, K., Prinz, M., Ramdohr, P. and El Goresy, A. (1970) Armalcolite: a new mineral from the Apollo 11 samples. *Proc. Apollo 11 Lunar Sci. Conf.*, Peragamon, New York, 55–63.

Anderson, D.L. (1981) Hot spots, basalts and the evolution of the mantle. *Science* **213**, 82–89.

Anderson, D.L. (1982) Hot spots, polar wander, Mesozoic convection and the geoid. *Nature* **297**, 391–393.

Andrews, A.J., Hugon, H., Durocher, M., Corfu, F., and Lavigne, M.J. (1986) The anatomy of a gold-bearing greenstone belt: Red Lake, Northwestern Ontario, Canada. In *Gold 86 Proc. Vol.*, ed. Macdonald, A.J., Toronto, Ont. 3–22.

Andrews, J.R., Bridgwater, D., Gormsen, K., Gilson, B., Keto, L. and Watterson, J. (1973) The Precambrian of South-East Greenland. In *The Precambrian of Scotland and Related Rocks of Greenland*, eds., Park, R.G. and Tarney, J., Keele Univ., 143–156.

Anhaeusser, C.R. (1971a) Cyclic volcanicity and sedimentation in the evolutionary development of Archaean greenstone belts of Shield areas. *Geol. Soc. Aust. Spec. Publ.* **3**, 57–70.

Anhaeusser, C.R. (1971b) The Barberton Moutain Land, South Africa – a guide to the understanding of the Archaean geology of Western Australia. *Geol. Soc. Aust. Spec. Publ.* **3**, 103–119.

Anhaeusser, C.R. (1972) The geology of the Jamestown Hills area of the Barberton Mountain Land, South Africa. *Trans. Geol. Soc. S. Afr.* **75**, 225–263.

Anhaeusser, C.R. (1973a) The evolution of the early Precambrian crust of southern Africa. *Phil. Trans. R. Soc. Lond.* **A273**, 359–388.

Anhaeusser, C.R. (1973b) The Archaean greenstone remnant west of Muldersdrif, Krugersdorp district. In *14th Ann. Rep. Econ. Geol. Res. Unit*, Pretorius, D.A. (comp.), Univ. Witwatersrand, 17–18.

Anhaeusser, C.R. (1974) The Archaean greenstone remnants north of Krugersdorp. In *15th Ann. Rep. Econ. Geol. Res. Unit*, Pretorius, D.A. (comp.), Univ. Witwatersrand, 8–9.

Anhaeusser, C.R. (1975) Precambrian tectonic environments. *Ann. Rev. Earth Planet. Sci.* **3**, 31–53.

Anhaeusser, C.R. (1977) Geological and geochemical investigations of the Roodekrans ultramafic complex and surrounding volcanic rocks, Krugersdrop district. *Trans. Geol. Soc. S. Afr.* **80**, 17–28.

Anhaeusser, C.R. (1978) The geological evolution of the primitive Earth: evidence from the Barberton Moutain Land. In *Evolution of the Earth's Crust*. ed. Tarling, D.H., Academic Press, London, 71–106.

Anhaeusser, C.R. (1981) The relation of mineral deposits to early crustal evolution. *Econ. Geol. 75th Anniv. Vol.* 42–62.

Anhaeusser, C.R. (1983) The geology of the Schapenburg greenstone remnant and surrounding Archaean granitic terrane south of Badplaas, Eastern Transvaal. *Spec. Publ. Geol. Soc. S. Afr.* **9**, 31–44.

Anhaeusser, C.R. (1985) Archaean layered ultramafic complexes in the Barberton mountain land, South Africa. In *Evolution of Archaean Supracrustal Sequences*, eds. Ayres, L.D., Thurston, P.C., Card, K.D., and Weber, W. *Geol. Assoc. Can. Spec. Pap.* **28**, 281–301.

Anhaeusser, C.R., Roering, C., Viljoen, M.J. and Viljoen, R.P. (1968) The Barberton Mountain

Land: a model of the elements and evolution of an Archaean fold belt. *Trans. Geol. Soc. S. Afr.* 7, annex, 225–254.

Anhaeusser, C.R., Mason R., Viljoen, M.J. and Viljoen, R.P. (1969) A reappraisal of some aspects of Precambrian shield geology. *Bull. Geol. Soc. Amer.* 80, 2175–2200.

Appel, P.W.U. and Secher, K. (1984) On a gold mineralization in the Precambrian Tartoq Group, South-West Greenland. *J. Geol. Soc. London* 141, 273–278.

Arculus, R.J. (1978) Mineralogy and petrology of Grenada, lesser Antilles island arc. *Contrib. Mineral. Petrol.* 65, 413–424.

Arculus, R.J. and Delano, J.W. 1982. Siderophile element abundances in the upper mantle: evidence for a sulfide signature and equilibirum with the core. *Geochim. Cosmochim. Acta* 45, 1331–1343.

Armbrustmacher, T.J. (1977) Geochemistry of Precambrian mafic dikes, central Bighorn Mountain, Wyoming, U.S.A. *Precamb. Res.* 4, 13–38.

Armbrustmacher, T.J. and Simons, F.S. (1977) Geochemistry of amphibolites from the central Beartooth Mountains, Montana, Wyoming. *U.S. Geol. Surv. J. Res.* 5, 53–60.

Armstrong, N.V., Hunter, D.R. and Wilson, A.H. (1982) Stratigraphy and petrology of the Archaean Nsuze Group, northern Natal and southeastern Transvaal, South Africa. *Precamb. Res.* 19, 75–107.

Armstrong, N.V., Wilson, A.H. and Hunter, D.R. (1986) The Nsuze Group Pongola Sequence, South Africa: geochemical evidence for Archaean volcanism in a continental setting. *Precamb. Res.* 34, 175–203.

Armstrong, R. (1968) A model for the evolution of strontium and lead isotopes in a dynamic Earth. *Rev. Geophysics* 6, 175–199.

Armstrong, R. (1981) Radiogenic isotopes: the case for crustal recycling on a near-steady-state no-continental growth Earth. *Phil. Trans. R. Soc. Lond.* A301, 443–472.

Armstrong, R.A., Compston, W., Retief, E.A. and Welke, H.J. (1986) Ages and isotopic evolution of the Ventesdorp volcanics. *Geocongress '86, Johannesburg, Ext. Abs.*, 89–92.

Armstrong, R.L. and Hills, F.A. (1967) Rb–Sr and K–Ar geochronologic studies of mantled gneiss domes, Albion Range, southern Idaho, U.S.A. *Earth Planet. Sci. Lett.* 3, 114–124.

Arndt, N.T. (1976) Melting relations of ultramafic lavas (komatiites) at one atmosphere and high pressure. *Carnegie Inst. Washington Yb.* 75, 555–561.

Arndt, N.T. (1977a) Thick, layered peridotite–gabbro lava flows in Munro Township, Ontario. *Can. J. Earth Sci.* 14, 2620–2637.

Arndt, N.T. (1977b) Ultrabasic magmas and high-degree melting of the mantle. *Contrib. Mineral. Petrol* 64, 205–221.

Arndt, N.T. (1983) Role of a thin, komatiite-rich oceanic crust in the Archaean plate-tectonic process. *Geology* 11, 372–375.

Arndt, N.T. (1986a) Komatiites: a dirty window to the Archaean mantle. *Terra Cogn.* 6, 59–66.

Arndt, N.T. (1986b) Differentiation of komatiite flows. *J. Petrol.* 27, 279–301.

Arndt, N.T. High Ni in Archaean tholeiites. *Tectonophysics* in press.

Arndt, N.T. and Fleet, M.E. (1979) Stable and metastable pyroxene crystallisation in layered komatiite lava flows. *Am. Mineral.* 64, 856–864.

Arndt, N.T. and Jenner, G.A. (1986) Crustally contaminated komatiites and basalts from Kambalda, Western Australia. *Chem. Geol.* 56, 229–255.

Arndt, N.T. and Nesbitt, R.W. (1982) Geochemistry of Munro Township basalts. In *Komatiites*, eds. Arndt, N.T. and Nisbet, E.G., George Allen and Unwin, London, 309–330.

Arndt, N.T. and Nesbitt, R.W. (1984) Magma mixing in komatiitic lavas from Munro Township, Ontario. In *Archaean Geochemistry*, ed. Kroner, A., Springer-Verlag, Berlin, 99–114.

Arndt, N.T. and Nisbet, E.G. (1982a) eds., *Komatiites*. George Allen and Unwin, London, 526 pp.

Arndt, N.T. and Nisbet, E.G. (1982b) What is a komatiite? In *Komatiites*. eds., Arndt, N.T. and Nisbet, E.G., George Allen and Unwin, London, 19–28.

Arndt, N.T., Naldrett, A.J. and Pyke, D.R. (1977) Komatiitic and iron-rich tholeiitic lavas of Munro Township, northeast Ontario. *J. Petrol.* 18, 319–369.

Arndt, N.T., Francis, D. and Hynes, A.J. (1979) The field characteristics and petrology of Archaean and Proterozoic komatiites. *Can. Mineral.* 17, 147–163.

Arndt, N.T., Bickle, M.J., Jenner, G.A., Nisbet, E.G. and Zindler, A. (1986) The nature and composition of the Archaean mantle. *Eos* 67, 172–174.

Arndt, N.T., Teixeira, N.A. and White, W.M. (1989) Bizarre geochemistry of komatiites from the

Crixas greenstone belt, Brazil. *Contrib. Mineral. Petrol.* **101**, 187–197.

Arth, J.G. and Hanson, G.N. (1975) Geochemistry and origin of the early Precambrian crust of northeastern Minnesota. *Geochim. Cosmochim. Acta* **39**, 325–362.

Arth, J.G., Arndt, N.T. and Naldrett, A.J. (1977) Genesis of Archaean komatiites from Munro Township, Ontario: trace-element evidence. *Geology* **5**, 590–594.

Arth, J.G., Barker, F. and Stern, T.W. (1980) Geochronology of Archaean gneisses in the Lake Helen area, southwestern Big Horn Mountains, Wyoming. *Precamb. Res.* **11**, 11–22.

Ashwal, L.D. and Card, K.D. eds. (1983) *Workshop on a Cross Section of Archaean Crust. Lun. Planet. Inst. Tech. Rep.* **83–03**, Lun. Planet. Inst., Huston.

Ashwal, L.D., Morrison, D.A., Phinney, W.C. and Wood, J. (1983) Origin of Archaean anorthosites: evidence from the Bad Vermilion Lake anorthosite complex, Ontario. *Contrib. Mineral Petrol.* **82**, 259–273.

Ashwal, L.D., Jacobsen, S.B., Myers, J.S., Kalsbeek, F. and Goldstein, S.J. (1989) Sm–Nd age of the Fiskenaesset Anorthosite Complex, West Greenland. *Earth Planet. Sci. Lett.* **91**, 261–270.

Auvray, B., Blais, S., Jahn, B.-M. and Piquet, D. (1982) Komatiites and the komatiitic series of the Finish greenstone belts. In *Komatiites.* eds. Arndt, N.T. and Nisbet, E.G., George Allen and Unwin, London, 131–146.

Ayres, L.D. and Cerny, P. (1982) Metallogeny of granitoid rocks in the Canadian Shield of Canada. *Can. Mineral.* **20**, 439–536.

Ayres L.D. and Thurston, P.C. (1985) Archaean supracrustal sequences in the Canadian Shield: an overview. In *Evolution of Archaean Supracrustal Sequences* eds. Ayres L.D., Thurston, P.C., Card, K.D. and Weber, W. 1985. *Geol. Assoc. Can. Spec. Paper* **28**, 343–380.

Ayres L.D., Thurston, P.C., Card, K.D. and Weber, W. eds. (1985) Evolution of Archaean Supracrustal Sequences. *Geol. Assoc. Can. Spec. Paper* **28**, 380 pp.

Baadsgaard, H., Nutman, A.P., Bridgwater, D., Rosing, M., McGregor, V.R. and Allaart, J.H. (1984) The zircon geochronology of the Akilia association and Isua supracrustal belt, West Greenland. *Earth Planet. Sci. Lett.* **68**, 221–228.

Baadsgaard, H., Nutman, A.P. and Bridgwater, D. (1986) Chronology and isotope geochemistry of the early Archaean Amîtsoq gneisses of the Isukasia area, southern West Greenland. *Geochim. Cosmochim. Acta* **50**, 2173–2183.

Baer, A.J. (1977) Speculations on the evolution of the lithosphere. *Precamb. Res.* **5**, 249–260.

Baglow, N. (1986) The Epoch nickel deposit, Zimbabwe. In *Mineral Deposits of Southern Africa,* Vol. 1, eds. Anhaeusser, C.R. and Maske, S. *Geol. Soc. S. Afr.*, Johannesburg, 255–262.

Balakrishnan, S., Hanson, G.N. and Rajamani, V. (1986) Nd and Pb isotopic constraints on the petrogenesis of komatiitic rocks from the Kolar schist belt, S. India. *Eos* **67**, 1266.

Ballhaus, G.G. and Stumpfl, E.F. (1986) Sulfide and platinum mineralisation in the Merensky Reef: evidence from hydrous silicates and fluid inclusions. *Contrib. Mineral. Petrol.* **94**, 193–204.

Banerjee, P.K. (1982) Stratigraphy, petrology and geochemistry of some basic volcanic and associated rocks of Singhbhum district, Bihar, and Mayurbhanj and Keonjhar Districts, Orissa. *Geol. Surv. India Mem.* **111**, 58 pp.

Barager, W.R.A., Coleman, L.C. and Hall, J.M. eds. (1977) Tectonic regimes in Canada. *Geol. Assoc. Canada, Spec. Paper 16*, 476 pp.

Barbey, P. and Martin, H. (1987) The role of komatiites in plate tectonics. Evidence from the Archaean and early Proterozoic crust in the eastern Baltic Shield. *Precamb. Res.* **35**, 1–14.

Barbey, P., Convert, J., Martin, H., Moreau, B., Capdevila, R. and Hameurt, J. (1980) Relationships between granite–gneiss terrains, greenstone belts and granulite belts in the Archaean crust of Lapland (Fennoscandia). *Geol. Rundsch.* **69**, 648–658.

Barbey, P., Capdevila, R. and Hameurt, J. (1982) Major and transition trace element abundances in the khondalite suite of the granulite belt of Lapland (Fennoscandia): evidence for an early Proterozoic flysch belt. *Precamb. Res.* **16**, 273–290.

Barbey, P., Convert, J., Moreau, B., Capdevila, R and Hameurt, J. (1984) Petrogenesis and evolution of an early Proterozoic collisional orogenic belt: the granulite belt of Lapland and the Belomorides (Fennoscandia). *Bull. Geol. Soc. Finland* **56**, 161–188.

Barley, M.E. (1981) Relations between volcanic rocks in the Warrawoona Group: continuous or cyclic evolution. In *Archaean Geology* eds. Glover, J.E. and Groves, D.I. *Geol. Soc. Aust. Spec. Publ.* **7**, 263–273.

Barley, M.E. (1986) Incompatible element enrichment in Archaean basalts: a consequence of contamination by older sialic crust rather than mantle heterogeneity. *Geology* **14**, 947–950.

Barley, M.E. (1987) The Archaean Whim Creek Belt, an ensialic fault-bounded basin in the Pilbara Block, Australia. *Precamb. Res.* **37**, 199–215.

Barley, M.E. and Bickle, M.J. 1982. Komatiites in the Pilbara Block, Western Australia. In *Komatiites.* eds. Arndt, N. and Nisbet, E., George Allen and Unwin, London, 97–104.

Barley, M.E., Sylvester, G.C., Groves, D.I., Borley, G.D. and Rogers, N. (1984) Archaean calc-alkaline volcanism in the Pilbara block, Western Australia. *Precamb. Res.* **24**, 285–319.

Barnes, S.J. (1989) Are Bushveld U-type parent magmas boninites or contaminated komatiites? *Contrib. Mineral. Petrol.* **101**, 447–457.

Barnes, S.J. and Naldrett, A.J. (1985) Geochemistry of the J–M (Howland) Reef of the Stillwater Complex, Minneapolis adit area. I. Sulfide chemistry and sulfide-olivine equilibrium. *Econ. Geol.* **80**, 627–645.

Barnes, S.J. and Naldrett, A.J. (1986) Geochemistry of the J–M Reef of the Stillwater Complex, Minneapolis adit area. II. Silicate mineral chemistry and petrogenesis. *J. Petrol.* **27**, 791–825.

Barnes, S.J. and Naldrett, A.J. (1987) Fractionation of the platinum-group elements and gold in some komatiites of the Abitibi Greenstone Belt, northern Ontario. *Econ. Geol.* **82**, 165–183.

Barnes, S.J., Naldrett, A.J. and Gorton, M.P. 1985. The origin of the fractionation of platinum group elements in terrestrial magmas. *Chem. Geol.* **53**, 303–323.

Barnes, S.J., Hill, R.E.T. and Gole, M.J. (1988) The Perseverence ultramafic complex, Western Australia: the product of a komatiite lava river. *J. Petrol.* **29**, 305–331.

Barrere, J. (1967) Le group Precambrien de l'Amsaga entre Atar et Akjoujt (Mauritanie). Etude d'un metamorphisme profond et de ses relations avec la migmatisation. *Mem. Bur. Rech. Geol. Min. Paris.* **42**.

Barton, E.S., Armstrong, R.A., Cornell, D.H. and Welke, H.J. (1986b) Feasibility of total-rock Pb–Pb dating of metamorphosed banded iron formation; the Marydale Group, southern Africa. *Chem. Geol. (Isot. Geosc. Sect.)* **59**, 255–271.

Barton, J.M. (1983) Isotopic constraints on possible tectonic models for crustal evolution in the Barberton granite-greenstone terrane, southern Africa. *Spec. Publ. Geol. Soc. S. Afr.* **9**, 73–79.

Barton, J.M. (1984) Timing of ore emplacement and deformation, Murchison and Sutherland greenstone belts, Kaapvaal craton. In *Gold '82: the Geology, Geochemistry and Genesis of Gold Deposit* ed. Foster, R.P., A.A. Balkema, Rotterdam, 629–644.

Barton, J.M. and McCourt, S. (1983) Rb–Sr age for the Palala granite, Limpopo Mobile Belt. *Geol. Soc. S. Afr. Spec. Publ.* **8**, 45–46.

Barton, J.M. and Ryan, B. (1977) A review of the geochronological framework of the Limpopo Mobile Belt *Bull. Geol. Surv. Botswana* **12**, 183–200.

Barton, J.M., Jr., Fripp, R.E.P., Horrocks, P. and McLean, N. (1979) The geology, age and tectonic setting of the Messina layered intrusion, Limpopo mobile belt, southern Africa. *Amer. J. Sci.* **279**, 1108–1134.

Barton, J.M., Hunter, D.R., Jackson, M.P.A. and Wilson, A.C. (1980) Rb–Sr age and source of the ·bimodal suite of the Ancient Gneiss Complex, Swaziland. *Nature* **283**, 756–758.

Barton, J.M., Robb, L.J., Anhaeusser, C.R. and van Nierop, D.A. (1983) Geochronologic and Sr isotopic studies of certain units in the Barberton granite–greenstone terrane, South Africa. *Spec. Publ. Geol. Soc. S. Afr.* **9**, 63–72.

Barton, J.M., Byron, C.L. and Klemd, R. (1986a) The setting of some gold mineralization on the farms Eerstelling and Rooderpoort, Pietersburg greenstone belt. *Geocongr. '86, Johannesburg, Ext. Abs.* 353–356.

Basu, A.R. and Pettingill, H.S. (1983) Origin and age of Adirondack anorthosites re-evaluated with Nd isotopes. *Geology* **11**, 514–518.

Basu, A.R., Ray, S.L., Sahu, A.K. and Sarkar, S.N. 1981. Eastern Indian 3800 million-year-old crust and early mantle differentiation. *Science* **212**, 1502–1506.

Bavington, D.A. (1981) The nature of sulphidic metasediments at Kambalda and their broad relationships with associated ultramafic rocks and nickel ores. *Econ. Geol.* **76**, 1606–1628.

Bavinton, D.A. and Keays, R.R. (1978) Precious metal values from interflow sedimentary rocks from the komatiite sequence at Kambalda, Western Australia. *Geochim. Cosmochim. Acta* **42**, 1151–1163.

Bayley, R.W., Proctor, P.D. and Condie, K.C. (1973) Geology of the South Pass area, Fremont County, Wyoming. *U.S. Geol. Surv. Prof. Pap.* **793**, 39pp.

Beckinsale, R.D., Gale, N.H., Pankhurst, R.J., McFarlane, A., Crow, M.J., Arthurs, J.W. and Wilkinson, A.F. (1980) Discordant Rb–Sr and Pb–Pb ages for the Archaean basement of Sierra Leone. *Precamb. Res.* **13**, 63–76.

Beisiegel, V.R. (1982) Distrito ferrífero da Serra dos Carajás. *I Simposio de Geologia da Amazonia, Belem. SBG/Nucleo Norte, Belem*, 21–26.

Bell, K. and Dodson, M.H. (1981) The geochronology of the Tanzanian Shield. *J. Geol.* **89**, 109–128.

Bence, A.E. and Papike, J.J. (1972) Pyroxenes as recorders of lunar basalt petrogenesis: chemical trends due to crystal-liquid interaction. *Proc. 3rd Lunar Sci. Conf.*, MIT Press, Cambridge, Mass., 431–469.

Bence, A.E., Grove, T.L. and Papike, J.J. (1980) Basalts as probes of planetary interiors: constraints on the chemistry and mineralogy of their source regions. *Precamb. Res.* **10**, 249–279.

Berger, B.R. (1989) Geology of Laval and Hartman Townships, district of Kenora. *Ont. Geol. Surv. Open File Rep.* **5703**. 209 pp.

Bernard-Griffiths, J., Peucat, J.J. Postaire, B., Vidal, P., Convert, J. and Moreau, B. (1984) Isotopic data (U–Pb, Rb–Sr, Rb–Rb and Sm–Nd) of mafic granulites from Finnish Lapland. *Precamb. Res.* **23**. 325–348.

Bernard-Griffiths, J., Jahn, B.M. and Sen, S.K. (1987) Sm–Nd isotopes and REE geochemistry of Madras granulites, India: an introductory statement. *Precomb. Res.* **37**, 343–355.

Bernardelli, A.L., Meireles, E. de M., Teixeira, I.T., Farias, N., Saueressing, R., Assad, R., Beisiegel, V. de R. and Hirata, W.K. (1982) Provincia mineral de Carajas-Para: depositos de ferro, manganes, cobre, ouro, niquel e bauxita. *I Simposio de Geologia da Amazonia, Belem. Anexo, SBG/Nucleo Norte*, 104.

Berthelsen, A. and Henriksen, N. (1975) Descriptive text to Geological map of Greenland 1:100 000, Ivigtut, 61 V 1 Syd. *Grønlands Geol Unders.* (also *Meddr. Grønland* **186**, 1), 169 pp.

Bessoles, B. (1977) Geologie de l'Afrique. Le craton Ouest Africain. *Mem. Bur, Rech. Geol. Min. Paris* 88.

Bhattacharya, A. and Sen, S.K. (1986) Granulite-metamorphism, fluid-buffering, and dehydration melting in the Madras charnockites and metapelites. *J. Petrol.* **27**, 1119–1141.

Bibikova, E.V. (1984) The most ancient rocks in the USSR territory by U–Pb data on accessory zircons. In *Archaean Geochemistry*. eds. Kröner, A., Hanson, G.A. and Goodwin, A.M., Springer-Verlag, Berlin, 235–250.

Bickford, M.E. (1988) The formation of continental crust: Part 1. A review of some principles; Part 2. An application to the Proterozoic evolution of southern North America. *Geol. Soc. Amer. Bull.* **100**, 1375–1391.

Bickford, M.E. and Boardman, S.J. (1984) A Proterozoic volcano-plutonic terrane, Gunnison and Salida areas, Colorado. *J. Geol.* **92**, 657–666.

Bickle, M.J. (1978) Heat loss from the Earth: a constraint on Archaean tectonics from the relation between geothermal gradients and the rate of plate production. *Earth Planet. Sci. Lett.* **40**, 301–315.

Bickle, M.J. (1982) The magnesium content of komatiitic liquids. In *Komatiites*, eds. Arndt, N.T. and Nisbet, E.G., George Allen and Unwin, London, 479–494.

Bickle, M.J. (1986) Implications of melting for stabilisation of the lithosphere and heat loss in the Archaean. *Earth Planet. Sci. Lett.* **80**, 314–324.

Bickle, M.J. and Eriksson, K.A. (1982) Evolution and subsidence of early Precambrian sedimentary basins. *Phil. Trans. R. Soc. Lond.* **A305**, 225–247.

Bickle, M.J. and Nisbet, E.G. (1986) Greenstone belt tectonics-thermal constraints. In *Workshop on tectonic evolution of greenstone belts*; eds. De Wit, M.J. and Ashwal, L.D. *Lun. Planet. Inst. Tech. Rep.* **86-LO**, 57–64.

Bickle, M.J., Martin, A. and Nisbet, E.G. (1975) Basaltic and peridotitic komatiites and stromatolites above a basal unconformity in the Belingwe greenstone belt. Rhodesia. *Earth Planet. Sci. Lett.* **27**, 155–162.

Bickle, M.J., Bettenay, L.F., Barley, M.E., Chapman, H.J., Groves, D.I., Campbell, I.H. and de Laeter, J.R. (1983) A 3500 Ma plutonic and volcanic calc-alkaline province in the Archaean East Pilbara Block. *Contrib. Mineral. Petrol.* **84**, 25–35.

Bickle, M.J., Morant, P., Bettenay, L.F., Boult, C.A., Blake, T.S. and Groves, D.I. (1985) Archaean tectonics of the Shaw batholith, Pilbara Block, Western Australia: structural and metamorphic tests of the batholith concept. *Geol. Assoc. Can. Spec. Pap.* **28**, 323–341.

Binder, A.B. (1980) On the origins of lunar pristine crustal rocks. *Proc. Conf. Lunar Highland Crust*, Pergamon, New York, 71–79.

Binder, A.B. (1982) The mare basalt magma source region and mare basalt magma genesis. *Proc. 13th Lunar Planet. Sci. Conf., J. Geophys. Res.* **87**, suppl., A37–A53.

Binder, A.B. (1985) Mare basalt genesis: modelling trace elements and isotopic ratios. *Proc. 16th Lunar Planet Sci. Conf., J. Geophys. Res.* **90**, suppl., D19–D30.

Binns, R.A., Gunthorpe, R.J. and Groves, D.I. (1976) Metamorphic patterns and development of Greenstone belts in Eastern Yilgarn Block, Western Australia. In *The Early History of Earth*. ed. Windley, B.F., Wiley, London, 303–313.

Binns, R.A., Hallberg, J.A. and Taplin, J.H. (1982) komatiites in the Yilgarn Block, Western Australia. In *Komatiites* eds. Arndt, N. and Nisbet, E., George Allen and Unwin, London, 117–130.

Blackburn, C.E. (1982) Geology of the Manitou Lake Area, District of Kenora. *Ont. Geol. Surv. Rep.* **223**, 61 pp.

Blais, S., Auvray, B., Capdevila, R., Jahn, B.M., Hameurt, J. and Bertrand, J.M. (1978) The Archaean greenstone belts of Karelia (Eastern Finland) and their komatiitic and tholeiitic series. in *Archaean Geochemistry*, eds. Windley, B.F. and Naqui, S.M., Elsevier, Amsterdam, 87–107.

Blake, T.S. and Groves, D.I. (1987) Continental rifting and the Archaean–Proterozoic transition. *Geology* **15**, 229–232.

Blanchard, D.P., Rhodes, J.M., Dungan, M.A., Rodgers, K.V., Donaldson, C.H., Brannon, J.C., Jacobs, J.W. and Gibson, E.K. (1976) The chemistry and petrology of basalts from Leg 37 of the Deep Sea Drilling Project. *J. Geophys. Res.* **81**, 4231–4246.

Bliss, N.W. and Stidolph, P.A. (1969) The Rhodesian basement complex: a review. *Spec. Publ. Geol. Soc. S. Afr.* **2**, 305–331.

Bloomer, S.H. and Hawkins, J.W. (1987) Petrology and geochemistry of boninite series volcanic rocks from the Marianas Trench. *Contrib. Mineral. Petrol.* **97**, 361–377.

Boak, J.L. and Dymek, R.F. (1982) Metamorphism of the ca. 3800 Ma supracrustal rocks at Isua, West Greenland: implications for early Archaean crustal evolution. *Earth Planet. Sci. Lett.* **59**, 155–176.

Boak, J.M. and Dymek, R.F. (1989) A systematic chemical stratigraphy for clastic metasedimentary and metavolcanic rocks from the 3800 Ma Isua supracrustal belt, West Greenland: implications for Archaean crustal evolution. *Geochim. Cosmochim. Acta* in press.

Bofinger, V.M. (1967) Geochronology in the East Kimberley area of Western Australia. Unpubl. Ph.D. thesis, Aust. Nat. Univ. Canberra.

Bondesen, E. (1970) The stratigraphy and deformation of the Precambrian rocks of the Graenseland area, South-West Greenland. *Bull. Grønlands Geol. Unders.* **86**, 210 pp.

Bosma, W., Kroonenberg, S.B. Maas, K. and de Roever, E.W.F. (1983) Igneous and metamorphic complexes of the Guiana Shield in Suriname. *Geol. Mijnb.* **62**, 241–254.

Bottinga, Y. and Weill, D.F. (1970) Densities of liquid silicate systems calculated from partial molar volumes of oxide components. *Amer. J. Sci.* **269**, 169–182.

Bottinga, Y., Weill, D.F., and Richet, P. (1982) Density calculations for silicate liquids I: revised method for aluminosilicate compositions. *Geochim. Cosmochim. Acta* **46**, 909–919.

Boudreau, A.E., Mathez, E.A. and McCallum, I.S. (1986) Halogen geochemistry of the Stillwater and Bushveld Complexes: evidence for transport of the platinum group elements by CO–rich fluids. *J. Petrol.* **27**, 967–987.

Bow, C., Wolfgram, D., Turner, A., Barnes, S., Evans, J., Zdepski, M. and Boudreau, A. (1982). Investigations of the Howland Reef of the Stillwater Complex, Minneapolis adit area – stratigraphy, structure and mineralization. *Econ. Geol.* **77**, 1481–1492.

Bowen, N.L. (1928) *The Evolution of the Igneous Rocks*. Princeton University Press (Repr. 1956 Dover Press), New York, 334 pp.

Bowen, T.B., Marsh, J.S., Bowen, M.P. and Eales, H.V. (1986) Volcanic rocks of the Witwatersrand Triad, South Africa. I. Description, classification and geochemical stratigraphy. *Precamb. Res.* **31**, 297–324.

Bowes, D.R. (1976) Archaean crustal history in North-western Britain. In *The Early History of the Earth*, ed. Windley, B.F., Wiley, London, 469–479.

Bowes, D.R., Barooah, B.C. and Khoury, S.G. (1971) Original nature of Archaean rocks of North-West Scotland. *Geol. Soc. Australia Spec. Publ.* **3**, 77–92.

Breaks, F.W., Bond, W.D. and Stone, D. (1978) Preliminary geological synthesis of the English River Subprovince Northwestern Ontario and its bearing upon mineral exploration. *Ont. Geol. Surv. Misc. Paper 72*, 55p.

Breaks, F.W., Bartlett, J.R., deKemp, E.A., Finamore, P.F., Jones, G.R., Macdonald, A.J., Shields, H.N. and Wallace, H. (1984) Precambrian geology, Quaternary geology and mineral deposits of the North Caribou Lake area, district of Kenora, Patricia Portion. *Ont. Geol. Surv. Misc. Paper* **119**, 258–273.

Breaks, F.W., Osmani, I.A. and DeKemp, E.A. (1986) Opapimiskan lake project: Precambrian geology of the Opapimiskan–Forester Lakes area, district of Kenora, Patricia portion. *Ont. Geol. Surv. Misc. Paper* **132**, 368–378.

Brevart, O., Dupre, B. and Allegre, C.J. 1986. Lead-lead age of komatiitic lavas and limitations on the structure and evolution of the Precambrian mantle. *Earth Planet. Sci. Lett.* **77**, 293–302.

Bridgwater, D. and Collerson, K.D. (1976) The major petrological and geochemical characters of the 3600 m.y. Uivak gneisses from Labrador. *Contrib. Mineral. Petrol.* **54**, 43–59.

Bridgwater, D., Watson, J. and Windley, B.F. (1973) The Archaean craton of the North Atlantic region. *Phil. Trans. R. Soc. Lond. A*, **273**, 493–512.

Bridgwater, D., McGregor, V.R. and Myers, J.S. (1974) A horizontal tectonic regime in the Archaean of Greenland and its implications for early crustal thickening. *Precamb. Res.* **1**, 179–197.

Bridgewater, D., Keto, L., McGregor, V.R. and Myers, J.S. (1976) Archaean gneiss complex of Greenland. In *Geology of Greenland*. eds. Escher, A. and Watt, W.S., Geol. Surv. Greenland, Copenhagen, 18–75.

Bridgwater, D., Collerson, K.D. and Myers, J.S. (1978) The development of the Archaean gneiss complex of the North Atlantic region. in *Evolution of the Earth's Crust*, ed. Tarling, D.H. Academic Press, London, 19–69.

Brigmann, G.E., Arndt, N.T., Hofmann, A.W. and Tobschall, H.J. (1987) Nobel metal abundances in komatiite suites from Alexo, Ontario and Gorgona Island, Columbia. *Geochim, Cosmochim. Acta* **51**, 2159–2169.

Brito, R.S.C. (1984) Geologia do sill estratificado do Rio Jacare–Maracas, Bahia. *Anais XXXIII Congresso Brasileiro de Geologia*, 4316–4331.

Brito Neves, B.B., Cordani, U.G. and Torquato, J.R.F. (1980) Evolucao geocronologica do precambriano do Estado da Bahia. Geologia e Recursos Minerais do Estado da Bahia. *Textos Basicos, Bahia, CPM/SME* **3**, 1–101.

Brooks, C. and Hart, S.R. (1974) On the significance of komatiite. *Geology* **2**, 107–110.

Brooks, C., Ludden, J. and Pigeon, Y. (1982) Volcanism of shoshonite to high-K andesite affinity in an Archaean arc environment, Oxford Lake, Manitoba. *Can. J. Earth Sci.* **19**, 55–67.

Brown, G.M. (1956) The layered ultrabasic rocks of Rhum, Inner Herbrides. *Phil. Trans. Roy. Soc. London*, **B240**, 1–53.

Brown, L., Klein, J., Middleton, R., Sacks, I.S. and Tera, F. (1982) [10]Be in island-arc volcanoes and implications for subduction. *Nature* **299**, 718–720.

Brozdowski, R.A. (1985) Cumulate xenoliths in the Lodgepole, Enos Mountain and Susie Peak intrusions: a guide. in *Stillwater Complex*; eds. Czamanske, G.K. and Zientek, M.L. *Mont. Bur. Mines Geol Spec. Publ.* **92**, 368–372.

Brugmann, G.E., Arndt, N.T., Hofmann, A.W. and Tobschall, H.J. (1987) Noble metal abundances in komatiite suites from Alexo, Ontario, and Gorgona Island, Colombia. *Geochim. Cosmochim. Acta.* **51**, 2159–2169.

Bryant, B. (1988) Geology of the Farmington Canyon complex, Wasatch Mountains, Utah. *U.S. Geol. Surv. Prof. Pap.* **1476**, in press.

Buchanan, D.L. (1975) The petrography of the Bushveld Complex intersected by boreholes in the Bethal area. *Trans. Geol. Soc. S. Afr.* **78**, 335–348.

Buchanan, D.L. and Jones, M.J. eds. (1984) *Sulphide deposits in mafic and ultramafic rocks*. Inst. Min. Metall., Stephen Austin, Hertford, 164 pp.

Buck, S.G. (1980) Stromatolite and ooid deposits within the fluvial and lacustrine sediments of the Precambrian Ventesdrop Supergroup of South Africa. *Precamb. Res.* **12**, 311–330.

Bultitude, R.J. and Wyborn, L.A.I. (1982) Distribution and geochemistry of volcanic rocks in the Duchess–Urandangi region Queensland. *BMR J. Aust. Geol. Geophys.* **7**, 99–112.

Buntin, T.J., Grandstaff, D.E., Ullmer, G.C. and Gold, D.P. (1985) A pilot study of geochemical and redox relationships between potholes and adjacent normal Merensky Reef of the Bushveld Complex. *Econ. Geol.* **80**, 975–987.

Burke, K. and Kidd, W.S.F (1978) Were Archaean continental geothermal gradients much steeper than those of today? *Nature* **272**, 240–241.

Burke, K., Kidd, W.S.F. and Kusky, T.M. (1986) Archean foreland basin tectonics in the Witwatersrand, South Africa. *Tectonics* **5**, 439–456.

Burger, A.J. and Coertze, J.F. (1973) Radiometric age measurements on rocks from southern Africa to the end of 1971. *Geol Surv. S. Afr. Bull.* **58**, 1–46.

Burger, A.J. and Walraven, F. (1979) Summary of age determinations carried out during the period April 1977 to March 1978. *Ann. Geol. Surv. S. Afr.* **12**, 209–218.

Burger, A.J., Nicolaysen, L.O. and De Villiers, J.W.L. (1962) Lead isotopic compositions of galenas from the Witwatersrand and Dominion Reef uraninites. *Geochim. Cosmochim. Acta* **26**, 25–59.

Cabri, L.J. (1987) The mineralogy of precious metals: new developments and metallurgical implications. *Can. Mineral.* **25**, 1–7.

Caen-Vachette, M., Vialette, Y., Bassot, J.-P. and Vidal, P. (1988) Rapport de la geochronologie isotopique à la connaissance de la geologie Gabonaise. *Chron. Rech. Min.* **491**, 35–54.

Cahen, L., Delhal, J. and Lavreau, J. (1976) The Archaean of Equatorial Africa: a review in *The Early History of the Earth.* ed. Windley, B.F. Wiley, London, 489–498.

Cahen, L., Snelling, N.J., Delhal, J. and Vail J.R. (1984). *The Geochronology and Evolution of Africa.* Clarendon Press, Oxford, 512 pp.

Cameron, E.M. (1988) Archaean gold: relation to granulite formation and redox zoning in the crust. *Geology* **16**, 109–112.

Cameron, E. M. and Hattori, K. (1987) Archaean gold mineralization and oxidized hydrothermal fluids. *Econ. Geol.* **82**, 1177–1191.

Cameron, E.N. (1977) Chromite in the central sector of the Eastern Bushveld Complex, South Africa. *Amer. Mineral.* **62**, 1082–1096.

Cameron, E.N. (1978) The zone of the eastern Bushveld Complex in the Olifants River trough. *J. Petrol.* **19**, 437–462.

Cameron, E.N. (1980) Evolution of the lower critical zone, central sector, eastern Bushveld Complex, and its chromitite deposits. *Econ. Geol.* **75**, 845–871.

Cameron, E.N. (1982) The upper critical zone of the eastern Bushveld Complex – precursor of the Merensky Reef. *Econ. Geol.* **77**, 1307–1327.

Cameron, E.N. and Desborough, G.A. (1969) Occurrence and characteristics of chromite deposits–eastern Bushveld Complex. *Econ. Geol. Man.* **4**, 23–40.

Cameron, E.N. and Emerson, M.E. (1959) The origin of certain chromite deposits in the eastern part of the Bushveld Complex. *Econ. Geol.* **54**, 1151–1213.

Cameron, W.E. (1985) Petrology and origin of primitive lavas from the Troodos ophiolite, Cyprus. *Contrib. Mineral. Petrol.* **89**, 239–255.

Cameron, W.E. and Nisbet, E.G. (1982) Phanerozoic analogues of komatiitic basalts in *Komatiites*, eds. Arndt, N.T. and Nisbet, E.G., George Allen and Unwin, London, 29–50.

Cameron, W.E., Nisbet, E.G. and Dietrich, V.J. (1979) Boninites, komatiites and ophiolitic basalts. *Nature* **280**, 550–553.

Cameron, W.E., McCulloch, M.T. and Walker, D.A. (1983) Boninite petrogenesis: chemical and Nd–Sr isotopic constraints. *Earth Planet. Sci. Lett.* **65**, 75–89.

Campbell, I.H. (1977) A study of macro-rhythmic layering and cumulate processes in the Jimberlana intrusion. *J. Petrol.* **18**, 183–215.

Campbell, I.H. (1985) The difference between oceanic and continental tholeiites: a fluid dynamic explanation. *Contrib. Mineral. Petrol.* **91**, 37–43.

Campbell, I.H. and Murck, B.W. (1986) The effects of temperature, oxygen fugacity and melt composition on the behaviour of chromium in basic and ultrabasic melts. *Geochim. Cosmochim. Acta* **50**, 1871–1888.

Campbell, I.H. and Naldrett, A.J. (1979) The influence of silicate:sulfide ratios on the geochemistry of magmatic sulfides. *Econ. Geol.* **74**, 1503–1505.

Campbell, I.H., McCall, G.J.H. and Tyrwhitt, D.S. (1970) The Jimberlana norite, Western Australia – a smaller analogue of the Great Dyke of Rhodesia. *Geol. Mag.* **107**, 1–11.

Campbell, I.H., Roeder, P.L. and Dixon, J.M. (1978) Crystal bouyancy in basaltic liquids and other experiments with a centrifuge furnace. *Contrib. Mineral. Petrol.* **67**, 369–377.

Campbell, I.H., Naldrett, A.J. and Barnes, S.J. (1983) A model for the origin of the platinum-rich sulfide horizons in the Bushveld and Stillwater Complexes. *J. Petrol.* **24**, 133–165.

Campbell, I.H., Lesher, C.M., Coàd, P.M., Franklin, J.M., Gorton, M.P. and Thurston, P.C. (1984) Rare-earth element mobility in alteration pipes beneath massive Cu–Zn sulfide deposits. *Chem. Geol.* **45**, 181–202.

Canil, D. (1987) The geochemistry of komatiites and basalts from the Deadman Hill area, Munro Township, Ontario Canada. *Can. J. Earth Sci.* **24**, 998–1008.

Card, K.D. and Ciesielski, A. (1986) DNAG 1. Subdivisions of the Superior Province of the Canadian Shield. *Geoscience Canada.* **13**, 5–14.

Carlson, R.W., Hunter, D.R. and Baker, F. (1983) Sm–Nd age and isotopic systematics of the bimodal suite, Ancient Gneiss Complex, Swaziland. *Nature* **305**, 701–704.

Carlson, R.W., Hunter, D.R. and Baker, F. (1985) Geochronological investigation of the Archaean Ancient Gneiss Complex, Swaziland. *Eos* **66**, 419.

Carter, A.H.C. (1988) The transport and deposition of gold within a major Archaean shear zone: Dalny Mine. Zimbabwe. *Rep. Univ. Southampton Dept. Geol.* **C12**, 51 pp.

Cartwright, I., Fitches, W.R., O'Hara, M.J., Barnicoat, A.C. and O'Hara, S. (1985) Archaean supracrustal rocks from the Lewisian near Stoer, Sutherland. *Scott. J. Geol.* **21**, 187–196.

Casella, C.J., Levay, J., Eble, E., Hirst, B., Huffman, K., Lahti, V. and Metzger, R. (1982) Precambrian geology of the southwestern Beartooth Mountains, Yellowstone National Park, Montana and Wyoming. In *Precambrian Geology of the Beartooth Mountains, Montana and Wyoming*: eds. Mueller, P.A. and Wooden, J.L. *Mont. Bur. Mines Geol. Spec. Publ.* **84**, 1–24.

Catanzaro, E.J. and Hanson, G.N. (1971) U–Pb ages for sphene in northeastern Minnesota-northwestern Ontario. *Can. J. Earth Sci.* **8**, 1319–1324.

Cattell, A.C. (1985) Geology and geochemistry of komatiites and associated basalts from Newton Township, Ontario, Canada. Unpub. PhD thesis, Univ. Southampton.

Cattell, A.C. (1987) Enriched komatiitic basalts from Newton Township, Ontario: their genesis by crustal contamination of depleted komatiite magma. *Geol. Mag.* **124**, 303–309.

Cattell, A.C. and Arndt, N.T. (1987) Low- and high-alumina komatiites from a late-Archaean sequence, Newton Township, Ontario. *Contrib. Mineral. Petrol.* **97**, 218–227.

Cattell, A., Krogh, T.E. and Arndt, N.T. (1984) Conflicting Sm–Nd whole rock and U–Pb zircon ages for Archaean lavas from Newton Township, Abitibi Belt, Ontario. *Earth Planet. Sci. Lett.* **70**, 280–290.

Cawthorn, R.G. and Davies, G. (1982) Possible komatiitic affinity of the Bushveld Complex, South Africa. In *Komatiites.* eds. Arndt, N.T. and Nisbet, E.G., George Allen and Unwin, London, 91–96.

Cawthorn, R.G. and Davies, G. (1983) Experimental data on the parental magmas to the Bushveld Complex. *Contrib. Mineral. Petrol.* **83**, 128–135.

Cawthorn, R.G. and McCarthy, T.S. (1980) Bottom crystallisation and diffusion control in layered complexes: evidence from Cr distribution in magnetite from the Bushveld Complex. *Trans. Geol. Soc. S. Afr.* **84**, 41–50.

Cawthorn, R.G. and Strong, D.F. (1974) The petrogenesis of komatiites and related rocks as evidence for a layered upper mantle. *Earth. Planet. Sci. Lett.* **23**, 369–375.

Cawthorn, R.G., Davies, G., Clubbley-Armstrong, A. and McCarthy, T.S. (1981) Sills associated with the Bushveld Complex, South Africa: an estimate of the parental magma composition. *Lithos* **14**, 1–15.

Chadwick, B. (1981) Field relations, petrography and geochemistry of Archaean amphibolite dykes and Malene supracrustal amphibolites, northwest Buksefjorden, southern West Greenland. *Precamb. Res.* **14**, 221–259.

Chadwick, B. (1985) Contrasting styles of tectonism and magmatism in the late Archaean crustal evolution of the northeastern part of the Ivisârtoq region, inner Godthåbsfjord, southern West Greenland, *Precamb. Res.* **27**, 215–238.

Chadwick, B. (1986) Malene stratigraphy and late Archaean structure: new data from Ivisârtoq, inner Godthåbsfjord, southern West Greenland. *Rapp. Grønlands Geol. Unders.* **130**, 74–85.

Chadwick, B. and Coe. K. (1984) Archaean crustal evolution in southern West Greenland: a review based on observations in the Buksefjorden region. *Tectonophysics* **105**, 121–130.

Chadwick, B. and Nutman, A.P. (1979) Archaean structural evolution in the northwest of the Buksefjorden region, southern West Greenland. *Precamb. Res.* **9**, 199–226.

Chadwick, B., Dawes, P.R., Escher, J.C., Friend, C.R.L., Hall, R.P., Kalsbeek, F., Nielsen, T.F.D.,

Nutman, A.P., Soper, N.J. and Vasudev, V.N. (1989) The Proterozoic mobile belt in the Ammassalik region, S.E. Greenland: an introduction and reappraisal. *Rapp. Grønlands Geol. Unders.* 146, 5–12.

Chakraborty, K.R. and Sen, S.K. (1983) A comparititive analysis of geochemical trends in basic granulites, greenstonstones and recent basalts based on three basic granulite suites from India. *Mem. Geol. Soc. India* 4, 462–487.

Chao, E.C.T., Minikin, J.A., Frondel, C., Klein, C, Drake, J.C., Fuchs, L., Tani, B., Smith, J.V., Anderson, A.T., Moore, P.B., Zechman, G.R. Jr., Traill, R.J., Plant, A.G., Douglas, J.A.V. and Dence, M.R. (1970) Pyroxferroite. a new calcium-bearing iron silicate from Tranquility Base. *Proc. Apollo 11th Lunar Sci. Conf.* Pergamon, New York, (*Geochim. Cosmochim. Acta suppl.* 14), 65–79.

Chapman, D.S. and Pollack, H.N. (1975) Global heat flow: A new look. *Earth Planet. Sci. Lett.* 28, 23–32.

Chapman, H.J. (1979) 2,390 Myr Rb–Sr whole rock for the Scourie dykes of northwest Scotland. *Nature* 277, 642–643.

Chapman, H.J. and Moorbath, S. (1977) Lead isotope measurements from the oldest recognised Lewisian gneisses of northwest Scotland. *Nature* 268, 41–42.

Chappell, B.W. and White, A.J.R. (1970) Further data on an 'ecologite' from the Sittampundi complex, India. *Mineralog. Mag.* 37, 555–560.

Chase, C.G. and Patchett, P.J. (1988) Stored mafic/ultramafic crust and early Archaean mantle depletion. *Earth Planet. Sci. Lett.* 91, 66–72.

Chauvel, C., Dupre, B. and Jenner, G.A. (1985) The Sm–Nd age of Kambalda volcanics is 500 Ma too old!. *Earth Planet. Sci. Lett.* 74, 315–321.

Chauvel, C., Arndt, N.T., Gruau, G. and Hofmann, A. Composition of the Archaean mantle: evidence fron Nd isotopes. *Earth Planet. Sci. Lett.* submitted 1987 in prep.

Chen, C.-Y. and Frey, F.A. (1985) Trace element and isotopic geochemistry of lavas from Haleakala volcano, East Maui, Hawaii: implications for the origin of Hawaiian basalts. *J. Geophys. Res.* 90, 8743–8768.

Cheng, Y. (1986) The Archaean. In *The Geology of China.* eds. Yang, Z.Y., Cheng, Y.Q. and Wang, H.Z. Clarendon Press, Oxford, 16–30.

Cheng, Y.Q., Shen, Q.H. and Wang, Z.J. (1982a) *Preliminary study of the metamorphosed basic volcano-sedimentary Yanlinguan Formation of the Taishan Group of Xintai, Shandong.* Geol. Publ. House, Beijing, 90 pp (with 5 pp of Engl. abs.).

Cheng, Y.Q., Bai, J. and Sun, D.Z. (1982b) The lower Precambrian of China. In *An outline of the stratigraphy of China.* Geol. Publ. House Beijing, 29 pp.

Cheng, Y.Q., Sun, D.Z. and Wu, J.S. (1984) Evolutionary mega-cycles of the early Precambrian proto-North China platform. *J. Geodynam.* 1, 251–277.

Chimimba, L.R. and Ncube, S.M.N. (1986) Nickel sulphide mineralization at Trojan mine, Zimbabwe. In *Mineral Deposits of Southern Africa, Vol. 1,* eds. Anhaeusser, C.R. and Maske, S. Geol. Soc. S. Afr., Johannesburg, 249–253.

Choubert, B. (1974) Le Precambrian des Guyanes. *Bur. Recherches Geol. Minieres, Mem.* 81.

Choudhuri, A. (1978) Geochemical trends in tholeiite dykes of different ages from Guiana. *Chem. Geol.* 22, 79–85.

Christensen, U. (1984) Convection with pressure- and temperature-dependent non-Newtonian rheology. *Geophys. J. Roy. Astron. Soc.* 77, 343–384.

Christensen, U. (1985) Thermal evolution models for the Earth. *J. Geophys. Res.* 90B, 2995–3007.

Claesson, L.Å. (1985) The geochemistry of early Proterozoic metavolcanic rocks hosting metal sulphide deposits in the Skellefte district, northern Sweden. *J. Geol. Soc. Lond.* 42, 899–909.

Claoué-Long, J.C. and Nesbitt, R.W. (1985) Contaminated komatiites. *Nature* 313, 247.

Claoué-Long J., Thirwall, M.F. and Nesbitt, R.W. (1984) Revised Sm–Nd systematics of Kambalda greenstones, Western Australia. *Nature* 307, 697–701.

Clarke, D.B. (1970) Tertiary basalts of Baffin Bay: possible primary magma from the mantle. *Contrib. Mineral. Petrol.* 25, 203–224.

Clarke, T. (1987) Platinum group element occurrences of the Labrador Trough. *Quebec Min. Ener. Res. Doc. Prom.* 18, 6 pp.

Cliff, R.A., Gray, C.M. and Huhma, H. (1983) A Sm–Nd isotopic study of the South Harris igneous complex. *Contrib. Mineral. Petrol.* 82, 91–98.

Clifford, T.N. (1970) The structural framework of Africa. In *African Magmatism and Tectonics.* eds.

Clifford, T.N. and Gass, I.G. Oliver and Boyd, Edinburgh, 1–26.

Coertze, F.J. (1970) The geology of the western part of the Bushveld Igneous Complex. *Geol. Soc. S. Afr. Spec. Publ.* **1**, 5–22.

Coertze, F.J. and Schumann, F.W. (1962) The basic portion and associated minerals of the Bushveld Igneous Complex north of Pilanesberg. *Bull. Geol. Surv. S. Afr.* **38**, 48 pp.

Coleman, R.G. (1971) Petrologic and Geophysical Nature of Serpentinites. *Bull. Geol. Soc. Amer.* **82**, 897–918.

Collender, F.D. (1964) The geology of the Cam and Motor Mine, southern Rhodesia. In *The Geology of Some Ore Deposits in Southern Africa*, 2. ed. Haughton, S.H. Geol. Soc. S. Afr., Johannesburg, 15–27.

Collerson, K.D. and Sheraton, J.W. (1986) Age and geochemical characteristics of a mafic dyke swarm in the Archaean Vestfold Block, Antarctica: inferences about Proterozoic dyke emplacement in Gondwanaland. *J. Petrol.* **27**, 853–886.

Collerson, K.D., Jesseau, C.W. and Bridgwater, D. (1976a) Crustal development of the Archaean gneiss complex: eastern Labrador. In *The Early History of the Earth*, ed. Windley, B.F. Wiley, London, 237–253.

Collerson, K.D., Jesseau, C.M. and Bridgwater, D. (1976b) Contrasting types of bladed olivine in ultramafic rocks from the Archaean of Labrador. *Can. J. Earth Sci.* **13**, 442–450.

Collerson, K.D., Kerr, A., Vocke, R.D. and Hanson, G. (1982) Reworking of sialic crust as represented in late Archaean age gneisses, northern Labrador. *Geology* **10**, 202–208.

Colvine, A.C., Andrews, A.J., Cherry, M.E., Durocher, M.E., Fyon, A.J., Lavigne, M.J., Jr., Macdonald, A.J., Marmont, S., Poulsen, K.H., Springer, J.S. and Troop, D.G. (1984) An integrated model for the origin of Archaean lode gold deposits. *Ontario Geol. Surv. Open File Rep.* **5524**, 98 pp.

Colvine, A.C., Fyon, J.A., Heather, K.B., Marmont, S., Smith, P.M. and Troop, D.G. (1988) Archaean lode gold deposits in Ontario. *Ontario Geol. Surv. Misc. Paper* **139**, 136 pp.

Compston, W. and Arriens, P. (1968) The Precambrian geochronology of Australia. *Can. J. Earth Sci.* **5**, 561–583.

Compston, W. and Kroner, A. (1988) Multiple zircon growth within early Archaean tonalitic gneiss from the Ancient Gneiss Complex, Swaziland. *Earth Planet. Sci. Lett.* **87**, 13–28.

Compston, W., Zhong, F.D. Foster, J.J., Collerson, K.D., Bai, J. and Sun, D.Z., (1983) Rb–Sr geochronology of Precambrian rocks from the Yenshan region, north China. *Precamb. Res.* **22**, 175–202.

Compston, W., Williams, I.S., Campbell, I.H. and Gresham, J.J. (1986) Zircon xenocrysts from the Kambalda volcanics: age constraints and direct evidence for order continental crust below the Kambalda-Norseman greenstones. *Earth Planet. Sci. Lett.* **76**, 299–311.

Compton, P. (1978) Rare earth evidence for the origin of the Nûk gneisses Buksefjorden region, southern West Greenland. *Contrib. Mineral. Petrol.* **66**, 283–293.

Condie, K.C. (1972) A plate tectonics evolutionary model of the South Pass Archaean greenstone belt, southwestern Wyoming. *Proc. 24th Int. Geol. Congr.* **1**, 104–112.

Condie, K.C. (1976) Trace element geochemistry of Archaean greenstone belts. *Earth Sci. Rev.* **12**, 393–417.

Condie, K.C. (1981) *Archaean Greenstone Belts*. Elsevier, Amsterdam, 434 pp.

Condie, K.C. (1982a) *Plate Tectonics and Crustal Evolution*. Pergamon, New York, 310 pp.

Condie, K.C. (1982b) A plate tectonic model for Proterozoic continental accretion in the southwestern United States. *Geology* **10**, 37–42.

Condie, K.C. (1982c) Early and Middle Proterozoic supracrustal successions and their tectonic setting. *Am. J. Sci.* **282**, 341–357.

Condie, K.C. (1985) Secular variation in the composition of basalts: an index to mantle evolution. *J. Petrol.* **26**, 545–563.

Condie, K.C. (1986a) Geochemistry and tectonic setting of early Proterozoic supracrustal rocks in the southwestern United States. *J. Geol.* **94**, 845–864.

Condie, K.C. (1986b) Origin and early growth rate of continents. *Precamb. Res.* **32**, 261–278.

Condie, K.C. (1987) Early Proterozoic volcanic regimes in southwestern North America. In *Geochemistry and Mineralization of Proterozoic Volcanic Suites*, eds. Pharaoh T.C., Beckinsale, R.D. and Rickard, D. *Geol. Soc. London Spec. Publ.* **33**, 211–218.

Condie, K.C. (1989a) Geochemical changes in basalts and andesites across the Archaean-

Proterozoic boundary: Identification and significance. *Lithos* in press.

Condie, K.C. (1989b) *Plate tectonics and crustal evolution* (3rd Ed.) Pergamon, Oxford, 476 pp.

Condie, K.C. and Allen, P. (1984) Origin of Archaean charnockites from southern India. In *Archaean Geochemistry*. eds. Kroner, A., Goodwin, A.M. and Hanson, G.N. Springer-Verlag, Berlin, 182–203.

Condie, K.C. and Crow, C. (1989) Early Precambrian within-plate basalts from the Kaapvaal Craton in southern Africa: a case for crustally contaminated komatiites. *J. Geol.* in press.

Condie, K.C. and Harrison, N.M. (1976) Geochemistry of the Archaean Bulawayan Group, Midlands greenstone belt, Rhodesia. *Precamb. Res.* 3, 253–271.

Condie, K.C. and Hunter, D.R. (1976) Trace element geochemistry of the Archaean granitic rocks from the Barberton region, South Africa. *Earth Planet. Sci. Lett.* 29, 389–400.

Condie, K.C. and Nuter, J.A. (1981) Geochemistry of the Dubois greenstone succession an early Proterozoic bimodal volcanic association in west central Colorado. *Precamb. Res.* 15, 131–156.

Condie, K.C. and Shadel, C. (1984) An early Proterozoic volcanic arc succession in southeastern Wyoming. *Can. J. Earth Sci.* 24, 415–427.

Condie, K.C., Barsky, C.K. and Mueller, P.A. (1969a) Geochemistry of Precambrian diabase dikes from Wyoming. *Geochim. Cosmochim. Acta* 33, 1371–1388.

Condie, K.C., Leech, A.P. and Baadsgaard, H. (1969b) Potassium–Argon ages of Precambrian mafic dikes in Wyoming. *Geol. Soc. Amer. Bull.* 8, 899–906.

Condie, K.C., Viljoen, M.J. and Kable, E.J.D. (1977) Effects of alteration on element distributions in Archaean tholeiites from the Barberton greenstone belt, South Africa. *Contrib. Mineral. Petrol.* 64, 75–89.

Condie, K.C., Allen, P. and Narayana, B.L. (1982) Geochemistry of the Archaean low- to high-grade transition zone, southern India. *Contrib. Mineral. Petrol.* 81, 157–167.

Condie, K.C., Bowling, G.P. and Vance, R.K. (1985) Geochemistry and origin of early Proterozoic supracrustal rocks. Dos Cabezas Mountains, southeastern Arizona. *Geol. Soc. Amer. Bull.* 96, 655–662.

Condie, K.D., Bobrow, D.J. and Card, K.D. (1987) Geochemistry of Precambrian mafic dykes from the southern Superior Province of the Canadian Shield. *Geol. Assoc. Can. Spec. Paper* 34, 95–108.

Conradie, J.A. and Schoch, A.E. (1986) Petrographical characteristics of the Koperberg suite, South Africa – An analogy to massif-type anorthosites? *Precamb. Res.* 31, 157–188.

Conway, C.M. and Silver, L.T. (1984) Extent and implications of silicic alkalic magmatism and quartz arenite sedimentation in the Proterozoic of central Arizona. *Geol. Soc. Amer. Abs. Prog.* 16, 219.

Cooke, D.L. and Moorhouse, W.W. (1969) Timiskamng volcanism in the Kirkland Lake Area, Ontario, Canada. *Can. J. Earth Sci.* 6, 117–132.

Cooper, J.R. (1936) Geology of the southern half of the Bay of Islands igneous complex. *Newfoundland Dept. Nat. Resour. Geol. Sect. Bull.* 4, 1–62.

Copper, J.A., Fanning, C.M. Flook, M.M. and Oliver, R.L. (1976) Archaean and Proterozoic metamorphic rocks on southern Eyre Peninsula, South Australia. *J. Geol. Soc. Aust.* 23, 287–292.

Cordani, U.G. and Hasui, Y. (1975) Comentarios disponiveis sobre os dados geocronologicos para a Folha de Goias. *Carta Geol. Brasil ao milionesimo, DNPM, Folha Goias,* 85–91.

Cordani, U.G. and Neves, B.B. de B. (1982) The geologic evolution of South America during the Archaean and Early Proterozoic. *Rev. Bras. Geoc.* 12, 78–88.

Cordani et al. (1988) The growth of the Brazilian Shield. *Episodes* 11, 163–167.

Corfu, F. (1988) Differential response of U–Pb systems in coexisting accessory minerals in the Winnipeg River Subprovince, Canadian Shield; implications for Archaean crustal growth and stabilization. *Contrib. Mineral. Petrol.* 98, 312–325.

Corfu, F. and Ayres, L.D. (1987) U–Pb geochronology of polycyclic volcanism and felsic plutonism, Favourable Lake belt, Northwestern Ontario. *Geol. Assoc. Can./Min. Assoc. Can. Prog. Abs.* 12, 35.

Corfu, F. and Wallace, H. (1986) U–Pb zircon ages for magmatism in the Red Lake greenstone belt, northwestern Ontario. *Can. J. Earth Sci.* 23, 27–42.

Corfu, F. and Wood, J. (1986) U–Pb zircon ages in supracrustal and plutonic rocks; North Spirit Lake area, northwestern Ontario. *Can. J. Earth Sci.* 23, 967–977.

Covello, L., Roscoe, S.M., Donaldson, J., Roach, D. and Fyson, W.K. (1988) Archaean quartz arenite and ultramafic rocks at Beniah lake, Slave Structural Province NWT. *Geol. Surv. Can. Paper* **88-1C**, 223–232.

Cotterill, P. (1979) The Selukwe schist belt and its chromitite deposits. *Geol. Soc. S. Afr. Spec. Publ.* **5**, 229–245.

Courtillot, V. and Besse, J. (1987) Magnetic field reversals, polar wander and core-mantle coupling. *Science* **237**, 1140–1147.

Courtney, R.C. and White, R.S. (1986) Anomalous heat flow and geoid across the Cape Verde Rise: evidence for dynamic support from a thermal plume in the mantle. *Geophys. J.R. Astr. Soc.* **87**, 815–868.

Coward, M.P. and Park, R.G. (1987) The role of mid-crustal shear zones in the early Proterozoic evolution of the Lewisian. In *Evolution of the Lewisian and Comparable Precambrian High Grade Terrains;* eds. Park, R.G. and Tarney, J. *Geol. Soc. Lond. Spec. Publ.,* **27**, 127–138.

Cowden, A. (1988) Emplacement of komatiite lava flows and associated nickel sulphides. *Econ. Geol.* **83**, 436–442.

Cowden, A. and Woolrich, P. (1987) Geochemistry of the Kambalda iron-nickel sulphides: implications for models of sulfide-silicate partitioning. *Can. Mineral.* **25**, 21–36.

Cowden, A., Donaldson, M.J., Naldrett, A.J. and Campbell, I.H. (1986) Platinum-group elements and gold in the komatiite-hosted Fe–Ni–Cu sulfide deposits at Kambalda, Western Australia. *Econ. Geol.* **81**, 1226–1235.

Cox, K.G. and Hawksworth, C.J. (1985) Geochemical stratigraphy of the Deccan Traps at Mahabaleshwar, western Ghats, India, with implications for open system magmatic processes. *J. Petrol.* **26**, 355–377.

Cox, K.G., Gurney, J.J. and Harte, A. (1973) Xenoliths from the Matsoku pipe In *Lesotho Kimberlites,* ed. Nixon, P.H. Lesotho Nat. Dev. Corp., Maseru, 76–100.

Cox, K.G., Bell, J.D. and Pankhurst, R.J. (1979) *The Interpretation of Igneous Rocks.* George Allen and Unwin, London, 450 pp.

Crawford, A.J. ed. (1989) *Boninites.* Unwin Hyman, London, 465 pp.

Crawford, A.J. and Cameron, W.E. (1985) Petrology and geochemistry of Cambrian boninites and low-Ti andesites from Heathcote, Victoria. *Contrib. Mineral. Petrol.* **91**, 93–104.

Crawford, A.J., Becculava, L., Serri, G. and Dostal, J. (1986) Petrology, geochemistry and tectonic implications of volcanics dredged from the intersection of the Yap and Mariana trenches. *Earth. Planet. Sci. Lett.* **80**, 265–280.

Crawford, A.R. and Compston, W. (1970) The age of the Vindhyan system of Peninsular India. *J. Geol. Soc. Lond.* **125**, 351–371.

Creager, K.C. and Jordan, T.H. (1984) Slab penetration into the lower mantle. *J. Geophys. Res.* **89**, 3031–3049.

Crewe, M.A. (1984) A textural study of Archaean peridotites, Ujaragssuit nunât, Ivisârtoq region, southern West Greenland. *Rapp. Grønlands Geol. Unders.* **120**, 70–74.

Crocket, J.H. (1974) Gold. In *Handbook of Geochemistry 4,* ed. Wedepohl, K.H., Springer-Verlag, Berlin, 79/1-79/0/1.

Crocket, J.H. and MacRae, W.E. (1986) Platinum-group element distribution in komatiitic and tholeiitic volcanic rocks from Munro Township, Ontario. *Econ. Geol.* **81**, 1242–1251.

Crocket, J.H. and Teruta, Y. (1977) Palladium, iridium, and gold contents of mafic and ultramafic rocks drilled from the Mid-Atlantic Ridge, Leg 37, Deep Sea Drilling Project. *Can. J. Earth Sci.* **14**, 777–784.

Crough, S.T. and Jurdy, D.M. (1980) Subducted lithosphere, hotspots, and the geoid. *Earth Planet. Sci. Lett.* **48**, 15–22.

Crow, C. and Condie, K.C. (1987) Geochemistry and origin of late Archaean volcanic rocks from the Rhenosterhoek Formation, Dominion Group, South Africa. *Precamb. Res.* **37**, 217–229.

Czamanske, G.K. and Zientek, M.L. eds. (1985) *Stillwater Complex.* Spec. Publ. Montana Bur. Mines Geol. **92**, 396 pp.

Daly, R.A. (1928) Bushveld Igneous Complex of the Transvaal. *Geol. Soc. Amer. Bull* **39**, 703–768.

Daly, S. (1986) The Mulgathing Complex. *Rep. Mines Energy S. Aust.* **86/41**, 16 pp.

Damon, P.E. (1983) Continental uplift, compensation and shunting during trench-spreading center collision. *Tectonophysics* **99**, T1–T8.

Danni, J.C.M., Fuck, R.A. and Leonardos, O.H. (1982) Archaean and Lower Proterozoic units in Central Brazil. *Geol. Rund.* **71**, 291–317.

Danni, J.C.M. and Teixeira, N.A. (1981) Caracteristicas e sistematizacao das associacoes de rochas maficas e ultramaficas pre-cambrianas do Estado de Goias. *Anais I Simp. Geol. Centro-*

Oeste, Geologia do Pre-Cambriano, Soc. Bras. Geol., Nucleo Centro-Oeste, 376–401.

Daoud, W.E.K. and Fuck, R.A. (1987) The stratigraphic position of the Proterozoic Urupi Formation, northern Brazil: A discussion. In *Proc. Precambrian Evolution of the Amazonian Region*, (1987) Carajás, Pará, Brazil, 71–78.

Dardenne, M.A., Ferreira Filho, C.F. and Meirelles, M.A. (1987) The role of shoshonitic and calc-alkaline suites in tectonic evolution of the Carajás District, Brazil. In *Proc. Precambrian Evolution of the Amazonian Region*, (1987) Carajás, Pará, Brazil, 40–50.

Dasch, E.J., Shih, C.-Y., Bansal, B.M., Weisman, H. and Nyquist, L.E. (1987) Isotopic analysis of basaltic fragments from lunar breccia 14321: Chronology and petrogenesis of pre-Imbrium mare volcanism. *Geochim. Cosmochim. Acta* **51**, 3241–3254.

Davidson, I., Teixeira, J.B.G., Silva, M.G., Rocha Neto, M.B. and Matos, F.M.V. (1988) The Rio Itapicuru Greenstone Belt, Bahia, Brazil: Structural and stratigraphical outline. *Precamb. Res.* **42**, 1–17.

Davis, D.W. and Jackson, M.C. 1988. Geochronology of the Lumby Lake greenstone belt: a 3 Ga complex within the Wabigoon subprovince, northwest Ontario. *Geol. Soc. Amer. Bull.* **100**, 818–824.

Davis, D.W., Corfu, F. and Krogh, T.E. (1986) High precision U–Pb Geochronology and implications for the tectonic evolution of the Superior Province. *LPI Tech. Rep.* **86–10**, 77–79.

Davis, D.W. and Edwards, G.R. 1985. The petrogenesis and metallogenesis of the Atikwa-Lawrence volcanic-plutonic terrain. *Ont. Geol. Surv. Misc. Paper* **127**, 101–111.

Davis, D.W., Sutcliffe, R.H. and Trowell, N.F. (1988) Geochronological constraints on the tectonic evolution of a late Archean greenstone belt. *Precamb. Res.* **39**, 171–191.

Davies, G.F. (1979) Thickness and thermal history of continental crust and root zones. *Earth Planet Sci. Lett.* **44**, 231–238.

Davies, G.F. (1981) Earth's neodymium budget and structure and evolution of the mantle. *Nature* **290**, 208–213.

Davies, G. and Tredoux, M. (1985) The platinum-group element and gold contents of the marginal rocks and sills of the Bushveld Complex. *Econ. Geol.* **80**, 838–48.

Davies, G., Cawthorn, R.G., Barton, J.M. and Morton, M. (1980). Parental magma to the Bushveld Complex. *Nature* **287**, 33–35.

Davis, P.A. and Condie, K.C. (1977) Trace element model studies of Nyanzian greenstone belts, western Kenya. *Geochim. Cosmochim. Acta* **41**, 271–277.

Davis, P.A., Stuber, A.M. and Potts, M.J. (1977) Rb and Sr concentrations and Sr–isotope ratios in the Preacher Creek ultramafic intrusion, Wyoming. *Univ. Wyo. Contrib. Geol.* **15**, 17–25.

Davis, R.D., Allsopp, H.L., Erlank, A.J. and Manton, W.I. (1970) Sr–isotope studies on various layered mafic intrusions in southern Africa. *Spec. Publ. Geol. Soc. S. Afr.* **1**, 576–593.

Dawes, P.R., Soper, J., Escher, J.C. and Hall, R.P. (1989) The northern boundary of the Proterozoic (Nagssugtoqidian) mobile belt of South-East Greenland. *Rapp. Grønlands Geol. Unders.* **146**, 54–65.

Dearnley, R. (1963) The Lewisian complex of South Harris; with some observations on the metamorphosed basic intrusions of the Outer Hebrides. *Quart. J. Geol. Soc. London* **118**, 143–166.

De Beer, J.H. and Stettler, E.H. 1986. The deep structures of South African greenstone belts. *Geocongress '86, Ext. Abs.*, Johannesburg, 361–364.

De Beer, J.H. and Stettler, E.H. (1988) Geophysical characteristics of southern Africa continental crust. In *Oceanic and Continental Lithosphere: Similarities and Differences*, eds. Menzies, M.A. and Cox, K.G., *Spec. Vol. J. Petrol.*, 163–184.

De Beer, J.H., Stettler, E.H., Duvenhage, A.W.A., Joubert, S.J. and Raath, C.J. (1984) Gravity and geoelectrical studies of the Murchison greenstone belt, South Africa. *Trans. Geol. Soc. S. Afr.* **87**, 347–359.

De Bruijn, H., van der Westhuizen, W.A., Grobler, N.J. and Botha, P.J. (1984) Lithogeochemistry and multivariate statistics as aids to the stratigraphic characterization of Proterozoic Ventesdorp lavas. *S. Afr. J. Sci.* **80**, 280–282.

DeKemp, E. (1987) Stratigraphy, Provenance, and Geochronology of Archaean Supracrustal Rocks of western Eyapamikama Lake Area, northwestern Ontario. Unpubl. MSc thesis, Carleton Univ., Ottawa, Ontario, 93 pp.

D'El Rey Silva, L.J.H. (1985) Geologia e controle estrutural do deposito cuprifero de Caraiba, Vale do Curaca – Bahia. *SME-CPM, Salvador, textos Basicos* **6**, 51–136.

De Laeter, J.R., Fletcher, I.R., Rosman, K.J.R., Williams, I.R., Gee, R.D. and Libby, W.G. (1983) Early Archean gneisses from the Yilgarn Block, Western Australia. *Nature* **292**, 322–324.

Delano, J.W. (1980) Chemistry and liquidus phase relations of Apollo 15 red glass: implications for the deep lunar interior. *Proc. 11th Lunar Planet. Sci. Conf.*, Pergamon, New York, 251–288.

Delano, J.W. (1986) Pristine lunar glasses: criteria, data, and implication. *Proc. 16th Lunar Planet. Sci. Conf.; J. Geophys. Res.* **91**, suppl., D201–D213.

Delany, P.T. (1987) Heat transfer during emplacement and cooling of mafic dykes. In *Mafic Dyke Swarms* eds. Halls, H.C. and Fahrig, W.F. *Geol. Assoc. Canada Spec. Paper* **34**, 31–46.

Delhal, J., Lendent, D. and Pasteels, P. (1975) L'age du complex granitique et migmatitique de Dibaya (region du Kasai, Zaire) par les methods Rb–Sr et U–Pb. *Ann. Soc. Geol. Belg.* **98**, 141–154.

Dence, M.R. (1972) Meteorite impace craters and the structure of the Sudbury basin. *Geol. Assoc. Can. Spec. Pap.* **10**, 7–18.

DePaolo, D.J. (1980) Crustal growth and mantle evolution: inferences from models of element transport and Nd and Sr isotopes. *Geochim. Cosmochim. Acta* **44**, 1185–1196.

DePaolo, D.J. (1981) Neodymium isotopes in the Colorado Front Range and crust-mantle evolution in the Proterozoic. *Nature* **291**, 193–196.

DePaolo, D.J. and Wasserberg, G.J. (1979) Sm–Nd age of Stillwater Complex and the mantle evolution curve for neodymium. *Geochim. Cosmochim. Acta* **43**, 998–1008.

De Roever, E.W.F. and Bosma, W. (1975) Precambrian magmatism and regional metamorphism in Suriname. In *Minist. Minas e Energia, D.N.P.M. Anais Decima Conf. Geol. Interguianas*, Belem, Brasil, **1**, 123–163.

Derrick, G.M. (1982) A Proterozoic rift zone at Mount Isa, Queensland, and implications for mineralisation. *BMR J. Aust. Geol. Geophys.* **7**, 81–92.

Desmaris, N.R. (1981) Metamorphosed Precambrian ultramafic rocks in the Ruby Mountain Range, Montana. *Precamb. Res.* **16**, 67–101.

Deus, P.B. and Viana, J.S. (1982) Jacurici Valley chromite district. *ISAP-Int. Sym. Arch. Early Prot. Geol. Evol. & Met., Abs. Excur.*, SME-Ba, 97–107.

Deus, P.B., Viana, J.S., Duarte, P.M. and Queiroz, W.J.A. (1982) Campo Formoso chromite district. *ISAP-Int. Sym. Arch. Early Prot. Geol. Evol. & Met., Abs. Excur.*, SME-Ba, 107–113.

Dewey, J. and Spall, H. (1975) Pre-Mesozoic plate tectonics: how far back in Earth history can the Wilson cycle be extended. *Geology* **3**, 422–424.

De Wit, M. (1982) Gliding and overthrust nappe tectonics in the Barberton greenstone belt. *J. Struct. Geol.* **4**, 117–136.

De Wit, M.J. and Ashwal, L., (1986) Workshop on tectonic evolution of greenstone belts. *Lun. Planet Inst. Tech. Rep.* **86–10**. Lunar Planet. Inst., Houston, 227 pp.

De Wit, M.J. and Stern, C.R. (1980) A 3500 Ma ophiolite complex from the Barberton greenstone belt, South Africa: Archaean oceanic crust and its geotectonic implications. *2nd Int. Arch. Symp.*, Perth, Abs., 85–87.

De Wit, M.J., Fripp, R.E.P. and Stanistreet, I.G. (1983) Tectonic and stratigraphic implications of new field observations along the southern part of the Barberton greenstone belt. *Spec. Publ. Geol. Soc. S. Afr.* **9**, 21–29.

De Wit, M.J., Armstrong, R., Hart, R.J. and Wilson, A.H. (1987a) Felsic igneous rocks within the 3.3 to 3.5 Ga Barberton greenstone belt: high crustal level equivalents of the surrounding tonalite-trondhjemite terrain, emplaced during thrusting. *Tectonics* **6**, 529–549.

De Wit, M.J., Hart, R.A. and Hart, R.J. (1987b) The Jamestown ophiolite complex, Barberton Mountain Belt: a section through 3.5 Ga oceanic crust. *J. Afr. Earth Sci.* **6**, 681–730.

Dewitt, E.J., Redden, J.A., Wilson, A.B. and Buscher, D. 1986. Mineral resource potential and geology of the Black Hills National Forest, South Dakota and Wyoming. *U.S. Geol. Surv. Bull.* **1580**, 135 pp.

Dick, H.J.B. and Bullen, T. (1984) Chromian spinel as petrogenetic indicator in abyssal and alpine-type peridotites and spatially associated lavas. *Contrib. Mineral. Petrol.* **86**, 54–76.

Dickinson, W.R. (1971) Plate tectonics in geologic history. *Science* **174**, 107–113.

Dietrich, V.J., Emmermann, R., Oberhansli, R. and Puchelt, H. (1978) Geochemistry of basaltic and gabbroic rocks from the west Mariana basin and the Mariana trench. *Earth Planet. Sci. Lett.* **39**, 127–144.

Dietrich, V.J., Gansser, A., Sommerauer, J. and Cameron, W.E. (1981) Palaeocene komatiites from Gorgona Island, East Pacific – a primary magma for ocean floor basalts? *Geochem. J.* **15**, 141–161.

Dietz, R.S. (1964) Sudbury structure as an astrobleme. *J. Geol.* **72**, 412–434.

Dillon-Leitch, H.C.H., Watkinson, D.H. and Coats, C.J.A. (1986) Distribution of platinum-group elements in the Donaldson West deposit, Cape Smith belt, Quebec. *Econ. Geol.* **81**, 1147–1158.

Dimroth, E., Cousineau, P., Leduc, M. and Sanschagrin, Y. (1978) Structure and organisation of Archean subaqueous basalt flows, Rouyn-Noranda area, Quebec, Canada. *Can. J. Earth Sci.* **15**, 902–918.

Dimroth, E., Imreh, L., Rocheleau, M. and Goulet, N. (1982) Evolution of the south-central part of the Archean Abitibi Belt, Quebec. 1: stratigraphy and paleogeographic model. *Can. J. Earth Sci.* **19**, 1729–1758.

Dimroth, E., Imreh, L., Cousineau, P., Leduc, M. and Sanschagrin, Y. (1985) Paleogeographic analysis of mafic submarine flows and its use in the exploration for massive sulphide deposits. *Geol. Assoc. Can. Spec. Paper* **28**, 203–222.

Divis, A.F. (1976) Geology and geochemistry of Sierra Madre Range, Wyoming. *Quart. J. Colorado Sch. Mines* **71**, 1–127.

Dixon, H.R. (1982) Archean granitic rocks in the Seminoe Mountains, Wyoming. Archaean Geochem. *Field Conf. Seminoe Mountains and Hartville Uplift, Wyoming*, **I**, 50–63.

Dixon, J.R. and Papike, J.J. (1975) Petrology of anorthosites from the Descartes region of the moon. *Apollo 16. Proc. Lunar Sci. Conf. 6th*, Pergamon, New York, 263–291.

Dixon, J.E., Fitton, J.G. and Frost, R.T.C. (1981) The tectonic significance of post-Carboniferous igneous activity in the North Sea basin. In *Petroleum Geology of the Continental Shelf of North-West Europe*, eds. Illing, L.V. and Hobson, G.D., Hayden and Son, London, 121–137.

Dodson, M.H., Gledhill, A.R., Shackleton, R.M. and Bell, K. (1975) Age differences between Archean cratons of east and southern Africa. *Nature* **254**, 315–318.

Donaldson, C.H. (1974) Olivine crystal types in harrisitic rocks of the Rhum pluton and in Archaean spinifex rocks. *Geol. Soc. Am. Bull.* **85**, 1721–1726.

Donaldson, C.H. (1976) An experimental investigation of olivine morphology. *Contrib. Mineral. Petrol.* **57**, 187–213.

Donaldson, C.H. (1982) Spinifex-textured komatiites: A review of textures, mineral compositions and layering. In *Komatiites* eds. Arndt, N.T. and Nisbet, E.G., George Allen and Unwin, London, 213–243.

Donaldson, M.J. (1974) Petrology of the Munni Munni Complex, Roebourne, Western Australia. *J. Geol. Soc. Aust* **21**, 1–16.

Dong, S.B., Shen, Q.H., Sun, D.Z. and Lu, L.Z. (1986) *Metamorphic map of China (1:4000000)*, *Explanatory text* (supervisor Cheng. Y.Q.). Geol. Publ. House, Beijing. 162 pp (233 pp in Chinese).

Dostal, J., Dupuy, C. and Liotard, J.M. (1982) Geochemistry and origin of basaltic lavas from Society islands. *Bull. Volc. Geotherm. Res.* **45**, 50–62.

Dougan, T.W. (1977) The Imataca Complex near Cerro Bolivar, Venezuela – a calc-alkaline Archaean protolith. *Precamb. Res.* **4**, 237–268.

Dowty, E., Prinz, M. and Keil, K. (1974) Ferroan anorthosite: a widespread and distinctive lunar rock type. *Earth Planet. Sci. Lett.* **24**, 15–25.

Dressler, B.O., Morrison, G.G., Peredery, W.V. and Rao, B.V. (1987) The Sudbury structure, Ontario, Canada–a review. In Research in Terrestial Impact Structures, ed. Pow, J., Frud Vieweg and Sohn, Braunschweig, 39–68.

Drury, S.A. (1983) The petrogenesis and setting of Archaean metavolcanics from Karnataka State, south India. *Geochim. Cosmochim. Acta* **47**, 317–329.

Drury, S.A., Naqvi, S.M. and Hussain, S.M. (1978) REE distributions in basaltic anorthosites from the Holenarsipur greenstone belt, Karnataka, South India. In *Archaean Geochemistry* eds. Windley, B.F. and Naqvi, S.M., Elsevier, Amsterdam, 363–374.

Duchac, K.C. and Hanor, J.S. (1987) Origin and timing of the metasomatic silicification of an early Archaean komatiite sequence, Barberton Mountain Land, South Africa. *Precamb. Res.* **37**, 125–146.

Dudas, F.O. and Egglar, D.H. (1986) Mantle xenoliths from the Crazy Mountains, Montana. *Geol. Soc. Amer. Abs. Prog.* **18**, 588.

Duncan, A.R. (1987) The Karoo igneous province – a problem area for inferring tectonic setting from basalt geochemistry. *J. Volc. Geotherm. Res.* **32**, 13–34.

Du Plessis, A. and Kleywegt, R.J. (1987) A dipping sheet model for the mafic lobes of the Bushveld Complex. *S. Afr. J. Geol.* **90**, 1–6.

Dupré, B. and Echeverria, L.M. (1984) Pb isotopes of Gorgona Island (Columbia): isotopic variations correlated with magma type. *Earth Planet. Sci. Lett.* **67**, 186–190.

Durney, D.W. (1972) A major unconformity in the Archaean Jones Creek area, Western Australia. *J. Geol. Soc. Aust.* **19**, 251–260.

Dymek, R.F. (1986) Characterization of the Apollo 15 feldpathic basalt suite. In *Workshop on the Geology and Petrology of the Apollo 15 Landing Site* eds. Spudis, P.D. and Ryder, G., *LPI Tech. Rep.* **86–03**. Lunar Planet. Inst., Houston, 52–57.

Dziewonski, A.M. (1984) Mapping the lower mantle: determination of lateral heterogeneity in P velocity up to degree and order 6. *J. Geophys. Res.* **89**, 5929–5952.

Dziewonski, A.M. and Anderson, D.L. (1981) Preliminary reference Earth model. *Phys. Earth Planet. Inter.* **25**, 297–356.

Eales, H.V. (1987) Upper critical zone chromitite layers at R.P.M. Union Section Mine, Western Bushveld Complex. In *Evolution of Chromian Ore Fields.* ed. Stowe, C.W. Van Nostrand Reinhold, 144–168.

Easton, R.M. (1985) The nature and significance of pre-Yellowknife supergroup rocks in the Point Lake area, Slave structural province, Canada. In *Evolution of Archaean Supracrustal Sequences;* eds. Ayres L.D., Thurston, P.C., Card, K.D. and Weber, W., *Geol. Assoc. Can. Spec. Paper* **28**, 153–167.

Echeverria, L.M. (1980) Tertiary or Mesozoic komatiites from Gorgona Island, Colombia: field relations and geochemistry. *Contrib. Mineral. Petrol.* **73**, 253–266.

Echeverria, L.M. and Aitken, B.G. (1986) Pyroclastic rocks: another manifestation of ultramafic volcanism on Gorgona Island, Columbia. *Contrib. Mineral. Petrol.* **92**, 428–436.

Eckelmann, F.D. and Poldervaart, A. (1957) Geologic evolution of the Beartooth Mountains, Montana and Wyoming. I. Archean history of the Quad Creek area. *Bull. Geol. Soc. Amer.* **68**, 1225–1262.

Edwards, J.S. (1981) The petrology and contact relationships of the southwestern portion of the Precambrian Mullen Creek mafic complex, Medicine Bow Mountains, Wyoming. *Geol. Soc. Amer. Abs. Prog.* **13**, 195.

Edwards, G.R. (1983) Geology of the Bethune Lake Area, Districts of Kenora and Rainy River. *Ont. Geol. Surv. Rep.* **201**, 59 pp.

Edwards, G.R. (1985) Geochemistry and evolution of an Archaean bimodal volcanic–plutonic complex, Wabigoon Subprovince, Ontario. Unpubl. PhD. thesis, Univ. West. Ontario, London, Ont. 344 p.

Edwards, G.R. and Nisbet, E.G. (1986) Hungry komatiites and indigestible zircons. *Nature* **322**, 771.

Eggler, D.H. (1987) Geochemistry of upper mantle and lower crust beneath Colorado and Wyoming. *Geol. Soc. Amer. Abs. Prog.* **19**, 272–273.

El Goersy, A. (1976) Oxide minerals in lunar rocks. In *Mineral. Soc. Amer. Short Course Notes, Vol. 3, Oxide Minerals.* ed. Rumble, D., Southern Printing Co., Blacksburg, EG1–46.

Ellam, R.M. and Hawkesworth, C.J. (1988) Is average continental crust generated at subduction zones? *Geology* **16**, 314–317.

Eller, J.A. and Friberg, L.M. (1982) Petrology of amphibolites on Casper Mountain, Wyoming. *Geol. Soc. Amer. Abs. Prog.* **19**, 272–273.

Elliott, J.E., Gaskill, D.L. and Raymond, W.H. (1983) Geological and geochemical investigations of the North Absaroka Wilderness study area, Park and Sweetgrass Counties, Montana. *U.S. Geol. Surv. Bull.* **1505A**, 251 pp.

Ellis, D.J. and Wyborn, L.A.I. (1984) Petrology and geochemistry of Proterozoic dolerites from the Mount Isa inlier. *BMR J. Aust. Geol. Geophys.* **9**, 19–32.

Elthon, D. (1986) Komatiite genesis in the Archaean mantle, with implications for the tectonics of Archaean greenstone belts. In *Workshop on Tectonic Evolution of Greenstone Belts. LPI Tech. Rep.* **86–10**, eds. M.J. de Wit and L.D. Ashwal, Lunar Planet. Inst. Houston, 97–99.

Engel, A.E.J., Nagy, B., Nagy, L.A., Engel, C.G., Kremp, G.O.W. and Drew, C. (1968) Algal-like fossils in the Onverwacht Series, South Africa: oldest recognised life-like forms on Earth. *Science* **161**, 1005–1008.

Engelbrecht, J.P. (1985) The chromites of the Bushveld Complex in the Nietverdiend area. *Econ. Geol.* **80**, 896–910.

England, P.C. and Bickle, M.J. (1984) Continental thermal and tectonic regimes during the Archaean. *J. Geol.* **92**, 353–367.

Ernst, W.G. (1989) Petrochemical comparison of the 3.5 Ga old mafic amphibolite inclusions from eastern Hebei Province, with Archaean mafic-ultramafic supracrustals of uncertain antiquity, southern Jilin/eastern Liaoning Provinces, China. *Chinese J. Geochem* **8**, 97–111.

Ernst, W.G., Cao, R.L. and Jiang, J.Y. (1988) Reconnaissance study of Precambrian metamorphic rocks, northeastern Sino-Korean Shield, People's Republic of China. *Geol. Soc. Amer. Bull.* **100**, 692–701.

Erslev, E.A. (1983) Pre-Beltian geology of the southern Madison Range, southwestern Montana. *Mont. Bur. Mines Geol. Mem.* **55**, 1–26.

Escher, A and Pulvertaft, T.C.R. (1976) Rinkian mobile belt of West Greenland. In *Geology of Greenland*. eds. Escher, A. and Watt, W.S., Geol. Surv. Greenland, Copenhagen, 105–119.

Escher, A., Escher, J.C. and Watterson, J. (1975) The reorientation of the Kangamiut dyke swarm, West Greenland. *Can. J. Earth Sci.* **12**, 158–173.

Escher, A., Sorensen, K. and Zeck, H.P. (1976) Nagssugtoqidian mobile belt in West Greenland. In *Geology of Greenland*, eds. Escher, A. and Watt, Geol. Surv. Greenland, W.S., Copenhagen, 77–95.

Escher, J.C. and Myers, J.S. (1975) New évidence concerning the original relationships between early Precambrian volcanics and anorthosites in the Fiskenaesset region, southern West Greenland. *Rapp. Grønlands Geol. Unders.* **75**, 72–76.

Escher, J.C., Friend, C.R.L. and Hall, R.P. (1989) The southern boundary of the East Greenland Proterozoic mobile belt: geology of the area between Umivik and Isortoq. *Rapp. Grølands Geol. Unders.* **146**, 70–78.

Espejo, A.C. (1974) Geologia de la region el Manteco-Guri, estado Bolivar. In Venezuela, Minis. Minas y Hidrocarburos, Direcion Geologia. Memor. Novena Conf. Geol. Inter-Guayanas, Ciudad Guayana, Venezuela, *Bol. Geol. Publ. Espec.* **6**, 207–250.

Etheridge, M.A., Rutland, R.W.R. and Wyborn, L.A.I. (1987) Orogenesis and tectonic processes in the early to middle Proterozoic of southern Australia. In *Proterozoic Lithospheric Evolution*. ed. Krøner, A., *Amer. Geophys. Un. Geodynamic Ser.* **17**, 131–147.

Evans, C.R. and Tarney, J. (1964) Isotopic ages of Assynt dykes. *Nature* **207**, 54–56.

Ewart, A., Bryan, W.B. and Gill, J.B. (1973) Mineralogy and geochemistry of the younger volcanic islands of Tonga, S.W. Pacific. *J. Petrol.* **14**, 429–465.

Fahrig, W.F. (1987) The tectonic settings of continental mafic dyke swarms: failed arm and early passive margin. In *Mafic Dyke Swarms*; eds. Halls, H.C. and Fahrig, *Geol. Assoc. Can. Spec. Pap.* W.F. **34**, 331–348.

Fahrig, W.F. and West, T.D. (1986) Diabase dyke swarms of the Canadian Shield. *Geol. Surv. Canada Map* **1627A**; also in (1987) *Mafic Dyke Swarms*. eds. Halls, H.C. and Fahrig, W.F. *Geol. Assoc. Can. Spec. Pap.* **34**.

Fanning, C.M., Oliver, R.L. and Cooper, J.A. (1986) The Carnot gneisses, southern Eyre Peninsula. *Geol. Surv. S. Aust. Quart. Geol. Notes* **80**, 7–12.

Ferguson, J. and Needham, R.S. (1978) The Zamu dolerite: a lower Proterozoic continental tholeiite suite from the Northern Territory, Australia. *J. Geol. Soc. Aust.* **25**, 309–322.

Ferreira Filho, C.F. and Danni, J.C.M. (1985) Petrologia e mineralizaçes sulfetadas do Prospecto Bahia-Carajás. In *II Simp. Amazônico, SBG/Nucleo Norte*, Belem, Brazil, 34–47.

Ferreira Filho, C.F., Cordani, U.G., Teixeira, W. and Danni, J.C.M. (1987) Geochronology of the Bahia Prospect copper deposit – Carajás Province – Brazil. In *Proc. Precamb. Evol. Amazonian Region, 1987*. Carajás, Pará, Brazil, 32–39.

Fettes, D.J. and Mendum, J.R. (1987) The evolution of the Lewisian complex in the Outer Hebrides. In *Evolution of the Lewisian and Comparable Precambrian High Grade Terrains* eds. Park, R.G. and Tarney, J. *Spec. Publ. Geol. Soc. Lond.* **27**, 27–44.

Figueiredo, A.N., Motta, J. and Marques, V.J. (1975) Estudo comparativo entre os complexos de Barro Alto e do Tocantins, Goias. *Rev. Bras. Geoc.* **5**, 15–29.

Fitton, M.J., Horowitz, R.C. and Sylvester, G. (1975) Stratigraphy of the early Precambrian in the West Pilbara, Western Australia. *Aust. Commonwealth Sci. Ind. Res. Org. Min. Res. Lab. Rep.* **FP11**.

Fletcher, I.R. and Rosman, K.J.R. (1982) Precise determination of initial ε_{Nd} from Sm–Nd isochron data. *Geochim. Cosmochim. Acta* **46**, 1983–1987.

Fletcher, I.R., Rosman, K.J.R., Williams, I.R., Hickman, A.H. and Baxter J.L. (1984) Sm–Nd geochronology of Greenstone belts in the Yilgarn Block, Western Australia. *Precamb. Res.* **26**, 333–361.

Fletcher, I.R., Libby, W.G. and Rosman, K.J.R. (1987) Sm–Nd dating of the 2411 Ma Jimberlana Dyke, Yilgarn Block, Western Australia. *Aust. J. Earth Sci.* **34**, 523–525.

Floyd, P.A. and Winchester, J.A. (1978) Identification and discrimination of altered and metamorphosed volcanic rocks using immobile elements. *Chem. Geol.* **21**, 291–306.

Foden, J.D., Buick, I.S. and Mortimer, G.E. (1988) The petrology and geochemistry of granulitic gneisses from the east Arunta inlier, Central Australia: implications for Proterozoic crustal development. *Precamb. Res.* **40–41**, 233–259.

Fodor, R.V., Corwin, C. and Roisenberg, A. (1985) Petrology of Serra Geral (Parana) continental flood basalts southern Brazil: crustal contamination, source material and S. Atlantic magmatism. *Contrib. Mineral. Petrol.* **91**, 54–65.

Foster, R.P. (1985) Major controls of Archaean gold mineralization in Zimbabwe. *Trans. Geol. Soc. S. Afr.* **88**, 109–133.

Foster, R.P. (1988) Archaean gold mineralization in Zimbabwe: implications for metallogenesis and exploration. *Bicent. Gold 88, Extend. Abs. Geol. Soc. Austr., Abs.* **22**, 62–72.

Foster, R.P. and Wilson, J.F. (1984) Geological setting of Archaean gold deposits in Zimbabwe. In *Gold '82: The Geology, Geochemistry and Genesis of Gold Deposits.* ed. Foster, R.P., A.A. Balkema, Rotterdam, 521–551.

Foster, R.P., Mann, A.G., Miller, R.G., and Smith, P.J.R. (1979) Genesis of gold mineralization in the Gatooma area, Rhodesia. In *Genesis of gold mineralization in the Gatooma area, Rhodesia* eds. Foster, R.P., Mann, A.G., Miller, R.G. and Smith, P.J.R. *Spec. Publ. Geol. Soc. S. Afr.* **5**, 25–38.

Foster, R.P., Mann, A.G., Stowe, C.W. and Wilson, J.F. (1986) Archaean gold mineralization in Zimbabwe. In *Mineral Deposits of Southern Africa Vol. 1.* eds. Anhaeusser, C.R. and Maske, S. *Geol. Soc. S. Afr.*, Johannesburg, 43–112.

Fraser, G.D., Waldrop, H.A. and Hyden, H.J. (1969) Geology of the Gardiner area, Park County, Montana. *U.S. Geol. Surv. Bull.* **1277**, 118 pp.

Friend, C.R.L. and Hughes, D.J. (1977) Archaean aluminous ultrabasic rocks with primary igneous textures from the Fiskenaesset region, southern West Greenland. *Earth Planet. Sci. Lett.* **36**, 157–167.

Friend, C.R.L. and Hughes, D.J. (1978) Relict plutonic textures in Archaean ultramafic rocks from the Fiskenaesset region, southern West Greenland: implications for crustal thickness. In *Archaean Geochemistry* eds. Windley, B.F. and Naqvi, S.M., Elsevier, Amsterdam, 375–392.

Friend, C.R.L., Hall, R.P. and Hughes, D.J. (1981) The geochemistry of the Malene (mid-Archaean) ultramafic-mafic amphibolite suite, southern West Greenland. In *Archaean Geology* eds. Glover, J.E. and Groves, D.I. *Spec. Publ. Geol. Soc. Aust.* **7**, 301–312.

Friend, C.R.L., Nutman, A.P. and McGregor, V.R. (1988) Late Archaean terrane accretion in the Godthåb region, southern West Greenland. *Nature* **335**, 535–538.

Froude, D.O., Ireland, T.R., Kinny, P.D., Williams, I.S., Compston, W., Williams, I.R. and Myers, J.S. (1983) Ion microprobe identification of 4100–4200 Myr-old terrestrial zircons. *Nature* **304**, 616–618.

Fryer, B.J. and Jenner, G.A. (1978) Geochemistry and origin of the Archean Prince Albert Group volcanics, western Melville Peninsula, Northwest Territories, Canada. *Geochim. Cosmochim. Acta* **42**, 1645–1654.

Fuck, R.A., Danni, J.C.M., Winge, M., Andrade, G.F., Barreira, C.F., Leonardos, O.H. and Kuyumjian, R.M. (1981) Geologia da regiao de Goianesia. *I. Simp. Geol. Centro-Oeste, Geologia do Pre-Cambriano, Soc. Bras. Geol., nucleo Centro-Oeste*, 447–467.

Fuck, R.A., Danni, J.C.M. and Leonardos, O.H. (1982) Outline of the Archaean and Lower Proterozoic of Central Goias. *ISAP-Int. Symp. Arch. Early Prot. Geol. Evol. and Met., Abs. Excur., SME-Ba*, 78–87.

Furgason, D.C. (1977) Petrology and geochemistry of part of the southern sector, southeastern Bushveld Complex. Unpubl. MSc thesis. Univ. Wisconsin-Madison, 129 pp.

Fyfe, W.S. and Kerrich, R. (1984) Gold: natural concentration processes, 99–127. In *Gold '82: The Geology, Geochemistry and Genesis of Gold Deposits.* ed. Foster, R.P., A.A. Balkema, Rotterdam, 753 pp.

Gaál, G. (1982) Proterozoic evolution and late Svecokarelian plate deformation of the Baltic Shield. *Geol. Rundsch.* **71**, 158–170.

Gaál, G. (1985) Nickel metallogeny related to tectonics. *Bull. Geol. Surv. Finland* **333**, 143–155.

Gaál, G. (1986) 2200 million years of crustal evolution: the Baltic Shield. *Bull. Geol. Soc. Finland* **58**, 149–168.

Gaál, G. and Gorbatschev, R. (1987) An outline of the Precambrian evolution of the Baltic Shield. *Precamb. Res.* **35**, 15–52.

Gaál, G., Koistinen, T. and Mattila, E. (1975) Tectonics and stratigraphy of the vicinity of

Outokumpu, north Karelia, Finland. *Bull. Geol. Surv. Finland.* **271**, 67 pp.

Gaál, G. Mikkola, A. and Sederholm, B. (1978) Evolution of the Archaean crust in Finland. *Precamb. Res.* **3**, 199–215.

Gable, D.J., Burford, A.E. and Corbett, R.G. (1988) The Precambrian geology of Casper Mountain, Natrona County, Wyoming. *U.S. Geol. Surv. Prof. Pap.* **1460**, 50 pp.

Gain, S.B. (1985) The geological setting of the platiniferous UG-2 chromitite layer on the farm Maandagshoek, eastern Bushveld Complex. *Econ. Geol.* **80**, 925–943.

Galer, S.J.G. and O'Nions, R.K. (1985) Residence time of thorium, uranium and lead in the upper mantle with implications for mantle convection. *Nature* **316**, 778–782.

Galer, S.J.G., Goldstein, S.L. and O'Nions, R.K. (1989) Limits on chemical and convective isolation in the Earth's interior. *Chem. Geol.* **75**, 257–290.

Galvão, C.F. (1980) Projeto Rio Jacare. *Relatorio Final, CBPM/SME, Bahia, vol. 1.*

Gariepy, C., Allègre, C.J. and Lajoie, J. (1984) U–Pb systematics in single zircons from the Pontiac sediments, Abitibi greenstone belt. *Can. J. Earth Sci.* **21**, 1296–1304.

Garihan, J.M. (1979) Geology and structure of the central Ruby Range, Madison County, Montana. *Geol. Soc. Amer. Bull.* **90**, 323–326.

Gee, R.D. (1975) Regional geology of the Archaean nuclei of the Western Australian Shield. In *Economic Geology of Australia and Papua New Guinea, 1* ed. Knight, G.C. Metals. Aust. Inst. Min. Metall., Melbourne, 43–55

Gee, R.D. (1979) Summary of the Precambrian stratigraphy of Western Australia. *Ann. Rep. Geol. Surv. West. Aust. 1979* 85–89.

Gee, R.D., Baxter, J.C., Wilde, S.A. and Williams, I.R. (1981) Crustal development in the Archaean Yilgarn Block, Western Australia. In *Archaean Geology*; eds. Glover, J.E. and Groves, D.I. *Spec. Publ. Geol. Soc. Aust.* **7**, 43–56.

Gelinas, L. and Brooks, C. (1974) Archaean quench-textured tholeiites. *Can. J. Earth Sci.* **11**, 324–340.

Gelinas, L., Brooks, C., Perreault, B., Carignan, J., Trudel, P. and Grasso, F. (1977) Chemostratigraphic divisions within the Abitibi volcanic Belt, Rouyn–Noranda district, Quebec. *Geol. Assoc. Can. Spec. Paper* **16**, 265–295. .

Gemuts, I. and Theron, A. (1975) The Archaean between Coolgardie and Norseman – stratigraphy and mineralisation. In *Economic Geology of Australia and Papua New Guinea, vol. 1,* ed. Knight, C.L. Aust. Inst. Mining Metall., Melbourne, 66–74.

Gibb, F.G.F. and Kanaris-Sotiriou, R. (1976) Jurassic igneous rocks of the Forties Field. *Nature* **260**, 23–25.

Gibbs, A.K. (1987a) Proterozoic volcanic rocks of the northern Guiana Shield, South America. In *Geochemistry and Mineralization of Proterozoic Volcanic Suites*; eds. Pharaoh, T.C., Beckinsale, R.D. and Rickard, D. *Geol. Soc. Lond. Spec. Publ.* **33**, 275–288.

Gibbs, A.K. (1987b) Contrasting styles of continental mafic intrusions in the Guiana Shield. In *Mafic Dyke Swarms* eds. Halls, H.C. and Fahrig, W.E. *Geol. Assoc. Can. Spec. Pap.* **34**, 457–465.

Gibbs, A.K. and Barron, C.N. (1983) The Guiana Shield reviewed. *Episodes* **2**, 7–14.

Gibbs, A.K. and Olszewski, W.J. Jr. (1982) Zircon U–Pb ages of Guyana greenstone-gneiss terrane. *Precamb. Res.* **17**, 199–214.

Gibbs, A.K., Wirth, K.R., Hirata, W.K. and Olszewski, W.J. Jr. (1986a) Age and composition of the Grão Pará Group volcanics, Serra dos Carajás, Pará, Brazil. *Rev. Bras. Geoc.* **16**, 201–211.

Gibbs, A.K., Montgomery, C.W., O'Day, P.A. and Erslev, E.A. (1986b) The Archaean-Proterozoic transition: Evidence from the geochemistry of metasedimentary rocks of Guyana and Montana. *Geochim. Cosmochim, Acta* **50**, 2125–2141.

Giles, C.W. and Hallberg, J.A. (1982) The genesis of the Archaean Welcome Well volcanic complex, Western Australia. *Contrib. Mineral. Petrol.* **80**, 307–318.

Gill, J.B. (1976) Composition and ages of Lau Basin and ridge volcanic rocks: implications for evolution of an inter-arc basin and remnant arc. *Geol. Soc. Am. Bull.* **87**, 1384–1395.

Gill, J.B. (1981) *Orogenic Andesites and Plate Tectonics,* Springer-Verlag, New York, 390 pp.

Gill, R.C.O. (1979) Comparative petrogenesis of Archaean and modern low-K tholeiites. A critical review of some geochemical aspects. In *Origin and Distribution of the Elements, 2,* ed. Ahrens, L.H. Pergamon, Oxford, (*Phys. Chem. Earth* **11**), 431–447.

Gill, R.C.O. and Bridgwater, D. (1979) Early Archaean basic magmatism in West Greenland: the geochemistry of the Ameralik dykes. *J. Petrol.* **20**, 695–726.

Gill, R.C.O., Bridgwater, D. and Allaart, J.H. (1981) The geochemistry of the earliest known basic

426   EARLY PRECAMBRIAN BASIC MAGMATISM

metavolcanic rocks, at Isua, West Greenland: a preliminary investigation. In *Archaean Geology* eds. Glover, J.E. and Groves, D.I. *Geol. Soc. Aust. Spec. Publ.* 7, 313–325.

Girardi, V.A.V. and Kurat, G. (1982) Precambrian mafic and ultramafic rocks of the Cana Brava Complex, Brazil – mineral compositions and evolution. *Rev. Bras. Geoc.* 12, 313–323.

Girardi, V.A.V., Kawashita, K., Basei, M.A.S. and Cordani, U.G. (1978) Algumas consideraçoes sobre a evolução geologica da região de Cana Brava a partir de dados geocronologicos. *Anais XXX Congr. Brasileiro Geol.* 1, 337–348.

Girardi, V.A.V., Rivalenti, G., Siena, F. and Sinigoi, S. (1981) Precambrian Barro Alto Complex of Goiás, Brazil: bulk geochemistry and phase equilibria. *Neus Jahr. Min Abh.* 142, 270–291.

Girardi, V.A.V., Rivalenti, G. and Sinigoi, S. (1986) The petrogenesis of the Niquelândia layered basic-ultrabasic complex, Central Goiás, Brazil. *J. Petrol.* 27, 715–744.

Glassley, W.E. and Sørensen, K. (1980) Constant $P_s - T$ amphibolite to granulite facies transition in Agto (West Greenland) metadolerites: implications and applications. *J. Petrol.* 21, 69–105.

Glikson, A.Y. (1971) Primitive Archaean element distribution patterns: chemical evidence and geotectonic significance. *Earth Planet. Sci. Lett.* 12, 309–320.

Glikson, A.Y. (1976) Archaean to early Proterozoic shield elements: relevance of plate tectonics *Geol. Assoc. Can. Spec. Pap.* 14, 489–516.

Glikson, A.Y. (1979) Early Precambrian tonalite-trondhjemite sialic nuclei. *Earth Sci. Rev.* 15, 1–73.

Glikson, A.Y. and Hickman, A. (1981) Geochemistry of Archaean volcanic successions, eastern Pilbara Block, Western Australia. *Bur. Min. Res. Geol. Geophys. Res.* 136, 83 pp.

Glikson, A.Y. and Jahn, B.M. (1985) REE and LIL elements, eastern Kaapvaal shield, South Africa: evidence of crustal evolution by 3-stage melting. In *Evolution of Archaean Supracrustal Sequences* eds. Ayres, L.D., Thurston, P.C., Card, K.D. and Weber, W. *Geol. Assoc. Can. Spec. Pap.* 28, 303–324.

Glikson, A.Y., Pride, C., Jahn, B., Davy, R. and Hickman, A.H. (1986) REE and HFS element evolution of Archaean mafic-ultramafic volcanic suites. Pilbara block, Western Australia. *Bur. Min. Res., Geol. Geophys. Res.* 1986/6.

Glover, J.E. (1971) ed., Symposium on Archaean Rocks. *Spec. Publ. Geol. Soc. Aust.* 3, 469 pp.

Glover, J.E. and Groves, D.I. (1981) eds. Archaean Geology. *Spec. Publ. Geol. Soc. Aust.* 7, 515 pp.

Goldich, S.S. and Wooden, J.L. (1980) Origin of the Morton gneiss, south-western Minnesota. Part III. Geochronology. *Geol. Soc. Am. Mem. Spec. Pap.* 182, 77–94.

Goldich, S.S., Wooden, J.L., Ankenbauer, G.A.Jr., Levy, T.M. and Suda, R.U. (1980) Origin of the Morton gneiss, south-western Minnesota. Part I. Lithology. *Geol. Soc. Am. Mem. Spec. Pap.* 182, 45–56.

Goldstein, S.L., O'Nions, R.K. and Hamilton, P.J. (1984) A Sm–Nd isotopic study of atmospheric dusts and particulates from major river systems. *Earth Planet. Sci. Lett.* 70, 221–236.

Gomes, C.B., Cordani, U.G. and Basei, M.A.S. (1975) Radiometric ages from the Serra dos Carajás area, northern Brazil. *Geol. Soc. Amer. Bull.* 86, 939–945.

Goodwin, A.M. (1977) Archaean volcanism in Superior Province, Canadian Shield. *Geol. Assoc. Can. Spec. Pap.* 16, 205–241.

Goodwin, A.M. (1981) Archaean plates and greenstone belts. In *Precambrian Plate Tectonics* ed. Kröner, A. Elsevier, Amsterdam, 105–135.

Gorbunov, G.I., Zagorodny, V.G. and Robonen, E.I. (1985) Main features of the geological history of the Baltic Shield and the epochs of ore formation. *Bull. Geol. Surv. Finland* 333, 3–41.

Gough, D.I. and van Niekerk, C.B. (1959) A study of the palaeomagnetism of the Bushveld gabbro. *Phil. Mag.* 4, 126–136.

Gould, D., Rathbone, P.A., Kimbell, G.S. and Burley, A.J. (1986) The Molopo Farms complex, Botswana – a possible target for Bushveld-type mineralisation? In *Metallogeny of Basic and Ultrabasic Rocks.* eds. Gallagher, M.J., Ixer, R.A., Neary, C.R. and Prichard, H.M. Inst. Min. Metall. London. 319–331.

Graff, P.J., Sears, J.W., Holden, G.S. and Hausel, W.D. (1982) Geology of the Elmers Rock greenstone belt, Laramie range, Wyoming. *Rep. Geol. Surv. Wyoming* 14, 23 pp.

Graham, R.H. (1980) The role of shear belts in the evolution of the South Harris igneous complex. *J. Struct. Geol.* 2, 29–37.

Granath, J.W. (1975) Wind River Canyon – an example of a greenstone belt in the Archaean of Wyoming, U.S.A. *Precamb. Res.* 2, 71–91.

Granger, H.C., McKay, E.J., Mattick, R.E., Patten, L.L. and McIlroy, P. (1971) Mineral resources of the Glacial primitive area Wyoming. *U.S. Geol. Surv. Bull.* **1319-F**, 113 pp.

Greeley, R. (1982) The Snake River plain, Idaho: representative of a new category of volcanism. *J. Geophys. Res.* **87**, 2705–2712.

Green, D.H. (1972) Archaean greenstone belts may include terrestrial equivalents of lunar maria? *Earth Planet. Sci. Lett.* **15**, 263–270.

Green, D.H. (1975) Genesis of Archaean peridotitic magmas and constraints on Archaean geothermal gradients and tectonics. *Geology* **3**, 15–18.

Green, N.L., (1975) Archaean glomeroporphyritic basalts. *Can. J. Earth Sci.* **12**, 1770–1784.

Green, A.G., Hajnal, Z., and Weber, W. (1985) An evolutionary model of the western Churchill Province and western margin of the Superior Province in Canada and the north-central United States. *Tectonophysics* **116**, 281–322.

Green, A.H. and Naldrett, A.J. (1984) The Langmuir volcanic peridotite-associated nickel deposit: Canadian equivalents of the Western Australian occurrences. *Econ. Geol.* **76**, 1503–1523.

Green, J.C. and Schulz, K.J. (1977) Iron-rich basaltic komatiites in the early Precambrian Vermilion district, Minnesota. *Can. J. Earth Sci.* **14**, 2181–2192.

Green, J.C. and Schulz, K.J. (1978) Iron-rich basaltic komatiites in the early Precambrian Vermilion district: Reply. *Can. J. Earth Sci.* **15**, 857–859.

Greenberg, J.K. and Brown, B.A. (1983) Lower Proterozoic volcanic rocks and their setting in the southern Lake Superior district. *Geol. Soc. Am. Mem.* **160**, 67–84.

Gresham, J.J. (1982) Kambalda nickel deposits. In *Regional geology and nickel deposits of the Norseman–Wiluna belt, Western Australia* eds. Groves, D.I. and Lesher, C.M. *Univ. West. Aust. Publ.* **7**, C18–C27.

Gresham, J.J. and Loftus Hills, G.D. (1981) The geology of the Kambalda Nickel field, Western Australia. *Econ. Geol.* **76**, 1373–1416.

Griffin, W.L., McGregor, V.R., Nutman, A., Taylor, P.N. and Bridgwater, D. (1980) Early Archaean granulite-facies metamorphism south of Ameralik, West Greenland. *Earth Planet. Sci. Lett.* **50**, 59–74.

Grobler, N.J. and Whitfield, G.G. (1970) The olivine-apatite magnetitites and related rocks in the Villa Nora occurrence of the Bushveld Igneous Complex. *Geol. Soc. S. Afr. Spec. Publ.* **1**, 208–227.

Grobler, N.J., Kleynhans, E.P.J., Botha, P.J. and De Bruiyn, H. (1982) Distinction between the lavas of the Allanridge andesites and Rietgat formations in the northern Cape and western Transvaal. *Trans. Geol. Soc. S. Afr.* **85**, 117–126.

Grobler, N.J., De Bruiyn, H., van der Westhuizen, W.A. and Schoch, A.E. (1986) Komatiitic affinities at the bottom and top of the late Archaean–early Proterozoic intracratonic Ventesdorp volcanic pile. *Geocongress '86. Johannesburg, Ext. Abs.*, 129–132.

Grout, F.F. (1918) The lopolith; an igneous form exemplified by the Duluth gabbro. *Amer. J. Sci.* **46**, 516–522.

Grove, T.L., Kinzler, R.J., Baker, M.B., Donnelly-Nolan, J.M. and Lescher, C.E. (1988) Assimilation of granite by basaltic magma at Burnt Lava flow, Medicine Lake volcano, northern California: decoupling of heat and mass tranfer. *Contrib. Mineral. Petrol.* **99**, 320–343.

Groves, D.I. (1988) Gold mineralization in the Yilgarn Block, Western Australia. *Bicent. Gold '88, Ext. Abs., Geol. Soc. Austr.* **22**, 13–18.

Groves, D.I. and Batt. W.D. (1984) Spatial and temporal variations of Archaean metallogenic associations in terms of evolution and granitoid–greenstone terrains with particular reference on the Western Australian Shield. In *Archaean Geochemistry*, eds. Kröner, A., Hanson, G.N. and Goodwin, A.M. Springer-Verlag, Berlin, 73–98.

Groves, D.I., Phillips. G.N., Ho, S.E. and Houstoun, S.M. (1985) The nature, genesis and regional controls of gold mineralization in Archaean greenstone belts of the Western Australian Shield: a brief review. *Trans. Geol. Soc. S. Afr.* **88**, 135–148.

Groves, D.I., Korkiakoski, E.A., McNaughton, N.J., Lesher, C.M. and Cowden, A. (1986) Thermal erosion by komatiites at Kambalda, Western Australia, and the genesis of nickel ores. *Nature* **319**, 136–139.

Groves, D.I., Philips, N., Ho, S.E., Houston, S.M. and Standing, C.A. (1987a) Craton scale distribution of Archaean greenstone belt gold deposits: predictive capacity of the metamorphic model. *Econ. Geol.* **82**, 2045–2058.

Groves, D.I., Korkiakoski, E.A., McNaughton, N.J., Lesher, C.M. and Cowden, A. (1987b) Thermal erosion by komatiites at Kambalda, Western Australia, and the genesis of nickel ores. *Nature* **319**, 136–139.

Gruau, G., Martin, H., Leveque, B. and Capdevila, R. (1985) Rb–Sr and Sm–Nd geochronology of Lower Proterozoic granite-greenstone terrains in French Guiana. *Precamb. Res.* **30**, 63–80.

Gruau, G., Arndt, N.T., Chauvel, C. and Jahn, B.M. (1986) Large scale compositional heterogeneity in the early Archaean mantle: Hf isotopic evidence. *Terra Cogn.* **6**, 245.

Gruau, G., Jahn, B.M., Glikson, A.Y., Davy, R., Hickman, A.H. and Chauvel, C. (1987) Age of the Archaean Talga–Talga Subgroup, Pilbara Block, Western Australia, and early evolution of the mantle: new Sm–Nd isotopic evidence. *Earth Planet. Sci. Lett.* **85**, 105–116.

Gruau, G., Tourpin, S., Jahn, B.M. and Anhaeusser, C.R. (1988) New geochemical and isotopic data for komatiites from the Onverwacht Group, southern Africa. *Chem. Geol.* **70**, 144.

Gubbins, D. and Richards, M. (1986) Coupling of the core dynamo and mantle; thermal or topographic? *Geophys. Res. Lett.* **13**, 1521–1524.

Hager, B.H. (1984) Subducted slabs and the geoid: constraints on mantle rheology and flow. *J. Geophys. Res.* **89**, 6003–6015.

Haggerty, S.E. (1981) Temperatures and gas fugacities of planetary basalts. In *Basaltic Volcanism on the Terrestrial Planets.* Pergamon, New York, 371–384.

Haggerty, S.E., Boyd, F.R., Bell, P.M., Finger, L.W. and Bryan, W.B. (1970) Opaque minerals and olivine in lavas and breccias from Mare Tranquillitatis. *Proc. Apollo 11 Lunar Sci. Conf.*, Pergamon, New York, (*Geochim. Cosmochim. Acta suppl.* 1), 513–538.

Hajash, Jr., A. (1984) Rare earth element abundances and distribution patterns in hydrothermally altered basalts: experimental results. *Contrib. Mineral. Petrol.* **85**, 409–412.

Halden, N.M. (1982) The structural, metamorphic and igneous history of migmatites in the deep levels of a wrench fault regime, Savonranta, central Finland. *Trans. R. Soc. Edin.* **73**, 17–30.

Hall, A.L. (1932) The Bushveld Igneous Complex of the Central Transvaal. *Geol. Surv. S. Afr. Mem.* **28**, 544 pp.

Hall, R.P. (1980) The tholeiitic and komatiitic affinities of the Malene metavolcanic amphibolites from Ivisârtoq, southern West Greenland. *Rapp. Grønlands Geol. Unders.* **97**, 20 pp.

Hall, R.P. (1982) Geochemistry of the Malene metavolcanic amphibolites from Ivisârtoq: significance to the Archaean stratigraphy of southern West Greenland. *Rapp. Grønlands Geol. Unders.* **110**, 68–72.

Hall, R.P. and Escher, J.C. (1989) A thrust contact between granulite and amphibolite facies gneisses, Niflheim, SE Greenland. *Rapp. Grønlands Geol. Unders.* **146**, 66–69.

Hall, R.P. and Friend, C.R.L. (1979) Structural evolution of the Archean rocks in Ivisârtoq and the neighboring inner Godthåbsfjord region, southern West Greenland. *Geology* **7**, 311–315.

Hall, R.P. and Hughes, D.J. (1986) A boninitic dyke in the eastern Sukkertoppen region: geochemistry of the boninitic-noritic dyke swarm of southern West Greenland. *Rapp. Grønlands Geol. Unders.* **130**, 44–52.

Hall, R.P. and Hughes, D.J. (1987) Noritic dykes of southern West Greenland: early Proterozoic boninitic magmatism. *Contrib. Mineral. Petrol.* **97**, 169–182.

Hall, R.P., Hughes, D.J. and Friend, C.R.L. (1985) Geochemical evolution and unusual pyroxene chemistry of the MD tholeiite dyke swarm from the Archaean craton of southern West Greenland. *J. Petrol.* **26**, 253–282.

Hall, R.P., Hughes, D.J. and Friend, C.R.L. (1986) Complex sequential pyroxene growth in tholeiitic hypabyssal rocks from southern West Greenland. *Mineralog. Mag.* **50**, 491–502.

Hall, R.P., Hughes, D.J. and Friend, C.R.L. (1987a) Mid-Archaean basic magmatism of southern West Greenland. In *Evolution of the Lewisian and Comparable Precambrian High Grade Terrains* eds. Park, R.G. and Tarney, J. *Spec. Publ. Geol. Soc. Lond.* **27**, 261–275.

Hall, R.P., Hughes, D.J., Friend, C.R.L. and Snyder, G.L. (1987b) Proterozoic mantle heterogeneity: geochemical evidence from contrasting basic dykes. In *Geochemistry and Mineralization of Proterozoic Volcanic Suites* eds. Pharaoh, T.C., Beckinsale, R.D. and Rickard, D. *Spec. Publ. Geol. Soc. Lond.* **33**, 9–21.

Hall, R.P., Hughes, D.J. and Friend. C.R.L. (1987c), Ti-rich plagioclase-phyric dykes of southern West Greenland. *Rapp. Grønlands Geol. Unders.* **135**, 46–52.

Hall. R.P., Hughes, D.J. and Joyner, L. (1988a) Complex pyroxene assemblages of Proterozoic dolerites, S.E. Greenland. *Mineralog. Mag.* **52**, 703–705.

Hall, R.P., Hughes, D.J. and Joyner, L. (1988b) Fe-enrichment in tholeiitic pyroxenes: complex two-pyroxene assemblages in Mesozoic dolerites, southern Tasmania. *Geol. Mag.* **125**, 573–582.

Hall, R.P., Escher, J.C., Chadwick, B. and Vasudev, V.N. (1989a) Supracrustal rocks in the Ammassalik region, South-East Greenland. *Rapp. Grønlands Geol. Unders.* **146**, 17–22.

Hall, R.P., Hughes, D.J. and Joyner, L. (1989b) Basic dykes of the southern Ammassalik region: preliminary mineralogical and geochemical results. *Rapp. Grønlands Geol. Unders.* **146**, 79–82.

Hallberg, J.A. (1972) Geochemistry of Archaean volcanic belts in the Eastern Goldfields region of Western Australia. *J. Petrol.* **13**, 45–56.

Hallberg, J.A. (1986) Archaean basin development and crustal extension in the northeastern Yilgarn Block, Western Australia. *Precamb. Res.* **31**, 133–156.

Hallberg, J.A. (1987) Postcratonization mafic and ultramafic dykes of the Yilgarn Block. *Aust. J. Earth Sci.* **34**, 135–149.

Hallberg, J.A. and Giles, C.W. (1986) Archaean felsic volcanism in the northeastern Yilgarn Block, Western Australia. *Aust. J. Earth Sci.* **33**, 413–428.

Hallberg, J.A. and Williams, D.A.C. (1972) Archaean mafic and ultramafic rock associations in the Eastern Goldfields region, Western Australia. *Earth Planet. Sci. Lett.*, **15**, 191–200.

Hallberg, J.A., Johnston, E. and Bye, S.M. (1976a) The Archaean Marda Complex, Western Australia. *Precamb. Res.* **3**, 111–136.

Hallberg, J.A., Carter, D.N. and West, K.N. (1976b) Archaean volcanism and sedimentation near Meekathara, Western Australia. *Precamb. Res.* **3**, 577–595.

Halls, H.C. and Fahrig, W.F. eds. (1987) Mafic Dyke Swarms. *Geol. Assoc. Can. Spec. Pap.* **34**, 503 pp.

Hamilton, P.J. (1977) Sr-isotope and trace element studies of the Great Dyke and Bushveld mafic phase and their relation to early Proterozoic magma genesis in southern Africa. *J. Petrol.* **18**, 24–52.

Hamilton, P.J., O'Nions, R.K. and Evensen, N.M. (1977) Sm–Nd dating of Archaean basic and ultrabasic volcanics. *Earth Planet. Sci. Lett.* **36**, 263–268.

Hamilton, P.J., Evensen, N.H., O'Nions, R.K., Smith, H.S. and Erlank, A.J. (1979) Sm–Nd dating of Onverwacht Group volcanics, southern Africa. *Nature* **279**, 298–300.

Hamilton, P.J., O'Nions, R.K., Bridgwater, D. and Nutman, A.P. (1983) Sm–Nd studies of Archaean metasediments and metavolcanics from West Greenland and their implications for the Earth's early history. *Earth Planet. Sci. Lett.* **62**, 263–272.

Hamlyn, P.R., Keays, R.R., Cameron, W.E., Crawford, A.J. and Waldron. H.M. (1985) Precious metals in magnesian low-Ti lavas: Implications for metallogenesis and sulfur saturation in primary magmas. *Geochim. Cosmochim. Acta* **49**, 1797–1811.

Hammond, R.L., Nisbet, B.W., Etheridge, M.A. and Wall, V.J. (1984) The Litchfield Complex, northwest Northern Territory: Archaean basement or Proterozoic cover. *Aust. J. Earth. Sci.* **31**, 485–496.

Hansen, W.R. (1965) Geology of the Flaming Gorge area, Utah–Colorado–Wyoming. *U.S. Geol. Surv. Prof. Pap.* **190**, 196 pp.

Hansen, B.T., Higgins, A.K. and Borchardt, B. (1987) Archaean U–Pb zircon ages from the Scorsby Sund region, East Greenland. *Rapp. Grønlands Geol. Unders*, **134**, 19–24.

Harding, R.R., Crocket, R.N. and Shelling, N.J. (1974) The Gaberone granite, Kanye volcanics and the Ventesdoro Plantation porphyry, Botswana: geochronology and review *Rep. Inst. Geol. Sci.* **75/5**, 1–26.

Harley, S.L. and Black, L.P. (1987) The Archaean geological evolution of Enderby Land, Antarctica. In *Evolution of the Lewisian and Comparable Precambrian High Grade Terrains* eds. Park, R.G. and Tarney, J. *Spec. Publ. Geol. Soc. Lond.* **27**, 285–296.

Harmer, R.E. and Sharpe, M.R. (1985) Field relations and strontium isotope systematics of the eastern Bushveld Complex. *Econ. Geol.* **80**, 813–837.

Harper, G.D. and Link, P.K. (1986) Geochemistry of Upper Proterozoic rift-related volcanics, northern Utah and southeastern Idaho. *Geology* **14**, 864–867.

Harris, N.B.W., Hawkesworth, C.J. and Ries, A.C. (1984) Crustal evolution in northeast and east Africa from model Nd ages. *Nature* **309**, 773–776.

Harrison, N.M. (1970) The geology of the country around Que Que. *Bull. Geol. Surv. Rhod.* **65**, 125 pp.

Harrison, P.H. (1984) The mineral potential of layered igneous complexes within the Western Gneiss Terrain. *Geol. Surv. W. Aust. Rep.* **19**, 37–54.

Harrison, J.E. and Peterman, Z.E. (1984) Introduction to correlation of Precambrian rock sequences: correlation of Precambrian rocks of the United States and Mexico. *U.S. Geol. Surv. Prof. Pap.* **1241-A**, 7pp.

Hartmann, W.K. (1980) Dropping stones in magma oceans: effects of early lunar cratering. *Proc. Conf. Lunar Highland Crust*, Pergamon, New York, 155–171.

Hartmann, W.K. (1986) Moon origin: the impact-trigger hypothesis. In *Origin of the Moon.* eds. Hartmann, W.K., Phillips, R.J. and Taylor, G.J. Lunar Planet. Inst., Houston, 579–608.

Hasui, Y and de Almeida, F.F.M. (1985) The Central Brazil Shield reviewed. *Episodes* **8**, 1, 29–37.

Hasui, Y., D'El Rey Silva, L.J.H., Lima e Silva, F.J., Mandetta, P., Moraes, J.A.C., Oliveira, J.G. and Miola, W. (1982) Geology and copper mineralization of Curaca River valley, Bahia. *Rev. Bras. Geoc.* **12**, 463–474.

Hattingh, P.J. (1980) The structure of the Bushveld Complex in the Groblersdal–Lydenburg–Belfast area of the eastern Transvaal as interpreted from a regional gravity survey. *Trans. Geol. Soc. S. Afr.* **83**, 125–133.

Hattingh, P.J. (1986) The palaeomagnetism of the main zone of the eastern Bushveld Complex. *Tectonophysics* **124**, 271–295.

Hatton, C.J. (1988) Densities and liquidus temperatures of Bushveld parental magmas as constraints on the origin of Merensky Reef. In *Magmatic Sulphides – the Zimbabwe Volume: Proc. 5th Magmat. Sulph. Field Conf.*, eds. Prendergast, M.D. and Jones, M.D. Inst Min. Metall. Lond., 87–93.

Hatton, C.J. and Sharpe, M.R. (1989) Significance of the origin of boninite-like rocks associated with the Bushveld Complex. In *Boninites.* ed. Crawford, A.J. Unwin Hyman, London, 174–207.

Hatton, C.J. and Von Gruenewaldt, G. (1987) The geological setting and petrogenesis of the Bushveld chromitite layers. In *Evolution of chromium ore fields*, ed. Stowe, C.W. Van Nostrand Rheinold, New York, 109–143.

Hatton, C.J., Harmer, R.E. and Sharpe, M.R. (1986) Petrogenesis of the middle group of chromitite layers. Doornvlei, eastern Bushveld Complex. In *Metallogeny of Basic and Ultrabasic Rocks.* eds. Gallagher, M.J., Ixer, R.A., Neary, C.R. and Prichard, H.M. Inst. Min. Metall. Lond., 241–247.

Haughton, D.R., Roeder, P.L. and Skinner, B.J. (1974) Solubility of sulphur in mafic magmas. *Econ. Geol.* **69**, 451–467.

Hausel, W.D., Graff, P.J. and Albert, K.G. (1985) Economic geology of the Copper Mountain supracrustal belt, Owl Creek Mountains, Fremont County, Wyoming. *Geol. Surv. Wyo. Rep. Invest.* **28**, 1–33.

Hawkes, D.D. (1966a) The petrology of the Guiana dolerites. *Geol. Mag.* **103**, 320–335.

Hawkes, D.D. (1966b) Differentiation of the Tumatumari–Kopinang dolerite intrusion, British Guiana. *Geol. Soc. Amer. Bull.* **77**, 1131–1158.

Hawkesworth, C.J. and O'Nions, R.K. (1977) The petrogenesis of some Archaean volcanic rocks from southern Africa. *J. Petrol.* **18**, 487–520.

Hawkesworth, C.J., Moorbath, S., O'Nions, R.K. and Wilson, J.F. (1975) Age relationships between greenstone belts and "granites" in the Rhodesian Archaean craton. *Earth Planet. Sci. Lett.* **25**, 251–262.

Hawkesworth, C.J., Gledhill, A.R., Wilson, J.F. and Orpen, J.L. (1979) A 2.9 b.y. event in the Rhodesian Archaean. *Earth Planet. Sci. Lett.* **43**, 285–297.

Hawkesworth, C.J., Erlank, A.J., Marsh, J.S., Menzies, M.A. and Calsterea, P.V. (1983) Evolution of the continental lithosphere: Evidence from volcanics and xenoliths in southern Africa. In *Continental Basalts and Mantle Xenoliths.* eds. Hawkesworth, C.J. and Norry, M.J., Shiva Press, Nantwich, 111–138.

Hay, W.W., Barron, E.J., Sloan, J.L. and Southam, J.R. (1981) Continental drift and the global pattern of sedimentation. *Geol. Runds.* **70**, 302–315.

Head, J.W.III. (1976) Lunar volcanism in space and time. *Rev. Geophys. Space Phys.* **14**, 265–300.

Head, J.W., Pieters, C., McCord, T., Adams, J. and Zisk, S. (1978) Definition and detailed characterization of lunar surface units using remote observations. *Icarus* **33**, 145–172.

Heaman, L.M. and Tarney, J. (1989) U–Pb baddeleyite ages for the Scourie dyke swarm, Scotland: evidence for two distinct intrusion events. *Nature* **340**, 705–708.

Hebeda, E.H., Boelrijk, N.A.I.M., Priem, H.N.A., Verdurmen, E.A.T. and Verschure, R.H. (1973) Excess radiogenic argon in the Precambrian Avanavero dolerite in western Surinam. *Earth Planet. Sci. Lett.* **20**, 189–200.

Hedge, C.E., Stacy, J.S. and Bryant, B. (1983) Geochronology of the Farmington Canyon complex,

Wasatch Mountains, Utah. In *Tectonic and stratigraphic studies in the northeastern Great Basin* eds. Miller, D.M., Howard, K.A. and Todd, V.R. *Geol. Soc. Amer. Mem.* **157**, 37–44.

Hegner, E., Tegtmeyer, A. and Kröner, A. (1981) Geochemie und petrogenese archäischer vulkanite der Pongola-Gruppe in Natal. *Chem. Erde* **40**, 23–57.

Hegner, E., Kröner, A. and Hofmann, A.W. (1984) Age and isotope geochemistry of the Archaean Pongola and Usushwana suites in Swaziland, southern Africa: a case for crustal contamination of mantle-derived magma. *Earth Planet. Sci. Lett.* **70**, 267–279.

Heimlich, R.A. and Banks, P.O. (1968) Radiometric age determinations, Bighorn Mountains, Wyoming. *Amer. J. Sci.* **266**, 180–192.

Heimlich, R.A. and Manzer, G.K. Jr. (1973) Flow differentiation within leopard rock dikes, Bighorn Mountains, Wyoming. *Earth Planet. Sci. Lett.* **17**, 350–356.

Heimlich, R.A., Nelson, G.C. and Gallagher, G.L. (1973) Metamorphosed mafic dikes from the southern Bighorn Mountains, Wyoming. *Geol. Soc. Amer. Bull.* **84**, 1439–1450.

Heimlich, R.A., Gallagher, G.L. and Shotwell, L.B. (1974) Quantitative petrography of mafic dikes from the central Bighorn Mountains, Wyoming. *Geol. Mag.* **111**, 97–107.

Helmstaedt, H., Padgham, W.A. and Brophy, J.A. (1986) Multiple dikes in lower Kam Group. Yellowknife greenstone belt: evidence for Archaean sea-floor spreading? *Geology* **14**, 562–566.

Helz, R.T. (1985) Composition of fine-grained mafic rocks from sills and dikes associated with the Stillwater Complex. In *Stillwater Complex* eds. Czamanske, G.K. and Zientek, M.L. *Mont. Bur. Mines Geol. Spec. Publ.* **92**, 97–117.

Helz, R.T. (1987) Evidence for melt extraction from the sills and dikes associated with the Stillwater Complex. *Geol. Soc. Amer. Abs. Prog.* **19**, 699.

Henderson, J.B. (1985) Geology of the Yellowknife-Hearne Lake area, District of Mackenzie: segment across an Archean basin. *Geol. Survey Can. Mem.* **414**, 135 pp.

Henderson, P., Fishlock, S.J., Laul, J.C., Cooper, T.D., Conard, R.L., Boynton, W.V. and Schmitt, R.A. (1976) Rare earth element abundances in rocks and minerals from the Fiskenaesset complex, West Greenland. *Earth Planet. Sci. Lett.* **30**, 37–49.

Henley, K.J. (1975) Gold-ore mineralogy and its relation to metallurgical treatment. *Minerals Sci. Eng.* **7**, **4**, 289–312.

Henley, R.W. (1984) Metals in hydrothermal fluids. In *Fluid-Mineral Equilibria in Hydrothermal Systems.* eds. Henley, R.W., Truesdell, A.H. and Barton, P.B. *Rev. Econ. Geol. 1, Soc. Econ. Geol.*, 115–127.

Henriksen, H. (1983) Komatiitic chlorite-amphibole rocks and mafic metavolcanics from the Karasjok greenstone belt, Finnmark, northern Norway. *Bull. Norges Geol. Unders.* **201**, 107pp.

Henriksen, N. (1969) Chemical relations between metabasaltic lavas and metadolerites in the Ivigtut area, South-West Greenland. *Meddr. Dansk Geol. Foren.* **19**, 27–50.

Henry, D.J., Mueller, P.A., Wooden, J.L., Warner, J.L. and Lee-Berman, R. (1982), Granulite grade supracrustal assemblages of the Quad Creek area, eastern Beartooth Mountains, Montana. In *Precambrian Geology of the Beartooth Mountains, Montana and Wyoming* eds. Mueller, P.A. and Wooden, J.L. *Mont. Bur. Mines Geol. Spec. Publ.* **84**, 147–159.

Hertogen, J., Janssens, M.J. and Palme, H. (1980) Trace elements in ocean ridge basalt glasses: implications for fractionation during mantle evolution and petrogenesis. *Geochim. Cosmochim. Acta* **44**, 2125–2143.

Herzberg, C.T. (1987) Magma density at high pressure Part 2: A test of the olivine flotation hypothesis. In *Magmatic Processes: Physicochemical Principles* ed. Mysen, B. *Geochem. Soc. Spec. Publ.* **1**, 47–58.

Herzberg, C.T. and O'Hara, M.J. (1985) Origin of mantle peridotite and komatiite by partial melting. *Geophys. Res. Lett.* **12**, 541–544.

Herzberg, C.T. and Ohtani, E. (1988) Origin of komatiite at high pressures. *Earth Planet. Sci. Lett.* **88**, 321–329.

Hess, H.H. (1960) Stillwater Igneous Complex, Montana – a quantitative mineralogical study. *Geol. Soc. Amer. Mem.* **80**, 230 pp.

Hewitt, J.M., McKenzie, D.P. and Weiss, N.O. (1980) Large aspect ratio cells in two-dimensional thermal convection. *Earth Planet. Sci. Lett.* **51**, 370–380.

Hickey, R.L. and Frey, F.A. (1982) Geochemical characteristics of boninite series volcanics: implications for their source. *Geochim. Cosmochim. Acta* **46**, 2099–2115.

Hickman, A.H. (1981) Crustal evolution of the Pilbara Block. In *Archaean Geology* eds. Glover, J.E. and Groves, D.I. *Spec. Publ. Geol. Soc. Aust.* **7**, 57–69.

Hickman, A.H. (1983) Geology of the Pilbara Block and its environs. *Bull. Geol. Surv. West. Aust.* **127**, 268pp.

Hickman, M.H. (1974) 3,500 Myr-old granite in southern Africa. *Nature* **251**, 295–296.

Hickman, M.H. (1976) Geochronological investigations in the Limpopo belt and part of the adjacent Rhodesian craton. *Ann. Rep. Res. Inst. Geol., Univ. Leeds* **20**, 30.

Hiemstra, S.A. (1979) The role of collectors in the formation of the platinum deposits in the Bushveld Complex. *Can. Mineral.* **17**, 469–482.

Hietanen, A. (1975) Generation of potassium-poor magmas in the northern Sierra Nevada and the Svecofennian in Finland. *U.S. Geol. Surv. J. Res.* **3**, 631–645.

Higgins, A.K. (1968) The Tartoq Group on Nuna qaqortoq and in the Iterdlak area, South-West Greenland. *Rapp. Grønlands Geol. Unders.* **17**, 17 pp.

Higgins, A.K. (1970) The stratigraphy and structure of the Ketilidian rocks of Midternaes, South-West Greenland. *Bull. Grønlands Geol. Unders.* **87**, 96 pp.

Hilde, T.W.C. (1983) Sediment subduction versus accretion around the Pacific. *Tectonophysics* **99**, 381–397.

Hill, R. and Roeder, P. (1974) The crystallisation of spinel from basaltic liquid as a function of oxygen fugacity. *J. Geol.* **82**, 709–729.

Hill, R.E.T., Cole, M.J. and Barnes, S.J. (1987) Physical volcanology of komatiites. *Geol. Soc. Aust. Excur. Guide Book* **1**, Perth, W. Aust., 74 pp.

Hills, F.A. and Armstrong, R.L. (1974) Geochronology of Precambrian rocks in the Laramine Range and implications for the tectonic framework of Precambrian southern Wyoming. *Precamb. Res.* **1**, 213–225.

Hirata, W.K., Rigon, J.C., Kadekaru, K., Cordeiro, A.A.C. and Meirelles, E.de M. (1982) Geologia regional da Provincia Mineral de Carajas. In *I Simposio de Geologia da Amazonia*, Belem. SBG/Nucleo Norte, 100–110.

Ho, S.E. (1987) Fluid inclusions: their potential as an exploration tool for Archaean gold deposits. In *Recent Advances in Understanding Precambrian Gold Deposits* eds. Ho, S.E. and Groves, D.I. *Geol. Dept. Univ. W. Austr. Publ.* **11**, 239–263.

Ho, S.E., Groves, D.I. and Philips, G.N. (1985) Fluid inclusions as indicators of the nature and source of ore fluids and ore depositional conditions for Archaean gold deposits of the Yilgarn Block, Western Australia. *Trans. Geol. Soc. S. Afr.* **88**, 149–158.

Hoatson, D.M. (1984) Potential for platinum group mineralisation in Australia – a review. *Bur. Min. Resour. Canberra Rec.* **1984/1**, 74 pp.

Hoatson, D.M. and Keays, R.R. (1987) Platinum group geochemistry of the Munni Munni layered intrusion, west Pilbara Block, Western Australia. *Proc. Symp. Platinif. Horizons in Lay. Intrus.*, Univ. New South Wales, (abs).

Hodgson, C.J. (1983) Preliminary report on a computer file of gold deposits of the Abitibi belt, Ontario. In *The Geology of Gold in Ontario*. ed. Colvine, A.C. *Ont. Geol. Surv. Misc. Pap.* **110**, 11–37.

Hodgson, C.J. (1986) Place of gold ore formation in the geological development of Abitibi greenstone belt. Ontario, Canada. *Trans. Inst. Min. Metall.* (**B**), *Appl. Earth Sci.* **95**, B183–B194.

Hodgson, C.J. and MacGeehan, P.J. (1982) Geological characteristics of gold deposits in the Superior Province of the Canadian Shield. In *Geology of Canadian Gold Deposits*, ed. Hodder, R.W. *CIM Spec. Vol.* **24**, 211–229.

Hoffman, P.F. (1988) United Plates of America, the birth of a craton: early Proterozoic assembly and growth of Laurentia. *Ann. Rev. Earth Planet. Sci.* **16**, 543–603.

Hofmann, A.W. (1986) Nb in Hawaiian magmas: constraints on source composition and evolution. *Chem. Geol.* **57**, 17–30.

Hofmann, A.W. and White, W.M. (1982) Mantle plumes from ancient oceanic crust. *Earth Planet. Sci. Lett.* **57**, 421–436.

Hofmann, A.W., Jochum, K.P., Seufert, M. and White, W.M. (1986) Nb and Pb in oceanic basalts: new constraints on mantle evolution. *Earth Planet. Sci. Lett.* **79**, 33–45.

Hofmann, H.J., Thurston, P.C. and Wallace, H. (1985) Archean stromatolites from Uchi Greenstone belt, northwestern Ontario. In *Evolution of Archaean Supracrustal Sequences* eds. Ayres, L.D., Thurston, P.C., Card, K.D. and Weber, W. *Geol. Assoc. Can. Spec. Paper* **28**, 125–132.

Holden, G.S. and Snyder. G.L. (1983) Compositional variation of mafic rocks from an Archean granite-greenstone terrane, central Laramie Range, Wyoming. *Geol. Soc. Amer. Abs. Prog.* **15**, 423.

Holst, T.B. (1984) Evidence for nappe development during the early Proterozoic Penokean orogeny, Minnesota. *Geology* **12**, 135–138.

Honkamo, M. (1985) On the Proterozoic metasedimentary rocks of the northern Pohjanmaa schist area, Finland. *Bull. Geol. Surv. Finland* **331**, 117–129.

Honkamo, M. (1987) Geochemistry and tectonic setting of early Proterozoic volcanic rocks in northern Ostrobothnia, Finland. In *Geochemistry and Mineralization of Proterozoic Volcanic Suites* eds. Pharaoh, T.C., Beckinsale, R.D. and Rickard, D. *Spec. Publ. Geol. Soc. Lond.* **33**, 59–88.

Hooper, P.R. (1985) A case of magma mixing in the Columbia River basalt group: the Wilbur Creek, Lapwai and Asotin flows, Saddle Mountain formation. *Contrib. Mineral. Petrol.* **91**, 66–73.

Hopwood, T. (1981) The significance of pyritic black shales in the genesis of Archaean nickel sulphide deposits. In *Handbook of strata-bound and stratiform ore deposits*, **9**. eds. Wolf, K.H. Elsevier, 411–467.

Horan, M.F., Hanson, G.N. and Spencer. K.J. (1987) Pb and Nd isotope and trace element constraints on the origin of basic rocks in an early Proterozoic igneous complex, Minnesota. *Precamb. Res.* **37**, 323–342.

Hörmann, P.K., Raith, M., Raase, P., Ackmand, D. and Seifert, F. (1980) The granulite complex of Finnish Lapland – petrology and metamorphic conditions in the Ivalojoki–Inarijarvi area. *Bull. Geol. Surv. Finland* **3**, 95 pp.

Houseman, G.A. and McKenzie, D.P. (1982) Numerical experiments on the onset of convective instability in the Earth's mantle. *Geophys. J.R. Astron. Soc.* **68**, 133–164.

Houston, R.S. and others. (1978) A regional study of rocks of Precambrian age in that part of the Medicine Bow Mountains lying in southeastern Wyoming – with a chapter on the relationship between Precambrian and Laramide structure. *Geol. Surv. Wyo. Mem.* **1** (2nd print.), 167 pp.

Houston, R.S., Reed, J.C., Karlstrom, K.E., Erslev, E.A., Snyder, G.L., Worl, R.G., Bryant, B., Reynolds, M.W., Peterman, Z.E., Page, N.J., Zientek, M.L. and Frost, C.D. (1988) The Wyoming Province. In *Precambrian, Conterminous United States; Vol. C-2, Geology of North America; Decade of North American Geology*, eds. Reed, J.C., Silver, L.T., Sims, P.K., Rankin, D.W., Houston, R.S. and Reynolds, M.W. Geol. Soc. Amer. in press.

Howie, R.A. (1955) The geochemistry of the charnockite series of Madras, India. *Trans. R. Soc. Edinb.* **62**, 725–768.

Huang, C.C. (1978) An outline of the tectonic characteristics of China. *Eclogae Geol. Helv.* **71**, 611–635.

Huang, X., Bi, Z and DePaolo, D.J. (1986) Sm–Nd isotope study of early Archaean rocks, Qianan, Hebei Province, China. *Geochim. Cosmochim. Acta* **50**, 625–631.

Hubbard, N.J. and Gast, P.W. (1971) Chemical composition and origin of non mare lunar basalts. *Proc. 2nd Lunar Sci. Conf.*, MIT Press, Cambridge, Mass., 999–1020.

Hubbard, N.J., Rhodes, J.M., Gast, P.W., Bansal, B.M., Shih, C.Y., Wiesmann, H. and Nyquist, L.E. (1973) Lunar rock types: the role of plagioclase in non-mare and highland rock types. *Proc. 4th Lunar Sci. Conf.*, Pergamon, New York, 1297–1312.

Hubbard, N.J., Rhodes, J.M., Wiesmann, H., Shih, C.Y. and Bansal, B.M., (1974) The chemical definition and interpretation of rock types returned from the non-mare regions of the Moon. *Proc. 5th Lunar Sci. Conf.*, Pergamon, New York, 1227–1246.

Hughes, C.J. (1976) Parental magma of the Great Dyke of Southern Rhodesia – voluminous late Archaean high magnesium basalt. *Trans. Geol. Soc. S. Afr.* **79**, 179–182.

Huhma, H. (1984) Nd-isotopic composition and age of Proterozoic basalts from Northern Finland. (Abstract). *Terra Cogn.* **4**, 192.

Hulbert, L.J. and Von Gruenewaldt, G. (1985) Textural and compositional features of chromite in the lower and critical zones of the Bushveld Complex south of Potgietersrus. *Econ. Geol.* **80**, 872–895.

Hulbert, L.J., Duke, J.M., Eckstrand, O.R., Lydon, J.W., Scoates, R.F.J., Cabri, L.J. and Irvine, T.N. (1988) Geological environments of the platinum group elements. *Geol. Surv. Canada Open File Rep.* **1440**, 148 pp.

Hunter, D.R. (1970a) The geology of the Usushwana Complex in Swaziland. *Geol. Soc. S. Afr. Spec. Publ.* **1**, 645–660.

Hunter, D.R. (1970b) The Ancient Gneiss complex in Swaziland. *Trans. Geol. Soc. S. Afr.* **73**, 107–153.

Hunter, D.R. (1974) Crustal development in the Kaapvaal craton. 1 The Archaean. *Precamb. Res.* **1**, 259–294.

Hunter, D.R. (1981) ed., *Precambrian of the Southern Hemisphere.* Elsevier, Amsterdam, 882 pp.

Hunter, D.R. and Wilson, A.H. (1988) A continuous record of Archaean evolution from 3.5 Ga to 2.6 Ga in Swaziland and northern Natal. *S. Afr. J. Geol.* **91**, 57–74.

Hunter, D.R., Barker, F. and Millard, H.T. (1984) Geochemical investigations of Archaean bimodal and Dwalile metamorphic suites, Ancient Gneiss Complex, Swaziland. *Precamb. Res.* **24**, 131–155.

Hunter, R.H. (1987) Textural equilibrium in layered igneous rocks. In *Origins of Igneous Layering.* ed. Parsons, I. Reidel, Dordrecht, 473–503.

Huppert, H.E. and Sparks, R.S.J. (1980) The fluid dynamics of a basaltic magma chamber replenished by influx of hot dense ultrabasic magma. *Contrib. Mineral. Petrol.* **75**, 279–289.

Huppert, H.E. and Sparks, R.S.J. (1985a) Cooling and contamination of mafic and ultramafic magmas during ascent through continental crust. *Earth Planet. Sci. Lett.* **74**, 371–386.

Huppert, H.E. and Sparks, R.S.J. (1985b) Komatiites I: eruption and flow. *J. Petrol.* **26**, 694–725.

Huppert, H.E., Sparks, R.S.J., Turner, J.S. and Arndt, N.T. (1984a) Emplacement and cooling of komatiite lavas. *Nature* **309**, 19–22.

Huppert, H.E., Sparks, R.S.J. and Turner, J.S. (1984b) Some effects of viscosity on the dynamics of magma chamhers. *J. Geophys. Res.* **89**, 6857–6877.

Huppert, H.E., Sparks, R.S.J. Wilson, R.J. and Hallworth, M.A. (1986) Cooling and crystallization at an inclined plane. *Earth Planet. Sci. Lett.* **72**, 319–328.

Huppert, H.E., Sparks, R.S.J., Wilson, J.R. Hallworth, M.A. and Leitch, A.M. (1987) Laboratory experiments with aqueous solutions modelling magma chamber processes. II. Cooling and crystallization along inclined planes. In *Origins of Igneous Layering.* ed. Parsons, I. Reidel, Dordrecht, 539–568.

Hurley, P.M., Melcher, G.C., Pinson, W.H. Jr. and Fairbairn, H.W. (1968) Some orogenic episodes in South America by K–Ar and whole-rock Rb–Sr dating. *Can. J. Earth Sci.* **5**, 633–638.

Inda, H.A.V. and Barbosa, J.F. (1978) Texto explicativo para o mapa geologico do Estado da Bahia, 1:1.000 000 *SME/CPM*, Salvador, Brazil, 122 pp.

Irvine, T.N. (1967) Chromium spinel as a petrogenetic indicator. Pt 2. Petrologic applications. *Can. J. Earth Sci.* **4**, 71–103.

Irvine, T.N. (1975) Crystallisation sequences in the Muskox intrusion and other layered intrusions—II. Origin of chromitite layers and similar deposits of other magmatic ores. *Geochim. Cosmochim, Acta* **39**, 991–1020.

Irvine, T.N. (1976) Chromite crystallization in the join Mg_2SiO_4–$CaMgSi_2O_6$–$CaAl_2Si_2O_8$–$MgCr_2O_4$–SiO_2. *Carnegie Inst. Wash. Yearb.* **76**, 465–472.

Irvine, T.N. (1977) Origin of chromitite layers in the Muskox intrusion and other layered intrusions: a new interpretation. *Geology* **5**, 273–277.

Irvine, T.N. (1979) Rocks whose composition is determined by crystal accumulation and sorting. In *The Evolution of Igneous Rocks.* ed. Yoder, H.S. Princeton Univ. Press, Princeton, 245–306.

Irvine, T.N. (1980) Magmatic infiltration metasomatism, double-diffusive fractional crystallisation, and adcumulus growth in the Muskox intrusion and other layered intrusions. In *Physics of Magmatic Processes* ed. Hargraves, R.B., Princeton Univ. Press, 325–383.

Irvine, T.N. (1982) Terminology for layered intrusions. *J. Petrol.* **23**, 127–162.

Irvine, T.N. and Baragar, W.R.A. (1971) A guide to the chemical classification of the common vocanic rocks. *Can. J. Earth Sci.* **8**, 523–548.

Irvine, T.N. and Sharpe, M.R. (1982) Source-rock compositions and depth of origin of Bushveld and Stillwater magmas. *Carnegie Inst. Wash. Yearb.* **81**, 294–303.

Irvine, T.N. and Sharpe, M.R. (1986) Magma mixing and the origin of stratiform oxide ore zones in the Bushveld and Stillwater Complexes. In *Metallogeny of Basic and Ultrabasic Rocks.* eds. Gallagher, M.J., Ixer, R.A., Neary, C.R. and Prichard, H.M., Inst. Min. Metall., 183–198.

Irvine, T.N. and Smith, C.H. (1967) The ultramafic rocks of the Muskox intrusion. In *Ultramafic and Related Rocks.* ed. Wyllie, P.J., Wiley, New York, 38–49.

Irvine, T.N. and Smith, C.H. (1969) Primary oxide minerals in the layered series of the Muskox intrusion. *Econ. Geol. Mon.* **4**, 76–94.

Irvine, T.N., Keith, D.W. and Todd, S.G. (1983) The J–M platinum-palladium reef of the

Stillwater Complex, Montana. II. Origin by double-diffusive convective magma mixing and implications for the Bushveld Complex. *Econ. Geol.* **78**, 1287–1334.

Irving, A.J. (1977) Chemical variation and fractionation of KREEP basalt magmas. *Proc. 8th Lunar Sci. Conf.*, Pergamon, New York, (*Geochim. Cosmochim. Acta suppl.* **8**), 2433–2488.

Irving, A.J. (1978) A review of experimental studies of crystal/liquid trace element partitioning. *Geochim. Cosmochim. Acta* **42**, 743–770.

Jackson, E.D. (1961) Primary textures and mineral associations in the ultramafic zone of the Stillwater Complex, Montana. *U.S. Geol. Surv. Prof. Pap.* **358**, 106 pp.

Jackson, E.D. (1963) Stratigraphic and lateral variation of chromite compositions in the Stillwater Complex. *Min. Soc. Amer. Spec. Pap.* **1**, 46–54.

Jackson, E.D. (1967) Ultramafic cumulates in the Stillwater, Great Dyke, and Bushveld intrusions. In *Ultramafic and Related Rocks.* ed. Wyllie, P.J., Wiley, New York, 20–38.

Jackson, E.D. (1969) Chemical variation in co-existing chromite and olivine in chromitite zones of the Stillwater Complex. *Econ. Geol. Mon.* **4**, 41–71.

Jackson, E.D. (1970) The cyclic unit in layered intrusions, a comparison of repetitive stratigraphy in the ultramafic parts of the Stillwater, Muskox, Great Dyke and Bushveld Complexes. *Geol. Soc. S. Afr. Spec. Publ.* **1**, 391–424.

Jackson, M.C. (1980) Geology and petrology of Archean basalts and associated rocks at Lava Flow Mountain, northern Ontario. Unpubl. M.Sc. thesis, Univ. Toronto, 287 pp.

Jackson, M.C. (1985) Geology of the Lumby Lake Area, Eastern part, districts of Kenora and Rainy River. *Ont. Geol. Surv. Open File Rept.* **5535**, 122 pp.

Jackson, M.P.A. (1984) Archaean structural styles in the Ancient Gneiss Complex, Swaziland, southern Africa. In *Precambrian Tectonics Illustrated.* eds. Kröner, A. and Greiling, R.E. Schweizerbart, Stuttgart, 1–18.

Jackson, M.P.A., Eriksson, K.A. and Harris, C.W. (1987) Early Archaean foredeep sedimentation related to crustal shortening: a reinterpretation of the Barberton Sequence. Southern Africa. *Tectonophysics* **136**, 197–221.

Jacobsen, S.B. (1988) Isotopic and chemical constraints on mantle-crust evolution. *Geochim. Cosmochim. Acta.* **52**, 1341–1350.

Jacobsen, S.B. and Pimentel-Klose, M.R. (1988) An Nd isotopic study of the Hamersley and Michicipoten banded iron formations: the source of Fe and Nd in Archaean oceans. *Earth Planet. Sci. Lett.* **87**, 29–44.

Jacobsen, S.B. and Wasserburg, G.J. (1979) The mean age of mantle and crustal reservoirs. *J. Geophys. Res.* **84**, 7411–7427.

Jahn, B.-M. and Condie, K.C. (1976) On the age of Rhodesian greenstone belts. *Contrib. Mineral. Petrol.* **57**, 317–330.

Jahn, B.-M. and Ernst, W.G. (1989) Late Archaean Sm–Nd isochron age for mafic–ultramafic supracrustal amphibolites from the northeastern Sino–Korean craton, China. *Precamb. Res.* in press.

Jahn, B.-M. and Murthy, V.R. (1975) Rb–Sr ages of the Archaean rocks from the Vermilion district, northeastern Minnesota. *Geochim. Cosmochim. Acta.* **39**, 1679–1689.

Jahn, B.-M. and Schrank, A. (1983) REE geochemistry of komatiites and associated rocks from Piumhi, southeastern Brazil. *Precamb. Res.* **21**, 1–20.

Jahn, B.-M. and Shih, C.Y. (1974) On the age of the Onverwacht Group. Swaziland–Sequence, South Africa. *Geochim. Cosmochim. Acta* **39**, 1679–1689.

Jahn, B.-M. and Zhang, Z.Q. (1984) Archaean granulite gneisses from eastern Hebei Province, China: rare earth geochemistry and tectonic implications. *Contrib. Mineral. Petrol.* **85**, 224–243.

Jahn, B.-M., Shih, C. and Murthy, V.R. (1974) Trace element geochemistry of Archaean volcanic rocks. *Geochim. Cosmochim. Acta* **38**, 611–627.

Jahn, B.-M., Auvray, B., Blais, S., Capdevila, R., Cornichet, J., Vidal, P. and Hameurt, J. (1980) Trace element geochemistry and petrogenesis of Finnish greenstone belts. *J. Petrol.* **21**, 201–244.

Jahn, B.-M., Glikson, A.Y., Peucat, J.J. and Hickman, A.H. (1981) REE geochemistry and isotopic data of Archean silicic volcanics and granitoids from the Pilbara Block, Western Australia: implications for early crustal evolution. *Geochim. Cosmochim. Acta* **45**, 1633–1652.

Jahn, B.-M., Gruau, G. and Glikson, A.Y. (1982) Komatiites of the Onverwacht Group, South Africa: REE geochemistry, Sm/Nd age and mantle evolution. *Contrib. Mineral. Petrol.* **80**, 25–40.

Jahn, B.-M., Vidal, P. and Kröner, A. (1984) Multi-chronometric ages and origin of tonalitic gneisses in Finnish Lapland: a case for long crustal residence time. *Contrib. Mineral. Petrol.* **86**, 398–408.

Jahn, B.-M., Auvray, B., Cornichet, J., Bai, Y.L., Shen, Q.H. and Liu, D.Y. (1987) 3.5 Ga old amphibolites from eastern Hebei Province, China: field occurrence, petrography, Sm–Nd isotochron age and REE geochemistry. *Precamb. Res.* **34**, 311–346.

Jahn, B.-M., Auvray, B., Shen, Q.H., Liu, D.Y., Zhang, Z.Q., Dong, Y.J., Ye, X.J., Zhang, Q.Z., Cornichet, J. and Mace, J. (1988) Archaean crustal evolution in China: the Taishan Complex and evidence for juvenile crustal addition from long-term depleted mantle. *Precamb. Res.* **38**, 381–403.

Jaques, A.L. (1976) An Archean tholeiitic layered sill from Mt Kilkenny, Western Australia. *J. Geol. Soc. Aust.* **23**, 157–168.

James, H.L. and Hedge, C.E. (1980) Age of the basement rocks of southwest Montana. *Geol. Soc. Amer. Bull.* **91**, 11–15.

James, O.B. (1980) Rocks of the early lunar crust. *Proc. 11th Lunar Sci. Conf.*, Pergamon, New York, (*Geochim. Cosmochim. Acta suppl.* **14**), 365–393.

James, O.B. and Fohr, M.K. (1983) Subdivision of the Mg-suite noritic rocks into Mg-gabbronorites and Mg-norites. *Proc. 13th Lunar Planet Sci. Conf. J. Geophys. Res.* **88**, A603–A614.

James, P.R. and Ding, P. (1988) 'Caterpillar Tectonics' in the Harts Range area: a kinship between two sequential extension-collision zone orogenic belts within the Arunta inlier of Central Australia. *Precamb. Res.* **40/41**, 199–216.

Janardhan, A.S. and Leake, B.E. (1975) The origin of the meta-anorthositic gabbros and garnetiferous granulites of the Sittampundi complex, Madras, India. *J. Geol. Soc. India.* **16**, 391–408.

Janardhan, A.S., Srikantappa, C. and Ramachandra, H.M. (1978) The Sargur schist complex – an Archaean high-grade terrain in southern India. In *Archaean Geochemistry* eds. Windley, B.F. and Naqvi, S.M., Elsevier, Amsterdam, 127–150.

Janardhan, A.S., Newton, R.C. and Smith, J.V. (1979) Ancient crustal metamorphism at low P_{H_2O}: Charnockite formation at Kabbaldurga, south India. *Nature* **278**, 511–514.

Janardhan, A.S., Newton, E.C. and Hansen, E.C. (1982) The transformation of amphibolite-facies gneiss to charnockite in southern Karnataka and northern Tamil Nadu, India. *Contrib. Mineral. Petrol.* **79**, 130–149.

Jaques, A.C. (1976) An Archaean tholeiitic layered sill from Mount Kilkenny, Western Australia. *J. Geol. Soc. Aust.* **23**, 157–168.

Jardim de Sa, E.F., Archanjo, C.J. and Legrand, J.M. (1982) Structural and metamorphic history of part of the high-grade terrain in the Curaca Valley, Bahia, Brazil. *Rev. Bras. Geoc.* **12**, 251–262.

Jardim de Sa, E.F., Souza, Z.S., Fonseca, V.P. and Legrand, J.M. (1984) Relacoes entre 'greenstone belts' e terrenos de alto grau: o caso da faixa Rio Capim, NE da Bahia. *Anais XXXIII Congr. Brasileiro Geol.*, 2615–2629.

Jenner, G.A. (1981) Geochemistry of high-Mg andesites from Cape Vogel, Papua New Guinea. *Chem. Geol.* **33**, 307–332.

Jenner, G.A., Fryer, B.J. and McLennan, S.M. (1981) Geochemistry of the Archaean Yellowknife Supergroup. *Geochim. Cosmochim. Acta* **45**, 1111–1129.

Jensen, L.S. (1976) A new cation plot for classifying subalkalic volcanic rocks. *Ont. Div. Mines Misc. Pap.* **66**, 22 pp.

Jensen, L.S. (1985) Stratigraphy and petrogenesis of Archaean metavolcanic sequences, southwestern Abitibi subprovince, Ontario. *Geol. Assoc. Can. Spec. Paper* **28**, 65–87.

Jensen, L.S. (1987) Geology of the Horseshoe Lake greenstone belt. *Ont. Geol. Surv. Misc. Paper* **137**, 104–108.

Jensen, L.S. and Langford, F.F. (1983) Geology and petrogenesis of the Archaean Abitibi Belt in the Kirkland lake Area. Ontario. *Ont. Geol. Surv. Open. File Rep.* **5455**, 512 pp.

Jensen, L.S. and Pyke, D.R. (1987) Komatiites in the Ontario portion of the Abitibi Belt. In *Komatiites* eds. Arndt, N.T. and Nisbet, E.G., George, Allen and Unwin, London, 147–158.

Jochum, K.P., Seufert, H.M., Spettel, B. and Palme, H. (1986) The Solar System abundances of Nb, Ta and Y and the relative abundances of refractory lithophile elements in differentiated planetary bodies. *Geochim. Cosmochim. Acta* **50**, 1173–1183.

Johnson, R.C. and Hills, F.A. (1976) Precambrian geochronology and geology of the Boxelder

Canyon area, northern Laramie Range, Wyoming. *Geol. Soc. Amer Bull.* **87**, 809–817.

Johnson, R.W., Jaques, A.L., Hickey, R.L., McKee, C.E. and Chappell, B.W. (1985) Manam Island, Papua New Guinea: petrology and geochemistry of a low-Ti basaltic island-arc volcano. *J. Petrol.* **26**, 283–323.

Johnson, Y.A., Park, R.G. and Winchester, J.A. (1987) Geochemistry, petrogenesis and tectonic significance of the early Proterozoic Loch Maree Group amphibolites of the Lewisian complex, NW Scotland. In *Evolution of the Lewisian and Comparable Precambrian High Grade Terrains* eds. Park, R.G. and Tarney, J. *Spec. Publ. Geol. Soc. Lond.* **27**, 255–269.

Jolly, W.T. (1977) Relations between Archaean lavas and intrusive bodies of the Abitibi Greenstone Belt. Ontario-Quebec. *Geol. Assoc. Can. Spec. Paper* **16**, 311–330.

Jones, D.L., Robertson, I.D.M. and McFadden, P.L. (1975) A palaeomagnetic study of the Precambrian dyke swarms associated with the Great Dyke of Rhodesia. *Trans. Geol. Soc. S. Afr.* **78**, 67–75.

Jones, G.M. (1977) Thermal interaction of the core and mantle and long term behaviour of the geomagnetic field. *J. Geophys. Res.* **82**, 1703–1709.

Jordon, T.H. (1978) Composition and development of the continental tectosphere. *Nature* **274**, 544–548.

Jorge João, X. da S., Frizzo, S.J., Marinho, P.A. da C., Carvalho, J.M.A., Neto, C.S.S., Souza, A.N. and Guimaraes, L.R. (1979) Geologia da regiao do sudoeste do Amapa e norte do Para. *Convenio DNPM/CPRM, Projeto Sudoeste do Amapa, Secao Geol. Basica,* **7**.

Kalliokoski, J.O. (1965) Geology of north-central Guyana Shield, Venezuela. *Geol. Soc. Amer. Bull.* **76**, 1027–1050.

Kalsbeek, F. (1981) The northward extent of the Archaean basement of Greenland – a review of Rb–Sr whole-rock ages. *Precamb. Res.* **14**, 203–219.

Kalsbeek, F. (1986) The tectonic framework of the Precambrian shield of Greenland. A review of new isotopic evidence. In *Developments in Greenland geology* eds. Kalsbeek, F. and Watt, W.S. *Rapp. Grønlands Geol. Unders.* **128**, 55–62.

Kalsbeek, F. and Leake, B.E. (1970) The chemistry and origin of some basement amphibolites between Ivigtut and Frederikshåb, South-West Greenland. *Bull. Grønlands Geol. Unders.* **90**, 36 pp.

Kalsbeek, F. and Taylor, P.N. (1985a) Pb isotopic studies of Proterozoic igneous rocks, West Greenland, with implications on the evolution of the Greenland shield. In *The Deep Proterozoic Crust in the North Atlantic Provinces* eds. Tobi, A.C. and Touret, J.L.R., Reidel, Amsterdam, 237–245.

Kalsbeek, F. and Taylor, P.N. (1985b) Isotopic and chemical variation across a Proterozoic continental margin – the Ketilidian mobile belt of South Greenland. *Earth Planet. Sci. Lett.* **73**, 65–80.

Kalsbeek, F. and Taylor, P.N. (1985c) Age and origin of early Proterozoic dolerite dykes in South-West Greenland. *Contrib. Mineral. Petrol.* **89**, 307–316.

Kalsbeek, F., Bridgwater, D. and Zeck, H.P. (1978) A 1950 ± 60 Ma Rb–Sr whole-rock age from two Kangamiut dykes and the timing of the Nagssugtoqidian (Hudsonian) orogeny in West Greenland. *Can. J. Earth Sci.* **15**, 1122–1128.

Kalsbeek, F., Taylor, P.N. and Henriksen, N. (1984) Age of rocks structures and metamorphism in the Nagssugtoqidian mobile belt, West Greenland – field and Pb-isotope evidence. *Can. J. Earth Sci.* **21**, 1126–1131.

Kanasewich, E.R. (1976) Plate tectonics and planetary convection. *Can. J. Earth Sci.* **13**, 331–340.

Karasevich, L.P., Garihan, J.M., Dahl, P.S. and Okuma, A.F. (1981) Summary of Precambrian metamorphic and structural history, Ruby Range, southwest Montana. *Montana Geol. Soc. Field. Conf. Guide. Butte*, 225–237.

Karlstrom, K.E. and Bowring, S.A. (1988) Early Proterozoic assembly of tectonostratigraphic terranes in southwestern North America. *J. Geol.* **96**, 561–576.

Karlstrom, K.E. and Houston, R.S. (1984) The Cheyenne belt: analysis of a Proterozoic suture in southern Wyoming. *Precamb. Res.* **25**, 415–446.

Karlstrom, K.E., Houston, R.S., Flurkey, A.J., Coolidge, C.M., Kratchovil, A.L. and Sever, C.K. (1981) A summary of the geology and uranium potential of Precambrian conglomerates in southeastern Wyoming. *U.S. Dept. Energy Rep.* **DJBX–139–81**, 541 pp.

Karlstrom, K.E., Flurkey, A.J. and Houston, R.S. (1983) Stratigraphy and depositional setting of Proterozoic sedimentary rocks in southeastern Wyoming: a record of an early Proterozoic

Atlantic-type cratonic margin. *Geol. Soc. Am. Bull.* **94**, 1257–1294.

Kato, T., Ringwood, A.E. and Irifune, T. (1988a) Experimental determination of element partitioning between silicate perovskites, garnets and liquids: constraints on early differentiation of the mantle. *Earth Planet. Sci. Lett.* **89**, 123–145.

Kato, T., Ringwood, A.E. and Irifune, T. (1988b) Constraints on element partition coefficients between $MgSiO_3$ perovskite and liquid determined by direct measurements. *Earth Planet. Sci. Lett.* **90**, 65–68.

Kazansky, V.I. and Moralev, V.M. (1981) Archaean geology and metallogeny of the Aldan Shield, USSR. In *Archaean Geology* eds. Glover, J.E. and Groves, D.I. *Spec. Publ. Geol. Soc. Aust.* **7**, 111–120.

Keats, W. (1974) The Roraima Formation in Guyana: a revised stratigraphy and a proposed environment of deposition. In *Memor. 2nd Congr. Latinoamericano Geol., Caracas. Venezuela Minis. Minas e Hidrocarburos, Bol. Geol., Publ. Espec.* **7**, (2) 901–940.

Keays, R.R. (1982) Palladium and iridium in komatiites and associated rocks: application to petrogenetic problems. In *Komatiites* eds. Arndt, N.T. and Nisbet, E.G., George, Allen and Unwin, London, 435–458.

Keays, R.R. (1984) Archaean gold deposits and their source rocks: the upper mantle connection. In *Gold '82: The Geology, Geochemistry and Genesis of Gold Deposits.* ed. Foster, R.P., A.A. Balkema, Rotterdam, 17–51.

Keays, R.R. and Campbell, I.A. (1981) Precious metals in the Jimberlana Intrusion, Western Australia: implications for the genesis of platiniferous ore in layered intrusions. *Econ. Geol.* **76**, 1118–1141.

Keays, R.R. and Scott, R.B. (1976) Precious metals in ocean-ridge basalts: implications for basalts as source rocks for gold mineralization. *Econ. Geol.* **71**, 705–720.

Keays, R.R., Ross, J.R. and Woolrich, P. (1981) Precious metals in volcanic peridotite-associated nickel sulfide deposits in Western Australia. II: distribution within the ores and lost rocks at Kambalda. *Econ. Geol.* **76**, 1645–1674.

Kerrich, R. and Fryer, B.J. (1979) Archaean precious metal hydrothermal systems, Dome mine, Abitibi greenstone belt. II. REE and oxygen isotope relations. *Can. J. Earth Sci.* **16**, 440–458.

Kerrich, R., Fryer, B.J., Milner, K.J. and Peirce, M.G. (1981) The geochemistry of gold-bearing chemical sediments, Dickenson Mine, Red Lake, Ontario: a reconnaissance study. *Can. J. Earth Sci.* **18**, 624–637.

Kesson, S.E. and Lindsley, D.H. (1976) Mare basalt petrogenesis – a review of experimental studies. *Rev. Geophys. Space Phys.* **14**, 361–373.

Kiilsgaard, T.H., Ericksen, G.E., Patten, L.L. and Bieniewski, C.L. (1972) Mineral resources of the Cloud Peak primitive area, Wyoming. *U.S. Geol. Surv. Bull.* **10**, 21–28.

Kinloch, E.D. (1982) Regional trends in the platinum-group mineralogy of the critical zone of the Bushveld Complex. *Econ. Geol.* **77**, 1328–1347.

Kishida, A. and Kerrich, R. (1987) Hydrothermal alteration zoning and gold concentration at the Kerr-Addison Archaean lode gold deposit, Kirkland Lake, Ontario. *Econ. Geol.* **82**, 649–690.

Kinzler, R.J. and Grove, T.L. (1985) Crystallization and differentiation of Archaean komatiite lavas from northeast Ontario: phase equilibrium and kinetic studies. *Amer. Mineral.* **80**, 40–51.

Kishida, A. and Riccio, L. (1980) Chemostratigraphy of lava sequences from the Rio Itapicuru Greenstone Belt. Bahia State, Brazil. *Precamb. Res.* **11**, 161–178.

Klein, T.L. (1982) Geology and geochemistry of the Seminoe metavolcanic sequence, Seminoe Mountains, Carbon Country, Wyoming. In *Precambrian Geology of the Beartooth Mountains* eds. Mueller, P.A. and Wooden, J.L. *Mont. Bur. Mines Geol. Spec. Publ.* **84**, 162–163.

Klein, E.M. and Langmuir, C.H. (1987) Global correlations of ocean ridge basalt chemistry with axial depth and crustal thickness. *J. Geophys. Res.* **92**, 8089–8115.

Kleinkopf, M.D. (1975) Regional gravity and magnetic anomalies of the Stillwater Complex area. *Montana Bur. Min. Geol. Spec. Publ.* **92**, 33–38.

Klemm, D.D., Schneider, W. and Wagner, B. (1984) The Precambrian metavolcano sedimentary sequence east of Ife and Ilesha, SW Nigeria. A Nigerian greenstone belt? *J. Afr. Earth Sci.* **2**, 161–176.

Klerkx, J. (1980) Age and metamorphic evolution of the basement complex around Jabal al 'Awaynat. In *The Geology of Libya. III.* eds. Salem, M.J. and Busrewil, M.T., Academic Press, New York, 901–906.

Klerkx, J. and Deutsch, S. (1977) Resultats preliminaires obtenus par la methode Rb/Sr sur l'age des formations Precambriennes de la region d'Uweinat (Libye). *Rapp. Ann. (1976) Mus. R. Afr. Centr. Tervuren (Belg), Dep. Geol. Min.* 83–94.

Kleywegt, R.J., De Beer, J.H., Stettler, E.H., Duvenhage, A.W.A. and Brandl, G. (1987). The structure of the Giyani greenstone belt as derived from geophysical studies. *S. Afr. J. Geol.* **90**, 282–295.

Knoper, M.W. and Condie, K.C. (1988) Geochemistry and petrogenesis of early Proterozoic amphibolites, west-central Colorado, USA. *Chem. Geol.* **67**, 209–225.

Kober, B. (1986) Whole-grain evaporation for $^{207}Pb/^{206}Pb$ age investigations on single zircons using a double filament thermal ion source. *Contrib. Mineral. Petrol.* **93**, 482–490.

Koehler, S.W. (1976) Petrology of the diabase dikes of the Tobacco Root Mountains, Montana. *Mont. Bur. Mines. Geol. Spec. Publ.* **73**, 27–36.

Koesterer, M.E., Frost, C.D., Frost, B.R., Hulsebosch, T.P., Bridgwater, D. and Worl, R.G. (1987) Development of the Archaean crust in the Medina Mountain area, Wind River Range, Wyoming (USA). *Precamb. Res.* **37**, 287–304.

Kogbe, C.A. (1983) Editorial. *J. Afr. Earth Sci.* **1**, i.

Koistinen, T.J. (1981) Structural evolution of an early Proterozoic stratabound Cu-Co-Zn deposit, Outokumpu, Finland. *Trans. R. Soc. Edin., Earth Sci.* **72**, 115–158.

Kontinen, A. (1987) An early Proterozoic ophiolite – the Jormua mafic-ultramafic complex. NE Finland. *Precamb. Res.* **35**, 313–341.

Korstgård, J.A. ed. (1979a) Nagssugtoqidian Geology. *Rapp. Grønlands. Geol. Unders.* **89**, 146 pp.

Korstgård, J.A. (1979b) Metamorphism of the Kangamiut dykes and the metamorphic and structural evolution of the southern Nagssugtoqidian boundary in the Itivdleg–Ikertoq region, West Greenland. In *Nagssugtoqidian Geology* ed. Korstgård, J.A. *Rapp. Grønlands Geol. Unders.* **89**, 63–75.

Korstgård, J.A., Ryan, B. and Wardle, R. (1987) The boundary between Proterozoic and Archaean crustal blocks in central West Greenland and northern Labrador. In *Evolution of the Lewisian and Comparable Precambrian High Grade Terrains* eds. Park, R.G. and Tarney, J. *Geol. Soc. Lond. Spec. Publ.* **27**, 247–259.

Kozlovsky, E.A. ed. (1984) *The Kola Superdeep Borehole.* Nedra, Moscow, 490 pp.

Kremenetsky, A.A. and Ovchinnikov, L.N. (1986) The Precambrian continental crust; its structure, composition and evolution as revealed by deep drilling in the USSR. *Precamb. Res.* **33**, 11–43.

Krill, A.G. (1985) Svecokarelian thrusting with thermal inversion in the Karasjok–Levajok area of the northern Baltic Shield. *Norges Geol. Unders.* **403**, 89–102.

Krill, A.G., Bergh, S., Lindal, I. and others, (1985) Rb–Sr, U–Pb and Sm–Nd isotopic dates from Precambrian rocks of Finmark. *Bull. Norges Geol. Unders.* **403**, 37–54.

Krogh, T.E., McNutt, R.H. and Davis, G.L. (1982) Two high precision U–Pb zircon ages for the Sudbury Nickel Irruptive. *Can. J. Earth Sci.* **19**, 723–728.

Kröner, A. (1977a) Precambrian mobile belts of southern and eastern Africa – ancient sutures or sites of ensialic mobility? A case for crustal evolution towards plate tectonics. *Tectonophysics* **40**, 101–135.

Kröner, A. (1977b) The Precambrian geotectonic evolution of Africa: plate tectonic accretion versus plate destruction. *Precamb. Res.* **4**, 163–213.

Kröner, A. ed. (1981) *Precambrian Plate Tectonics.* Elsevier, Amsterdam, 781 pp.

Kröner, A. (1984) Evolution, growth and stabilisation of the Precambrian lithosphere. *Phys. Chem. Earth* **15**, 69–106.

Kröner, A. (1985) Evolution of the Archaean continental crust. *Ann. Rev. Earth Planet. Sci.* **13**, 49–74.

Kröner, A. and Compston, W. (1988) Ion microprobe ages of zircons from Archaean granite pebbles and greywacke, Barberton greenstone belt, southern Africa. *Precamb. Res.* **38**, 367–380.

Kröner, A. and Greiling R. eds. (1984) *Precambrian Tectonics Illustrated.* Schweizerbart'sche Verlagsbuchhandlung, Stuttgart.

Kröner, A. and Todt, W. (1988) Single zircon dating constraining the maximum age of the Barberton greenstone belt, Southern Africa. *J. Geophys. Res.* **93**, 329–337.

Kröner, A., Puustinen, K. and Hickman, M. (1981) Geochronology of an Archaean tonalitic gneiss dome in northern Finland and its relation with an unusual overlying conglomerate and komatiitic greenstone. *Contrib. Mineral. Petrol.* **76**, 33–41.

Kröner, A., Hanson, G.N. and Goodwin, A.M. (1984) *Archaean Geochemistry.* Springer-Verlag, Berlin, 286 pp.

Kröner, A., Compston, W., Zhang, G.W., Guo, A.L. and Todt, W. (1988). Age and tectonic setting of late Archaean greenstone-gneiss terrain in Henan Province, China, as revealed by single-grain zircon dating. *Geology* **16**, 211–215.

Kruger, F.J. and Marsh, J.S. (1982) The significance of $^{87}Sr/^{86}Sr$ ratios in the Merensky cycle of the Bushveld Complex. *Nature* **298**, 53–55.

Kuehner, S.M. (1989) Petrology and geochemistry of early Proterozóic high-Mg dykes from the Vestfold Hills, Antarctica. In *Boninites.* ed. Crawford, A.J., Unwin Hyman, London, 208–231.

Kuo, H.Y. and Crocket, J.H. (1979) Rare earth elements in the Sudbury Nickel Irruptive: comparison with layered gabbros and implications for nickel irruptive petrogenesis. *Econ. Geol.* **74**, 590–605.

Kusky, T.M. (1989) Accretion of the Slave Province. *Geology* **17**, 63–67.

Kutty, T.R.N., Anantha Iyer, G.V., Ramakrishnan, M. and Verma, S.P. (1984) Geochemistry of meta-anorthosites from Holenarsipur, Karnataka, South India. *Lithos* **17**, 317–328.

Kuyumjian, R.M. and Dardenne, M.A. (1982) Geochemical characteristics of the Crixas greenstone belt, Goiás, Brazil. *Rev. Bras. Geoc.* **12**, 324–330.

Kvenvolden, K.A. and Von Huene, R. (1986) Natural gas generation in sediments of the convergent margin of the eastern Aleutian trench area. In *Tectonostratigraphic terranes of the circum-Pacific region. Circum-Pacific Coun. Ener. Min. Resour.* ed. Howell, D.G., *Earth Sci. Ser.* **1**, 31–49.

Kwong, Y.T.J. and Crocket, J.H. (1978) Background and anomalous gold in rocks of an Archean greenstone assemblage, Kakagi Lake area, northwestern Ontario. *Econ. Geol.* **73**, 50–63.

Kyser, T.K., Cameron, W.E. and Nisbet, E.G. (1986) Boninite petrogenesis and alteration history: constraints from stable isotope compositions of boninites from Cape Vogel, New Caledonia and Cyprus. *Contrib. Mineral. Petrol.* **93**, 222–226.

Laajoki, K. (1983) On the geology of the south Puolanka area, Finland. *Bull. Geol. Surv. Finland* **263**, 1–54.

Laajoki, K. (1986) The Precambrian supracrustals of Finland and their tectonic-exogenic evolution. *Precamb. Res.* **33**, 67–85.

Labotka, T.C. (1985) Petrogenesis of the metamorphic rocks beneath the Stillwater Complex: assemblages and conditions of metamorphism. In *Stillwater Complex* eds. Czamanske, G.K. and Zientek, M.L. *Montana Bur. Min. Geol. Spec. Publ.* **92**, 70–76.

De Laeter, J.R., Fletcher, I.R., Rosman, K.J.R., Williams, I.R., Gee, R.D. and Libby, W.A. (1981) Early Archaean gneisses from the Yilgarn Block, Western Australia. *Nature* **292**, 322–323.

Lahtinen, J. (1985) PGE-bearing copper-nickel occurrences in the marginal series of the early Proterozoic Koillismaa layered intrusion, northern Finland. *Bull. Geol. Surv. Finland* **333**, 161–178.

Laing, W.P. and Beardsmore, T.J. (1986) Stratigraphic rationalisation of the eastern Mount Isa Block, recognition of key correlations with the Georgetown and Broken Hill Blocks, and an eastern Australian Proterozoic terrain, and their metallogenic implications. *Proc. 8th Aust. Geol. Conv., Adelaide,* **15**, 112–113.

Lajoie, J. and Gelinas, L. (1978) Emplacement of Archaean peridotitic komatiites in La Motte Township, Quebec. *Can. J. Earth Sci.* **15**, 672–677.

Lambert, R. St. J. (1980) The thermal history of the Earth in the Archaean. *Precamb. Res.* **11**, 199–214.

Lambert, D.D., Unruh, D.M. and Simmons, E.C., (1985) Isotopic investigations of the Stillwater Complex: a review. In *Stillwater Complex* eds. Czamanske, G.K. and Zientek, M.L. *Mont. Bur. Mines Geol. Spec. Publ.* **92**, 46–54.

Lamothe, D., Giovennazzo, D. and Picard, C. (1987) Platinum group element occurrences in the Ungava trough, New Quebec. *Quebec Min. Ener. Resour. Doc. Prom.* **15**, 14 pp.

Langford, F.F. and Morin J.A. (1976) The development of the Superior Province of northwestern Ontario by merging island arcs. *Amer. J. Sci.* **276**, 1023–1034.

Larue, D.K. (1983) Early Proterozoic tectonics of the Lake Superior region: tectonostratigraphic terranes near the purported collision zone. *Geol. Soc. Amer. Mem.* **160**, 33–48.

Larue, D.K. and Sloss, L.L. (1980) Early Proterozoic sedimentary basins of the Lake Superior region: summary. *Geol. Soc. Am. Bull.* **91**, 450–452.

Laul, J.C., Smith, M.R. and Schmitt, R.A. (1983) ALHA81005 meteorite: chemical evidence for lunar highland origin. *Geophys. Res. Lett.* **10**, 825–828.

Lavreau, J. (1980) Etude geologique du Haute-Zaire-Genese et evolution d'un segment lithospherique. Unpubl. thesis, Univ. Brussels.

Lawrence, R.W., James, P.R. and Oliver, R.L. (1987) Relative timing of folding and metamorphism in the Ruby Mine area of the Harts Range, central Australia. *Aust. J. Earth Sci.* **34**, 293–312.

Lee, C.A. and Fesq, H.W. (1986) Au, Ir, Ni and Co in some chromitites of the eastern Bushveld Complex, South Africa. *Chem. Geol.* **62**, 227–237.

Le Pichon, X. and Huchon, P. (1984), Geoid, Pangea and convection. *Earth Planet. Sci. Lett.* **67**, 123–135.

Le Roex, A.P. (1983) Geochemistry, mineralogy and magmatic evolution of the basaltic and trachytic lavas from Gough Island, south Atlantic. *J. Petrol.* **6**, 149–186.

Lesher, C.M., Arndt, N.T. and Groves, D.I. (1984) Genesis of komatiite-associated nickel sulphide deposits at Kambalda, Western Australia: a distal volcanic model. In *Sulphide deposits in mafic and ultramafic rocks.* eds. Buchannan, D.L. and Jones, M.J. Inst. Min. Metall., 132–140.

Lewis, J.D., Rosman, K.J.R. and de Laeter, J.R. (1975) The age and metamorphic effects of the Black Range dolerite dyke. *Ann. Rep. W. Aust. Geol. Surv.* **1974**, 80–88.

Lewry, J.F., Sibbald, T.I.I. and Schledewitz, D.C.P. (1985) Variation in character of Archean rocks in the Western Churchill Province and its significance. In *Evolution of Archaean Supracrustal Sequences* eds. Ayres. L.D., Thurston, P.C., Card, K.D. and Weber, W. *Geol. Assoc. Can. Spec. Paper* **8**, 239–261.

Li, S.G., Hart, S.R., Guo, A.L. and Todt, W. (1988) Whole-rock Sm–Nd isotopic age of the Dengfeng Group, central Henan, and its tectonic significance. *Ke Xue Tong Bao (Science News)*, **22**, 178–1731.

Lindenmayer, Z.G. (1980) Evolucão geologica do Vale do Curaca e dos corpos mafico/ultramaficos mineralizados a cobre. Unpubl. MSc thesis, UFBa., 140 pp.

Lindsley, D.H. (1983) Pyroxene thermometry. *Amer. Mineral.* **68**, 477–493.

Lindsley, D.H. and Munoz, J.L. (1969) Subsolidus relations along the join hedenbergite-ferrosilite. *Amer. J. Sci.* **267A**. 295–324.

Lindstrom, M.M. (1984) Alkali gabbronorite, ultra–KREEPy melt rock and the diverse suite of clasts in North Ray Crater feldspathic fragmental breccia 67975. *Proc. 15th Lunar Planet. Sci. Conf. J. Geophys. Res.* **89**, C50–C62.

Lindstrom, M.M., Knapp, S.A., Shervais, J.W. and Taylor, L.A. (1984) Magnesian anorthosites and associated troctolites and dunite in Apollo 14 breccias. *Proc. 15th Lunar Planet. Sci. Conf.; J. Geophys. Res.* **89**, C41-C49.

Liotard, J.M., Barsczus, H.G., Dupuy, C. and Dostal, J. (1986) Geochemistry and origin of basaltic lavas from Marquesas Archipelago, French Polynesia. *Contrib. Mineral. Petrol.* **92**, 260–268.

Lipin, B.R. and Loferski, P.J. (1983) The origin of chromite deposits in the Stillwater Complex, Montana. (abst.). *Eos* **64**, 884.

Liu, D.Y., Page, R.W., Compston, W. and Wu, J.S. (1985) U–Pb zircon geochronology of late Archaean metamorphic rocks in the Taihangshan–Wutaishan area, north China. *Precamb. Res.* **27**, 85–109.

Liu, D.Y., Shen, Q.H., Jahn, B.-M., Zhang, Z.Q., Auvray, B., Zhang, Q.Z. and Ye, X.J. (1986) Early crustal evolution in China I. U–Pb geochronology of intrusive granitoids from the Qianxi Group. *Terra Cogn.* **6**, 238.

Livingstone, A. (1976) The paragenesis of spinel- and garnet-amphibole lherzolite in the Rodel area, South Harris. *Scott. J. Geol.* **12**, 293–300.

Lobach-Zhuchenko, S.B., Levchenkov, O.A., Chekulaev, V.P. and Krylov, I.N. (1986) Geological evolution of the Karelian granite–greenstone terrain. *Precamb. Res.* **33**, 45–66.

Loferski, P.J. (1986) Petrology of metamorphosed chromite-bearing ultramafic rocks from the Red Lodge district, Montana. *U.S. Geol. Surv. Bull.* **1626-B**, 34 pp.

Lofgren, G.E. (1981) Comparative petrography and cooling rates. In *Basaltic Volcanism on the Terrestrial Planets.* Pergamon Press, New York, 364–370.

Longhi, J. (1977) Magma oceanography 2: chemical evolution and crustal formation. *Proc. 8th*

Lunar Sci. Conf., Pergamon, New York, (*Geochim. Cosmochim Acta suppl.* **8**), 601–621.

Longhi, J. (1980) A model of early lunar differentiation. *Proc. 11th Lunar Planet. Sci. Conf.*, Pergamon, New York, (*Geochim. Cosmochim Acta suppl.* **14**), 289–315.

Longhi, J., Wooden, J.L. and Coppinger, K.D. (1983) The petrology of high-Mg dikes from the Beartooth Mountains, Montana: a search for the parent magma of the Stillwater Complex. *Proc. 14th Lunar Planet. Sci. Conf.*, 1, *J. Geophys. Res.* **88**, suppl., **B53–B69**.

Loper, D.E. (1985) A simple model of whole-mantle convection. *J. Geophys. Res.* **90**, 1809–1836.

Loper, D.E. and McCartney, K. (1986) Mantle plumes and the periodicity of magnetic field reversals. *Geophysical Research Letters* **13**, 1525–1528.

Loucks, R.R., Premo, W.R. and Snyder, G.L. (1988) Petrology, structure, and age of the Mullen Creek layered mafic complex and age of arc accretion, Medicine Bow Mountains, Wyoming. *Geol. Soc. Am. Abs. Prog.* **20**, A73.

Louthean, R. (1983) *Register of Australian Mining*. Lodestone Publ. Co., Leederville, W. Austr., 357 pp.

Lowe, D.R., Byerly, G.R., Ransom, B.L. and Nocita, B.W. (1985) Stratigraphic and sedimentological evidence bearing on structural repetition in early Archaean rocks of the Barberton greenstone belt., South Africa. *Precamb. Res.* **27**, 165–186.

Ludden, J., Gelinas, L. and Trudel, P. (1982) Archean metavolcanics from the Rouyn-Noranda district, Abitibi Greenstone Belt, Quebec. 2: Mobility of trace elements and petrogenetic constraints, *Can. J. Earth Sci.* **19**, 2276–2287.

Lugmair, G.W. and Carlson, R.W. (1978) The Sm–Nd history of KREEP. *Proc. 9th Lunar Planet. Sci. Conf.*, Pergamon, New York, (*Geochim. Cosmochim Acta suppl.* **10**), 689–704.

Luhr, J.F. and Carmichael, I.S.E. (1985) Jorullo volcano, Michoacan, Mexico (1759–1774): The earliest stages of fractionation in calc-alkaline magmas. *Contrib. Mineral. Petrol.* **90**, 142–161.

Lundberg, B. (1980) Aspects of the geology of the Skellefte field, northern Sweden. *Geol. Fören, Stockholm Förh.* **102**, 156–166.

Lundstrom, I. (1987) Lateral variation in supracrustal geology within the Swedish part of the southern Svecofennian volcanic belt. *Precamb. Res.* **35**, 353–365.

Lush, A.P., McGrew, A.J., Snoke, A.W. and Wright, J.E. (1988) Allochthonous Archean basement in the northern East Humboldt Range, Nevada. *Geology* **16**, 349–353.

Lusk, J. (1976) A possible volcanic exhalitive origin for lenticular nickel sulfide deposits of volcanic association, with special reference to those in Western Australia. *Can. J. Earth Sci.* **13**, 451–458.

Ma, X.Y. *et al.* (1957) *Tectonic Features of the Wutaishan Area, Shanxi Province, China*. (in Chinese). Geol. Publ. House, Beijing, 75 pp.

MacDonald, G.A. and Katsura, T. (1964) Chemical composition of Hawaiian lavas. *J. Petrol.* **5**, 82–133.

MacFarlane, A., Crow, M.J., Arthurs, J.W., Wilkinson, J.W. and Aucott, A.F. (1981) The geology and mineral resources of northern Sierra Leone. *Inst. Geol. Sci. U.K. Overseas. Mem.* **7**, HMSO, 103 pp.

MacGregor, A.M. (1932) The geology of the country around Que Que, Gwelo district. *Bull. Geol. Surv. Rhod.* **20**, 113 pp.

MacGregor, A.M. (1951) Some milestones in the Precambrian of southern Rhodesia. *Proc. Geol. Soc. S. Afr.* **54**, 27–71.

MacGregor, A.M. (1953) Precambrian formations of tropical southern Africa. *19th Int. Geol. Congr. Algeria*, **1**, 39–52.

Machado, N., Brooks, C. and Hart, S.R. (1986) Determination of initial $^{87}Sr/^{86}Sr$ and $^{143}Nd/^{144}Nd$ in primary minerals from mafic and ultramafic rocks: experimental procedure and implications for the isotopic characteristics of the Archean mantle under the Abitibi greenstone belt, Canada. *Geochim. Cosmochim. Acta* **50**, 2335–2348.

MacKenzie, W.S., Donaldson, C.H. and Guilford, C. (1982) *Atlas of Igneous Rocks and Their Textures*. Longman, Harlow, 148 pp.

Macrae, N.D. (1969) Ultramafic intrusions of the Abitibi area, Ontario. *Can. J. Earth Sci.* **6**, 281–304.

Makovicky, M., Makovicky, E. and Rose-Hannen, J. (1986) Experimental studies on the solubility and distribution of platinum-group elements in base-metal sulphides in platinum deposits. In *Proc. Int.-Symp. Metallogeny of Basic and Ultrabasic Rocks, Edinburgh*, eds. Gallagher, M.J., Ixer, R.A., Neary, C.R. and Pritchard, H.M. Inst. Min. Metall., London, 415–425.

Mann, A.C. (1983) Trace element geochemistry of high alumina basalt–andesite–dacite–

rhyodacite lavas of the main volcanic series of Santorini volcano, Greece. *Contrib. Mineral. Petrol.* **84**, 43–57.

Mantovani, S.M., Marques, L.S., DeSousa, M.A., Civetta, L., Atalla, L. and Innocenti, F. (1985) Trace element and strontium isotope constraints on the origin and evolution of Parana continental flood basalts of Santa Catarina State (Southern Brazil). *J. Petrol.* **26**, 187–209.

Manzer, G.K. Jr. and Heimlich, R.A. (1974) Petrology and geochemistry of mafic and ultramafic rocks from the northern Bighorn Mountains, Wyoming. *Geol. Soc. Amer. Bull.* **85**, 703–708.

Marinho, M.M., Rocha, G.M.F., Deus, P.B. and Viana, J.S. (1986) Geologia e potential cromitifero do Vale do Jacurici-Bahia. *Proc. XXX Congresso Brasileiro de Geologia.*

Marini, O.J., Fuck, R.A., Danni, J.C.M., Dardenne, M.A., Loguercio, S.O.C. and Ramalho, R. (1984) As faixas de dobramentos Brasilia, Uruacu e Paraguai–Araguaia e o Macico Mediano de Goiás. In *Geologia do Brasil.* eds. Schobbenhaus, C., Campus, D.A., Derze, G.R. and Asmus, H.E. DNPM, Brazil, 251–303.

Marmo, J.-S. and Ojakangas, R.W. (1984) Early Proterozoic glaciogenic deposits, eastern Finland. *Bull. Geol. Soc. Amer.* **95**, 1055–1062.

Marot, A., Capdevila, R., Leveque, B., Gruau, G., Martin, G., Charlot, R. and Hocquard, C. (1984) Le "Synclinorium du sud" de Guyane Française: un ceinture de roches vertes d'age Proterozoique Inferieur. In *10e Reun. Ann. Sciences de la Terre, Bordeaux, 1984.* Soc. Geol. Fr. Edit. Paris.

Marsh, J.S. (1987) Basalt geochemistry and tectonic discrimination within continental flood basalt provinces. *J. Volc. Geotherm. Res.* **32**, 35–49.

Marston, R.J. and Kay, B.D. (1980) The distribution petrology and genesis of nickel ores at the Juan Complex, Kambalda, Western Australia. *Econ. Geol.* **75**, 546–565.

Marston, R.J. and Travis, G.A. (1976) Stratigraphic implications of heterogeneous deformation in the Jones Creek conglomerate (Archaean), Kathleen Valley, Western Australia. *J. Geol. Soc. Aust.* **23**, 141–156.

Marston, R.J., Groves, D.I., Hudson, D.R. and Ross, J.R. (1981) Nickel sulfides in Western Australia: A review. *Econ. Geol.* **76**, 1330–1363.

Martin, A., Nisbet, E.G. and Bickle, M.J. (1980) Archean stromatolites of the Belingwe greenstone belt, Zimbabwe (Rhodesia). *Precamb. Res.* **13**, 337–362.

Martin, H. (1986) Effect of steeper Archaean geothermal gradient on geochemistry of subduction-zone magmas. *Geology* **14**, 753–756.

Martin, H. (1987) Evolution in composition of granitic rocks controlled by time-dependent changes in petrogenetic processes: examples from the Archaean of eastern Finland. *Precamb. Res.* **35**, 257–276.

Martin, H. and Quérré, G. (1984) A 2.5 Ga reworked sialic crust: Rb–Sr ages and isotopic geochemistry of the late Archaean volcanic and plutonic rocks from eastern Finland. *Contrib. Mineral. Petrol.* **85**, 292–299.

Martin, H., Chauvel, C., Jahn, B.-M. and Vidal, P. (1983) Rb–Sr and Sm–Nd ages and isotopic geochemistry of Archaean granodioritic gneisses from eastern Finland. *Precamb. Res.* **20**, 79–91.

Martin, H., Auvray, B., Blais, S., Capdevila, R., Hameurt, J., Jahn, B.-M., Piquet, D., Quérré, G. and Vidal, P. (1984) Origin and geodynamic evolution of the Archaean crust of eastern Finland. *Geol. Soc. Finl. Bull.* **56**, 135–160.

Martinez, M.L., York, D., Hall, C.M. and Hanes, J.A. (1984) Oldest reliable $^{40}Ar/^{39}Ar$ ages for terrestrial rocks, Barberton Mountain komatiites. *Nature* **307**, 352–354.

Mascarenhas, J.F. (1976) Estruturas do tipo 'greenstone belt' not leste da Bahia. *XXIX Congr. Brasileiro Geol., abs.*, 185.

Mathison, C.I. and Marshall, A.E. (1981) Ni–Cu sulfides and their host ultramafic rocks in the Mt. Scholl intrusion, Pilbara region, Western Australia. *Econ. Geol.* **76**, 1581–1596.

Maurel, C. and Maurel, P. (1982a) Etude experimentale de la distribution de l'aluminium entre basin silicate et spinelle chromifère. Implications petrogenetiques: teneur en chrome des spinelles. *Bull. Mineral* **105**, 197–202.

Maurel, C. and Maurel, P. (1982b) Etude experimentale de la distribution du fer ferrique entre spinelle chromifère et basin silicate basique. *Bull. Mineral* **197**, 25–33.

McBirney, A.R. (1985) Further considerations of double-diffusive stratification and layering in the Skaergaard intrusion. *J. Petrol.* **26**, 993–1001.

McBirney, A.R. and Noyes, R.M. (1979) Crystallisation and layering of the Skaergaard intrusion. *J. Petrol.* **20**, 487–554.

McCall, G.J.H. ed. (1977) *The Archaean – Search for the Beginning*. Dowden, Hutchinson and Ross, Stroudsburg, 505 pp.

McCall, G.J.H. and Peers, R. (1971) Geology of the Binneringie Dyke, Western Australia. *Geol. Runds.* **60**, 1174–1263.

McCallum, I.S., Raedeke, L.D. and Mathez, E.A. (1980) Investigations of the Stillwater Complex: Part 1. Stratigraphy and structure of the banded zone. *Amer. J. Sci.* **280A**, 59–87.

McCarthy, T.S., Lee, C.A., Fesq, H.W., Kable, E.J.D. and Erasmus, C.S. (1984) Sulphur saturation in the lower and critical zones of Eastern Bushveld Complex. *Geochim. Cosmochim. Acta* **48**, 1005–1019.

McClay, K.R. and Campbell, I.H. (1976) The structure and shape of the Jimberlana Intrusion, Western Australia as indicated by a combined geological and geophysical investigation of the Bronzite Complex. *Geol. Mag.* **96**, 75–80.

McCulloch, M.T. and Cameron, W.E. (1983) Nd–Sr isotopic study of primitive lavas from the Troodos ophiolite, Cyprus: evidence for a subduction-related setting. *Geology* **11**, 727–731.

McCulloch, M.T. and Compston, W. (1981) The Sm–Nd age of Kambalda and Kanowna greenstones, and heterogeneity in the Archaean mantle. *Nature* **294**, 322–327.

McCulloch, W.R. and Cummings, M. L. (1987) Metasomatism between Archaean age metamorphosed mafic and ultramafic rock, Tobacco Root Mountains, Montana. *Geol. Soc. Amer. Abs. Prog.* **19**, 320.

McCulloch, M.T., Collerson, K.D. and Compston, W. (1983a) Growth of Western Australia. *J. Geol. Soc. Aust.* **30**, 155–160.

McCulloch, M.T., Compston, W. and Froude, D. (1983b) Sm–Nd dating of Archaean gneisses, eastern Yilgarn Block, Western Australia. *J. Geol. Soc. Aust.* **30**, 149–153.

McDougall, I. (1976) Geochemistry and origin of basalts of the Columbia River group, Oregon and Washington. *Geol. Soc. Am. Bull.* **87**, 777–792.

McDougall, J.D., Gopalan, K., Lugmair, G.W. and Roy, A.B. (1983) An ancient depleted mantle source for Archaean crust in Rajasthan, India. *LPI Tech. Rep.* **83–03**, Lunar Planet. Inst., Houston, 55–56.

McDougall, J.D., Willis, R., Lugmair, G.W., Roy, A.B. and Gopalan, K. (1985) The Aravalli sequence of Rajasthan, India: A Precambrian continental margin? *LPI Tech. Rep.* **85–01**, Lunar Planet. Inst., Houston, 57–58.

McGregor, V.R. (1973) The early Precambrian gneisses of the Godthåb district, West Greenland. *Phil. Trans. R. Soc. Lond.* **A273**, 343–358.

McGregor, V.R. and Mason, B. (1977) Petrogenesis and geochemistry of metabasaltic and metasedimentary enclaves in the Amîtsoq gneisses, West Greenland. *Amer. Mineral.* **62**, 887–904.

McGregor, V.R., Nutman, A.P. and Friend, C.R.L. (1986) The Archaean geology of the Godthåbsfjord region, southern West Greenland. In *Workshop on early crustal genesis: the world's oldest rocks*. ed. Ashwal, L.D. *LPI Tech. Rep.* **86–04**, Lunar Planet. Inst., Houston, 113–169.

McIver, J.R., Cawthorne, R.G. and Wyatt, B.A. (1982) The Ventersdorp Supergroup: the youngest komatiitic sequence in South Africa. In *Komatiites*, eds. Arndt, N.T. and Nisbet, E.G., George Allen and Unwin. London, 81–90.

McKenzie, D.P. (1967) Some remarks on heat flow and gravity anomalies. *J. Geophys. Res.* **72**, 6261–6273.

McKenzie, D.P. (1978) Some remarks on the development of sedimentary basins. *Earth Planet. Sci. Lett.* **40**, 25–32.

McKenzie, D.P. (1984a) The generation and compaction of partially molten rock. *J. Petrol.* **25**, 713–765.

McKenzie, D.P. (1984b) A possible mechanism for epeirogenic uplift. *Nature* **307**, 616–618.

McKenzie, D. (1985) The extraction of magma from the crust and mantle. *Earth Planet. Sci. Lett.* **74**, 81–91.

McKenzie, D. (1986) Mantle mixing still a mystery. *Nature* **323**, 297.

McKenzie, D. and Bickle, M.J. (1988) The volume and composition of melt generated by extension of the lithosphere. *J. Petrol.* **29**, 625–679.

McKenzie D.P. and O'Nions, R.K. (1983) Mantle reservoirs and ocean island basalts *Nature* **301**, 229–231.

McKenzie, D.P. and Weiss, N. (1980) The thermal history of the Earth. In *The Continental Crust and its Mineral Deposits*. ed. D.W. Strangway *Geol. Assoc. Can. Spec. Paper* **20**, 575–590.

McKenzie, D.P., Nisbet, E.G. and Sclater, J. (1980) Sedimentary basin development in the Archaean. *Earth Planet. Sci. Lett.* **48**, 35–41.

McLaren, C.H. and De Villiers, J.P.R. (1982) The platinum-group chemistry and mineralogy of the UG-2 chromite layer of the Bushveld Complex. *Econ Geol.* **77**, 1348–1366.

McLennan, S.M. and Taylor, S.R. (1982) Geochemical constraints on the growth of the continental crust. *J. Geol.* **90**, 342–361.

McNeil, A.M. and Kerrich, R. (1986) Archaean lamprophyre dykes and gold mineralization, Matheson, Ontario: the conjunction of LILE-enriched mafic magmas, deep crustal structures, and Au concentration. *Can. J. Earth Sci.* **23**, 324–343.

McRae, N.D. (1969) Ultramafic intrusions of the Abitibi area, Ontario. *Can. J. Earth Sci.* **6**, 281–303.

McQueen, (1981) Volcanic-associated nickel deposits from around the Widgiemooltha Dome, Western Australia. *Econ. Geol.* **76**, 1417–1443.

Meijer, A. (1980) Primitive arc volcanism and a boninite series: example from western Pacific island arcs. *Amer. Geophys. Union Monog.* **23**, 269–282.

Menendez, A. (1972) Geologia de la region de Guasipati, Guayana Venezolana. *Memor. IV Congr. Geol. Venezolano, Bol. Geol., Publ. Espec.* **5**, 2001–2043.

Mengel, F.C. (1983a) Chemistry of coexisting mafic minerals in granulite facies amphibolites from West Greenland: clues to conditions of metamorphism. *Neues Jahrb. Miner. Abh.* **147**, 315–340.

Mengel, F.C. (1983b) Petrography and geochemistry of amphibolites from the Norde Strømfjord area in the central part of the Nagssugtoqidian of West Greenland. *Rapp. Grønlands Geol. Unders.* **113**, 20 pp.

Meriläinen, K. (1976) The granulite complex and adjacent rocks in Lapland, northern Finland. *Bull. Geol. Surv. Finland* **281**, 129 pp.

Meyer, C. Jr. (1977) Petrology, mineralogy and chemistry of KREEP basalt. *Phys. Chem. Earth* **10**, 239–260.

Michard-Vitrac, A., Lancelot, J. and Allègre, C.J. (1977) U–Pb ages on single zircons from the early Precambrian rocks of West Greenland and the Minnesota River Valley. *Earth Planet. Sci. Lett.* **35**, 449–453.

Miller, D.M. (1980) Structural geology of the northern Albion Mountains, southcentral Idaho. *Geol. Soc. Amer. Mem.* **153**, 399–423.

Mogk, D.E. and Geissman, J.W. (1984) The Stillwater Complex is allocthonous. *Geol. Soc. Amer. Abs. Prog.* **16**, 598.

Mogk, D.W., Mueller, P.A. and Wooden, J.L. (1988) Archaean tectonics of the North Snowy Block, Beartooth Mountains, Montana. *J. Geol.* **96**, 125–141.

Moeskops, P.G. (1977) New type of layering due to fractionation of Archaean ultimafic magma. *Nature* **267**, 508–509.

Molyneux, T.G. (1974) A geological investigation of the Bushveld Complex in Sekhukhuneland and part of the Steelpoort valley. *Trans. Geol. Soc. S. Afr.* **77**, 329–338.

Molyneux, T.G. and Klinkert, P.S. (1978) A structural interpretation of part of the eastern mafic lobe of the Bushveld Complex and its surroundings. *Trans. Geol. Soc. S. Afr.* **81**, 359–368.

Montalvão, R.M.G., Hildred, P.D., Bezerra, P.E.L, Prado, P. and Silva, S.J. (1982) Petrographic and chemical aspects of the mafic-ultramafic rocks of the Crixás, Guarinos, Pilar de Goiás–Hidrolina and Goiás greenstone belts, Central Brazil. *Rev. Bras. Geoc.* **12**, 331–347.

Montalvão, R.M.G., Muniz, M.D., Issler, R.S., Dall'Agnol, R., Lima, M.I.C., de Gernandes, P.E.C.A. and Da Silva, G.G. (1975) Geologia. In *Dep. Nacion. Prod. Mineral. Brasil. Projecto RADAMBRASIL, Folha NA. 20, Boa Viasta e Parte da Folhas NA. 21, Tumucumacque, NB. 20, Roraima, e NB. 21. Levantamento de Recursos Naturais*, **8**, 15–136.

Montgomery, C. and Hurley, P.M. (1978) Total rock U–Pb and Rb–Sr systematics in the Imataca Series, Guyana Shield, Venezuela. *Earth Planet. Sci. Lett.* **39**, 281–290.

Moorbath, S. (1983) The most ancient rocks. *Nature* **304**, 585–586.

Moorbath, S. and Taylor, P.N. (1981) Isotopic evidence for continental growth in the Precambrian. In *Precambrian Plate Tectonics* ed. Kröner, A. Elsevier, Amsterdam, 491–526.

Moorbath, S., Wilson, J.F. and Cotterill, P. (1976) Early Archaean age for the Sebakwian group at Sulukwe, Rhodesia. *Nature* **264**, 536–538.

Moores, E.M. and Jackson E.D. (1974) Ophiolites and oceanic crust. *Nature* **250**, 136–139.

Morey, G.B. (1980) A brief review of the geology of the western Vermilion district, northeastern Minnesota. *Precamb. Res.* **11**, 247–265.

Morgan, P. (1985) Crustal radiogenic heat production and the selective survival of ancient continental crust. *J. Geophys. Res.* **90**, C561–C570.

Morgan, P. (1986) Thermal implications of metamorphism in greenstone belts and the hot asthenosphere-thick continental lithosphere paradox. In *Workshop on Tectonic Evolution of Greenstone Belts* eds. de Wit, M.J. and Ashwal, L.D. *Lunar Planet. Inst. Houston Tech. Rep.* **86–10**, 157–159.

Morgan, J.W., Wandless, G.A., Petrie, R.K. and Irving, A.J. (1981) Composition of the Earth's upper mantle. I. Siderophile trace elements in ultramafic nodules. *Tectonophysics* **75**, 47–67.

Morrice, M. (1988) Geology of the Northwest Angle Inlet Area, Lake of the Woods, District of Kenora. *Ont. Geol. Surv. Open File Rept.* **5683**, 139 pp.

Mose, S.A. (1982) Adcumulus growth of anorthosite at the base of the lunar crust. *Proc. 13th Lunar Planet. Sci. Conf.; J. Geophys. Res.* **87**, A10–A18.

Mortimer, G.E., Cooper, J.A. and Oliver, R.L. (1988) Proterozoic mafic dykes near Port Lincoln, South Australia: composition, age and origin. *Aust. J. Earth Sci.* **35**, 93–110.

Mueller, P.A. and Rogers, J.J.W. (1973) Secular chemical variation in a series of Precambrian mafic rocks, Beartooth Mountains, Montana and Wyoming. *Geol. Soc. Amer. Bull.* **84**, 3645–3652.

Mueller, P.A., Wooden, J.L., Odom, A.L. and Bowes, D.R. (1982) Geochemistry of the Archaean rocks of the Quad Creek and Hellroaring plateau areas of the eastern Beartooth Mountains. In *Precambrian Geology of the Beartooth Mountains, Montana and Wyoming* eds. Mueller, P.A. and Wooden, J.L. *Mont. Bur. Mines Geol. Spec. Publ.* **84**, 69–82.

Mueller, P.A., Wooden, J.L., Schulz, K. and Bowes, D.R. (1983) Incompatible-element-rich andesitic amphibolites from the Archaean of Montana and Wyoming: evidence for mantle metasomatism. *Geology* **11**, 203–206.

Mueller, P.A., Wooden, J.L., Henry, D.J. and Bowes, D.R. (1985) Archaean crustal evolution of the eastern Beartooth Mountains, Montana and Wyoming. In *Stillwater Complex* eds. Czamanske, G.K. and Zientek, M.L. *Montana Bur. Mines Geol. Spec. Publ.* **92**, 9–20.

Mullan, H.S. and Bussel, M.A. (1977) The basic rock series in batholithic associations. *Geol. Mag.* **114**, 265–280.

Murck, B.W. and Campbell, I.H. (1986) The effects of temperature, oxygen fugacity and melt composition on the behaviour of chromium in basic and ultrabasic melts. *Geochim. Cosmochim. Acta* **50**, 1871–1888.

Murthy, N.G.K. (1987) Mafic dyke swarms of the Indian Shield. In *Mafic Dyke Swarms* eds. Halls, H.C. and Fahrig, W.F., *Geol. Ass. Can. Spec. Pap.* **34**, 393–400.

Mutanen, T., Tornroos, R. and Johanson, B. (1987) The significance of cumulus chlorapatite and hight-temperature dashkesanite to the genesis of PGE mineralizations in the Koitelainen and Keivitsa-Satovaara complexes, northern Finland. *Proc. Geo-platinum 87 symp.* (*abs*). Open University, Milton Keynes.

Myers, J.D., Marsh, B.D. and Sinha, A.K. (1986) Geochemical and strontium isotopic characteristics of parental Aleutian arc magmas: evidence from the basaltic lavas of Atka. *Contrib. Mineral. Petrol.* **94**, 1–11.

Myers, J.D., Patchen, A.D. and Houston, R.S. (1987) The Lake Owen mafic complex, SE Wyoming: 1. geologic field relations. *Eos* **68**, 430.

Myers, J.S. (1975a) Igneous stratigraphy of Archaean anorthosite at Majorqap qâva, near Fiskenaesset, south-west Greenland. *Rapp. Grønlands Geol. Unders.* **74**, 27 pp.

Myers, J.S. (1975b) Stratigraphy of the Fiskenaesset anorthosite complex, southern West Greenland, and comparison with the Bushveld and Stillwater complexes. *Rapp. Grønlands Geol. Unders.* **75**, 72–77.

Myers, J.S. (1975c) Pseudo-fractionation trend of the Fiskenaesset anorthosite complex, southern West Greenland. *Rapp. Grønlands Geol. Unders.* **75**, 77–80.

Myers, J.S. (1976) The early Precambrian gneiss complex of Greenland. In *The Early History of the Earth*, ed. Windley, B.F., Wiley, London, 165–176.

Myers, J.S. (1981) The Fiskenaesset anorthosite complex – a stratigraphic key to the tectonic evolution of the West Greenland gneiss complex 3000–2800 m.y. ago. In *Archaean Geology*, eds. Glover, J.E. and Groves, D.I., *Geol. Soc. Aust. Spec. Publ.* **7**, 351–360.

Myers, J.S. (1984) The Nagssugtoqidian mobile belt of Greenland. In *Precambrian Tectonics Illustrated*. eds. Kroner, A. and Greiling, R., Schweizerbart'sche Verlagsbuchhandlung, Stuttgart, 237–250.

Myers, J.S. (1985) Stratigraphy and structure of the Fiskenaesset Complex, southern West Greenland. *Bull. Grønlands Geol. Unders.* **150**, 72 pp.

Myers, J.S. and Platt, R.G. (1977) Mineral chemistry of layered Archaean anorthosite at Majorqap qava, near Fiskenaesset, southwest Greenland. *Lithos* **10**, 59–72.

Myers, J.S. (1988) Oldest known terrestrial anorthosite at Mount Narryer, Western Australia. *Precamb. Res.* **38**, 309–323.

Nakamura, E., Campbell, I.H. and Sun, S.-S. (1985) The influence of subduction processes on the geochemistry of Japanese alkaline basalts. *Nature* **316**, 55–58.

Naldrett, A.J. (1973) Nickle sulphide deposits – their classification and genesis with special emphasis on deposits of volcanic association. *Can. Inst. Min. Bull.* **66**, 739, 45–63.

Naldrett, A.J. (1981) Nickle sulfide deposits: classification, composition and genesis. *Econ. Geol. 75th Anniv. Vol.*, 628–685.

Naldrett, A.J. and Mason, G.D. (1968) Contrasting Archean ultramafic igneous bodies in Dundonald and Clergue Townships, Ontario. *Can. J. Earth Sci.* **5**, 111–143.

Naldrett, A.J. and Turner, A.R. (1977) The geology and petrogenesis of a greenstone belt and related nickel sulphide mineralisation at Yakabindie, Western Australia. *Precamb. Res.* **5**, 43–103.

Naldrett, A.J., Innes, D.G., Sowa, J. and Gorton, M.P. (1982) Compositional variations within and between five Sudbury ore deposits. *Econ. Geol.* **77**, 1519–1534.

Naldrett, A.J., Gasparrini, E.C., Barnes, S.J., Von Gruenewaldt, G. and Sharpe, M.R. (1986) The upper critical zone of the Bushveld Complex and the origin of Merensky-type ores. *Econ. Geol.* **81**, 1105–1117.

Natland, J.H. (1982) Crystal morphologies and pyroxene compositions in boninites and tholeiitic basalts from DSDP holes 458 and 459B in the Mariana fore-arc region *Init. Rep. Deep Sea Drill. Proj.* **60**, 681–707.

Naqvi, S.M. and Hussain, S.M. (1979) Geochemistry of meta-anorthosites from a greenstone belt in Karnataka, India. *Can. J. Earth Sci.* **16**, 1254–1264.

Naqvi, S.M. and Rogers, J.J.W. (1987) *Precambrian Geology of India*. Oxford University Press, Oxford, 223 pp.

Naqvi, S.M., Divakara Rao, V. and Narain, H. (1974) The protocontinental growth of the Indian shield and the antiquity of its rift valleys. *Precamb. Res.* **1**, 345–398.

Neall, F.B. (1987) Sulphidation of iron-rich rocks as a precipitation mechanism for large Archaean gold deposits in Western Australia: thermodynamic confirmation. In *Recent Advances in Understanding Precambrian Gold Deposits* eds. Ho, S.E. and Groves, D.I. *Geol. Dept. and Univ. Ext., Univ. W. Aust., Publ.* **11**, 265–269.

Nelson, B.K. and DePaolo, D.J. (1984) 1700 Ma greenstone volcanic successions in southwestern North America and isotopic evolution of Proterozoic mantle. *Nature* **312**, 143–146.

Nelson, B.K. and DePaolo, D.J. (1985) Rapid production of continental crust 1.7 to 1.9 b.y. ago: Nd isotopic evidence from the basement of the north American mid-continent. *Geol. Soc. Amer. Bull.* **96**, 746–754.

Nelson, D.R., Crawford, A.J. and McCulloch, M.T. (1984) Nd–Sr isotopic and geochemical systematics in Cambrian boninites and tholeiites from Victoria, Australia. *Contrib. Mineral. Petrol.* **88**, 164–172.

Nesbitt, B.E., Murowchick, J.B. and Muehlenbachs, K. (1986) Dual origins of lode gold deposits in the Canadian Cordillera. *Geology* **14**, 506–509.

Nesbitt, R.W. (1971) Skeletal crystal forms in the ultramafic rocks of the Yilgarn Block, Western Australia: evidence for an Archaean ultramafic liquid. *Spec. Publ. Geol. Soc. Aust.* **3**, 331–348.

Nesbitt, R.W. and Sun, S.-S. (1976) Geochemistry of Archaean spinifex-textured peridotites and magnesian and low magnesian tholeiites. *Earth Planet. Sci. Lett.* **31**, 433–453.

Nesbitt, R.W., Sun, S.-S. and Purvis, A.C. (1979) Komatiites, geochemistry and genesis. *Can. Mineral.* **17**, 165–186.

Nesbitt, R.W., Jahn, B.M. and Purvis, A.C. (1982) Komatiites: An early Precambrian phenomenon. *J. Volc. Geotherm. Res.* **14**, 31–45.

Nesbitt, R.W., Walker, I.W. and Blight, D.F. (1984) Geochemistry of Archaean metabasaltic lavas, Diemals, Western Australia. *Geol. Surv. W. Aust. Rep.* **12**, *Prof. Pap. for 1982*, 15–26.

Newsom, H.E., White, W.M., Jochum, K.P. and Hofmann, A.W. (1986) Siderophile and chalcophile element abundances in oceanic basalts, Pb isotope evolution and growth of the Earth's core. *Earth Planet. Sci. Lett.* **80**, 299–313.

Nicolaysen, L.O., Burger, A.J. and Liebenberg, W.R. (1962) Evidence for the extreme age of certain minerals from the Dominion Reef conglomerates and the underlying granite in western Transvaal. *Geochim. Cosmochim. Acta* **26**, 15–23.

Niewland, D.A. and Compston, W. (1981) Crustal evolution in the Yilgarn Block, near Perth, Western Australia. In *Archaean Geology* eds. Glover, J.E. and Groves, D.I. *Spec. Publ. Geol. Soc. Aust.* **7**, 159–171.

Nilson, A.A. (1984) O atual estagio de conhecimento dos complexos mafico-ultraficos precambrianos do Brasil-Uma avaliacão preliminar. *XXXIII Congr. Brasileiro Geol.*, IV, 4166–4203.

Nisbet, E.G. (1982) Definition of 'Archaean' – comment and a proposal on the recommendations of the international subcommision on Precambrian stratigraphy. *Precamb. Res.* **19**, 111–118.

Nisbet, E.G. (1984) Turbulence in petrology – the behaviour of komatiites. *Nature* **309**, 14–15.

Nisbet, E.G. (1987) *The Young Earth*. Allen and Unwin, London, 402 pp.

Nisbet, E.G. and Chinner, G.A. (1981) Controls on the eruption of mafic and ultramafic flows, Ruth Well Cu–Ni prospect, Western Australia. *Econ. Geol.* **76**, 1729–1735.

Nisbet, E.G. and Fowler, C.M.R. (1983) Model for Archaean plate tectonics. *Geology* **11**, 376–379.

Nisbet, E.G. and Walker, D. (1982) Komatiites and the structure of the Archaean mantle. *Earth Planet. Sci. Lett.* **60**, 105–113.

Nisbet, E.G., Bickle, M.J. and Martin, A. (1977) The mafic and ultramafic lavas of the Belingwe greenstone belt, Rhodesia. *J. Petrol.* **4**, 521–566.

Nisbet, E.G., Wilson J.F. and Bickle M.J. (1981) The evolution of the Rhodesian craton and adjacent Archaean terrain: tectonic models. In *Precambrian Plate Tectonics*. ed. Kröner, A., Elsevier, Amsterdam, 161–183.

Nisbet, E.G., Arndt, N.T., Bickle, M.J., Cameron, W.E., Chauvel, C., Cheadle, M., Henger, E., Kyser, T.K., Martin, A., Renner, R. and Roeder, E. (1987) Uniquely fresh 2.7 Ga komatiites from the Belingwe greenstone belt, Zimbabwe. *Geology* **15**, 1147–1150.

Nixon, G.T. (1988) Petrology of the younger andesites and dacites of Iztaccihuatl volcano, Mexico: II. Chemical stratigraphy, magma mixing, and the composition of basalt magma influx. *J. Petrol.* **29**, 265–303.

Norman, M.D. and Ryder, G. (1979) A summary of the petrology and geochemistry of pristine highland rocks. *Proc. 10th Lunar Planet. Sci. Conf.* Pergamon, New York, 531–559.

Nunes, P.D. and Tilton, G.R. (1971) Uranium-lead ages of minerals from the Stillwater igneous complex and associated rocks, Montana. *Geol. Soc. Amer. Bull.* **82**, 2231–2249.

Nutman, A.P. (1984) Precambrian gneisses and intrusive anorthosite of Smithson Bjerge, Thule district, North-west Greenland. *Rapp. Grønlands Geol. Unders.* **119**, 31pp.

Nutman, A.P. (1986) The early Archaean to Proterozoic history of the Isukasia area, southern West Greenland. *Bull. Grønlands Geol. Unders.* **154**, 80 pp.

Nutman, A.P. and Bridgwater, D. (1983) Deposition of Malene supracrustal rocks on an Amîtsoq basement in outer Ameralik, southern West Greenland. *Rapp. Grønlands Geol. Unders* **112**, 43–51.

Nutman, A.P., Bridgewater, D., Dimroth, E., Gill, R.C.O. and Rosing, M. (1983) Early (3700 Ma) Archaean rocks of the Isua supracrustal belt and adjacent gneisses. *Rapp. Grønlands Geol. Unders.* **112**, 5–22.

Nutman, A.P., Allaart, J.H., Bridgwater, D., Dimroth, E. and Rosing, M. (1984) Stratigraphic and geochemical evidence for the depositional environment of the early Archaean Isua supracrustal belt, southern West Greenland. *Precamb. Res.* **25**, 365–396.

Nwe, Y.Y. (1975) Two different pyroxene crystallization trends in the trough bands of the Skaergaard intrusion, East Greenland. *Contrib. Mineral. Petrol.* **49**, 285–300.

Nyquist, L.E., Bansal, B. and Weismann, H. (1975) Rb–Sr ages and initial $^{87}Sr/^{86}Sr$ for Apollo 17 basalts and KREEP basalt 15386. *Proc. 6th Lunar Sci. Conf.*, Pergamon, New York, 1445–1465.

Obradovich, J.D., Zartman, R.E. and Peterman, Z.E. (1984) Update of the geochronology of the Belt Supergroup. *Mont. Bur. Mines Geol. Spec. Publ.* **90**, 82–84.

Oen, I.S. (1987) Rift related igneous activity and metallogeneis in SW Bergslagen, Sweden. *Precamb. Res.* **35**, 367–382.

OGS/MERQ. (1983) Lithostratigraphic map of the Abitibi subprovince. *Ontar. Geol. Surv. Map* **2484**, DV 83–36.

O'Hara, M.J. (1977) Geochemical evolution during fractional crystallisation of a periodically refilled magma chamber. *Nature* **243**, 507–508.

O'Hara, M.J. (1985) Importance of the 'shape' of the melting regime during partial melting of the mantle. *Nature* **314**, 58–62.

O'Hara, M.J., Saunders, M.J. and Mercy, E.L.P. (1975) Garnet-peridotite, primary ultrabasic magma and eclogite; interpretation of upper mantle processes in kimberlite. *Phys. Chem. Earth.* **9**, 571–604.

Ohnenstetter, D., Watkinson, D.H., Jones, P.C. and Talkington, R. (1986) Cryptic compositional variation in laurite and enclosing chromite from the Bird River Sill, Manitoba. *Econ. Geol.* **81**, 1159–1168.

Ohtani, E. (1984a) Generation of komatiite magma and gravitational differentiation in the deep upper mantle. *Earth Planet. Sci. Lett.* **67**, 261–272.

Ohtani, E. (1984b) The primordial terrestrial magma ocean and its implication for stratification of the mantle. *Phys. Earth Planet. Interiors* **38**, 70–80.

Okeke, P.O., Borley, G.D. and Watson, J. (1983) A geochemical study of metasedimentary granulites and gneisses in the Scourie-Laxford area of the north-west Scotland. *Mineralog. Mag.* **47**, 1–9.

Old, R.A. and Rex, D.C. (1971) Rubidium and strontium age determinations of some Precambrian granitic rocks, S.E. Uganda. *Geol. Mag.* **108**, 353–360.

Oldham, J.W. (1968) A short note an recent geological mapping of the Shabani area. *Trans. Geol. Soc. S. Afr. (ann.)* **71**, 189–194.

Oliveira, E.P. and Montes, M.L. (1984) Os enxames de diques maficos do Brasil. *XXXIII Congr. Brasileiro Geol.*, 4137–4154.

Oliver, R.L. and Purvis, A.C. (1986) High pressure, high temperature metamorphism, northern Eyre Peninsula. *Proc. 8th Aust. Geol. Conv. Adelaide*, **15**, 216 (abstr).

Oliver, R.L., James, P.R. and Jago, J.B. eds. (1983) *Antarctic Earth Science*. Cambridge Univ. Press, 697 pp.

Olszewski, W.J., Jr., Wirth, K.R., Gibbs, A.K. and Gaudette, H.E. (1989) The age, origin and tectonics of the Grão Pará Group and associated rocks, Serra dos Carajás, Brazil: Archaean continental volcanism and rifting. *Precamb. Res.* **42**, 229–254.

O'Neill, J.M., Duncan, M.S. and Zartman, R.E. (1988) An early Proterozoic gneiss dome in the Highland Mountains of southwestern Montana. *Mont. Bur. Mines Geol. Spec. Paper (8th Int. Conf. Basement Tect.)*.

O'Nions, R.K. (1987) Relationships between chemical and convective layering in the Earth. *J. Geol. Soc. Land.* **144**, 259–274.

O'Nions R.K. and Oxburgh, E.R. (1983) Heat and helium in the Earth. *Nature* **306**, 429–431.

O'Nions, R.K., Evensen, N.M., Hamilton, P.J. and Carter, S.R. (1978) Melting of the mantle past and present. *Phil. Trans. R. Soc. Lond.* **288A**, 547–559.

O'Nions, R.K., Evensen, N.M. and Hamilton, P.J. (1979) Geochemical modelling of mantle differentiation and crustal growth. *J. Geophys. Res.* **84**, 6091–6101.

O'Nions, R.K., Hamilton, P.J. and Hooker, P.J. (1983) A Nd isotope investigation of sediments related to crustal development in the British Isles. *Earth Planet. Sci. Lett.* **63**, 229–240.

Onstott, T.C., York, D. and Hargraves, R.B. (1984) Dating of Precambrian diabase dykes of Venezuela using paleomagnetic and $^{40}Ar/^{39}Ar$ methods. *II symp. Amazonico, Manaus*, 513–518.

Orpen, J.L. and Wilson, J.F. (1981) Stromatolites at 3500 Myr and a greenstone–granite unconformity in the Zimbabwean Archaean. *Nature* **291**, 443–445.

Osborne, E.F. (1978) Changes in phase relations in response to change in pressure from 1 atm to 10Kb for the system Mg_2SiO_4–Fe_3O_4–$CaAl_2Si_2O_8$–SiO_2. *Carnegie Inst. Wash. Yearb.* **77**, 784–790.

Oxburgh, E.R. and Parmentier, E.M. (1977) Compositional density stratification in oceanic lithosphere–causes and consequences. *J. Geol. Soc. Lond.* **133**, 343–355.

Page, M.L. and Schmulian, M.L. (1981) The proximal volcanic environment of the Scotia Nickel deposit. *Econ. Geol.* **76**, 1469–1479.

Page, N.J., Zientek, M.L., Czamanske, G.K. and Foose, M.P. (1985) Sulfide mineralisation in the Stillwater Complex and underlying rocks. In *Stillwater Complex* eds. Czamanske, G.K. and Zientek, M.L. *Montana Bur. Mines & Geol. Spec. Publ.* **92**, 93–96.

Page, N.J. and Zientek, M.L. (1985) Geologic and structural setting of the Stillwater Complex. In *Stillwater Complex* eds. Czamanske, G.K. and Zientek, M.L. *Montana Bur. Mines & Geol. Spec. Publ.* **92**, 1–8.

Page, R.W. (1985) Isotope record of Proterozoic crustal events in Northern Australia. In *Tectonics and Geochemistry of Early to Middle Proterozoic Fold Belts.*, BMR Aust. Res. 1985/28, abs.

Palme, H. and Nickel, K.G. (1985) Ca/Al ratio and composition of the Earth's upper mantle. *Geochim. Cosmochim. Acta* **310**, 2123–3132.

Pankhurst, R.J. (1977) Open system crystal fractionation and incompatible element variation in basalts. *Nature* **268**, 36–38.

Papike, J.J. (1981) Silicate mineralogy of planetary basalts. In *Basaltic Volcanism on the Terrestrial Planets.* Pergamon Press, New York, 340–363.

Papike, J.J. and Bence, A.E. (1978) Lunar mare versus terrestrial mid-ocean ridge basalts: planetary constraints on basaltic volcanism. *Geophys. Res. Lett.* **5**, 803–806.

Papike, J.J., Hodges, F.N., Bence, A.E., Cameron, M. and Rhodes, J.M. (1976) Mare basalts: crystal chemistry, mineralogy and petrology. *Rev. Geophys. Space Phys.* **14**, 475–540.

Papike, J.J., Taylor, L.A. and Simon, S.B. (1990) Lunar minerals. In *The Lunar Sourcebook* eds. Heiken, G. and Vaniman, D.T. Cambridge Univ. Press, Cambridge in press.

Papike, J.J. and Simon, S.B. (1984) Petrology of lunar materials: highland, mare, and regiolith. *Proc. 27th Int. Geol. Congress, Vol.* **11**, VNU Science Press, Utrecht 91–136.

Papike, J.J. and Vaniman, D.T. (1978) Luna 24 ferrobasalts and the Mare basalt suite: comparative chemistry, mineralogy and petrology. In *Mare Crisium: the view from Luna 24*, eds. Merrill, R.B. and Papike, J.J., Pergamon, New York, 371–401.

Papike, J.J., Vaniman, D.T. and Taylor, S.R. (1981) Lunar mare basalts. In *Basaltic Volcanism on the Terrestial Planets.* Pergamon Press, New York, 236–267.

Paris, I., Stanistreet, I.G. and Hughes, M.J. (1985) Cherts of the Barberton greenstone belt interpreted as products of submarine exhalative activity. *J. Geol.* **93**, 111–129.

Park, A.F. (1983) Sequential development of metamorphic fabric and structural elements in polyphase deformed serpentinites in the Svecokarelides of eastern Finland. *Trans. R. Soc. Edin. Earth Sci.* **74**, 33–60.

Park, A.F. (1984) Nature, affinities and significance of metavolcanic rocks in the Outokumpu assemblage of eastern Finland. *Bull. Geol. Soc. Finland* **56**, 25–52.

Park, A.F. (1985) Accretion tectonism in the Proterozoic Svecokarelides of the Baltic Shield. *Geology* **13**, 725–729.

Park, A.F. (1987) Nature of the early Proterozoic Outokumpu assemblage, eastern Finland. *Precamb. Res.* **38**, 131–146.

Park, A.F. and Bowes, D.R. (1981) Metamorphosed and deformed pillows from Losomaki: evidence of sub-aqueous volcanism in the Outokumpu association, eastern Finland. *Bull. Geol. Soc. Finland* **53**, 135–145.

Park, R.G. and Tarney, J. (1987a) The Lewisian complex: a typical Precambrian high-grade terrain? In *Evolution of the Lewisian and Comparable High Grade Terrains* eds. Park, R.G. and Tarney, J. *Geol. Soc. Lond. Spec. Publ.* **27**, 13–25.

Park, R.G. and Tarney, J. (1987b) eds. *Evolution of the Lewisian and Comparable High Grade Terrains; Geol. Soc. Lond. Spec. Publ.* **27**, 315 pp.

Park, R.G., Crane, A. and Niamatullah, M. (1987) Early Proterozoic structure and kinematic evolution of the southern mainland Lewisian. In *Evolution of the Lewisian and Comparable High Grade Terrains* eds. Park, R.G. and Tarney, J. *Geol. Soc. Lond. Spec. Publ.* **27**, 139–151.

Parker, A.J. and Lemon, N.M. (1981) Reconstruction of the early Proterozoic stratigraphy of the Gawler Craton, South Australia. *J. Geol. Soc. Aust.* **29**, 221–238.

Parker, A.J., Rickwood, P.C., Baillie, P.W., Boyd, D.M., McClenaghan, M.P., Freeman, M.J., Pietsch, B.A., Murray, C.G. and Myers, J.S. (1987) Mafic dyke swarms of Australia. In *Mafic*

Dyke Swarms eds. Halls, H.C. and Fahrig, W.F. *Geol. Assoc. Can. Spec. Pap.* **34**, 401–417.

Parks, J. and Hill, R.E.T. (1985) Phase compositions and cryptic variations in a 2.2 km section of the Windimurra layered gabbroic intrusion, Yilgarn block, Western Australia – a comparison with the Stillwater Complex. *Econ. Geol.* **81**, 1196–1202.

Parr, J. and Rickard, D. (1987) Early Proterozoic subaerial volcanism and its relationship to Broken Hill-type mineralization in central Sweden. In *Geochemistry and Mineralization of Proterozoic Volcanic Suites* eds. Pharaoh, T.C., Beckinsale, R.D. and Rickard, D. *Spec. Publ. Geol. Soc. Lond.* **33**, 81–93.

Parsons, I. (1987) ed. *Origin of Igneous Layering.* Reidel, Dordrecht, 666 pp.

Parsons, B.E. and McKenzie, D.P. (1978) Mantle convection and the thermal structure of plates. *J. Geophys. Res.* **83**, 4485–4496.

Parsons, B.E. and Sclater, J.G. (1977) An analysis of the variation of ocean floor bathymetry and heat flow with age. *J. Geophys. Res.* **82**, 803–827.

Patchen, A.D. and Myers, J.D. (1987) The Lake Owen mafic complex, SE Wyoming: II. Mineralogy and compositional characteristics. *Eos* **68**, 430.

Patchett, P.J. and Bridgwater, D. (1984) Origin of continental crust of 1.9–1.7 Ga defined by Nd isotopes in the Ketilidian terrain of South Greenland. *Contrib. Mineral. Petrol.* **87**, 311–318.

Pavoni, N. (1970) Zones lateraler horizontaler Verschiebung in der Erdkruste und daraus ableitbare Aussagen zur globalen Tektonik. *Geol. Runds.* **59**, 56–77.

Pavoni, N. (1981) A global geotectonic reference system inferred from Cenozoic tectonics. *Geol. Runds.* **70**, 189–206.

Pavoni, N. (1985) Die pazifisch-antipazifische Bipolarität im Strukturbild der Erde und ihre geodynamische Deutung. *Geol. Runds.* **74**, 251–266.

Pearce, J.A. (1982) Trace element characteristics of lavas from destructive plate boundaries. In *Andesites*. ed. Thorpe, R.S. John Wiley, New York, 525–548.

Pearce, J.A. (1983) Role of sub-continental lithosphere in magma genesis at active continental margins. In *Continental Basalts and Mantle Xenoliths*. eds. Hawkesworth, C.J. and Norry, M.J. Shiva, Nantwich, 230–249.

Pearce, J.A. and Cann, J.R. (1973) Tectonic setting of basic volcanic rocks determined using trace element analyses. *Earth Planet Sci. Lett.* **19**, 290–300.

Pearce, J.A. and Norry, M.J. (1979) Petrogenetic implications of Ti, Zr, Y and Nb variations in volcanic rocks. *Contrib. Mineral. Petrol.* **69**, 33–47.

Pearce, T.H., Gorman, B.E. and Birkett, T.C. (1975) The $TiO_2-K_2O-P_2O_5$ diagram: a method of discriminating between oceanic and non-oceanic basalts. *Earth Planet. Sci. Lett.* **24**, 419–426.

Pearson, R.C., Kiilsgaard, T.H., Patten, L.L. and Mattick, R.E. (1971) Mineral resources of the Popo Agie primitive area, Fremont and Sublette Counties, Wyoming. *U.S. Geol. Surv. Bull.* **1353-B**, 55 pp.

Pekkarinen, L.J. (1979) The Karelian formations and their depositional basement in the Kiihtelysvaara-Värtsilä area, East Finland. *Bull. Geol. Surv. Finland* **301**, 141 pp.

Percival, J.A. and Card, K.D. (1985) Structure and evolution of Archaean Crust in Central Superior Province, Canada. *Geol. Assoc. Can. Spec. Pap.* **28**, 179–192.

Percival, J.A. and Williams, H.R. (1989), The late Archaean Quetico accretionary complex, Superior province, Canada. *Geology* **17**, 23–25.

Perfit, M.R., Gust, D.A., Bence, A.E., Arculus, R.J. and Taylor, S.R. (1980) Chemical characteristics of island-arc basalts: implications for mantle sources. *Chem. Geol.* **30**, 227–256.

Perkins, D.P. and Newton, R.C. (1981) Charnockite geobarometers based on coexisiting garnet–pyroxene–plagioclase–quartz. *Nature* **292**, 144–146.

Perttunen, V. (1985) On the Proterozoic stratigraphy and exogenic evolution of the Peräpohja area, Finland. *Bull. Geol. Surv. Finland* **331**, 131–141.

Peterman, Z.E. (1979) Geochronology and the Archaean of the United States. *Econ. Geol.* **74**, 1544–1562.

Peterman, Z.E. (1981) Dating of Archaean basement in northeastern Montana. *Geol. Soc. Amer. Bull.* **92**, 139–146.

Peterman, Z.E. and Futa, K. (1987) Is the Archaean Wyoming province exotic to the Superior craton? Evidence from Sm–Nd model ages of basement cores. *Geol. Soc. Amer. Abs. Prog.* **19**, 803.

Peterman, Z.E. and Hildreth, R.A. (1978) Reconnaissance geology and geochronology of the Precambrian of the Granite Mountains, Wyoming. *U.S. Geol. Surv. Prof. Pap.* **1055**, 22 pp.

Peterman, Z.E., Zartman, R.E. and Sims, P.K. (1980) Early Archaean tonalitic gneiss from northern Michigan, U.S.A. *Geol. Soc. Am. Mem. Spec. Pap.* **182**, 125–134.

Peucat, J.J., Jahn, B.-M. and Cornichet, J. (1986) High precision zircon U–Pb age of a tonalite from the Archean granite greenstone terrain, Quigyuan, NE China. *Proc. Int. Symp. Precamb. Crust. Evol.* **3**, Geol. Publ. House, Beijing. 222–229.

Pharaoh, T.C. (1980) Geological history of the Komagfjord Tectonic Window, Finnmark, nothern Norway. Unpubl. PhD thesis, Univ. Dundee.

Pharaoh, T.C. (1984) The Precambrian geology of the south-eastern part of mapsheet Skoganvarre 2034 IV. *Geol. Surv. Norway Unpub. Rep.*, 13 pp.

Pharaoh, T.C. (1985) Volcanic and geochemical stratigraphy of the Nussir Group of Arctic Norway – an early Proterozoic greenstone suite. *J. Geol. Soc. Lond.* **142**, 259–278.

Pharaoh, T.C. and Pearce, J.A. (1984) Geochemical evidence for the geotectonic setting of early Proterozoic metavolcanic sequences in Lapland. *Precamb. Res.* **10**, 283–309.

Pharaoh, T.C., Ramsay, D.M. and Jansen, O. (1983) Stratigraphy and structure of the northern part of the Repparfjord-Komagfjord window, Finnmark, northern Norway. *Bull. Norges Geol. Unders.* **377**, 1–45.

Pharaoh, T.C., Beckinsale, R.D. and Rickard, D. eds. (1987). Geochemistry and Mineralization of Proterozoic Volcanic Suites. *Spec. Publ. Geol. Soc. Lond.* **33**, 575 pp.

Pharaoh, T.C., Warren, A and Walsh, N.J. (1987) Early Proterozoic metavolcanic suites of the northernmost part of the Baltic shield. In *Geochemistry and Mineralization of Proterozoic Volcanic Suites*, eds. Pharaoh, T.C., Beckinsale, R.D. and Rickard, D. *Spec. Publ. Geol. Soc. Lond.* **33**, 41–58.

Phaup, A.E. (1932) The geology of the Antelope gold belt. *Bull. Geol. Surv. S. Rhod.* **21**, 119 pp.

Phillips, G.N. (1985) Archaean gold deposits of Australia. *Econ. Geol. Res. Unit Inf. Circ.* **175**, 41 pp.

Phillips, G.N. (1986) Geology and alteration in the Golden Mile, Kalgoorlie. *Econ. Geol.* **81**, 779–808.

Phillips, G.N. and Groves, D.F. (1984) Fluid access and fluid-wallrock interaction in the genesis of the Archaean gold-quartz vein deposit at the Hunt Mine, Kambalda, Western Australia. In *Gold '82: The Geology, Geochemistry and Genesis of Gold Deposits*. ed. Foster R.P. Balkena, Rotterdam, 389–416.

Phinney, W.C., Morrison, D. and Maczuga, D.E. (1988) Anorthosites and related megacrystic units in the evolution of Archaean crust. *J. Petrol.* **29**, 1283–1323.

Pichamuthu, C.S. (1960) Charnockite in the making. *Nature* **188**, 135–136.

Pidgeon, R.T. (1980) 2480 Ma old zircons from granulite facies rocks from east Hebei Province, North China. *Geol. Rev.* **26**, 198–207.

Pidgeon, R.T. (1984) Geochronological constraints on early volcanic evolution of the Pilbara Block, Western Australia. *Aust. J. Earth Sci.* **31**, 237–242.

Pieters, C.M. (1978) Mare basalt types on the front side of the moon: a summary of spectral reflectance data. *Proc. 9th Lunar Planet. Sci. Conf.*, Pergamon, New York, (*Geochim. Cosmochim. Acta suppl.* **10**), 2825–2849.

Plumb, K.A. and James, H.L. (1985) Subdivision of Precambrian time: recommendations and suggestions by the Subcommission on Precambrian Stratigraphy. *Precamb. Res.* **32**, 65–92.

Plumb, K.A., Derrick, G.M., Needham, R.S. and Shaw, R.P. (1981) The Proterozoic of Northern Australia. In *Precambrian of the Southern Hemisphere* ed. Hunter, D.R. Elsevier, Amsterdam, 205–307.

Podmore, F. and Wilson, A.H. (1987) A reappraisal of the structure, geology and emplacement of the Great Dyke, Zimbabwe. In *Mafic Dyke Swarms* eds. Halls, H.C. and Fahrig, W.E. *Geol. Assoc. Can. Spec. Pap.* **34**, 317–330.

Poidevin, J.L., Dostal, J. and Dupuy, C. (1981) Archaean greenstone belt from the Central African Republic (Equatorial Africa). *Precamb. Res.* **16**, 157–170.

Pollack, H.N. and Chapman, D.S. (1977) On the regional variation of heat flow, geotherms, and lithospheric thickness. *Tectonophysics* **38**, 279–296.

Porter, D.J. and McKay, K.G. (1981) The nickel sulfide mineralization and metamorphic setting of the Forrestania area, Western Australia. *Econ. Geol.* **76**, 1524–1549.

Premo, W.R. and Van Schmus, W.R. (1989) Zircon geochronology of Precambrian rocks in southeastern Wyoming and northern Colorado. *Geol. Soc. Amer. Spec. Pap.* **235**, 13–32.

Premo, W.R., Tatsumo, M., Helz, R.T. and Zientek, M.L. (1986) A zircon age for Stillwater complex-related rocks. *Geol. Soc. Amer. Abs. Prog.* **18**, 722.

Prendergast, M.D. (1987) The chromite ore field of the Great Dyke, Zimbabwe. In *Evolution of Chromian Ore Fields.* ed. Stowe, C.W. Van Nostrand Reinhold, New York, 89–108.

Prian, J.-P., Simeon, Y., Johan, V., Ledru, P., Piantone, P., Coste, B. and Eko N'Dong, J. (1988) Valorisation geologique de l'inventaire minier de l'Archaean et du Proterozoique inferieur des feuilles Mitzic, Booue et Mouila, a 1/200000 (Gabon central). *Chron. Rech. Min.* **491**, 67–104.

Priem, H.N.A., Boelrijk, N.A.I.M., Hebeda, E.H., Verdurmen, E.A. and Verschure, R.H. (1973) Age of the Precambrian Roraima Formation in northeastern South America: Evidence from isotopic dating of Roraima pyroclastic volcanic rocks in Suriname. *Geol. Soc. Amer. Bull.* **84**, 1677–1684.

Prinz, M. (1964) Geologic evolution of the Beartooth Mountains, Montana and Wyoming. Part 5. Mafic dike swarms of the southern Beartooth Mountains. *Geol. Soc. Amer. Bull.* **75**, 1217–1248.

Prinz, M. and Keil, K. (1977) Mineralogy, petrology and chemistry of ANT-suite rocks from the lunar highlands. *Phys. Chem. Earth* **10**, 215–237.

Pulfrey, W. (1938) Geological survey of No. 2 mining area. Interim report and map of the south west quadrant. *Rep. Min. Geol. Depot. Kenya* **7**.

Purvis, A.C. (1980) The geochemistry and metamorphic petrology of the Southern Cross-Forrestania greenstone belt at Diggers Rocks, Western Australia. Unpubl. Ph.D. thesis, Univ. Adelaide.

Purvis, A.C. (1983) Metamorphosed altered komatiites at Mount Martin, Western Australia. Archaean weathering products metamorphosed at the alumino–silicate triple point. *Aust. J. Earth Sci.* **31**, 91–106.

Purvis, A.C. and Moeskops, P.G., (1981) Nickel-copper sulfide rich Proterozoic dikes at Cowarna Rocks, Western Australia. *Econ. Geol.* **76**, 1597–1605.

Purvis, A.C., Nesbitt, R.W. and Hallberg, J.A. (1972) The geology of part of the Carr Boyd Rocks Complex, Western Australia, and its associated nickel mineralization. *Econ. Geol.* **67**, 1093–1113.

Pyke, D.R., Naldrett, A.J. and Eckstrand, O.R. (1973) Archaean ultramafic flows in Munro Township. Ontario. *Geol. Soc. Am. Bull.* **84**, 955–977.

Qian, X.L., Cui, W.Y., Wang, S.Q. and Wang, G.Y. (1985) *Geology of Precambrian Iron Formations in Eastern Hebei Province, China.* (in Chinese). Hebei Sci. Tech. Press, China. 273 pp.

Raase, P., Raith, M., Ackermand, D. and Lal, R.K. (1986) Progressive metamorphism of mafic rocks from greenschist to granulite facies in the Dharwar craton of South India. *J. Geol.* **94**, 261–282.

Radhakrishna, B.P. and Naqvi, S.M. (1986) Precambrian continental crust of India and its evolution. *J. Geol.* **94**, 145–166.

Raedeke, L.D. and McCallum, I.S. (1980) Lunar fractionation trends are not anomalous. *Lunar Planet. Sci.* **XI**, 908–910.

Raedeke, L.D. and McCallum, I.S. (1984) Investigations in the Stillwater Complex. II. Petrology and petrogenesis of the ultramafic series. *J. Petrol.* **25**, 395–420.

Raedeke, L.D. and McCallum, I.S.E. (1985) Guide to the Chrome Mountain area. Czamanske, in *Stillwater Complex*; G.K. and Zientek, M.L. eds. *Montana Bur. Min. Geol. Spec. Publ.* **92**, 277–285.

Raedeke, L.D., McCallum, I.S.E., Mathez, E.A. and Criscenti, L.J. (1985) The Contact Mountain section of the Stillwater Complex. In *Stillwater Complex;* Czamanske, G.K. and Zientek, M.L. eds. *Montana Bur. Min. Geol. Spec. Publ.* **92**, 286–292.

Raith, M., Raase, P., Ackermand, D. and Lal, R.K. (1983) Regional geothermobarometry in the granulite-facies terrane of South India. *Trans. R. Soc. Edinb., Earth Sci.* **73**, 221–244.

Rajamani, V., Shivkumar, K., Hanson, G.N. and Shirey, S.B. (1985) Geochemistry and petrogenesis of amphibolites, Kolar schist belt, South India: evidence for komatiitic magma derived by low percentages of melting of the mantle. *J. Petrol.* **26**, 92–123.

Ramadurai, S., Sankaran, M., Selven, T.A. and Windley, B.F. (1975) The stratigraphy and structures of the Sittampundi complex, Tamil Nadu, India. *J. Geol. Soc. India* **16**, 409–414.

Ramsay, W.R.H., Crawford, A.J. and Foden, J.D. (1984) Field setting, mineralogy, chemistry and genesis of arc picrites, New Georgia, Solomon Islands. *Contrib. Mineral Petrol* **88**, 386–402.

Redden, J.A. (1981) Summary of the geology of the Nemo area. In *Geology of the Black Hills, South Dakota and Wyoming*; ed. Rich, F.J., *Geol. Soc. Amer. Guidebook. Rocky Mountains* (1981). *Am.*

Geol. Inst., Falls Church, Va., 193–210.

Redden, J.A., Peterman, Z.E., Zartman, R.E. and Dewitt, E.J. (1988) U–Th–Pb geochronology and preliminary interpretation of Precambrian tectonic events in the Black Hills, South Dakota. Geol. Assoc. Can.; Min. Assoc. Can. Trans-Hudson Orog. Symp. in press.

Redman, B.A. and Keays, R.R. (1985) Archaean basic volcanism in the Eastern Goldfields Province, Western Australia. Precamb. Res. 30, 113–152.

Reed, J.C. (1987) Precambrian geology of the USA. Episodes 10, 243–247.

Reed, J.C. and Zartman, R.E. (1973) Geochronology of Precambrian rocks of the Teton Range, Wyoming. Geol. Soc. Amer. Bull. 84, 561–582.

Reed, J.C., Bickford, M.E., Premo, W.R., Aleinkoff, J.N. and Pallister, J.S. (1987) Evolution of the early Proterozoic Colorado province: constraints from U–Pb geochronology. Geology 15, 861–865.

Rehkopff, A. (1984) Origin of quartzo-feldspathic supracrustal rocks from the central part of the Nagssugtoqidian mobile belt in West Greenland. Rapp. Grønlands Geol. Unders. 117, 26 pp.

Reid, D.L. (1981) Sm–Nd ages from the Namaqua and Richtersveld provinces. Geocongr. '81, Geol. Soc. S. Afr., Ext. Abs., 173–174.

Reid, R.R., McMannis, W.J. and Palmquist, J.C. (1975) Precambrian geology of the North Snowy block, Beartooth Mountains, Montana. Geol. Soc. Amer. Spec. Pap. 157, 135 pp.

Renner, R. and Gibbs, A.K. (1987) Geochemistry and petrology of metavolcanic rocks of the early Proterozoic Mazaruni greenstone belt, northern Guyana. In Geochemistry and Mineralization of Proterozoic Volcanic Suites; eds. Pharaoh, T.C., Beckinsale, R.D. and Rickard, D., Geol. Soc. Lond. Spec. Publ. 33, 289–309.

Richardson, S.H., Gurney, J.J., Erlank, A.J. and Harris, J.W. (1984) Origin of diamonds in old enriched mantle. Nature 310, 198–202.

Richter, F.M. (1984) Time and space scales mantle convection. In Pattern of change of Earth Evolution, eds. Holland, H.D. and Trendall, A.F., Dahlem Konferenzen 1984, Springer-Verlag, Berlin, 271–289.

Richter, F.M. (1984) Regionalised models for the thermal evolution of the Earth. Earth Planet. Sci. Lett. 68, 471–484.

Richter, F.M. (1985) Models for the Archaean thermal regime. Earth Planet. Sci. Lett. 73, 350–360.

Richter, F.M. (1988) A major change in the thermal state of the earth at the Archaean–Proterozoic boundary: consequences for the nature and preservation of continental lithosphere. In Oceanic and Continental Lithosphere: Similarities and differences. eds. Menzies, M.A. and Cox, K.G., Spec. Vol. J. Petrol., 39–52.

Rickard, D. (1979) Scandinavian metallogenesis. Geojournal 3, 235–252.

Ringwood, A.E. (1982) Phase transformations and differentiation in subducted lithosphere: implications for mantle dynamics, basalt petrogenesis and crustal evolution. J. Geol. 90, 611–643.

Rivalenti, G. (1975) Chemistry and differentiation of mafic dykes in an area near Fiskenaesset, West Greenland. Can. J. Earth Sci. 12, 721–730.

Rivalenti, G. (1976) Geochemistry of metavolcanic amphibolites from South-West Greenland. In The Early History of the Earth. ed. Windley, B.F., Wiley, London, 213–223.

Rivalenti, G. and Rossi, A. (1972) The geology and petrology of the Precambrian rocks to the north-east of the fjord Qagssit, Frederikåb district, South-West Greenland. Bull. Grønlands Geol. Unders 103, 98 pp.

Rivalenti, G. and Rossi, A. (1975) Geochemistry of Precambrian amphibolites in an area near Fiskenaesset, South-West Greenland. Boll. Soc. Geol. Ital. 94, 27–49.

Rivalenti, G., Girardi, V.A.V., Sinigoi, S., Rossi, A. and Siena, F. (1982) The Niquelândia mafic-ultramafic complex of Central Goiás, Brazil: petrological considerations. Rev. Bras. Geoc. 12, 380–391.

Robert, F. and Brown, A.C. (1986a) Archaean gold-bearing quartz veins at the Sigma Mine, Abitibi Greenstone Belt, Quebec. I. Geologic relations and formation of the vein system. Econ. Geol. 81, 578–592.

Robert, F. and Brown, A.C. (1986b) Archaean gold-bearing quartz veins at the Sigma Mine, Abitibi Greenstone Belt, Quebec. II. vein paragenesis and hydrothermal alteration. Econ. Geol. 81, 593–616.

Rock, N.M.S., Duller, P., Haszeldine, R.S. and Groves, D.I. (1987) Lamprophyres as potential gold exploration targets: some preliminary observations and speculations. In *Recent Advances in Understanding Precambrian Gold Deposits.* eds. Ho, S.E. and Groves, D.I., *Geol. Dept. and Univ. Ext., Univ. W. Austr. Publ.* **11**, 271–286.

Rock, N.M.S., Groves, D.I. and Perring, C.S. (1988) Gold, porphyries and lamprophyres: a new genetic model. In *Gold mineralization in the Yilgarn Block, Western Australia.* ed. Groves, D.I., *Bicent. Gold 88, Ext. Abs. 22, Geol. Soc. Austr.*, **13–18** (307–312).

Roeder, P.L., Campbell, I.H. and Jamieson, H.E. (1979) A re-evaluation of the olivine-spinel geothermometer. *Contrib. Mineral. Petrol.* **68**, 325–335.

Rogers, J.J.W. (1986) The Dharwar craton and the assembly of Peninsular India. *J. Geol.* **94**, 129–143.

Rollinson, H.R. (1978) Zonation of supracrustal relics in the Archaean of Sierra Leone, Liberia, Guinea and Ivory Coast. *Nature* **272**, 440–442.

Rollinson, H.R. (1982) P-T conditions in coeval greenstone belts and granulites from the Archaean of Sierra Leone. *Earth Planet. Sci. Lett.* **59**, 177–191.

Rollinson, H.R. (1987) Early basic magmatism in the evolution of Archaean high-grade gneiss terrains: an example from the Lewisian of NW Scotland. *Mineralog. Mag.* **51**, 345–355.

Rollinson, H.R. and Cliff, R.A. (1982) New Rb–Sr age determinations of the Archaean basement of Sierra Leone. *Precamb. Res.* **17**, 63–72.

Rollinson, H.R. and Fowler, M.B. (1987) The magmatic evolution of the Scourian complex at Gruinard Bay. In *Evolution of the Lewisian and Comparable High Grade Terrains*; eds. Park, R.G. and Tarney, J., *Geol. Soc. Lond. Spec. Publ.* **27**, 57–71.

Ross, J.R. and Keays, R.R. (1979) Precious metals in volcanic-type nickel sulfide deposits in Western Australia. I. Relationship with the composition of ores and their host rocks. *Can. Mineral.* **17**, 417–436.

Ross, J.R. and Hopkins, G.M.F. (1975) Kambalda nickel sulfide deposits. In *Economic Geology of Australia and Papua New Guinea. I.* ed. Knight, C.L., *Metals. Austr. Inst. Min. Metall. Mon* **5**, 100–121.

Ross, J.R. and Travis, G.A. (1981) The nickel sulphide deposits of western Australia in global perspective. *Econ. Geol.* **76**, 1291–1329.

Ross, M.E. and Heimlich, R.A. (1972) Petrology of Precambrian mafic dikes from the Bald Mountain area, Bighorn Mountains, Wyoming. *Geol. Soc. Amer. Bull.* **83**, 1117–1124.

Runcorn, S.K. (1962) Convection currents in the Earth's mantle. *Nature* **195**, 1248–1249.

Runcorn, S.K. (1965) Changes in the convection pattern of the earth's mantle and continental drift; evidence for a cold origin of the earth. *Phil. Trans. R. Soc. Lond.* **A258**, 228–251.

Russell, R.D. and Birney, D.J. (1974) A bi-directional mixing model for lead isotype evolution. *Phys. Earth Planet. Int.* **8**, 158–166.

Ryabchikov, I.D., Suddaby, P., Girnis, A.V., Kulikov, V.S., Kulikova, V.V. and Bogatikov, O.A. (1988) Trace element geochemistry of Archaean and Proterozoic rocks from eastern Karelia, USSR. *Lithos* **21**, 183–194.

Ryder, G. (1977a) The Moon. *Rev. Geophys.* **25**, 277–284.

Ryder, G. (1977b) Petrographic evidence for nonlinear cooling rates and a volcanic origin for Apollo 15 KREEP basalts. *Proc. 17th Lunar Planet. Sci. Conf.; J. Geophys. Res.* **92**, E331–E339.

Ryder, G. and Spettel, B. (1985) The parental magma for some rocks from the Norite 1 subzone of the Stillwater Complex – a lunar analog study. *Proc. 15th Lunar Planet. Sci. Conf.* (2): *J. Geophys. Res.* **90**, (Sup.), C591–600.

Saager, R. and Koppel, V. (1976) Lead isotopes and trace elements from sulphides of Archaean greenstone belts in South Africa – a contribution to the knowledge of the oldest known mineralization. *Econ. Geol.* **71**, 44–57.

Saager, R. and Meyer, M. (1984) Gold distribution in Archaean granitoids and supracrustal rocks from Southern Africa: a comparison. In *Gold '82: The Geology, Geochemistry and Genesis of Gold Deposits.* ed. Foster, R.P., Balkema, A.A. Rotterdam, 53–70.

Saager, R., Meyer, M. and Muff, R. (1982) Gold distribution in supracrustal rocks from Archaean greenstone belts of southern Africa and from Palaeozoic ultramafic complexes of the European Alps. *Econ. Geol.* **77**, 1–24.

Saboia, L.A. (1979) Os 'greenstone-belts' de Crixas e Goias, GO. *Soc. Bras. Geol., N. Centro-Oeste, Bol. Inf.* **9**, 43–72.

Saboia, L.A. and Teixeira, N.A. (1980) Lavas ultrabasicas da unidade basal do 'greenstone-belt' de

Crixas, GO: Uma nova classe de rochas ultrabasicas no Estado de Goias. *Rev. Bras. Geoc.* **10**, 21–42.

SACS (South African Committee for Stratigraphy) (1980) Stratigraphy of South Africa. I. Lithostratigraphy of the Republic of South Africa, South West Africa/Namibia, and the republics of Bophuthatswana, Transkei and Venda (comp. L.E. Kent). *Handb. Geol. Surv. S. Afr.* **8**, *Govt. Print.*, Pretoria.

Saha, A.K. and Ray, S.L. (1984) The structural and geochemical evolution of the Singhbhum granite batholithic complex, India. *Tectonophysics* **105**, 163–176.

Sakko, M. (1971) Radiometric zircon ages on the Early Karelian metadiabases. *Geology* **23**, 117–118. (English summary).

Sakuyama, M. and Nesbitt, R.W. (1986) Geochemistry of the Quarternary volcanic rocks of the northeast Japan arc. *J. Volc. Geotherm. Res.* **29**, 413–450.

Sang, J. and Ho, S.E. (1987) A review of gold deposits in China. In *Recent Advances in Understanding Precambrian Gold Deposits.* eds. Ho, S.E. and Groves, D.I., *Geol. Dept. and Univ. Ext., Univ. W. Austr. public.* **11**, 307–329.

Santos, B.A. (1981) *Amazonia: Potential Minerale Perspectivas de Desenvolvimento.* Ed. Univ. São Paulo, São Paulo, Brazil.

Santos, J.O. and Loguercio, S.O.C. (1984) A parte meridional do Craton Amazonico (Escudo Brasil Central) e as bacias do Alto Tapajos e Parecis-Alto Xingu. In *Geologia do Brasi – Texto Explicativo do Mapa Geologico do Brasil e da Area Oceanica Adjacente, Incluindo Depositos Minerais, 1:2,500,000.* eds. Schobbenhaus, C., *et al.* Dept. Nacion. Prod. Mineral, Brasilia, Brazil, 93–127.

Sarkar, S.N. and Saha, A.K. (1983) Structure and tectonics of the Singhbhum–Orissa–Iron Ore craton, eastern India. *Recent. Res. Geol.* **10**, 1–25.

Sarkar, S.N., Saha, A.K., Boelrijk, A.I.M. and Hebeda, E.H. (1979) New data on the geochronology of the Older Metamorphic Group and the Singhbhum Granite of Singhbum–Koenjhar–Mayurbhanj region, eastern India. *Ind. J. Earth Sci.* **6**, 32–51.

Sarkar, S.N., Gopalan, K. and Trivedi, J.R. (1981) New data on the geochronology of the Precambrian of Bhandara-Drug, Central India. *Ind. J. Earth Sci.* **8**, 131–151.

Saunders, A.D., Tarney, J. and Weaver, S.D. (1980) Transverse geochemical variations across the Antarctic peninsula: implications for the genesis of calc-alkaline magmas. *Earth Planet. Sci. Lett.* **46**, 344–360.

Saunders, A.D., Norry, M.J. and Tarney, J. (1988) Origin of MORB and chemically-depleted mantle reservoirs: trace element constraints. In *Oceanic and Continental Lithosphere: Similarities and Differences.* eds. Menzies, M.A. and Cox, K.G., *Spec. Vol. J. Petrol.*, 415–445.

Saverikko, M. (1983a) Explosive komatiitic volcanism in Finnish Lapland. *Geology* **35**, 21–23.

Saverikko, M. (1983b) The Kummitsoiva komatiite complex and its satellites in northern Finland. *Bull. Geol. Soc. Finland* **55**, 111–139.

Saverikko, M. (1985) The pyroclastic komatiite complex at Sattasvaara in northern Finland. *Bull. Geol. Soc. Finland* **57**, 55–87.

Saverikko, M. (1988) The Oraniemi arkose–slate–quartzite association: an Archaean aulacogen fill in northern Finland. In *Sedimentoloav of Precambrian Iron Formations in Eastern and Northern Finland;* eds. Laajoki, K. and Paakkola, J., *Geol. Surv. Finland, Spec. Pap.* **5**, 189–212.

Schiotte, L., Compston, W. and Bridgwater, D. (1988) Late Archaean ages for the deposition of clastic sediments belonging to the Malene supracrustals, southern West Greenland: evidence fron an ion probe U–Pb study. *Earth Planet. Sci. Lett.* **87**, 45–58.

Schmidt, C.J. and Garihan, J.M. (1986) Middle Proterozoic and Laramide tectonic activity along the southern margin of the Belt Basin. *Montana Bur. Mines Geol. Spec. Publ.* **94**, 217–235.

Schorscher, H.D. (1978) Komatiitos na estrutura 'greenstone belt' serie Rio das Velhas, Quadrilatero Ferrifero, Minas Gerais, Brasil. *XXX Congr. Brasileiro Geol.*, abs., 292–293.

Schorscher, H.D., Santana, F.C., Polonia, J.C. and Moreira, J.M.P. (1982) Quadrilatero Ferrifero-Minas Gerais State: Rio das Velhas greenstone belt and Proterozoic rocks. *ISAP-Int. Sym. Arch. Early Prot. Geol. Evol. and Met., SMR-Ba, Excur. Ann.*, 44 p.

Schreiber, H.D. and Haskin, L.A. (1976) Chromium in basalts: experimental determination of redox states and partitioning among synthetic silicate phases. *Proc. 7th Lunar Sci. Conf.*, Pergamon, New York, 1221–1259.

Schulz, K.J. (1980) The magmatic evolution of the Vermilion greenstone belt, NE Minnesota. *Precamb. Res.* **11**, 215–245.

Schulz, K.J. (1982) Magnesian basalts from the Archaean terrains of Minnesota. In *Komatiites*. eds. Arndt, N.T. and Nisbet, E.G., George Allen and Unwin, London, 171–186.

Sclater, J.G., Jaupart, C. and Galson, D. (1980) The heat flow through oceanic and continental crust and the heat lost of the Earth. *Rev. Geophys. Space Phys.* **18**, 269–311.

Scoates, R.F.J. (1983) A preliminary stratigraphic examination of the ultramafic zone of the Bird River sill. *Manitoba Min. Resour. Div. Rep. Field Act.*, 70–83.

Scoates, R.F.J. and Eckstrand, O.R. (1986) Platinum-group elements in the upper central layered zone of the Fox River Sill, northeastern Manitoba. *Econ. Geol.* **81**, 1137–1146.

Sen, S.K. and Ray, S. (1971) Hornblende-pyroxene granulites versus pyroxene granulites: A study from the type charnockite area. *Neues Jahrb. Mineral. Monatash.* **115**, 291–314.

Sen, S.K. and Sun, S.-S. (1980) Some petrological and geochemical aspects of basic granulites from India. *Proc. 26 Int. Geol. Congr.*, Paris.

Seward, T.M. (1973) Thio complexes of gold and the transport of gold in hydrothermal ore solutions. *Geochim. Cosmochim. Acta* **37**, 379–399.

Seward, T.M. (1984) The transport and deposition of gold in hydrothermal systems. In *Gold '82: The Geology, Geochemistry and Genesis of Gold Deposits*. ed. Foster, R.P., A.A., Balkema, Rotterdam, 165–181.

Shackleton, R.M. (1946) Geology of the Migori gold belt and adjoining areas. *Rep. Geol. Surv. Kenya* **10**.

Sharma, R.P. (1982) Lithostratigraphy, structure and petrology of the Bundalkhand Group. In *Geology of Vindhyanchal*. eds. Valdiya, K.S., Bhatia, S.B. and Gaur, V.K., Hindustan Publ. Co., New Delhi, 30–46.

Sharma, R.P. (1983) Structure and tectonics of the Bundalkhand complex, central India. *Recent. Res. Geol.* **10**, 198–210.

Sharpe, M.R. (1980) Metasomatic zontion of an ultramafic lens at Ikatoq, near Faeringhavn, southern West Greenland. *Bull. Grønlands Geol. Unders.* **135**, 32 pp.

Sharpe, M.R. (1981) The chronology of magma influxes to the eastern compartment of the Bushveld Complex as exemplified by its marginal border groups. *J. Geol. Soc. Lond.* **138**, 307–326.

Sharpe, M.R. (1982) Noble metals in the marginal rocks of the Bushveld Complex. *Econ. Geol.* **77**, 1286–1295.

Sharpe, M.R. (1984) Petrography, classification and chronology of mafic sill intrusions beneath the eastern Bushveld Complex. *Geol. Surv. S. Afr. Bull.* **77**, 40 pp.

Sharpe, M.R. (1985) Strontium isotope evidence for preserved density stratification in the main zone of the Bushveld Complex. *Nature* **316**, 119–126.

Sharpe, M.R. and Hulbert, L.J. (1985) Ultramafic sills beneath the eastern Bushveld Complex: mobilized suspensions of early lower zone cumulates in a parental magma with boninitic affinities. *Econ. Geol.* **80**, 849–871.

Sharpe, M.R. and Irvine, T.N. (1983) Melting relations of two Bushveld chilled margin rocks and implications for the origin of chromitite. *Carnegie Inst. Wash. Yearb.* **82**, 295–300.

Sharpe, M.R. and Snyman, J.A. (1980) A model for the emplacement of the eastern compartment of the Bushveld Complex. *Tectonophysics.* **65**, 85–110.

Sharpe, M.R., Evensen, N.K. and Naldrett, A.J. (1986) Sm/Nd and Rb/Sr evidence for liquid mixing, magma generation and contamination in the Eastern Bushveld Complex. *Geocongress, Univ. Witwatersrand*, 621–624.

Shau, M. (1977) 'Komatiites' and quartzites in the Archean Prince Albert Group. *Geol. Assoc. Can. Spec. Pap.* **16**, 341–354.

Shaw, H.R. (1985) Links between magma-tectonic rate balances, plutonism and volcanism. *J. Geophys. Res.* **90**, 11275–11288.

Shaw, H.F., Tracey, R.J., Niemeyer, S. and Colodner, D. (1986) Age and Nd–Sr systematics of the Spring Creek Lake body, Sierra Madre Mtns., WY. *Eos* **67**, 1266.

Shegelski, R. (1980) Archaean cratonization, emergence and red bed development, Lake Shebandowan area, Canada. *Precamb. Res.* **12**, 331–347.

Shen, Q.H., Liu, D.Y., Wang, P., Gao, J.F. and Zhang, Y.F. (1987) U–Pb and Rb–Sr isotope age study of the Jining Group from nel Mongol of China. *Bull. Chin. Acad. Geol. Sci.* **16**, 165–178.

Sheraton, J.W., Thomson, J.W. and Collerson, K.D. (1987) Mafic dyke swarms of Antarctica. In *Mafic Dyke Swarms*, eds. Halls, H.C. and Fahrig, W.F., *Geol. Ass. Can. Spec. Pap.* **34**, 419–432.

Shervais, J.W. (1982) Ti–V plots and the petrogenesis of modern and ophiolitic lavas. *Earth Planet. Sci. Lett.* **59**, 101–118.

Shirey, S.B. and Hanson, G.N. (1984) Mantle derived Archaean monzodiorites and trachyandesites. *Nature* **310**, 222–224.

Shirey, S.B. and Hansen, G.N. (1986) Mantle heterogeneity and crustal recycling in Archaean granite-greenstone belts: evidence from Nd isotopes and trace elements in the Rainy Lake area Superior Province, Ontario, Canada. *Geochim. Cosmochim. Acta* **50**, 2631–2651.

Shirley, D.N. (1983) A partially molten magma ocean model. *Proc. 13th Lunar Planet. Sci. Conf.*; *J. Geophys. Res.* **88**, A519–A527.

Sial, A.N., Oliveira, E.P. and Choudhuri, A. (1987) Mafic dyke swarms of Brazil. In *Mafic Dyke Swarms*; eds. Halls, H.C. and Fahrig, W.F., *Geol. Assoc. Can. Spec. Pap.* **34**, 467–481.

Siedlecka, A., Krill, A.G., Often, M., Sandstad, J.S., Solli, A. and Iversen, E. (1985) Lithostratigraphy and correlation of the Archaean and Early Proterozoic rocks of the Finnmarksvidda and Sorvaranger district. *Norges Geol. Unders. Bull.* **403**, 7–36.

Sills, J.D. and Rollinson, H.R. (1987) Metamorphic evolution of the mainland Lewisian complex. In *Evolution of the Lewisian and Comparable High Grade Terrains*; eds. Park, R.G. and Tarney, J., *Spec. Publ. Geol. Soc. Lond.* **27**, 81–92.

Sills, J.D. and Windley, B.F. (1982) Petrogenesis of Lewisian amphibolite dykes from Clashnessie Bay, near Stoer, Sutherland. *Scott. J. Geol.* **18**, 167–176.

Sills, J.D., Savage, D., Watson, J.V. and Windley, B.F. (1982) Layered ultramafic-gabbro bodies in the Lewisian of northwest Scotland: geochemistry and petrogenesis. *Earth Planet. Sci. Lett.* **58**, 345–360.

Sills, J.D., Wang, K.Y., Yan, Y.H. and Windley, B.F. (1987) The Archaean granulite gneiss terrain in E. Hebei Province, NE China: geological framework and conditions of metamorphism. In *Evolution of the Lewisian and Comparable High Grade Terrains*; eds. Park, R.G. and Tarney, J., *Spec. Publ. Geol. Soc. London* **27**, 297–305.

Silva, M.G. (1987) Geochemie, petrologie und tektonische entwicklung eines proterozoischen gruensteingurtels: Rio Itapicuru, Bahia, Brasilien. Unpubl. Dr. Ret. Nat. thesis, Univ. Freiburg, Germany, 180 pp.

Silvennoinen, A. (1985) On the Proterozoic stratigraphy of northern Finland. *Bull. Geol. Surv. Finland* **331**, 107–116.

Simmons, E.C. and Lambert, D.D. (1982) Magma evolution in the Stillwater Complex, Montana: a preliminary evaluation using REE data for whole rocks and cumulate feldspars. In *Precambrian Geology of the Beartooth Mountains, Montana and Wyoming*; eds. Mueller, P.A. and Wooden, J.L., *Mont. Bur. Mines. Geol. Spec. Publ.* **84**, 91–106.

Simmons, E.C., Hanson, G.N. and Lumbers, S.B. (1980) Geochemistry of the Shawmere anorthosite complex, Kapuskasing structural zone, Ontario. *Precamb. Res.* **11**, 43–71.

Simon, S.B., Papike, J.J. and Laul, J.C. (1988) Chemistry and petrology of the Appenine Front, Apollo 15, Part I: KREEP basalt and plutonic rocks. *Proc. 18th Lunar Planet. Sci. Conf.*, Cambridge Univ. Press, Cambridge and Lunar Planet Inst., Houston, 187–201.

Simonen, A. (1980) The Precambrian in Finland. *Bull. Geol. Surv. Finland* **304**, 58 pp.

Simons, F.S., Armbrustmacher, T.J., Van Noy, R.M., Zilka, N.T., Federspiel, F.E., Ridenour, J. and Anderson, L.A. (1979) Mineral resources of the Beartooth primitive area and vicinity, Carbon, Park, Stillwater and Sweet Grass. *U.S. Geol. Surv. Bull.* **1391–F**, 125 pp.

Sims, P.K. (1972) Vermilion district and adjacent areas. In *Geology of Minnesota – a centennial volume*. eds. Sims, P.K. and Morey, G.B., Minnesota Geol. Surv., 49–62.

Sims, P.K. (1976) Early Precambrian tectonic–igneous evolution in the Vermilion district, northeastern Minnesota. *Geol. Soc. Amer. Bull.* **87**, 378–389.

Sims, P.K., Card, K.D., Morey, G.B. and Peterman, Z.E. (1980) The Great Lakes tectonic zone – a major crustal structure in central North America. *Geol. Soc. Amer. Bull.* **91**, 690–698.

Sims, P.K., Peterman, Z.E. and Schulz, K.J. (1985) The Dunbar gneiss-granitoid dome: implications for early Proterozoic tectonic evolution of northern Wisconsin. *Geol. Soc. Amer. Bull.* **96**, 1101–1112.

Sivell, W.J. and Foden, J.D. (1985) Banded amphibolites of the Harts range meta-igneous complex, central Australia: an early Proterozoic basalt-tonalite suite. *Precamb. Res.* **28**, 223–252.

Sivell, W.J. and Foden, J.D. (1988) Amphibolites from the Entia Gneiss Complex, eastern Arunta inlier: geochemical evidence for a Proterozoic transition from extensional to compressional tectonics. *Precamb. Res.* **38**, 235–255.

Sivell, W.J., Foden, J.D. and Lawrence, R.W. (1985) The entire anorthosite gneiss, eastern Arunta inlier, Central Australia, geochemistry and genesis. *Aust. J. Earth Sci.* **32**, 449–466.

Skiöld, T. (1981) Radiometric ages of plutonic rocks from the Vittangi-Karesuando area, northern Sweden. *Förh. Geol. Fören. Stockholm* **103**, 317–329.

Skiöld, T. (1987) Aspects of the Proterozoic geochronology of northern Sweden. *Precamb. Res.* **35**, 161–168.

Skiöld, T. and Cliff, R.S. (1984) Sm–Nd and U–Pb dating of early Proterozoic mafic-felsic volcanism in northernmost Sweden. *Precamb. Res.* **26**, 1–13.

Skinner, W.R. (1969) Geologic evolution of the Beartooth Mountains, Montana and Wyoming: part 8. Ultramafic rocks in the Highline Trail Lakes area, Wyoming. *Geol. Soc. Amer. Mem.* **115**, 19–52.

Sleep, N.H. (1979) Thermal history and degassing of the Earth: some simple calculations. *J. Geol.* **87**, 671–686.

Smewing, J.D., Simonian, K.O. and Gass, I.G. (1975) Metabasalts from the Troodos Massif, Cyprus: genetic implication deduced from petrography and trace element geochemistry. *Contrib. Mineral. Petrol.* **57**, 49–64.

Smith, A.R. and Sutcliffe, R.H. (1988) Plutonic rock of the Abitibi subprovince. *Ont. Geol. Surv. Misc. Paper* **141**, 188–196.

Smith, H.S. (1988) Silicification of basalts in the Barberton greenstone belt: major element, trace element and oxygen isotope evidence. *Geol. Soc. Amer. Barberton Mem.*

Smith, H.S. and Erlank, A.J. (1982) Geochemistry and petrogenesis of komatiites from the Barberton greenstone belt, South Africa. In *Komatiites*. eds. Arndt, N.T. and Nisbet E.G. George Allen and Unwin, London, 347–397.

Smith, H.S. and Erlank, A.J. (1984) Komatiite-hosted Ni–sulphide deposits: assessment of potential in volcanics and sulphur-rich 'dykes' from the Barberton greenstone belt. *Arch. Gold. Barberton Cent. Symp., Abs. and Guide bk.* 20.

Smith, H.S., Erlank, A.J. and Duncan, A.R. (1980) Geochemistry of some ultramafic komatiite lava flows from the Barberton Mountain Land, South Africa. *Precamb. Res.* **11**, 399–415.

Smith, H.S., Anhaeusser, C.R., Kimber, B., Jardine, R., Harris, C. and Erlank, A.J. (1988) Komatiite flows, Barberton greenstone belt: geochemical comparison of GI and GII types. *Chem. Geol.* **70**, 148.

Smith, R.L. (1979) Ash-flow magmatism. *Geol. Soc. Amer. Spec. Paper* **180**, 1–27.

Smith, P.E., Perdix, J.L. and Parks, T.C. (1982) Burial metamorphism in the Hammersley Basin, Western Australia. *J. Petrol.* **23**, 75–102.

Smith, P.E., Tatsumoto, M. and Farquhar, R.M. (1987) Zircon Lu–Hf systematics and the evolution of the Archaean crust in the southern Superior Province, Canada. *Contrib. Mineral. Petrol.* **97**, 93–104.

Smith, T.E. and Longstaffe, F.J. (1974) Archaean rocks of Shoshonitic affinities at Bijou Point, Northwestern Ontario. *Can. J. Earth Sci.* **11**, 1407–1413.

Smith, T.J., Cloke, P.L. and Kesler, S.E. (1984) Geochemistry of fluid inclusions from the McIntyre-Hollinger gold deposit, Timmins, Ontario, Canada. *Econ. Geol.* **79**, 1265–1285.

Snyder, G.A., Simmons, E.C., Holden, G.S. and Snyder, G.L. (1985) A Nd, Sr and REE geochemical study of early Proterozoic rifting of the Archaean Wyoming craton: evidence for two distinct mantle source regions. *Eos* **66**, 415.

Snyder, G.L. (1980a) Geologic map of the northernmost Park Range and southernmost Sierra Madre, Jackson and Routt Counties, Colorado. *U.S. Geol. Surv. Misc. Inv. Ser.* Map **I-1113**.

Snyder, G.L. (1980b) Map of Precambrian and adjacent Phanerozoic rocks of the Hartville Uplift, Goshen, Niobrara and Platte Counties, Wyoming. *U.S. Geol. Surv. Open-File Rep.* **80–779**.

Snyder, G.L. (1984) Preliminary geologic maps of the central Laramie Mountains. Albany and Platte Counties, Wyoming. *U.S. Geol. Surv. Open-File Rep.* **84–358**.

Snyder, G.L. (1986) Preliminary geologic maps of the Reese Mountain and part of the Hightower SW 7.5 minute quadrangles (part A) and parts of the Fletcher Park and Johnson Mountain 7.5 minute quadrangles (part B), Albany and Platte Counties, Wyoming. *U.S. Geol. Surv. Open-File Rep.* **86–201**, 16 pp.

Snyder, G.L. and Hedge, C.E. (1978) Intrusive rocks northeast of Steamboat Springs, Park Range, Colorado. *U.S. Geol. Surv. Prof. Paper* **1041**, 42 pp.

Snyder, G.L. and Peterman, Z.E. (1982) Precambrian geology and geochronology of the Hartville Uplift, Wyoming. *Proc. 1982 Geochem. Field Conf., Seminoe Mtns. and Hartville Uplift, Wyoming, pt. 1, Guide to Field Trips*, 64–94.

Snyder, G.L., Hughes, D.J., Hall, R.P. and Ludwig, K.R. (1989) Distribution of Precambrian mafic intrusives penetrating some Archaean rocks of western North America. *U.S. Geol. Surv. Open File Rep.* **89–125**, 36 pp.

Solomon, S.C. and Longhi, J. (1977) Magma oceanography: 1. Thermal evolution. *Proc. 8th Lunar Sci. Conf.*, Pergamon, New York, 583–599.

Sohnge, P.G., le Roex, H.D. and Nel, H.J. (1948) The geology of the country around Messina. *Geol. Surv. S. Afr., Explan. sheet* **46**, 82 pp.

Sokolov, V.A. and Heiskanen, K.J. (1985) Evolution of the Precambrian volcanogenic-sedimentary lithologies in the south-eastern part of the Baltic Shield. *Bull. Geol. Surv. Finland* **331**, 91–106.

Spall, H. (1971) Palaeomagnetism and K–Ar age of mafic dikes from the Wind River Range, Wyoming. *Geol. Soc. Amer. Bull.* **82**, 2457–2472.

Sparks, R.S.J. (1986) The role of crustal contamination in magma evolution through geological time. *Earth Planet. Sci. Lett.* **78**, 211–223.

Sparks, R.S.J. and Huppert, H.E. (1984) Density changes during the fractional crystallization of basaltic liquid: Fluid dynamic implications. *Contrib. Mineral. Petrol.* **85**, 300–309.

Sparks, R.S.J., Huppert, H.E., Kerr, R.C., McKenzie, D.P. and Tait, S.R. (1985) Post-cumulus processes in layered intrusions. *Geol. Mag.* **122**, 555–568.

Spencer, E.W. and Kozak, S.J. (1975) Precambrian evolution of the Spanish Peaks area, Montana. *Geol. Soc. Amer. Bull.* **86**, 785–792.

Srikantappa, C., Hormann, P.K. and Raith, M. (1984) Petrology and geochemistry of layered ultramafic to mafic complexes from the Archaean craton of Karnataka, southern India. In *Archaean Geochemistry* eds. Kröner, A., Goodwin, A.M. and Hanson, G.N. Springer-Verlag, Berlin, 138–160.

Stacey, F.D. and Loper, D.E. (1983) The thermal boundary-layer interpretation of D″ and its role as a plume source. *Phys. Earth Planet. Int.* **33**, 45–55.

Stagman, J.G. (1978) An outline of the geology of Rhodesia. *Rhod. Geol. Surv. Bull.* **80**, 126 pp.

Stahle, H.J., Raith, M., Hoernes, S. and Delf, A. (1987) Element mobility during incipient granulite formation at Kabbaldurga, southern India. *J. Petrol.* **28**, 803–834.

Starmer, I.C. (1985) The evolution of the southern Norwegian Proterozoic as revealed by the major and megatectonics of the Kongsberg and Bamble sectors. In *The Deep Proterozoic Crust in the North Atlantic Provinces.* eds. Tobi, A.C. and Touret, J.L.R. Reidel, 259–290.

Staudacher, T. and Allegre, C.J. (1982) Terrestrial xenology. *Earth Planet. Sci. Lett.* **60**, 389–406.

Stecher, O. (1981) Geokemien af et Arkaeisk amfibolit baelte fra Kangerdluarssungup Taserssua-Iortuarssup Tasia området, Vestgrønland. Spec. Afhandling (thesis), Univ. Arhus, Denmark, 128 pp.

Stecher, O., Carlson, R.W., Shirey, S.B., Bridgwater, D. and Nielsen, T. (1986) Nd-isotope evidence for the evolution of metavolcanic rocks from the Archaean block of Greenland and Labrador. *Terra Cogn.* **6**, 236.

Steel, I.M., Bishop, F.C., Smith, J.V. and Windley, B.F. (1977) The Fiskenaesset complex, West Greenland, III: Chemistry of silicate and oxide minerals from oxide-bearing rocks, mostly from Qeqertarssuatsiaq. *Bull. Grønlands Geol. Unders.* **124**, 38 pp.

Stettler, E.H., du Plessis, J.G. and de Beer, J.H. (1988) The structure of the Pieterseberg greenstone belt, South Africa as derived from geophysics. *S. Afr. J. Geol.* **91**, 292–303.

Stewart, A.J., Shaw, R.D. and Black, L.P. (1984) The Arunta inlier: a complex ensialic mobile belt in central Australia. Part 1: stratigraphy, correlations and origin. *Aust. J. Earth Sci.* **31**, 445–456.

Stöffler, D., Knoll, H.-D., Marvin, U.B., Simonds, C.H. and Warren, P.H. (1980) Recommended classification and nomenclature of lunar highland rocks. *Proc. Conf. Lunar Highland Crust.* Pergamon, New York, 51–70.

Stolper, E., Walker, D., Hager, B.H. and Hays, J.F. (1981) Melt segregation from partially molten source regions: the importance of melt density and source region size. *J. Geophys. Res.* **86**, 6261–6271.

Stolz, G.W. and Nesbitt, R.W. (1981) The komatiite-nickel sulfide association at Scotia: a petrochemical investigation of the ore environment. *Econ. Geol.* **76**, 1480–1502.

Stott, G.M., Sanborn-Barrie, M. and Corfu, F. (1987) Major transpression events recorded across Archaean subprovince boundaries in northwestern Ontario. *Geol. Assoc. Can. Yellowknife '87, Prog. Abs.*

Stowe, C.W. (1968a) The geology of the country south and west of Selukwe. *Rhod. Geol. Surv. Bull.* **59**, 209 pp.

Stowe, C.W. (1968b) Intersecting fold trends in the Rhodesian basement complex south and west of Selukwe. *Trans. Geol. Soc. S. Afr.* **71**, 53–78.

Stowe, C.W. (1971) Summary of the tectonic development of the Rhodesian Archaean craton. In *Symp. on Archaean Rocks*; ed. Glover, J.E., *Spec. Publ. Geol. Soc. Aust.* **3**, 377–383.

Stowe, C.W. (1980) Wrench tectonics in the Archaean Rhodesian craton. *Trans. Geol. Soc. S. Afr.* **83**, 193–205.

Stowe, C.W. (1987a) Chromitite deposits of the Shurugwi greenstone belt, Zimbabwe. In *Evolution of chromium ore fields.* ed. Stowe, E.W., Van Nostrand Reinhold, New York, 71–88.

Stowe, C.W. (1987b) Extended tabulation of chromitite ore fields. In *Evolution of chromium ore fields.* ed. Stowe, C.W., Van Nostrand Reinhold, New York, 297–320.

Stuckless, J.S., Nkomo, L.T. and Doe, B.R. (1981) U–Th–Pb systematics in hydrothermally altered granites from the Granite Mountains, Wyoming. *Geochim. Cosmochim. Acta* **45**, 635–645.

Stuckless, J.S., Hedge, C.E., Worl, R.G., Simmons, K.R., Nkomo, L.T. and Wenner, D.B. (1985) Isotopic studies of the late Archaean plutonic rocks of the Wind River Range, Wyoming. *Geol. Soc. Amer. Bull.* **96**, 850–860.

Stueber, A.M., Heimlich, R.A. and Ikramuddin, M. (1976) Rb–Sr ages of Precambrian mafic dikes from the Bighorn Mountains, Wyoming. *Geol. Soc. Amer. Bull.* **87**, 909–914.

Subramaniam, A.P. (1956) Mineralogy and petrology of the Sittampundi complex, Salem District, Madras State, India. *Geol. Soc. Amer. Bull.* **67**, 317–390.

Sun, S.-S. (1980) Lead isotopic study of young volcanic rocks from mid-ocean ridges, ocean islands and island arcs. *Phil. Trans. R. Soc. Lond.* **A297**, 409–445.

Sun, S.-S. (1982) Chemical composition and origin of the earth's primitive mantle. *Geochim. Cosmochim. Acta* **46**, 179–192.

Sun, S.-S. (1984a) Geochemical characteristics of Archaean ultramafic and mafic volcanic rocks: implications for mantle composition and evolution. In *Archaean Geochemistry.* eds. Kröner, A. Hansen, G.N. and Goodwin A.M., Springer-Verlag, Berlin, 25–46.

Sun. S.-S. (1984b) Some geochemical constraints on mantle evolution models. *Proc. 27th Int. Geol. Congr.* Vol. 9, VNU Sci. Press, 475–508.

Sun, S.-S. (1987) Chemical composition of Archaean komatiites: implications for the early history of the Earth and mantle evolution. *J. Volc. Geotherm. Res.* **32**, 67–82.

Sun, S.-S. and McCulloch, M.T. (1985) Chemical and Nd isotope study of late Archaean to middle Proterozoic mafic volcanic rocks in Northern Australia. *Tectonics and Geochem. of early to mid. Prot. Fold Belts Conf., Darwin, Australia, 1985, Abs.*

Sun, S.-S. and McDonough, W.F. (1988) Chemical and isotopic systematics of oceanic basalts: implications for mantle composition and processes. In *Magmatism in Oceans Basins.* eds. Saunders, A.D. and Norry, M.J., *Spec. Publ. Geol. Soc. Lond.* **42**, 313–345.

Sun, S.-S. and Nesbitt, R.W. (1977) Chemical heterogeneity of the Archaean mantle, composition of the Earth and mantle evolution. *Earth. Planet Sci. Lett.* **35**, 429–448.

Sun, S.-S. and Nesbitt, R.W. (1978a) Petrogenesis of Archaean ultrabasic and basic volcanic rocks: evidence from rare earth elements. *Contrib. Mineral. Petrol.* **65**, 301–325.

Sun, S.-S. and Nesbitt, R.W. (1978b) Geochemical regularities and genetic significance of ophiolitic basalts. *Geology.* **6**, 689–693.

Sun, S.-S., Nesbitt, R.W. and Sharaskin, A.Y. (1979) Geochemical characteristics of mid-ocean ridge basalts. *Earth Planet. Sci. Lett.* **44**, 119–138.

Sun, S.-S., Bai, J., Jin, W.S., Jiang, Y.N., Gao, F., Wang, W.Y., Wang, J.L., Gao, Y.D. and Yang, C.L. (1984) *The early Precambrian Geology of Eastern Hebei.* Tianjin Sci. Tech. Press, Tianjin, 275 pp (in Chinese, 20 pp English summary).

Sun, S.-S., Nesbitt, R.W. and McCulloch, M.T. (1989) Geochemistry and petrogenesis of Archaean and early Proterozoic siliceous high magnesian basalts. In *Boninites.* ed. Crawford, A.J., Unwin Hyman, London, 149–173.

Sutcliffe, R.H. (1984) Geology of the Mulcahy Lake Gabbro Intrusion. *Ont. Geol. Surv. Misc.* Paper **119**, 33–37.

Sutcliffe, R.H. (1988) Magma Mixing in Late Archaean Tonalitic and mafic rocks of the Lac des Iles area, Western Superior Province. *Precamb. Res.* **44**, 81–101.

Sutcliffe, R.H. and Sweeny, J.H. (1985) Geology of the Lac des Iles Complex, District of Thunder

Bay. *Ont. Geol. Survey Misc.* Paper **126**, 47–53.

Sutton, J. (1963) Long-term cycles in the evolution of the continents. *Nature* **198**, 731–735.

Takahashi, E. and Scarfe, C.M. (1985) Melting of peridotite to 14 GPa and the genesis of komatiite. *Nature* **315**, 566–568.

Talbot, (1979) A klippe of Nagssugtoqidian suprascrustal rocks at Sarfartûp nunâ, central West Greenland. In *Nagssugtoqidian Geology*, ed. Korstgård, J.A., *Rapp. Grønlands Geol. Unders.* **89**, 23–42.

Talkington, R.W. and Lippin, B. (1986) Platinum-group minerals in chromite seams of the Stillwater complex, Montana. *Econ. Geol.* **81**, 1179–1186.

Tankard, A.J., Jackson, M.P.A., Eriksson, K.A., Hobday, D.K., Hunter, D.R. and Minter, W.E.L. (1982) *Crustal Evolution of Southern Africa*. Sprinter-Verlag, New York, 523 pp.

Tarney, J. (1969) Epitaxic relations between coexisting pyroxenes. *Mineralog. Mag.* **37**, 115–122.

Tarney, J. (1978) Achmelvic Bay, Assynt. In *The Lewisian and Torridonian Rocks of North-West Scotland*, eds. Barner, A.J., Beach, A., Park, R.G., Tarney, J. and Stewart, A.D. *Geol. Assoc. Guide* **21**, 35–50.

Tarney, J. and Weaver, B.L. (1987a) Geochemistry and petrogenesis of early Proterozoic dyke swarms. In *Mafic Dyke Swarms*; eds. Halls, H.C. and Fahrig, W.F. *Geol. Assoc. Can. Spec. Pap.* **34**, 81–94.

Tarney, J. and Weaver, B.L. (1987b) Mineralogy, petrology and geochemistry of the Scourie dykes: petrogenesis and crystallization processes in dykes intruded at depth. In *Evolution of the Lewisian and Comparable High Grade Terrains*; eds. Park, R.G. and Tarney, J., *Publ. Geol. Soc. Lond.* **27**, 217–233.

Tarney, J. and Weaver, B.L. (1987c) Geochemistry of the Scourian complex; petrogenesis and tectonic models. In *Evolution of the Lewisian and Comparable High Grade Terrains*; eds. Park, R.G. and Tarney, J., *Spec. Publ. Geol. Soc. Lond.* **27**, 45–56.

Tarney, J. and Windley, B.F. (1977) Chemistry, thermal gradients and evolution of the lower crust. *J. Geol. Soc. Lond.* **134**, 153–172.

Tarney, J., Dalziel, I.W.D. and de Wit, M.J. (1976) Marginal basin "Rocas Verdes" complex from S. Chile: a model for Archaean greenstone belt formation. In *The Early History of the Earth* ed. Windley, B.F., Wiley, London, 131–146.

Tarney, J., Saunders, A.D., Mattey, D.P., Wood, D.A. and Marsh, N.G. (1981) Geochemical aspects of back-arc spreading in the Scotia Sea and western Pacific. *Phil. Trans. R. Soc. Lond.* **A300**, 263–285.

Tarney, J., Weaver, S.D., Saunders, A.D., Pankhurst, R.J, and Barker, P.F. (1982) Volcanic evolution of the northern Antarctic Peninsula and the Scotia arc. In *Andesites*. ed. Thorpe, R.S. Wiley, New York, 371–400.

Tassinari, C.C.G., Hirata, W.K. and Kawashita, K. (1982) Geologic evolution of the Serra dos Carajas, Para, Brazil. *Rev. Bras. Geoc.* **12**, 263–267.

Tatsumi, Y. and Ishizaka, K. (1982) Origin of high-Mg andesites in the Setouchi volcanic belt, southwest Japan, I. Petrographical and chemical characteristics. *Earth Planet. Sci. Lett.* **60**, 293–304.

Taylor, L.A. (1989) Oxide minerals. In *The Lunar Sourcebook*. eds. Heiken, G. and Vaniman, D.T., Cambridge University Press, in press.

Taylor, L.A., Shervais, J.W., Hunter, R.H., Shih, C.-Y., Bansal, B.M., Wooden, J., Nyquist, L.E. and Laul, J.C. (1983) Pre-4.2 AE mare-basalt volcanism in the lunar highlands. *Earth Planet. Sci. Lett.* **66**, 33–47.

Taylor, P.N., Moorbath, S., Goodwin, R. and Petrykowski, A.C. (1980) Crustal contamination as an indicator of the extent of early Archaean crust: Pb isotopic evidence from the late Archaean gneisses of West Greenland. *Geochim. Cosmochim Acta* **44**, 1437–1453.

Taylor, P.N., Jones, N.W. and Moorbath, S. (1984) Isotopic assessment of relative contributions from crust and mantle sources to the magma genesis of Precambrian granitoid rocks. *Phil. Trans. R. Soc. London* **A310**, 605–625.

Taylor, R.N. (1987) Lava stratigraphy and volcanism in the upper lavas of the Troodos opholite: Margi area. In *Troodos '87 ophiolites and oceanic lithosphere*, eds. Xenophontos, C. and Malpas, J.G., *Field excur. guidebook*. 130–157.

Taylor, R.N. (1988) The stratigraphy, geochemistry and petrogenesis of the Troodos extrusive sequence, Cyprus. Unpubl. Ph.D thesis, Southampton, 318 pp.

Taylor, R.N: and Nesbitt, R.W. (1988) Light rare-earth enrichment of supra subduction-zone mantle: evidence from the Troodos ophiolite, Cyprus. *Geology* **16**, 448–451.

Taylor, R.N. (1989) Geochemical stratigraphy of the Troodos extrusive sequence: temporal developments of a spreading centre magma chamber. In *Ophiolites and Oceanic Lithosphere.* eds. Moores, E.M. and Malpas, J., *Geol. Surv. Dept., Nicosia, Cyprus.* (in press).

Taylor, S.R. (1982) *Planetary Sciences: A Lunar Perspective.* Lunar Planet. Inst., Houston, 481 pp.

Taylor, S.R. and McLennan, S.M. (1985) *The Continental Crust: its Composition and Evolution.* Blackwell, Oxford, 312 pp.

Tegtmeyer, A.R. and Kröner, A. (1987) U–Pb zircon ages bearing on the nature of early Archaean greenstone belt evolution, Barberton Mountainland, southern Africa. *Precamb. Res.* **36**, 1–20.

Teixeira, N.A. and Danni, J.C.M. (1979) Petrologia das lavas ultrabasicas e basicas da sequencia vulcano-sedimentar Morro do Ferro, Fortaleza de Minas (MG). *Rev. Bras. Geoc.* **9**, 151–158.

Teixeira, J.B.G., Silva, M.G., Costa, U.R., Oliveira, M.A., Fratin, O., Teles, P.J., Vianna, I.A. and Gama, H.B. (1982) Rio Itapicuru Greenstone Belt (Serrinha region) with Faixa Weber gold deposits and Serra de Jacobina gold-bearing metasedimentary sequence. *ISAP-Int. Sym. Arch. Early Prot. Geol. Evol. and Met., Abs.,* SME-Ba, 118–138.

Tera, F., Brown, L., Morris, J. and Sacks, I.S. (1986) Sediment incorporation in island-arc magmas: Inferences from [10]Be. *Geochim. Cosmochim. Acta* **50**, 535–550.

Thompson, R.N. (1987) Phase-equilibria constraints on the genesis and magmatic evolution of oceanic basalts. *Earth Sci. Rev.* **24**, 161–210.

Thompson, R.N., Dickin, A.P., Gibson, I.L. and Morrison, M.A. (1982) Elemental fingerprints of isotopic contamination of Hebridean Palaeocene mantle-derived magmas by Archaean sial. *Contrib. Mineral. Petrol.* **79**, 159–168.

Thompson, R.N., Morrison, M.A., Dickin, A.P. and Hendry, G.L. (1983) Continental flood basalts... Arachnids rule OK? In *Continental Basalts and Mantle Xenoliths.* eds. Hawkesworth, C.L. and Norry, M.J. Shiva, Nantwich, 158–185.

Thurston, P.C. (1981) Economic evaluation of Archaean felsic volcanic rocks using REE geochemistry. In *Archaean Geology;* eds. Glover, J.E. and Groves, D.I. *Geol. Soc. Aust. Spec. Publ.* **7**, 439–450.

Thurston, P.C. (1986) Volcanic cyclicity in mineral exploration; the caldera cycle and zoned magma chambers. *Ont. Geol. Surv. Misc.* Paper **129**, 104–123.

Thurston, P.C. and Fryer, B.J. (1983) The geochemistry of repetitive cyclical volcanism from basalt through rhyolite in the Uchi-Confederation greenstone belt, Canada. *Contrib. Mineral. Petrol.* **83**, 204–226.

Thurston, P.C., Ayres, L.D., Edwards, G.R., Gelinas, L., Ludden, J.N.V. and Verpaelst, P. (1985) Archean bimodal volcanism. In *Evolution of Archaean Supracrustal Sequences;* eds. Ayres, L.D., Thurston, P.C., Card, K.D. and Weber, W. *Geol. Assoc. Can. Spec.* Paper **28**, 7–22.

Thurston, P.C., Ayres, L.D., Dimroth, E., Easton, R.M. and Johns, G.W., (1986) Archean volcanology – progress to date. *Int. Volc. Cong. Abs.* 358.

Thurston, P.C. and Sutcliffe, R.H. (1986) The Archean crust as a density filter. *Geol. Assoc. Can.; Min. Assoc. Can. Prog. Abs.* **11**, 136.

Thurston, P.C., Cortis, A.L. and Chivers, K.M. (1987) A reconnaissance re-evaluation of a number of northwestern greenstone belts: evidence for an early Archean sialic crust. *Ont. Geol. Survey Misc.* Paper **137**, 4–24.

Thurston, P.C. and Chivers, K.M. (1989) Secular variation in greenstone sequence development emphasizing Superior Province, Canada, *Precamb. Res.* (in press).

Tilling, R.I., Gottfried, D. and Rowe, J.J. (1973) Gold abundance in igneous rocks: bearing on gold mineralization. *Econ. Geol.* **68**, 168–186.

Todd, S.G., Keith, D.W., LeRoy, L.W., Schissel, D.J., Mann, E.L. and Irvine, T.N. (1982) The J–M platinum-palladium reef of the Stillwater Complex, Montana: I. Stratigraphy and petrology. *Econ. Geol.* **77**, 1454–1480.

Tolbert, G.E., Tremaine, J.W., Melcher, G.C. and Comes, C.B. (1971) The recently-discovered Serra dos Carajas iron deposits, northern Brazil. *Econ. Geol.* **66**, 985–994.

Torske, T. (1978) En Proterozoisk aulakogen i Nord-Norges grunnfjell. *Abstracts, 13, Nordisk Geol. Vintermøte, København.*

Townend, R., Ferreira. P.M. and Franke, N.D. (1980) Caraiba, a new copper deposit in Brazil. *Trans. Inst. Min. Met.* **89**, B159–B164.

Trendall, A.F. (1983) The Hamersley Basin. In *Iron-formation, Facts and Problems.* eds. Trendall, A.F. and Morris, R.C. Elsevier, Amsterdam, 69–129.

Trowell, N.F., Blackburn, C.E. and Edwards, G.R. (1980) Preliminary geological synthesis of the Savant lake–Crow lake metavolcanic-metasedimentary belt Northwestern Ontario and its bearing upon mineral exploration. *Ont. Geol. Surv. Misc.* Paper **89**, 30 pp.

Tuniz, C., Pal, D.K., Moniot, R.K., Savin, W., Kruse, T.H., Hergoz, G.F. and Evans, J.C. (1983) Recent cosmic ray exposure history of ALHA 81005. *Geophys. Res. Lett.* **10**, 804–806.

Turcotte, D.L. and Oxburgh, E.R. (1967) Finite amplitude convection cells and continental drift. *J. Fluid Mech.* **28**, 29–42.

Turner, J.S. and Chen, C.F. (1974) Two-dimensional effects of double-diffusive convection. *J. Fluid. Mech.* **63**, 577–593.

Turner, J.S., Huppert, H.E. and Sparks, R.S.J. (1986) Komatiites II. Experimental and theoretical investigation of post-emplacement cooling and crystallization. *J. Petrol.* **27**, 397–438.

Tweto, O. (1987) Rock units of the Precambrian basement in Colorado. *U.S. Geol. Surv. Prof. Pap.* **1321-A**, 54 pp.

Twist, D. (1985) Geochemical evolution of the Rooiberg silicic lavas in the Loskop Dam area, southeastern Bushveld. *Econ. Geol.* **80**, 1153–1165.

Tyler, N. (1979) Stratigraphy, geochemistry and correlation of the Ventesdorp Supergroup in the Derdepoort area, west-central Transvaal. *Trans. Geol. Soc. S. Afr.* **82**, 133–147.

Ujike, O. and Goodwin, A.M. (1987) Geochemistry and origin of Archean felsic metavolcanic rocks, central Noranda area, Quebec, Canada. *Can. J. Earth Sci.* **24**, 2551–2567.

Ulmer, G.C. (1969) Experimental investigations of chrome-spinels. In *Magmatic Ore Deposits*; ed. Wilson, H.D.B. *Econ. Geol. Mon.* **4**, 114–131.

Umeji, A.C. (1983) Archaean greenstone belts of Sierra Leone with comments on the stratigraphy and metallogeny. *J. Afr. Earth Sci.* **1**, 1–8.

Usselman, T.M., Hodge, D.S., Naldrett, A.J. and Campbell, I.H. (1979) Physical constraints on the characteristics of nickel sulfide ore in ultramafic lavas. *Can. Mineral.* **17**, 361–372.

Vaasajoki, M. (1977) Rapakivi granites and other postorogenic rocks in Finland: their age and the lead isotopic composition of certain associated galena mineralisations. *Geol. Surv. Finland Bull.* **294**, 1–66.

Van der Merwe, M.J. (1976) The layered sequence of the Potgietersrus limb of the Bushveld Complex. *Econ. Geol.* **71**, 1337–1351.

Van Niekerk, C.B. and Burger, A.J. (1964) The age of the Ventesdorp system. *Ann. Geol. Surv. S. Afr.* **3**, 75–86.

Van Niekerk, C.B. and Burger, A.J. (1969) Lead isotopic data relating to the age of the Dominion Reef lavas. *Trans. Geol. Soc. S. Afr.* **72**, 37–45.

Van Niekerk, C.B. and Burger, A.J. (1978) A new age for the Ventesdorp acid lavas. *Trans. Geol. Soc. S. Afr.* **81**, 155–163.

Van Schmus, W.R. and Bickford, M.E. (1981) Proterozoic chronology and evolution of the midcontinent region, North America. In *Precambrian Plate Tectonics.* ed. Kröner, A. Elsevier, Oxford, 261–296.

Vearncome, J.R. (1986) Structure of veins in a gold-pyrite deposit in a banded iron formation, Amalia belt, South Africa. *Geol. Mag.* **123**, 601–609.

Vearncome, J.R., Barton, J.M. and Walsh, K.L. (1987) The Rooiwater Complex and associated rocks, Murchison granitoid-greenstone terrane, Kaapvaal craton. *S. Afr. J. Geol.* **90**, 361–377.

Vearncombe, J.R., Barley, M.E., Eisenlohr, B., Grigson, M.W., Groves, D.I., Houston, M.S., Partington, G.A. and Swarnecki, M.S. (1988) Structural controls on gold mineralization – examples from the Archaean terrains of Western Australia and southern Africa. *Bicent. Gold 88, Ext. Abs., Geol. Soc. Austr. Abstr.* **22**, 19–23.

Veizer, J. (1988) The evolving exogenic cycle. In *Chemical Cycles in the Evolution of the Earth.* eds. Gregor, C.B., Garrelo, R.M., Mackenzie, F.T. and Maynard, J.B. Wiley, New York, 175–219.

Veizer, J. and Jansen, S.L. (1979) Basement and sedimentary recycling and continental evolution. *J. Geol.* **87**, 341–370.

Venkatesh, A.S., Poornachandra Rao, G.V.S., Prasada Rao, N.T.V. and Bhalla, M.S. (1987) Palaeomagnetic and geochemical studies on dolerite dykes from Tamil Nadu, India. *Precamb. Res.* **34**, 291–310.

Vermaak, C.F. (1970) The geology of the lower portion of the Bushveld Complex and its relationship to the floor rocks in the area west of the Pilanesberg, Western Transvaal. *Geol. Soc. S. Afr. Spec. Publ.* **1**, 242–262.

Vermaak, C.F. (1976) The Merensky Reef – thoughts on its environment and genesis. *Econ. Geol.* **71**, 1270–1298.

Vermaak, C.F. and Hendriks, L.P. (1976) A review of the mineralogy of the Merensky Reef with specific reference to new data on the precious metal mineralogy. *Econ. Geol.* **71**, 1244–1269.

Verpaelst, P. (1985) Geologie de la sequence volcanique Archeenne du complexe de Duprat Abitibi, Quebec. Unpubl. D. Sci. thesis. Univ. Montreal, 262 pp.

Vidal, P., Blais, S., Jahn, B.-M., Capdevila, R. and Tilton, G.R. (1980) U–Pb and Rb–Sr systematics of the Sudmussalmi Archaean greenstone belt (eastern Finland). *Geochim. Cosmochim. Acta* **44**, 2033–2044.

Viljoen, M.J. and Bernasconi, A. (1979) The geochemistry, regional setting and genesis of the Shangari–Danba nickel deposits, Rhodesia. In *Genesis of gold mineralization in the Gatooma area, Rhodesia*; eds. Foster, R.P., Mann, A.G., Miller, R.G. and Smith, P.J.R. *Spec. Publ. Geol. Soc. S. Afr.* **5**, 25–38 (67–98).

Viljoen, M.J. and Viljoen, R.P. (1969a) The geology and geochemistry of the Lower Ultramafic Unit of the Onverwacht Group and a proposed new class of igneous rocks. *Spec. Publ. Geol. Soc. S. Afr.* **2**, 55–86.

Viljoen, M.J. and Viljoen, R.P. (1969b) Evidence for the existence of mobile extrusive peridotitic magma from the Komati Formation of the Onverwacht Group. *Spec. Publ. Geol. Soc. S. Afr.* **2**, 87–112.

Viljoen, R.P. and Viljoen, M.J. (1969c) Evidence for the composition of the primitive mantle and its products of partial melting from a study of the rocks of the Barberton Mountain land. *Spec. Publ. Geol. Soc. S. Afr.* **2**, 275–298.

Viljoen, M.J. and Viljoen, R.P. (1969d) An introduction to the geology of the Barberton granite-greenstone terrain. *Spec. Publ. Geol. Soc. S. Afr.* **2**, 9–28.

Viljoen, R.P. and Viljoen, M.J. (1969e) The effects of metamorphism and serpentinization on the volcanic and associated rocks of the Barberton region. *Spec. Publ. Geol. Soc. S. Afr.* **2**, 29–54.

Viljoen, R.P. and Viljoen, M.J. (1969f) The geological and geochemical significance of the upper formations of the Onverwacht Group. *Spec. Publ. Geol. Soc. S. Afr.* **2**, 113–152.

Viljoen, R.P. and Viljoen, M.J. (1969g) The relationship between mafic and ultramafic material derived from the upper mantle and ore deposits of the Barberton region. *Spec. Publ. Geol. Soc. S. Afr.* **2**, 221–244.

Viljoen, R.P. and Viljoen, M.J. (1970) The geology and geochemistry of the layered ultramafic bodies of the Kaapmuiden area, Barberton Mountain Land. *Spec. Publ. Geol. Soc. S. Afr.* **1**, 661–688.

Viljoen, M.J. and Viljoen, R.P. (1979) Reassessment of part of the Barberton type area. South Africa – a comment. *Precamb. Res.* **9**, 349–352.

Viljoen, R.P. and Viljoen, M.J. (1982) Komatiites – an historical review. In *Komatiites*. eds. Arndt, N.T. and Nisbet, E.G. George Allen and Unwin, London, 5–17.

Viljoen, M.J., Viljoen, R.P. and Pearton, T.N. (1982) The nature and distribution of Archaean komatiite volcanics in South Africa. In *Komatiites*. eds. Arndt. N.T. and Nisbet, E.G., George Allen and Unwin, London, 53–79.

Viljoen, M.J., Viljoen, R.P., Smith, H.S. and Erlank, A.J. (1983) Geological, textural and geochemical features of komatiitic flows from the Komati Formation. *Spec. Publ. Geol. Soc. S. Afr.* **9**, 1–20.

Viljoen, M.J., De Klerk, W.J., Coetzar, P.M., Hatch, N.P., Kinloch, E. and Peyerl, W. (1986) The Union section of the Rustenburg Platinum mines, with reference to the Merensky Reef. In *Mineral Deposits of Southern Africa II*. eds. Anhaeusser, C.R. and Maske, S. Geol. Soc. S. Africa. 1061–1090.

Vinogradov, A.P. and Tugarinov, A.I. (1961) The geologic age of pre-Cambrian rocks of the Ukrainian and Baltic shields. *Annal. New York Acad. Sci.* **91**, 500–513.

Vitaliano, C.J., Cordua, W.S., Burger, H.R., Hanley, T.B., Hess, D.F. and Root, F.K. (1979) Geology and structure of the southern part of the Tabacco Root Mountains, southwestern Montana. *Geol. Soc. Amer. Bull.* **90**, 712–715.

Vivallo, W. and Claesson, L.-A. (1987) Intra-arc rifting, and massive sulphide mineralization in an early Proterozoic volcanic arc, Skellefte district, northern Sweden. In *Geochemistry and Mineralization of Proterozoic Volcanic Suites*; eds. Pharaoh, T.C., Beckinsale, R.D. and Rickard, D. *Spec. Publ. Geol. Soc. Lond.* **33**, 69–79.

Vivallo, W. and Rickard, D. (1984) Early Proterozoic ensialic spreading-subsidence: evidence from the Garpenberg enclave, central Sweden. *Precamb. Res.* **26**, 203–221.

Volpe, A.M., MacDougall, J.D. and Hawkins, J.W. (1987) Mariana Trough basalts (MTB): trace element and Sr–Nd isotopic evidence for mixing between MORB-like and arc-like melts. *Earth. Planet. Sci. Lett.* **82**, 241–254.

Von Gruenewaldt, G. (1973) The main and upper zones of the Bushveld Complex in the Roossenekal area, eastern Transvaal. *Trans. Geol. Soc. S. Afr.* **76**, 207–227.

Von Gruenewaldt, G. (1977) The mineral resources of the Bushveld Complex. *Min. Sci. Eng.* **9**, 83–95.

Von Gruenewaldt, G., Sharpe, M.R. and Hatton, C.J. (1985) The Bushveld Complex: introduction and review. *Econ. Geol.* **80**, 803–812.

Von Gruenewaldt, G., Behr, S. and Wilhelm, H.J. (1988) Some preliminary investigations of the Molopo Farms Complex, Botswana and its Ni–Cu sulphide mineralization. In *Proc. 5th Magmatic Sulphides Field Conf.*; ed., Prendergast, M., Inst. Min. Metall., London (in press).

Vuorelainen, Y., Hakli, T., Hanninen, E., Papunen, H., Reino, J. and Tornroos, R. (1982) Isomertiete and other platinum-group minerals from the Konttijarvi layered mafic intrusion, northern Finland. *Econ. Geol.* **77**, 1511–1518.

Wager, L.R. and Deer, W.A. (1939) Geol. investigations in East Greenland, III. The petrology of the Skaergaard intrusion, Kangerdlugssuaq, East Greenland. *Medd. om Grønland* **105**, 352 pp.

Wager, L.R. and Brown, G.M. (1968) *Layered Igneous Rocks.* Oliver and Boyd, Edinburgh, 588 pp.

Wager, L.R., Brown, G.K. and Wadsworth, W.J. (1960) Types of igneous cumulates. *J. Petrol.* **1**, 73–85.

Wahlstrom, E.E. (1956) Petrology and weathering of the Iron Dike, Boulder and Laramie Counties, Colorado. *Geol. Soc. Amer. Bull.* **67**, 147–163.

Walker, D. (1983) Lunar and terrestrial crust formation. *Proc. 14th Lunar Planet. Sci. Conf.; J. Geophys. Res.* **88**, B17–B25.

Walker, D. (1986) Melting equilibrium in multicomponent systems and liquidus/solidus convergence in mantle peridotite. *Contrib. Mineral. Petrol.* **92**, 303–307.

Walker, D.A. and Cameron, W.E. (1983) Boninite primary magmas: evidence from the Cape Vogel peninsula. *Contrib. Mineral. Petrol.* **83**, 150–158.

Walker, R.J., Hanson, G.N., Papike, J.J. and O'Neil, J.R. (1986). Nd, O and Sr isotopic constraints on the origin of Precambrian rocks, southern Black Hills, South Dakota. *Geochim. Cosmochim. Acta* **50**, 2833–2846.

Walker, R.J., Shirey, S.B. and Stecher, O. (1988) Comparitive Re–Os, Sm–Nd and Rb–Sr isotope and trace element systematics for Archaean komatiite flows from Munro Township, Abitibi Belt, Ontario. *Earth Planet. Sci. Lett.* **87**, 1–12.

Wallace, H. (1985) Geology of the Slate Falls area. *Ont. Geol. Surv. Geol. Rep.* **232**, 85 pp.

Wallace, H., Thurston, P.C. and Corfu, F. (1986) Developments in stratigraphic correlation. *Ont. Geol. Surv. Misc. Paper* **129**, 88–102.

Wang, K.Y., Yan, Y.H., Yang, R.Y. and Chen, Z.Z. (1985) REE geochemistry of early Precambrian charnockites and tonalitic-granodioritic gneisses of the Qianan region, eastern Hebei, north China. *Precamb. Res.* **27**, 63–84.

Wang, K.Y., Sills, J.D., Yan, Y.H. and Windley, B.F. (1989) The Archaean gneiss complex in eastern Hebei Province, North China: geochemistry and evolution. *Precamb. Res.* in press.

Wang, R.M., He, S.Y., Chen, Z.Z., Li, P.F. and Dai, F.Y. (1985). Geochemical evolution and metamorphic development of the early Precambrian in eastern Hebei, China. *Precamb. Res.* **27**, 111–129.

Wang, S.S., Zhai, M.G., Hu, S.L., Sang, H.Q. and Qiu, J. (1986) $^{40}Ar/^{39}Ar$ age spectrum for biotite separated from Qingyuan tonalite, NE China. *Scientia Geol. Sinica* **1**, 97–99.

Wang, S.S., Hu, S.L., Zhai, M.G., Sang, H.Q. and Qiu, J. (1987) An application of the $^{40}Ar/^{39}Ar$ dating technique to the formation time of qingyuan granite–greenstone terrain in NE China. *Acta Petrol. Sinica* **4**, 55–62.

Wang, Z.J., Shen, Q.H. and Jin, S.W. (1987) Petrology, geochemistry and U–Pb isotopic dating of the Shipaihe 'metadiorite mass' in Dengfeng Country, Henan Province, China. *Bull. Chin. Acad. Geol. Sci.* **16**, 215–225.

Ward, H.J. (1975) Barrambie iron–titanium–vanadium deposit. In *Economic Geology of Australia and Papua New Guinea.* ed. Knight, C.C. Aust. Inst. Min. Metal., Melbourne, 207–211.

Ward, P. (1987) Proterozoic deposition and deformation at the Karelian craton margin in southeastern Finland. *Precamb. Res.* **35**, 71–93.

Warner, J.L., Simonds, C.H. and Phinney, W.C. (1976) Genetic distinction between anorthosites and Mg-rich plutonic rocks: new data from 76255 (abs.). *Lunar Sci.* **VII**, 915–917.

Warner, R.D., Snipes, D.S., Hughes, S.S., Steiner, J.C., Davis, M.W., Manoogian, P.R. and Schmitt, R.A. (1985) Olivine-normative dolerite dykes from western South Carolina: Mineralogy, chemical composition and petrogenesis. *Contrib. Mineral. Petrol.* **90**, 386–400.

Warren, P.H. (1984) Primordial degassing, lithosphere thickness, and the origin of komatiites. *Geology* **12**, 335–338.

Warren, P.H. (1985) The magma ocean concept and lunar evolution. *Ann. Rev. Earth Planet. Sci.* **13**, 201–240.

Watkeys, M.K. (1983) Brief explanatory notes on the provisional geological map of the Limpopo Belt and its environs. *Spec. Publ. Geol. Soc. S. Afr.* **8**, 5–8.

Watkeys, M.K. (1988) The geology of the Tjakastad Subgroup: a review. *Geol. Soc. Amer. Barberton Mem.*

Watters, B.R. and Armstrong, R.L. (1985) Rb–Sr study of metavolcanic rocks from the La Ronge and Flin Flon domains, northern Saskatchewan. *Can. J. Earth Sci.* **22**, 452–463.

Watts, A.B. and Davies, S.F. (1981) Long wavelenth gravity and topography anomalies. *Ann. Rev. Earth Planet. Sci.* **9**, 415–458.

Watts, A.B., Mckenzie, D.P., Parsons, B.E. and Roufosse, M. (1985) The relationships between gravity and bathymetry in the Pacific Ocean. *Geophys. J. R. Astron. Soc.* **83**, 263–298.

Weaver, B.L. (1980) Rare earth element geochemistry of Madras granulites. *Contrib. Mineral. Petrol.* **71**, 271–279.

Weaver, B.L. and Tarney, J. (1979) Thermal aspects of komatiite generation and greenstone belt models. *Nature* **79**, 689–692.

Weaver, B.L. and Tarney, J. (1980) Rare earth geochemistry of Lewisian granulite facies gneisses, northwest Scotland: implications for the petrogenesis of the Archaean lower continental crust. *Earth Planet. Sci. Lett.* **51**, 279–296.

Weaver, B.L. and Tarney, J. (1981a) Lewisian gneiss geochemistry and Archaean crustal development models. *Earth Planet. Sci. Lett.* **55**, 171–180.

Weaver, B.L. and Tarney, J. (1981b) The Scourie dyke suite: petrogenesis and geochemical nature of the Proterozoic sub-continental mantle. *Contrib. Mineral. Petrol.* **78**, 175–188.

Weaver, B.L. and Tarney, J. (1983) Chemistry of the subcontinental mantle: inferences from Archaean and Proterozoic dykes and continental flood basalts. In *Continental Basalts and Mantle Xenoliths.* eds. Hawkesworth, C.L. and Norry, M.J. Shiva, Nantwich, 209–229.

Weaver, B.L., Tarney, J., Windley, B.F., Sugavanem, E.B. and Venkata Rao, V. (1978) Madras granulites: geochemistry and P–T conditions of crystallisation. In: *Archaean Geochemistry.* eds. Windley, B.F. and Naqvi, S.M. Elsevier, Amsterdam, 177–204.

Weaver, B.L., Tarney, J. and Windley, B.F. (1981) Geochemistry and petrogenesis of the Fiskenasset anorthosite complex, southern West Greenland; nature of the parent magma. *Geochim. Cosmochim. Acta* **45**, 711–725.

Weaver, B.L., Tarney, J., Windley, B.F. and Leake, B.E. (1982) Geochemistry and petrogenesis of Archaean metavolcanic amphibolites from Fiskenaesset, S.W. Greenland. *Geochim. Cosmochim. Acta* **46**, 2203–2215.

Wedow, H.Jr., Gaskill, D.J., Banister, D.P., Pattee, E.C. and Peterson, D.L. (1975) Mineral resources of the Absaroka primitive area and vicinity, Parkand Sweet Grass Counties, Montana. *U.S. Geol. Surv. Bull.* **1391-B**, 115 pp.

Welin, E. (1987) The depositional evolution of the Svecofennian supracrustal sequences in Finland and Sweden. *Precamb. Res.* **35**, 95–114.

Wernick, E. (1981) The Archaean of Brazil. *Earth Sci. Rev.* **17**, 31–48.

White, R.S., Spence, G.D., Fowler, S.R., McKenzie, D.P., Westbrook, G.K. and Bowen, A.N. (1987) Magmatism at rifted continental margins. *Nature* **330**, 439–444.

Wilde, S.A. and Pidgeon, R.T. (1986) Geology and geochronology of the Saddleback greenstone belt in the Archaean Yilgarn block, southwestern Western Australia. *Aust. J. Earth Sci.* **33**, 491–501.

Wiles, J.W. (1957) The geology of the eastern portion of the Hartley gold belt (pt 1); Gold deposits and mines (pt 2). *Bull. Geol. Surv. Rhod.* **44**, 128pp (pt 1); 180pp (pt 2).

Wiles, J.W. (1968) Some aspects of the metamorphism of the Basement Complex in the Sipolilo district. *Trans. Geol. Soc. S. Afr.* **71**, *Ann.* 71–88.

Wilks, M.E. (1988) The Himalayas, a modern analogue for Archaean crustal evolution. *Earth Planet. Sci. Lett.* **87**, 127–136.

Wilks, M.E. and Nisbet, E.G. (1988) Stratigraphy of the Steep Rock Group, northwest Ontario: a major Archaean unconformity and Archaean stromatolites. *Can. J. Earth Sci.* **25**, 370–391.

Willemse, J. and Viljoen, E.A. (1970) The fate of argillaceous material in the gabbroic magma of the Bushveld Complex. *Spec. Publ. Geol. Soc. S. Afr.* **1**, 336–366.

Williams, D.A.C. (1972) Archaean ultramafic, mafic and associated rocks. Mt. Monger, Western Australia. *J. Geol. Soc. Aust.* **19**, 163–188.

Williams, D.A.C. (1979) The associations of some nickel sulphide deposits with komatiite volcanism in Rhodesia. *Can. Mineral.* **17**, 337–349.

Williams, D.A.C. and Hallberg, J.A. (1973) Archaean layered intrusions of the Eastern Goldfields region, Western Australia. *Contrib. Mineral. Petrol.* **38**, 45–70.

Williams, D.A.C. and Furnell, R.G. (1979a) Reassessment of part of the Barberton type area, South Africa. *Precamb. Res.* **9**, 325–347.

Williams, D.A.C. and Furnell, R.G. (1979b) Reassessment of part of the Barberton type area, South Africa – a reply. *Precamb. Res.* **9**, 353–357.

Williams, G.E. (1973) Geotectonic cycles, lunar evolution, and the dynamics of the earth-moon system. *Modern Geol.* **4**, 159–183.

Williams, H.R. (1978) The Archaean geology of Sierra Leone. *Precamb. Res.* **6**, 251–268.

Williams, H.R. (1988) The Archaean Kasila Group of western Sierra Leone: geology and relations with adjacent granite-greenstone terrane. *Precamb. Res.* **38**, 201–213.

Williams, H.R. and Williams, R.A. (1976) The Kasila Group, Sierra Leone, an interpretation of new data. *Precamb. Res.* **3**, 505–508.

Willis, R.E., Brown, R.E., Stroud, W.J. and Stevens, B.P.J. (1983) The early Proterozoic Willyama Supergroup: stratigraphic subdivision and interpretation of high to low grade metamorphic rocks in the Broken Hill Block, New South Wales. *J. Geol. Soc. Aust.* **30**, 195–224.

Wilson, A.F. (1978) Comparison of some of the geochemical features and tectonic setting of Archaean and Proterozoic granulites, with particular reference to Australia. In *Archaean Geochemistry.* eds. Windley, B.F. and Naqvi, S.M. Elsevier, Amsterdam, 241–267.

Wilson, A.H. (1982) The Geology of the Great "Dyke", Zimbabwe: The ultramafic rocks. *J. Petrol.* **23**, 240–292.

Wilson, A.H. and Prendergast, M.D. (1987) The Great Dyke, Zimbabwe: an overview. *Abs. and Field Guide, 5th magmatic sulphides field conference, IGCP Proj.* **161**, Harare, Zimbabwe, 55 pp.

Wilson, J.F. (1968) The Mashaba igneous complex and its subsequent deformation. *Trans. Geol. Soc. S. Afr.* **71**, *Ann.*, 175–188.

Wilson, J.F. (1973) The Rhodesian Archaean craton – an essay in cratonic evolution. *Phil. Trans. R. Soc. Lond.* **A273**, 389–411.

Wilson, J.F. (1979) A preliminary re-appraisal of the Rhodesian basement complex. *Spec. Publ. Geol. Soc. S. Afr.* **5**, 1–23.

Wilson, J.F. (1981) The granitic gneiss – greenstone shield of Zimbabwe. In *Precambrian of the Southern Hemisphere* ed. Hunter, D.R. Elsevier, Amsterdam, 459–499.

Wilson, J.F., Bickle, M.J., Hawkesworth, C.J., Nisbet, E.G., Martin, A and Orpen, J.L. (1978) Granite-greenstone terrains of the Rhodesian Archaean craton. *Nature* **271**, 23–27.

Winchester, J.A. and Floyd, P.A. (1977) Geochemical discrimination of different magma series and their differentiation products using immobile elements. *Chem. Geol.* **20**, 325–343.

Windley, B.F. (1973) Archaean anorthosites: a review with the Fiskenaesset Complex, West Greenland as a model for interpretation. In *Symp. on Archaean Rocks; Geol. Soc. S. Afr. Spec. Publ.* **33**, 319–332.

Windley, B.F. ed. (1976) *The Early History of the Earth.* Wiley, London, 619 pp.

Windley, B.F. (1984) *The Evolving Continents* (2nd edition). Wiley, London, 399 pp.

Windley, B.F. (1981) Precambrian rocks in the light of the plate-tectonic concept. In *Precambrian Plate Tectonics,* ed. Kröner, A. Elsevier, Amsterdam, 1–16.

Windley, B.F. and Naqvi, S.M. eds. (1978) *Archaean Geochemistry.* Elsevier, Amsterdam, 406 pp.

Windley, B.F. and Selvan, T.A. (1975) Anorthosites and associated rocks of Tamil Nadu, southern India. *J. Geol. Soc. India* **16**, 209–215.

Windley, B.F. and Smith, J.V. (1974) The Fiskenaesset complex, West Greenland, part II. General mineral chemistry from Qeqertarssuatsiaq. *Bull Grønlands Geol. Unders* **108**, 54 pp.

Windley, B.F., Herd, R.K. and Bowden, A.A. (1973) The Fiskenaesset complex, West Greenland, part I. A preliminary study of stratigraphy, petrology and wholerock chemistry from Qeqertarssuatsiaq. *Bull. Grønlands Geol. Unders.* **106**, 80 pp.

Windley, B.F., Bishop, F.C. and Smith, J.V. (1981) Metamorphosed layered igneous complexes in Archaean granulite-gneiss belts. *Ann. Rev. Earth Planet. Sci.* **9**, 175–198.

Winge, M. and Danni, J.C.M. (1980) Compartimentos geotectonicos pre-brasilianos entre Carataca e Bendengo–Municipio de Uaua–Ba. *Anais XXXI Congr. Brasileiro Geol.* **5**, 2785–2795.

Winter, H. de la Rey (1976) A lithostratigraphic classification of the Ventesdorp succession. *Trans. Geol. Soc. S. Afr.* **79**, 31–48.

Winter, H. de la Rey (1978) A cratonic foreland model for the Witwatersrand basin development in a continental back-arc, plate tectonic setting. *S. Afr. J. Geol.* **90**, 409–427.

Wirth, K.R., Gibbs, A.K. and Olszewski, W.M. Jr. (1986) U–Pb ages zircons from the Grão Para Group and Serra dos Carajás Granite, Para, Brazil. *Rev. Bras. Geoc.* **16**, 195–200.

Withnall, I.W., Bain, J.H.C. and Rubenach, M.J. (1980) The Precambrian geology of Northeastern Queensland. In *The Geology and Geophysics of Northeastern Australia.* eds. Henderson, R.A. and Stephenson, P.J. Geol. Soc. Aust., Queensland Div., Brisbane, 109–127.

Witschard, F. (1984) The geological and tectonic evolution of the Precambrian of Northern Sweden – a case of basement reactivation? *Precamb. Res.* **23**, 275–315.

Wood, B.J. (1975) The influence of pressure, temperature and bulk composition on the apperance of garnet in orthogneisses – an example from South Harris, Scotland. *Earth Planet. Sci. Lett.* **26**, 299–311.

Wood, C.P. (1980) Boninite at a continental margin. *Nature* **288**, 692–694.

Wood, D.A. (1979) A variably veined and sub-oceanic mantle: genetic significance for mid-ocean ridge basalts from geochemical evidence. *Geology* **7**, 499–503.

Wood, D.A. (1980) The application of a Hf–Th–Ta diagram to problems of tectonomagmatic classification and to establishing the nature of crustal contamination of basaltic lavas of the Britich Teritary province. *Earth Planet. Sci. Lett.* **50**, 11–30.

Wood, J.A. (1986) Moon over Mauna Loa: a review of hypotheses of formation of Earth's Moon. In *Origin of the Moon.* eds. Hatmann, W.K., Phillips, R.J. and Taylor, G.J. Lunar Planet. Inst., Houston, 17–55.

Wood, D.A., Joron, J.L. and Treuil, M. (1979) A re-appraisal of the use of trace elements to classify and discriminate between magma series erupted in different tectonic setting. *Earth Planet. Sci. Lett.* **45**, 326–336.

Wood, D.A., Joron, J.L., Treuil, M., Norry, M.J. and Tarney, J. (1979) Elemental and Sr isotope variations in basic lavas from Iceland and the surrounding ocean floor. The nature of mantle source inhomogeneities. *Contrib. Mineral. Petrol.* **70**, 319–339.

Woodall, G.A. and Travis, R. (1969) The Kambalda nickel deposits, Western Australia. *9th Commonwealth Mining Metal. Cong., London*, **2**, 517–533.

Wooden, J.L. (1975) Geochemistry and Rb–Sr geochronology of Precambrian mafic dikes from the Beartooth, Ruby Range, and Tobacco Root Mountains, Montana. Ph.D. thesis, Univ. North Carolina, Chapel Hill, 194 pp.

Wooden, J.L. and Mueller, P.A. (1979) Mafic dikes of the Beartooth Mountains. IGCP Archaean Geochem. *Work. Gp. Field Conf. (1979); Guide to the Precambrian rocks of the Beartooth mountains*, 50–61.

Wooden, J.L. and Mueller, P.A. (1988) Pb, Sr and Nd isotopic compositions of a suite of late Archaean, igeneous rocks, eastern Beartooth Mountains: implications for crust-mantle evolution. *Earth Planet. Sci. Lett.* **87**, 59–72.

Wooden, J.L., Vitaliano, C.J., Koehler, S.W. and Ragland, P.C. (1978) The late Precambrian mafic dikes of the Tobacco Root Mountains, Montana: geochemistry, Rb–Sr geochronology and relationship to Belt tectonics. *Can. J. Earth Sci.* **15**, 467–479.

Wooden, J.L., Goldich, S.S. and Suhr, N.H. (1980) Origin of the Morton gneiss, southwestern Minnesota: part 2. Geochemistry. *Geol. Soc. Amer. Spec. Pap.* **182**, 57–76.

Wooden, J.L., Mueller, P.A., Hunt, D.K. and Bowes, D.R. (1982) Geochemistry and Rb–Sr geochronology of Archaean rocks from the interior of the southeastern Beartooth Mountains,

Montana and Wyoming. In *Precambrian Geology of the Beartooth Mountains, Montana and Wyoming*; eds. Mueller, P.A. and Wooden, J.L. *Mont. Bur. Mines Geol. Spec. Publ.* **84**, 45–56.

Worl, R.G. (1969) Migmatites in the Three Waters area, northern Wind River Mountains, Wyoming. *Geol. Soc. Amer. Spec. Paper* **121**, 647–648.

Worl, R.G. (1972) Layered paratectonic migmatites of the Three Waters area, Wind River Range, Wyoming, U.S.A. *Proc. 24th Int. Geol. Congr.* **2**, 135–143.

Worst, B.G. (1960) The Great Dyke of Southern Rhodesia. *S. Rhod. Geol. Surv. Bull.* **47**, 234 pp.

Wyatt, B.A. (1981) The geology and geochemistry of Klipriviersberg volcanics, Ventersdorp Supergroup, south of Johannesburg. MSc thesis, Univ. Witwatersrand.

Wyborn, L.A.I. (1987) Extensional magmatism in the Australian Proterozoic and its metallogenic implications. *BMR J. Aust. Geol. Geophys.* **12**, 189–193.

Wyborn, L.A.I., Wyborn, D., Chappell, B.W., Sheraton, J., Tarney, J., Collins, W. and Drummond, D.J. (1987a) Geological evolution of granite compositions with time in the Australian continent – implications for tectonic and mantle processes. *Proc. 9th Geol. Con.* **8**, 434–435.

Wyborn, L.A.I., Page, R.W. and Parker, A.J. (1987b) Geochemical and geochronological signatures in Australian Proterozoic igneous rocks. In *Geochemistry and Mineralization of Proterozoic Volcanic Suites*; eds. Pharaoh, T.C., Beckinsale, R.D. and Rickard, D. *Spec. Publ. Geol. Soc. Lond.* **33**, 377–394.

Wylie, P.J., Donaldson, C.H., Irving, A.J., Kesson, S.E., Merrill, R.B., Presnall, D.C., Stopler, E.M., Usselman, T.M. and Walker, D. (1981) Experimental petrology of basalts and their source rocks. In *Basaltic Volcanism on the Terrestrial Planets*. Pergamon Press, New York, 493–630.

Wyman, D. and Kerrich, R. (1988) Alkaline magmatism, major structures, and gold deposits: implications for greenstone belt gold metallogeny. *Econ. Geol.* **83**, 454–461.

Wyman, D.A. and Kerrich, R. (1989) Archean lamprophyre dikes of the Superior Province, Canada: distribution, petrology, and geochemical characteristics. *J. Geophys. Res.* **94**, 4667–4696.

Ying, S.H. (1980) *The Taishan Complex*. Science Publ. Co., Beijing, 83 pp (in Chinese).

Zartman, R.E. and Doe, B.R. (1981) Plumbotectonics – the model. *Tectonophysics* **75**, 135–162.

Zartman, R.E., Peterman, Z.E., Obradovitch, J.D., Gallego, M.D. and Bishop, D.T. (1982) Age of the Crossport Sill near Eastport, Idaho. *Idaho Bur. Mines Geol. Bull.* **24**, 61–69.

Zeck, H.P. and Kalsbeek, F. (1981) Geochemistry of amphibolite facies metamorphism of a suite of basic dykes, Precambrian basement, Greenland. *Chem. Erde* **40**, 1–22.

Zeck, H.P., Morthorst, J.R. and Kalsbeek, F. (1983) Metasomatic control of K/Rb ratios in amphibolites. *Chem. Geol.* **40**, 313–321.

Zhai, M.G., Yang, R.Y., Lu, W.Z. and Zhou, J. (1985) Geochemistry and evolution of the Qingyuan Archaean granite–greenstone terrain, NE China. *Precamb. Res.* **27**, 37–62.

Zhang, G.W., Bai, Y.B., Sun, Y., Guo, A.L., Zhou, D.W. and Li, T.H. (1985) Composition and evolution of the Archaean crust in central Henan, China. *Precamb. Res.* **27**, 7–35.

Zhong, F.D. (1984) Geochronology study of Archaean granite–gneisses in Anshan area, Northeast China. *Geochimica* **3**, 195–205.

Zhu, B.Q., Zhai, M., Stacey, J.S. and Lanphere, M.A. (1986) Ages of petrogenesis and metamorphism in Archaean rocks, Huadian region, NE China. *Terra Cogn.* **6**. 145.

Zientek, M.L., Czamanske, G.K. and Irvine, T.N. (1985) Stratigraphy and nomenclature for the Stillwater Complex. In *Stillwater Complex*; eds. Czamanske, G.K. and Zientek, M.L. *Mont. Bur. Mines Geol. Spec. Publ.* **92**, 21–32.

Zientek, M.L., Foose, M.P. and Mei, L. (1986) Palladium, platinum and rhodium contents of rocks near the lower margin of the Stillwater Complex, Montana. *Econ. Geol.* **81**, 1169–1178.

Zindler, A. (1982) Nd and Sr isotope studies of komatiites and related rocks. In *Komatiites*. eds. Arndt, N.T. and Nisbet, E.G. George Allen and Unwin, London, 399–420.

Zindler, A and Hart, S. (1986) Chemical geodynamics. *Ann. Rev. Earth Planet. Sci.* **14**, 493–571.

Index